PRINCIPLES AND APPLICATIONS OF ELECTROMAGNETIC FIELDS

PRINCIPLES AND APPLICATIONS OF

Electromagnetic Fields

ROBERT PLONSEY

ROBERT E. COLLIN

Associate Professors of Electrical Engineering
Case Institute of Technology

McGRAW-HILL BOOK COMPANY, INC.

New York Toronto London

1961

PRINCIPLES AND APPLICATIONS OF
ELECTROMAGNETIC FIELDS

 Library of Congress Catalog Card Number 61-11131

THE MAPLE PRESS COMPANY, YORK, PA.

50340

TO VIVIAN AND KATHLEEN

WHO DID SO MUCH
TO MAKE THE WRITING OF THIS BOOK
A PLEASANT TASK

PREFACE

In view of the number of texts published in the area of electromagnetic fields for undergraduate students in recent years, another book in this area needs some explanation. Generally, we find that the available books suffer from one or more of the following deficiencies:

1. Inadequate coverage of topics
   a. Very little or no material beyond Maxwell's equations
   b. Omission of important derivations
   c. Lack of depth in treatment
2. Lack of separation or distinction between purely mathematical and physical concepts
3. *Ad hoc* treatment of the subject of fields in the presence of material bodies
4. Insufficient number of examples worked out
5. A lack of problems designed to give the student ability and confidence in developing analytical solutions and extensions to the theory in contrast with routine drill problems

The current text represents the authors' efforts to overcome these shortcomings. We have endeavored to produce a book in electromagnetic theory suitable for undergraduate use that is sophisticated enough to establish a firm basis for advanced study in this area. Furthermore, a sufficient number of applications treated in adequate depth are included to illustrate the basic concepts and also provide a nontrivial background in a diverse number of areas of current interest.

We feel that an adequate text should be sufficiently complete and have enough scope to warrant a place on a personal bookshelf of the undergraduate student after he leaves school. In many cases it is desirable to avoid presentation of a lengthy derivation of certain formulas in class. On the other hand, many students wish to know how a given result is obtained, and in this case an outline of the steps to be followed suffices, provided a detailed derivation is included in the text. We have therefore included some material in the text of a more advanced and sophisticated nature for the advanced or curious student and as a supplement to the main course. The chapters have been organized so that this material is reserved to the last sections. These are set in smaller type and may be omitted by the instructor without detriment to the main continuity of the text.

The first nine chapters of the book constitute the basic principles of electromagnetic theory and are more than ample for a junior- or senior-level course of one-semester duration. The level can be varied somewhat by inclusion or elimination of certain of the topics. The organization of these chapters is fairly standard and includes vector analysis, electrostatics, mathematical techniques in the solution of Laplace's equation, current fields, magnetostatics, and time-varying fields (Maxwell's equations), in that order. The material on vector analysis gives greater emphasis to the relationship between fields and their sources. The climax of this development is in the presentation of the Helmholtz theorem. The flux concept is introduced in the same chapter so that the student will understand its general usefulness in representing any type of vector field. Vector-analysis techniques are freely utilized in the main body of the text.

The theory of electrostatics is developed in Chapter 2 for free-space conditions. In this way the electric field and its relationship to charge sources are developed as a basic formulation. The atomic properties of dielectrics are considered in some detail in Chapter 3. The effect of the presence of dielectrics in an electric field is next explained in terms of the induced dipole sources, from which the equivalent charge source is then determined. This procedure is also followed in Chapters 6 and 7, where the subject of magnetostatics is first developed for currents under free-space conditions. The effect of magnetic materials is then introduced in terms of the induced current sources.

Chapter 4 discusses the method of separation of variables and conformal-mapping techniques for obtaining solutions to Laplace's equation with specified boundary conditions. This chapter includes a discussion of cylindrical and spherical functions and sufficient elementary material on functions of a complex variable to be essentially self-contained. Since this chapter presents no new physical concepts, it may be omitted without detriment to the continuity of the text. Simple solutions to Laplace's equation, including image theory, are included in Chapter 3. The uniqueness theorem is developed in Chapter 2.

Chapter 5 discusses currents and Ohm's law. It presents a formal solution to the computation of the resistance of an arbitrary conducting body. The graphical techniques of flux plotting and the use of the electrolytic tank as auxiliary techniques for solution of Laplace's equation are included here.

Energy storage in electric and magnetic fields is discussed in Chapters 3 and 8, respectively. The calculation of the electric or magnetic force by the principle of virtual work is developed. Finally, the Maxwell stress tensor is introduced, and its significance for the field concept discussed.

In Chapter 9 the displacement current is postulated and Maxwell's equations are formulated. The development of vector and scalar potentials, their relationship to the sources, and retardation effects are then discussed. The chapter concludes with a fairly complete description of the relationship between circuit theory and field theory.

Many fields books written for the undergraduate terminate after introducing Maxwell's equations. Since Maxwell's equations form a climax and the bulk of the practical applications involve time-varying fields, it can hardly be considered a wise choice to terminate the book just when the door to a large variety of interesting and important applications has been opened. We feel that the student can proceed to the application of Maxwell's equations most expediently when he does not have to change texts, since this eliminates the necessity of getting acquainted with another author's notation and way of doing things. For this reason we have included three rather complete chapters on applications. These are on wave guiding, radiation, and interaction of fields with charged particles.

The material on wave guiding is included in Chapter 10. This chapter also discusses plane waves, refraction at a plane interface, and cavity resonators. The technique for accounting for losses at good conductors is considered so that attenuation in waveguides and Q of cavities may be computed. The chapter includes a general treatment of the resolution of fields into TEM, TM, and TE modes. The rectangular and circular waveguides are considered.

The subject of radiation is considered in Chapter 11. The simple linear antenna is described, and the fundamental properties of arrays developed. A full description of the receiving antenna and reciprocity is also included.

The final chapter discusses the interaction of fields with charged particles. This includes the subject of electron ballistics and parallel-plane vacuum-tube theory, including the effect of space charge and transit time. Space-charge-wave theory is developed and applied to the klystron and the traveling-wave tube. In the latter case the helix as a slow-wave structure is described. Propagation in gyrotropic media is considered, with specific application to the ionosphere and ferrites. Finally, an introduction to magnetohydrodynamics is given.

The greatest utility of the book is seen as a full-year course which would substantially cover the entire text. This would provide a very firm foundation in electromagnetic theory and also a good insight into a variety of applications. Chapters 9 to 12 may also be used in a one-semester course on applications of electromagnetic fields if the students have a sufficient background in the fundamentals. As with the full course, this would serve as an introduction to advanced graduate study in special topics. The level of this material may in some instances be deemed suitable for an introductory graduate course.

Objections are at times raised as regards presenting field theory in a layered package, that is, electrostatics, stationary currents, magnetostatics, etc. It certainly is feasible to begin with Maxwell's equations and specialize to static fields and then return to time-varying fields. This is essentially a direct-analysis point of view. By beginning with the experimental laws for static fields and building up and generalizing to time-varying fields the whole approach becomes more of a synthesis procedure. This we feel makes the subject material more acceptable from the student's point of view. It also permits the concept of a vector field to be firmly developed in connection with fields that have a relatively simple behavior, for example, electrostatic field vs. time-varying electromagnetic fields.

We do not particularly feel that relativistic electrodynamics should be introduced at the undergraduate level. The practical applications are few in number, and usually insufficient time is available to give anything more than a brief introduction, which probably leaves the student confused rather than informed in the subject. We have adopted the conventional formulation of considering **E** and **B** as the fundamental force vectors and treat magnetization on the Amperian-current basis. This is in contrast to the view adopted by Professor Chu of the Massachusetts Institute of Technology. Although Professor Chu's formulation has certain features to recommend it when considering the four-vector formulation of electrodynamics, we feel it wise to adhere to the conventional formulation. Not only does this keep our presentation similar to that in most other reference books the student will consult; it is also in keeping with the scope of our presentation.

Since this book is based, to a large extent, on the combined work of many earlier contributors, it is impossible to acknowledge this on an individual basis. We should like, however, to express our thanks to those who assisted us during the preparation of the manuscript. We are greatly appreciative of the help received from colleagues at Case Institute of Technology and in particular from Professors Forest E. Brammer and Robert D. Chenoweth. Many fruitful ideas were also received from students who made use of a preliminary version of this book. We are also grateful to Professor John R. Martin, Acting Chairman of the Department of Electrical Engineering, for making available facilities for the preparation of the manuscript. And finally, we are indebted to Misses Florence Alaimo, Ruth Hudak, and Janet Leonard, who expertly deciphered the original handwritten manuscript and typed the final copy.

It may be worth noting that a toss of a coin determined the order of the authors' names for the book.

Robert Plonsey
Robert E. Collin

CONTENTS

Preface . . . vii

Chapter 1. Vector Analysis . . . 1

1.1. Scalars and vectors . . . 1
1.2. Addition and subtraction of vectors . . . 2
1.3. Unit vectors and vector components . . . 3
1.4. Vector representation of surfaces . . . 4
1.5. The vector product of two vectors . . . 5
1.6. The scalar product of two vectors . . . 6
1.7. Product of three vectors . . . 7
1.8. Scalar and vector fields . . . 8
1.9. Gradient . . . 11
1.10. The divergence . . . 13
1.11. Line integral . . . 15
1.12. The curl . . . 16
1.13. Successive application of ∇ . . . 19
1.14. Gauss' law . . . 21
1.15. Stokes' theorem . . . 22
1.16. Orthogonal curvilinear coordinates . . . 23
1.17. Point sources . . . 26
1.18. Helmholtz's theorem . . . 29
1.19. Vector summary . . . 36

Chapter 2. Electrostatics . . . 39

2.1. Introduction to electrostatics . . . 39
2.2. Coulomb's law . . . 40
2.3. Units . . . 41
2.4. Electric field . . . 42
2.5. Gauss' flux theorem . . . 47
2.6. Electrostatic potential . . . 50
2.7. Conducting boundaries . . . 54
2.8. Poisson's equation . . . 59
2.9. Uniqueness theorem . . . 61
2.10. Method of images . . . 63
2.11. Dipoles and volume discontinuities . . . 72
2.12. Field behavior at a charged surface . . . 76

Chapter 3. Electrostatic Fields in Material Bodies, Energy, and Forces . . . 79

3.1. Polarizability . . . 79
3.2. Electric flux density **D** . . . 85

3.3. Boundary conditions . . . 88
3.4. Dielectric strength . . . 93
3.5. Capacitance . . . 94
3.6. Electrostatic energy . . . 101
3.7. Electrostatic forces . . . 106
3.8. Electrostatic forces (Maxwell's stress tensor) . . . 112
3.9. Molecular fields . . . 115

Chapter 4. Solution of Boundary-value Problems . . . 119

4.1. Rectangular coordinates . . . 120
4.2. Cylindrical coordinates . . . 130
4.3. Spherical coordinates . . . 138
4.4. Solution of two-dimensional problems by conformal mapping . . . 146
4.5. The Schwarz-Christoffel transformation . . . 158

Chapter 5. Stationary Currents . . . 164

5.1. Ohm's law . . . 164
5.2. Nonconservative fields—emf . . . 167
5.3. Conservation of charge . . . 169
5.4. Relaxation time . . . 170
5.5. Resistance of arbitrary shaped conductors . . . 171
5.6. Boundary conditions and refraction of current flow lines . . . 177
5.7. Duality between **J** and **D** . . . 183
5.8. Joule's law . . . 186
5.9. Convection current . . . 188
5.10. Flux plotting . . . 189
5.11. Electrolytic tank . . . 192

Chapter 6. Static Magnetic Field in Vacuum . . . 198

6.1. Ampere's law of force . . . 199
6.2. The magnetic field **B** . . . 201
6.3. The vector potential . . . 204
6.4. The magnetic dipole . . . 211
6.5. Magnetic flux and divergence of **B** . . . 216
6.6. Ampere's circuital law . . . 218
6.7. Differential equation for vector potential . . . 224

Chapter 7. Magnetic Field in Material Bodies . . . 226

7.1. Magnetic dipole polarization per unit volume . . . 227
7.2. Equivalent volume and surface polarization currents . . . 228
7.3. The magnetic field intensity **H** . . . 231
7.4. *B-H* curve . . . 233
7.5. Boundary condition for **B** and **H** . . . 235
7.6. Scalar potential for **H** . . . 243
7.7. The magnetic circuit . . . 252
7.8. Physical properties of magnetic materials . . . 257

Chapter 8. Quasi-stationary Magnetic Field . . . 265

8.1. Faraday's law . . . 266
8.2. Induced electric field due to motion . . . 268

8.3. Inductance . . . 273
8.4. Energy of a system of current loops . . . 280
8.5. Energy as a field integral . . . 283
8.6. Forces as derivatives of coefficients of inductance . . . 286
8.7. Lifting force of magnets . . . 290
8.8. Magnetic stress tensor . . . 294
8.9. Hysteresis loss . . . 297

Chapter 9. Time-dependent Fields . . . 299

9.1. Modification of static field equations under time-varying conditions . . . 299
9.2. Maxwell's equations . . . 301
9.3. Source-free wave equation . . . 303
9.4. Power flow and energy . . . 306
9.5. Sinusoidal time-varying fields . . . 310
9.6. Helmholtz's equation . . . 312
9.7. Complex Poynting vector . . . 315
9.8. Boundary conditions . . . 317
9.9. Scalar and vector potentials . . . 321
9.10. Relation between field theory and circuit theory . . . 326

Chapter 10. Plane Waves, Waveguides, and Resonators . . . 342

10.1. Classification of wave solutions . . . 342
10.2. Plane waves . . . 348
10.3. Reflection from a conducting plane . . . 356
10.4. Transmission lines . . . 361
10.5. Transmission-line parameters . . . 363
10.6. Rectangular waveguide . . . 370
10.7. Circular waveguides . . . 380
10.8. Electromagnetic cavities . . . 385

Chapter 11. Radiation and Antennas . . . 393

11.1. Radiation from a linear-current element . . . 393
11.2. The half-wave antenna . . . 398
11.3. Introduction to arrays . . . 401
11.4. Linear arrays . . . 407
11.5. Two-dimensional arrays . . . 409
11.6. Continuous distributions . . . 412
11.7. Network formulation for transmitting-receiving system . . . 413
11.8. Antenna equivalent circuits . . . 417
11.9. Receiving antennas . . . 419

Chapter 12. Interaction of Charged Particles with Fields . . . 425

12.1. Charged particles in static fields—fundamental equations . . . 425
12.2. Motion of charged particles in an electrostatic field . . . 427
12.3. Motion of charged particles in a magnetostatic field . . . 431
12.4. Motion of charged particles in a static electric and magnetic field . . . 434
12.5. Fields in the presence of space charge . . . 437
12.6. Llewellyn-Peterson equations . . . 445
12.7. Space-charge waves . . . 449
12.8. The klystron . . . 456

12.9. Traveling-wave-tube amplifier 463
12.10. Electromagnetic-wave propagation in gyrotropic media 473
12.11. Magnetohydrodynamics 486

Table of Numerical Constants 493

Problems 495

General Bibliography 543

Index 545

CHAPTER 1

VECTOR ANALYSIS

This chapter develops the mathematics of vector analysis that will be needed in the succeeding chapters of the book. Based on the work of this chapter, it is possible to considerably simplify the formulation of the physical laws of electromagnetic theory. Furthermore, manipulation of the equations with the goal of solving physical problems is also greatly facilitated. One of the purposes of this chapter is to lay the necessary groundwork in the use of vector algebra and vector calculus.

Another purpose is also sought in this chapter. For along with the mathematical simplifications in the use of vector analysis, there go certain concomitant philosophical concepts. This chapter, consequently, contains a discussion of fields, the flux representation of vector fields, and some general remarks concerning sources of fields. The definition of the divergence and curl of a field can then be understood as measures of the strength of the sources and vortices of a field. When in the succeeding chapters the specific nature of the electric and magnetic fields is considered, the student will have an appropriate framework into which to fit them.

Although much effort has been directed to the development of a physical basis for the mathematical definitions of this chapter, they may still seem somewhat artificial. The full justification of their utility, and a deepening of their meaning, will become apparent when the physical laws of electromagnetics are considered in the later chapters.

1.1. Scalars and Vectors

In this book we deal with physical quantities that can be measured. The measurement tells how many times a given unit is contained in the quantity measured. The simplest physical quantities are those that are completely specified by a singe number, along with a known unit. Such quantities are called scalars. Volume, density, and mass are examples of scalars.

Another group of physical quantities are called vectors. We may see how the vector arises if we consider as an example a linear displacement of a point from a given initial position. It is true that the final position of the point could be described in terms of three scalars, e.g., the cartesian coordinates of the final point with respect to axes chosen through the initial point. But this obscures the fact that the concept of displacement is a single idea and does not depend on a coordinate system. Consequently, we introduce displacements as quantities of a new type and establish a system of rules for their use. All physical quantities which can be represented by such displacements and which obey their respective rules are called vectors.

The vector can be represented graphically by a straight line drawn in the direction of the vector, the sense being indicated by an arrowhead and its length made proportional to the magnitude of the vector. Examples of vector quantities include displacement, acceleration, and force. In this book all vector quantities are designated by boldface type, while their magnitudes only are indicated through the use of italics.

1.2. Addition and Subtraction of Vectors

From the definition of a vector, just given, it is possible to deduce the rule for addition of vectors. Thus, consider two vectors **A** and **B** as illustrated in Fig. 1.1. Vector **A** represents the displacement of a movable point from point 1 to point 2. Vector **B** represents a displacement from point 2 to point 3. The result is equivalent to a total displacement of a point from 1 to 3. This linear displacement from 1 to 3 is called the resultant, or geometric sum of the two displacements (1,2) and (2,3). It is represented by the vector **C**, which we call the sum of the vectors **A** and **B**:

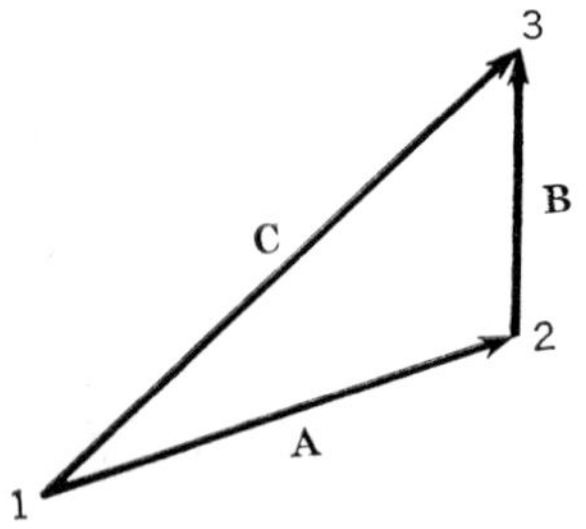

FIG. 1.1. Vector addition.

$$\mathbf{C} = \mathbf{A} + \mathbf{B} \tag{1.1}$$

Note that vectors **A** and **B** are of the same dimensions and type and that the geometric construction of Fig. 1.1 requires that the origin of one be placed at the head of the other. We may inquire whether the order of addition is of significance.

Consider that the displacement **B** is made first and then the displacement **A**. In this case the movable point describes the path 143 as in Fig. 1.2 and consequently produces the same resultant. Vector addition thus obeys the commutative law; i.e., the geometric sum of two vectors is independent of the order of addition so that

$$\mathbf{A} + \mathbf{B} = \mathbf{B} + \mathbf{A} \tag{1.2}$$

The path 143 and 123 together make up a parallelogram whose diagonal is the resultant of the displacement represented by the two adjacent sides.

Accordingly, the law of vector addition is often referred to as the parallelogram law. This law of addition is characteristic of the quantities called vectors. Thus it is proved in statics that forces acting on a rigid body follow the parallelogram law of addition; consequently, such forces are vectors.

It is easy to show that vectors satisfy the associative law of addition, which states that the order of adding any number of vectors is immaterial. Thus the sum of three vectors **A**, **B**, **C** can be expressed as

$$(\mathbf{A} + \mathbf{B}) + \mathbf{C} = \mathbf{A} + (\mathbf{B} + \mathbf{C}) \tag{1.3}$$

The proof can be established by considering Fig. 1.3, in which the same resultant (1,4) is arrived at by carrying out the summation indicated by

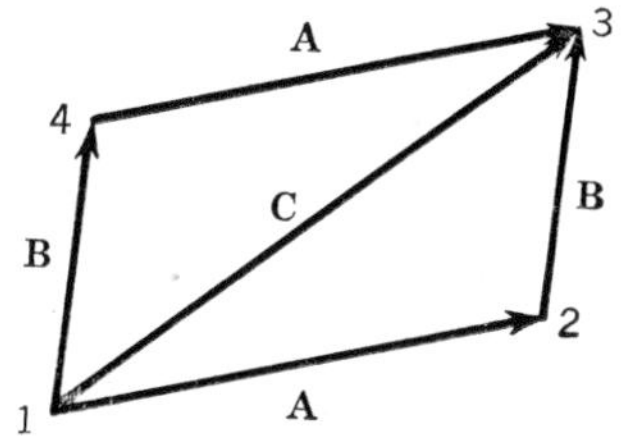

FIG. 1.2. Illustration of parallelogram of vector addition.

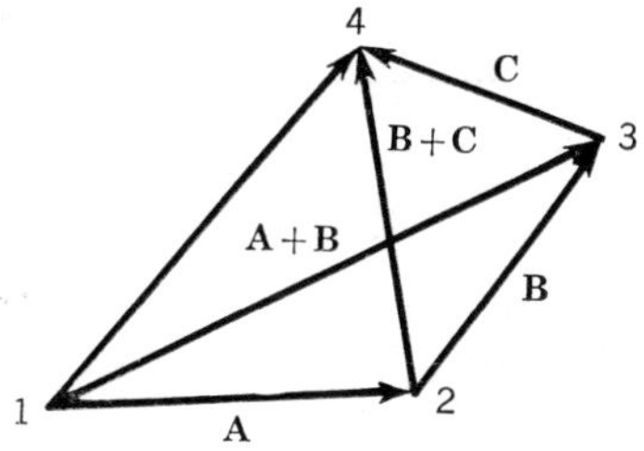

FIG. 1.3. Illustration of associative law of vector addition.

either the left- or right-hand side of Eq. (1.3). The former path is (1,3,4); the latter is (1,2,4).

To obtain the difference of two vectors **A** − **B**, it becomes necessary to define the negative of a vector. This is taken to mean a vector of the same magnitude but of opposite direction to the original vector. Thus

$$\mathbf{A} - \mathbf{B} = \mathbf{A} + (-\mathbf{B}) \tag{1.4}$$

We may therefore define vector subtraction as follows: A vector **B** is subtracted from a vector **A** by adding to **A** a vector of the same magnitude as **B** but in the opposite direction. In the parallelogram of Fig. 1.2 a diagonal from 4 to 2 would represent the geometric difference **A** − **B**.

1.3. Unit Vectors and Vector Components

The result of multiplying a vector **A** by a positive scalar m is to produce a new vector in the same direction as **A** but whose magnitude is that of **A** times m. The resultant **P** is thus related to **A** and m by the following:

$$\mathbf{P} = m\mathbf{A} \tag{1.5}$$

$$|\mathbf{P}| = m|\mathbf{A}| \quad \text{or} \quad P = mA \tag{1.6}$$

A unit vector is one whose magnitude is unity. It is often convenient to express a vector as the product of its magnitude and a unit vector having the same direction. Thus if **a** is a unit vector having the direction of **A**, then $\mathbf{A} = A\mathbf{a}$. The result expressed by (1.5) follows immediately, since $m\mathbf{A} = mA\mathbf{a}$. The three unit vectors $\mathbf{a}_x$, $\mathbf{a}_y$, $\mathbf{a}_z$ parallel to the right-hand rectangular axes x, y, z, respectively, are of particular importance.

The components of a vector are any vectors whose sum is the given vector. We shall often find it convenient to choose as components the three rectangular components of cartesian coordinates. Thus, if A_x, A_y, A_z are the magnitude of the projections of vector **A** on the x, y, z axes, its rectangular components are $\mathbf{a}_xA_x$, $\mathbf{a}_yA_y$, $\mathbf{a}_zA_z$. The vector **A** is completely determined by its components since the magnitude is given by

$$A^2 = A_x^2 + A_y^2 + A_z^2 \qquad (1.7)$$

and the direction cosines l, m, n are given by

$$l = \frac{A_x}{A} \qquad m = \frac{A_y}{A} \qquad n = \frac{A_z}{A} \qquad (1.8)$$

For brevity, we shall usually designate A_x, A_y, A_z, without the associated unit vectors, as the components of **A**.

Equal vectors have the same magnitude and direction; consequently, their respective rectangular components are equal. Therefore a vector equation can always be reduced, in general, to three scalar equations. For example, $\mathbf{A} + \mathbf{B} = \mathbf{C}$ can be expressed as

$$\mathbf{a}_x(A_x + B_x) + \mathbf{a}_y(A_y + B_y) + \mathbf{a}_z(A_z + B_z) = \mathbf{a}_xC_x + \mathbf{a}_yC_y + \mathbf{a}_zC_z \qquad (1.9)$$

i.e., addition is commutative and associative. Since the vector represented by the left-hand side of (1.9) equals that of the right-hand side, we are led to the result

$$A_x + B_x = C_x \qquad A_y + B_y = C_y \qquad A_z + B_z = C_z \qquad (1.10)$$

1.4. Vector Representation of Surfaces

Figure 1.4 illustrates a plane surface of arbitrary shape. We may represent this surface by a vector **S** whose length corresponds to the magnitude of the surface area and whose direction is specified by the normal to the surface. To avoid ambiguity, however, some convention must be adopted which establishes the positive sense of the normal.

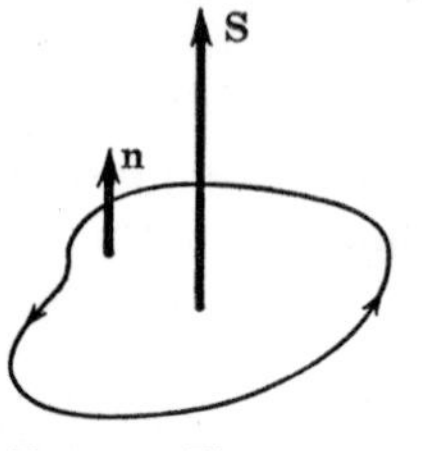

FIG. 1.4. Vector surface area. **n** is a unit surface normal.

When the surface forms part of a closed surface, the positive normal is usually taken as directed outward. For an open surface the positive normal can be associated with the positive sense of describing the periphery. This relationship is defined by taking the positive normal in the direction that a right-hand screw would advance when turned so as to describe the positive periphery. This definition actually arises out of a mathematical description of certain physical phenomena which will be discussed in later chapters. One can choose either positive periphery or positive normal arbitrarily.

If the surface is not plane, it is subdivided into elements which are sufficiently small so that they may be considered plane. The vector representing the surface is then found by vector addition of these components. This means that an infinite number of surfaces correspond to a given surface vector. The unit surface normal is designated by **n**.

1.5. The Vector Product of Two Vectors

Certain rules have been set up governing multiplication of vectors. The vector or cross product $\mathbf{A} \times \mathbf{B}$ of two vectors **A** and **B** is, by definition,

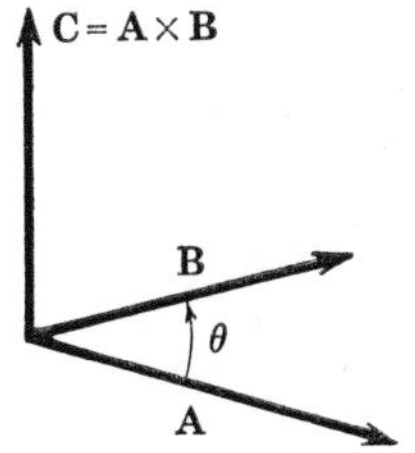

FIG. 1.5. Vector cross product.

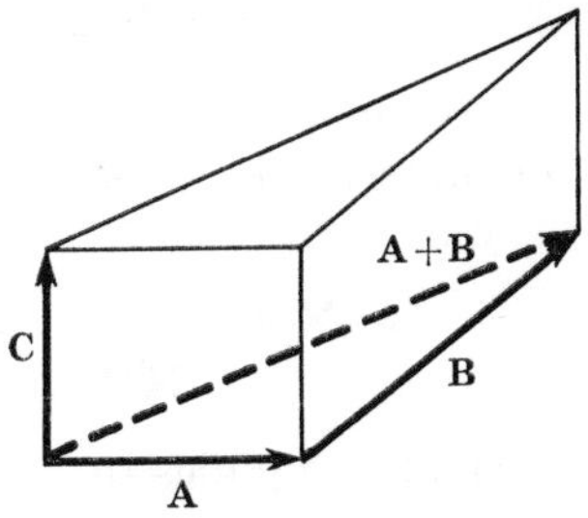

FIG. 1.6. Illustration for the distributive law of vector multiplication.

a vector of magnitude $AB \sin \theta$ in the direction of the normal to the plane determined by **A** and **B**. Its sense is that of advance of a right-hand screw rotated from the first vector to the second through the angle θ between them, as in Fig. 1.5. Since the direction reverses if the order of multiplication is interchanged, the commutative law of multiplication does not hold. Actually, we have

$$\mathbf{A} \times \mathbf{B} = -\mathbf{B} \times \mathbf{A} \tag{1.11}$$

This definition of vector product was chosen because it corresponds to a class of physically related quantities. Geometrically, the magnitude $|\mathbf{A} \times \mathbf{B}|$ is the area of a parallelogram formed by **A** and **B** as the sides. If we think of the periphery of the parallelogram as described from the origin to head of **A** followed by the origin to head of **B**, then, in accordance with the definitions in the last section, $\mathbf{A} \times \mathbf{B}$ represents the vector area of the parallelogram.

The preceding geometric interpretation is the basis for a proof that vector multiplication follows the distributive law. Thus, consider the prism described in Fig. 1.6, whose sides are **A**, **B**, **A** + **B**, and **C**. Since the total surface is closed, the vector representing the total surface of the prism is zero (see Prob. 1.6). Consequently, taking the positive normal as directed outward, the sum of the component surface areas may be set equal to zero, giving

$$\tfrac{1}{2}(\mathbf{A} \times \mathbf{B}) + \tfrac{1}{2}(\mathbf{B} \times \mathbf{A}) + \mathbf{A} \times \mathbf{C} + \mathbf{B} \times \mathbf{C} + \mathbf{C} \times (\mathbf{A} + \mathbf{B}) = 0 \tag{1.12}$$

from which we obtain

$$\mathbf{C} \times (\mathbf{A} + \mathbf{B}) = \mathbf{C} \times \mathbf{A} + \mathbf{C} \times \mathbf{B} \tag{1.13}$$

Equation (1.13) expresses the distributive law of multiplication.

The vector product of two vectors can be expressed in terms of the rectangular components of each vector. Since the distributive law holds,

$$\begin{aligned} \mathbf{A} \times \mathbf{B} &= (\mathbf{a}_x A_x + \mathbf{a}_y A_y + \mathbf{a}_z A_z) \times (\mathbf{a}_x B_x + \mathbf{a}_y B_y + \mathbf{a}_z B_z) \\ &= \mathbf{a}_x \times \mathbf{a}_x A_x B_x + \mathbf{a}_x \times \mathbf{a}_y A_x B_y + \mathbf{a}_x \times \mathbf{a}_z A_x B_z \\ &\quad + \mathbf{a}_y \times \mathbf{a}_x A_y B_x + \mathbf{a}_y \times \mathbf{a}_y A_y B_y + \mathbf{a}_y \times \mathbf{a}_z A_y B_z \\ &\quad + \mathbf{a}_z \times \mathbf{a}_x A_z B_x + \mathbf{a}_z \times \mathbf{a}_y A_z B_y + \mathbf{a}_z \times \mathbf{a}_z A_z B_z \end{aligned} \tag{1.14}$$

The sine of the angle between two vectors is zero when they are in the same or opposite directions and is ± 1 when they are orthogonal. It is thus easy to verify that

$$\begin{gathered} \mathbf{a}_x \times \mathbf{a}_y = \mathbf{a}_z \qquad \mathbf{a}_y \times \mathbf{a}_z = \mathbf{a}_x \qquad \mathbf{a}_z \times \mathbf{a}_x = \mathbf{a}_y \\ \mathbf{a}_x \times \mathbf{a}_x = \mathbf{a}_y \times \mathbf{a}_y = \mathbf{a}_z \times \mathbf{a}_z = 0 \end{gathered} \tag{1.15}$$

so that (1.14) simplifies to

$$\mathbf{A} \times \mathbf{B} = \mathbf{a}_x(A_y B_z - A_z B_y) + \mathbf{a}_y(A_z B_x - A_x B_z) + \mathbf{a}_z(A_x B_y - A_y B_x) \tag{1.16}$$

A convenient way of remembering the formula given by (1.16) is to note that it is obtained from the formal expansion of the following determinant:

$$\mathbf{A} \times \mathbf{B} = \begin{vmatrix} \mathbf{a}_x & \mathbf{a}_y & \mathbf{a}_z \\ A_x & A_y & A_z \\ B_x & B_y & B_z \end{vmatrix} \tag{1.17}$$

Once one term of the expansion is found, the remaining can be obtained by cyclical permutation; that is, replace x by y, y by z, and z by x. For example, the first term in (1.17) is $\mathbf{a}_x(A_y B_z - A_z B_y)$, from which the second term is found to be $\mathbf{a}_y(A_z B_x - A_x B_z)$ by replacing x, y, z by y, z, x, respectively.

1.6. The Scalar Product of Two Vectors

As mentioned, vector multiplication is useful in mathematically describing the relationship between vectors that arise out of a class of physical problems. In handling another class of physically related quantities, it will be desirable to define a scalar product of two vectors.

The scalar, or dot, product of two vectors **A** and **B**, written $\mathbf{A} \cdot \mathbf{B}$, is a scalar and has the magnitude $AB \cos \theta$, where θ is the angle between the two vectors, as illustrated in Fig. 1.5. From the definition it is clear that

$$\mathbf{A} \cdot \mathbf{B} = \mathbf{B} \cdot \mathbf{A} \tag{1.18}$$

and consequently the commutative law of multiplication applies.

The scalar product of **A** and **B** may be interpreted as the algebraic product of the magnitude of one of them (say **A**) by the component of the other (**B**) in the direction of the first. Referring to Fig. 1.7, we apply this concept to establish a basic property of the scalar product. We have

$$\mathbf{A} \cdot \mathbf{D} + \mathbf{B} \cdot \mathbf{D} = OP|\mathbf{D}| + PQ|\mathbf{D}| = OQ|\mathbf{D}| \tag{1.19}$$

But $(\mathbf{A} + \mathbf{B}) \cdot \mathbf{D} = OQ|\mathbf{D}|$; consequently,

$$(\mathbf{A} + \mathbf{B}) \cdot \mathbf{D} = \mathbf{A} \cdot \mathbf{D} + \mathbf{B} \cdot \mathbf{D} \tag{1.20}$$

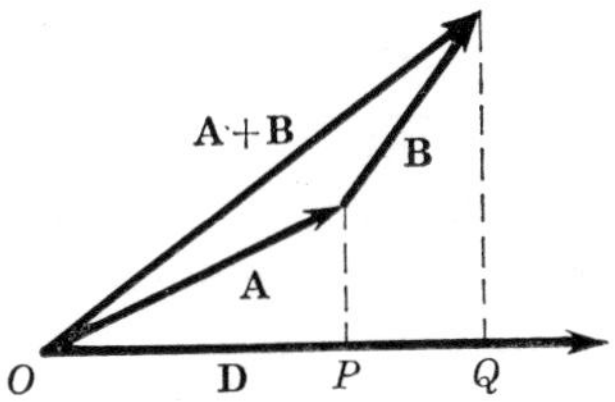

FIG. 1.7. Illustration of distributive law of scalar multiplication.

In other words, the distributive law applies to scalar multiplication.

The cosine of the angle between the directions of two vectors is +1 when the directions are the same, −1 when they are opposite, and 0 when they are at right angles. Consequently, the following vector relationships involving the unit vectors $\mathbf{a}_x$, $\mathbf{a}_y$, $\mathbf{a}_z$ must be true:

$$\begin{aligned} \mathbf{a}_x \cdot \mathbf{a}_x &= \mathbf{a}_y \cdot \mathbf{a}_y = \mathbf{a}_z \cdot \mathbf{a}_z = 1 \\ \mathbf{a}_x \cdot \mathbf{a}_y &= \mathbf{a}_y \cdot \mathbf{a}_z = \mathbf{a}_z \cdot \mathbf{a}_x = 0 \end{aligned} \tag{1.21}$$

Since the commutative and distributive laws hold, it follows that scalar multiplication of vectors is carried out by the rules of ordinary algebra. In particular, we may expand $\mathbf{A} \cdot \mathbf{B}$ in terms of rectangular components.

$$\begin{aligned} \mathbf{A} \cdot \mathbf{B} &= (\mathbf{a}_x A_x + \mathbf{a}_y A_y + \mathbf{a}_z A_z) \cdot (\mathbf{a}_x B_x + \mathbf{a}_y B_y + \mathbf{a}_z B_z) \\ &= \mathbf{a}_x \cdot \mathbf{a}_x A_x B_x + \mathbf{a}_x \cdot \mathbf{a}_y A_x B_y + \mathbf{a}_x \cdot \mathbf{a}_z A_x B_z + \mathbf{a}_y \cdot \mathbf{a}_x A_y B_x \\ &\quad + \mathbf{a}_y \cdot \mathbf{a}_y A_y B_y + \mathbf{a}_y \cdot \mathbf{a}_z A_y B_z + \mathbf{a}_z \cdot \mathbf{a}_x A_z B_x + \mathbf{a}_z \cdot \mathbf{a}_y A_z B_y \\ &\quad + \mathbf{a}_z \cdot \mathbf{a}_z A_z B_z \end{aligned} \tag{1.22}$$

With the aid of (1.21) this simplifies to

$$\mathbf{A} \cdot \mathbf{B} = A_x B_x + A_y B_y + A_z B_z \tag{1.23}$$

A special case is the dot product of **A** with itself, which gives

$$\mathbf{A} \cdot \mathbf{A} = A^2 = A_x A_x + A_y A_y + A_z A_z \tag{1.24}$$

1.7. Product of Three Vectors

Three vectors can be multiplied in three different ways. As an example of one such possibility we consider $\mathbf{A}(\mathbf{B} \cdot \mathbf{C})$. This is nothing more than the product of a scalar $(\mathbf{B} \cdot \mathbf{C})$ times a vector $\mathbf{A}$, and the resultant may be evaluated by the rules given in Sec. 1.3.

A second arrangement is known as the triple scalar product. An example is $\mathbf{A} \cdot \mathbf{B} \times \mathbf{C}$. The vector product is necessarily formed before taking the scalar product, and the resultant is a scalar. This product has a simple geometrical interpretation that is evident in Fig. 1.8. For $\mathbf{B} \times \mathbf{C}$ is clearly the vector area of the base, whereupon $\mathbf{A} \cdot \mathbf{B} \times \mathbf{C}$ is the volume of the parallelepiped. But this volume is calculated equally well by the expressions $\mathbf{B} \cdot \mathbf{C} \times \mathbf{A}$ and $\mathbf{C} \cdot \mathbf{A} \times \mathbf{B}$. Consequently,

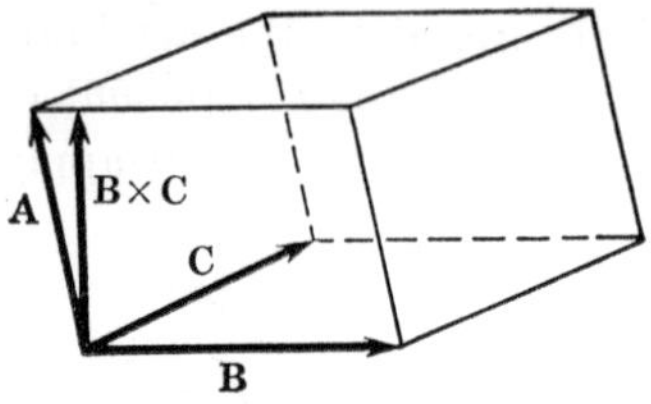

FIG. 1.8. Geometric interpretation of triple scalar product.

$$\mathbf{A} \cdot \mathbf{B} \times \mathbf{C} = \mathbf{A} \times \mathbf{B} \cdot \mathbf{C} = \mathbf{C} \cdot \mathbf{A} \times \mathbf{B} = \mathbf{C} \times \mathbf{A} \cdot \mathbf{B}$$

This result is often expressed by the statement that in the triple scalar product the dot and cross may be interchanged and/or the order of the vectors altered by a cyclic rearrangement.

The third arrangement involving the product of three vectors is the triple vector product, exemplified by $\mathbf{A} \times (\mathbf{B} \times \mathbf{C})$. The parentheses indicate which product is taken first, since the result depends on the order of forming the product; that is, the associative law does not hold and $\mathbf{A} \times (\mathbf{B} \times \mathbf{C})$ is not the same as $(\mathbf{A} \times \mathbf{B}) \times \mathbf{C}$. By inspection we note that the resultant of $\mathbf{A} \times (\mathbf{B} \times \mathbf{C})$ lies in the plane of $\mathbf{B}$ and $\mathbf{C}$ and is orthogonal to $\mathbf{A}$. If each vector is expressed in terms of its rectangular components and the indicated operations carried out, it can be verified that

$$\mathbf{A} \times (\mathbf{B} \times \mathbf{C}) = \mathbf{B}(\mathbf{A} \cdot \mathbf{C}) - \mathbf{C}(\mathbf{A} \cdot \mathbf{B}) \tag{1.25}$$

The details are left as an exercise for the student (see Prob. 1.7).

1.8. Scalar and Vector Fields

If a particle is in motion, then at any instant its velocity can be designated by a vector; i.e., the velocity possesses the properties characteristic of a vector. But if we examine the state of motion of a fluid filling space, then the velocities of the different particles will not be the same, in general. In this case every point has its own velocity vector and the moving continuous fluid can be represented by what is called a vector field.

In mathematical physics the field of a physical quantity refers to nothing more than the dependence of that quantity on position in a region of space. It is assumed that the variation is, ordinarily, a continuous one. The field may be a vector field, as illustrated above, or a scalar field.

The scalar field is simply a scalar function of position in space; that is, there is associated with each point in space a definite scalar magnitude. For example, the barometric pressure at each point on the earth's surface constitutes a scalar field. The field is a scalar field because pressure is a scalar quantity.

Since a field is a function of x, y, z, say, it can also be expressed as a function of a new set of coordinates x', y', z' by an appropriate transformation. Ordinarily, such a transformation brings in the direction cosines of the new axes measured relative to the old axes. But the presence of direction cosines would make any physical law involving the scalar field depend on the choice of axes, which is contrary to the character of the laws of nature. Consequently, the only scalar functions of the coordinates which can enter physical laws are those in which the direction cosines do not appear in an arbitrary transformation of axes. Such a function is called a proper scalar function of the coordinates.

To each point on the surface of the earth a temperature can be measured and a temperature field established. It is convenient to organize this information graphically (or conceptually) by connecting points that are at the same temperature, choosing certain specific values of temperature. In this way the isotherms of a weather map give a rough idea of the temperature field. Scalar fields are sometimes called potential fields, and lines or surfaces over which the field has a constant magnitude are referred to as equipotentials. For example, points on the same contour line of a topological map correspond to points on the earth with the same potential energy, that is, equipotentials.

A vector function of position associates a definite vector with each point of a special region, the aggregate of these vectors constituting a vector field. A simple example is the position vector $\mathbf{r} = \mathbf{a}_x x + \mathbf{a}_y y + \mathbf{a}_z z$, which is a function of position of the point (x,y,z) relative to the origin of the axes chosen. There is thus associated with every point a vector having the magnitude and direction of the line drawn from the origin to the point in question. The field strength at a few points of the field is illustrated in Fig. 1.9.

A vector field may be described in terms of its components at every point in space. In this way one can form three scalar fields from the given vector field. If the component fields are proper scalar functions, then the vector function is a proper vector function.

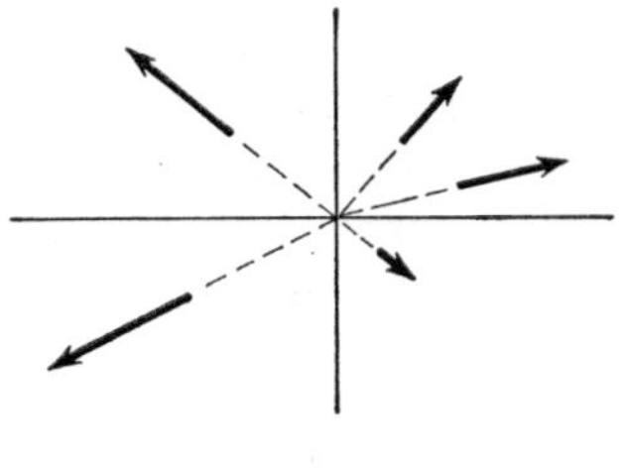

FIG. 1.9. The vector field $\mathbf{r}$.

Among the vector fields that arise in physics are those involving the flow of some quantity. In fluid flow, if $\mathbf{v}$ is the fluid velocity and ρ is the density at (x,y,z), then $\rho\mathbf{v}$ is the vector representing the flow of mass per unit area at each point (x,y,z). Similarly, the electric current density $\mathbf{J}(x,y,z)$ represents the flow of charge per unit area in a current flow field.

As we have already said, a vector field is defined by specifying a vector at each point in space, for example, $\mathbf{J}(x,y,z)$. In most cases of interest this vector is a continuous function of (x,y,z), except possibly at isolated

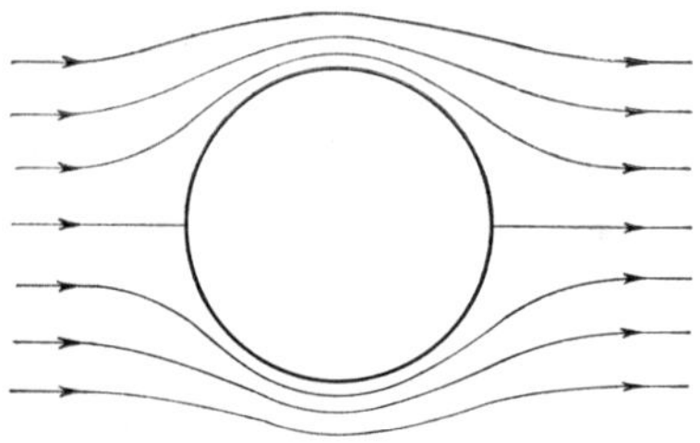

FIG. 1.10. Illustration of flow lines around a sphere.

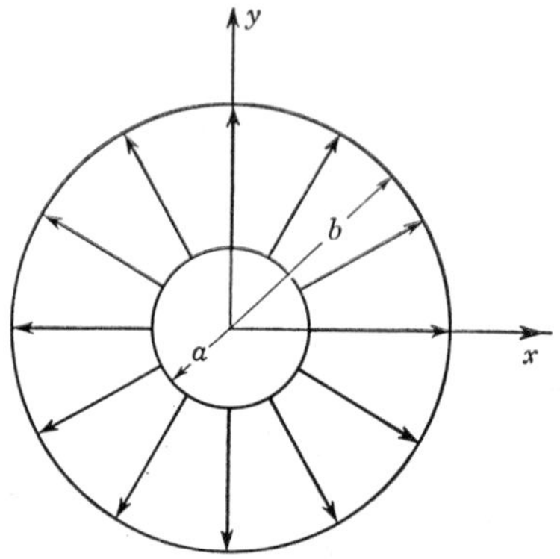

FIG. 1.11. Flow lines between two coaxial cylinders.

points or singularities or along isolated lines or singular lines. Where the vector field is continuous we can define lines of flow of the field, which are lines at every point tangent to the vector at that point. The differential equation for the line can be found by forming a proportionality between the components of displacement along the line dx, dy, dz and the corresponding components of the vector field at the same point:

$$\frac{dx}{J_x} = \frac{dy}{J_y} = \frac{dz}{J_z} \tag{1.26}$$

Just as equipotential surfaces proved convenient in visualizing a scalar field, flow lines are useful in "mapping" vector fields. Obviously, only a few of the infinite number of lines would be drawn in such a map. As a matter of fact, in a two-dimensional field we can represent the field intensity by taking equal quantity of flow between adjacent lines. In this way the density of lines transverse to the direction of flow is proportional to the magnitude of the flow vector. Figure 1.10 illustrates fluid flow through a figure of revolution; adjacent lines actually symbolize concentric tubes, each of which carries the same quantity of fluid.

As a simple example of the evaluation of flow lines, consider the radial flow of current between two coaxial cylinders as illustrated in Fig. 1.11. The current density is given by

$$\mathbf{J} = \frac{I}{2\pi}\frac{\mathbf{a}_r}{r} = \frac{I}{2\pi}\left(\frac{x\mathbf{a}_x}{r^2} + \frac{y\mathbf{a}_y}{r^2}\right)$$

and hence the differential equation for the flow lines is

$$\frac{dx}{J_x} = \frac{dy}{J_y} \qquad \text{or} \qquad \frac{dx}{x} = \frac{dy}{y}$$

Integration gives the result

$$\ln y = \ln x + \ln c = \ln cx$$

or

$$y = cx$$

where c is a constant of integration. Every different value of the constant c gives a new flow line. A complete flow map as illustrated in Fig. 1.11 is obtained by considering a range of values of c. Note that with the values selected, the relative density of the lines graphically corresponds to the relative current density. Figure 1.11 is therefore a particularly appropriate representation of the vector field **J**. A more complete discussion of flux plotting is given in Chap. 5.

The representation of a vector field by means of flow lines seems fairly obvious where the field represents the flow of some quantity. But the technique for construction of flow lines as given in the previous paragraph contains no restrictions on such fields. It may, indeed, be used to represent any vector field. We shall see that the representation of vector fields through the concept of flow or flux (which means the same thing) will prove very useful in our future work.

1.9. Gradient

Let us suppose that $\Phi(x,y,z)$ represents a scalar field and that Φ is a single-valued, continuous, and differentiable function of position. These properties will always be true of the physical fields that we shall encounter. An equipotential surface, then, has the equation

$$\Phi(x,y,z) = C \tag{1.27}$$

where C is a constant. By assigning to C a succession of values, a family of equivalued surfaces is obtained. No two such surfaces will intersect since we have taken Φ to be single-valued.

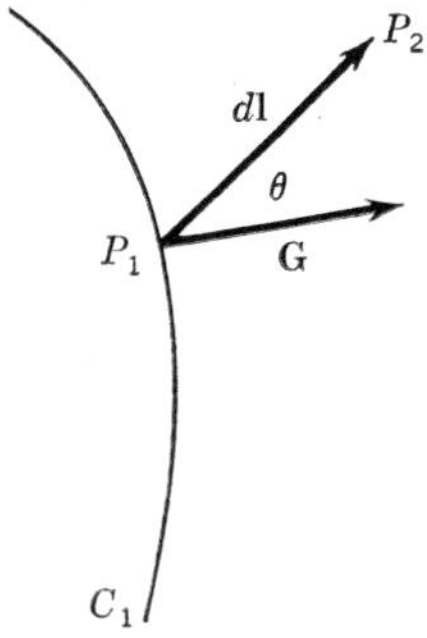

FIG. 1.12

Consider two closely spaced points P_1 and P_2, where P_1 lies on the equipotential surface C_1, and P_2 may or may not be on this surface, as in Fig. 1.12. Let the coordinates of P_1 be (x,y,z); then P_2 is located at $(x + dx,\ y + dy,\ z + dz)$. The displacement $d\mathbf{l}$ from P_1 to P_2 can be expressed in terms of its rectangular components:

$$d\mathbf{l} = \mathbf{a}_x\, dx + \mathbf{a}_y\, dy + \mathbf{a}_z\, dz \tag{1.28}$$

We desire to evaluate the difference in potential $d\Phi$ between P_1 and P_2. From the rules of calculus this can be written

$$d\Phi = \frac{\partial \Phi}{\partial x} dx + \frac{\partial \Phi}{\partial y} dy + \frac{\partial \Phi}{\partial z} dz \tag{1.29}$$

It will prove very useful to define the following vector $\mathbf{G}$:

$$\mathbf{G} = \mathbf{a}_x \frac{\partial \Phi}{\partial x} + \mathbf{a}_y \frac{\partial \Phi}{\partial y} + \mathbf{a}_z \frac{\partial \Phi}{\partial z} \tag{1.30}$$

for we may now combine (1.28) and (1.30) to express $d\Phi$ as

$$d\Phi = \mathbf{G} \cdot d\mathbf{l} \tag{1.31}$$

Let us deduce some of the properties of the vector $\mathbf{G}$, as defined above. Suppose that P_2 lies on surface C_1 (see Fig. 1.12). Since P_2 is a differential distance from P_1, $d\mathbf{l}$ will be tangent to the surface C_1 at P_1. Furthermore, $d\Phi$ will, of course, be zero, since Φ is constant on C_1. In order for (1.31) to be satisfied, it is clear that $\mathbf{G}$ is normal to the equipotential surface at P_1.

To determine the magnitude of $\mathbf{G}$, let P_2 be chosen in such a way that $d\mathbf{l}$ makes an angle θ with the normal to C, as illustrated in Fig. 1.12. Since $\mathbf{G}$ is in the direction of the normal, then

$$d\Phi = \mathbf{G} \cdot d\mathbf{l} = G \cos \theta \, dl \tag{1.32}$$

and consequently

$$\frac{d\Phi}{dl} = G \cos \theta \tag{1.33}$$

In words, the component of $\mathbf{G}$ in the direction $d\mathbf{l}$ is the rate of increase of Φ in that direction. The latter is also termed the directional derivative of Φ in the direction $d\mathbf{l}$. If $\theta = 0$, $d\mathbf{l}$ becomes an element normal to the equipotential surface, written dn, and the directional derivative is maximum and equal to the magnitude G; that is,

$$G = \frac{\partial \Phi}{\partial n} \tag{1.34}$$

and is the maximum rate of increase of Φ. Because $\mathbf{G}$ thus coincides with the maximum space rate of increase of Φ, in both direction and magnitude, it is called the gradient of Φ.

The vector $\mathbf{G}$, for the reason just expressed, is often written grad Φ. Another more useful notation involves the ∇ (read del) operator, which is defined as

$$\nabla \equiv \mathbf{a}_x \frac{\partial}{\partial x} + \mathbf{a}_y \frac{\partial}{\partial y} + \mathbf{a}_z \frac{\partial}{\partial z} \tag{1.35}$$

We shall have much more to say about this operator, but for the present it is sufficient to note that $\mathbf{G} = \text{grad}\ \Phi = \nabla\Phi$.

The gradient operation may be viewed formally as converting a scalar field into a vector field. We shall discover later that certain vector fields can be specified as the gradient of some fictitious scalar field. The scalar field may then play a simplifying role in the ensuing mathematical analysis.

1.10. The Divergence

We have noted that it may be convenient to think of any vector field as representing the flow of a fluid. It is desirable to specify that the fluid is incompressible. But this restriction would ordinarily prevent us from representing any arbitrary vector field. For example, a consequence of incompressibility is the requirement that as much fluid enter as leave a fixed region. Only special vector fields could be represented by fluid motion of this type (solenoidal fields).

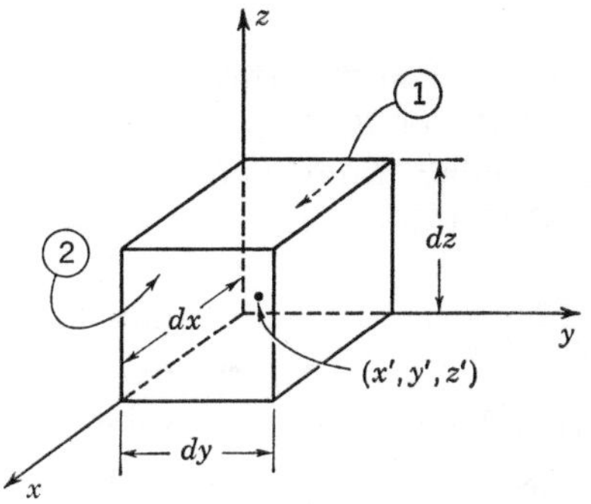

FIG. 1.13. Illustration for evaluation of divergence.

To avoid this limitation we also suppose that fluid at certain points may be created or destroyed. Points of the first sort will be called sources; points of the second kind, sinks, or negative sources. By setting up a suitable source system (including vortices to be discussed later), we may represent any arbitrary vector field by a steady motion of an incompressible fluid. In general, the sources will be found to be continuously distributed in space. We shall now describe how the source strength can be calculated.

To find the source magnitude within a volume V we may measure the volume of fluid leaving in a unit time. Since the fluid is incompressible and in steady motion, this volume of fluid represents the algebraic sum of all the sources contained within V.

Now this same evaluation can be performed by calculating the fluid flow through the bounding surface S of V. If we actually desire the source strength at a point (x',y',z'), we take V to be a differential volume element $dx\,dy\,dz$ enclosing this point as in Fig. 1.13. We assume the existence of a vector field $\mathbf{F}$. A Taylor expansion of F_x (the x component of $\mathbf{F}$) in the vicinity of the point (x',y',z') can be made. The leading terms are

$$F_x = F_x(x',y',z') + \frac{\partial F_x}{\partial x}(x - x') + \frac{\partial F_x}{\partial y}(y - y') + \frac{\partial F_x}{\partial z}(z - z') \quad (1.36)$$

where the partial derivatives are evaluated at (x',y',z'). If (x',y',z') is taken at the center of the volume element, then the outflow through surface 2, shown in Fig. 1.13, is

$$\int_2 F_x\, dS = dy\, dz \left[F_x(x',y',z') + \frac{1}{2}\frac{\partial F_x}{\partial x}\, dx \right] + \text{higher-order terms} \tag{1.37}$$

Notice that the vector is in the positive x direction and hence represents an outflow from the volume V. The outflow through surface 1 involves, in this case, an x component $-F_x$; consequently,

$$-\int_1 F_x\, dS = -dy\, dz \left[F_x(x',y',z') - \frac{1}{2}\frac{\partial F_x}{\partial x}\, dx \right] + \text{higher-order terms} \tag{1.38}$$

The total outflow through surfaces 1 and 2, neglecting higher-order terms, is $(\partial F_x/\partial x)\, dx\, dy\, dz$. In a similar way the contributions from the remaining two pairs of surfaces will be found to be $(\partial F_y/\partial y)\, dx\, dy\, dz$ and $(\partial F_z/\partial z)\, dx\, dy\, dz$. The net outflow is therefore

$$\oint_S \mathbf{F} \cdot d\mathbf{S} = \left(\frac{\partial F_x}{\partial x} + \frac{\partial F_y}{\partial y} + \frac{\partial F_z}{\partial z} \right) dx\, dy\, dz \tag{1.39}$$

Here $\mathbf{F} \cdot d\mathbf{S}$ is the rate of flow of material through the surface element $d\mathbf{S}$. For example, $\int_\sigma \mathbf{F} \cdot d\mathbf{S}$ evaluates the total flux through a surface σ. In (1.39) the notation $\oint$ signifies that the integral is over a closed surface. It therefore correctly expresses, in vector notation, the net flow of flux through the surface S which bounds V. Integrals of this type are known as surface integrals.

If we divide (1.39) by $dx\, dy\, dz$, the left-hand side becomes the net outflow per unit volume at (x',y',z'). This is a scalar quantity and is called the divergence of the vector $\mathbf{F}$ at the point (x',y',z') and is written

$$\operatorname{div} \mathbf{F} = \lim_{V \to 0} \frac{\oint \mathbf{F} \cdot d\mathbf{S}}{V} = \frac{\partial F_x}{\partial x} + \frac{\partial F_y}{\partial y} + \frac{\partial F_z}{\partial z} \tag{1.40}$$

The divergence may be thought of as equal to the rate of increase of lines of flow per unit volume. The div $\mathbf{F}$ is a scalar field which at each point is a measure of the strength of the source of the vector field $\mathbf{F}$ at that point.

Since the operator ∇ defined earlier has the formal properties of a vector, we may form its product with any vector according to the usual rules. Thus if we write the scalar product

$$\nabla \cdot \mathbf{F} = \frac{\partial F_x}{\partial x} + \frac{\partial F_y}{\partial y} + \frac{\partial F_z}{\partial z} \tag{1.41}$$

we discover a convenient representation for div $\mathbf{F}$ which leads to the correct expansion in cartesian coordinates. Accordingly, we adopt the notation

$$\nabla \cdot \mathbf{F} = \operatorname{div} \mathbf{F} \tag{1.42}$$

1.11. Line Integral

There are several properties of vector fields, exhibited through their representation by flux lines, that will concern us. One has to do with the "spreading out," or "outflow," of lines of flow in a given region. As we have already noted, this is a measure of the net strength of the sources within that region. The net outflow is most readily evaluated by means of a surface integral $\oint_S \mathbf{F} \cdot d\mathbf{S}$ over the bounding surface of the region. The magnitude of the resultant is also sometimes referred to as "the number of flux lines crossing S."

The outflow per unit volume at a point measures the divergence of the field at that point and consequently the strength of the source located there. An evaluation of the divergence of a vector field in rectangular coordinates is given in (1.40).

Now, instead of integrating the normal component of a vector over a surface, we can integrate its component along a line. Thus if $d\mathbf{l}$ is a vector element of length along an arbitrary path C, the integral $\int_1^2 \mathbf{F} \cdot d\mathbf{l}$ taken over the extent of the path C is called the line integral. The value of the integral, in general, depends on the path, although for certain vector functions it depends only on the end points. A simple problem that leads to a line integral is the calculation of the work done in moving a particle over the path C, where $\mathbf{F}$ represents the force applied at each point along the contour.

When the path is a closed one, the line integral is denoted by $\oint_C \mathbf{F} \cdot d\mathbf{l}$ and the quantity is called the "net circulation integral" for $\mathbf{F}$ around the chosen path. It is a measure of another vector field property, namely, the curling up of the field lines. For example, if the flux lines are closed loops, then the circulation integral over any such loop will obviously yield a nonzero result. For the fluid-flow analog, $\oint_C \mathbf{F} \cdot d\mathbf{l}$ measures the circulation of fluid around the path C; hence the expression "circulation integral."

For a special class of vector fields, $\mathbf{F}$ can be derived from the gradient of a scalar Φ. The line integral over a path C joining points 1 and 2 can then be written

$$\int_1^2 \mathbf{F} \cdot d\mathbf{l} = \int_1^2 \nabla\Phi \cdot d\mathbf{l} = \int_1^2 \frac{d\Phi}{dl}\, dl \tag{1.43}$$

The last expression arises out of the definition of the directional derivative. Consequently,

$$\int_1^2 \mathbf{F} \cdot d\mathbf{l} = \Phi_2 - \Phi_1 \tag{1.44}$$

where Φ_2 is the value of the scalar potential at point 2, while Φ_1 is the value at point 1. What is most important about this result is that it depends only on the location of the end points and is independent of the path C. Consequently, if the path is a closed one, the line integral vanishes; i.e., we may state that the line integral of a gradient over any closed path vanishes. The absence of circulation in such functions is expressed by designating them irrotational.

We have shown that any vector function that can be derived as the gradient of a scalar potential function is irrotational. It follows, conversely, that if the line integral of a vector function around any closed loop is zero, then the vector function is derivable as the gradient of a scalar function.

1.12. The Curl

We have already defined the divergence operator and noted how it evaluates a certain property of a vector field, namely, its rate of increase of lines of flow. Another very important property of a vector field, just noted, is the amount of "curl" in the flow lines. The latter is related to the magnitude of the circulation. By analogy with fluid flow, the region that "produces" the circulation is a vortex region; hence the "curling" of flow lines may be thought of as related to the "vorticity" of the field. We should like to specify a differential operator that measures the "vorticity." Following our experience with the divergence operator, one might suggest that the vorticity of a field at some point P be found by computing the net circulation integral around an element of area at P and dividing by the area of the element, taking the limit as the area approaches zero.

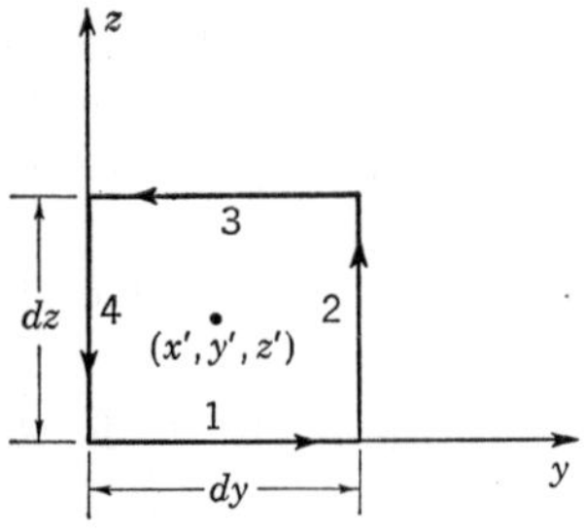

FIG. 1.14. Contour for evaluation of x component of the curl of a vector.

The above definition is not quite satisfactory because the orientation of the differential area is not specified, and the result may be expected to depend on that choice. The process of evaluating the vorticity is thus more complicated than that for finding the divergence. Let us, however, carry out the suggested operation for a differential area in the yz plane as illustrated in Fig. 1.14. If the path shown by the arrows (counterclockwise) is followed, then the net circulation integral is

$$\oint_C \mathbf{F} \cdot d\mathbf{l} = \int_1 F_y\,dy + \int_2 F_z\,dz - \int_3 F_y\,dy - \int_4 F_z\,dz \tag{1.45}$$

where the total line integral is expressed in terms of its component parts along paths 1, 2, 3, 4. If we let the point (x',y',z') be at the center of the path chosen and expand the field components in a Taylor series expansion about that point [as in (1.36)], then (1.45) can be written

$$\begin{aligned}\int_C \mathbf{F} \cdot d\mathbf{l} &= F_y(x',y',z')\,dy - \frac{\partial F_y}{\partial z}\frac{dz}{2}\,dy + F_z(x',y',z')\,dz + \frac{\partial F_z}{\partial y}\frac{dy}{2}\,dz \\ &\quad - F_y(x',y',z')\,dy - \frac{\partial F_y}{\partial z}\frac{dz}{2}\,dy - F_z(x',y',z')\,dz + \frac{\partial F_z}{\partial y}\frac{dy}{2}\,dz \\ &= \left(\frac{\partial F_z}{\partial y} - \frac{\partial F_y}{\partial z}\right) dy\,dz \end{aligned} \tag{1.46}$$

where the partial derivatives are evaluated at the point (x',y',z'). If we divide by the area $dy\,dz$, we obtain $\partial F_z/\partial y - \partial F_y/\partial z$, which is a measure of the circulation per unit area in the yz plane at the point (x',y',z'). This quantity is defined as the x component of the curl of **F**. The direction of the curl is thus that associated with the direction of advance of a right-hand screw when rotated in the direction in which the fluid circulates.

A similar derivation may be applied to elementary contours in the xy and xz planes to obtain the z and y components of the curl of **F**. We may obtain the other components of the curl of **F** by cyclic permutation of the variables x, y, z as well. The final result is

$$\text{curl}\,\mathbf{F} = \mathbf{a}_x\left(\frac{\partial F_z}{\partial y} - \frac{\partial F_y}{\partial z}\right) + \mathbf{a}_y\left(\frac{\partial F_x}{\partial z} - \frac{\partial F_z}{\partial x}\right) + \mathbf{a}_z\left(\frac{\partial F_y}{\partial x} - \frac{\partial F_x}{\partial y}\right) \tag{1.47}$$

The curl of **F** is a vector quantity. Its component along an arbitrary direction equals the circulation per unit area in the plane normal to that direction. This is clearly true for the direction of the coordinate axes, and will be true for any direction once it is shown that the curl **F** is a proper vector function. Proof of the latter is left to the student (see Prob. 1.10).

The vector curl **F** clearly is a measure of what we have called the vorticity of the field. It corresponds to the maximum circulation per unit area at a point, the maximum being obtained when the area $d\mathbf{S}$ is so oriented that curl **F** is normal to it. When **F** represents a fluid velocity, the direction of curl **F** at a point P is along the axis of rotation of the fluid close to P, using the right-hand rule.

If we take the vector product $\nabla \times \mathbf{F}$, treating ∇ as a vector according to the rules of vector multiplication, then

$$\nabla \times \mathbf{F} = \begin{vmatrix} \mathbf{a}_x & \mathbf{a}_y & \mathbf{a}_z \\ \dfrac{\partial}{\partial x} & \dfrac{\partial}{\partial y} & \dfrac{\partial}{\partial z} \\ F_x & F_y & F_z \end{vmatrix}$$

$$= \mathbf{a}_x\left(\frac{\partial F_z}{\partial y} - \frac{\partial F_y}{\partial z}\right) + \mathbf{a}_y\left(\frac{\partial F_x}{\partial z} - \frac{\partial F_z}{\partial x}\right) + \mathbf{a}_z\left(\frac{\partial F_y}{\partial x} - \frac{\partial F_x}{\partial y}\right) \quad (1.48)$$

But this result is nothing more than curl $\mathbf{F}$ in rectangular coordinates. Because of this, the usual notation for designating the curl of $\mathbf{F}$ is $\nabla \times \mathbf{F}$.

The importance of the curl (and divergence) property will be clear later, when we show that an arbitrary vector field is specified by giving its divergence and curl.

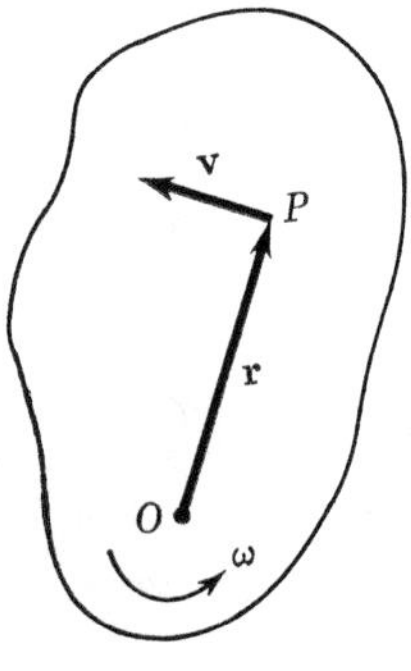

FIG. 1.15. A rotating rigid body.

Example 1.1. We wish to calculate the curl of the vector field $\mathbf{v}$, where $\mathbf{v}$ is found in the following way. Consider an arbitrary rigid body, as illustrated in Fig. 1.15, rotating about an axis through 0 with an angular velocity ω. Then, as we know, the linear velocity at any point P in the body is the product of the angular velocity by the moment arm of the point about the axis. The latter is the perpendicular distance from P to the axis. In vector notation the relation between the linear velocity $\mathbf{v}$, the moment arm vector $\mathbf{r}$, and the angular velocity $\boldsymbol{\omega}$ (the vector angular velocity $\boldsymbol{\omega}$ is defined to be directed along the axis of rotation with a sense established by the advance of a right-hand screw) is given by

$$\mathbf{v} = \boldsymbol{\omega} \times \mathbf{r} \quad (1.49)$$

The velocity $\mathbf{v}$ as given above defines a vector field whose curl we desire.

If (1.49) is reduced to rectangular coordinates, the curl operation may be carried out. Thus

$$(\nabla \times \mathbf{v})_x = \frac{\partial(\omega_x y - x\omega_y)}{\partial y} - \frac{\partial(\omega_z x - \omega_x z)}{\partial z} \quad (1.50)$$

Note that $\boldsymbol{\omega}$ is a fixed vector and consequently a constant in the differentiation. Thus

$$(\nabla \times \mathbf{v})_x = 2\omega_x \quad (1.51)$$

Similarly, $(\nabla \times \mathbf{v})_y = 2\omega_y$, $(\nabla \times \mathbf{v})_z = 2\omega_z$, and hence

$$\nabla \times \mathbf{v} = 2\boldsymbol{\omega} \quad (1.52)$$

This result verifies the concept of the curl operation as a measure of the vorticity. Intuitively, $\boldsymbol{\omega}$ measures the vorticity of the $\mathbf{v}$ field. Note that the result is independent of position within the rotating body.

1.13. Successive Application of ∇

It is possible to form scalar and vector products in which the operator ∇ appears more than once. For example, since the gradient is a vector, it is possible to take the divergence of the gradient. If this is expanded in rectangular coordinates, there results

$$\nabla \cdot \nabla\Phi = \nabla \cdot \left(\mathbf{a}_x \frac{\partial \Phi}{\partial x} + \mathbf{a}_y \frac{\partial \Phi}{\partial y} + \mathbf{a}_z \frac{\partial \Phi}{\partial z}\right) = \frac{\partial^2 \Phi}{\partial x^2} + \frac{\partial^2 \Phi}{\partial y^2} + \frac{\partial^2 \Phi}{\partial z^2} \tag{1.53}$$

The same result is obtained if we think of $\nabla \cdot \nabla$ as a new operator ∇^2 with properties

$$\nabla \cdot \nabla = \nabla^2 = \frac{\partial^2}{\partial x^2} + \frac{\partial^2}{\partial y^2} + \frac{\partial^2}{\partial z^2} \tag{1.54}$$

The operator ∇^2 is called the Laplacian and is a scalar. It may also be applied to a vector, with the result

$$\nabla^2 \mathbf{F} = \frac{\partial^2 \mathbf{F}}{\partial x^2} + \frac{\partial^2 \mathbf{F}}{\partial y^2} + \frac{\partial^2 \mathbf{F}}{\partial z^2} \tag{1.55}$$

This result is interpreted as three scalar equations; e.g., the x component is

$$\nabla^2 F_x = \frac{\partial^2 F_x}{\partial x^2} + \frac{\partial^2 F_x}{\partial y^2} + \frac{\partial^2 F_x}{\partial z^2} \tag{1.56}$$

In addition to taking the divergence of the gradient, it is also possible to form the curl of the gradient. The reader should verify that

$$\nabla \times \nabla\Phi = \nabla \times \left(\mathbf{a}_x \frac{\partial \Phi}{\partial x} + \mathbf{a}_y \frac{\partial \Phi}{\partial y} + \mathbf{a}_z \frac{\partial \Phi}{\partial z}\right) = 0 \tag{1.57}$$

by expanding the curl of $\nabla\Phi$ by the determinant rule. The result is not unexpected, since it has already been noted that the gradient is irrotational. Consequently, its curl must vanish everywhere, as is verified by (1.57).

The divergence of the curl of a vector is also identically zero. This can be verified by expansion in rectangular components. Thus

$$\nabla \cdot \nabla \times \mathbf{F} = 0 \tag{1.58}$$

Any vector field that has zero divergence is called solenoidal. This describes the fact that flux lines are closed on themselves since there are no sources or sinks in the field for the lines of flux to terminate on. Equation (1.58) specifies that the curl of any vector field is solenoidal.

An important identity for work involving the wave equation is

$$\nabla \times (\nabla \times \mathbf{F}) \equiv \nabla(\nabla \cdot \mathbf{F}) - \nabla^2\mathbf{F} \tag{1.59}$$

This can be readily verified by expansion in rectangular coordinates. Equation (1.59) can also be considered as a defining equation for $\nabla^2\mathbf{F}$, when the vectors are expanded in other than rectangular coordinates.

A number of additional vector identities involving composite vectors are summarized at the end of this chapter. They can be verified by expansion in rectangular coordinates, carrying out the indicated operation. As an example, let us consider $\nabla \cdot (\mathbf{A} \times \mathbf{B})$.

Example 1.2. Consider

$$\nabla \cdot (\mathbf{A} \times \mathbf{B}) = \nabla \cdot \Sigma \mathbf{a}_x(A_yB_z - B_yA_z) \tag{1.60}$$

The summation, here, represents cyclical permutation; that is, the second term is obtained by replacing x by y, y by z, z by x. The third term is obtained from the second in a similar fashion. This notation avoids the need to write what is essentially repetitious material. Expanding (1.60) gives

$$\begin{aligned}\nabla \cdot (\mathbf{A} \times \mathbf{B}) &= \sum \frac{\partial}{\partial x}(A_yB_z - B_yA_z) \\ &= \sum \left(A_y\frac{\partial B_z}{\partial x} + B_z\frac{\partial A_y}{\partial x} - B_y\frac{\partial A_z}{\partial x} - A_z\frac{\partial B_y}{\partial x}\right) \\ &= \sum A_y\frac{\partial B_z}{\partial x} + \sum B_z\frac{\partial A_y}{\partial x} - \sum B_y\frac{\partial A_z}{\partial x} - \sum A_z\frac{\partial B_y}{\partial x}\end{aligned} \tag{1.61}$$

Since it does not matter which of the three terms indicated by the summation is chosen to represent the summation, we can write

$$\begin{aligned}\nabla \cdot (\mathbf{A} \times \mathbf{B}) &= \sum A_z\frac{\partial B_x}{\partial y} + \sum B_z\frac{\partial A_y}{\partial x} - \sum B_z\frac{\partial A_x}{\partial y} - \sum A_z\frac{\partial B_y}{\partial x} \\ &= \sum B_z\left(\frac{\partial A_y}{\partial x} - \frac{\partial A_x}{\partial y}\right) - \sum A_z\left(\frac{\partial B_y}{\partial x} - \frac{\partial B_x}{\partial y}\right)\end{aligned} \tag{1.62}$$

But $\partial B_y/\partial x - \partial B_x/\partial y$ is the z component of the curl of $\mathbf{B}$, and $\partial A_y/\partial x - \partial A_x/\partial y$ is the z component of the curl of $\mathbf{A}$. Accordingly,

$$\nabla \cdot (\mathbf{A} \times \mathbf{B}) = \Sigma B_z(\nabla \times A)_z - \Sigma A_z(\nabla \times B)_z \tag{1.63}$$

Expanding the indicated summation yields three terms which are clearly the three terms of a scalar product. Accordingly, the right-hand side of (1.63) can be written in vector notation, and the following identity results:

$$\nabla \cdot (\mathbf{A} \times \mathbf{B}) = \mathbf{B} \cdot \nabla \times \mathbf{A} - \mathbf{A} \cdot \nabla \times \mathbf{B} \tag{1.64}$$

1.14. Gauss' Law

In Sec. 1.9 we noted that the net outflow of "fluid" from a given volume can be found by integrating the scalar product of the vector function (considered as a flow function) and an element of area over the boundary surface. This same quantity can also be found by integrating the divergence throughout the volume since the divergence evaluates the net outflow per unit volume. Expressed mathematically,

$$\int_V \nabla \cdot \mathbf{F}\, dV = \oint_S \mathbf{F} \cdot d\mathbf{S} \tag{1.65}$$

In this equation S is the boundary surface for the volume V, and $d\mathbf{S}$ a surface element with positive normal drawn outward. Equation (1.65) is known as Gauss' law, or sometimes as the divergence theorem.

It is worthwhile pointing out that (1.65) applies even when the bounding surface is not simply connected. For example, in Fig. 1.16 the volume is bounded by the closed surfaces S_1, S_2, and S_3, where S_3 contains S_1 and S_2. In applying Gauss' theorem the volume integral of $\nabla \cdot \mathbf{F}$ is taken throughout the designated volume, while the surface integral of $\mathbf{F} \cdot \mathbf{n}\, dS$ is taken over the bounding surface, in this case the separate component surfaces S_1, S_2, S_3. We conclude that Gauss' theorem is applicable to a volume enclosed by a multiply connected surface; the surface integral in (1.65) then designates an integral over each separate surface involved. It is important to remember that positive surface area is outward from the volume; this accounts for the direction of $\mathbf{n}$ as shown in Fig. 1.16.

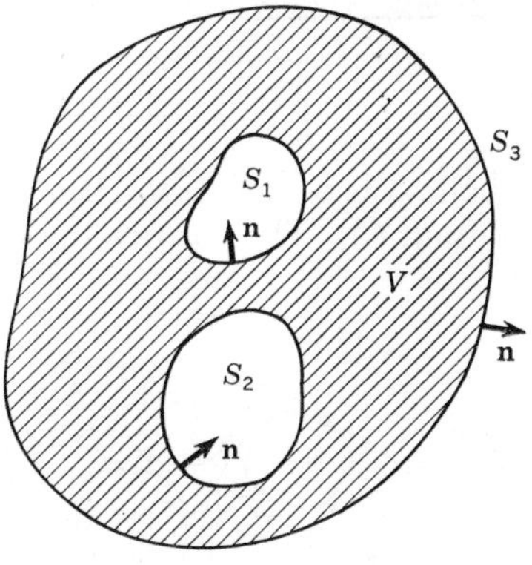

FIG. 1.16. Volume bounded by multiply connected surfaces.

A useful corollary of Gauss' law is known as Green's theorem. To derive this we let $\mathbf{F}$ be the product of a scalar Φ and a vector $\nabla\psi$. Then, using the vector identity (1.118) given at the end of the chapter,

$$\nabla \cdot \mathbf{F} = \nabla \cdot \Phi \nabla\psi = \Phi \nabla^2\psi + \nabla\Phi \cdot \nabla\psi \tag{1.66}$$

Integrating (1.66) over an arbitrary volume and making use of Gauss' law leads to Green's first theorem:

$$\int_V \nabla \cdot \mathbf{F}\, dV = \oint_S \Phi \nabla\psi \cdot d\mathbf{S} = \int_V (\Phi \nabla^2\psi + \nabla\Phi \cdot \nabla\psi)\, dV \tag{1.67}$$

The result expressed by (1.67) holds for any two scalar functions Φ and ψ which are finite and continuous and can be differentiated twice in the region V. If, from (1.67), we subtract the equation obtained by interchanging ψ and Φ, we obtain

$$\oint_S \left(\Phi \frac{\partial \psi}{\partial n} - \psi \frac{\partial \Phi}{\partial n}\right) dS = \int_V (\Phi \nabla^2 \psi - \psi \nabla^2 \Phi)\, dV \qquad (1.68)$$

Note that $\nabla\psi \cdot d\mathbf{S}$ is replaced by $(\partial\psi/\partial n)\, dS$ in (1.68), where $\mathbf{n}$ is the outward unit normal to S. Equation (1.68) is known as Green's second theorem and is very important in the solution of boundary-value problems in electromagnetic theory.

1.15. Stokes' Theorem

Stokes' theorem may be stated in the form

$$\oint_C \mathbf{F} \cdot d\mathbf{l} = \int_S \nabla \times \mathbf{F} \cdot d\mathbf{S} \qquad (1.69)$$

where S is an arbitrary surface (not necessarily plane) bounded by the contour C. The positive direction of $d\mathbf{S}$ is related to the positive sense of describing C according to the right-hand rule, as discussed in Sec. 1.4.

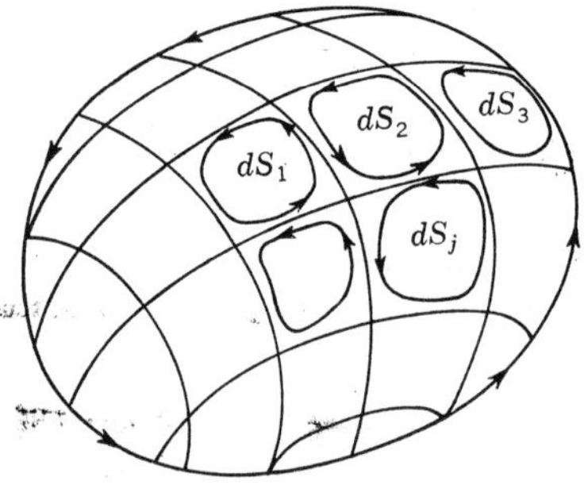

FIG. 1.17. Illustration for proof of Stokes' theorem.

To establish Stokes' theorem, consider the arbitrary surface S illustrated in Fig. 1.17. Let us divide the surface into differential elements of area $d\mathbf{S}_1$, $d\mathbf{S}_2$, $d\mathbf{S}_3$, etc. For each such area element we form $\oint \mathbf{F} \cdot d\mathbf{l}$, taking the contour direction to correspond with the positive sense of S. If now all such integrals are summed, the contributions arising from the common boundary of any two elements (e.g., $d\mathbf{S}_1$ and $d\mathbf{S}_2$) exactly cancel each other, since they are described in opposite directions in the adjoining differential areas. Thus after addition there is left only the integral over the original bounding contour. Consequently,

$$\oint_C \mathbf{F} \cdot d\mathbf{l} = \oint_{dS_1} \mathbf{F} \cdot d\mathbf{l} + \oint_{dS_2} \mathbf{F} \cdot d\mathbf{l} + \cdots \qquad (1.70)$$

Now for each integral on the right-hand side of (1.70) the definition of $\nabla \times \mathbf{F}$ can be applied. Thus we have

$$\oint_C \mathbf{F} \cdot d\mathbf{l} = (\nabla \times \mathbf{F}) \cdot d\mathbf{S}_1 + (\nabla \times \mathbf{F}) \cdot d\mathbf{S}_2 + \cdots \qquad (1.71)$$

Note that in view of the choice of describing the contour around each element, each differential area element $d\mathbf{S}_1$, $d\mathbf{S}_2$, . . . corresponds in the same way to the positive contour C. Consequently, the summation indicated on the right-hand side of (1.71) can be expressed as a surface integral. We are thus led to Stokes' theorem as postulated earlier; i.e.,

$$\oint_C \mathbf{F} \cdot d\mathbf{l} = \int_S \nabla \times \mathbf{F} \cdot d\mathbf{S} \tag{1.72}$$

If we consider two surfaces S_1 and S_2 which have the same contour C, then from Stokes' theorem,

$$\int_{S_1} \nabla \times \mathbf{F} \cdot d\mathbf{S}_1 = \int_{S_2} \nabla \times \mathbf{F} \cdot d\mathbf{S}_2 \tag{1.73}$$

By reversing the direction of the normal to one of the surfaces, say surface 1, then the direction of positive area is outward (or inward) to $S_1 + S_2$ considered as a closed surface, and we have

$$-\int_{S_1} \nabla \times \mathbf{F} \cdot d\mathbf{S}_1 + \int_{S_2} \nabla \times \mathbf{F} \cdot d\mathbf{S}_2 = \oint_S \nabla \times \mathbf{F} \cdot d\mathbf{S} = 0 \tag{1.74}$$

In other words, the vector function $\mathbf{A} = \nabla \times \mathbf{F}$ has no net outflow from an arbitrary region. It is therefore solenoidal, a result that we have already noted in Sec. 1.13.

1.16. Orthogonal Curvilinear Coordinates

It should be noted that the fundamental definitions of gradient, divergence, and curl do not involve a particular coordinate system. That we have expressed them, so far, in rectangular coordinates reflects merely that it is easiest to do so. But in a wide variety of problems other coordinate systems will be more appropriate. Accordingly, it is desirable to develop expansions for the preceding differential operations in other such systems. The easiest way of doing this is to work out general formulas in orthogonal curvilinear coordinates. Then the expansions for a specific system (e.g., spherical, cylindrical) can be obtained by substitution of appropriate parameters.

A generalized coordinate system consists of three families of surfaces whose equations in terms of rectangular coordinates are

$$u_1(x,y,z) = \text{constant} \qquad u_2(x,y,z) = \text{constant} \qquad u_3(x,y,z) = \text{constant} \tag{1.75}$$

We are interested only in the case where these three families of surfaces are orthogonal to each other (problems requiring nonorthogonal coordinates practically never can be solved exactly, and approximate techniques usually involve use of orthogonal coordinate systems). Equation (1.75) can be inverted so that (x,y,z) are expressed in terms of (u_1,u_2,u_3).

The lines of intersection of the coordinate surfaces constitute three families of lines, in general curved. At the point (x,y,z) or (u_1,u_2,u_3) we assign three unit vectors $\mathbf{a}_1$, $\mathbf{a}_2$, $\mathbf{a}_3$ tangent to the corresponding coordinate line at the point. The vector field $\mathbf{F}$ may be expressed in terms of components along these unit vectors, as we have been doing for rectangular coordinates. For the assumed orthogonal system the unit vectors are mutually perpendicular at any point.

Consider the infinitesimal parallelepiped, illustrated in Fig. 1.18, whose faces coincide with the surfaces u_1 or u_2 or u_3 = constant. Since the coordinates need not express a distance directly (e.g., the angles of spherical coordinates), the differential elements of length must be expressed as $dl_1 = h_1\,du_1$, $dl_2 = h_2\,du_2$, $dl_3 = h_3\,du_3$, where h_1, h_2, h_3 are suitable scale factors and may be functions of u_1, u_2, u_3. As an illustration, in cylindrical coordinates (r,ϕ,z), $h_1 = 1$, $h_2 = r$, $h_3 = 1$, since the elements of length along the coordinate curves r, ϕ, z are dr, $r\,d\phi$, and dz. The square of the diagonal dl of the parallelepiped may be written

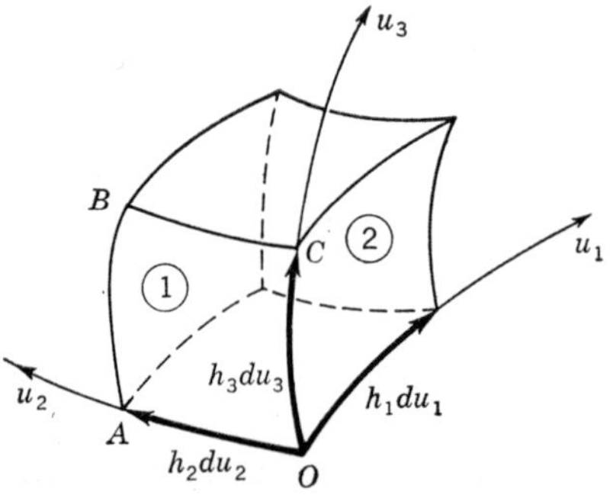

FIG. 1.18. Orthogonal curvilinear coordinates.

$$dl^2 = h_1^2\,du_1^2 + h_2^2\,du_2^2 + h_3^2\,du_3^2 \tag{1.76}$$

and its volume is $h_1h_2h_3\,du_1\,du_2\,du_3$.

Gradient

Let $\Phi(u_1,u_2,u_3)$ be a scalar function. Then, according to the properties of $\nabla\Phi$, it is a vector whose component in any direction is given by the directional derivative of Φ in that direction. Thus, for the u_1 component, we have

$$(\nabla\Phi)_1 = \frac{\partial\Phi}{\partial l_1} = \frac{1}{h_1}\frac{\partial\Phi}{\partial u_1} \tag{1.77}$$

and similarly for directions 2 and 3. The resultant vector expansion is

$$\nabla\Phi = \frac{\mathbf{a}_1}{h_1}\frac{\partial\Phi}{\partial u_1} + \frac{\mathbf{a}_2}{h_2}\frac{\partial\Phi}{\partial u_2} + \frac{\mathbf{a}_3}{h_3}\frac{\partial\Phi}{\partial u_3} \tag{1.78}$$

Divergence

To calculate the divergence of a vector $\mathbf{A}$ it is necessary to evaluate the net outflow per unit volume in the limit as the volume approaches zero.

If we refer to the differential volume of Fig. 1.18, it is possible to proceed as we did in rectangular coordinates. Let the components of **A** be $(\mathbf{a}_1A_1,\mathbf{a}_2A_2,\mathbf{a}_3A_3)$. Then the flux through surface 1 ($OABC$) taking the outward normal is

$$\text{Flux}_{S_1} = -A_1h_2h_3\,du_2\,du_3 + \frac{1}{2}\frac{\partial}{\partial u_1}(A_1h_2h_3)\,du_1\,du_2\,du_3$$

while the flux through surface 2 is

$$\text{Flux}_{S_2} = A_1h_2h_3\,du_2\,du_3 + \frac{1}{2}\frac{\partial}{\partial u_1}(A_1h_2h_3)\,du_1\,du_2\,du_3$$

If we add the outflow for the remaining two surface pairs, the net flux will be found to be

$$\left[\frac{\partial}{\partial u_1}(h_2h_3A_1) + \frac{\partial}{\partial u_2}(h_3h_1A_2) + \frac{\partial}{\partial u_3}(h_1h_2A_3)\right]du_1\,du_2\,du_3$$

The second and third terms can also be written down by cyclical permutation of the first. From the definition of divergence we can now write

$$\nabla\cdot\mathbf{A} = \frac{1}{h_1h_2h_3}\left[\frac{\partial}{\partial u_1}(h_2h_3A_1) + \frac{\partial}{\partial u_2}(h_3h_1A_2) + \frac{\partial}{\partial u_3}(h_1h_2A_3)\right] \tag{1.79}$$

Curl

The component 1 of the curl can be found by calculating the circulation around contour $OABC$ and dividing by the enclosed surface area. Thus

$$\int_0^A A_2\,dl_2 + \int_B^C A_2\,dl_2 = -\frac{\partial}{\partial u_3}(A_2h_2)\,du_2\,du_3$$

and

$$\int_A^B A_3\,dl_3 + \int_C^0 A_3\,dl_3 = \frac{\partial}{\partial u_2}(A_3h_3)\,du_2\,du_3$$

The details are analogous to those in Sec. 1.12. In vector notation the above result, by definition of the curl, leads to

$$(\nabla\times\mathbf{A})_1 = \frac{1}{h_2h_3}\left[\frac{\partial}{\partial u_2}(A_3h_3) - \frac{\partial}{\partial u_3}(A_2h_2)\right] \tag{1.80}$$

By cyclic changes in the indices the remaining components are obtained. Consequently,

$$\begin{aligned}(\nabla\times\mathbf{A}) = {} & \frac{\mathbf{a}_1}{h_2h_3}\left[\frac{\partial}{\partial u_2}(A_3h_3) - \frac{\partial}{\partial u_3}(A_2h_2)\right] \\ & + \frac{\mathbf{a}_2}{h_3h_1}\left[\frac{\partial}{\partial u_3}(A_1h_1) - \frac{\partial}{\partial u_1}(A_3h_3)\right] \\ & + \frac{\mathbf{a}_3}{h_1h_2}\left[\frac{\partial}{\partial u_1}(A_2h_2) - \frac{\partial}{\partial u_2}(A_1h_1)\right]\end{aligned} \tag{1.81}$$

which may be written in determinant form as

$$\nabla \times \mathbf{F} = \frac{1}{h_1h_2h_3}\begin{vmatrix} h_1\mathbf{a}_1 & h_2\mathbf{a}_2 & h_3\mathbf{a}_3 \\ \dfrac{\partial}{\partial u_1} & \dfrac{\partial}{\partial u_2} & \dfrac{\partial}{\partial u_3} \\ h_1A_1 & h_2A_2 & h_3A_3 \end{vmatrix} \tag{1.82}$$

Laplacian

The Laplacian of a scalar is defined as the divergence of the gradient of the scalar and may be formed by combining (1.79) and (1.78). The result is

$$\nabla^2\Phi = \nabla \cdot \nabla\Phi = \frac{1}{h_1h_2h_3}\left[\frac{\partial}{\partial u_1}\left(\frac{h_2h_3}{h_1}\frac{\partial\Phi}{\partial u_1}\right) + \frac{\partial}{\partial u_2}\left(\frac{h_3h_1}{h_2}\frac{\partial\Phi}{\partial u_2}\right) + \frac{\partial}{\partial u_3}\left(\frac{h_1h_2}{h_3}\frac{\partial\Phi}{\partial u_3}\right)\right] \tag{1.83}$$

The results of this section are used in Sec. 1.19 to evaluate the aforementioned vector operations in rectangular, cylindrical, and spherical coordinates.

1.17. Point Sources

In physical problems vector fields arise from source distributions which are continuous in space. Nevertheless, it is convenient, mostly from a

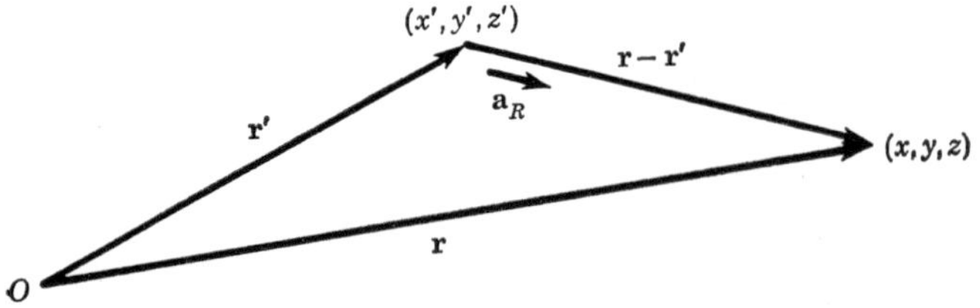

FIG. 1.19. Illustration of notation for source and field points.

mathematical standpoint, to assume the source distribution discontinuous. We shall consider here the characteristics of fields set up by point sources. It should be noted that by properly superposing such point sources, an arbitrary distribution can be represented.

In field theory it is necessary to clearly distinguish between the coordinates that determine the location of the source and the coordinates that designate the point at which the field is being evaluated. In this book primed coordinates x', y', z' will be used to designate the source point while unprimed coordinates x, y, z will be used to designate the field point. The vector $\mathbf{r}' = x'\mathbf{a}_x + y'\mathbf{a}_y + z'\mathbf{a}_z$ is a vector from the origin to the source point, while $\mathbf{r} = x\mathbf{a}_x + y\mathbf{a}_y + z\mathbf{a}_z$ is a vector from the origin to the field point, as in Fig. 1.19. The vector from the source point to the field point is

$$\mathbf{r} - \mathbf{r}' = (x - x')\mathbf{a}_x + (y - y')\mathbf{a}_y + (z - z')\mathbf{a}_z \tag{1.84}$$

with the magnitude of the distance given by

$$|\mathbf{r} - \mathbf{r}'| = [(x - x')^2 + (y - y')^2 + (z - z')^2]^{1/2} \tag{1.85}$$

We shall also use an abbreviated notation, described below, when there is no danger of confusion. In this notation, the magnitude $|\mathbf{r} - \mathbf{r}'|$ will be designated by R and a unit vector directed from the source point to the field point by $\mathbf{a}_R$; thus

$$|\mathbf{r} - \mathbf{r}'| = R \tag{1.86a}$$

$$\frac{\mathbf{r} - \mathbf{r}'}{|\mathbf{r} - \mathbf{r}'|} = \mathbf{a}_R \tag{1.86b}$$

The capital letter R is used in order to avoid confusion with the usual notation for spherical coordinates.

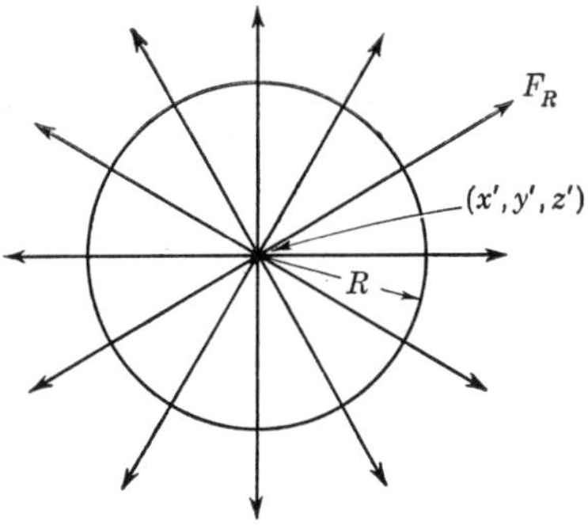

FIG. 1.20. Flow lines from a point source.

For a single point source located at (x',y',z') symmetry requires that the flow lines be radial and diverge uniformly. If we choose any spherical surface whose center is at the point source, as illustrated in Fig. 1.20, the total flux crossing the surface will be independent of the radius. In particular, the total flux computed is a measure of the total outflow from the source—hence is a measure of the strength of the source. If we call this quantity Q and let $\mathbf{F}$ be the vector field, then

$$Q = k \oint_S \mathbf{F} \cdot d\mathbf{S} = k4\pi R^2 F_R \tag{1.87}$$

The surface integral is over a spherical surface of radius R, and it can be evaluated because $\mathbf{F}$ is everywhere radial and of the same magnitude on S. In (1.87) k is a constant of proportionality to be determined on the basis of our definition of source strength. We shall choose $k = 1$, so that

$$Q = 4\pi R^2 F_R \tag{1.88}$$

and hence we must have

$$\mathbf{F} = \frac{Q}{4\pi R^2}\mathbf{a}_R \tag{1.89}$$

where $\mathbf{a}_R$ is a unit vector in the radial direction.

The vector field is irrotational, a fact readily established by demonstrating that $\mathbf{F}$ can be derived as the gradient of a scalar Φ. By inspection it is clear that if $\Phi = Q/4\pi R$, then

$$\mathbf{F} = -\nabla\Phi = -\frac{Q}{4\pi}\mathbf{a}_R\frac{\partial R^{-1}}{\partial R} = \frac{Q}{4\pi R^2}\mathbf{a}_R \tag{1.90}$$

One advantage of representing the vector field $\mathbf{F}$ in terms of a scalar potential Φ becomes apparent when the field due to a series of point sources is required. We could superpose the vector fields $\mathbf{F}_1$, $\mathbf{F}_2$, $\mathbf{F}_3$, . . . , $\mathbf{F}_n$ due to Q_1, Q_2, Q_3, . . . , Q_n vectorially, but it is much easier to add algebraically Φ_1, Φ_2, Φ_3, . . . , Φ_n. Then, if $\mathbf{F}_S$ is the resultant field, it can be expressed as

$$\mathbf{F}_S = \sum_{i=1}^{n} \mathbf{F}_i = -\nabla\Phi \qquad \Phi = \sum_{i=1}^{n} \frac{Q_i}{4\pi R_i} \tag{1.91}$$

where R_i is the distance from the source at (x'_i,y'_i,z'_i) to the field point at (x,y,z). This equation shows how any irrotational field can be calculated when its sources are given.

If x'_i, y'_i, z'_i are the coordinates of the ith source, then the potential $\Phi(x,y,z)$ can be expressed by the following, according to (1.91):

$$\Phi(x,y,z) = \sum_{i=1}^{n} \frac{Q_i}{4\pi[(x - x'_i)^2 + (y - y'_i)^2 + (z - z'_i)^2]^{1/2}} \tag{1.92}$$

In any physical problem the sources will be confined to some finite region. It is often of interest to compute the potential set up at distances which are very large compared with the extent of the source region. To find this potential, take the origin of coordinates in the neighborhood of the source system and expand (1.92) in powers of x'_i, y'_i, z'_i which are small compared with x, y, z. Then by Taylor's theorem,

$$\Phi = \Phi_0 + \sum_{i=1}^{n} \left(\frac{\partial\Phi}{\partial x'_i}\bigg|_0 x'_i + \frac{\partial\Phi}{\partial y'_i}\bigg|_0 y'_i + \frac{\partial\Phi}{\partial z'_i}\bigg|_0 z'_i + \cdots \right) \tag{1.93}$$

where the index 0 means that the quantities in the parentheses are evaluated for $x'_i = y'_i = z'_i = 0$. Thus

$$\Phi_0 = \sum_{i=1}^{n} \frac{Q_i}{4\pi(x^2 + y^2 + z^2)^{1/2}} = \sum_{i=1}^{n} \frac{Q_i}{4\pi r} \qquad r = (x^2 + y^2 + z^2)^{1/2}$$

Carrying out the remaining operations indicated in (1.93) with respect to (1.92) leads to

$$\Phi(x,y,z) = \frac{1}{4\pi r} \sum_{i=1}^{n} Q_i + \frac{1}{4\pi r^2} \left(\frac{x}{r} \sum_{i=1}^{n} Q_i x'_i + \frac{y}{r} \sum_{i=1}^{n} Q_i y'_i + \frac{z}{r} \sum_{i=1}^{n} Q_i z'_i + \cdots \right) \tag{1.94}$$

The total source strength is obviously

$$Q = \sum_{i=1}^{n} Q_i$$

We define the moment of the source system as

$$\mathbf{m} = \sum_{i=1}^{n} Q_i\mathbf{r}_i'$$

where the vector $\mathbf{r}_i' = \mathbf{a}_x x_i' + \mathbf{a}_y y_i' + \mathbf{a}_z z_i'$. The first two terms may now be written as

$$\Phi = \frac{Q}{4\pi r} + \frac{\mathbf{m} \cdot \mathbf{r}}{4\pi r^3} \tag{1.95}$$

Note that to a first approximation the system of sources acts at great distances like a point source of strength

$$Q = \sum_{i=1}^{n} Q_i$$

The second term is a dipole term, about which more will be said in a later chapter.

1.18. Helmholtz's Theorem

All vector fields will be found to be made up of one or both of two fundamental types: solenoidal fields that have identically zero divergence everywhere and irrotational fields that have zero curl everywhere. The most general vector field will have both a nonzero divergence and a nonzero curl. We shall show that this field can always be considered as the sum of a solenoidal and an irrotational field. This statement is essentially the content of Helmholtz's theorem. Another way of stating the Helmholtz theorem is that a vector field is completely specified by its divergence and curl. The latter constitute the source and vortex source of the field, respectively. Before proceeding to the general case we shall treat the two special cases mentioned above first. Many of the properties of a vector field, whether it be an electric, magnetic, velocity, etc., field, stem directly from its solenoidal or irrotational characteristic. When we consider the electric and magnetic fields in later chapters, it will be seen that they fit into the general framework presented in this section.

Case 1. *Irrotational Field*

A vector field $\mathbf{F}$ that has zero curl or rotation everywhere is called an irrotational field. Thus $\nabla \times \mathbf{F} = 0$, but at the same time the divergence of $\mathbf{F}$ cannot be identically zero or else the field $\mathbf{F}$ would vanish everywhere.† Hence let

$$\nabla \cdot \mathbf{F} = \rho(x,y,z) \tag{1.96}$$

where ρ is now interpreted as the source function for the field $\mathbf{F}$.

The gradient of any scalar function Φ has zero curl, as was noted in Sec. 1.13, and hence the condition $\nabla \times \mathbf{F} = 0$ is satisfied if we take

$$\mathbf{F} = -\nabla\Phi \tag{1.97}$$

since $\nabla \times \nabla\Phi = 0$. The minus sign is chosen arbitrarily so that these results compare directly with later work in the book; a positive sign would be an equally correct choice. Substituting into (1.96) we get

$$-\nabla \cdot \mathbf{F} = \nabla^2\Phi = -\rho \tag{1.98}$$

Thus the scalar function Φ, which is called the scalar potential, is a solution of (1.98), a partial differential equation known as Poisson's equation. Once a solution for Φ has been found, we may obtain our vector field $\mathbf{F}$ at once from (1.97).

Case 2. Solenoidal Fields

A vector field for which $\nabla \cdot \mathbf{F} = 0$ is called a solenoidal field. In a field of this type all the flow lines are continuous and close upon themselves. If $\nabla \cdot \mathbf{F} = 0$, we cannot have an identically vanishing curl or again our field $\mathbf{F}$ would vanish.† Thus let

$$\nabla \times \mathbf{F} = \mathbf{J}(x,y,z) \tag{1.99}$$

The vector function $\mathbf{J}$ is the vortex source for the field $\mathbf{F}$. It must be a vector source function, since $\nabla \times \mathbf{F}$ is a vector.

A mathematical identity that has been established is $\nabla \cdot \nabla \times \mathbf{A} = 0$, where $\mathbf{A}$ is any vector function. Thus $\nabla \times \mathbf{A}$ is a solenoidal field, and hence we may take

$$\mathbf{F} = \nabla \times \mathbf{A} \tag{1.100}$$

The vector $\mathbf{A}$ is called the vector potential since it plays a role similar to that of the scalar potential Φ. Whether $\mathbf{A}$ has any significant physical properties is usually of little importance since the use of a vector potential is mainly to facilitate the integration of (1.99).

If we substitute (1.100) into (1.99), we obtain

$$\nabla \times \nabla \times \mathbf{A} = \nabla\nabla \cdot \mathbf{A} - \nabla^2\mathbf{A} = \mathbf{J} \tag{1.101}$$

after expanding the curl-curl operation. If $\nabla \cdot \mathbf{A}$ could be taken as zero,

† Although this result is plausible in that $\nabla \times \mathbf{F} = \nabla \cdot \mathbf{F} = 0$ signifies that there are no sources or vortex sources, we have yet to demonstrate the conclusion. It is a consequence of the Helmholtz theorem.

(1.101) would simplify to

$$\nabla^2 \mathbf{A} = -\mathbf{J} \tag{1.102}$$

and **A** would be a solution of the vector Poisson equation; i.e., each component of **A** satisfies the scalar Poisson equation. For example,

$$\nabla^2 A_x = -J_x$$

On the basis of the Helmholtz theorem the divergence of **A** is at our disposal since only its curl has been specified thus far. Consequently, we can always choose **A** so that $\nabla \cdot \mathbf{A} = 0$. This can also be demonstrated in the following way. In place of the potential **A** we could use a potential

$$\mathbf{A}' = \mathbf{A} + \nabla\psi$$

where ψ is an arbitrary scalar function. This will not change the value of **F** obtained from (1.100) since

$$\nabla \times \mathbf{A}' = \nabla \times \mathbf{A} + \nabla \times \nabla\psi = \nabla \times \mathbf{A}$$

If **A** does not have a zero divergence, then we use the potential **A**′ and choose ψ so that $\nabla \cdot \mathbf{A}' = 0$, that is, so that $\nabla \cdot \mathbf{A} + \nabla^2\psi = 0$. Since a function ψ can always be found that satisfies this (Poisson's) equation, a function **A**′ with zero divergence and with curl equal to **F** can always be obtained.

Case 3. General Vector Field

Helmholtz's theorem states that the most general vector field will have both a nonzero divergence and a nonzero curl and, furthermore, can be derived from the negative gradient of a scalar potential Φ and the curl of a vector potential **A**. In view of our discussion above, this statement is fairly obvious, since a general field would be simply a superposition of the two types of fields discussed separately. It will, nevertheless, be instructive to examine the mathematical statement of Helmholtz's theorem. A proof of the theorem is to be found in the following section.

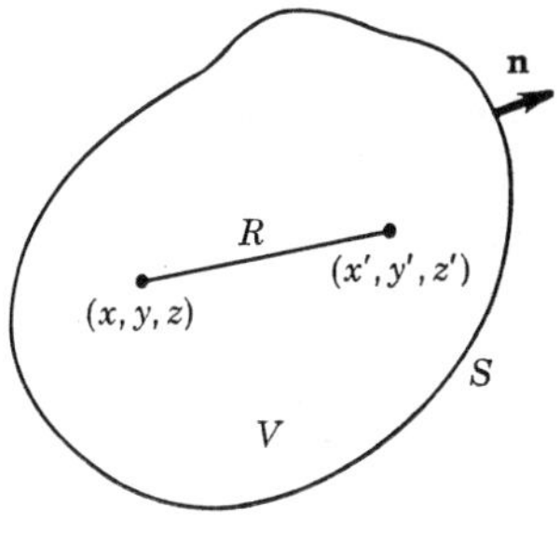

FIG. 1.21. Illustration of Helmholtz's theorem.

Consider a volume V bounded by a closed surface S, as in Fig. 1.21. A mathematical identity (proved later) states that the vector field **F** at the point (x,y,z) is given by

$$\mathbf{F}(x,y,z) = -\nabla\left[\int_V \frac{\nabla' \cdot \mathbf{F}(x',y',z')}{4\pi R}\,dV' - \oint_S \frac{\mathbf{F}(x',y',z')\cdot\mathbf{n}}{4\pi R}\,dS'\right] + \nabla \times \left[\int_V \frac{\nabla' \times \mathbf{F}(x',y',z')}{4\pi R}\,dV' + \oint_S \frac{\mathbf{F}(x',y',z')\times\mathbf{n}}{4\pi R}\,dS'\right] \tag{1.103}$$

where $\nabla' \equiv \mathbf{a}_x(\partial/\partial x') + \mathbf{a}_y(\partial/\partial y') + \mathbf{a}_z(\partial/\partial z')$ operates on the source coordinates, the integration is over the source coordinates (x',y',z'), and $\mathbf{n}$ is a unit normal directed out of the volume V. This is the mathematical statement of Helmholtz's theorem.

The term $\nabla' \cdot \mathbf{F}(x',y',z')$ gives the source function $\rho(x',y',z')$, while the term $\nabla' \times \mathbf{F}(x',y',z')$ determines the (vortex)† source function $\mathbf{J}(x',y',z')$.

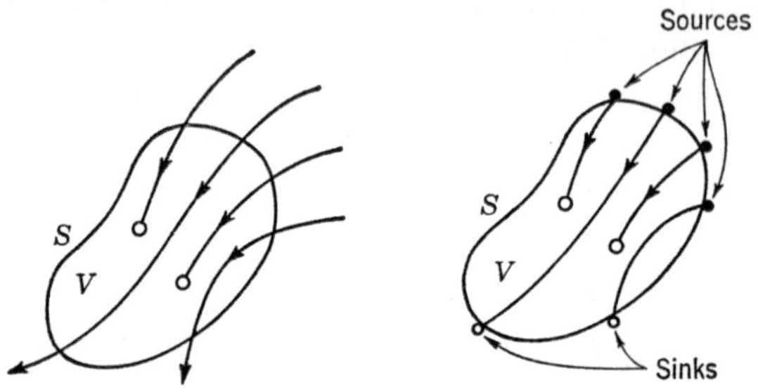

FIG. 1.22. Illustration of need for surface sources in Helmholtz's theorem.

The surface integrals represent integration over the surface sources on S. If S recedes to infinity, the field $\mathbf{F}$ will generally vanish at infinity, and hence the surface sources will vanish also. If V is finite, however, sources will occur on the surface S in general.

The physical significance of the surface sources may be seen as follows. Consider the situation where the flow lines of $\mathbf{F}$ extend into the volume V from outside the surface S, as in Fig. 1.22*a*. If we sever or cut off these flow lines at the surface S, then the field inside V can be maintained at its original value only if we place an equivalent source on the surface S to produce the same flow into the volume V as was produced by the original sources outside V. This situation is illustrated in Fig. 1.22*b*. The strength of the surface source must be equal to the original flow per unit area across S and hence equal to $-\mathbf{F} \cdot \mathbf{n}$. The minus sign arises since $\mathbf{F} \cdot \mathbf{n}$ is a measure of the outward flow, whereas the source strength must equal the inward flow. The other surface source $\mathbf{F} \times \mathbf{n}$ arises for similar reasons and is the equivalent vortex source that must be placed on S in order to maintain the proper circulation for the field $\mathbf{F}$ in V.

We now let $-\mathbf{F} \cdot \mathbf{n} = \sigma$ and $\mathbf{F} \times \mathbf{n} = \mathbf{K}$, where σ and $\mathbf{K}$ are the equivalent surface sources. The scalar and vector potentials are next

† In the future little effort will be made to distinguish between the two types of sources; both will be referred to as sources, and the context will clarify whether it is a source or vortex source.

defined to be

$$\Phi(x,y,z) = \int_V \frac{\rho(x',y',z')}{4\pi R}\, dV' + \oint_S \frac{\sigma(x',y',z')}{4\pi R}\, dS' \tag{1.104a}$$

$$\mathbf{A}(x,y,z) = \int_V \frac{\mathbf{J}(x',y',z')}{4\pi R}\, dV' + \oint_S \frac{\mathbf{K}(x',y',z')}{4\pi R}\, dS' \tag{1.104b}$$

Thus in place of (1.103) we have

$$\mathbf{F}(x,y,z) = -\nabla\Phi(x,y,z) + \nabla \times \mathbf{A}(x,y,z) \tag{1.105}$$

which is the mathematical statement of the second part of Helmholtz's theorem.

In summary, we can thus state:

1. If the curl of **F** is identically zero, then **F** is an irrotational field and can be obtained from the gradient of a scalar potential function.
2. If the divergence of **F** is identically zero, then **F** is a solenoidal field and may be derived from the curl of a vector potential function.
3. A general vector field can be derived from the negative gradient of a scalar potential and the curl of a vector potential.
4. The potentials are determined by the volume and surface source functions ρ, **J** and σ, **K**.

Integration of Poisson's Equation

Let us return to a consideration of the integrals stated in (1.104) and show that the potentials are, indeed, solutions of the Poisson equation. From our discussion on point sources at the beginning of this section we obtained the result that for a point source Q

$$\Phi = \frac{Q}{4\pi R}$$

If instead of a point source Q we have a distribution of point sources with a volume density $\rho(x',y',z')$, it follows, by superposition, that the potential is given by

$$\Phi(x,y,z) = \int_V \frac{\rho(x',y',z')}{4\pi R}\, dV' \tag{1.106}$$

Consequently, from (1.90) and (1.98), Φ defined by (1.106) satisfies Poisson's equation. Although this proof is probably satisfactory from an intuitive point of view, it is worthwhile to show mathematically that Φ as given by (1.106) is a solution of the Poisson equation

$$\nabla^2\Phi(x,y,z) = -\rho(x,y,z) \tag{1.107}$$

The mathematical details involved are themselves of great importance.

The Laplacian of (1.106) is

$$\nabla^2\Phi(x,y,z) = \int_V \frac{\rho(x',y',z')}{4\pi}\, \nabla^2\left(\frac{1}{R}\right) dV'$$

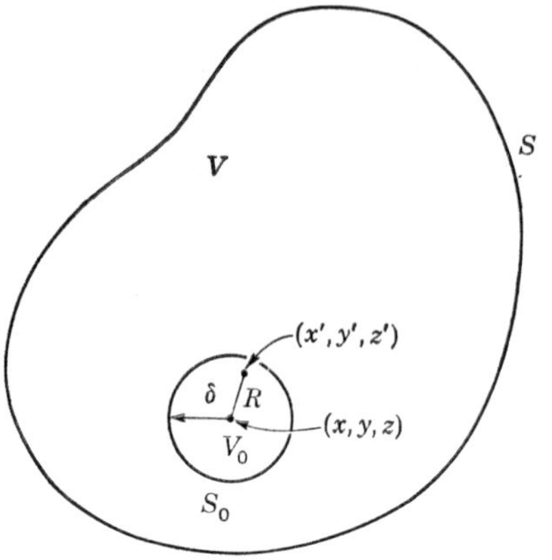

FIG. 1.23. Sphere surrounding singularity point (x,y,z).

We may bring the ∇^2 operator inside the integral because ∇^2 affects the x, y, z variables only while the integration is over the x', y', z' variables. Next we note that $\nabla^2(1/R)$ equals zero at all points except at the singularity point $R = 0$ (see Prob. 1.11). Thus the volume integral is zero except, possibly, for a contribution from the singular point $R = 0$. As x', y', z' approach x, y, z, R tends toward zero. Our procedure is to surround the singular point (x,y,z) by a small sphere of radius δ, surface S_0, and volume V_0, as in Fig. 1.23. Since $\rho(x',y',z')$ is a continuous function, we may choose δ so small that for all values of x', y', z' inside the sphere, ρ is essentially equal to its value $\rho(x,y,z)$ at the singular point.

Our integral now becomes

$$\int_V \frac{\rho(x',y',z')}{4\pi} \nabla^2 \left(\frac{1}{R}\right) dV' = \frac{\rho(x,y,z)}{4\pi} \int_{V_0} \nabla^2 \left(\frac{1}{R}\right) dV'$$

We designate by ∇' the del operator, which has been defined as

$$\nabla' \equiv \mathbf{a}_x \frac{\partial}{\partial x'} + \mathbf{a}_y \frac{\partial}{\partial y'} + \mathbf{a}_z \frac{\partial}{\partial z'}$$

so that ∇' operates on the source coordinates. Similarly,

$$\nabla'^2 \equiv \frac{\partial^2}{\partial x'^2} + \frac{\partial^2}{\partial y'^2} + \frac{\partial^2}{\partial z'^2}$$

Since $R = [(x - x')^2 + (y - y')^2 + (z - z')^2]^{1/2}$, then we can confirm by direct expansion that $\nabla(1/R) = -\nabla'(1/R)$, and $\nabla^2(1/R) = \nabla'^2(1/R)$. Using the latter identity and the divergence theorem in the above volume integral, we obtain

$$\frac{\rho}{4\pi} \int_{V_0} \nabla'^2 \left(\frac{1}{R}\right) dV' = \frac{\rho}{4\pi} \oint_{S_0} \nabla' \left(\frac{1}{R}\right) \cdot d\mathbf{S}'_0$$

Now $\nabla'(1/R) = -\mathbf{a}'_R/R^2$ and $d\mathbf{S}'_0 = \mathbf{a}'_R R^2\, d\Omega$, where $\mathbf{a}'_R$ is a unit vector directed outward from the point (x,y,z) and $d\Omega$ is an element of solid angle. Substitution of these relations now shows finally that

$$\nabla^2\Phi(x,y,z) = -\frac{\rho(x,y,z)}{4\pi} \oint_{S_0} d\Omega = -\rho(x,y,z)$$

and hence verifies that Φ as given by (1.106) is a solution of Poisson's equation.

For the vector potential **A** each component is a solution of the scalar Poisson equation; so it follows by vector addition that the solution to $\nabla^2\mathbf{A} = -\mathbf{J}$ is given by

$$\mathbf{A} = \int_V \frac{\mathbf{J}(x',y',z')}{4\pi R}\, dV'$$

For surface sources the solutions are the same, with the exception that the integration is now over a surface instead of throughout a volume.

Proof of Helmholtz's Theorem

In view of the properties of the function $\nabla^2(1/R)$ as discussed in connection with the integration of Poisson's equation, it is clear that the vector function $\mathbf{F}(x,y,z)$ can be represented as

$$\begin{aligned}\mathbf{F}(x,y,z) &= -\int_V \frac{\mathbf{F}(x',y',z')}{4\pi} \nabla^2 \left(\frac{1}{R}\right) dV' \\ &= -\nabla^2 \int_V \frac{\mathbf{F}(x',y',z')}{4\pi R}\, dV'\end{aligned}$$

Using the vector identity $\nabla \times \nabla \times = \nabla\nabla \cdot - \nabla^2$, we may rewrite the above as

$$\mathbf{F}(x,y,z) = \nabla \times \nabla \times \int_V \frac{\mathbf{F}(x',y',z')}{4\pi R}\, dV' - \nabla\nabla \cdot \int_V \frac{\mathbf{F}(x',y',z')}{4\pi R}\, dV' \qquad (1.108)$$

Consider the divergence term first. We have

$$\nabla \cdot \int_V \frac{\mathbf{F}(x',y',z')}{4\pi R}\, dV' = \frac{1}{4\pi} \int_V \mathbf{F}(x',y',z') \cdot \nabla \left(\frac{1}{R}\right) dV'$$

since ∇ does not operate on the primed variables. Next we note that

$$\begin{aligned}\mathbf{F}(x',y',z') \cdot \nabla \left(\frac{1}{R}\right) &= -\mathbf{F}(x',y',z') \cdot \nabla' \left(\frac{1}{R}\right) \\ &= -\nabla' \cdot \frac{\mathbf{F}(x',y',z')}{R} + \frac{\nabla' \cdot \mathbf{F}(x',y',z')}{R}\end{aligned}$$

Hence

$$\begin{aligned}\nabla \cdot \int_V \frac{\mathbf{F}(x',y',z')}{4\pi R}\, dV' &= -\int_V \nabla' \cdot \frac{\mathbf{F}(x',y',z')}{4\pi R}\, dV' + \int_V \frac{\nabla' \cdot \mathbf{F}(x',y',z')}{4\pi R}\, dV' \\ &= -\oint_S \frac{\mathbf{F}(x',y',z') \cdot \mathbf{n}}{4\pi R}\, dS' + \int_V \frac{\nabla' \cdot \mathbf{F}(x',y',z')}{4\pi R}\, dV' = \Phi \qquad (1.109)\end{aligned}$$

which is the desired form for the scalar potential Φ.

We now return to the curl term in (1.108) and note that

$$\begin{aligned}\nabla \times \int_V \frac{\mathbf{F}(x',y',z')}{4\pi R}\, dV' &= -\frac{1}{4\pi} \int_V \mathbf{F}(x',y',z') \times \nabla \left(\frac{1}{R}\right) dV' \\ &= \frac{1}{4\pi} \int_V \mathbf{F}(x',y',z') \times \nabla' \left(\frac{1}{R}\right) dV'\end{aligned}$$

We next use the relation

$$\nabla' \times \frac{\mathbf{F}(x',y',z')}{R} = -\mathbf{F}(x',y',z') \times \nabla'\left(\frac{1}{R}\right) + \frac{\nabla' \times \mathbf{F}(x',y',z')}{R}$$

to get

$$\nabla \times \int_V \frac{\mathbf{F}(x',y',z')\,dV'}{4\pi R} = \int_V \frac{\nabla' \times \mathbf{F}(x',y',z')}{4\pi R}\,dV' - \frac{1}{4\pi}\int_V \nabla' \times \frac{\mathbf{F}(x',y',z')}{R}\,dV' \tag{1.110}$$

The first integral on the right-hand side is in the desired form. The remaining step is to show that

$$-\frac{1}{4\pi}\int_V \nabla' \times \frac{\mathbf{F}(x',y',z')}{R}\,dV' = \oint_S \frac{\mathbf{F}(x',y',z') \times \mathbf{n}}{4\pi R}\,dS' \tag{1.111}$$

To prove this result let $\mathbf{C}$ be a constant vector and apply the divergence theorem to the quantity $\nabla' \cdot \mathbf{C} \times \mathbf{F}/R$ to obtain

$$\begin{aligned}\int_V \nabla' \cdot \mathbf{C} \times \frac{\mathbf{F}}{R}\,dV' &= -\int_V \mathbf{C} \cdot \nabla' \times \frac{\mathbf{F}}{R}\,dV' \\ &= \oint_S \mathbf{C} \times \frac{\mathbf{F}}{R} \cdot \mathbf{n}\,dS'\end{aligned} \tag{1.112}$$

In the surface integral we have

$$\mathbf{C} \times \frac{\mathbf{F}}{R} \cdot \mathbf{n} = \mathbf{C} \cdot \frac{\mathbf{F} \times \mathbf{n}}{R}$$

and (1.112) becomes

$$-\mathbf{C} \cdot \int_V \nabla' \times \frac{\mathbf{F}}{R}\,dV' = \mathbf{C} \cdot \oint_S \frac{\mathbf{F} \times \mathbf{n}}{R}\,dS'$$

Since $\mathbf{C}$ is an arbitrary vector, the two integrals are equal and the relation (1.111) is verified. Thus we have

$$\nabla \times \int_V \frac{\mathbf{F}(x',y',z')}{4\pi R}\,dV' = \int_V \frac{\nabla' \times \mathbf{F}(x',y',z')}{4\pi R}\,dV' + \oint_S \frac{\mathbf{F}(x',y',z') \times \mathbf{n}}{4\pi R}\,dS' = \mathbf{A} \tag{1.113}$$

Consequently, it now follows that

$$\underline{\mathbf{F} = -\nabla\Phi + \nabla \times \mathbf{A}}$$

when (1.109) and (1.113) are used in (1.108). This completes the proof of Helmholtz's theorem.

1.19. Vector Summary

From the general equations in orthogonal curvilinear coordinates the vector operations in the three most common systems are found to be as follows.

Rectangular Coordinates

$$u_1 = x \qquad u_2 = y \qquad u_3 = z$$
$$h_1 = 1 \qquad h_2 = 1 \qquad h_3 = 1$$

$$\nabla\Phi = \mathbf{a}_x \frac{\partial \Phi}{\partial x} + \mathbf{a}_y \frac{\partial \Phi}{\partial y} + \mathbf{a}_z \frac{\partial \Phi}{\partial z}$$

$$\nabla \cdot \mathbf{F} = \frac{\partial F_x}{\partial x} + \frac{\partial F_y}{\partial y} + \frac{\partial F_z}{\partial z}$$

$$\nabla \times \mathbf{F} = \mathbf{a}_x \left(\frac{\partial F_z}{\partial y} - \frac{\partial F_y}{\partial z} \right) + \mathbf{a}_y \left(\frac{\partial F_x}{\partial z} - \frac{\partial F_z}{\partial x} \right) + \mathbf{a}_z \left(\frac{\partial F_y}{\partial x} - \frac{\partial F_x}{\partial y} \right)$$

$$\nabla^2\Phi = \frac{\partial^2 \Phi}{\partial x^2} + \frac{\partial^2 \Phi}{\partial y^2} + \frac{\partial^2 \Phi}{\partial z^2}$$

Cylindrical Coordinates

$$u_1 = r \qquad u_2 = \phi \qquad u_3 = z$$
$$h_1 = 1 \qquad h_2 = r \qquad h_3 = 1$$

$$\nabla\Phi = \mathbf{a}_r \frac{\partial \Phi}{\partial r} + \mathbf{a}_\phi \frac{1}{r} \frac{\partial \Phi}{\partial \phi} + \mathbf{a}_z \frac{\partial \Phi}{\partial z}$$

$$\nabla \cdot \mathbf{F} = \frac{1}{r} \frac{\partial}{\partial r} (rF_r) + \frac{1}{r} \frac{\partial F_\phi}{\partial \phi} + \frac{\partial F_z}{\partial z}$$

$$\nabla \times \mathbf{F} = \mathbf{a}_r \left(\frac{1}{r} \frac{\partial F_z}{\partial \phi} - \frac{\partial F_\phi}{\partial z} \right) + \mathbf{a}_\phi \left(\frac{\partial F_r}{\partial z} - \frac{\partial F_z}{\partial r} \right) + \mathbf{a}_z \left[\frac{1}{r} \frac{\partial (rF_\phi)}{\partial r} - \frac{1}{r} \frac{\partial F_r}{\partial \phi} \right]$$

$$\nabla^2\Phi = \frac{1}{r} \frac{\partial}{\partial r} \left(r \frac{\partial \Phi}{\partial r} \right) + \frac{1}{r^2} \frac{\partial^2 \Phi}{\partial \phi^2} + \frac{\partial^2 \Phi}{\partial z^2}$$

Spherical Coordinates

$$u_1 = r \qquad u_2 = \theta \qquad u_3 = \phi$$
$$h_1 = 1 \qquad h_2 = r \qquad h_3 = r \sin \theta$$

$$\nabla\Phi = \mathbf{a}_r \frac{\partial \Phi}{\partial r} + \mathbf{a}_\theta \frac{1}{r} \frac{\partial \Phi}{\partial \theta} + \frac{\mathbf{a}_\phi}{r \sin \theta} \frac{\partial \Phi}{\partial \phi}$$

$$\nabla \cdot \mathbf{F} = \frac{1}{r^2} \frac{\partial}{\partial r} (r^2 F_r) + \frac{1}{r \sin \theta} \frac{\partial}{\partial \theta} (\sin \theta \, F_\theta) + \frac{1}{r \sin \theta} \frac{\partial F_\phi}{\partial \phi}$$

$$\nabla \times \mathbf{F} = \frac{\mathbf{a}_r}{r \sin \theta} \left[\frac{\partial}{\partial \theta} (F_\phi \sin \theta) - \frac{\partial F_\theta}{\partial \phi} \right] + \frac{\mathbf{a}_\theta}{r} \left[\frac{1}{\sin \theta} \frac{\partial F_r}{\partial \phi} - \frac{\partial}{\partial r} (rF_\phi) \right] + \frac{\mathbf{a}_\phi}{r} \left[\frac{\partial}{\partial r} (rF_\theta) - \frac{\partial F_r}{\partial \theta} \right]$$

$$\nabla^2\Phi = \frac{1}{r^2} \frac{\partial}{\partial r} \left(r^2 \frac{\partial \Phi}{\partial r} \right) + \frac{1}{r^2 \sin \theta} \frac{\partial}{\partial \theta} \left(\sin \theta \frac{\partial \Phi}{\partial \theta} \right) + \frac{1}{r^2 \sin^2 \theta} \frac{\partial^2 \Phi}{\partial \phi^2}$$

Vector Identities

$$\nabla(\Phi + \psi) = \nabla\Phi + \nabla\psi \tag{1.114}$$

$$\nabla \cdot (\mathbf{A} + \mathbf{B}) = \nabla \cdot \mathbf{A} + \nabla \cdot \mathbf{B} \tag{1.115}$$

$$\nabla \times (\mathbf{A} + \mathbf{B}) = \nabla \times \mathbf{A} + \nabla \times \mathbf{B} \tag{1.116}$$

$$\nabla(\Phi\psi) = \Phi\, \nabla\psi + \psi\, \nabla\Phi \tag{1.117}$$

$$\nabla \cdot (\psi\mathbf{A}) = \mathbf{A} \cdot \nabla\psi + \psi\nabla \cdot \mathbf{A} \tag{1.118}$$

$$\nabla \cdot \mathbf{A} \times \mathbf{B} = \mathbf{B} \cdot \nabla \times \mathbf{A} - \mathbf{A} \cdot \nabla \times \mathbf{B} \tag{1.119}$$

$$\nabla \times (\Phi\mathbf{A}) = \nabla\Phi \times \mathbf{A} + \Phi\nabla \times \mathbf{A} \tag{1.120}$$

$$\nabla \times (\mathbf{A} \times \mathbf{B}) = \mathbf{A}\nabla \cdot \mathbf{B} - \mathbf{B}\nabla \cdot \mathbf{A} + (\mathbf{B} \cdot \nabla)\mathbf{A} - (\mathbf{A} \cdot \nabla)\mathbf{B} \tag{1.121}$$

$$\nabla(\mathbf{A} \cdot \mathbf{B}) = (\mathbf{A} \cdot \nabla)\mathbf{B} + (\mathbf{B} \cdot \nabla)\mathbf{A} + \mathbf{A} \times (\nabla \times \mathbf{B}) + \mathbf{B} \times (\nabla \times \mathbf{A}) \tag{1.122}$$

$$\nabla \cdot \nabla\Phi = \nabla^2\Phi \tag{1.123}$$

$$\nabla \cdot \nabla \times \mathbf{A} = 0 \tag{1.124}$$

$$\nabla \times \nabla\Phi = 0 \tag{1.125}$$

$$\nabla \times \nabla \times \mathbf{A} = \nabla(\nabla \cdot \mathbf{A}) - \nabla^2\mathbf{A} \tag{1.126}$$

$$\int_V \nabla\Phi\, dV = \oint_S \Phi\, d\mathbf{S} \tag{1.127}$$

$$\int_V \nabla \cdot \mathbf{A}\, dV = \oint_S \mathbf{A} \cdot d\mathbf{S} \tag{1.128}$$

$$\int_V \nabla \times \mathbf{A}\, dV = \oint_S \mathbf{n} \times \mathbf{A}\, dS \tag{1.129}$$

$$\int_S \mathbf{n} \times \nabla\Phi\, dS = \oint_C \Phi\, d\mathbf{l} \tag{1.130}$$

$$\int_S \nabla \times \mathbf{A} \cdot d\mathbf{S} = \oint_C \mathbf{A} \cdot d\mathbf{l} \tag{1.131}$$

$$\int_V [Q\cdot(\nabla\times\nabla\times P) - P\cdot(\nabla\times\nabla\times Q)]\, dV$$

$$= \oint_S [\nabla\times Q]\cdot[\mathbf{u}_n\times P] - [\nabla\times P]\cdot[\mathbf{u}_n\times Q]\, dS$$

$$= \oint_S \mathbf{u}_n\cdot[P\times(\nabla\times Q) - Q\times(\nabla\times P)]\, dS$$

CHAPTER 2

ELECTROSTATICS

This chapter develops the basic properties of the electrostatic field in vacuum. The law of Coulomb for the force between two point charges is the experimental basis for the work of this chapter. The electric field is defined as the force exerted on a unit positive charge and leads to its establishment as a fundamental entity. From Coulomb's law the electric field due to a point charge is readily evaluated. The principle of superposition is next used to establish the law for the field produced by a volume or surface distribution of charge. The total flux of the electrostatic field is then related to the charge by means of Gauss' law.

From an investigation of the electrostatic field its nature is discovered, which permits the field to be derived from the negative gradient of a scalar potential function. The scalar potential is related to the work done against the field in moving a unit charge to an arbitrary field point from a given reference. The differential form of Gauss' law is used to show that the potential function is a solution of Poisson's equation, the charge density being the source function.

With the concept of a potential established, the treatment leads naturally into a discussion of conductors, the behavior of the electric field at conductor surfaces, and the constant potential nature of conducting bodies. Several elementary boundary-value problems involving conductors are solved by application of Gauss' law and image techniques.

The last section introduces the electric dipole, the dipole potential, and the dipole field. The field from a volume distribution of dipoles is shown to be the same as that from an equivalent volume and surface charge distribution. This sets the stage for the theory of the behavior of insulating materials in electric fields, which is presented in Chap. 3.

2.1. Introduction to Electrostatics

The phenomenon that underlies the study of electrostatics was known to man since very early times. Thales of Miletus is credited with first observing that amber when rubbed attracted light objects to itself, a discovery that dates back to 600 B.C. Subsequent experimenters found that most substances when rubbed possessed this property to some extent. In particular, if a glass rod is rubbed with a silk cloth, both rod and cloth will attract small bits of paper. We say that they are electrified (a word derived from *ēlektron*, the Greek name for amber).

If an electrified glass rod is brought into contact with a gilded pith ball suspended by a silk string, the ball becomes electrified. Substances, such as the gold covering of the ball, which have the property of removing electrification from electrified objects are called conductors. Other substances, such as the silk thread, which remove the electrification very slowly are insulators. If two balls are electrified by the glass rod, they will be found to repel each other. But if one is electrified by the rod and the other by the silk cloth, they attract each other. A hypothesis to explain this assumed the existence of two kinds of electricity. Arbitrarily, that on the glass rod is taken as positive and that on the silk negative.

The connection between the static electricity noted above and electric currents such as are caused to flow by means of a battery was established by Faraday (1833) and Rowland (1876). They showed that electric current was the flow of electric charge, of the same nature as the charges of electrostatics. Developments in the field of atomic and nuclear physics have deepened our knowledge concerning the nature of positive and negative charged particles. We shall assume that the student is familiar with at least a qualitative picture of atomic structure.

2.2. Coulomb's Law

By means of a torsion balance which he developed, Coulomb, in 1785, investigated the nature of the force between charged bodies. The following conclusions were drawn from the results of his experiments as they relate to the force between two charged bodies which are very small compared with their separation, i.e., point charges:

1. The magnitude of the force is proportional to the product of the charge magnitudes.
2. The magnitude of the force is inversely proportional to the square of the distance between the charges.
3. The direction of the force is along the line connecting the charges.
4. The force is attractive if charges are unlike, repulsive if they are alike.
5. The force depends on the medium in which the charges are placed.

Coulomb's experiments have subsequently been repeated with much greater precision; the inverse-square-law behavior is known to be true to at least 1 part in 10^9. It should be noted that the nature of Coulomb's and subsequent experiments is such as to provide a basis for a macroscopic theory. The above conclusions can be expected to hold only so long as the charged bodies are small compared with the distance separating them.

The information obtained by Coulomb can be formulated mathematically in what is known as Coulomb's law. Using vector notation, we have

$$\mathbf{F}_{12} = \frac{kq_1q_2}{\epsilon R^2}\mathbf{a}_R \tag{2.1}$$

In this equation $\mathbf{F}_{12}$ is the vector force acting on charge q_2 due to charge q_1. Its direction is governed by $\mathbf{a}_R$, a unit radius vector in the direction from q_1 to q_2. The symbols q_1 and q_2 specify both the magnitude and sign of the charges involved. The parameter ϵ is a property of the medium called the electric permittivity, R is the distance between the charges, and k is a constant of proportionality.

2.3. Units

In order to measure physical quantities, a standard of reference, or unit, must be defined so that the quantity can be expressed numerically. Because we deal with many physically related quantities, we seek a self-consistent system of units where every quantity can be defined in terms of a minimum number of basic, independent units consistent with a need for convenience and precision.† Many competing systems have been employed in the past for use in the area of electric and magnetic fields. A discussion of their development and conversion from one to another can be found elsewhere.‡

The system of units that has been almost universally adopted for use in applied electromagnetic theory is the meter-kilogram-second (mks, for short) system introduced by Giorgi in 1901. In this system length is measured in meters, mass in kilograms, time in seconds. For electric and magnetic quantities, a further basic unit must be defined, and this is usually chosen to be the coulomb for unit of charge or ampere for unit of current. Any electric or magnetic quantity can be expressed in terms of these fundamental units. The mks system has the advantage over systems used earlier in that primary electrical quantities are in the practical system; that is, potential is measured in volts, resistance in ohms, power in watts, etc.

We may illustrate how the unit of force is derived from the fundamental units by recalling that force equals mass times acceleration. Consequently, the unit of force in the mks system is kilogram-meters per

† For a discussion of optimization of the fundamental units, see A. G. McNish, The Basis of Our Measuring System, *Proc. IRE*, vol. 47, pp. 636–643, May, 1959.

‡ J. A. Stratton, "Electromagnetic Theory," sec. 1.8, Appendix I, McGraw-Hill Book Company, Inc., New York, 1941.

second squared and is called a newton. We may write

$$1 \text{ newton} = \frac{1 \text{ kilogram-meter}}{\text{second}^2} \tag{2.2}$$

The unit of energy is called the joule, and since it is given by the product of force times distance,

$$1 \text{ joule} = 1 \text{ newton-meter} \tag{2.3}$$

It is possible to determine the units of ϵ by inspection of the force equation (2.1). We easily verify that

$$\text{Units of } \epsilon = \frac{\text{coulombs}^2}{\text{newton–square meter}} = \frac{\text{coulombs}^2}{\text{joule-meter}} \tag{2.4}$$

We shall show in a later chapter that for a charged capacitor the energy in joules can be expressed in terms of the charge in coulombs and capacitance in farads as

$$\text{Energy in joules} = \frac{\frac{1}{2} \text{ coulombs}^2}{\text{farad}} \tag{2.5}$$

Consequently, (2.4) becomes

$$\text{Units of } \epsilon = \frac{\text{farads}}{\text{meter}} \tag{2.6}$$

The dimensionless constant k in (2.1) is chosen as either $1/4\pi$ or unity, depending on whether a "rational" or "irrational" system of units is desired. The choice that is usually made is the rational system, and such a system will be used in this text. Thus the force equation may be written in rationalized mks units as

$$\mathbf{F}_{12} = \frac{q_1 q_2}{4\pi\epsilon R^2}\,\mathbf{a}_R \tag{2.7}$$

The value of ϵ depends on the medium. For the balance of this chapter we assume the charges to lie in a free-space medium (vacuum). We adopt the notation ϵ_0 as the electric permittivity of free space. This has been measured to be 8.854×10^{-12} farad per meter, or very closely $(1/36\pi) \times 10^{-9}$ farad per meter. In (2.7) we may now remark that $\mathbf{F}_{12}$ is measured in newtons, q_1 and q_2 in coulombs, and R in meters.

2.4. Electric Field

The force on a charge q_0 due to a system of point charges q_1, q_2, q_3, as illustrated in Fig. 2.1, can be found by successive application of Coulomb's law and the principle of superposition; that is, the force due to each charge is found as if it alone were present; then the vector sum of these forces is calculated to give the resultant force. A graphical solution is shown in Fig. 2.1.

Before giving the expression for the force caused by several charges we digress briefly to explain the notation that will be adopted. In field theory we must often consider simultaneously two sets of points, the source point (x',y',z'), which specifies the location of the source, and the field point (x,y,z), which specifies the point at which the field is measured. The primes are used to clearly distinguish the source coordinates from the unprimed field coordinates. We shall find this distinction to be very

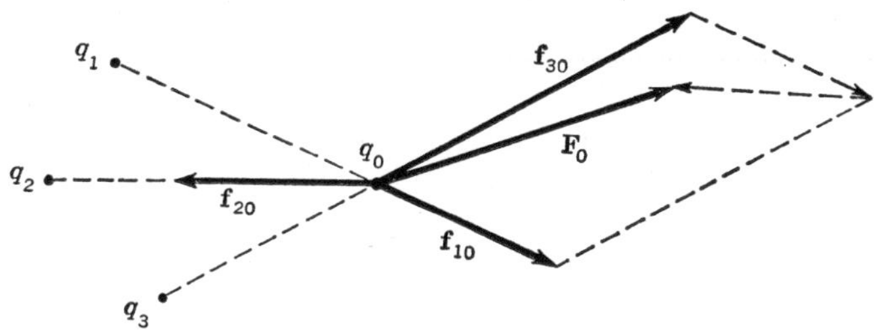

FIG. 2.1. Graphical solution for force on a single charge q_0 due to other charges.

helpful on many occasions. We shall follow the convention outlined in Sec. 1.17 and let $\mathbf{r}$ be a vector from the origin to the field point and let $\mathbf{r}'$ be a vector from the same origin to the source point. The vector $\mathbf{r} - \mathbf{r}'$ is then the vector distance from the source point to the field point. For brevity we denote the magnitude of $\mathbf{r} - \mathbf{r}'$ by the capital letter R and a unit vector in the direction of $\mathbf{r} - \mathbf{r}'$ by $\mathbf{a}_R$; thus

$$R = |\mathbf{r} - \mathbf{r}'| = [(x - x')^2 + (y - y')^2 + (z - z')^2]^{1/2}$$

$$\mathbf{a}_R = \frac{\mathbf{r} - \mathbf{r}'}{|\mathbf{r} - \mathbf{r}'|}$$

If we have several source points, we denote the vector distance from the origin to the ith source point (x_i',y_i',z_i') by $\mathbf{r}_i'$ and let $R_i = |\mathbf{r} - \mathbf{r}_i'|$ and $\mathbf{a}_{R_i} = (\mathbf{r} - \mathbf{r}_i')/|\mathbf{r} - \mathbf{r}_i'|$.

For the force acting on a charge q_0 at (x,y,z) due to point charges $q_1, q_2, \ldots, q_n$ at $(x_1',y_1',z_1'), \ldots, (x_n',y_n',z_n')$, we can now write

$$\mathbf{F} = \frac{q_0}{4\pi\epsilon_0} \sum_{i=1}^{n} \frac{q_i}{|\mathbf{r} - \mathbf{r}_i'|^2} \frac{\mathbf{r} - \mathbf{r}_i'}{|\mathbf{r} - \mathbf{r}_i'|} = \frac{q_0}{4\pi\epsilon_0} \sum_{i=1}^{n} \frac{q_i \mathbf{a}_{R_i}}{R_i^2} \tag{2.8}$$

In applying Coulomb's law, as in the above example, we have conceptually implied that the charge q_i acts through the intervening medium, in some way, on charge q_0. The formulation of Coulomb is thus an "action-at-a-distance law." In other words, q_i acts through the intervening space directly on charge q_0, causing the observable force, and vice versa, of course.

The same final result can be obtained on the basis of another principle, known as the field concept. Here the view is taken that the charge q_i sets up a field which pervades all space and in particular the point at which q_0 lies. In this view the force exerted by q_i on q_0 is communicated by means of this field. In other words, q_i sets up a field; the field in turn has the property of causing a force to be exerted on a charge in the field. The force depends only on the strength of the field and not on the origin of the field; that is, the field has an independent existence; consequently, it does not even depend on whether a charge q_0 is present to detect it.

When a number of charges are present, each sets up a partial field and the total field is obtained by superposition. The net force on q_0 is the vector sum of the partial forces due to each component field or, equivalently, is the force due to the total field. The symbol **E** is used to represent the electric field.

The electrostatic field **E** that fulfills the property discussed above must consequently be defined by the force that is exerted on a unit charge in the field. It is a vector quantity in the same direction as the force. As we shall see, this definition is not quite satisfactory operationally, and so we modify it to read

$$\mathbf{E} = \lim_{\Delta q \to 0} \frac{\mathbf{F}}{\Delta q} \tag{2.9}$$

The formulation of (2.9) arises from an awareness that in making a measurement of **E** there is always the possibility that the measuring process itself may seriously disturb the conditions existing prior to measurement. The limiting process is introduced in (2.9) for the purpose of ensuring that the introduction of the exploratory charge Δq into the field will not perturb the value of that field, i.e., will not affect the sources of that field. By letting Δq go to zero, the value calculated by (2.9) should approach, as a limit, the field strength prior to measurement.

A fundamental difficulty arises in carrying out the limiting process $\Delta q \to 0$, because charge cannot be subdivided indefinitely. The smallest unit of charge, an electron or positron, is 1.60×10^{-19} coulomb. In practice, so long as Δq is very small compared with the sources producing the field, its introduction into the field may be presumed to have little influence on the behavior of the sources. In this case the ratio of force to charge Δq will satisfactorily evaluate **E**. The definition embodied in (2.9) must thus be qualified by the afore-mentioned restriction.

Although the presence of an electrostatic field is revealed only if a force is exerted on a test charge, the field concept postulates the existence of the field anyway. In the study of electrostatics the field approach and the action-at-a-distance concept are indistinguishable. With time-varying sources, one is forced to ascribe a physical reality to the field because of the finite velocity of propagation of the interaction. Consequently, we shall emphasize the field concept under static conditions as well.

On the basis of the definition of **E** and by using Coulomb's law, we can calculate the electric field set up by a point charge q. In free space the electric field at the point (x,y,z) caused by a point charge q located at (x',y',z') is given by

$$\mathbf{E}(x,y,z) = \frac{\mathbf{F}}{\Delta q} = \frac{q}{4\pi\epsilon_0|\mathbf{r} - \mathbf{r}'|^2} \frac{\mathbf{r} - \mathbf{r}'}{|\mathbf{r} - \mathbf{r}'|} \frac{\Delta q}{\Delta q} = \frac{q\mathbf{a}_R}{4\pi\epsilon_0 R^2} \tag{2.10}$$

For a series of point charges $q_1, q_2, \ldots, q_n$, the total electric field is found by superposing the field from each individual charge; thus

$$\mathbf{E} = \frac{1}{4\pi\epsilon_0} \sum_{i=1}^{n} \frac{q_i}{R_i^2} \mathbf{a}_{R_i} \tag{2.11}$$

where $R_i^2 = (x - x_i')^2 + (y - y_i')^2 + (z - z_i')^2$. The sum indicated in (2.11) must be performed vectorially. The simplest way to proceed is to evaluate R_i and $\mathbf{a}_{R_i} = (\mathbf{r} - \mathbf{r}_i')/R_i$ from their defining relations and combine the x, y, and z components for all values of the summation index i.

When we encounter a large number of point charges in a finite volume, it is convenient to describe the source in terms of a charge density ρ. By conventional concepts we define the charge density as

$$\rho = \lim_{\Delta V \to 0} \frac{\Delta q}{\Delta V} \tag{2.12}$$

where Δq is the algebraic sum of the charge in the volume ΔV. The result of the limiting process is the charge density at the point in question. We have already noted that such a limiting process cannot be completely carried out because the charge cannot be subdivided indefinitely. However, so long as ΔV is small enough so that further decrease in ΔV does not substantially affect the value computed for ρ, yet Δq is large compared with 1.60×10^{-19} coulomb, there is no difficulty in defining ρ. We shall assume that as a result of this process the charge density ρ can be represented by a continuous function of position. Clearly, the total charge Q in a volume V is given by

$$Q = \int_V \rho \, dV \tag{2.13}$$

The electric field set up by an arbitrary volume charge density can be found by superposition since each element $\rho \, dV$ behaves like a point source. Consequently, (2.11) generalizes to

$$\mathbf{E}(x,y,z) = \frac{1}{4\pi\epsilon_0} \int_V \frac{\rho(x',y',z')}{R^2} \mathbf{a}_R \, dV' \tag{2.14}$$

where the integration is taken over the source coordinates (x',y',z'). In this equation R is the distance from the source point dV' to the field point and $\mathbf{a}_R$ is a unit vector in that direction.

Similar remarks can be made concerning the mathematically convenient concept of a surface charge density ρ_s coulombs per square meter and a line charge density ρ_l coulombs per meter as is presented above for a volume charge density. The calculation of electric field from a surface charge distribution ρ_s, for example, would be given by

$$\mathbf{E} = \frac{1}{4\pi\epsilon_0} \int_S \frac{\rho_s}{R^2} \mathbf{a}_R \, dS \tag{2.15}$$

with R and $\mathbf{a}_R$ defined as in (2.14).

Example 2.1. Field from an Infinite Charged Plane. Through application of Coulomb's law and the principle of superposition, we wish to evaluate the field above a uniformly charged infinite plane surface. The charge density is ρ_s coulombs per square meter. Let the point P be an arbitrary field point at a fixed distance h from the plane of charge. We first consider the contribution to the total field from an annular ring of charge centered about the foot of the perpendicular from P, as in Fig. 2.2. This ring contributes to the field in a direction normal to the plane (z direction) only. This becomes clear if we divide the ring into a sum of pairs of charge elements on diametrically opposite sides of the ring, as illustrated. The partial field from each pair is in the z direction. Consequently, the net contribution from the ring of area $dS = 2\pi r \, dr$ is

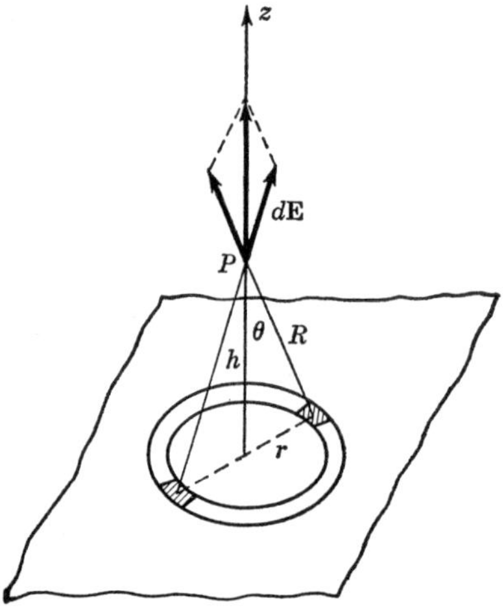

FIG. 2.2. Evaluation of field from an infinite charged plane.

$$dE_z = \frac{\rho_s 2\pi r \, dr}{4\pi\epsilon_0 R^2} \cos\theta \tag{2.16}$$

In this equation we made use of the obvious fact that each element of the ring is the same distance R from P and produces a partial field, making the same angle θ with the z axis. From Fig. 2.2 it is possible to verify that $r = R \sin\theta = h \tan\theta$ and $dr = h \sec^2\theta \, d\theta$. Thus

$$E_z = \frac{\rho_s}{2\epsilon_0} \int_0^{\pi/2} \sin\theta \, d\theta = \frac{\rho_s}{2\epsilon_0} \tag{2.17}$$

Written vectorially, with **n** a unit vector normal to the charge surface and directed away from the surface,

$$\mathbf{E} = \frac{\rho_s}{2\epsilon_0}\mathbf{n} \tag{2.18}$$

Note that the magnitude of **E** is independent of the position of the point P. The field is uniform everywhere above the plane. Below the plane the field is also uniform but pointing in the negative z direction.

2.5. Gauss' Flux Theorem

The electric field defined in (2.9) is an example of a vector field as discussed in Sec. 1.8. In particular, the field may be represented by means of the flux concept. The total flux of **E** from a point source q may be readily calculated by integrating $\mathbf{E} \cdot d\mathbf{S}$ over a surface enclosing q.

Thus consider an element of surface $d\mathbf{S}$ as shown in Fig. 2.3 at a vector distance $r\mathbf{a}_r$ from a charge q, taken as the origin of a spherical coordinate system. The flux through an element of surface $d\mathbf{S}$ is

$$\mathbf{E} \cdot d\mathbf{S} = \frac{1}{4\pi\epsilon_0}\frac{q}{r^2}\mathbf{a}_r \cdot d\mathbf{S} \tag{2.19}$$

But the solid angle $d\Omega$ subtended by $d\mathbf{S}$ at q is

$$d\Omega = \frac{\mathbf{a}_r \cdot d\mathbf{S}}{r^2}$$

and consequently,

$$\mathbf{E} \cdot d\mathbf{S} = \frac{q\,d\Omega}{4\pi\epsilon_0} \tag{2.20}$$

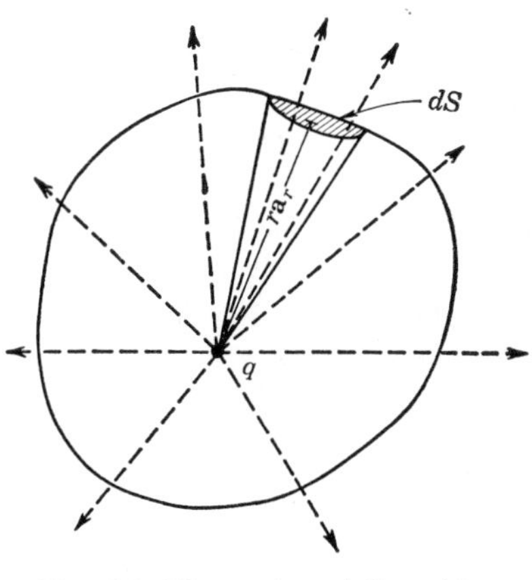

FIG. 2.3. Illustration of Gauss' law.

Let us integrate both sides of (2.20) over any closed surface S containing q. In this case, since the total solid angle is 4π,

$$\oint_S \mathbf{E} \cdot d\mathbf{S} = \frac{q}{\epsilon_0} \tag{2.21}$$

If the charge q lies outside the surface S, then the surface integral vanishes since the total solid angle subtended at q by the surface is zero. The physical interpretation of this result is that flux lines originating from an external charge q and entering the surface S must also leave this surface. If the enclosed volume contains a number of point charges $q_1, q_2, \ldots, q_n$, then (2.21) holds for the partial fields due to each charge separately.

By superposition,

$$\oint_S \mathbf{E} \cdot d\mathbf{S} = \sum_{i=1}^{n} \frac{q_i}{\epsilon_0} = \frac{Q}{\epsilon_0} \tag{2.22}$$

where **E** is now the total field, and Q is the algebraic sum of all charges contained in S. Equation (2.22) is a statement of Gauss' flux theorem. In words we say that the net outflow of flux of **E** through a closed surface

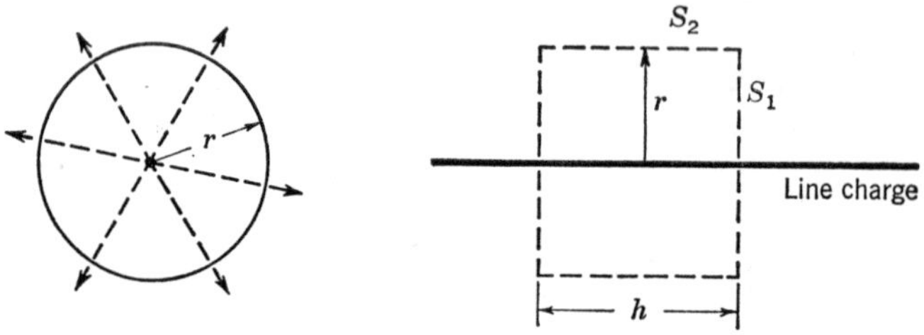

FIG. 2.4. Field from a line charge.

S equals $1/\epsilon_0$ times the total charge (i.e., the algebraic sum of the sources) contained within S. Flux lines must therefore originate on positive charge and terminate on negative charge, the relative number of lines depending on the source strength. This theorem is particularly useful in simplifying the calculation of the electric field from symmetrical distributions of charge. We illustrate this point with the following examples.

Example 2.2. Field from an Infinite Line Charge. An infinite line charge has a strength of ρ_l coulombs per meter. It is desired to find the electric field which it sets up. We note that because of symmetry, the electric field must be a function of r only and, furthermore, can be in the r direction only. Here r is the distance from the line charge to any field point; i.e., it is the radial variable in cylindrical coordinates.

Let us surround the wire by a concentric cylindrical surface of axial length h and radius r, as illustrated in Fig. 2.4. Application of Gauss' flux theorem to this surface leads to

$$\oint_S \mathbf{E} \cdot d\mathbf{S} = \int_{S_1} \mathbf{E} \cdot d\mathbf{S} + \int_{S_2} \mathbf{E} \cdot d\mathbf{S} = \frac{Q}{\epsilon_0} \tag{2.23}$$

where S_1 is the end surfaces and S_2 the cylindrical surface. Since **E** has no component normal to the end surfaces, there is no contribution from that integral. Furthermore, because **E** is normal to the cylindrical surface and uniform over the fixed radius, we can integrate (2.23), giving

$$2\pi r h E_r = \frac{h}{\epsilon_0} \rho_l$$

and hence

$$E_r = \frac{\rho_l}{2\pi\epsilon_0 r} \tag{2.24}$$

In the flux concept, lines of flow of **E** are directed radially outward. Their divergent nature implies a continual reduction in the strength of the field. Indeed, this is borne out by the $1/r$ variation in (2.24).

For contrast, let us calculate the field by direct application of Coulomb's law. Let the line charge lie along the x axis, and take the origin at the foot of the perpendicular from the field point. Consider the field due to

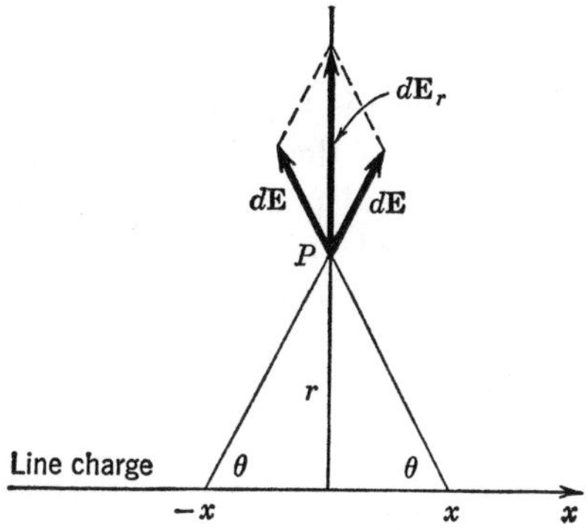

FIG. 2.5

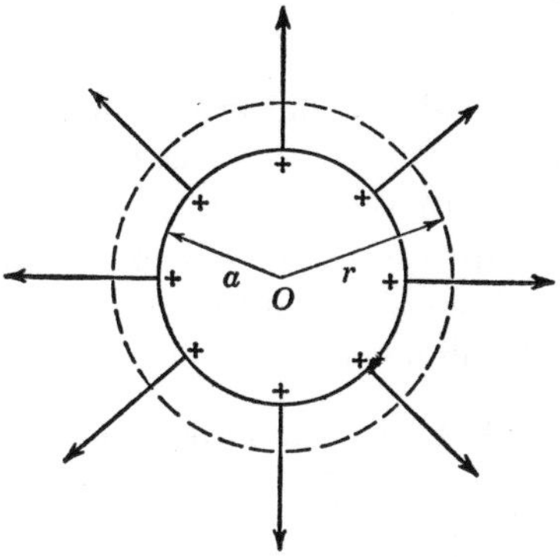

FIG. 2.6. Field from a charged sphere.

two differential charge elements of magnitude $\rho_l\, dx$, with one at x and the other at $-x$. Using the geometry in Fig. 2.5, it is clear that the resultant is in the radial direction and of magnitude

$$dE_r = \frac{\rho_l\, dx}{4\pi\epsilon_0 r^2/\sin^2\theta}\, 2\sin\theta$$

Hence

$$E_r = \frac{\rho_l}{2\pi\epsilon_0 r^2}\int_0^\infty \sin^3\theta\, dx \tag{2.25}$$

Since $x = r\cot\theta$, and $dx = -r\csc^2\theta\, d\theta$, the previous integral becomes

$$E_r = \frac{\rho_l}{2\pi\epsilon_0 r}\int_0^{\pi/2}\sin\theta\, d\theta = \frac{\rho_l}{2\pi\epsilon_0 r} \tag{2.26}$$

The result is the same as that obtained by the simpler Gauss-theorem technique.

Example 2.3. Field from a Charged Sphere. A spherical surface of radius a carries a total charge Q which is uniformly distributed over the surface. It is required to calculate the field from the charge.

From the symmetry of the charge distribution it is clear that the field depends only on the spherical coordinate r, where the origin is at the center of the spherical charge distribution. Furthermore, the electric field can have only a radial component; any other possibility violates either the symmetry or Gauss' flux theorem.

We choose the Gaussian surface to be that of a concentric sphere of radius r, as in Fig. 2.6. Applying Gauss' theorem and noting that E_r is everywhere normal to the spherical surface and uniform thereon leads to

$$\oint_S \mathbf{E} \cdot d\mathbf{S} = 4\pi r^2 E_r = \begin{cases} \dfrac{Q}{\epsilon_0} & r > a \\ 0 & r < a \end{cases}$$

and hence

$$E_r = \begin{cases} \dfrac{Q}{4\pi\epsilon_0 r^2} & r > a \\ 0 & r < a \end{cases} \tag{2.27}$$

The field of a uniformly charged spherical surface is thus identical with that due to a point charge located at the center of the spherical surface and with the same total charge, provided the field point is external to the sphere. No field exists within the spherical surface. Note how much more difficult it would be to calculate the field by direct application of Coulomb's law.

2.6. Electrostatic Potential

In this section we shall show that the electrostatic field may be derived from the negative gradient of a scalar potential function. The scalar potential will be found to be equal to the work done against the field in moving a point charge from infinity up to its final position. The work done in moving a charge around a closed path turns out to be zero. This property classifies the electrostatic field as a conservative field (a field with zero rotation or curl).

Consider the electric field set up by a number of point charges q_1, q_2, . . . , q_n, as illustrated in Fig. 2.7. We desire to evaluate the integral I:

$$I = -\int_C \mathbf{E} \cdot d\mathbf{l} \tag{2.28}$$

This is a contour integral between the points P_1 and P_2 along some arbitrary path C, and $d\mathbf{l}$ is a displacement along this contour. Since $\mathbf{E}$ represents the force on a unit charge, the integral expressed by (2.28) evaluates the total external work (against the field, hence absorbing the minus sign) required to move a unit charge from P_1 to P_2 along C. The integral can be evaluated by using the principle of superposition. Let us first determine the result due to the single charge q_i. If the origin of coordinates is chosen at q_i, then the partial contribution becomes

$$I_i = -\int_C \frac{q_i \mathbf{a}_r \cdot d\mathbf{l}}{4\pi\epsilon_0 r^2} \tag{2.29}$$

An element of the path C is illustrated in Fig. 2.8. From the geometry we note that $\mathbf{a}_r \cdot d\mathbf{l} = dl \sin\theta = dr$. Thus

$$I_i = -\int_C \frac{q_i\,dr}{4\pi\epsilon_0 r^2} = \frac{q_i}{4\pi\epsilon_0}\left(\frac{1}{R_{2i}} - \frac{1}{R_{1i}}\right) \tag{2.30}$$

where R_{2i} is the magnitude of the distance from point P_2 to the position of the ith charge, and R_{1i} is the magnitude of the distance from P_1 to the position of the ith charge. If the effect of all charges is now considered, the result is a summation over i as follows:

$$I = -\int_C \mathbf{E} \cdot d\mathbf{l} = \sum_{i=1}^{n} \frac{q_i}{4\pi\epsilon_0}\left(\frac{1}{R_{2i}} - \frac{1}{R_{1i}}\right) \tag{2.31}$$

A very important conclusion follows from the result expressed by (2.31): the work done in moving a test charge between any two points in an

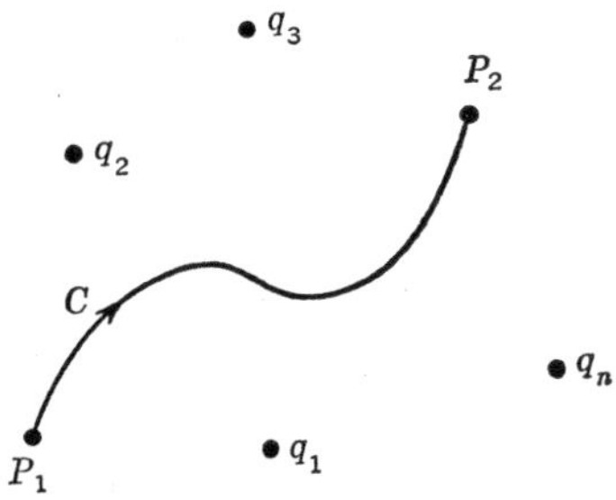

Fig. 2.7. Contour C along which a unit charge is moved.

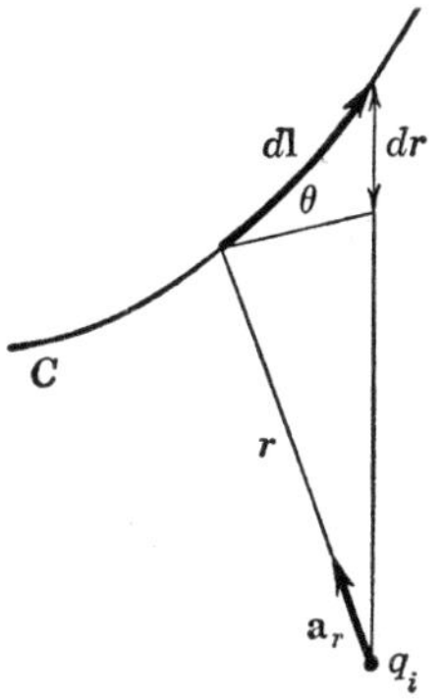

Fig. 2.8

electrostatic field depends only on the position of the end points and not on the path. It follows, for example, that the energy required to move a unit charge from some arbitrary reference point to some other point in the field is unique. Consequently, a relative potential may be assigned to that point, hence to every point, in the field. The "potential" at a point is nothing more than a scalar quantity which designates the energy required to move a unit charge from the reference point to the given point. The aggregate of potentials is a scalar field. Note the close analogy to potential energy in mechanics.

The difference in potential $\Phi_{12} = \Phi(P_2) - \Phi(P_1)$ between the points P_2 and P_1 may be defined as the work required to move a unit positive charge from P_1 to P_2 and is given by (2.31). The difference of potential is independent of the reference potential, of course. When the reference point is chosen as the point at infinity, then the relative potential of all other points under these conditions is known as the absolute potential. From (2.31) it is seen that the absolute potential $\Phi(P_2)$ at the point P_2 is given by

$$\Phi(P_2) = \frac{1}{4\pi\epsilon_0}\sum_{i=1}^{n} \frac{q_i}{R_{2i}} \tag{2.32}$$

From what has been said so far it is clear that the line integral of electric field around any closed path must be zero. For if (see Fig. 2.9)

$$\Phi(P_2) - \Phi(P_1) = -\int_{C_1} \mathbf{E} \cdot d\mathbf{l} \quad \text{and} \quad \Phi(P_1) - \Phi(P_2) = -\int_{C_2} \mathbf{E} \cdot d\mathbf{l}$$

then

$$\oint_{C_1+C_2} \mathbf{E} \cdot d\mathbf{l} = 0$$

In this discussion C_1 and C_2 are completely arbitrary, as is the location of P_1 and P_2. But this means that the electric field is irrotational and consequently can be derived from a scalar potential. This fact has, however, already been noted. Actually, we could have proceeded along more analytic lines, as was done in Sec. 1.17. Specifically, (2.10) can be rewritten by inspection as

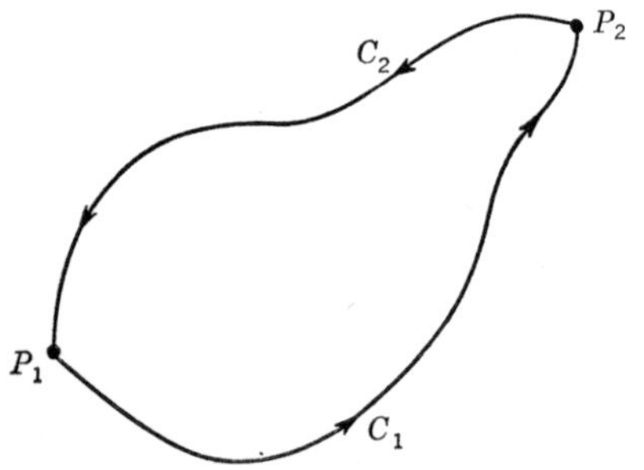

FIG. 2.9

$$\mathbf{E} = -\nabla \frac{q}{4\pi\epsilon_0 R} \tag{2.33}$$

This means that in general we can write

$$\mathbf{E} = -\nabla\Phi \tag{2.34}$$

The negative sign can be understood physically from the fact that **E** is in the direction that a positive charge moves, hence in the direction of *decreasing* potential. The scalar potential, by integration, is

$$\Phi = \frac{q}{4\pi\epsilon_0 R} + C \tag{2.35}$$

for a single point source q. Furthermore, for a volume source density, by superposition,

$$\Phi = \int_V \frac{\rho\, dV'}{4\pi\epsilon_0 R} + C \tag{2.36}$$

where R is the distance from the source to the field point. The value of **E** is unaffected by the choice of the integration constant C, which is determined arbitrarily by assigning a potential to some reference point, as already noted. Since the curl of a gradient is zero, it follows immediately from (2.34) that

$$\nabla \times \mathbf{E} = 0 \tag{2.37}$$

which is further confirmation of the irrotational (conservative) nature of the electrostatic field.

The unit of potential is the volt. A difference of potential $\Phi(P_2) - \Phi(P_1)$ of 1 volt means that 1 joule of work is required to move a coulomb of charge from P_1 to P_2. From (2.34) we note that the electric field **E** may be expressed in units of volts per meter, as well as joules per coulomb.

We have shown that, corresponding to any electrostatic field **E**, there exists a scalar potential field Φ such that the electric field is equal to the negative gradient of the potential field. As a consequence, **E** will be normal to the equipotential surface Φ = constant; it points in the direction of the maximum rate of decrease of potential. If a charge is moved around any closed path, no net energy is required since the electrostatic field is conservative.

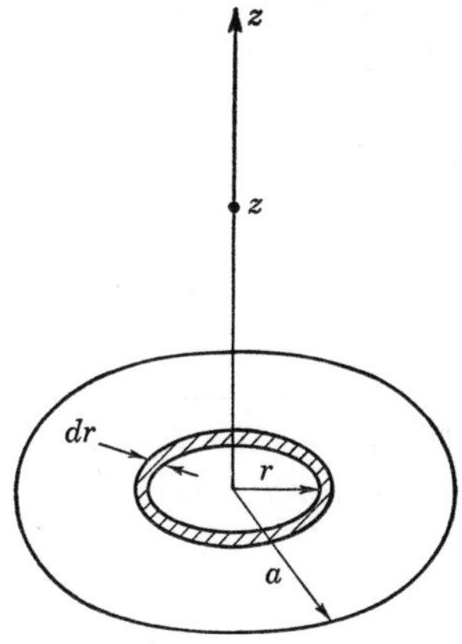

FIG. 2.10. A uniformly charged disk.

Example 2.4. Potential on Axis of Charged Disk. We wish to find the potential along the axis of a uniformly charged disk of radius a, as illustrated in Fig. 2.10. The surface charge density is ρ_s coulombs per square meter. The polar axis is designated z, and the origin is its intersection with the disk.

For an annular ring of radius r and width dr, the contribution to the potential at a point z along the polar axis, using (2.35), is

$$d\Phi = \frac{\rho_s 2\pi r\, dr}{4\pi\epsilon_0 \sqrt{z^2 + r^2}}$$

If we now integrate over all the charge distribution, we obtain

$$\Phi_{\text{axis}} = \frac{\rho_s}{2\epsilon_0} \int_0^a \frac{r\, dr}{\sqrt{z^2 + r^2}} + C$$

$$= \frac{\rho_s}{2\epsilon_0} [(a^2 + z^2)^{1/2} - |z|] + C \tag{2.38}$$

The sign of $(z^2)^{1/2}$ has been chosen so that $\Phi(z) = \Phi(-z)$ and Φ decreases as $|z|$ becomes very large. This is necessitated by the symmetry in z and the requirement that at great distances from the disk the potential behaves like that due to a point source. If the point z recedes to infinity, then the potential can be made equal to zero by taking $C = 0$. Consequently, the absolute potential is

$$\Phi_{\text{axis}} = \frac{\rho_s}{2\epsilon_0} [(a^2 + z^2)^{1/2} - |z|] \tag{2.39}$$

Because of axial symmetry, the electric field on the axis will be in the z direction. As a consequence, the electric field on the axis is related only to the variation in potential along the axis in accordance with (2.34). Note that for a calculation of the **E** field in general it is necessary to know Φ as a function of r, ϕ, z. It is only because of the symmetry here that one can say, a priori, that $\partial\Phi/\partial r = \partial\Phi/\partial\phi = 0$ along the axis. Thus

$$(E_z)_{\text{axis}} = -\frac{\partial\Phi}{\partial z} = \frac{\rho_s}{2\epsilon_0}[1 - z(a^2 + z^2)^{-1/2}] \qquad \text{for } z > 0$$
$$(E_z)_{\text{axis}} = -\frac{\rho_s}{2\epsilon_0}[1 + z(a^2 + z^2)^{-1/2}] \qquad \text{for } z < 0 \tag{2.40}$$

We note that $E_z(z) = -E_z(-z)$, as is clear on physical grounds. The result for $z = 0$ is independent of a, hence corresponds to the earlier solution (2.18) for an infinite sheet of charge. The electric field is discontinuous by an amount ρ_s/ϵ_0 in crossing the charged surface, a result which will be shown (Sec. 2.12) to be independent of the geometry of the surface.

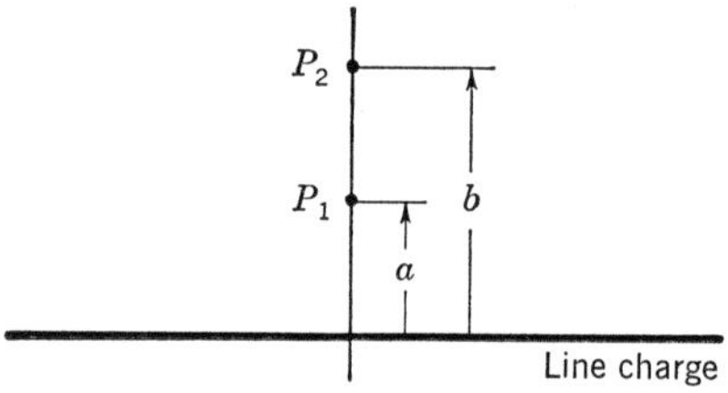

FIG. 2.11. An infinite line charge.

Example 2.5. Potential from a Line Charge. We wish to evaluate the difference in potential between the points P_1 and P_2 in the field of a line charge of infinite extent and of strength ρ_l coulombs per meter. The point P_1 is at a radius a, and P_2 is at a radius b along the same radial line from the line charge, as in Fig. 2.11. From (2.24) the electric field is $E_r = \rho_l/2\pi\epsilon_0 r$; consequently,

$$\Phi(b) - \Phi(a) = -\int_a^b \frac{\rho_l\,dr}{2\pi\epsilon_0 r} = \frac{\rho_l}{2\pi\epsilon_0}\ln\frac{a}{b} \tag{2.41}$$

Since the field is in the radial direction only, equipotential surfaces are concentric cylinders. Consequently, the result in (2.41) is true even if P_1 and P_2 are not on the same radial line. That is, for any $P'(a,\phi,z)$ then $\Phi(P_2) - \Phi(P') = \Phi(P_2) - \Phi(P_1) + [\Phi(P_1) - \Phi(P')]$. But

$$\Phi(P_1) - \Phi(P') = 0$$

since P_1 and P' lie on an equipotential surface $r = a$.

If $b \to \infty$, then the potential relative to infinity as a reference results. Unfortunately, the expression given by (2.41) becomes infinite. The difficulty arises because in this case the infinite line charge itself extends to infinity. Consequently, we cannot express an absolute potential for this problem.

2.7. Conducting Boundaries

So far we have considered the problem of evaluating the electric field from given charge sources. To facilitate this process the electrostatic scalar potential was defined and related to the sources. This relationship is a scalar one and hence much simpler than the vector relationship. The electric field is readily found once the potential function is known.

Usually, however, the distribution of charge is not known. What is given are configurations of metallic bodies which are connected to primary sources such as a battery. The charge distribution will then be a consequence of these conditions, and in a sense, the problem will be to determine the resultant source distribution. We first need to know a little more about the characteristics of metal substances.

A property of metal bodies that has already been noted is that they are good "conductors of electricity"; that is, they readily permit a current flow or motion of charge. As a consequence, if charges are placed on or in conductors, they will move about as long as there is the slightest electric field producing a force on them. The charges move until they reach an equilibrium configuration. This is obviously characterized by the necessity that no field exists within or tangent to the surface of the conductor. This can happen only when all the charges reside on the surface of the conductor, for if any charge remained in the body, then by Gauss' flux theorem a nonzero field in the vicinity of such charges would have to exist. The surface charge at equilibrium must be so distributed that the total electric field inside the conductor and tangential to its surface is zero.

A finite time is required for equilibrium to be essentially achieved if some charge is suddenly placed in a body. This time is designated as the relaxation time and is the basis for distinguishing conductors from insulators. For a good conductor, such as copper, the relaxation time is of the order of 10^{-19} second, while for a good insulator, such as fused quartz, it is 10^6 seconds. For all purposes we may assume the electrostatic field within a conductor to be identically zero.

The above discussion holds also in the case where a conductor is placed in an external electric field. In this case, current, consisting of free charge in the conductor, flows until a surface charge distribution is built up so that the field it produces within the conductor and tangent to its surface just cancels the external field. The body must remain electrically neutral, and hence the algebraic sum of the surface charge is zero in this case. Since the tangential component of electric field is zero on the surface of a conductor, this surface must be an equipotential one. Furthermore, the region within the body has no field; so it too is at the same electrostatic potential. The electrostatic field at the surface of a conductor has a direction normal to the surface.

By applying Gauss' flux theorem it is possible to determine a relationship between the surface charge density on a conductor at equilibrium and the electrostatic field at the surface. Consider a very small portion of any charged conducting surface, as illustrated in Fig. 2.12. An infinitesimal coin-shaped surface is visualized with one broad face parallel to and just outside the conducting surface and the opposite face inside.

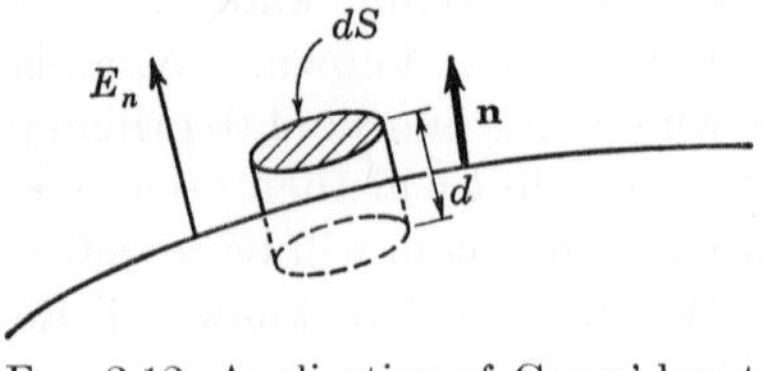

FIG. 2.12. Application of Gauss' law to relate normal **E** to surface charge on a conductor.

No flux crosses the lower surface which is within the conductor since **E** is zero in this region. No flux leaves through the sides since this would require that **E** have a component tangential to the surface. Furthermore, we can let $d \to 0$, so that the area of the sides is of lower order compared with that of the broad faces. Above the surface, however, a normal component of **E** exists. Thus the net outflow of flux of **E** from the closed coin surface is given by $\mathbf{E} \cdot d\mathbf{S}$, and this must equal the net charge within. If we let ρ_s be the surface charge density and denote by E_n the electric field which is in the direction of the outward normal at the surface, we have

$$E_n\, dS = \frac{\rho_s}{\epsilon_0}\, dS$$

or

$$\underline{E_n = \mathbf{n} \cdot \mathbf{E} = \frac{\rho_s}{\epsilon_0}} \tag{2.42}$$

If Fig. 2.12 represents a portion of an infinite plane conductor, then (2.42) seems, superficially, to contradict the result $E_n = \rho_s/2\epsilon_0$ obtained for a plane charge layer [Eq. (2.18)]. The reason for the difference is that (2.42) is a relationship that applies at a conducting surface where the total field on one side of the charge layer is zero. Thus all the lines of

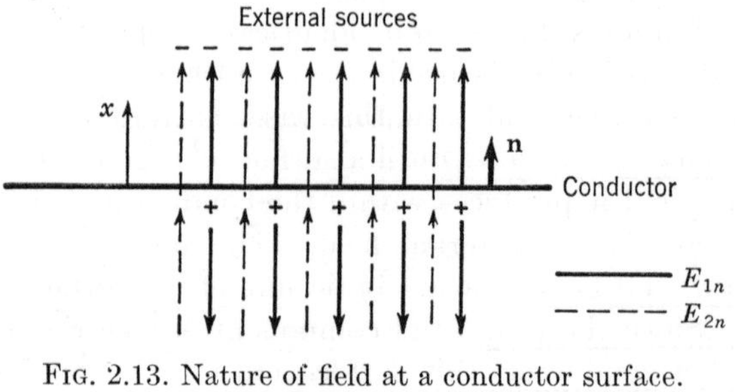

FIG. 2.13. Nature of field at a conductor surface.

flux are directed away from the surface in one direction only. Actually, the field at the surface arises from two systems of charges, the local surface charge ρ_s and the charges that are remote from the conductor and which we could think of as lying uniformly on a distant parallel plane surface. The remote charges are of opposite sign and are physically required in order that the total charge be zero. (They may not be ignored on the basis of their remoteness since the magnitude of the charge is infinite. This is also clear when we note that the field of an infinite surface charge is independent of the distance from the surface.) We realize now that the problem solved in Example 2.1 resulted only in the partial field caused by the sources under consideration and that the total field can be found only if the opposite polarity sources are included. For

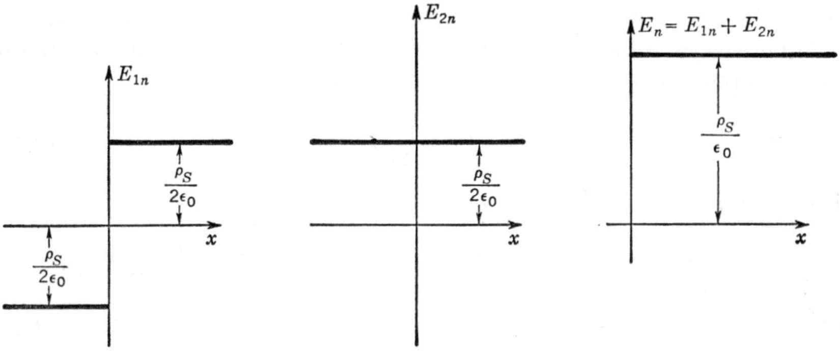

FIG. 2.14. Plot of normal fields as a function of distance x away from conductor surface.

the conducting plane the induced surface charge can be maintained only if these other sources are present. The local surface charge will give a field $\rho_s/2\epsilon_0$ directed in both directions normal to the surface, as explained in Example 2.1. Let this field be denoted by E_{1n}, as in Fig. 2.13. The remote charges contribute a field E_{2n}, which is continuous across the surface charge layer. The combination of the fields E_{1n} and E_{2n} results in a zero field in the conductor and a field $E_n = \rho_s/\epsilon_0$ on the air side of the surface. In the case of both the single plane of charge and a surface charge layer on a conductor, the normal electric field changes discontinuously by an amount ρ_s/ϵ_0 as the charge layer is crossed. A sketch of the fields E_{1n}, E_{2n}, and E_n is given in Fig. 2.14.

Example 2.6. Field between Coaxial Cylinders. A typical example of a problem involving conducting boundaries and known potentials, where the charge and field distribution is to be determined, will now be considered. Figure 2.15 illustrates the cross section of a coaxial cable with air dielectric. The inner cylinder has a radius a; the outer cylindrical conductor has a radius b. By connecting a battery between inner and outer conductors a potential difference V can be established. For definiteness, let the inner

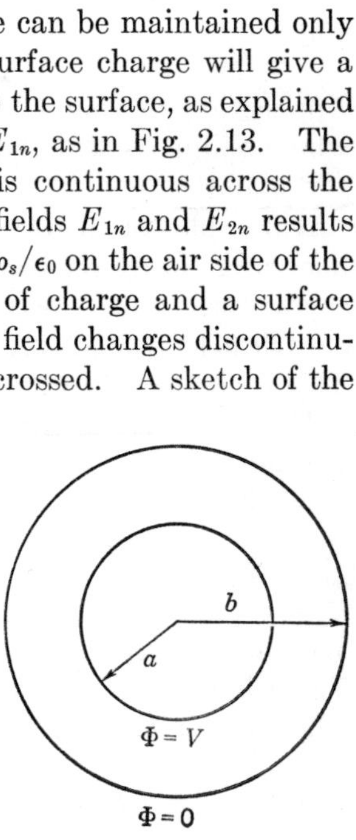

FIG. 2.15. Cross section of coaxial cylinders.

conductor be at a potential V (the entire cylinder is at this potential since conducting surfaces are equipotential surfaces) and the outer conductor at zero potential. We wish to know the electric field everywhere. This is of practical importance since excessive values will cause dielectric breakdown, as will be explained in Chap. 3.

If we can determine the surface charge, then the field can be found. Let us assume ρ_l coulombs per meter on the inner conductor and $-\rho_l$ coulombs per meter on the outer. From symmetry considerations the field between the inner and outer conductor is in the radial direction and by an application of Gauss' law is found to be

$$E_r = \frac{\rho_l}{2\pi\epsilon_0 r} \tag{2.43}$$

From the information given,

$$\Phi(a) - \Phi(b) = V = \int_a^b E_r\,dr = \frac{\rho_l}{2\pi\epsilon_0}\int_a^b \frac{dr}{r}$$

or

$$V = \frac{\rho_l}{2\pi\epsilon_0}\ln\frac{b}{a} \tag{2.44}$$

Consequently,

$$\rho_l = \frac{2\pi\epsilon_0 V}{\ln(b/a)} \tag{2.45}$$

and

$$E_r = \frac{V}{r\ln(b/a)} \qquad a \le r \le b \tag{2.46}$$

Application of Gauss' flux theorem to a concentric cylindrical surface whose radius is less than a or greater than b reveals that $E_r = 0$ for $r < a$ and $r > b$.

A flux plot of the electric field in this problem would show radial flux lines originating on the inner conductor and terminating at the outer conductor. Since the charge density on the inner conductor is $\rho_l/2\pi a$, the electric field, from (2.42), should be of magnitude $V/[a\ln(b/a)] = \rho_l/2\pi a\epsilon_0$. This is confirmed by (2.46) with $r = a$.

If the electric field were uniform in the region $a < r < b$, instead of varying with r, then it would be directly related to the difference of potential and the spacing $b - a$ and given by

$$E \approx \frac{V}{b-a} \tag{2.47}$$

We should expect this approximation to be very good if the spacing $b - a$ is very small compared with b or a. Thus, if we let $b = a + \epsilon$, where $\epsilon \ll a$, then (2.46) can be approximated by

$$E_r = \frac{V}{r\ln(1+\epsilon/a)} \approx \frac{V}{a\epsilon/a} = \frac{V}{\epsilon} = \frac{V}{b-a} \tag{2.48}$$

which is equivalent to (2.47).

2.8. Poisson's Equation

For a volume charge distribution Gauss' flux theorem as expressed in (2.21) can be written

$$\oint_S \mathbf{E} \cdot d\mathbf{S} = \frac{1}{\epsilon_0} \int_V \rho \, dV \tag{2.49}$$

In this equation the surface S encloses the volume V. If, now, Gauss' theorem is used, (2.49) transforms to

$$\oint_S \mathbf{E} \cdot d\mathbf{S} = \int_V \nabla \cdot \mathbf{E} \, dV = \frac{1}{\epsilon_0} \int_V \rho \, dV \tag{2.50}$$

By letting V become very small, it is seen that the integrands in (2.50) must be equal, and a differential relationship between the field and source at the same point results. Thus, at any point,

$$\nabla \cdot \mathbf{E} = \frac{\rho}{\epsilon_0} \tag{2.51}$$

As was noted in Sec. 1.10, the divergence of a vector field at a point P is a measure of the strength of the source at P. In (2.51) we note the physically satisfying interpretation of ρ/ϵ_0 as being the source of the electrostatic field.

If $\mathbf{E}$ is expressed as the negative of the gradient of a scalar potential Φ, the following partial differential equation results:

$$\nabla^2 \Phi = -\frac{\rho}{\epsilon_0} \tag{2.52}$$

This is known as Poisson's equation. For a source-free region of space, Laplace's equation results; i.e.,

$$\nabla^2 \Phi = 0 \tag{2.53}$$

If the $\mathbf{E}$ field is known everywhere in a space in which conducting bodies are present, then by (2.42) the charges on the surfaces of the conductors are known. Conversely, given the surface charge on the conductors, the $\mathbf{E}$ field can be calculated by using (2.15). As we noted earlier, neither kind of information is likely to be available in the typical problem of electrostatics. The fundamental problem in a space free of charge (except for the charged conductors) is to solve for an electrostatic potential that satisfies Laplace's equation and also the boundary conditions on the conducting bodies, namely, Φ_i = constant, on the surface S_i. In addition, the interior of the conductors must have the same potential as the surface. The data that are available are the geometry of the conducting bodies and either the potential or the total charge on each. We shall show, in Sec. 2.9, that the electrostatic field is uniquely determined from such information, while the scalar potential is determined to within some arbitrary constant.

The fundamental problem of electrostatics can be solved in only a relatively few cases. For simple symmetrical geometry the technique using Gauss' law as in Example 2.6 can be employed. This works for parallel-plane and concentric spherical boundaries as well as concentric cylinders. Somewhat more elaborate boundary-value problems can be handled by methods of mathematical physics as exemplified by use of cylindrical and spherical harmonics. This approach will be considered in detail in Chap. 4. For the moment we shall content ourselves with the following example, which illustrates the method of direct solution of Laplace's equation. Its simplicity in the present case arises from the choice of boundaries such that Φ is a function of a single variable only.

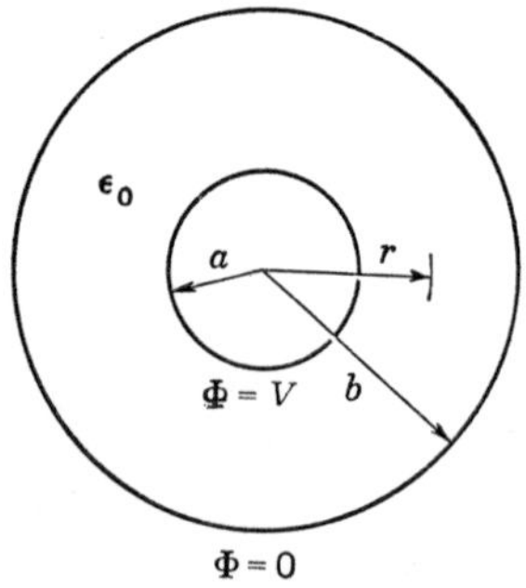

FIG. 2.16. Cross section of coaxial cable.

Example 2.7. Solution of Laplace's Equation. Let us consider here the problem presented in Example 2.6, which was solved by means of Gauss' flux theorem. The geometry is repeated, for convenience, in Fig. 2.16. We wish to determine the field within the coaxial cable subject to the boundary conditions that

$$\Phi(a) = V \qquad \Phi(b) = 0$$

Our purpose is to do this by direct solution of Laplace's equation, thereby illustrating that technique. The cylindrical nature of the boundaries suggests that Laplace's equation be written in cylindrical coordinates (Sec. 1.19).

$$\nabla^2\Phi = \frac{1}{r}\frac{\partial}{\partial r}\left(r\frac{\partial\Phi}{\partial r}\right) + \frac{1}{r^2}\frac{\partial^2\Phi}{\partial\phi^2} + \frac{\partial^2\Phi}{\partial z^2} = 0 \tag{2.54}$$

If the cable is extremely long, then except near the ends, no variation with z is to be expected. Furthermore, because of cylindrical symmetry, the potential cannot be a function of ϕ. Accordingly, (2.54) becomes

$$\frac{1}{r}\frac{d}{dr}\left(r\frac{d\Phi}{dr}\right) = 0 \tag{2.55}$$

It is possible to solve this differential equation by integrating twice; thus

$$\Phi = C_1 \ln r + C_2 \tag{2.56}$$

The integration constants may now be determined from the boundary conditions specified earlier. We have

$$V = C_1 \ln a + C_2 \tag{2.57a}$$
$$0 = C_1 \ln b + C_2 \tag{2.57b}$$

From this pair of equations it is easy to determine that

$$C_1 = \frac{V}{\ln (a/b)} \qquad C_2 = -\frac{V \ln b}{\ln (a/b)}$$

so that

$$\Phi = \frac{V}{\ln (a/b)} (\ln r - \ln b)$$

$$= \frac{V}{\ln (a/b)} \ln \frac{r}{b} \tag{2.58}$$

The electric field is obtained by taking the negative of the gradient of Φ. In cylindrical coordinates, and because Φ is a function of r only,

$$E_r = -\frac{\partial \Phi}{\partial r} = \frac{V}{r \ln (b/a)} \tag{2.59}$$

This result checks with that found in Example 2.6 and given by (2.46).

2.9. Uniqueness Theorem

Consider an arbitrary distribution of conducting bodies in a space free of charge, as in Fig. 2.17. According to the uniqueness theorem, the field is uniquely specified everywhere by giving the potential at the surface of each conductor, or by giving the total surface charge on each, or by giving the potential of some of the conductors and the total charge on the remainder. We proceed, now, to develop this theorem.

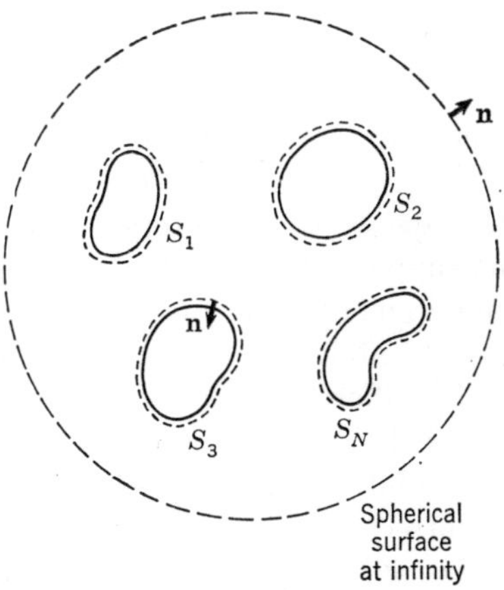

FIG. 2.17. Illustration of volume V enclosed by the surface S (dashed lines).

Let Ψ represent any scalar function which is a solution of Laplace's equation $\nabla^2\Psi = 0$. Expanding $\nabla \cdot \Psi \nabla\Psi$ gives

$$\nabla \cdot \Psi \nabla\Psi = \nabla\Psi \cdot \nabla\Psi + \Psi \nabla^2\Psi = |\nabla\Psi|^2$$

since $\nabla^2\Psi$ is assumed equal to zero. If we integrate both sides throughout a volume V and use Gauss' law (divergence theorem), we obtain, as in Sec. 1.14,

$$\int_V \nabla \cdot \nabla\Psi \, dV = \oint_S \Psi \nabla\Psi \cdot d\mathbf{S} = \oint_S \Psi \frac{\partial\Psi}{\partial n} \, dS = \int_V |\nabla\Psi|^2 \, dV \tag{2.60}$$

where $\mathbf{n}$ is the unit outward normal to the closed surface S surrounding V. This result may be used to prove the uniqueness theorem. In (2.60) the surface S must enclose the volume V. This requirement may be met if we surround our conductors by the surface of a sphere of infinite radius. The total closed surface so obtained is illustrated by the dashed lines in Fig. 2.17. On the surface of the infinite sphere of radius r the potential Ψ decreases at least as fast as $1/r$, $\nabla\Psi$ decreases as $1/r^2$ or faster, and the surface area increases as r^2 only so that

$$\lim_{r\to\infty} \oint_{\text{sphere}} \Psi \nabla\Psi \cdot d\mathbf{S} \to 0$$

With a closed surface constructed in the above manner we can apply (2.60) to obtain

$$\oint_{\text{sphere}} \Psi \frac{\partial\Psi}{\partial n}\, dS + \sum_{i=1}^{N} \oint_{S_i} \Psi \frac{\partial\Psi}{\partial n}\, dS = \sum_{i=1}^{N} \oint_{S_i} \Psi \frac{\partial\Psi}{\partial n}\, dS = \int_V |\nabla\Psi|^2\, dV \tag{2.61}$$

since the integral over the surface at infinity vanishes.

The requirement of the boundary-value problem is to find a potential function Ψ which is a solution of Laplace's equation and which reduces to a specified constant value

$$\Psi = \Phi_i \text{ on conductor } S_i \qquad i = 1, 2, 3, \ldots, K-1$$

where $K - 1$ is less than N. On the remaining conductors S_j ($j = K$, $K + 1, \ldots, N$) the potential Ψ is to be compatible with the condition that on these conductors a total charge Q_j ($j = K, \ldots, N$) exists. Let us suppose that you have found a solution Ψ_1 which satisfies the above requirements and your classmate has found a solution Ψ_2 that also satisfies the above requirements. The two solutions appear to be mathematically different since yours is in a closed form while your classmate's solution is in the form of a series. The question to be settled is whether the two solutions are identical or whether, perhaps, only one of the solutions is correct.

Let Ψ in (2.61) be the difference between your solution and your classmate's solution; that is, $\Psi = \Psi_1 - \Psi_2$. In view of the fact that Ψ_1 and Ψ_2 are known to be solutions of Laplace's equation, then by superposition $\Psi = \Psi_1 - \Psi_2$ is also a solution, so that

$$\sum_{i=1}^{K-1} \oint_{S_i} (\Psi_1 - \Psi_2) \frac{\partial(\Psi_1 - \Psi_2)}{\partial n}\, dS + \sum_{j=K}^{N} \oint_{S_j} (\Psi_1 - \Psi_2) \frac{\partial(\Psi_1 - \Psi_2)}{\partial n}\, dS = \int_V |\nabla(\Psi_1 - \Psi_2)|^2\, dV \tag{2.62}$$

Now on S_i ($i = 1, 2, \ldots, K - 1$), both Ψ_1 and Ψ_2 reduce to the constant values Φ_i. Hence the first set of terms on the left-hand side of

(2.62) vanish. On the remaining conductors Ψ_1 and Ψ_2 are, of course, constant (conductors always have constant potential surfaces) and $\partial\Psi_1/\partial n = \rho_{s1}/\epsilon_0$, $\partial\Psi_2/\partial n = \rho_{s2}/\epsilon_0$, where ρ_{s1} and ρ_{s2} are the surface density of charge for the two solutions; that is, $\partial\Psi/\partial n$ gives the normal electric field at the surface, noting that the outward normal to S_i is the negative of the outward normal at the conducting surface. We may now replace (2.62) by

$$\sum_{j=K}^{N} (\Psi_1 - \Psi_2) \oint_{S_j} \frac{\rho_{s1} - \rho_{s2}}{\epsilon_0} \, dS = \int_V |\nabla(\Psi_1 - \Psi_2)|^2 \, dV \qquad (2.63)$$

But $\oint_{S_j} \rho_{s1} \, dS = \oint_{S_j} \rho_{s2} \, dS = Q_j$, since both solutions are compatible with the condition that the total charge on S_j is Q_j. Hence the left-hand side of (2.63) vanishes, and we must have

$$\int_V |\nabla(\Psi_1 - \Psi_2)|^2 \, dV = 0$$

The integrand is always positive, and therefore the volume integral vanishes only if

$$\nabla(\Psi_1 - \Psi_2) = 0$$

which integrates to

$$\Psi_1 = \Psi_2 + C \qquad (2.64)$$

where C is a constant. However, $\Psi_1 = \Psi_2$ on the surfaces S_i ($i = 1, \ldots, K - 1$); so C must equal zero. Thus the answer to the question posed earlier is that the two solutions are identical. In other words, if a solution can be found that satisfies all the conditions of the problem, then this solution is unique.

In the special case where the total charge on each conducting body is specified, we cannot say that the constant C in (2.64) is zero. For this situation the potential is unique to within an arbitrary constant since no point has been chosen as a reference potential point. The electric field is, however, unique since its value does not depend on the constant C. The previous theorem obviously also includes the case where the potential on all conducting bodies is specified.

2.10. Method of Images

A certain class of boundary-value problems involving infinite conducting planes and wedges, conducting spheres, and conducting cylinders may be solved by a special technique known as the "method of images." When applicable, the technique is very powerful and leads to the solution in a direct manner. However, the technique is not a general one and applies only to a narrow class of problems. On the other hand, some of the concepts involved in the image technique are of a very fundamental nature and applicable to both static and time-varying fields.

Consider the boundary-value problem which consists of a point source q placed a distance h in front of an infinite conducting plane of negligible thickness, as in Fig. 2.18. Now in view of the uniqueness proof of the preceding section, if a potential field can be found such that it is an equipotential over the plane surface and such that its behavior in the immediate vicinity of q is that appropriate to a point source of strength q, and

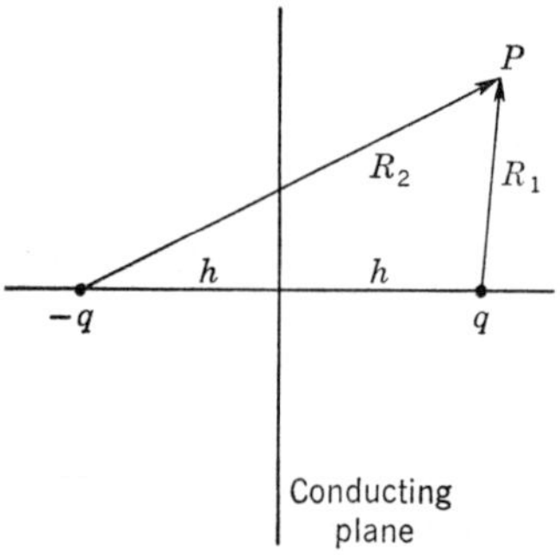

FIG. 2.18. Solution of the problem of a point charge and a conducting plane by the image technique.

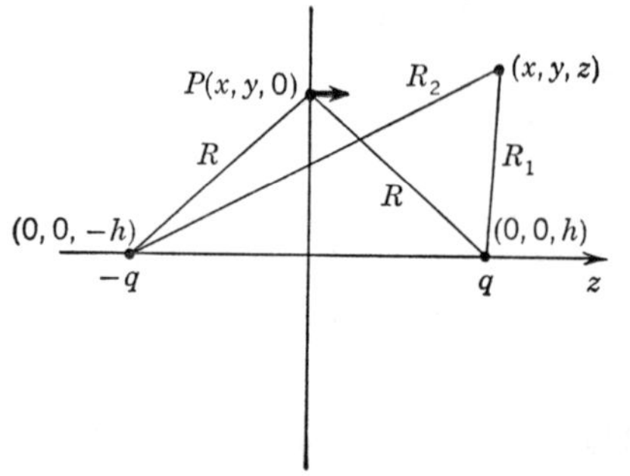

FIG. 2.19

also satisfies Laplace's equation, then it is unique.† This point is being emphasized because sometimes it is possible by intuition, inspection, and/or experience to construct a potential field which meets the necessary requirements. That we know the solution so obtained is unique is, of course, vital.

In the problem at hand all requirements can be met by a scalar potential field which arises from the charge q and a charge $-q$ located at the mirror image of q. The potential at any point P (see Fig. 2.18) is given by

$$\Phi = \frac{q}{4\pi\epsilon_0}\left(\frac{1}{R_1} - \frac{1}{R_2}\right) \tag{2.65}$$

This function clearly satisfies Laplace's equation, since it arises from the superposition of the potentials due to point sources which, individually, are solutions. Furthermore, the behavior of the field in the neighborhood of q corresponds to the field of a point source of strength q. Finally, the potential of points located on the conducting plane, for which $R_1 = R_2$, is constant (arbitrarily zero). Accordingly, (2.65) is the solution to the problem. The charge $-q$ is known as the image charge, and the technique involved is known as the method of images.

† This problem corresponds to the case where both charge and potential are specified. The point charge is a limiting form.

The electric field at the conducting plane can be calculated since the potential field is known. Consequently, we are in a position to determine the charge on the plane. It will be helpful to consider Fig. 2.19 where coordinate axes are set up. The field at $P(x,y,0)$ is desired. At any point (x,y,z) the field in the z direction is given by

$$E_z(x,y,z) = \frac{-q}{4\pi\epsilon_0}\frac{\partial}{\partial z}\left(\frac{1}{R_1} - \frac{1}{R_2}\right)$$

where $R_1 = [x^2 + y^2 + (z - h)^2]^{1/2}$ and $R_2 = [x^2 + y^2 + (z + h)^2]^{1/2}$. After differentiating, we obtain

$$E_z(x,y,z) = \frac{-q}{4\pi\epsilon_0}\left\{\frac{h - z}{[x^2 + y^2 + (z - h)^2]^{3/2}} + \frac{h + z}{[x^2 + y^2 + (z + h)^2]^{3/2}}\right\}$$

The normal field E_n at the surface of the conducting plane is $E_z(x,y,0)$ and is given by

$$E_n = -\frac{qh}{2\pi\epsilon_0 R^3} = \frac{\rho_s}{\epsilon_0} \tag{2.66}$$

where $R^3 = (x^2 + y^2 + h^2)^{3/2}$. The surface charge density is, consequently, inversely proportional to the cube of the distance from the point charge. The total charge on the plane is evaluated by the following integral:

$$\int_S \rho_s\, dS = -\frac{qh}{2\pi}\int_S \frac{1}{R^3}\, dS \tag{2.67}$$

Using polar coordinates r, θ, then $R^2 = h^2 + r^2$ and $dS = r\, dr\, d\theta$. Thus

$$\int_S \rho_s\, dS = -\frac{qh}{2\pi}\int_0^\infty\int_0^{2\pi}\frac{r\, dr\, d\theta}{(h^2 + r^2)^{3/2}} = -qh\int_0^\infty \frac{r\, dr}{(h^2 + r^2)^{3/2}}$$

$$= \frac{qh}{(h^2 + r^2)^{1/2}}\Bigg|_0^\infty = -q \tag{2.68}$$

It is apparent that all lines of electric flux emanating from the point charge q terminate on the plane conductor. An illustration of a portion of these flux lines is given in Fig. 2.20. For interest, equipotential lines are drawn in as well.

Although the combination of the image charge $-q$ and the point charge q gives the correct field in the half space containing q, the image charge is, of course, fictitious. The component of field evaluated as due to the image charge is actually caused by the surface distribution ρ_s given in (2.66). The fields due to either $-q$ or ρ_s are fully equivalent in the half space containing q. For field points in the region $z < 0$, the sources continue to be ρ_s on the plane and q at $(0,0,h)$. Now if, for the points in the region $z > 0$, ρ_s on the plane is equivalent to the point charge $-q$ at $(0,0,-h)$, then by symmetry the field due to ρ_s in the region $z < 0$ is equivalent to $-q$ but at $(0,0,h)$. This means that the total field is due to $+q$ and $-q$, which are both at $(0,0,h)$, and hence the field in the space $z < 0$ is zero. The conducting sheet may be thought of as shielding the region $z < 0$ from electrostatic field sources in the region $z > 0$.

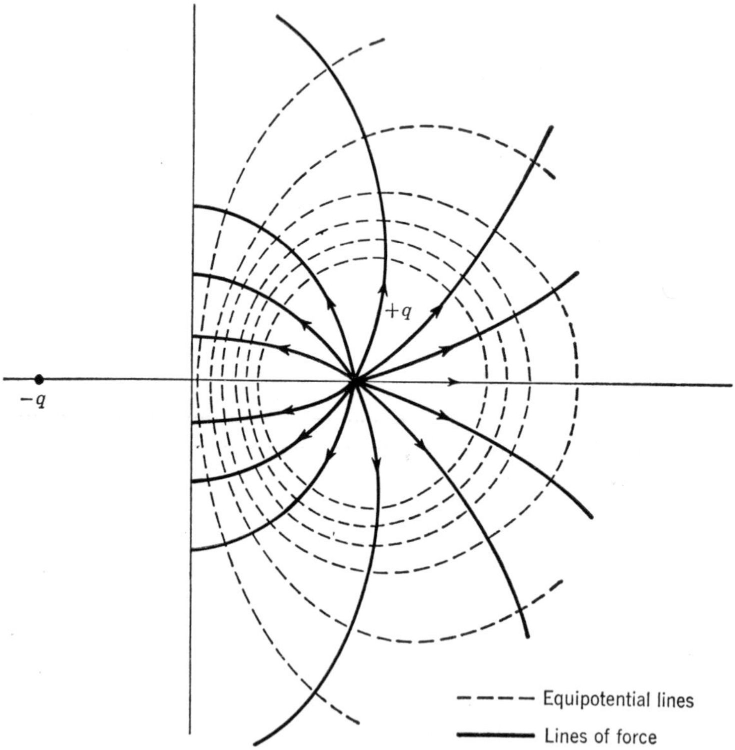

FIG. 2.20. Flux lines from a point charge in front of a conducting plane.

The problem considered here shows how charges may be induced in conducting bodies in the presence of other charges. This illustrates the reason why in measuring the electric field by taking the ratio $\mathbf{F}/\Delta q$ the test charge Δq must be made very small if it is not to disturb a priori conditions. Thus suppose q in Fig. 2.19 is actually introduced at $(0,0,h)$ to measure the field that exists prior to making the measurement (in this case zero, of course). Note that a nonzero force of strength $q^2/[4\pi\epsilon_0(2h)^2]$ is measured and an erroneous field of $q/16\pi\epsilon_0 h^2$ is presumed to have existed. The error in the measurement depends on the size of q.

A simple extension of the image technique involving a point source and a conducting plane is the case where a point source is located within two intersecting planes, such as q in the right angle AOB of Fig. 2.21. The image of q in OA is $-q$ at P_1, while the image of q in OB is $-q$ at P_2. But the combination of all three charges makes neither OA nor OB an equipotential surface. What has been neglected is imaging the images, a process that must be continued repeatedly. In this example, q at P_3 satisfies both the requirement of $-q$ at P_1 and $-q$ at P_2 for proper imaging, and the group of four charges satisfies the boundary conditions and gives the correct solution for the field in AOB. The boundary requirements can be satisfied with a finite number of images only if AOB is an exact submultiple of 180°, as in the 90° case illustrated. For the purpose of evaluating the position of the image charges it should be noted that they all lie on a circle.

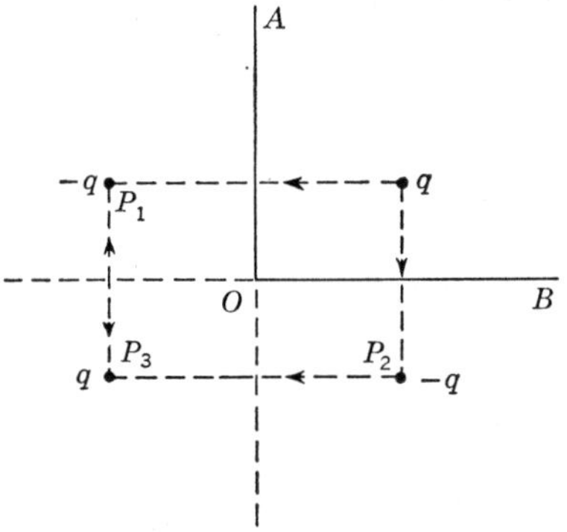

FIG. 2.21. A point charge located inside a 90° corner.

Example 2.8. Inversion in a Sphere. Another type of boundary-value problem that can be solved by the method of images is considered in this example. The charge q_1 is given as being located at a point P_1, a distance R_1 from the center of a conducting sphere of radius a, where $R_1 > a$. The field external to the sphere is desired.

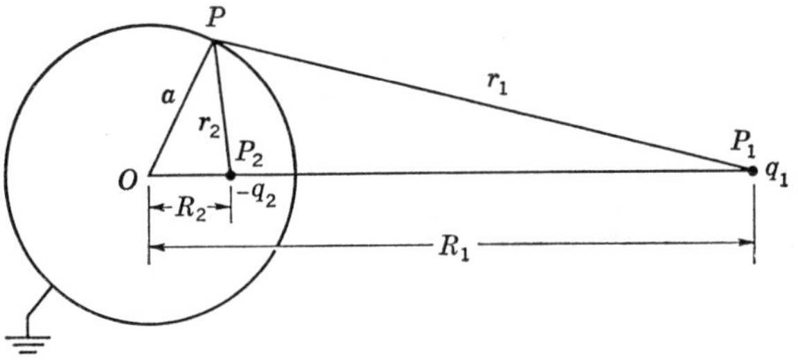

FIG. 2.22. Imaging of point charge q_1 in a sphere.

Let us consider an image charge $-q_2$ located at P_2 which is a distance R_2 from the center of the sphere along the line OP_1, as in Fig. 2.22. It is necessary that the combination of q_1 and $-q_2$ make the spherical surface a zero potential surface (the sphere is grounded). If we take any arbitrary point P on the sphere, then it is required that

$$\frac{q_1}{4\pi\epsilon_0 r_1} - \frac{q_2}{4\pi\epsilon_0 r_2} = 0 \qquad (2.69)$$

This will always be satisfied if we take

$$\frac{q_1}{q_2} = \frac{r_1}{r_2} \tag{2.70}$$

provided that an R_2 can be found so that r_1/r_2 is a constant, independent of the position P. Since any arbitrary point P and the line OP_1 determine a plane, points in Fig. 2.22 can be thought of as located in that plane. If $OP_2 = R_2$ is chosen so that

$$\frac{OP_2}{a} = \frac{a}{OP_1} \tag{2.71}$$

then the triangle OP_2P is similar to triangle OP_1P, and consequently

$$\frac{r_1}{r_2} = \frac{a}{OP_2} \tag{2.72}$$

will be a constant, as is necessary. To summarize, the field due to a charge q, a distance R_1 from the center of a grounded sphere of radius a, is found from the charge q_1 and an image charge $-q_2$ whose magnitude is

$$q_2 = \frac{R_2}{a} q_1 = \frac{a}{R_1} q_1 \tag{2.73}$$

and which is located on the line joining the center of the sphere and q_1 and at a distance

$$R_2 = \frac{a^2}{R_1} \tag{2.74}$$

from the center. Because of this relation the image technique in connection with a sphere is referred to as inversion in a sphere.

If the sphere were not grounded, then to maintain electrical neutrality an additional charge $+q_2$ must be placed inside the sphere.† (For the grounded sphere, $+q_2$ is essentially removed to infinity.) The location of $+q_2$ must be such as not to destroy the surface of the sphere as an equipotential. This is achieved by locating it at the center. Then the potential of the combination at any external field point is given as

$$\Phi = \frac{1}{4\pi\epsilon_0}\left(\frac{q_1}{r_1} - \frac{q_2}{r_2} + \frac{q_2}{r_0}\right) \tag{2.75}$$

† The total charge induced on the grounded sphere must be $-q_2$ because the combination $-q_2$ and q_1 correctly describes the electric field at the surface of the sphere according to the image theory just developed. Thus it would surely show the lines of flux as if they would terminate on $-q_2$. The location of a $+q_2$ within the sphere would lead to no net flux terminating on the sphere, as is necessary in the nongrounded case.

and the geometry is as given in Fig. 2.23. Actually, the charge $+q_2$ at the center is only an image charge that produces the same external effect as the uniform charge distribution $\rho_s = q_2/4\pi a^2$ residing on the outer surface of the sphere. The total surface charge on the sphere is zero

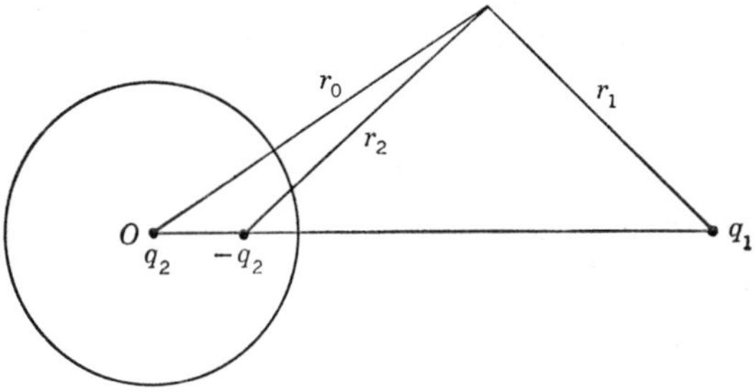

FIG. 2.23. Image charges $+q_2$ and $-q_2$ inside an ungrounded sphere.

since it equals the sum of the uniform distribution $q_2/4\pi a^2$ and a non-uniform distribution of total amount $-q_2$ which sets up the same external field as the $-q_2$ image charge. The total surface charge density could, of course, be found from (2.75) in the manner illustrated in connection with the infinite plane conductor.

Example 2.9. Inversion in a Cylinder. The problem of a line charge parallel with and outside a conducting cylinder of radius a may be solved in a manner similar to the sphere problem considered in the previous

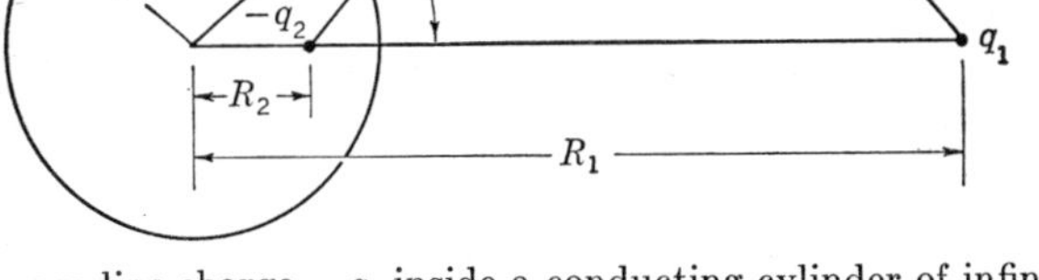

FIG. 2.24. Image line charge $-q_2$ inside a conducting cylinder of infinite length.

example. Consider a line charge of density q_1 coulombs per meter located a distance R_1 from the center of the cylinder and also an image line charge of density $-q_2$ at a distance R_2 from the center, as in Fig. 2.24.

The potential at any point P is, from (2.41),

$$\Phi(P) = -\frac{q_1}{2\pi\epsilon_0}\ln r_1 + \frac{q_2}{2\pi\epsilon_0}\ln r_2 + C$$

where C is an arbitrary constant, depending on the reference potential point. If we choose $q_2 = q_1$, we obtain

$$\Phi(P) = -\frac{q_1}{4\pi\epsilon_0} \ln \left(\frac{r_1}{r_2}\right)^2 + C \tag{2.76}$$

The constant potential surfaces are given by the condition

$$\frac{r_1}{r_2} = \text{constant} \tag{2.77}$$

The geometry here is the same as in the sphere problem, and condition

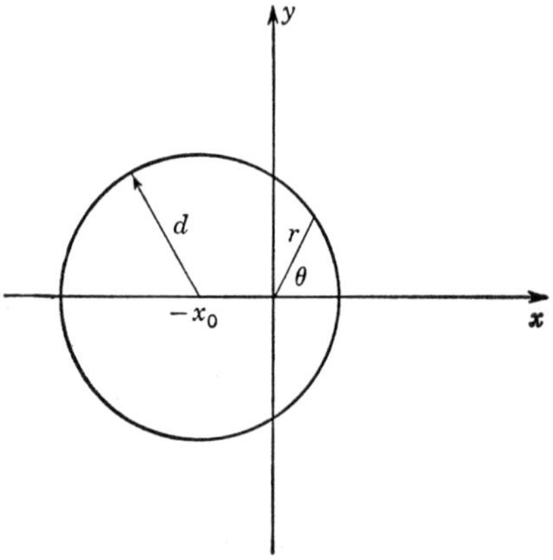

FIG. 2.25. Circle with center at $x = -x_0$, $y = 0$.

(2.77) is also the same; hence if we choose

$$R_2 = \frac{a^2}{R_1} \tag{2.78}$$

condition (2.77) will hold.

It will be of interest to derive equations for the entire family of equipotential surfaces specified by (2.77) and (2.78). From the law of cosines we get

$$r_2^2 = r^2 + R_2^2 - 2rR_2 \cos\theta$$
$$r_1^2 = r^2 + R_1^2 - 2rR_1 \cos\theta$$

To satisfy (2.77), let $r_1^2 = kr_2^2$, and then

$$r_1^2 - kr_2^2 = 0 = (1-k)r^2 + R_1^2 - kR_2^2 - 2r(R_1 - kR_2)\cos\theta \tag{2.79}$$

A circle with center at $x = -x_0$, $y = 0$ and radius d, as in Fig. 2.25, has the equation

$$r^2 + 2rx_0 \cos\theta + x_0^2 - d^2 = 0 \tag{2.80}$$

Comparing with (2.79) shows that

$$x_0 = \frac{R_1 - kR_2}{k - 1} \tag{2.81a}$$

$$d^2 - x_0^2 = \frac{R_1^2 - kR_2^2}{k - 1} \tag{2.81b}$$

In the example we are considering, the radius of the cylinder is a, and hence $d = a$. Also the center of the cylinder is located at the origin of coordinates so that $x_0 = 0$. Thus from (2.81a) we find that

$$k = \frac{R_1}{R_2} \tag{2.82}$$

and from (2.81b) we confirm the relationship given in (2.78). The family of equipotential surfaces depends on the parameter k, and when k is given

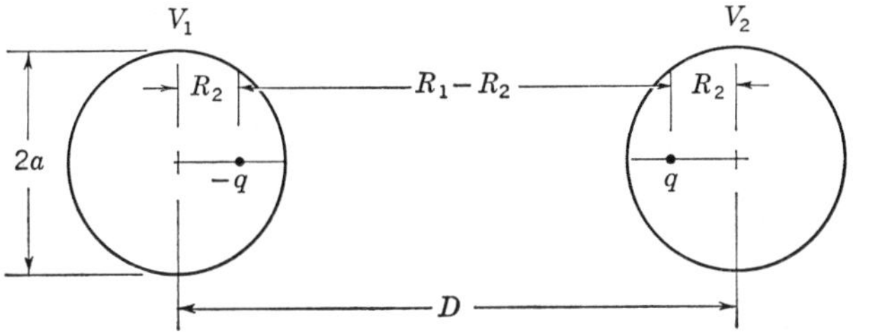

FIG. 2.26. Two-conductor transmission line.

by (2.82), the equipotential surface that coincides with the conducting cylinder is specified. The potential of the cylindrical conducting surface is

$$\Phi(a) = -\frac{q_1}{4\pi\epsilon_0} \ln k + C = -\frac{q_1}{4\pi\epsilon_0} \ln \frac{R_1}{R_2} + C \tag{2.83}$$

If the cylinder is grounded, then (2.83) determines C, since $\Phi(a)$ will be zero. The potential at any other point is then given by

$$\Phi(P) = -\frac{q_1}{4\pi\epsilon_0} \ln \left(\frac{r_1}{r_2}\right)^2 + \frac{q_1}{4\pi\epsilon_0} \ln \frac{R_1}{R_2} = -\frac{q_1}{4\pi\epsilon_0} \ln \frac{r_1^2 R_2}{r_2^2 R_1} \tag{2.84}$$

The above results may be used to solve the two-wire transmission-line problem. Consider two infinitely long parallel cylinders of radius a and center-to-center separation D, as in Fig. 2.26. If we choose the constant C to be zero in (2.83) and hypothesize that the distributed surface charge on the cylinders is equivalent to the line charges $-q$ and $+q$, as illustrated in Fig. 2.26, then the potential of the left-side cylindrical surface will be, as for Fig. 2.24,

$$V_1 = \frac{-q}{4\pi\epsilon_0} \ln \frac{R_1}{R_2} \tag{2.85a}$$

By symmetry, a similar constant potential surface surrounds the positive line charge and the potential of the right-side cylinder is

$$V_2 = \frac{q}{4\pi\epsilon_0} \ln \frac{R_1}{R_2} = -V_1 \tag{2.85b}$$

The fact that the cylindrical surfaces can be made equipotential ones is the basis of the original hypothesis. We must still determine R_1 and R_2. From the geometry of Fig. 2.26 it is seen that $D = R_1 + R_2$. Also $R_1R_2 = a^2$, and hence

$$R_1 = \frac{D}{2} + \left(\frac{D^2}{4} - a^2\right)^{1/2} \tag{2.86a}$$

$$R_2 = \frac{D}{2} - \left(\frac{D^2}{4} - a^2\right)^{1/2} \tag{2.86b}$$

Since R_1 and R_2 are uniquely specified, satisfaction of (2.86) will always lead to a valid solution of the boundary-value problem. The potential difference between the cylinders is

$$V_2 - V_1 = \frac{q}{2\pi\epsilon_0} \ln \frac{D + (D^2 - 4a^2)^{1/2}}{D - (D^2 - 4a^2)^{1/2}} \tag{2.87}$$

and may be adjusted to any given value by a proper choice of the line charge density q.

The equipotential surfaces are circles with centers located a distance x_0 from the cylinder center and away from the other cylinder and having a radius d. Equations (2.81) and (2.83) relate the parameters x_0 and d to the particular constant potential surface (value of k) being considered. The magnitude of the total charge on each cylinder is q coulombs per meter and could be determined from (2.87) if the difference in potential between the conductors were given.

2.11. Dipoles and Volume Discontinuities

The field due to two point charges of equal magnitude but opposite sign at distances which are large compared with the separation of the charges is called a dipole field. The product of the charge times the separation is the dipole moment. The dipole field is of particular importance, as we shall see, in the discussion of dielectrics.

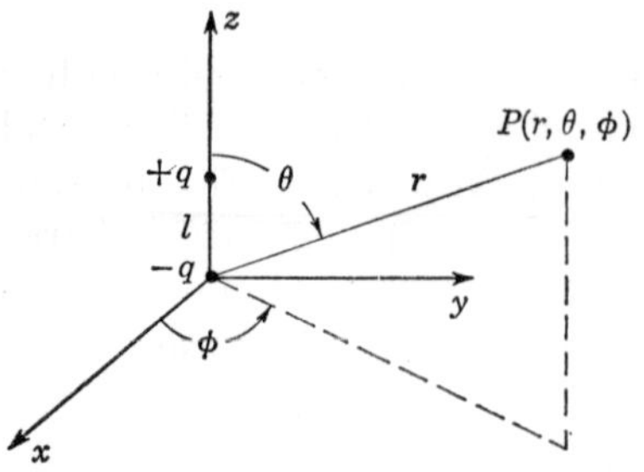

FIG. 2.27. The electric dipole.

Consider Fig. 2.27, where a dipole of charge magnitude q and separation l is illustrated. The origin has been chosen at the charge $-q$, and the polar axis is defined by the line joining $-q$ and q. An obvious and correct way of finding the field at an arbitrary point P is to superpose the field due to each point charge separately. We are interested in the potential field as well as the electric field and will determine the former first.

Referring to Fig. 2.27, the potential at P due to $-q$ alone is

$$\Phi(P)_- = \frac{-q}{4\pi\epsilon_0 r} \tag{2.88}$$

Since the charge $+q$ is displaced a distance l along the positive z axis, its contribution to the potential at P is the same as what would be found by displacing P a distance l in the negative z direction with $+q$ located at the origin. By Taylor's theorem, the latter comes out to be

$$\Phi(P)_+ = \frac{q}{4\pi\epsilon_0 r} + \frac{\partial}{\partial z}\left(\frac{q}{4\pi\epsilon_0 r}\right)\bigg|_P (-l) + \frac{1}{2!}\frac{\partial^2}{\partial z^2}\left(\frac{q}{4\pi\epsilon_0 r}\right)\bigg|_P (-l)^2 + \cdots \tag{2.89}$$

The total dipole potential at P, by superposition, is then

$$\Phi(P) = -\frac{\partial}{\partial z}\left(\frac{q}{4\pi\epsilon_0 r}\right)\bigg|_P l + \frac{1}{2!}\frac{\partial^2}{\partial z^2}\left(\frac{q}{4\pi\epsilon_0 r}\right)\bigg|_P l^2 + \cdots \tag{2.90}$$

Since

$$r = \sqrt{x^2 + y^2 + z^2}$$

$$\frac{\partial}{\partial z}\left(\frac{1}{r}\right) = \frac{-z}{r^3}$$

$$\frac{\partial^2}{\partial z^2}\left(\frac{1}{r}\right) = -\frac{1}{r^3}\left(1 - \frac{3z^2}{r^2}\right) \qquad \text{etc.}$$

For $l \ll r$, as originally specified, all terms except the first can be neglected and the dipole potential reduces to

$$\Phi(P) = \frac{qzl}{4\pi\epsilon_0 r^3} \tag{2.91}$$

where z and r correspond to the point P. Now

$$\cos\theta = \frac{z}{r}$$

and hence

$$\Phi(P) = \frac{ql\cos\theta}{4\pi\epsilon_0 r^2} \tag{2.92}$$

The product ql is the dipole moment p. It can be written as a vector, with a direction defined as from $-q$ to $+q$. In this case (2.92) has the simple form

$$\Phi(P) = \frac{\mathbf{p}\cdot\mathbf{a}_r}{4\pi\epsilon_0 r^2} \tag{2.93}$$

The electric field of the dipole can be readily found by taking the negative gradient of (2.93), using spherical coordinates. We get

$$\mathbf{E} = \frac{ql}{4\pi\epsilon_0 r^3}(\mathbf{a}_r 2\cos\theta + \mathbf{a}_\theta \sin\theta) \tag{2.94}$$

A sketch of the lines of force in a dipole field is given in Fig. 2.28. The three-dimensional picture of the lines of force is obtained by revolving the pattern in Fig. 2.28 about the dipole axis. In the r, θ plane the differential equation for the lines of force is

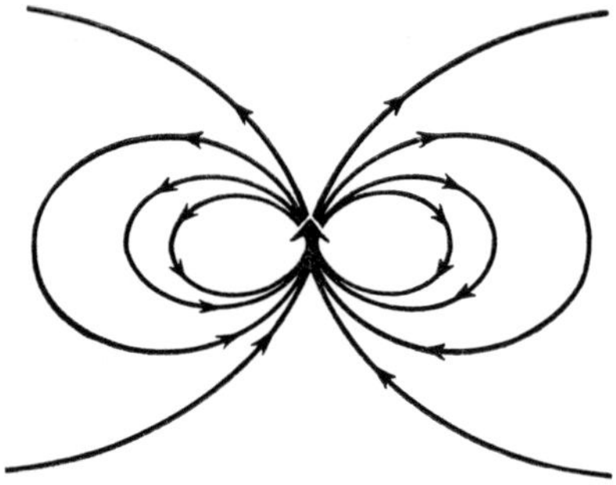

FIG. 2.28. Lines of force in an electric dipole field.

$$\frac{dr}{E_r} = \frac{r\,d\theta}{E_\theta} \qquad \text{or} \qquad \frac{dr}{r} = 2\cot\theta\,d\theta$$

after substituting from (2.94). This equation is readily integrated to give

$$\ln r = 2\ln\sin\theta + \ln C$$

or

$$r = C\sin^2\theta$$

For each value of the integration constant C, a particular line of force is obtained. The constant potential contours are orthogonal to the lines of force and hence have a slope which is the negative reciprocal of the slope of the lines of force. The differential equation is therefore

$$\frac{dr}{r\,d\theta} = -\tfrac{1}{2}\tan\theta$$

This may be integrated to give

$$\ln r = \tfrac{1}{2}\ln\cos\theta + \tfrac{1}{2}\ln C$$

or

$$r^2 = C\cos\theta$$

That this equation is the equation for a constant potential contour is readily seen from (2.92) by equating Φ to a constant. Each value of the constant C determines a particular constant potential surface.

For a volume distribution of charge it was convenient to specify a charge density ρ as a mathematically well-defined function. Its definition and the restrictions required were discussed in Sec. 2.4. In the same way a volume distribution of dipoles can be represented by a vector function $\mathbf{P}$, which gives the dipole moment per unit volume at a point. Specifically,

$$\mathbf{P} = \lim_{\Delta V \to 0} \frac{\Sigma \mathbf{p}_i}{\Delta V} \tag{2.95}$$

where $\Sigma\mathbf{p}_i$ is the vector sum of the dipole moments in the volume ΔV. It is again necessary that ΔV be sufficiently large so that individual dipole characteristics do not affect the result but small enough to get a true limit. We shall assume that the **P** that results is a continuous function.

The potential set up by an arbitrary volume distribution of dipoles will now be computed by using the above definition of **P**. Let $\mathbf{P}(x',y',z')$ be the dipole density at the point (x',y',z'), and let dV' be an element of volume. Since the distance to a field point is inherently large compared with the extent of the differential volume element dV', (2.93) applies at any field point. Consequently, by superposition,

$$\Phi(x,y,z) = \int_V \frac{\mathbf{P}(x',y',z') \cdot \mathbf{a}_R}{4\pi\epsilon_0 R^2} \, dV' \tag{2.96}$$

where $R^2 = (x - x')^2 + (y - y')^2 + (z - z')^2$ and is the distance between the volume element dV' and the field point (x,y,z). This result can be transformed into one that has an interesting physical interpretation. Noting that

$$\nabla' \left(\frac{1}{R}\right) = \frac{\mathbf{a}_R}{R^2}$$

we have

$$\frac{\mathbf{P} \cdot \mathbf{a}_R}{R^2} = \mathbf{P} \cdot \nabla' \left(\frac{1}{R}\right) = \nabla' \cdot \frac{\mathbf{P}}{R} - \frac{\nabla' \cdot \mathbf{P}}{R}$$

Thus (2.96) becomes

$$\Phi(x,y,z) = \frac{1}{4\pi\epsilon_0} \int_V \nabla' \cdot \frac{\mathbf{P}}{R} \, dV' - \frac{1}{4\pi\epsilon_0} \int_V \frac{\nabla' \cdot \mathbf{P}}{R} \, dV'$$
$$= \oint_S \frac{\mathbf{P} \cdot \mathbf{n}}{4\pi\epsilon_0 R} \, dS' + \int_V \frac{-\nabla' \cdot \mathbf{P}}{4\pi\epsilon_0 R} \, dV \tag{2.97}$$

after using Gauss' theorem on the first volume integral. From a comparison of (2.97) with (2.36) we deduce that the field due to a volume dipole distribution **P** is the same as that from an equivalent volume and surface charge distribution such that

$$\rho_p = -\nabla' \cdot \mathbf{P} \qquad \text{in } V \tag{2.98a}$$

and

$$\rho_{sp} = \mathbf{P} \cdot \mathbf{n} \qquad \text{on } S \tag{2.98b}$$

Actually, if the volume be taken to include the surface, then the surface charge distribution is included in the expression for ρ_p; that is, just as we can let ρ contain ρ_s as a limiting case, we let $-\nabla' \cdot \mathbf{P}$ include $\mathbf{P} \cdot \mathbf{n}$ as a limiting case and

$$\Phi = \int_V \frac{-\nabla' \cdot \mathbf{P}}{4\pi\epsilon_0 R} \, dV' \tag{2.99}$$

The results expressed by (2.98) can be understood on a purely physical basis. Where $\nabla' \cdot \mathbf{P} \neq 0$, this means a net creation of dipole moment per unit volume, hence an incomplete cancellation of charge density from adjacent dipoles. Similarly, the surface charge density occurs because the dipoles ending on the surface cannot be neutralized for want of an adjacent dipole layer, as illustrated in Fig. 2.29.

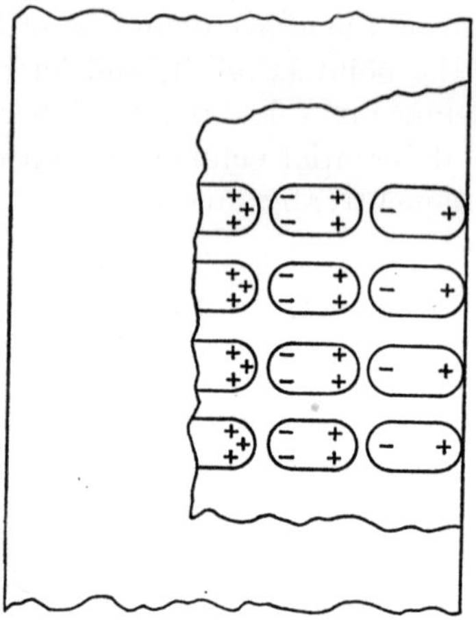

FIG. 2.29. Illustration of creation of volume and surface polarization charge because of incomplete cancellation of charge between adjacent dipoles.

2.12. Field Behavior at a Charged Surface

A general shaped open surface on which an arbitrary surface charge density $\rho_s(x',y',z')$ exists is illustrated in Fig. 2.30. In solving certain types of problems in electrostatics it is necessary to know the behavior of the scalar potential and the electric field in crossing such a surface. This can be determined from the fundamental relationships. Thus the scalar potential Φ due to the surface charge distribution is given by

$$\Phi = \frac{1}{4\pi\epsilon_0} \int_S \frac{\rho_s}{R} \, dS' \tag{2.100}$$

where R is the distance between the charge element $\rho_s \, dS'$ and the field point; that is, $R = |\mathbf{r} - \mathbf{r}'|$. Consider the variation in Φ as the field point moves along a path Γ that crosses the charged surface. In the vicinity of the surface the path followed may be considered linear and a small element of surface (ΔS_0) surrounding the intersection of the path with the surface may be considered plane. Figure 2.30 illustrates the over-all geometry, and Fig. 2.31 is an enlargement of the region near the surface. We can resolve (2.100) into the following components:

$$\Phi = \frac{1}{4\pi\epsilon_0} \int_{S-\Delta S_0} \frac{\rho_s \, dS'}{R} + \frac{1}{4\pi\epsilon_0} \int_{\Delta S_0} \frac{\rho_s \, dS'}{R} \tag{2.101}$$

In the first integral of (2.101), R is always finite and the contribution to Φ that is produced is clearly continuous as the field point crosses the surface. It will be easier to evaluate the second integral if we consider ΔS_0 to be a very small circular area of radius r_0 centered about the point of crossing the surface, that is, O of Fig. 2.31. If the field point is within a very small distance Δd of the surface, and recognizing that over ΔS_0, ρ_s is essentially constant, we have

$$\frac{1}{4\pi\epsilon_0}\int_{\Delta S_0}\frac{\rho_s\,dS'}{R} = \frac{\rho_s}{2\epsilon_0}\int_0^{r_0}\frac{r'\,dr'}{\sqrt{r'^2+(\Delta d)^2}} = \frac{\rho_s}{2\epsilon_0}\left[\sqrt{r_0^2+(\Delta d)^2} - |\Delta d|\right] \tag{2.102}$$

For a point passing through the surface, Δd decreases to zero and then increases negatively. But under these conditions the contribution to the total potential from (2.102) will also be continuous in crossing the surface; consequently, the total electrostatic potential is continuous across an arbitrary charged surface.

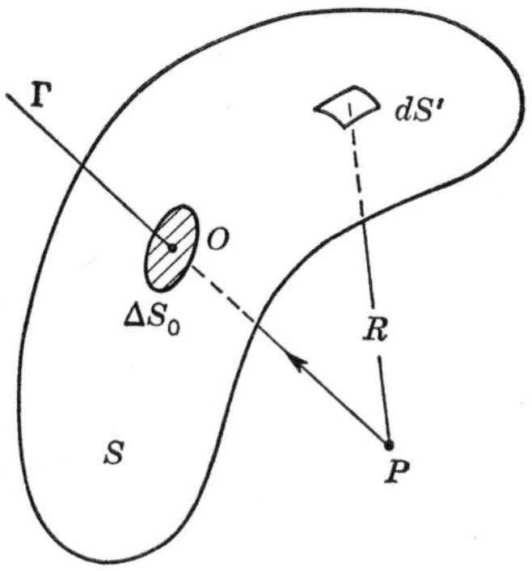

FIG. 2.30. An arbitrary surface charge distribution.

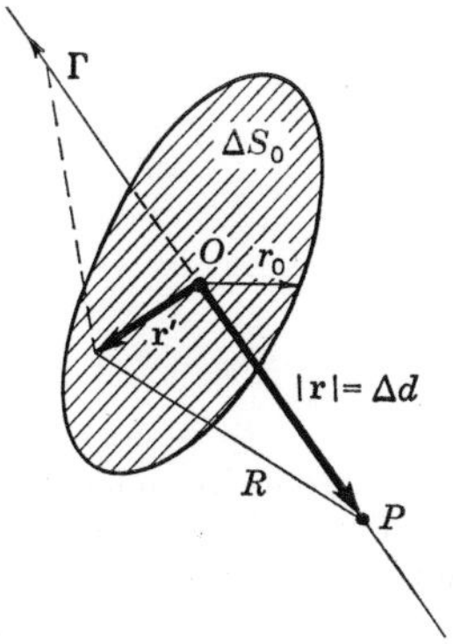

FIG. 2.31. Geometry of surface charge layer close to the surface.

A somewhat different situation arises if we consider the behavior of the electric field. In this case the normal component of **E** suffers a discontinuity equal to ρ_s/ϵ_0 in crossing a charged surface. This can be established very easily by application of Gauss' flux theorem, and this will be presented in Sec. 3.3. This same result can also be verified in a direct way as follows. We have

$$E_n = -\nabla\Phi\cdot\mathbf{n} = \frac{-1}{4\pi\epsilon_0}\int_S \rho_s\nabla\left(\frac{1}{|\mathbf{r}-\mathbf{r}'|}\right)\cdot\mathbf{n}\,dS' \tag{2.103}$$

The total surface can be broken up into two parts as before. Then

$$E_n = \frac{-1}{4\pi\epsilon_0}\int_{S-\Delta S_0}\rho_s\nabla\left(\frac{1}{|\mathbf{r}-\mathbf{r}'|}\right)\cdot\mathbf{n}\,dS' - \frac{1}{4\pi\epsilon_0}\int_{\Delta S_0}\rho_s\nabla\left(\frac{1}{|\mathbf{r}-\mathbf{r}'|}\right)\cdot\mathbf{n}\,dS' \tag{2.104}$$

The field contributed by the first integral in (2.104) must be continuous along a path passing through the charge surface since the only variable, $|\mathbf{r}-\mathbf{r}'|$, is finite and well-behaved. In this case, however, a discontinuity is introduced by the second integral of (2.104). Again, for simplicity, ΔS_0 is chosen sufficiently small so that it may be considered plane and so that ρ_s may be taken out as a constant. If we let the origin be located at the field point, then with reference to Fig. 2.32,

$$\begin{aligned}\frac{1}{4\pi\epsilon_0}\int_{\Delta S_0}\rho_s\nabla\left(\frac{1}{|\mathbf{r}-\mathbf{r}'|}\right)\cdot\mathbf{n}\,dS' \\ &= -\frac{\rho_s}{4\pi\epsilon_0}\int_{\Delta S_0}\nabla'\left(\frac{1}{R}\right)\cdot\mathbf{n}\,dS' = \frac{\rho_s}{4\pi\epsilon_0}\int_{\Delta S_0}\frac{\mathbf{a}_R\cdot\mathbf{n}}{R^2}\,dS' \\ &= \frac{\rho_s}{4\pi\epsilon_0}\int_{\Delta S_0}d\Omega = \frac{\rho_s\Omega_0}{4\pi\epsilon_0}\end{aligned} \tag{2.105}$$

where Ω_0 is the solid angle subtended at the field point by ΔS_0. As the field point moves across ΔS_0, as illustrated in Fig. 2.32, Ω_0 increases to a value of 2π, and then in

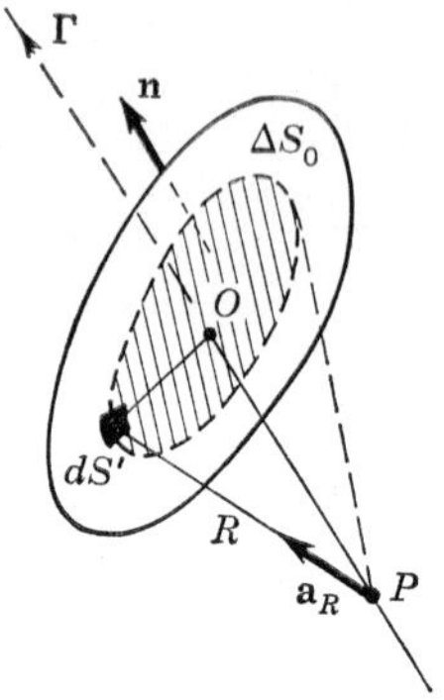

FIG. 2.32

crossing the surface it suddenly changes to -2π. Consequently, the normal component of **E** suffers a discontinuity, in crossing a charged surface, which we have found to be

$$E_{n1} - E_{n2} = \frac{\rho_s}{\epsilon_0} \tag{2.106}$$

where ρ_s is the surface charge density at the point of discontinuity. This result corresponds to what was found in Example 2.4, except that we have now assured it for any surface geometry.

We show later, in Sec. 3.3, that the tangential component of **E** is continuous across a charge surface.

CHAPTER 3

ELECTROSTATIC FIELDS IN MATERIAL BODIES, ENERGY, AND FORCES

In the previous chapter the properties of conducting materials in electrostatic fields were discussed. We turn now to a consideration of the properties of insulators in an electric field. In order to accomplish this aim, a short discussion of the microscopic (atomic) properties of insulators will be required. For this purpose simplified physical models of the atom will be used. Despite this simplicity it will be possible to satisfactorily predict the macroscopic behavior of dielectrics.

After discussing the basic properties of dielectrics, a study is made of capacitance and capacitors. This is followed by an evaluation of the work required to assemble a charge configuration. The work done is next related to the energy stored in the field. The latter part of the chapter introduces the principle of virtual work to evaluate the force acting on a body in an electrostatic field. The virtual-work principle of evaluating forces is a very powerful technique which greatly simplifies the solution of some otherwise difficult problems.

3.1. Polarizability

Electronic Polarizability

Let us begin by considering a monatomic gas in an electric field. This choice takes advantage of the fact that the spacing between the molecules of a gas is very much greater than the size of the molecule, so that, as we shall confirm, interaction between molecules can be neglected. This means that the effect of the field on any molecule is substantially the same as if it were the only particle present. A further geometric simplification arises in the choice of a single atom molecule.

Figure 3.1*a* illustrates a simplified model of the atom consisting of a positively charged nucleus surrounded by a spherically symmetric cloud of electrons. Since the nucleus has a diameter of the order of 10^{-15} meter while that of the electron cloud is of the order of 10^{-9} meter, the nucleus is essentially a point source. If, now, an external **E** field is applied, then a relative displacement of the nucleus from the center will occur. The relative displacement of the nucleus z can be computed by equating the force exerted by the external field to the coulomb restoring force.

Let R be the radius of the electron cloud, as in Fig. 3.1. The electron cloud is considered to be equivalent to a uniform sphere of charge with a density $-3q/4\pi R^3$. The field due to this charge at a distance z from the

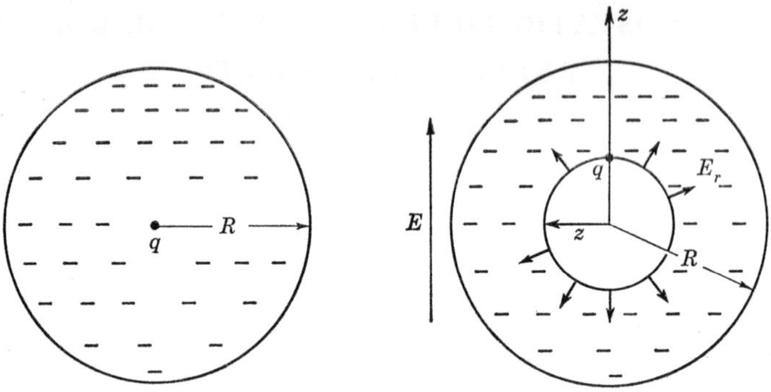

FIG. 3.1. (a) Model of atom; (b) displacement of nucleus by external field **E**.

center may be found by applying Gauss' law to the region bounded by a sphere of radius z, as in Fig. 3.1b. We have

$$\epsilon_0 4\pi z^2 E_r = -\frac{4\pi z^3}{3}\frac{3q}{4\pi R^3}$$

or

$$E_r = \frac{-qz}{4\pi R^3 \epsilon_0}$$

Since the charge on the nucleus is q, the coulomb restoring force F acting on the nucleus when displaced an amount z is

$$F = E_r q = -\frac{zq^2}{4\pi R^3 \epsilon_0} \tag{3.1}$$

Note that the restoring force, given by (3.1), is proportional to the displacement. This consequence is confirmed in atomic physics for small displacements z. Equating the external force qE to $-F$ gives

$$qz = p = 4\pi\epsilon_0 R^3 E \tag{3.2a}$$

or in vector form,

$$\mathbf{p} = 4\pi\epsilon_0 R^3 \mathbf{E} \tag{3.2b}$$

The dipole moment per atom, $\mathbf{p}$, is proportional to $\mathbf{E}$, according to (3.2). If N is the number of atoms per cubic meter, then $\mathbf{P} = N\mathbf{p}$, where $\mathbf{P}$ is the dipole moment per unit volume. In view of the proportionality expressed in (3.2), we may relate $\mathbf{p}$ to $\mathbf{E}$ as $\mathbf{p} = \alpha\mathbf{E}$, where α is called the polarizability. When the polarization arises in the way just described, α is designated the electronic polarizability α_e. In terms of the above parameters,

$$\alpha_e = 4\pi R^3\epsilon_0 \tag{3.3}$$

The quantity α_e/ϵ_0 is seen, from (3.3), to be of the order of magnitude of the atomic volume. In Table 3.1 the ratio $\alpha_e/4\pi R^3\epsilon_0$ is given for several of the inert gases. Despite the crude model used, this ratio comes out in the order of unity.

TABLE 3.1. POLARIZABILITIES AND RADII FOR SOME COMMON ELEMENTS

	Gas				
Parameter	He	Ne	A	Kr	Xe
$\alpha_e \times 10^{40}$ cu m (measured)	0.222	0.43	1.8	2.7	4.4
R, A	0.95	1.15	1.4	1.6	1.75
$\alpha_e/4\pi R^3\epsilon_0$	0.28	0.29	0.66	0.67	0.83

In summary, we note that the effect of an applied electric field on the molecules of matter may be to create electrostatic dipoles. These in turn will set up a secondary (induced) field so that the net field in the presence of matter is modified from its free-space value. Before proceeding to a consideration of the fields, we mention several other polarization-producing mechanisms.

Ionic Polarizability

In a molecule characterized essentially by ionic bonds, we can think of that molecule as composed of positively and negatively charged ions. It is the coulomb forces between these ions which mainly account for the binding force. The application of an electric field to any such molecule will tend to displace the positive ions relative to the negative ones. This process will induce a dipole moment in the molecule. Note that it is quite distinct from electronic polarizability, where the displacement between the nucleus and electron cloud accounts for the polarization. In a polyatomic gas one can expect both processes to occur. We shall designate ionic polarizability by α_i.

Orientational Polarizability

In certain polyatomic molecules where the atomic bond is at least partially ionic, the individual atoms tend to be either positively or negatively charged. A two-atom molecule will thus have a permanent dipole moment, the magnitude of which depends on the time-average transfer of charge between atoms and the internuclear distance. For polyatomic molecules several bonds may have a permanent dipole moment; the dipole moment of the entire molecule is the vector sum of the component moments. For certain kinds of symmetry the latter may come out zero.

The application of an electric field to molecules with a permanent dipole moment would ordinarily cause all molecules to align themselves with the applied field. This orienting tendency is opposed by the random thermal agitation of the molecule, and in solids and some liquids by mutual interactions of the molecules. Assuming that the molecules are free to move, we can calculate their effective orientational or dipolar polarizability.

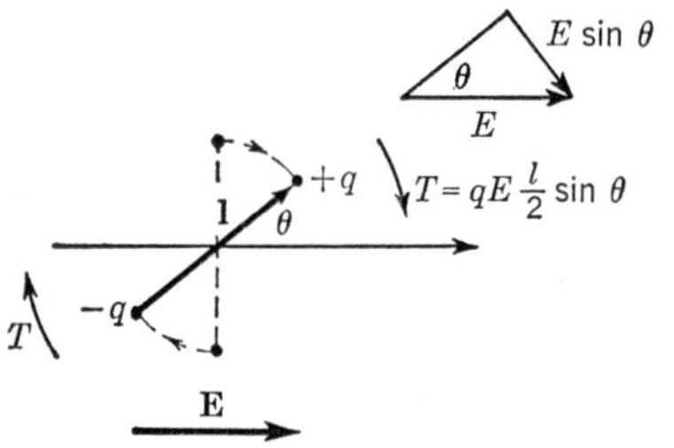

FIG. 3.2. Dipole in a field **E**.

Consider a system of N molecules per unit volume, each possessing a permanent dipole moment of $\mathbf{p}$. In the absence of an electric field, their orientation is completely random, so that there is no net time-average dipole moment. Now let an electric field be applied in the z direction. We represent in Fig. 3.2 a typical molecule, with moment $\mathbf{p} = q\mathbf{l}$, making an angle θ with the z direction.

For the potential energy of the dipole, let the zero reference be taken to be when $\mathbf{p} \cdot \mathbf{E} = 0$, that is, $\theta = 90°$. If the dipole is allowed to rotate into alignment with the field, then the decrease in potential energy of $+q$, from the position shown in Fig. 3.2, would be

$$W_+ = \int_{\pi/2}^{\theta} \frac{qlE}{2} \sin\theta \, d\theta = -\frac{qlE\cos\theta}{2} \tag{3.4a}$$

and similarly for the negative charge,

$$W_- = -\frac{qlE\cos\theta}{2} \tag{3.4b}$$

since the torque acting on each charge of the dipole is $T = (qlE/2)\sin\theta$. Consequently, the potential energy of the dipole in any position can be written as

$$W = -\mathbf{p} \cdot \mathbf{E} = -pE\cos\theta \tag{3.5}$$

The contribution to the total dipole moment in the z direction from the afore-mentioned molecule is $\mathbf{p} \cdot \mathbf{a}_z = p\cos\theta$. To find the total moment it is necessary to know the relative number of molecules making different angles θ with z. With an applied field, the distribution is no longer uniform, but it can be found from the Boltzmann distribution law. The latter states that the probability that the direction of $\mathbf{p}$ lies between θ

and $\theta + d\theta$ is proportional to $e^{-W/kT}\, d\omega$, where $d\omega$ is an element of solid angle corresponding to $d\theta$, k is Boltzmann's constant and equals 1.38×10^{-23} joule per degree Kelvin, while T is absolute temperature. From Fig. 3.3 we note that

$$d\omega = 2\pi \sin\theta\, d\theta$$

The number of dipoles, per unit volume, whose moment lies between θ and $\theta + d\theta$, is now given by

$$dN = Ae^{(pE\cos\theta)/kT} \sin\theta\, d\theta \qquad (3.6)$$

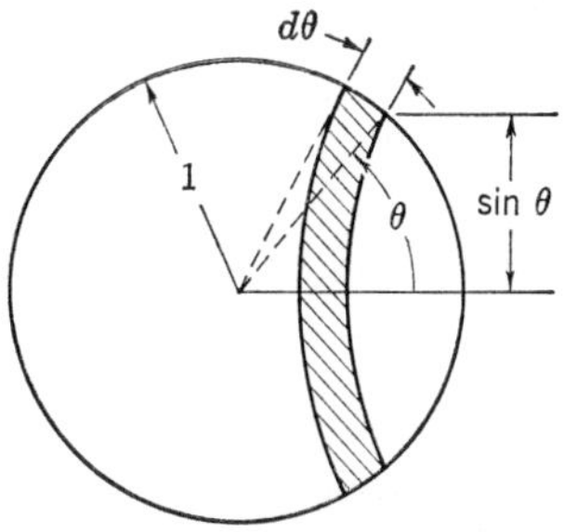

FIG. 3.3. Element of solid angle on a unit sphere.

where A is a constant of proportionality. The latter can be determined from the requirement that the total number of molecules per unit volume be given by the integral of (3.6) from $\theta = 0$ to π; that is,

$$A = \frac{N}{\int_0^\pi e^{(pE\cos\theta)/kT} \sin\theta\, d\theta} \qquad (3.7)$$

The partial contribution to the polarization from dipoles lying between θ and $\theta + d\theta$ is

$$dP = dN\, p\cos\theta$$

The total polarization P is given by

$$\begin{aligned} P &= Ap\int_0^\pi e^{(pE\cos\theta)/kT} \sin\theta\cos\theta\, d\theta \\ &= pN\frac{\int_0^\pi e^{(pE\cos\theta)/kT} \sin\theta\cos\theta\, d\theta}{\int_0^\pi e^{(pE\cos\theta)/kT} \sin\theta\, d\theta} \end{aligned} \qquad (3.8)$$

Let $x = \cos\theta$, $a = pE/kT$; then

$$P = pN\frac{\int_{-1}^{1} e^{ax}x\, dx}{\int_{-1}^{1} e^{ax}\, dx} = Np\left(\coth a - \frac{1}{a}\right) = NpL(a) \qquad (3.9)$$

where integration by parts was used to evaluate the numerator. The function $L(a)$ was first introduced in connection with a similar study of magnetic dipoles by Langevin (1905), and it is called the Langevin function. A plot of $L(a)$ is shown in Fig. 3.4. The curve depicts the saturation property of the orientational polarization. At room temperatures laboratory fields are too weak to approach saturation and $pE \ll kT$ can be assumed. As a consequence,

$$L(a) \approx \frac{a}{3} = \frac{pE}{3kT}$$

and the polarization per unit volume is

$$P = \frac{Np^2E}{3kT} \tag{3.10}$$

The orientational polarizability per molecule, α_0, is thus given by

$$\alpha_0 = \frac{p^2}{3kT} \tag{3.11}$$

Typical values of p are around 10^{-30} coulomb-meter, so that at room temperatures α_0 comes out around 10^{-40} cubic meter. This is the same order of magnitude as the electronic polarizability. Note, however, that while both electronic and ionic polarizability depend only on atomic configurations, and hence are essentially independent of temperature, the orientational polarizability is inversely proportional to temperature. This corresponds to the observation that at elevated temperatures it becomes more difficult to align the dipoles against the thermal motion.

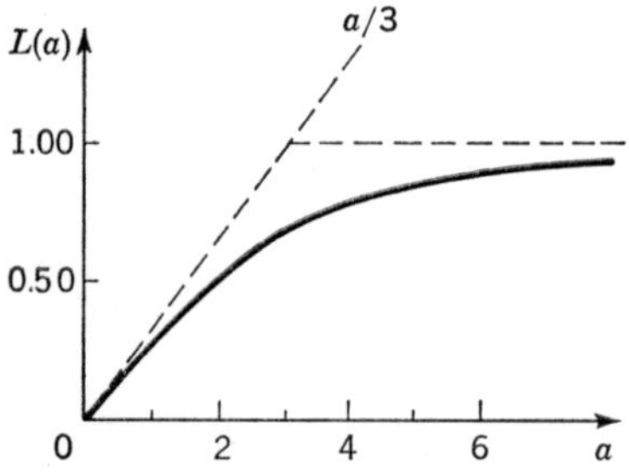

FIG. 3.4. The Langevin function.

The total polarization of a polyatomic gas may arise as a result of electronic, ionic, and orientational polarizability. Therefore, in general, we have

$$\mathbf{P} = N\left(\alpha_e + \alpha_i + \frac{p^2}{3kT}\right)\mathbf{E} \tag{3.12}$$

where the vector notation, which is implicit in the previous work, has been restored.

If the foregoing analysis is applied to solid dielectrics, then the field $\mathbf{E}$ in (3.12) cannot be taken as that which exists prior to the introduction of the dielectric. This is because adjacent dipoles become sufficiently influential in modifying the field at a point within the dielectric. The field $\mathbf{E}$ in (3.12) must be interpreted as being the molecular field in the dielectric. In a later section we shall evaluate the first-order interaction, and hence relate the molecular field to the external field.

We turn now to a consideration of the macroscopic effects of the polarizability of dielectric materials. It will be useful to lump together the various contributions to the polarization in the following expression:

$$\mathbf{P} = \epsilon_0 \chi_e \mathbf{E} \tag{3.13}$$

where χ_e is a dimensionless quantity called the electric susceptibility, and $\mathbf{E}$ is the total field; that is, χ_e is assumed to absorb the factor relating the molecular field to the total field. Equation (3.13) assumes that the polarization is proportional to $\mathbf{E}$, a relationship usually confirmed in practice. Note that, in part, it is based on an assumption of being far from saturation for dipolar molecules. Equation (3.13) fails to describe the polarization of certain substances which exhibit spontaneous polarization. The latter class are known as ferroelectrics. A description of their properties can be found in the references on solid-state physics.

3.2. Electric Flux Density D

In Chap. 2 we learned how to calculate the electric field from a given distribution of charge. This can be accomplished by first determining the scalar potential field that is set up. The electric field is then derived from the negative gradient of the scalar potential. The pertinent formulas follow (the reader is reminded that they apply under free-space conditions):

$$\Phi = \frac{1}{4\pi\epsilon_0} \int_V \frac{\rho \, dV'}{R} \tag{3.14}$$

$$\mathbf{E} = -\nabla\Phi \tag{3.15}$$

where $R = |\mathbf{r} - \mathbf{r}'|$.

Let us consider what happens if a dielectric is introduced into the electric field set up by an arbitrary charge distribution, as described above. As a consequence of its polarizability, the volume occupied by the dielectric now contains a dipole moment distribution $\mathbf{P}$. This distribution constitutes a secondary source for the electric field. Thus, in addition to the original charges that set up the field, the dipole moment $\mathbf{P}$ must also be included as a source for the complete field, i.e., the field in the presence of the dielectric.

The scalar potential due to $\mathbf{P}$ has already been determined and is given by (2.97). The total potential, by superposition, must then be

$$\begin{aligned}\Phi &= \frac{1}{4\pi\epsilon_0}\left(\int_V \frac{\rho - \nabla' \cdot \mathbf{P}}{R} \, dV' + \oint_S \frac{\mathbf{P} \cdot \mathbf{n}}{R} \, dS'\right) \\ &= \frac{1}{4\pi\epsilon_0}\left(\int_V \frac{\rho + \rho_p}{R} \, dV' + \oint_S \frac{\rho_{sp}}{R} \, dS'\right)\end{aligned} \tag{3.16}$$

In the last expression, $\rho_p = -\nabla' \cdot \mathbf{P}$ is the equivalent volume polarization charge, ρ_{sp} the surface polarization charge, while ρ is called the "true" charge. Note that ρ in (3.16) is, in general, not exactly the same as in

(3.14). This comes about because ρ_p may react back on the original distribution of charge, thus affecting that distribution. In many cases, however, such effects will be small.

The polarization charge does not only arise as a physical interpretation of the mathematical equivalence expressed in (2.97). If one could actually measure the excess charge in a small volume within a dielectric, one would confirm that $\rho_p = -\nabla \cdot \mathbf{P}$. [See remarks following (2.97).] Nevertheless, a distinction between true and polarization charge is made, and this difference is related to the origin of the volume charge density. True charge is essentially accessible for measurement; it is a free charge. Polarization charge would also be accessible if one could make measurements on an atomic scale within dielectric materials. However, we ordinarily consider the polarization charge as arising from constituent dipoles so that the "bound" nature of the charges (hence "inaccessibility" of individual charges) is evident. We consequently distinguish the polarization charge as a separate entity.

As a consequence of (3.16), the sources of electric field must be generalized to include the polarization charge. In the dielectric we must consider ρ_p as fully equivalent to the true charge ρ. Thus the divergence of $\mathbf{E}$ must be related to the sum of $\rho + \rho_p$ as follows:

$$\nabla \cdot \mathbf{E} = \frac{\rho + \rho_p}{\epsilon_0} \tag{3.17}$$

This result is readily verified by taking the Laplacian of (3.16) and using the singularity property of $\nabla^2(1/R)$. Since (3.17) relates the divergence of $\mathbf{E}$ to the charge density at a point, the surface charge term does not enter as long as we are in the interior of the dielectric. Later on we shall show that the surface polarization charge is readily taken into account in practice by imposing a discontinuity condition on the normal component of $\mathbf{E}$ at the surface of the dielectric body. If we express ρ_p in terms of $\mathbf{P}$, we have

$$\nabla \cdot (\epsilon_0 \mathbf{E} + \mathbf{P}) = \rho \tag{3.18}$$

In practice, it is inconvenient to take explicit account of the polarization $\mathbf{P}$. We can avoid this by introducing a new field vector $\mathbf{D}$ defined as

$$\mathbf{D} = \epsilon_0 \mathbf{E} + \mathbf{P} \tag{3.19}$$

Equation (3.18) now becomes

$$\nabla \cdot \mathbf{D} = \rho \tag{3.20}$$

The vector $\mathbf{D}$ has the dimensions of coulombs per square meter and is called the electric displacement, or electric-flux-density vector. The source for the vector $\mathbf{D}$ is the true charge density ρ.

With materials for which (3.13) holds we have

$$\mathbf{D} = \epsilon_0\mathbf{E} + \epsilon_0\chi_e\mathbf{E} = \epsilon_0(1 + \chi_e)\mathbf{E} = \epsilon\mathbf{E} \tag{3.21}$$

The parameter $\epsilon = \epsilon_0(1 + \chi_e)$ is called the permittivity of the dielectric. The permittivity relative to that of free space is $\epsilon/\epsilon_0 = 1 + \chi_e$ and is called the relative dielectric constant and will be designated by the symbol κ. If ϵ is known, we can solve for the polarization $\mathbf{P}$ since

$$\mathbf{P} = \epsilon_0\chi_e\mathbf{E} = (\epsilon - \epsilon_0)\mathbf{E} \tag{3.22}$$

Values of the relative dielectric constant of several typical materials are given in Table 3.2.

TABLE 3.2. RELATIVE DIELECTRIC CONSTANT OF SEVERAL MATERIALS

Material	Relative dielectric constant	Dielectric strength, kv/m
Air	1.00	3,000
Oil	2.3	15,000
Paper	1.5–4	15,000
Polystyrene	2.7	20,000
Glass	6.0	30,000
Paraffin	2.1	30,000
Quartz	5.0	30,000
Mica	6.0	200,000

Sometimes (3.21) is taken as a definition of $\mathbf{D}$. It is clear that such a relationship holds only for a class of dielectrics under certain conditions. It depends primarily on the linear relation between the polarization and the electric field as expressed in (3.13). In addition to this, it also requires that the material be isotropic, that is, χ_e should be independent of the direction of $\mathbf{E}$. When this is not the case, the relation between $\mathbf{P}$ and $\mathbf{E}$, and hence between $\mathbf{D}$ and $\mathbf{E}$, becomes a matrix, i.e., a tensor, relation. For example, if the molecule is not symmetrical, its dipole moment will in general not be collinear with the field $\mathbf{E}$ and each component of $\mathbf{P}$ will be related to each component of $\mathbf{E}$, so that

$$\begin{bmatrix} P_x \\ P_y \\ P_z \end{bmatrix} = \epsilon_0 \begin{bmatrix} \alpha_{xx} & \alpha_{xy} & \alpha_{xz} \\ \alpha_{yx} & \alpha_{yy} & \alpha_{yz} \\ \alpha_{zx} & \alpha_{zy} & \alpha_{zz} \end{bmatrix} \begin{bmatrix} E_x \\ E_y \\ E_z \end{bmatrix}$$

It is clear that in this case χ_e and ϵ are matrix (tensors of rank 2) quantities. Materials characterized by a tensor permittivity are called anisotropic materials. For our purposes in the remainder of this book we shall assume that ϵ is a scalar constant. Most materials fall into this class.

If (3.20) is integrated throughout a volume V and Gauss' law is used, then an integral relationship between $\mathbf{D}$ and the total true charge results; i.e.,

$$\oint_S \mathbf{D} \cdot d\mathbf{S} = \int_V \rho \, dV = Q \tag{3.23}$$

where Q is the total charge within V. Utilizing the flux concept, we see that the number of flux lines originating within V is proportional to the quantity of charge Q.

3.3. Boundary Conditions

In solving problems in electrostatics it is necessary to relate the inaccessible (polarization) charges to the accessible (true) charges, or to the

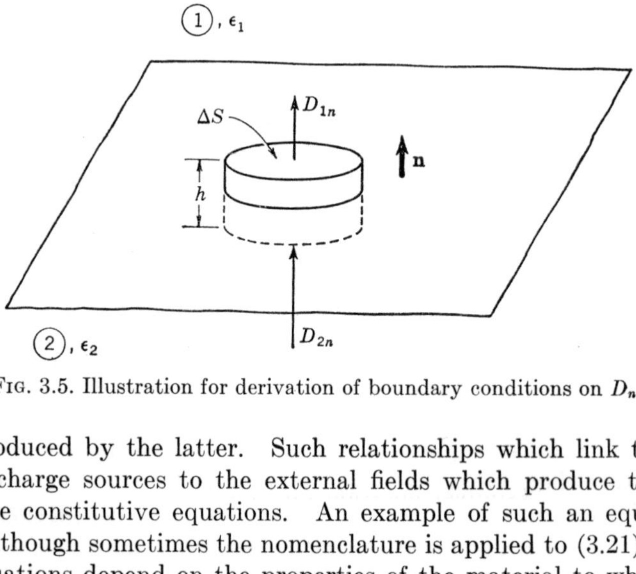

FIG. 3.5. Illustration for derivation of boundary conditions on D_n.

fields produced by the latter. Such relationships which link the inaccessible charge sources to the external fields which produce them are called the constitutive equations. An example of such an equation is (3.13), although sometimes the nomenclature is applied to (3.21) as well. Such equations depend on the properties of the material to which they apply. As was noted earlier, (3.21) is restricted to linear, isotropic materials. However, the material need not be homogeneous; that is, ϵ may be a function of position.

One very common case of nonhomogeneity occurs when the dielectric constant varies discontinuously as between two different homogeneous media. The way in which $\mathbf{D}$ and $\mathbf{E}$ behave in crossing the boundary between two dielectrics is of much interest and will be discussed now.

Figure 3.5 illustrates a very small element of the interface between dielectrics 1 and 2 whose permittivities are ϵ_1 and ϵ_2, respectively. Since the element of surface is of differential extent, it may be considered to be plane. A coin-shaped surface is placed with its broad face parallel to the interface and so that one surface is in region 1 and the other in region 2. The area of the broad face is ΔS, and the thickness is h. Let us apply Gauss' flux theorem to the volume of the coin. If we make use of (3.23), then

$$(\mathbf{n} \cdot \mathbf{D}_1)\, \Delta S - (\mathbf{n} \cdot \mathbf{D}_2)\, \Delta S = \rho_s\, \Delta S \tag{3.24}$$

where ρ_s is the true surface charge density on the interface. Equation (3.24) does not include the outflow of flux of $\mathbf{D}$ through the sides, because this flow can be made negligible by letting $h \to 0$, while at the same time the terms in (3.24) remain unaffected.

For the simple case of a surface charge ρ_s in a free-space medium, we can set $\mathbf{D}_1 = \epsilon_0 \mathbf{E}_1$ and $\mathbf{D}_2 = \epsilon_0 \mathbf{E}_2$ in (3.24), with the result that

$$\mathbf{n} \cdot (\mathbf{E}_1 - \mathbf{E}_2) = \frac{\rho_s}{\epsilon_0} \tag{3.25a}$$

In other words, the normal component of electric field is discontinuous through a charged surface, the magnitude of the discontinuity being given by (3.25*a*). For a dielectric interface ρ_s is ordinarily equal to zero unless a surface charge is actually placed at the interface. Taking $\rho_s = 0$, (3.24) becomes

$$\mathbf{n} \cdot \mathbf{D}_1 = \mathbf{n} \cdot \mathbf{D}_2 \tag{3.25b}$$

That is, the normal component of $\mathbf{D}$ is continuous across a dielectric boundary. The normal component of $\mathbf{E}$, on the other hand, is discontinuous. This is clear if (3.21) is substituted into (3.25*b*) to give

$$\frac{\mathbf{n} \cdot \mathbf{E}_1}{\mathbf{n} \cdot \mathbf{E}_2} = \frac{\epsilon_2}{\epsilon_1} \tag{3.26a}$$

Using a somewhat simpler notation we have

$$\epsilon_1 E_{1n} = \epsilon_2 E_{2n} \tag{3.26b}$$

The discontinuity in $\mathbf{n} \cdot \mathbf{E}$ is readily explained physically. The field $\mathbf{E}$ arises from the total effective charge consisting of the true charge ρ, the volume polarization charge $\rho_p = -\nabla \cdot \mathbf{P}$, and the surface polarization charge ρ_{sp}. At the surface of a dielectric the normal component of $\mathbf{E}$ is discontinuous by an amount equal to ρ_{sp}/ϵ_0, just as it would be if we considered a surface layer of true charge equal to ρ_{sp}. Equation (3.26*b*) is readily shown to verify this result. From (3.22) the normal component of $\mathbf{P}$ at the surface is seen to be given by

$$P_{1n} = (\epsilon_1 - \epsilon_0) E_{1n} \qquad \text{medium 1}$$
$$P_{2n} = (\epsilon_2 - \epsilon_0) E_{2n} \qquad \text{medium 2}$$

The surface polarization charge is given by $P_{2n} - P_{1n}$ since this represents the amount of charge on the positive ends of the dipoles in medium 2 that is not canceled by the opposite charge on the negative ends of the dipoles in medium 1. From the above relations we see that

$$P_{2n} - P_{1n} = \rho_{sp} = \epsilon_0(E_{1n} - E_{2n}) + \epsilon_2 E_{2n} - \epsilon_1 E_{1n} \tag{3.27}$$

We can now see that a discontinuity of E_n by the amount ρ_{sp}/ϵ_0 corresponds to the requirement that $\epsilon_2 E_{2n} = \epsilon_1 E_{1n}$. In other words, satisfaction of (3.26) is consistent with the necessity that E_n be discontinuous by an amount ρ_{sp}/ϵ_0.

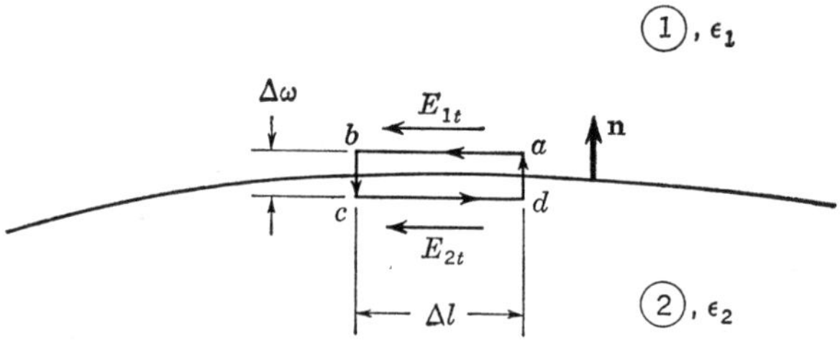

FIG. 3.6. Illustration for derivation of boundary conditions on E_t.

In practice, we usually find a suitable solution for **E** and **D** in the two dielectric regions. We then adjust the magnitudes of these solutions so that (3.26) holds at the boundary. By this means we avoid the necessity of taking the surface polarization charge into account explicitly.

Boundary conditions on the tangential components of the field can be found in the following manner. Figure 3.6 is a cross section normal to the interface separating two media of different permittivity. Considering the small rectangular path of length Δl and width $\Delta\omega$, where opposite sides of the long dimension lie in the separate media, we have

$$\int_a^b \mathbf{E} \cdot d\mathbf{l} + \int_b^c \mathbf{E} \cdot d\mathbf{l} + \int_c^d \mathbf{E} \cdot d\mathbf{l} + \int_d^a \mathbf{E} \cdot d\mathbf{l} = 0 \tag{3.28}$$

since the line integral of **E** around any closed path is zero. Without affecting the remaining two integrals,

$$\int_b^c \mathbf{E} \cdot d\mathbf{l} = \int_d^a \mathbf{E} \cdot d\mathbf{l} = 0$$

by letting $\Delta\omega \to 0$, while keeping Δl fixed. If we symbolize the tangential component of **E** in region 1 by E_{1t} and that in region 2 by E_{2t}, then, with the choice of directions given in Fig. 3.6,

$$E_{1t}\,\Delta l - E_{2t}\,\Delta l = 0$$

or

$$\underline{E_{1t} = E_{2t}} \tag{3.29a}$$

This may be written in vector form as

$$\mathbf{n} \times \mathbf{E}_1 = \mathbf{n} \times \mathbf{E}_2 \qquad (3.29b)$$

For the tangential components of **D** we must now have

$$\frac{D_{1t}}{D_{2t}} = \frac{\epsilon_1}{\epsilon_2} \qquad (3.30)$$

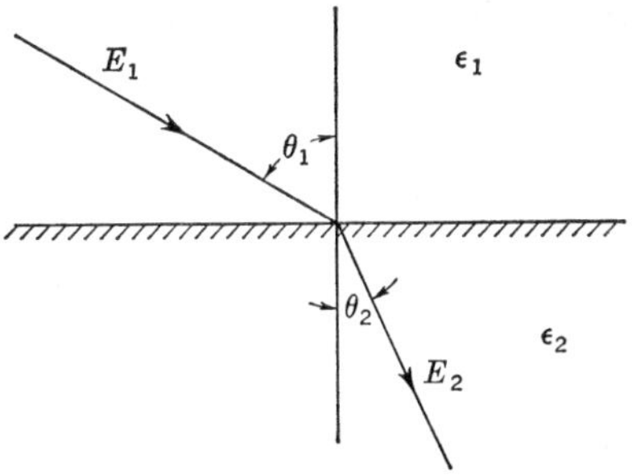

FIG. 3.7. Refraction of **E** lines of force across a boundary.

Thus the tangential component of electric field is continuous across a boundary between two dielectrics, while tangential **D** is discontinuous. Note that this result would not be affected by the presence of a surface charge layer at the interface.

The total change in electric field in crossing an interface may be found from the above equations. The net result is analogous to the refraction of a light ray in passing from one medium to another. Thus in terms of the geometry of Fig. 3.7 and using (3.26) and (3.29), we have

$$E_1 \sin \theta_1 = E_2 \sin \theta_2 \qquad (3.31a)$$

$$\epsilon_1 E_1 \cos \theta_1 = \epsilon_2 E_2 \cos \theta_2 \qquad (3.31b)$$

Dividing (3.31*a*) by (3.31*b*) yields an electrostatic "Snell's law," relating the angle of incidence to the angle of refraction in terms of the properties of the media; i.e.,

$$\frac{\tan \theta_1}{\tan \theta_2} = \frac{\epsilon_1}{\epsilon_2} \qquad (3.32)$$

Let us consider a problem that illustrates some of the concepts just considered. Figure 3.8 depicts two very large (essentially infinite) parallel conducting planes which are maintained at a potential difference V by means of a battery. Because of the uniformity (the very large size is chosen so that fringing of the field at the edges can be neglected), the

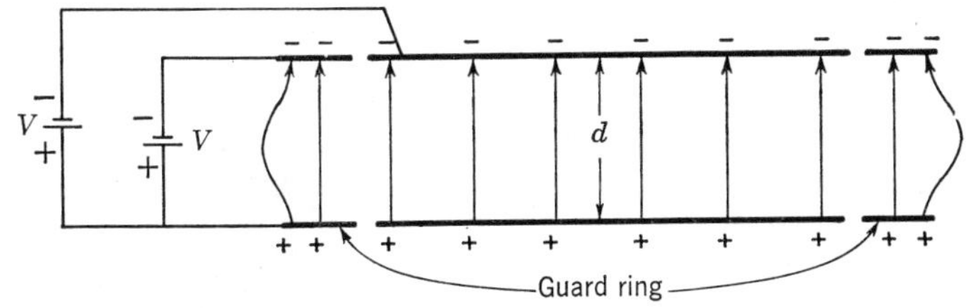

FIG. 3.8. Electric field between two parallel plates.

electric field must be given by

$$E = \frac{V}{d} \tag{3.33}$$

The uniformity of the electric field is represented in the flux plot in Fig. 3.8.†

Suppose we now insert a uniform slab of dielectric, of relative permittivity κ, between the plates, as shown in Fig. 3.9. Let the thickness of

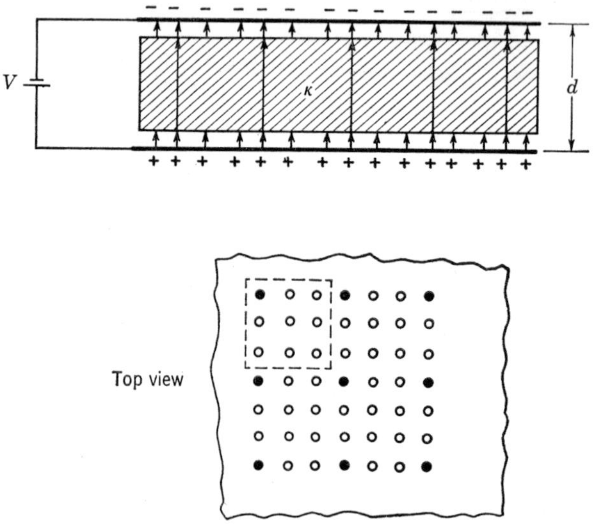

FIG. 3.9. Dielectric slab between parallel plates.

the slab be slightly less than d, leaving a small air gap. Since V is unchanged, E within the dielectric is still given by V/d; that is, $\int_0^d \mathbf{E} \cdot d\mathbf{l}$ must equal V as before. (We are neglecting the small error that the air gap introduces in this argument.)

In passing from the dielectric into the air gap, the electric field increases by virtue of (3.26). Thus, letting E_a be the field in the air gap and E_d that in the dielectric and noting that both are normal to the interface and the conducting plates, we have

$$E_a = \kappa E_d \tag{3.34}$$

where, of course, $\kappa > 1$. In Fig. 3.9, $\kappa = 9$, since the field strength is proportional to the number of lines per unit area.

† Fringing of the field at the sides of the plates may be avoided by increasing the size of the bottom plate and surrounding the upper plate by a guard ring, insulated from the upper plate and kept at the same potential $-V$ as in Fig. 3.8.

The discontinuity in $\mathbf{E}$ is explained by the polarization charge which terminates the flux lines. The polarization of the dielectric is represented in Fig. 3.9, and it should be clear that an equivalent surface charge layer is available at the dielectric surface. If the charge density on the conductor is ρ_s, then $E_a = \rho_s/\epsilon_0$. Now since $\kappa = 9$, $\chi_e = \kappa - 1 = 8$, and $P = \epsilon_0\chi_e E_d = 8\rho_s/9$. In view of (3.27),

$$\rho_{sp} = \mathbf{n} \cdot \mathbf{P} = \begin{cases} -\frac{8}{9}\rho_s & \text{lower interface} \\ \frac{8}{9}\rho_s & \text{upper interface} \end{cases}$$

where $\mathbf{n}$ is the outward normal to the dielectric surface. From Fig. 3.9 we see that this is precisely the charge density needed to terminate the lines of flux of E_a at the surface of the dielectric.

3.4. Dielectric Strength

As we have noted, the process of induced polarization involves the displacement, from an equilibrium position, of the nucleus relative to the surrounding electron cloud. In our discussion of this phenomenon we tacitly assumed that an elastic process was involved. It might be anticipated, though, that if the field strength is increased sufficiently, new phenomena would be involved, including the possibility of a permanent change in the dielectric. This does, indeed, happen. We say that dielectric breakdown occurs, and the field strength at which this takes place is called the dielectric strength of the material. Table 3.2 lists the breakdown field strength for several common materials.

In general, breakdown begins as a result of the movement of electrons within the insulator under the influence of the applied field. The source of the electrons comes from impurities in the insulator, or crystal-lattice defects, or by field emission due to the applied field itself. If the energy gained by the electron from the accelerating field is greater than that lost in collision with the lattice structure, the electron will accumulate sufficient energy to create a hole-electron pair. The latter will then also be accelerated, so that a continuous discharge through the material can result.

If the solid dielectric contains gas bubbles or layers of gas, breakdown may first occur in that region because the breakdown field strength in a gas is lower than in the solid dielectric, while at the same time the field strength in the gas will be larger than in the dielectric [e.g., see (3.34)]. As a consequence of the gas discharge so initiated, the solid dielectric will be subjected to ionic bombardment, leading, finally, to complete breakdown.

It is not the intention to provide here a detailed physical description and analysis of dielectric breakdown. What is important is that it serves as a motivation in the solution of potential problems where excessive field strengths that can cause breakdown must be avoided. In problems involving composite dielectrics one is faced with a need to reconcile the different dielectric constants and dielectric strengths. The following is a simple example of this.

Example 3.1. Dielectric Breakdown. For the parallel conducting planes as in Fig. 3.10, but with an air dielectric, the field is approximately

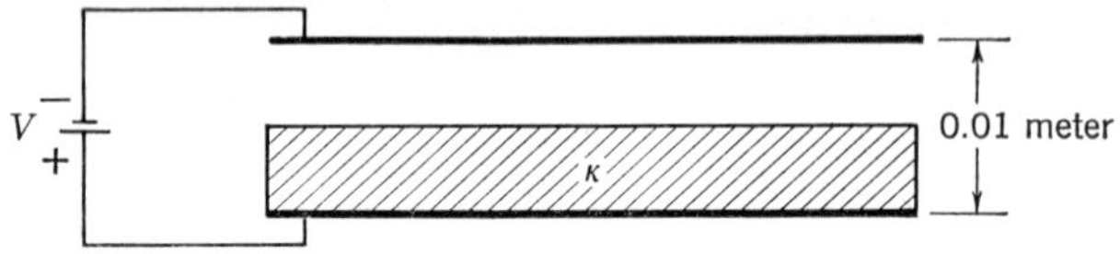

FIG. 3.10. Partially filled parallel-plate capacitor.

uniform and of magnitude V/d. Consequently, for $d = 0.01$ meter, the maximum potential is 30 kilovolts, as this results in a field of 3,000 kilovolts per meter, which is the breakdown stress for air.

If the region between the plates were filled with glass, then a voltage of 150 kilovolts would be permitted, since the dielectric strength of glass is approximately five times greater than that of air. Suppose, however, that the glass is only half the thickness of the air gap. Then since $\kappa = 6$ for glass, the field in the air is six times that in the glass, so that $\frac{6}{7}V$ exists across the air gap. In this case

$$\frac{\frac{6}{7}V_{\text{max}}}{0.005} = 3{,}000 \text{ kv/m}$$

and

$$V_{\text{max}} = 17.5 \text{ kv}$$

Thus breakdown in the air will occur at a lower voltage than it would in the absence of the glass. Once breakdown occurred, the entire voltage would appear across the glass, but since the field strength would be only $17.5/0.005 = 35{,}000$ volts per meter, the glass would not break down.

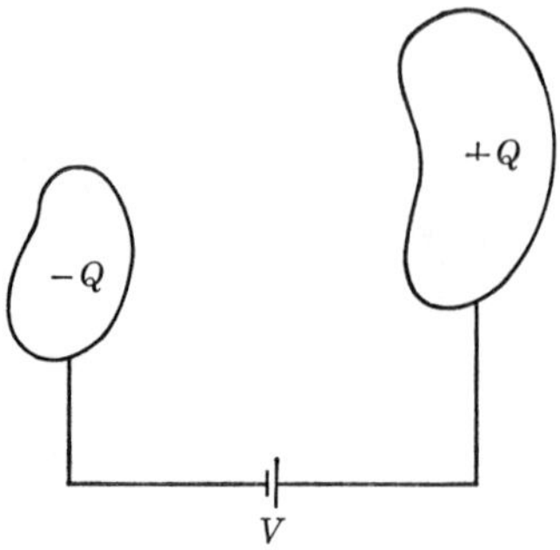

FIG. 3.11. Two charged bodies forming a capacitor.

3.5. Capacitance

Consider two perfectly conducting bodies of arbitrary shape, as illustrated in Fig. 3.11, and let a quantity of charge be transferred from one to the other, such as would be accomplished by connecting a battery between the two. The charge on one is Q and on the other $-Q$. According to the uniqueness theorem, the difference in potential between the bodies is uniquely determined by the charge Q and the geometry of the conducting bodies. In view of the linear dependence of potential on charge, as revealed for example in (2.32), the difference of potential between the bodies, V, can be expressed as

$$V = \frac{1}{C} Q \tag{3.35}$$

The parameter C depends only on the geometry and for a given geometry is a constant. It is called the capacitance and is measured in units of farads or coulombs per volt.

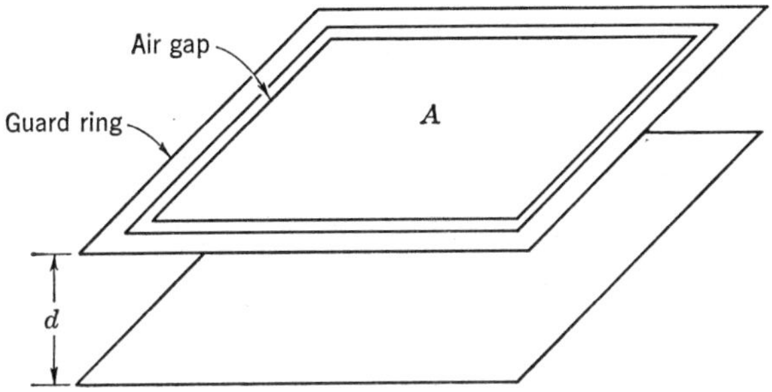

FIG. 3.12. A parallel-plate capacitor with guard ring.

The parallel-plate capacitor shown in Fig. 3.12 presents a simple geometry for illustrating the calculation of capacitance. If the smallest linear dimension of the plates is large compared with the spacing, then the field may be assumed to be uniform and the fringing at the edges neglected. Alternatively, a guard ring may be used as suggested in Fig. 3.12. In this case the relationship expressed by (3.33) applies, that is, $E = V/d$, where V is an assumed potential difference between the plates. For an air dielectric, $D = \epsilon_0 E$, and $\rho_s = D$; consequently, the total charge Q on each plate, of area A, is

$$Q = \rho_s A = \frac{\epsilon_0 V A}{d} \tag{3.36}$$

and

$$C = \frac{Q}{V} = \frac{\epsilon_0 A}{d} \tag{3.37}$$

Example 3.2. Capacitance between Concentric Spheres. It is required to find the capacitance between the two conducting concentric spherical shells illustrated in Fig. 3.13. If we assume a charge $+Q$ on the inner sphere and $-Q$ on the outer, then the electric field between the two shells is

$$E_r = \frac{Q}{4\pi\epsilon_0 r^2} \tag{3.38}$$

This result can be verified by applying Gauss' flux theorem to the concentric spherical surface of radius r, where $a < r < b$. The potential difference between the spheres, V, is given by

$$V = \int_a^b E_r \, dr = \frac{Q}{4\pi\epsilon_0}\left(\frac{1}{a} - \frac{1}{b}\right) \tag{3.39}$$

Consequently, the capacitance between the two spheres is

$$C = \frac{Q}{V} = 4\pi\epsilon_0 \frac{ab}{b - a} \tag{3.40}$$

If $b \to \infty$, the capacitance of an isolated sphere results; that is, we have a uniformly charged sphere with lines of flux extending radially outward to

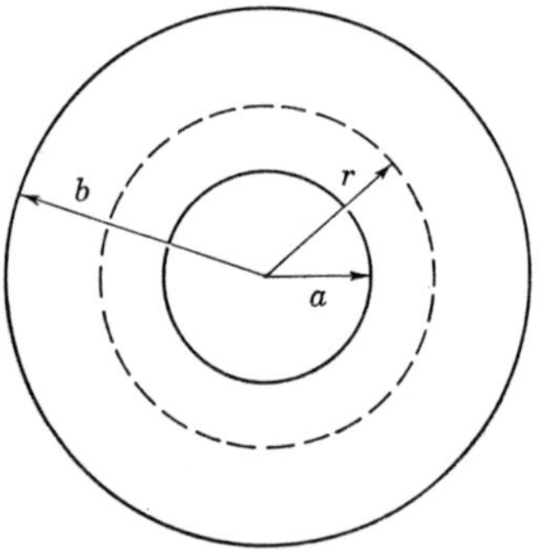

FIG. 3.13. A spherical capacitor.

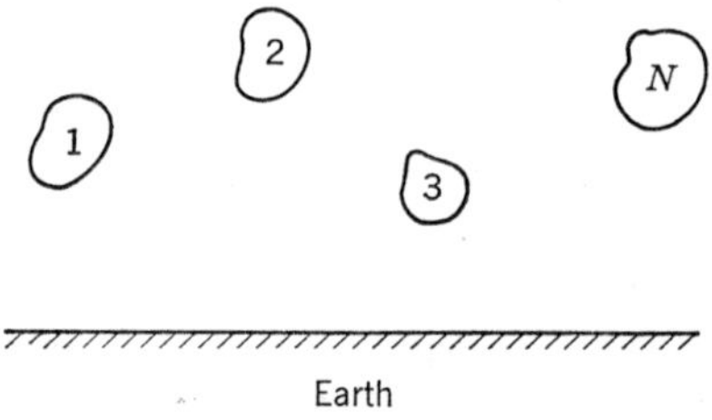

FIG. 3.14. A general arrangement of conductors.

terminate at "infinity." Such a condition is essentially obtained in the case of a small charged sphere in a relatively large laboratory room, where the flux tends to terminate on the distant walls. From (3.40) the capacitance of the isolated sphere is seen to be

$$C = 4\pi\epsilon_0 a \tag{3.41}$$

Multicapacitor System

The concept of capacitance can be extended to a region containing more than two conducting bodies. In Fig. 3.14 we show N arbitrary conducting bodies and the earth. The latter may or may not figure in a practical problem. We can simply consider it as just another conducting body, chosen as the reference, in the following analysis.

On the basis of the uniqueness theorem in Sec. 2.9, specification of the charge on each body plus specification that the earth be a zero reference potential uniquely determines the potential everywhere. Furthermore, because of the linear dependence of potential on charge, the following equations result:

$$\begin{aligned}
\Phi_1 &= p_{11}q_1 + p_{12}q_2 + \cdots + p_{1N}q_N \\
\Phi_2 &= p_{21}q_1 + p_{22}q_2 + \cdots + p_{2N}q_N \\
&\cdots\cdots\cdots\cdots\cdots \\
\Phi_N &= p_{N1}q_1 + p_{N2}q_2 + \cdots + p_{NN}q_N
\end{aligned} \tag{3.42}$$

In the above set of equations, the p_{ij}'s are constants, called the coefficients of potential, and they depend only on the geometry. We note that the potential on each body is properly a function of the total charge on each and that the relationship is linear.

The N equations can be solved to give the charges as functions of the potentials. The result will be in the form

$$\begin{aligned}
q_1 &= c_{11}\Phi_1 + c_{12}\Phi_2 + \cdots + c_{1N}\Phi_N \\
q_2 &= c_{21}\Phi_1 + c_{22}\Phi_2 + \cdots + c_{2N}\Phi_N \\
&\cdots\cdots\cdots\cdots\cdots \\
q_N &= c_{N1}\Phi_1 + c_{N2}\Phi_2 + \cdots + c_{NN}\Phi_N
\end{aligned} \tag{3.43}$$

Since the c_{ij}'s depend only on the p_{ij}'s of (3.42), they also depend only on the geometry. The terms $c_{11}, c_{22}, \ldots, c_{NN}$ are called coefficients of capacitance, while the c_{12}, c_{13}, etc., are coefficients of induction. The coefficient c_{ii} can be obtained by evaluating the ratio q_i/Φ_i of the ith body, with all others grounded. Since a positive charge on i produces a positive potential on i, the c_{ii}'s are all positive. The c_{ji} can be measured by grounding all but the ith body and evaluating the ratio q_j/Φ_i. Note that if q_i is positive, then Φ_i is positive, but the charge induced on j (q_j for $j \neq i$) will be negative. Accordingly, c_{ji} ($i \neq j$) is negative.

The coefficients of induction must satisfy a condition of reciprocity; that is, $c_{ij} = c_{ji}$. We can show this in the following way. Let all bodies be grounded but the first and second, which we take to be initially uncharged. If we begin charging body 1 to a final value of q_1 by adding very small increments of charge, then at some intermediate point the accumulated charge is q and the potential [from (3.42)] is $\Phi_1 = p_{11}q$. The energy required to add a charge dq is then, by definition of potential,

$$dW = p_{11}q\,dq$$

Accordingly, the total energy required to place a charge of q_1 on 1 is

$$W = \int_0^{q_1} p_{11}q\,dq = \frac{p_{11}q_1^2}{2} \tag{3.44}$$

If we proceed now to charge body 2 until q_2 is accumulated, then, corresponding to the presence of a charge q ($0 < q < q_2$), the potential is $\Phi_2 = p_{21}q_1 + p_{22}q$ and the work done in adding an increment of charge dq

to body 2 is

$$dW = (p_{21}q_1 + p_{22}q)\,dq$$

Altogether, the following amount of work is required to charge body 2 with q_2 coulombs:

$$W = \int_0^{q_2} (p_{21}q_1 + p_{22}q)\,dq = p_{21}q_1q_2 + \frac{p_{22}q_2^2}{2} \tag{3.45}$$

The total energy required to put q_1 on 1 and q_2 on 2 is

$$W_T = p_{21}q_1q_2 + \frac{p_{11}q_1^2}{2} + \frac{p_{22}q_2^2}{2} \tag{3.46}$$

If, however, we had first charged body 2 with q_2 and then body 1 with q_1, the total energy required is expressed by [we have only to interchange the subscripts 1 and 2 in (3.46)]

$$W_T = p_{12}q_2q_1 + \frac{p_{22}q_2^2}{2} + \frac{p_{11}q_1^2}{2} \tag{3.47}$$

Since the final result is the same, the energies expressed by (3.46) and (3.47) are equal; consequently,

$$p_{12} = p_{21} \tag{3.48}$$

The above proof can be repeated between any two bodies, say the ith and jth, so that

$$p_{ij} = p_{ji} \tag{3.49}$$

From purely algebraic considerations it then follows that

$$c_{ij} = c_{ji} \tag{3.50}$$

Equation (3.43) can now be rewritten in a way that is more informative. We let $C_{ij} = -c_{ij}$, noting that C_{ij} will be positive ($i \neq j$), and let

$$C_{ii} = c_{i1} + c_{i2} + c_{i3} + \cdots + c_{iN} \tag{3.51}$$

Then by adding and subtracting additional terms that will be clear upon examination, we can write in place of (3.43) the following:

$$\begin{aligned}
q_1 &= C_{11}\Phi_1 + C_{12}(\Phi_1 - \Phi_2) + C_{13}(\Phi_1 - \Phi_3) + \cdots + C_{1N}(\Phi_1 - \Phi_N) \\
q_2 &= C_{21}(\Phi_2 - \Phi_1) + C_{22}\Phi_2 + C_{23}(\Phi_2 - \Phi_3) + \cdots + C_{2N}(\Phi_2 - \Phi_N) \\
&\cdots\cdots\cdots\cdots\cdots\cdots\cdots\cdots\cdots\cdots\cdots\cdots \\
q_N &= C_{N1}(\Phi_N - \Phi_1) + C_{N2}(\Phi_N - \Phi_2) + C_{N3}(\Phi_N - \Phi_3) + \cdots + C_{NN}\Phi_N
\end{aligned} \tag{3.52}$$

Of the total charge q_1 on body 1, a portion $C_{12}(\Phi_1 - \Phi_2)$ depends on the difference of potential with body 2. Correspondingly, on body 2, we note that since $C_{12} = C_{21}$, an equal but opposite charge is bound. But this corresponds precisely to the view that body 1 and body 2 are connected by a capacitance C_{12}. Thus, of the total charge on 1, portions are bound on capacitors connecting 1 with every other body, and what remains lies on a capacitor (C_{11}) with potential Φ_1 to ground. The entire

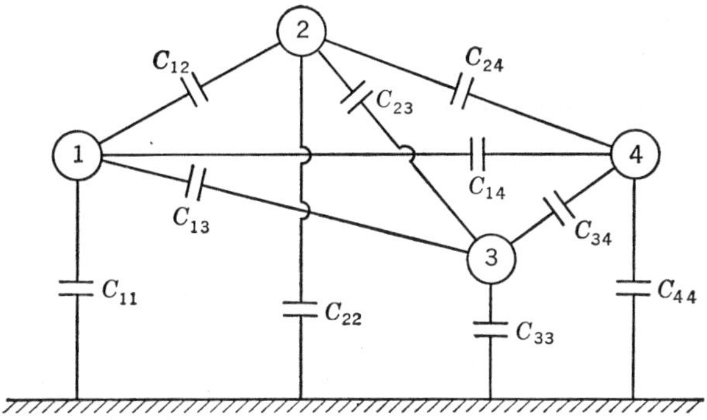

FIG. 3.15. Equivalent capacitor circuit for four arbitrary bodies in the presence of the earth.

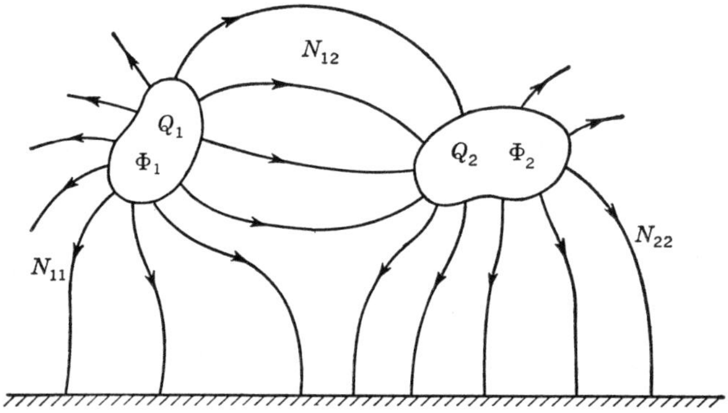

FIG. 3.16. Lines of flux surrounding two charged bodies.

set of equations can therefore be interpreted as representing the connection of the ith with the jth body by means of a capacitance C_{ij} and with capacitances C_{ii} from the ith body to ground. Figure 3.15 illustrates this for a four-body system.

A more detailed explanation of the significance of the capacitances C_{ij} will be given in the following discussion. With reference to Fig. 3.16, which illustrates two conductors in the presence of a ground plane, let us choose a scale of N lines per coulomb in a flux plot of the **E** field surrounding the bodies. Then, if the charge on body 1 is Q_1, $N_1 = NQ_1$ lines of flux leave body 1. Of these a number N_{12} terminate on body 2 and the remainder $N_{11} = N_1 - N_{12}$ terminate on the ground plane. The number of lines N_{12} terminates on an amount of charge $-q_2 = -Q_1N_{12}/N_1$ on body 2, while N_{11} corresponds to a total charge

$$-q_1 = \frac{-Q_1N_{11}}{N_1} = -(Q_1 - q_2)$$

residing on the ground plane. If the total charge on body 2 is Q_2, then since N_{12} lines of flux from body 1 terminate in a charge $-q_2$ on body 2, there must be $N(Q_2 + q_2) = N_{22}$ lines of flux leaving body 2 and terminating on the ground plane. If body 1 is at a potential Φ_1 and body 2 is at a potential Φ_2, then the capacitances C_{ij} are given by

$$C_{11} = \frac{N_{11}}{N\Phi_1} \qquad C_{12} = \frac{N_{12}}{N(\Phi_1 - \Phi_2)} \qquad C_{22} = \frac{N_{22}}{N\Phi_2}$$

and are just proportional to the fraction of the number of lines of flux leaving the body and terminating on adjacent bodies, divided by the potential difference. These results follow from (3.52) as follows. We have

$$Q_1 = C_{11}\Phi_1 + C_{12}(\Phi_1 - \Phi_2) = q_1 + q_2$$

and thus $q_1 = C_{11}\Phi_1 = N_{11}/N$, $q_2 = C_{12}(\Phi_1 - \Phi_2) = N_{12}/N$. Hence, $C_{11} = N_{11}/N\Phi_1$, etc., as stated above.

Electrostatic Shielding

The electric field within a closed conducting surface must be zero if no charge is placed within. For suppose that a field were present; then just inside the conducting surface, which is an equipotential, we could define another equipotential surface but at a different potential. Then over this surface $\mathbf{E}$ is everywhere outward (or inward). However, by Gauss' law, this requires the presence of an interior charge. But this contradicts the original assumption.

This characteristic leads to the use of hollow shells for electrostatic screening. If a body is placed within, it is uninfluenced by external fields. If the internal body is charged, the interior electric field is independent of external effects. The potential difference to the shell is also independent of external sources, but the potential with respect to ground depends on the potential of the shell with respect to ground. This variability is eliminated by grounding the shell.

Some of these remarks are more rigorously formulated through the use of (3.52). For example, based on Fig. 3.17 we have the following:

$$\begin{aligned} q_1 &= C_{11}\Phi_1 + C_{12}\Phi_1 + C_{13}(\Phi_1 - \Phi_3) \\ q_2 &= -C_{12}\Phi_1 - C_{23}\Phi_3 \\ q_3 &= C_{13}(\Phi_3 - \Phi_1) + C_{23}\Phi_3 + C_{33}\Phi_3 \end{aligned} \tag{3.53}$$

since $\Phi_2 = 0$ because body 2 is grounded. If we choose $q_1 = 0$, then the field inside body 2 is zero since there is no charge contained within. Therefore the potential Φ_1 must equal Φ_2, that is, equal zero. From the first equation in the set (3.53), we see this requires that $C_{13} = 0$. Thus in place of (3.53), we have

$$\begin{aligned} q_1 &= (C_{11} + C_{12})\Phi_1 \\ q_2 &= -C_{12}\Phi_1 - C_{23}\Phi_3 \\ q_3 &= (C_{23} + C_{33})\Phi_3 \end{aligned} \tag{3.54}$$

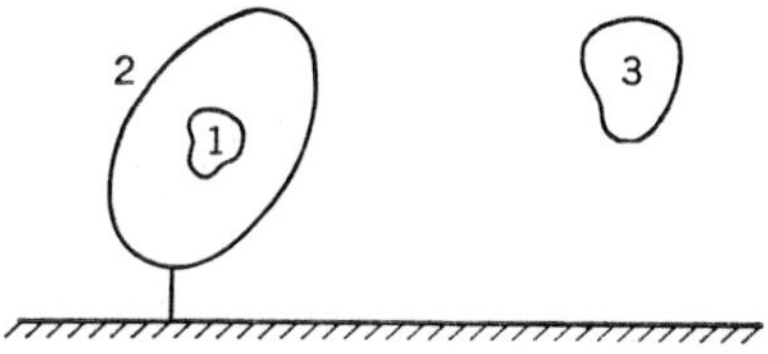

FIG. 3.17. Electrostatic shielding of bodies 1 and 3 by a grounded closed conducting shell 2.

Since body 2 is grounded, we may consider it as part of the ground plane, and then $C_{11} + C_{12}$ is the capacitance of body 1 with respect to ground while $C_{23} + C_{33}$ is the capacitance of body 3 with respect to ground. Since the potentials of body 1 and 3 depend only on their own charges, there is no interaction between them. Consequently, body 2 behaves as an electrostatic shield.

3.6. Electrostatic Energy

The absolute potential due to a single charge q has been found to be

$$\Phi = \frac{q}{4\pi\epsilon_0 R}$$

where R is the distance between the field point and the position of q. Consequently, the energy required to bring a charge q' from infinity to a distance R from q is

$$W_e = \frac{qq'}{4\pi\epsilon_0 R} \tag{3.55}$$

This energy is what is meant by the potential energy of the charges and is a function of their final position only.

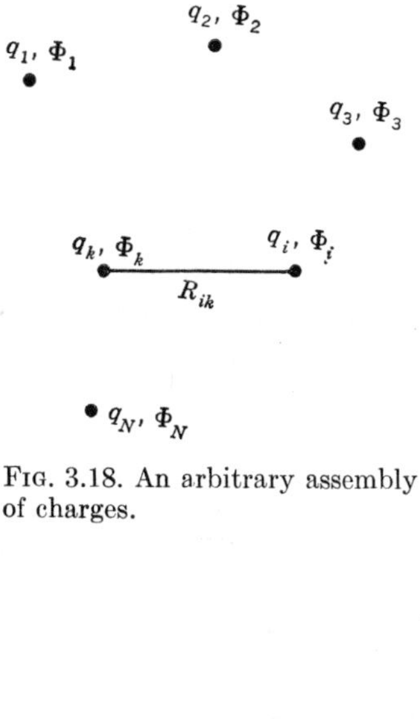

FIG. 3.18. An arbitrary assembly of charges.

Let us evaluate the energy required to assemble an arbitrary distribution of charges of arbitrary magnitudes, such as illustrated in Fig. 3.18. We can think of bringing each charge in from infinity in succession and evaluating the energy required. Thus no energy is required to introduce the first charge. The second charge brought in requires an energy as given by (3.55). For the third charge we must consider the net energy arising from interaction with the first and second charges, and so on.

The total energy to establish N charges can be formulated in a relatively simple way. Consider that the kth charge is the last one to be brought in from infinity. Since all others are present, then by superposition the energy required would be

$$(W_e)_k = \frac{1}{4\pi\epsilon_0} \sum_{i=1}^{N}{}' \frac{q_k q_i}{R_{ik}} \tag{3.56}$$

where R_{ik} is the distance from the ith to kth charge, and the primed summation symbolizes that $i = k$ is not included. The total energy required to assemble all N charges would not be correctly given if (3.56) were summed from $k = 1$ to $k = N$ since that equation is valid only for the last charge brought in; that is, no matter what order the charges are assembled in, the partial energy associated with any two charges, say, kth and jth, equals $q_k q_j / 4\pi\epsilon_0 R_{jk}$, and this term occurs only once during the assembly process. But if (3.56) is summed over k, each interaction occurs twice; for example, a partial energy contribution due to charges q_3 and q_5 is included not only when $k = 3$, $i = 5$, but also when $k = 5$, $i = 3$. This means that a summation of (3.56) over k yields exactly twice the desired value of energy. Then the energy W_e to assemble the charges is simply

$$W_e = \frac{1}{2} \sum_{k=1}^{N} \sum_{i=1}^{N}{}' \frac{q_k q_i}{4\pi\epsilon_0 R_{ik}} \tag{3.57}$$

If Φ_k designates the potential at the kth charge due to all other charges, i.e.,

$$\Phi_k = \frac{1}{4\pi\epsilon_0} \sum_{i=1}^{N}{}' \frac{q_i}{R_{ik}} \tag{3.58}$$

then (3.57) can be written as

$$W_e = \frac{1}{2} \sum_{k=1}^{N} q_k \Phi_k \tag{3.59}$$

We may inquire where the energy associated with a charge distribution is stored. An analogous question can be raised in mechanics if we are given two masses that are attached to the opposite ends of a compressed spring. The seat of the stored energy may be considered to be in the masses. We say that initially they lie in a region of higher potential energy; after expansion of the spring, potential energy is converted into kinetic energy. But one could also consider that the stored energy reposed in the stressed state of the spring. The first viewpoint coincides with the expression (3.57), where the stored energy is linked to the charges and their positions. This view is the counterpart of the action-at-a-distance concept, in that it ascribes physical reality to the charges and their spatial distribution alone. On the other hand, the field concept should be capable of expressing the stored energy without recourse to a description of the charges causing the field, if the concept is to be fully complete. The energy would then be described in terms of the "elastic" quality of the electric field, in similarity with the viewpoint that the spring stores the energy in the mechanical system considered. It turns out that the energy associated with a charge distribution can be expressed in terms of the fields alone, and we now proceed to show this.

For a continuous charge distribution, (3.59) can be generalized to read

$$W_e = \tfrac{1}{2} \int_V \rho\Phi \, dV \tag{3.60}$$

where Φ is now a continuous function of position and is the potential distribution due to all charges. It is no longer necessary to exclude the contribution due to $\rho \, \Delta V$ at the point where Φ is evaluated, since in the limit $\Delta V \to 0$ the contribution is zero anyway. Since $\nabla \cdot \mathbf{D} = \rho$, we have

$$W_e = \frac{\epsilon_0}{2} \int_V (\nabla \cdot \mathbf{E})\Phi \, dV$$

Now $\nabla \cdot (\Phi\mathbf{E}) = \Phi\nabla \cdot \mathbf{E} + \mathbf{E} \cdot \nabla\Phi$, and hence

$$W_e = \frac{\epsilon_0}{2} \int_V \nabla \cdot (\Phi\mathbf{E}) \, dV - \frac{\epsilon_0}{2} \int_V \nabla\Phi \cdot \mathbf{E} \, dV$$

Using the divergence (Gauss') theorem and noting that $\mathbf{E} = -\nabla\Phi$,

$$W_e = \frac{\epsilon_0}{2} \oint_S \Phi\mathbf{E} \cdot d\mathbf{S} + \frac{\epsilon_0}{2} \int_V E^2 \, dV \tag{3.61}$$

If the original volume extends indefinitely, as it should to include all possible charge, then the surface S in (3.61) extends to infinity. Considering S as the surface of a sphere of radius R, where $R \to \infty$, we note that $\Phi \propto 1/R$, $E \propto 1/R^2$, so that even though $S \propto R^2$, the integral is of order $1/R$ and vanishes in the limit $R \to \infty$. We are then left with

$$W_e = \frac{\epsilon_0}{2} \int_V E^2 \, dV \tag{3.62}$$

We have consequently found a way of expressing the stored energy of a charge distribution in terms of the field alone.

The previous analysis, which applies to charges in free space, can be extended to regions involving dielectrics. In this derivation we shall again consider the true charges as continuously distributed. If an amount of charge $\Delta\rho$ is introduced into an existing field Φ, then the energy is increased by an amount ΔW_e, where

$$\Delta W_e = \int_V \Phi\, \Delta\rho\, dV$$

This can also be written as

$$\Delta W_e = \int_V \Phi\Delta(\nabla \cdot \mathbf{D})\, dV = \int_V \Phi(\nabla \cdot \Delta\mathbf{D})\, dV \qquad (3.63)$$

Using the identity $\nabla \cdot \Phi\, \Delta\mathbf{D} = \Phi\, \nabla \cdot \Delta\mathbf{D} + \Delta\mathbf{D} \cdot \nabla\Phi$ and the divergence theorem gives

$$\Delta W_e = \int_V [\nabla \cdot (\Phi\, \Delta\mathbf{D}) - \Delta\mathbf{D} \cdot \nabla\Phi]\, dV = \oint_S \Phi\, \Delta\mathbf{D} \cdot d\mathbf{S} - \int_V \Delta\mathbf{D} \cdot \nabla\Phi\, dV$$

The surface integral term goes out for the same reasons as given in discussing (3.61); thus

$$\Delta W_e = \int_V \Delta\mathbf{D} \cdot \mathbf{E}\, dV$$

To obtain the total energy in the field, the above equation must be integrated. This involves the unknown functional relationship between **D** and **E**. For the case where $\mathbf{D} = \epsilon\mathbf{E}$ and ϵ is a constant, we have

$$W_e = \int_0^D \Delta W_e = \int_0^E \int_V \mathbf{E} \cdot \epsilon\, \Delta\mathbf{E}\, dV = \int_V \int_0^E \frac{\epsilon\, dE^2}{2}\, dV$$

or

$$W_e = \frac{\epsilon}{2} \int_V E^2\, dV = \tfrac{1}{2} \int_V \mathbf{E} \cdot \mathbf{D}\, dV \qquad (3.64)$$

The above expression reduces to the earlier one if free-space conditions are supposed.

In (3.64), as in (3.62), we have shown that the potential energy of a charge distribution can be calculated from the field produced by the charges. In the expressions themselves, the quantity $U_e = \mathbf{E} \cdot \mathbf{D}/2$ has the dimensions of energy density. It is often referred to as if it were truly the energy density of the field. However, just as it was impossible to localize the energy as being associated with the charge or the field, it is impossible to relate the total energy of the field with any particular distribution of component energy. Thus $\mathbf{E} \cdot \mathbf{D}/2$ is an energy density only to the extent that its volume integral over all space correctly evaluates the total potential energy.

Example 3.3. Energy Storage in a Capacitor. The parallel-plate capacitor provides a simple example for the application of the results of this section. Conditions are as illustrated in Fig. 3.19, where the plates have an area A and spacing d. For simplicity, one plate may be chosen to have a potential zero; the other is then at potential V. By an extension of (3.60) we may write

$$W_e = \tfrac{1}{2} \int_S \Phi \rho_s \, dS \tag{3.65}$$

This is zero over the lower surface, while for the upper surface we have $W_e = \tfrac{1}{2}VQ$. Since $C = Q/V$, this may also be expressed as

$$W_e = \tfrac{1}{2}CV^2 \tag{3.66}$$

From the method of derivation it is clear that (3.66) is valid for any two-body capacitor.

We may also calculate this result from the fields. We note that

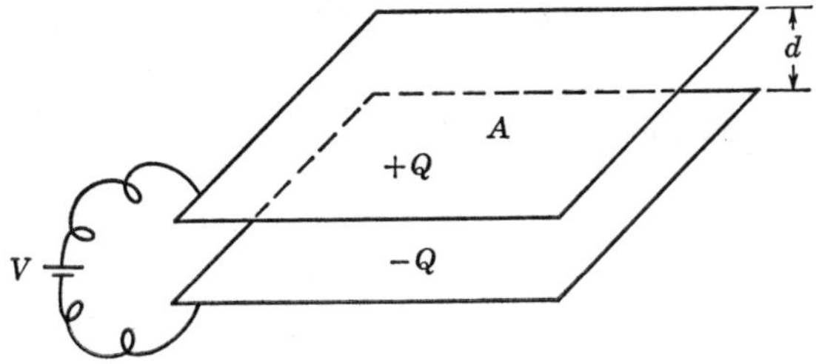

FIG. 3.19. A parallel-plate capacitor.

$D = Q/A$ and is uniform within the capacitor and zero outside, neglecting fringing. Hence,

$$W_e = \tfrac{1}{2} \int_V \mathbf{D} \cdot \mathbf{E} \, dV = \frac{1}{2\epsilon_0} \left(\frac{Q}{A}\right)^2 Ad = \frac{Q^2 d}{2\epsilon_0 A} \tag{3.67}$$

But $C = \epsilon_0 A/d$; hence $W_e = Q^2/2C = CV^2/2$ as before.

This same result can also be obtained in a more familiar way. The total charge Q on the upper plate is considered as having been transferred from the lower plate in differential increments. At an intermediate stage the charge is q, and the potential is therefore $\Phi = q/C$. The energy required to add a charge dq is simply $\Phi \, dq = q \, dq/C$. Then the work to build up to a charge Q is

$$W_e = \frac{1}{C} \int_0^Q q \, dq = \frac{Q^2}{2C}$$

thus confirming (3.66) once again.

Energy in a Multicapacitor System

The method used to derive (3.59) is also valid for a system of N conducting bodies as illustrated in Fig. 3.14. The energy stored in a system of N charged conducting bodies can be determined from (3.65). We note that over each separate conductor Φ is a constant, and hence the surface integral, if expressed as the sum of surface integrals over each body separately, becomes

$$W_e = \tfrac{1}{2}\int_S \rho_s\Phi\, dS = \tfrac{1}{2}\Phi_1\int_{S_1}\rho_s\, dS + \tfrac{1}{2}\Phi_2\int_{S_2}\rho_s\, dS + \cdots + \tfrac{1}{2}\Phi_j\int_{S_j}\rho_s\, dS + \cdots \tag{3.68}$$

But each integral is just the total charge on each conducting surface. Thus, if we let

$$Q_j = \int_{S_j}\rho_s\, dS \tag{3.69}$$

we have

$$\underline{W_e = \tfrac{1}{2}\sum_j \Phi_j Q_j} \tag{3.70}$$

If (3.43) is used to express Q_i in terms of the potentials Φ_j, we can rewrite (3.70) as

$$\underline{W_e = \tfrac{1}{2}\sum_{i=1}^{N}\sum_{j=1}^{N} c_{ij}\Phi_i\Phi_j} \tag{3.71}$$

This expression gives the energy stored in a system of N conducting bodies in terms of the potentials of the bodies.

For a two-body problem the capacitance C between the two bodies may be defined in terms of the stored energy in the electric field surrounding the bodies. From (3.66) we have

$$C = \frac{2W_e}{V^2} \tag{3.72}$$

This alternative definition of capacitance C is often easier to evaluate than some of the other formulations.

3.7. Electrostatic Forces

Figure 3.20 shows a parallel-plate capacitor in a free-space medium with a total charge $+Q$ on one plate and $-Q$ on the other. Because of the presence of charge of opposite sign, one would expect that a force of attraction exists between the plates. Such a force does, of course, exist; our present purpose is to calculate its magnitude. One of the simplest ways of doing this makes use of the principle of virtual work. As applied to this problem one of the capacitor

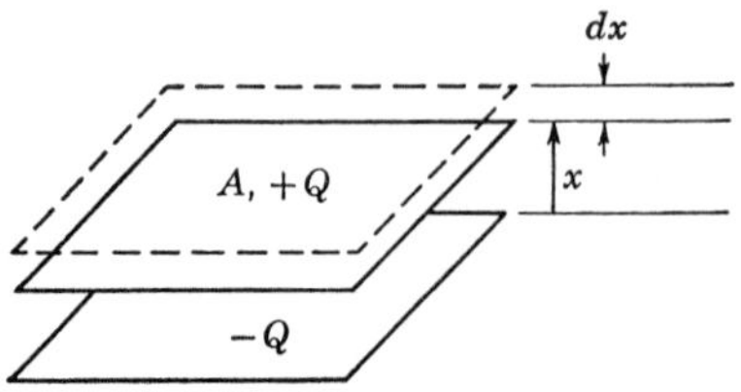

FIG. 3.20. Virtual displacement of one plate in a parallel-plate capacitor.

plates, say the upper one, is visualized as being displaced by an amount Δx away from the lower plate (x is the plate separation and is positive from lower to upper plate, as in Fig. 3.20). As a consequence, work must be performed, and the amount is simply $-F_x \Delta x$, where F_x is the electrostatic force that must be overcome by the external agent. From conservation of energy, this mechanical work must reappear as energy elsewhere, and in this case the stored energy of the field is the only other energy term that could possibly be involved. It must have increased, and the amount of increase is readily established from (3.67) to be

$$\Delta W_e = \frac{Q^2}{2\epsilon_0 A} \Delta x = \frac{\epsilon_0}{2} E^2 A \,\Delta x$$

where A is the area of the plates, and fringing of the field near the edges of the plates is neglected. Consequently,

$$-F_x \,\Delta x = \frac{\epsilon_0}{2} E^2 A \,\Delta x \tag{3.73a}$$

or in vector form,

$$\frac{\mathbf{F}}{A} = -\frac{\epsilon_0 E^2}{2} \mathbf{a}_x \tag{3.73b}$$

A generalization of this technique can be used to determine the force acting on any conducting body in the presence of an arbitrary distribution of charged conductors in free space. The x component of electrostatic force $\mathbf{F}_i \cdot \mathbf{a}_x$ on the ith body can be found by giving the ith body a virtual displacement $\Delta x_i \mathbf{a}_x$ and equating the work done, $(-\mathbf{F}_i \cdot \mathbf{a}_x) \,\Delta x_i$, to the net increase in the algebraic sum of all other energy changes that are produced. It is necessary that all energy terms that might be affected by the virtual displacement be included in the energy balance. In general, this includes a change in stored energy that results from the new position of the displaced body. Furthermore, if the potential between two or more bodies is held constant by the presence of a battery, then charge may be transferred by the battery in the virtual process, hence requiring the inclusion of a battery energy term also. If ΔW represents the work done by the batteries, then

$$-(F_i)_x \,\Delta x_i + \Delta W = \Delta W_e \tag{3.74}$$

or

$$(F_i)_x = \frac{\Delta W}{\Delta x_i} - \frac{\Delta W_e}{\Delta x_i} \tag{3.75}$$

The remaining force components are found by replacing x by y and z in turn.

Certain constraints are evident in a system of conducting charged bodies during the processes of displacing one or several of them. Those charged bodies that are isolated are constrained by the requirement that the total charge be kept constant in the process; their potential may vary, however. Those bodies linked together by batteries will be constrained to maintain a constant relative potential; in this case the total charge on each body may change. If one or several bodies are actually moved through a finite distance and it is desired to calculate the total work done on the body by setting up an energy balance, then obviously the true constraints of the system must be known and utilized. However, if it is only required that the force, for a particular configuration, be evaluated, then the virtual work process is conceptual only. Consequently, we may set up hypothetical constraints in the knowledge that so long as the resulting energy balance is correctly described, the correct value of force will be determined. The force that exists under a given set of conditions has a definite value, and this value cannot be affected by the constraints that are assumed under a purely hypothetical displacement of one of the bodies. However, if a constraint is assumed, then the correct consequences of that constraint must be included in the energy calculations. We consider now two cases, that of constant charge and that of constant potential.

Constant Charge

Since the constraint requires that each conducting body be isolated, the possibility that energy be supplied by a battery source does not occur. In the limit $\Delta x_i \to 0$, (3.75) may be written as

$$(F_i)_x = -\left.\frac{\partial W_e}{\partial x_i}\right|_{\text{constant charge}} \tag{3.76}$$

where x_i is the x variable at the ith body. Equation (3.70) can be used to evaluate W_e. When this equation is substituted into (3.76) and the fact that Q_j is constant is made use of, we find that

$$(F_i)_x = -\tfrac{1}{2} \sum_j Q_j \frac{\partial \Phi_j}{\partial x_i} \tag{3.77}$$

Constant Potential

In order to maintain each body at a constant potential when one body suffers a virtual displacement, it may be necessary to transfer charge to any or all of the conducting bodies in the system. This is performed by a battery, and the work involved must be included in an energy balance. If dQ_j is added to the jth body as a consequence of the virtual displacement, the energy supplied is $\Phi_j \, dQ_j$, and the total of such energy is

$$dW = \sum_j \Phi_j \, dQ_j \tag{3.78}$$

Combining this equation with (3.70) and (3.75), we have in the limit

$$(F_i)_x = \sum_j \Phi_j \frac{\partial Q_j}{\partial x_i} - \frac{1}{2} \sum_j \Phi_j \frac{\partial Q_j}{\partial x_i}$$

$$= \frac{1}{2} \sum_j \Phi_j \frac{\partial Q_j}{\partial x_i} = \frac{\partial W_e}{\partial x_i}\bigg|_{\text{constant potential}} \tag{3.79}$$

Note that the second summation makes use of the constant-potential constraint. Equation (3.79) expresses the very interesting fact that in a virtual displacement under a constant-potential constraint the work done by the field is equal to the increase in stored energy. The energy balance is explained by the fact that the battery supplies an energy equal to the sum of the mechanical and field terms.

The above results are applicable to the evaluation of torques also. It is only necessary to interpret F_i as a torque and dx_i as a small rotation $d\theta_i$.

Example 3.4. Force on a Capacitor Plate. The force between plates of a parallel-plate capacitor is given by (3.73). The method used to obtain this result involved the principle of virtual work together with a constraint of constant charge. Note that the result can be obtained immediately from (3.77). As pointed out above, the constraint can also be chosen as that of constant potential. Such a constraint can be interpreted physically as being due to the connection of a battery across the capacitor. Let us calculate the force under such assumed conditions. For our present purpose Fig. 3.19 correctly describes the problem. We assume that the upper plate is pushed away from the lower plate by an amount dx, where x is the plate separation. Since $W_e = \frac{1}{2}CV^2$,

$$dW_e = \tfrac{1}{2}V^2\, dC = \tfrac{1}{2}V^2\, d\left(\frac{\epsilon_0 A}{x}\right) = -\tfrac{1}{2}V^2 \frac{\epsilon_0 A}{x^2}\bigg|_{x=d} dx$$

$$= -\frac{1}{2}\frac{V^2 C}{d}\, dx$$

Since V remains constant, Q must vary in view of the change in C. Thus

$$Q = CV$$

$$dQ = V\, dC = -\frac{V}{d} C\, dx$$

The change in Q is accomplished by the transfer of the charge dQ by the battery. The battery therefore does an amount of work $V\,dQ$. This plus the work done by an external force acting against the field, $-F_x\, dx$, must equal the increase in stored energy. Consequently,

$$\frac{-V^2C}{d}\,dx - F_x\,dx = -\frac{1}{2}\frac{V^2C}{d}\,dx$$

or

$$F_x = -\frac{1}{2}\frac{V^2C}{d} = -\frac{\epsilon_0 AE^2}{2}$$

The result confirms that given by (3.73). Note that it is obtained almost at once by using (3.79).

The above results are easily generalized to the case of an arbitrary two-body capacitor with capacitance C. For an arbitrary virtual displacement dr, the electrostatic force acting in the direction of dr is readily found if the rate of change $\partial C/\partial r$ can be found. For a constant-charge constraint we write $W_e = Q^2/2C$, and hence from (3.76),

$$F_r = -\frac{\partial W_e}{\partial r} = -\frac{Q^2}{2}\frac{\partial C^{-1}}{\partial r} = \frac{Q^2}{2C^2}\frac{\partial C}{\partial r} \tag{3.80}$$

For a constant-potential constraint we may write $W_e = CV^2/2$ in place of $Q^2/2C$, and since V is constant, we have, from (3.79),

$$F_r = \frac{\partial W_e}{\partial r} = \frac{V^2}{2}\frac{\partial C}{\partial r} = \frac{Q^2}{2C^2}\frac{\partial C}{\partial r} \tag{3.81}$$

since $V = Q/C$. The force F_r is, of course, the same in both cases, since for a given charge on the capacitor a unique field exists around the conductors, and hence the force acting on the conductors is also unique. The existence of the force does not depend on the virtual displacement or constraints that are assumed. The virtual displacement is conceptual and, together with assumed constraints and an energy balance, permits the force exerted by the field to be evaluated. Any change in the assumed constraints cannot change the force acting; it only changes the details of the energy-balance equation.

Forces in the Presence of Dielectrics

In the previous paragraphs we showed how the force on a charged body in an electrostatic field could be found. The equations that were derived assumed that the conducting bodies lie in a free-space medium. The results obtained, however, would still apply in the presence of dielectrics; that is, the process whereby a virtual displacement is given a particular body, and the consequent mechanical work equated to the increase in field energy less the work performed by the batteries, still applies. However, the change in energy stored in the field must now include energy storage in the dielectrics.

If a dielectric body which lies in an electric field is given a virtual displacement, then the energy stored in the field may be expected to change in view of its dependence on the geometry. As a consequence it follows that dielectric bodies in electrostatic fields will experience a net force. This force may be calculated from the principle of virtual work, just as in the case for conducting bodies. The following example illustrates this technique.

Example 3.5. Force on a Dielectric Slab. A parallel-plate capacitor of width W, separation d, and length L is partially filled with a uniform

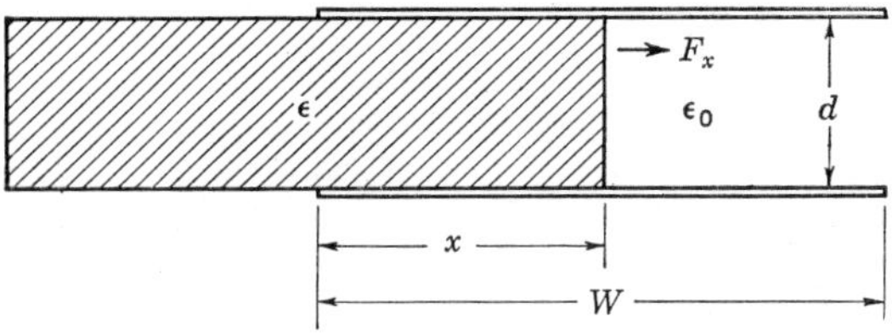

FIG. 3.21. Force on a dielectric slab in a parallel-plate capacitor.

dielectric slab, of permittivity ϵ, as illustrated in Fig. 3.21. The force acting on the dielectric slab is desired.

With the geometry as given, the total stored energy is calculated to be

$$W_e = \tfrac{1}{2}\epsilon_0 \left(\frac{V}{d}\right)^2 (W - x)Ld + \tfrac{1}{2}\epsilon \left(\frac{V}{d}\right)^2 xLd + W_1$$

where W_1 is a correction term that takes account of the fringing of the field at the sides. Let us postulate a virtual displacement Δx of the dielectric body in the positive x direction. We shall arbitrarily consider this to occur under the condition of constant potential. The work done by the field, $F_x \, \Delta x$, equals the increase in stored energy. We have

$$\begin{aligned} F_x \, \Delta x = \Delta W_e &= \tfrac{1}{2}\epsilon \left(\frac{V}{d}\right)^2 Ld \, \Delta x - \tfrac{1}{2}\epsilon_0 \left(\frac{V}{d}\right)^2 Ld \, \Delta x + \frac{\partial W_1}{\partial x} \Delta x \\ &= \tfrac{1}{2}E^2(\epsilon - \epsilon_0)Ld \, \Delta x \end{aligned} \tag{3.82}$$

since the correction term W_1 does not change if the ends of the dielectric slab are not too close to the plate edges, i.e., if the fringing field remains constant. The direction of the force is such as to draw the dielectric slab farther into the air gap of the capacitor since $\epsilon > \epsilon_0$ always. The pressure P_x, in the x direction, is

$$P_x = \frac{F_x}{Ld} = \tfrac{1}{2}(\epsilon - \epsilon_0)E^2 \tag{3.83}$$

and is the force per unit area exerted by the field on the plane end of the slab.

3.8. Electrostatic Forces (Maxwell Stress Tensor)

The total force acting on a volume distribution of charge ρ, which lies in an external electric field $\mathbf{E}$, is given by

$$\mathbf{F} = \int \mathbf{F}_v \, dV \tag{3.84}$$

where the volume density of force is

$$\mathbf{F}_v = \rho \mathbf{E} \tag{3.85}$$

As expressed by (3.84) and (3.85), the point of view that the force arises from an action at a distance is still manifest, even though the sources of the $\mathbf{E}$ field do not enter the calculation explicitly. The formulation is actually hybrid since it involves both fields and charge. According to the Faraday-Maxwell field theory, it should be possible to evaluate forces completely in terms of the field alone; that is, if the field is the means whereby forces are transmitted, one should be able to determine these forces by considering only the field. One can think of the transmission of forces via the field in the same way as one views that transmitted by a stretched rubber band. Indeed, one may view the flux lines as being elastic, the force action being a consequence of their state of stress. This implies that if a charge distribution in an external field is surrounded by an arbitrary surface area, then the total force acting on the charges contained must, in a sense, cross this area. But then it should be possible to calculate the total force by integrating over the arbitrary surface a "stress function" that depends, at each point, only on the field at that point and the direction of the surface area. Put mathematically, it suggests the existence of a force $\mathbf{T}\,dS$ which acts on an element of an arbitrary closed surface that contains charge and which satisfies the relation

$$\int_V \mathbf{F}_v \, dV = \int_S \mathbf{T} \, dS \tag{3.86}$$

The existence of a function $\mathbf{T}$ was asserted by Maxwell and forms an important part of the field concept.

We proceed now to show that the surface force $\mathbf{T}$ does exist and to derive an expression for it. This will require some vector manipulation so that the volume integral of (3.86) may be transformed into surface integrals. We begin by substituting for $\mathbf{F}_v$ the value given in (3.85) and note that $\nabla \cdot \mathbf{D} = \rho$; thus

$$\int_V \mathbf{F}_v \, dV = \int_V \rho \mathbf{E} \, dV = \int_V \mathbf{E} \nabla \cdot \mathbf{D} \, dV$$

If $\mathbf{E}$ is now expanded into rectangular components, we obtain

$$\int_V \mathbf{F}_v \, dV = \mathbf{a}_x \int_V E_x \nabla \cdot \mathbf{D} \, dV + \mathbf{a}_y \int_V E_y \nabla \cdot \mathbf{D} \, dV + \mathbf{a}_z \int_V E_z \nabla \cdot \mathbf{D} \, dV$$

Using vector identity (1.118), each term may be written as

$$\begin{aligned}\int_V \mathbf{F}_v \, dV = \mathbf{a}_x \int_V \nabla \cdot E_x \mathbf{D} \, dV + \mathbf{a}_y \int_V \nabla \cdot E_y \mathbf{D} \, dV + \mathbf{a}_z \int_V \nabla \cdot E_z \mathbf{D} \, dV \\ - \mathbf{a}_x \int_V \mathbf{D} \cdot \nabla E_x \, dV - \mathbf{a}_y \int_V \mathbf{D} \cdot \nabla E_y \, dV - \mathbf{a}_z \int_V \mathbf{D} \cdot \nabla E_z \, dV\end{aligned} \tag{3.87}$$

The first three right-hand terms of (3.87) can be integrated by Gauss' theorem, and then summed, whereupon the sum becomes equal to $\oint \mathbf{E}(\mathbf{n} \cdot \mathbf{D}) \, dS$. Concerning the remaining three terms, let us consider the coefficient of $\mathbf{a}_x$ in some detail. Expanding in rectangular coordinates we obtain

$$\mathbf{a}_x \int_V \mathbf{D} \cdot \nabla E_x \, dV = \mathbf{a}_x \int_V \left(D_x \frac{\partial E_x}{\partial x} + D_y \frac{\partial E_x}{\partial y} + D_z \frac{\partial E_x}{\partial z} \right) dV$$

But since $\nabla \times \mathbf{E} = 0$, it follows that $\partial E_x/\partial y = \partial E_y/\partial x$ and $\partial E_x/\partial z = \partial E_z/\partial x$. Accordingly,

$$\mathbf{a}_x \int_V \mathbf{D} \cdot \nabla E_x \, dV = \mathbf{a}_x \int_V \left(D_x \frac{\partial E_x}{\partial x} + D_y \frac{\partial E_y}{\partial x} + D_z \frac{\partial E_z}{\partial x} \right) dV$$
$$= \mathbf{a}_x \frac{\epsilon_0}{2} \int_V \frac{\partial}{\partial x} (E_x^2 + E_y^2 + E_z^2) \, dV \tag{3.88}$$

A similar expression is obtained for the $\mathbf{a}_y$ coefficient and the $\mathbf{a}_z$ coefficient, except that $\partial/\partial x$ is replaced by $\partial/\partial y$ and $\partial/\partial z$, respectively. Equation (3.87) may now be written as

$$\int_V \mathbf{F}_v \, dV = \oint_S \mathbf{E}(\mathbf{n} \cdot \mathbf{D}) \, dS - \frac{\epsilon_0}{2} \int_V \nabla(E^2) \, dV \tag{3.89}$$

The volume integral on the right-hand side can be converted to a surface integral using vector identity (1.127) (also see Prob. 3.10). As a result,

$$\int_V \mathbf{F}_v \, dV = \oint_S \epsilon_0 \left[(\mathbf{n} \cdot \mathbf{E})\mathbf{E} - \frac{E^2}{2} \mathbf{n} \right] dS \tag{3.90}$$

We can now identify (3.86) with (3.90), and therefore

$$\underline{\mathbf{T} = \epsilon_0 \left[(\mathbf{n} \cdot \mathbf{E})\mathbf{E} - \frac{E^2}{2} \mathbf{n} \right]} \tag{3.91}$$

Some idea of the vector relationships involved between **T**, **E**, and **n** can be obtained from Fig. 3.22. An element of surface is illustrated, and this determines the direction of **n**, of course. The field **E** is illustrated as making an arbitrary angle θ with **n**. The surface force **T** is described by (3.91). From (3.91) we see that **T** has a component in the direction of **n** which is either a compression or tension force, while that in the direction **E** involves a shearing force, in general. Furthermore, **T** must lie in the plane determined by **E** and **n**. If the components of **T** in the direction normal (T_n) and tangential (T_t) to the surface are computed, we find

$$T_n = \epsilon_0 \left(E^2 \cos^2 \theta - \frac{E^2}{2} \right) = \frac{\epsilon_0 E^2}{2} \left(2 \cos^2 \theta - 1 \right)$$
$$= \underline{\frac{\epsilon_0 E^2}{2} \cos 2\theta} \tag{3.92a}$$

$$\underline{T_t = \epsilon_0 E^2 \cos \theta \sin \theta = \frac{\epsilon_0 E^2}{2} \sin 2\theta} \tag{3.92b}$$

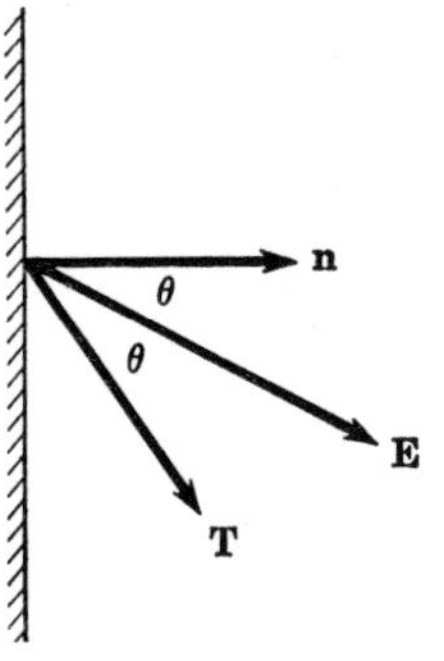

FIG. 3.22. Relation between **n**, **E**, and **T** at a plane surface.

This clearly requires that **T** have a magnitude $\epsilon_0 E^2/2$ and make an angle 2θ with **n**, as illustrated in Fig. 3.22.

To summarize, the force exerted on a charged region can be computed by integrating the surface force **T** over any bounding surface. At each surface element, **E** bisects the angle between **T** and the surface normal. Where **E** and **n** are in the same direction, the force is pure tension. For $\theta = 45°$, the stress transmitted is pure shearing stress. As

θ increases further, the shearing stress diminishes until at $\theta = 90°$ it is pure pressure. The magnitude of **T** remains equal to $\epsilon_0 E^2/2$ and is independent of the angle between **E** and **n**.

It is also possible to write

$$\int_S \mathbf{T}\, dS = \int_S \boldsymbol{\tau} \cdot d\mathbf{S} \tag{3.93}$$

in which case $\boldsymbol{\tau}$ has the properties of a tensor. Maxwell actually expressed his results in this form; consequently $\boldsymbol{\tau}$ is referred to as the Maxwell stress tensor.

The result expressed by (3.91) has been derived only for forces involving charges in free space. If dielectric bodies are also present, then the volumetric force is no longer

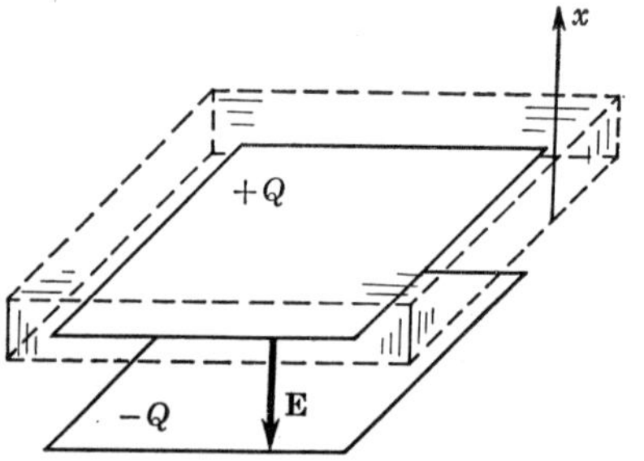

FIG. 3.23. Closed surface for evaluating force on a capacitor plate.

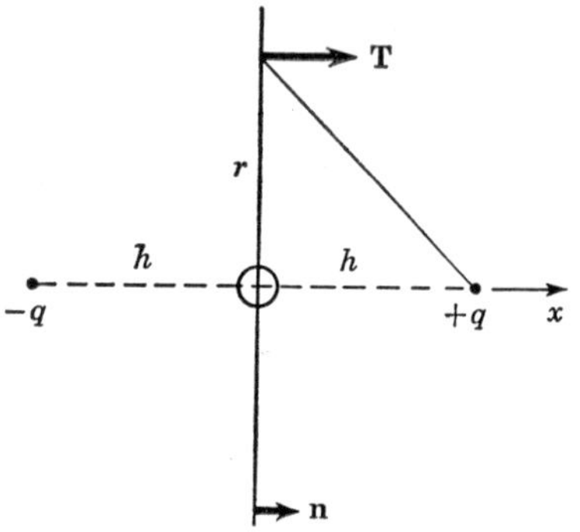

FIG. 3.24. Evaluation of force between two point charges.

given by the simple expression (3.85) at every point. It turns out, however, that if the surface of integration lies completely in free space, then (3.90) correctly evaluates the net force exerted on enclosed dielectric bodies in addition to that exerted on enclosed true charge. This result is not surprising since we know that the external behavior of a dielectric body can be interpreted in terms of an equivalent true charge distribution (polarization charge).

The full meaning of the surface force **T** may be more apparent if we consider some examples. First let us note how it could have been used to determine the force between the charged parallel plates illustrated in Fig. 3.20. To find the force on the upper plate, a rectangular parallelepiped may be envisioned surrounding that plate, as shown in Fig. 3.23. Since **E** lies only between the plates, a nonzero value of **E** appears only at the lower surface of the parallelepiped. In view of the uniformity of **E**,

$$\int_S \mathbf{T}\, dS = \frac{-\epsilon_0 E^2 \mathbf{a}_x}{2} \int_S dS = \frac{-\epsilon_0 E^2}{2} A \mathbf{a}_x$$

and the result of (3.73) is given at once.

In Fig. 3.24 we have two equal and opposite charges separated by a distance $2h$. The force exerted on the negative charge will be found by integrating $\mathbf{T}\, dS$ over an infinite plane surface (closed at infinity). The line joining $-q$ and $+q$ is bisected by the surface and is perpendicular to it. The geometry is then precisely that in Fig. 2.18; consequently, the electric field over the plane is given by (2.66). Since **E** is always normal to the surface and in the negative x direction, the force exerted on $-q$

is in the positive x direction. Its magnitude is given by

$$\begin{aligned} F_{-q} &= \frac{\epsilon_0}{2}\left(\frac{q2h}{4\pi\epsilon_0}\right)^2 \int_0^\infty \int_0^{2\pi} \frac{r\,dr\,d\theta}{(h^2+r^2)^3} \\ &= -\pi\epsilon_0\left(\frac{q2h}{4\pi\epsilon_0}\right)^2 \frac{1}{4(h^2+r^2)^2}\Bigg]_0^\infty \\ &= \frac{q^2}{4\pi\epsilon_0(2h)^2} \end{aligned} \tag{3.94}$$

We note that the final result correctly corresponds to Coulomb's law of force.

3.9. Molecular Fields

In Sec. 3.1 some of the details of the polarization of dielectric materials were discussed. Basically, we noted that under the influence of an applied electric field the molecules of a dielectric would become polarized. Further, the dipole polarization was noted to be, in general, proportional to the applied field for field strengths well below the saturation level.

In order to actually calculate the polarization of a given material, it is necessary to know the polarizability of the material and the field. The latter must be the field that actually exists at the molecule. For gases, this is to a good approximation the applied field, i.e., the field that exists in the absence of the gas. Actually, the field at a particular molecule is the superposition of the applied field and that due to its neighboring polarized molecules. But for a gas, the separation of the molecules is sufficiently great so that the effect of adjacent dipoles on one another can be neglected.

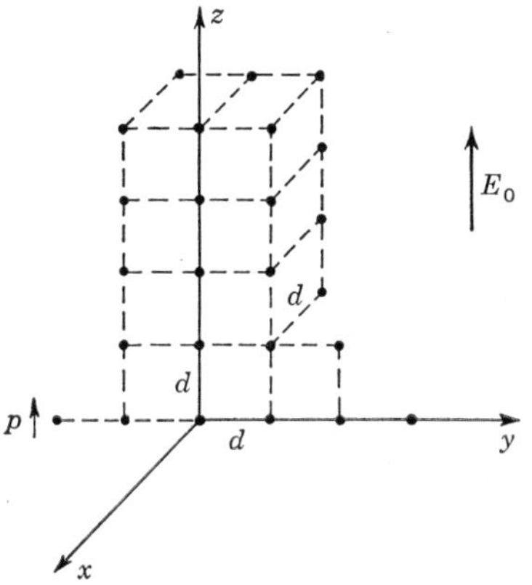

FIG. 3.25. An infinite cubical array of atoms.

For a solid dielectric, dipole-dipole interaction must be considered. In general, this may be fairly difficult to express analytically. We shall consider in some detail a specific, and common, configuration to illustrate the principle. We shall assume that the dielectric consists of a cubical array of atoms (or molecules) with spacing d between each atom along the x, y, z axis, as in Fig. 3.25. The medium is assumed infinite in extent, and a field $\mathbf{E}_0$ is applied in the z direction. Under the influence of the field $\mathbf{E}_0$, each atom becomes polarized with a dipole moment $\mathbf{p}$. Since there are $N = 1/d^3$ atoms per unit volume, the dipole moment per unit volume is $\mathbf{p}/d^3 = \mathbf{P}$. The field acting to polarize the atom at the origin (or any other atom) is the sum of the applied field $\mathbf{E}_0$ plus an interaction field $\mathbf{E}_i$. The interaction field is the field produced by all the neighboring dipoles. The polarizing field $\mathbf{E}_p$ may be written as

$$\mathbf{E}_p = \mathbf{E}_0 + \mathbf{E}_i \tag{3.95}$$

and hence the induced dipole moment in each atom is

$$\mathbf{p} = \alpha_e \mathbf{E}_p \tag{3.96}$$

where α_e is the polarizability of the atom. When the material behaves linearly, $\mathbf{p}$, $\mathbf{E}_i$, and $\mathbf{E}_p$ are all proportional to the applied field $\mathbf{E}_0$, and hence we may take $\mathbf{E}_i$ proportional to $\mathbf{p}$. Thus

$$\mathbf{E}_i = C\mathbf{p} \tag{3.97}$$

where C is called the field interaction constant. Equation (3.96) now becomes

$$\mathbf{p} = \alpha_e(\mathbf{E}_0 + \mathbf{E}_i) = \alpha_e\mathbf{E}_0 + \alpha_e C\mathbf{p}$$

or

$$\mathbf{p} = \frac{\alpha_e\mathbf{E}_0}{1 - \alpha_e C} \tag{3.98}$$

During the course of the analysis it will be shown that the average total electric field $\mathbf{E}$ in the dielectric is equal to the applied field $\mathbf{E}_0$. Hence the displacement flux density $\mathbf{D}$ is given by

$$\mathbf{D} = \epsilon_0\mathbf{E} + \mathbf{P} = \epsilon_0\mathbf{E}_0 + \frac{N\alpha_e\mathbf{E}_0}{1 - \alpha_e C} = \epsilon\mathbf{E} \tag{3.99}$$

where $N = d^{-3}$ is the number of dipoles per unit volume. Solving for the permittivity ϵ gives

$$\epsilon = \epsilon_0 + \frac{N\alpha_e}{1 - \alpha_e C} \tag{3.100}$$

a result known as the Clausius-Mossotti equation. Assuming that α_e is known and that the interaction constant C can be found, we can compute ϵ from (3.100). The interaction constant C is a function of the lattice geometry only.

Evaluation of Interaction Constant

The interaction constant C may be found by evaluating the field at the origin as produced by all the neighboring dipoles in the lattice illustrated in Fig. 3.25. From symmetry considerations it is readily seen that this field has a z component only. The scalar potential due to a dipole of moment p and located at (x',y',z') is

$$\Phi(x,y,z) = \frac{p}{4\pi\epsilon_0}\frac{\cos\theta}{R^2} = \frac{p(z - z')}{4\pi\epsilon_0 R^3}$$

since $z - z' = R\cos\theta$, and $R^2 = (x - x')^2 + (y - y')^2 + (z - z')^2$. The z component of electric field is

$$\begin{aligned} E_z = -\frac{\partial\Phi}{\partial z} &= \frac{-p}{4\pi\epsilon_0}\left[\frac{1}{R^3} - \frac{3(z - z')^2}{R^5}\right] \\ &= \frac{p}{4\pi\epsilon_0}\frac{2(z - z')^2 - (x - x')^2 - (y - y')^2}{[(x - x')^2 + (y - y')^2 + (z - z')^2]^{5/2}} \end{aligned} \tag{3.101}$$

Now the dipoles are located at $x' = nd$, $y' = md$, $z' = sd$ ($n, m, s = 0, \pm1, \pm2, \ldots,$ but $n = m = s = 0$ excluded). Summing (3.101) over all the dipoles and placing $x = y = z = 0$, we obtain the following expression for the interaction field at the origin:

$$E_i = \frac{p}{4\pi\epsilon_0}\sum_{n=-\infty}^{\infty}\sum_{m=-\infty}^{\infty}\sum_{s=-\infty}^{\infty}{}' \frac{2(sd)^2 - (nd)^2 - (md)^2}{[(sd)^2 + (nd)^2 + (md)^2]^{5/2}} \tag{3.102}$$

where the prime means omission of the term $n = m = s = 0$, which corresponds to the dipole at the origin.

The above series may be transformed into a rapidly converging series by use of the Poisson summation formula, and the sum may then be evaluated. For our purpose we shall approximate the sum by a triple integral. Let us introduce the variables

$x = n$, $y = m$, $z = s$ and the differentials $dx = dn$, $dy = dm$, $dz = ds$. The integration extends from $|x|$, $|y|$, $|z|$ equal to unity out to infinity. Since all quantities are squared, we may integrate over one octant only and multiply the result by 8. We now have

$$E_i = \frac{8p}{4\pi\epsilon_0 d^3}\int_1^\infty\int_1^\infty\int_1^\infty \frac{2z^2 - x^2 - y^2}{(x^2 + y^2 + z^2)^{5/2}}\,dx\,dy\,dz$$

Since the integrand was originally obtained by differentiating with respect to z, we may integrate with respect to z at once to get

$$E_i = \frac{2p}{\pi\epsilon_0 d^3}\int_1^\infty\int_1^\infty \left.\frac{-z}{(x^2 + y^2 + z^2)^{3/2}}\right|_1^\infty dx\,dy$$
$$= \frac{2p}{\pi\epsilon_0 d^3}\int_1^\infty\int_1^\infty \frac{dx\,dy}{(x^2 + y^2 + 1)^{3/2}}$$

With the substitution $y = (x^2 + 1)^{1/2}\tan\theta$ and the limits $\sin^{-1}(2 + x^2)^{-1/2}$ to $\pi/2$ on θ, the integration over y is readily performed (the integrand becomes simply $\cos\theta\,d\theta$) to give

$$E_i = \frac{2p}{\pi\epsilon_0 d^3}\int_1^\infty \left[\frac{1}{x^2 + 1} - \frac{1}{(x^2 + 1)(x^2 + 2)^{1/2}}\right] dx$$

The first term gives

$$\int_1^\infty \frac{dx}{1 + x^2} = \left.\tan^{-1} x\right|_1^\infty = \frac{\pi}{2} - \frac{\pi}{4} = \frac{\pi}{4}$$

In the second term replace $x^2 + 1$ by u^2 to get

$$-\int_{\sqrt{2}}^\infty \frac{du}{u(u^4 - 1)^{1/2}} = \left.-\tfrac{1}{2}\sec^{-1} u^2\right|_{\sqrt{2}}^\infty = -\frac{\pi}{4} + \frac{\pi}{6}$$

The final value for E_i is thus

$$E_i = \frac{2p}{\pi\epsilon_0 d^3}\left(\frac{\pi}{4} + \frac{\pi}{6} - \frac{\pi}{4}\right) = \frac{p}{3\epsilon_0 d^3} \tag{3.103}$$

From (3.97) we find that the interaction constant C is

$$C = \frac{E_i}{p} = \frac{1}{3\epsilon_0 d^3} = \frac{N}{3\epsilon_0} \tag{3.104}$$

If the series (3.102) is summed, it is found that the interaction constant C is equal to $0.34/d^3\epsilon_0$, and hence we can conclude that the triple integral is a good approximation to the triple sum.

If we use the above value of C in (3.100), we find that the permittivity is given by

$$\epsilon = \epsilon_0 + \frac{N\alpha_e}{1 - N\alpha_e/3\epsilon_0}$$

or

$$\frac{\epsilon - \epsilon_0}{\epsilon + 2\epsilon_0} = \frac{N\alpha_e}{3\epsilon_0} \tag{3.105}$$

For a gas, ϵ is only slightly greater than ϵ_0, so that (3.105) becomes

$$\frac{\epsilon - \epsilon_0}{3\epsilon_0} = \frac{N\alpha_e}{3\epsilon_0}$$

or

$$\epsilon = \epsilon_0 + N\alpha_e \tag{3.106}$$

which, upon comparison with (3.100), is seen to be equivalent to neglecting the interaction field. By combining (3.104) and (3.95) we find that the polarizing field is given by

$$\mathbf{E}_p = \mathbf{E}_0 + C\mathbf{p} = \mathbf{E}_0 + \frac{\mathbf{P}}{3\epsilon_0} \tag{3.107}$$

As a final step we shall show that the average electric field in the dielectric is equal to the applied field $\mathbf{E}_0$. To show this we must evaluate the average field produced by all the dipoles. The average is to be taken over the volume of a unit cell. We have already found a value for the interaction field, which is the field set up by all the dipoles except one. If we add to this the average field set up by the individual dipole, we shall obtain the average field produced by all the dipoles in the lattice. At this point it should be noted that replacing the triple sum in (3.102) by a triple integral is equivalent to determining an average value E_i for the interaction field.

The z component of the field produced by a single dipole, say the dipole at the origin, is given by (3.101) with x', y', and z' equal to zero. The average value of this field is E_{1a}, where

$$E_{1a} = \frac{8p}{4\pi\epsilon_0 d^3} \int_0^{d/2} \int_0^{d/2} \int_0^{d/2} \frac{2z^2 - x^2 - y^2}{(x^2 + y^2 + z^2)^{5/2}}\, dx\, dy\, dz \tag{3.108}$$

In (3.108) the integration is taken over one octant of the unit cell, which has a volume of $d^3/8$. This is equivalent to averaging over the unit cell because of the symmetry of the integrand. The above may be integrated in the same manner as the integral for E_i was performed. The reader may readily verify by carrying out the details that the end result is simply

$$E_{1a} = -\frac{p}{3\epsilon_0 d^3} \tag{3.109}$$

This result is interesting in that it shows that the average field produced by a single dipole in its own unit cell is the negative of the average field produced by all the neighboring dipoles. Consequently, the average field produced by all the dipoles in the lattice is $(E_d)_{\text{av}}$ and vanishes since

$$(E_d)_{\text{av}} = E_i + E_{1a} = \frac{p}{3\epsilon_0 d^3} - \frac{p}{3\epsilon_0 d^3} = 0$$

This completes the proof that the average field in the dielectric is just equal to the applied field E_0.

The above analysis clearly shows that the effective field acting to polarize a given molecule is not equal to E_0, for this is the average field in the dielectric and includes the contribution from the molecule itself. The field acting to polarize the molecule is equal to E_0 minus the contribution to the total average field from the molecule itself. This gives $E_0 - (-p/3\epsilon_0 d^3) = E_0 + p/3\epsilon_0 d^3$ for the polarizing field, a result in agreement with (3.107).

CHAPTER 4

SOLUTION OF BOUNDARY-VALUE PROBLEMS

In the previous two chapters a number of elementary boundary-value problems were solved by means of certain specialized techniques such as the method of images, inversion in a sphere or cylinder, and the use of Gauss' law. These methods, when applicable, lead to the required solutions for the scalar potential Φ and the electric field in a very direct manner. However, for many problems that occur in practice, these special techniques are too restricted in their scope. In this chapter we shall examine a much more general approach, that is, the method of separation of variables. For this method it is necessary that the boundaries over which the potential or its normal derivative is specified coincide, at least piecewise, with the constant coordinate surfaces in a suitable orthogonal curvilinear coordinate system. Furthermore, it is necessary that the partial differential equation in question be separable in the appropriate system of coordinates so that the solution can be represented as a product of three functions which individually are a function of one coordinate variable only. When the solution can be represented in such a product form, the partial differential equation is said to be separable.

In three dimensions Laplace's equation is separable in 11 different coordinate systems. Among these systems are rectangular, spherical, and cylindrical coordinates, all of which will be examined in this chapter. In two dimensions Laplace's equation is separable in virtually an infinite number of different two-dimensional coordinate systems. We shall show that any coordinate system which is generated by an analytic function $W = F(Z)$, where Z is the complex variable $x + jy$, is a coordinate system in which Laplace's equation is separable.

Boundary-value problems are usually classified into three types. If the value of the potential is specified everywhere on the whole boundary, this boundary condition is referred to as a Dirichlet boundary condition. If, on the other hand, the normal derivative of Φ (this is proportional to the charge density) is specified on the whole boundary, we refer to this as Neumann boundary conditions. Finally, if the potential is specified on part of the boundary and the normal derivative $\partial\Phi/\partial n$ on the remainder, we have a mixed boundary-value problem. The method of solution in all three cases is essentially the same. It is not possible to specify both Φ and its normal derivative in an arbitrary manner over a common portion of the boundary since this overspecifies the problem. For example, if Φ is specified on the whole boundary, then since the solution to Laplace's equation is unique, $\partial\Phi/\partial n$ is completely determined. Thus, in this case, there is no choice in the values that $\partial\Phi/\partial n$ can have on the boundary. Similar remarks also apply to the Neumann and mixed boundary-value problems.

As was demonstrated in Sec. 2.9, if we can find a solution to Laplace's equation that satisfies all the required boundary conditions, then this solution is unique. Similar uniqueness theorems are readily established for other types of partial differential equations also. Likewise, the method of separation of variables, as discussed in this chapter in connection with Laplace's equation, is applicable to all partial differential equations.

If the boundaries do not coincide with constant coordinate surfaces in a coordinate system for which the partial differential equation is separable, then the method of separation of variables is of little use. In these cases (which do occur very often in practice) approximate methods of analysis or experimental techniques must be used.

4.1. Rectangular Coordinates

In rectangular coordinates Laplace's equation for the scalar potential Φ is

$$\frac{\partial^2\Phi}{\partial x^2} + \frac{\partial^2\Phi}{\partial y^2} + \frac{\partial^2\Phi}{\partial z^2} = 0 \tag{4.1}$$

To determine if this equation is separable in rectangular coordinates, we assume a product solution for Φ of the form

$$\Phi = f(x)g(y)h(z) \tag{4.2}$$

where f, g, and h are functions of x, y, and z, respectively, only. Substituting into (4.1) gives

$$ghf'' + fhg'' + fgh'' = 0$$

where $f'' = d^2f/dx^2$, $g'' = d^2g/dy^2$, and $h'' = d^2h/dz^2$. Dividing by fgh gives

$$\frac{f''}{f} + \frac{g''}{g} + \frac{h''}{h} = 0 \tag{4.3}$$

Each term is a function of one variable only; for example, f''/f depends on x only. If we keep y and z constant and vary x only, the term f''/f can possibly vary. However, the sum of the three terms must equal zero, and therefore each term must be equal to a constant in order for (4.3) to hold for all arbitrary values of x, y, and z. Thus we must have

$$\frac{d^2f}{dx^2} + k_x^2 f = 0 \tag{4.4a}$$

$$\frac{d^2g}{dy^2} + k_y^2 g = 0 \tag{4.4b}$$

$$\frac{d^2h}{dz^2} + k_z^2 h = 0 \tag{4.4c}$$

where k_x, k_y, and k_z are constants (as yet arbitrary) called separation constants. The only restriction on the separation constants so far is that the sum be equal to zero, so that (4.3) will hold; i.e.,

$$k_x^2 + k_y^2 + k_z^2 = 0 \tag{4.5}$$

By means of the above procedure we have reduced the solution of the partial differential equation to that of solving three ordinary differential equations. It is this property that leads to the designation "method of separation of variables" for the technique used. In practice, the separation constants are determined by the boundary conditions which the potential or its normal derivative must satisfy. The details will become clear from the examples to be discussed, but first we shall consider some special cases.

Case 1

If $k_x = 0$ but k_y and k_z are not zero, then from (4.5) it is seen that $k_z = \pm jk_y$. When $k_x = 0$, the solution to (4.4*a*) is $f = A_1x + A_2$, where A_1 and A_2 are arbitrary constants. The solution to (4.4*b*), with k_y^2 chosen to be positive, is then

$$g = B_1 \sin k_y y + B_2 \cos k_y y$$

where B_1 and B_2 are arbitrary constants. This solution is readily verified by substitution in (4.4*b*). Since $k_z^2 = -k_y^2$, (4.4*c*) becomes

$$\frac{d^2h}{dz^2} - k_y^2 h = 0$$

with a general solution of the form

$$h = C_1 \sinh k_y z + C_2 \cosh k_y z$$

where C_1 and C_2 are constants. In the above solutions we could equally well choose the exponential forms

$$g = B_1 e^{jk_y y} + B_2 e^{-jk_y y}$$
$$h = C_1 e^{k_y z} + C_2 e^{-k_y z}$$

According to the mathematical theory, any two independent solutions constitute a general solution for a second-order differential equation. In the present case the solution for Φ is

$$\Phi = fgh = (A_1x + A_2)(B_1 \sin k_y y + B_2 \cos k_y y)(C_1 \sinh k_y z + C_2 \cosh k_y z) \tag{4.6}$$

The constants in this solution would normally be determined by the boundary conditions. If k_y^2 had been chosen as a negative constant, the hyperbolic and trigonometric functions in (4.6) would be interchanged.

Case 2

If $k_x = k_y = 0$, it is necessary that $k_z = 0$ also in order for (4.5) to hold. In this case the solution is of the form

$$\Phi = (A_1x + A_2)(B_1y + B_2)(C_1z + C_2) \tag{4.7}$$

Case 3

If k_x and k_y are both positive real constants, then from (4.5) we have

$$k_z = \pm j(k_x^2 + k_y^2)^{1/2} \tag{4.8}$$

In this case the general solution for Φ is

$$\Phi = (A_1 \sin k_x x + A_2 \cos k_x x)(B_1 \sin k_y y + B_2 \cos k_y y)(C_1 \sinh |k_z|z + C_2 \cosh |k_z|z) \tag{4.9}$$

where $|k_z| = (k_x^2 + k_y^2)^{1/2}$. Other variations of (4.9) can be derived by cyclic permutation of the variables.

In many problems it is found that a combination of the various solutions discussed above must be used in order to satisfy all the required boundary conditions. Also, it is usually found that the separation constants can take on an infinite sequence of values. The general solution for Φ is then given by a summation over all the possible individual solutions. This particular property makes the general solution extremely flexible in that it can be made to satisfy any arbitrary boundary condition. These points will be further elaborated in the course of solutions of the following examples.

Example 4.1. A Rectangular Dirichlet Boundary-value Problem. We wish to find a solution to Laplace's equation in the interior of a rectangular enclosure such that the potential Φ reduces to zero on all sides except the side at $z = c$. On the surface $z = c, 0 \leq x \leq a, 0 \leq y \leq b$, the potential Φ is equal to the specified value $V(x,y)$. With reference to Fig. 4.1, it should be noted that the boundaries coincide with constant coordinate surfaces.

The boundary condition $\Phi = 0$ on the two faces $x = 0, a$ must be satisfied for all values of y and z on these faces. This condition is met if $f(x)$

vanishes at $x = 0, a$. Since the solution for f is of the form

$$f = A_1 x + A_2 \qquad k_x = 0$$
$$f = A_1 \sin k_x x + A_2 \cos k_x x \qquad k_x \neq 0$$

we see that the first solution for $k_x = 0$ is valid only if A_1 and A_2 are zero.

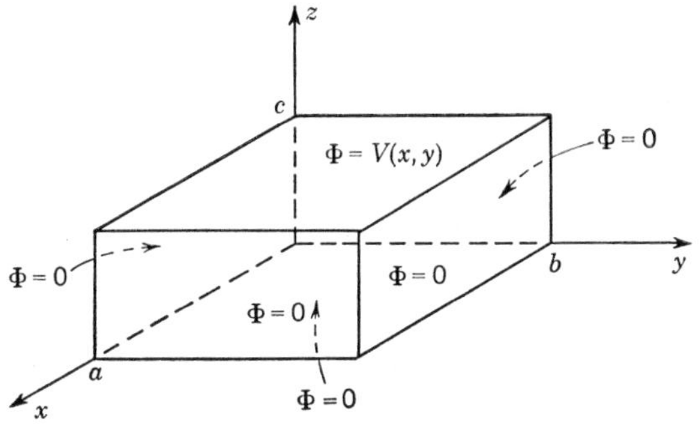

FIG. 4.1. A rectangular parallelepiped with specified boundary conditions.

This is a trivial solution; so the second form must be chosen. For f to equal zero at $x = 0$, we must choose $A_2 = 0$. Hence we have

$$f = A_1 \sin k_x x$$

Now f must equal zero at $x = a$ also, and hence $\sin k_x a = 0$. From this result we see that the separation constant k_x is given by

$$k_x = \frac{n\pi}{a} \qquad n = 1, 2, 3, \ldots$$

The function $\sin (n\pi x/a)$ is called an eigenfunction (proper function) and $n\pi/a$ an eigenvalue. The most general solution for f is

$$f = \sum_{n=1}^{\infty} A_n \sin \frac{n\pi x}{a} \tag{4.10a}$$

where the A_n are as yet arbitrary constants.

What we have said about the function $f(x)$ is applicable to the function $g(y)$ also. It is therefore not difficult to see that a general solution for $g(y)$ that vanishes at $y = 0, b$ is

$$g(y) = \sum_{m=1}^{\infty} B_m \sin \frac{m\pi y}{b} \qquad m = 1, 2, 3, \ldots \tag{4.10b}$$

For the nmth solution for fg we have $k_x{}^2 = (n\pi/a)^2$ and $k_y{}^2 = (m\pi/b)^2$.

The corresponding value of k_z must be

$$k_z = \pm j\left[\left(\frac{n\pi}{a}\right)^2 + \left(\frac{m\pi}{b}\right)^2\right]^{1/2} = \pm j\Gamma_{nm}$$

Of the two possible solutions $\sinh |k_z|z$ and $\cosh |k_z|z$ for $h(z)$, the hyperbolic sine function must be chosen, since this is the only solution that vanishes at $z = 0$. The general solution for Φ is thus of the form

$$\Phi = \sum_{n=1}^{\infty}\sum_{m=1}^{\infty} A_nB_m \sin\frac{n\pi x}{a}\sin\frac{m\pi y}{b}\sinh\left[\left(\frac{n\pi}{a}\right)^2 + \left(\frac{m\pi}{b}\right)^2\right]^{1/2} z \tag{4.11}$$

We note that the nature of the boundary conditions has led to a solution such as was discussed under case 3.

In order to determine the coefficients A_n and B_m we must impose the final boundary condition at $z = c$. From (4.11) we obtain

$$V(x,y) = \sum_{n=1}^{\infty}\sum_{m=1}^{\infty} C_{nm}\sin\frac{n\pi x}{a}\sin\frac{m\pi y}{b} \tag{4.12}$$

where for convenience we have defined

$$\begin{aligned} C_{nm} &= A_nB_m \sinh\left[\left(\frac{n\pi}{a}\right)^2 + \left(\frac{m\pi}{b}\right)^2\right]^{1/2} c \\ &= A_nB_n \sinh \Gamma_{nm}\, c \end{aligned}$$

The eigenfunctions $\sin (n\pi x/a)$ and $\sin (m\pi y/b)$ occurring in (4.12) have an orthogonality property that enables C_{nm} to be determined. This orthogonality property is the vanishing of the following integrals:

$$\int_0^a \sin\frac{n\pi x}{a}\sin\frac{s\pi x}{a}\,dx = 0 \qquad n \neq s \tag{4.13a}$$

$$\int_0^b \sin\frac{m\pi y}{b}\sin\frac{s\pi y}{b}\,dy = 0 \qquad m \neq s \tag{4.13b}$$

When $n = s$ or $m = s$ we have

$$\int_0^a \sin^2\frac{n\pi x}{a}\,dx = \frac{a}{2} \tag{4.14a}$$

$$\int_0^b \sin^2\frac{m\pi y}{b}\,dy = \frac{b}{2} \tag{4.14b}$$

Orthogonality properties similar to these are found to apply to the eigenfunctions that occur in other coordinate systems as well.

The above orthogonality properties (4.13) are readily proved by direct integration. However, since we shall be dealing with more complicated functions later on, it will be instructive to prove these properties by a more general method. Let $f_n = \sin(n\pi x/a)$ and $f_s = \sin(s\pi x/a)$. Multiplying the equation satisfied by f_n by f_s and similarly the equation for f_s by f_n gives

$$f_s \frac{d^2 f_n}{dx^2} + \left(\frac{n\pi}{a}\right)^2 f_n f_s = 0$$

$$f_n \frac{d^2 f_s}{dx^2} + \left(\frac{s\pi}{a}\right)^2 f_n f_s = 0$$

Subtracting the two equations and integrating over $0 \leq x \leq a$ gives

$$\left[\left(\frac{n\pi}{a}\right)^2 - \left(\frac{s\pi}{a}\right)^2\right] \int_0^a f_n f_s\, dx = \int_0^a \left(f_n \frac{d^2 f_s}{dx^2} - f_s \frac{d^2 f_n}{dx^2}\right) dx$$

If we integrate the right-hand side by parts once, we obtain

$$\int_0^a \left(f_n \frac{d^2 f_s}{dx^2} - f_s \frac{d^2 f_n}{dx^2}\right) dx = \left(f_n \frac{df_s}{dx} - f_s \frac{df_n}{dx}\right)\Bigg|_0^a - \int_0^a \left(\frac{df_n}{dx}\frac{df_s}{dx} - \frac{df_s}{dx}\frac{df_n}{dx}\right) dx$$

Since both f_n and f_s vanish at $x = 0, a$ and the integrand in the integral on the right-hand side vanishes, it follows that the integral on the left-hand side is equal to zero. Hence, since $n \neq s$, it follows that

$$\int_0^a f_n f_s\, dx = 0 \qquad n \neq s$$

In a similar way (4.13*b*) may be proved.

Returning to (4.12) and multiplying both sides by $\sin(s\pi x/a)$ and $\sin(t\pi y/b)$, where s and t are integers, we obtain

$$\int_0^a \int_0^b V(x,y) \sin\frac{s\pi x}{a} \sin\frac{t\pi y}{b}\, dx\, dy = \frac{ab}{4} C_{st} \tag{4.15}$$

by virtue of (4.13) and (4.14). Equation (4.12) represents a double Fourier series for $V(x,y)$, and by virtue of the orthogonal properties of the eigenfunctions an equation for C_{st}, that is, (4.15), is readily obtained. If $V(x,y)$ is known, (4.15) can be evaluated.

Let us choose for $V(x,y)$ the form

$$V(x,y) = V_0 \sin\frac{\pi x}{a} \sin\frac{\pi y}{b}$$

In this case all $C_{st} = 0$ except C_{11}, which from (4.15) has the value V_0. Thus the solution for Φ is

$$\begin{aligned}\Phi &= C_{11} \sin\frac{\pi x}{a} \sin\frac{\pi y}{b} \frac{\sinh \Gamma_{11} z}{\sinh \Gamma_{11} c} \\ &= \frac{V_0}{\sinh \Gamma_{11} c} \sin\frac{\pi x}{a} \sin\frac{\pi y}{b} \sinh \Gamma_{11} z \end{aligned} \tag{4.16}$$

where $$\Gamma_{11}^2 = \left(\frac{\pi}{a}\right)^2 + \left(\frac{\pi}{b}\right)^2$$

If $V(x,y)$ was of the form

$$V(x,y) = V_0 \sin \frac{\pi y}{b}$$

instead, then all C_{st} for $t \neq 1$ are zero. From (4.15) the values of C_{s1} are

$$\begin{aligned}
\frac{ab}{4} C_{s1} &= \int_0^a \int_0^b V_0 \sin \frac{s\pi x}{a} \sin^2 \frac{\pi y}{b}\, dx\, dy \\
&= \frac{V_0 b}{2} \int_0^a \sin \frac{s\pi x}{a}\, dx \\
&= -\frac{V_0 b}{2} \frac{a}{s\pi} \cos \frac{s\pi x}{a} \Big|_0^a = \frac{V_0 b}{2} \frac{a}{s\pi} (1 - \cos s\pi) \\
&= \frac{V_0 ba}{s\pi} \qquad s = 1, 3, 5, \ldots
\end{aligned}$$

Hence $C_{s1} = 4V_0/s\pi$, and the solution for Φ becomes

$$\Phi = \sum_{s=1,3,5,\ldots}^{\infty} \frac{4V_0}{s\pi} \sin \frac{s\pi x}{a} \sin \frac{\pi y}{b} \frac{\sinh \Gamma_{s1} z}{\sinh \Gamma_{s1} c} \tag{4.17}$$

If the potential Φ is specified different from zero on some of the other faces as well, the complete solution to the problem is readily constructed by finding a solution for Φ similar to the above that vanishes on all sides but one. On the latter side, Φ is made to satisfy the required boundary condition. A superposition of these potential functions will then satisfy all the boundary conditions. For example, if the side $z = c$ is kept at a potential $V_1(x,y)$, side $x = a$ at a potential $V_2(y,z)$, and the side $y = b$ at a potential $V_3(x,z)$, we construct three potential functions Φ_1, Φ_2, and Φ_3 with the following properties. All Φ_i $(i = 1, 2, 3)$ satisfy Laplace's equation. In addition, the Φ_i are determined so that the following boundary conditions are satisfied:

$$\begin{array}{ll}
\Phi_1 = 0 & \text{on all sides except } z = c \\
\Phi_1 = V_1(x,y) & \text{at } z = c \\
\Phi_2 = 0 & \text{on all sides except } x = a \\
\Phi_2 = V_2(y,z) & \text{at } x = a \\
\Phi_3 = 0 & \text{on all sides except } y = b \\
\Phi_3 = V_3(x,z) & \text{at } y = b
\end{array}$$

The potential $\Phi = \Phi_1 + \Phi_2 + \Phi_3$ is then a solution of Laplace's equation and in addition satisfies all the required boundary conditions. The solution for each Φ_i is similar to that used to obtain the solutions (4.16) and (4.17).

Example 4.2. A Two-dimensional Problem. As a second example consider a two-dimensional region with boundaries at $x = 0, a$, $y = 0, b$, as in Fig. 4.2. Let the boundary conditions be

$$\frac{\partial \Phi}{\partial x} = 0 \qquad x = 0 \tag{4.18a}$$

$$\Phi = 0 \qquad y = 0, b \tag{4.18b}$$

$$\Phi = V(y) \qquad x = a \tag{4.18c}$$

The potential Φ is independent of z (the structure is of infinite extent in the z direction), and hence is a solution of the two-dimensional Laplace equation

$$\frac{\partial^2 \Phi}{\partial x^2} + \frac{\partial^2 \Phi}{\partial y^2} = 0 \tag{4.19}$$

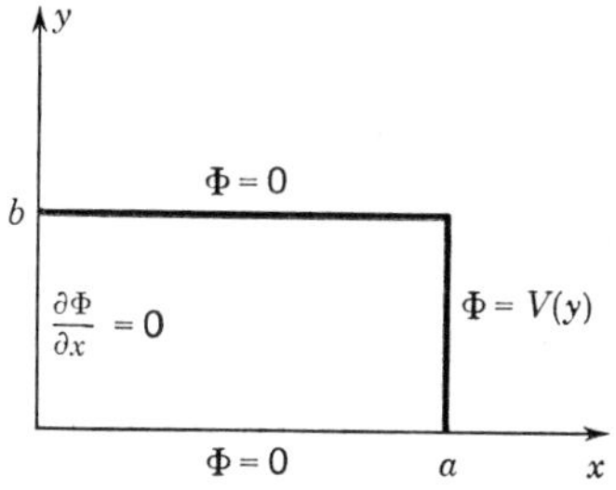

FIG. 4.2. A two-dimensional boundary-value problem.

Two-dimensional structures of the above form usually do not occur in practice; i.e., we do not have structures of infinite extent. However, we could in the present case be dealing equally well with a current flow problem where the current is confined to flow in a very thin sheet of conducting material. If we had a sheet of resistance paper and painted silver strips along the three sides $x = a$, $y = 0, b$, as in Fig. 4.3, and kept the side at $x = a$ at a constant potential V_0, we should have a two-dimensional problem similar to that illustrated in Fig. 4.2. The current density $\mathbf{J}$ is equal to the conductivity σ of the resistance sheet times the electric field and hence given by $\mathbf{J} = -\sigma \nabla \Phi$. The current flow lines would be similar to those sketched in Fig. 4.3. Since the sheet terminates along $x = 0$, the

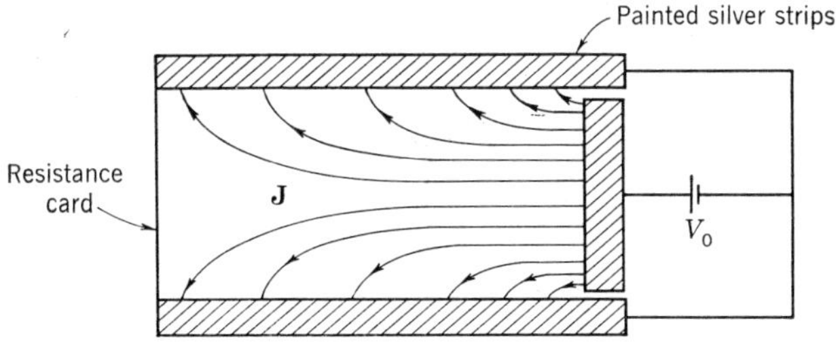

FIG. 4.3. A two-dimensional current flow problem.

current cannot flow normal to this edge and the boundary condition $\partial \Phi / \partial x = 0$ is satisfied.

To solve (4.19) subject to the boundary conditions (4.18), we assume a product solution $f(x)g(y)$. As before, we find that f and g satisfy (4.4a) and (4.4b), respectively. In addition, since $k_z = 0$, we must have $k_x = jk_y$, so that $k_x^2 + k_y^2 = 0$. In order to satisfy the boundary conditions at $y = 0, b$, each possible solution for $g(y)$ must be chosen as $\sin (n\pi y/b)$, where n is an integer. The corresponding solution for f must be either $\cosh (n\pi x/b)$ or $\sinh (n\pi x/b)$. Only the $\cosh (n\pi x/b)$ solution satisfies the boundary condition (4.18a) and hence is the solution chosen. The most general solution for Φ that satisfies the boundary conditions at $x = 0$, $y = 0$, and $y = b$ is thus

$$\Phi = \sum_{n=1}^{\infty} A_n \cosh \frac{n\pi x}{b} \sin \frac{n\pi y}{b} \tag{4.20}$$

At $x = a$ we must have

$$\Phi = V(y) = \sum_{n=1}^{\infty} A_n \cosh \frac{n\pi a}{b} \sin \frac{n\pi y}{b} \tag{4.21}$$

To determine the coefficients A_n we use the orthogonality property (4.13b) and also the result (4.14b) to expand $V(y)$ into a Fourier series. Multiplying both sides of (4.21) by $\sin (m\pi y/b)$ and integrating over y gives

$$A_m \frac{b}{2} \cosh \frac{m\pi a}{b} = \int_0^b V(y) \sin \frac{m\pi y}{b}\, dy \tag{4.22}$$

This equation determines A_m and hence completely specifies the potential function Φ.

If $V(y)$ is equal to a constant V_0, we have

$$\begin{aligned} A_m &= \frac{2V_0}{b \cosh (m\pi a/b)} \frac{b}{m\pi} (1 - \cos m\pi) \\ &= \frac{4V_0}{m\pi \cosh (m\pi a/b)} \qquad m = 1, 3, 5, \ldots \end{aligned} \tag{4.23}$$

Thus
$$\Phi = \frac{4V_0}{\pi} \sum_{m=1,3,\ldots}^{\infty} \frac{1}{m} \frac{\cosh (m\pi x/b)}{\cosh (m\pi a/b)} \sin \frac{m\pi y}{b} \tag{4.24}$$

This is the solution for the potential in the current flow problem illustrated in Fig. 4.3. The current density $\mathbf{J}$ is given by $-\sigma\, \nabla\Phi$.

If we modify the boundary conditions (4.18) and require that the side $y = b$ be kept at a constant potential V_1, we may satisfy this boundary condition by means of a partial solution $\Phi_1 = V_1 y/b$ corresponding to a choice $k_x = k_y = 0$ for the separation constants. For Φ we now choose

$$\Phi = V_1 \frac{y}{b} + \sum_{n=1}^{\infty} B_n \cosh \frac{n\pi x}{b} \sin \frac{n\pi y}{b} \tag{4.25}$$

At $x = 0$, clearly, $\partial\Phi/\partial x = 0$, since $V_1 y/b$ is not a function of x. Also at $y = 0$, $\Phi = 0$, while at $y = b$, $\Phi = V_1$. At $x = a$ we must have $\Phi = V(y)$, and hence

$$V(y) = V_1 \frac{y}{b} + \sum_{n=1}^{\infty} B_n \cosh \frac{n\pi a}{b} \sin \frac{n\pi y}{b} \tag{4.26}$$

In this case the coefficients B_n are given by

$$B_n = \frac{2}{b}\left(\cosh \frac{n\pi a}{b}\right)^{-1} \int_0^b \left[-V_1 \frac{y}{b} + V(y)\right] \sin \frac{n\pi y}{b}\, dy \tag{4.27}$$

As a third modification consider the same rectangular region illustrated in Fig. 4.2 but with the following boundary conditions:

$$\begin{aligned} \frac{\partial\Phi}{\partial x} &= 0 \qquad & x &= 0 \\ \Phi &= 0 & y &= 0 \\ \Phi &= V(y) & x &= a \\ \frac{\partial\Phi}{\partial y} &= p(x) & y &= b \end{aligned}$$

The function $p(x)$ is equal to $\rho(x)/\epsilon_0$, where ρ is the charge density on the side $y = b$. To solve this problem we construct two partial solutions Φ_1 and Φ_2 with the following properties:

For Φ_1

$$\begin{aligned} \frac{\partial\Phi_1}{\partial x} &= 0 \qquad & x &= 0 \\ \Phi_1 &= 0 & y &= 0 \\ \frac{\partial\Phi_1}{\partial y} &= 0 & y &= b \\ \Phi_1 &= V(y) & x &= a \end{aligned}$$

For Φ_2

$$\begin{aligned} \frac{\partial\Phi_2}{\partial x} &= 0 \qquad & x &= 0 \\ \Phi_2 &= 0 & y &= 0 \\ \Phi_2 &= 0 & x &= a \\ \frac{\partial\Phi_2}{\partial y} &= p(x) & y &= b \end{aligned}$$

A superposition of Φ_1 and Φ_2 gives a potential $\Phi = \Phi_1 + \Phi_2$, which satisfies all the boundary conditions. The reader may readily verify that appropriate solutions for Φ_1 and Φ_2 are

$$\Phi_1 = \sum_{n=1}^{\infty} A_n \cosh\left(\frac{2n-1}{2b}\pi x\right)\sin\left(\frac{2n-1}{2b}\pi y\right) \tag{4.28a}$$

where $$A_n \frac{b}{2}\cosh\left(\frac{2n-1}{2b}\pi a\right) = \int_0^b V(y)\sin\left(\frac{2n-1}{2b}\pi y\right)dy$$

and $$\Phi_2 = \sum_{m=1}^{\infty} B_m \cos\left(\frac{2m-1}{2a}\pi x\right)\sinh\left(\frac{2m-1}{2a}\pi y\right) \tag{4.28b}$$

where

$$B_m \frac{2m-1}{2a}\pi\frac{a}{2}\cosh\left(\frac{2m-1}{2a}\pi b\right) = \int_0^a p(x)\cos\left(\frac{2m-1}{2a}\pi x\right)dx$$

In (4.28) the functions

$$\sin\frac{2n-1}{2b}\pi y \qquad \text{and} \qquad \cos\frac{2m-1}{2a}\pi x$$

are orthogonal over the respective ranges $0 \le y \le b$ and $0 \le x \le a$, and hence the usual Fourier series analysis could be used to determine the coefficients A_n and B_m.

The above technique of superimposing partial solutions in order to satisfy arbitrary boundary conditions on a number of sides is also applicable to problems occurring in other coordinate systems as well.

4.2. Cylindrical Coordinates

In a cylindrical coordinate system r, ϕ, z, Laplace's equation is

$$\frac{1}{r}\frac{\partial}{\partial r} r \frac{\partial \Phi}{\partial r} + \frac{1}{r^2}\frac{\partial^2 \Phi}{\partial \phi^2} + \frac{\partial^2 \Phi}{\partial z^2} = 0 \tag{4.29}$$

This equation is separable; so solutions of the form $\Phi = f(r)g(\phi)h(z)$ exist. Substituting into (4.29) gives

$$gh\frac{1}{r}\frac{d}{dr} r \frac{df}{dr} + \frac{fh}{r^2}\frac{d^2 g}{d\phi^2} + fg\frac{d^2 h}{dz^2} = 0$$

Dividing by fgh and multiplying by r^2 gives

$$\left(\frac{r}{f}\frac{d}{dr} r \frac{df}{dr}\right) + \left(\frac{1}{g}\frac{d^2 g}{d\phi^2}\right) + r^2\left(\frac{1}{h}\frac{d^2 h}{dz^2}\right) = 0 \tag{4.30}$$

The second term is a function of ϕ only, and hence (4.30) can hold for all values of r, ϕ, and z only if this term is constant. Thus we must have

$$\frac{d^2 g}{d\phi^2} + \nu^2 g = 0 \tag{4.31}$$

where ν^2 is a separation constant. In many practical problems the whole range $0 \le \phi \le 2\pi$ is involved, and since Φ must be single-valued, that is, $\Phi(2\pi) = \Phi(0)$, ν must be equal to an integer n. The solutions to (4.31) are clearly

$$g = B_1 \sin n\phi + B_2 \cos n\phi \tag{4.32a}$$

or

$$g = B_1 e^{jn\phi} + B_2 e^{-jn\phi} \tag{4.32b}$$

when $\nu = n$. Of course, (4.32) is still the appropriate solution even if n is not an integer.

The second term in (4.30) may now be replaced by $-\nu^2$. Making this substitution and dividing by r^2 gives

$$\left(\frac{1}{rf}\frac{d}{dr}r\frac{df}{dr} - \frac{\nu^2}{r^2}\right) + \frac{1}{h}\frac{d^2h}{dz^2} = 0 \tag{4.33}$$

Each term in this equation is a function of one variable only and must equal a constant if the equation is to hold for all values of r and z. Consequently, we have

$$\frac{d^2h}{dz^2} + k_z^2 h = 0 \tag{4.34}$$

$$\frac{1}{r}\frac{d}{dr}r\frac{df}{dr} - \left(\frac{\nu^2}{r^2} + k_z^2\right)f = 0 \tag{4.35}$$

Equation (4.34) is of the type already considered and has solutions of the form

$$h = C_1 \sin k_z z + C_2 \cos k_z z \tag{4.36a}$$

or if $k_z = j\Gamma$ and Γ is real,

$$h = C_1 \sinh \Gamma z + C_2 \cosh \Gamma z \tag{4.36b}$$

Equation (4.35) is Bessel's equation, and the two independent solutions are called Bessel's functions of the first and second kinds and of order ν. In the special case when $k_z = 0$, the solution reduces to a simple power of r. We shall consider this special case first.

Solution When $k_z = 0$

When the potential has no variation with z, the separation constant $k_z = 0$ and (4.25) becomes

$$\frac{1}{r}\frac{d}{dr}r\frac{df}{dr} - \frac{\nu^2}{r^2}f = 0 \tag{4.37}$$

Let us see if a simple function such as $f = r^\alpha$ will be a solution. Substituting into (4.37) gives

$$\frac{1}{r}\frac{d}{dr}r\frac{dr^\alpha}{dr} - \nu^2 r^{\alpha-2} = (\alpha^2 - \nu^2)r^{\alpha-2} = 0$$

Hence r^α is a possible solution provided $\alpha = \pm\nu$. A general solution for Φ in this case is

$$\Phi = \sum_{n=1}^{\infty} [r^n(A_n \sin n\phi + B_n \cos n\phi) + r^{-n}(C_n \sin n\phi + D_n \cos n\phi)] \quad (4.38)$$

where we have chosen $\nu = n$, and A_n, B_n, C_n, D_n are amplitude constants.

Example 4.3. Dielectric Cylinder in a Uniform Applied Field. Consider a dielectric cylinder of radius r_0 and permittivity ϵ infinitely long and parallel to the z axis and placed in a uniform electrostatic field $\mathbf{E}_0$ directed along the x axis, as in Fig. 4.4. We wish to determine the induced potential and field for all values of r and ϕ.

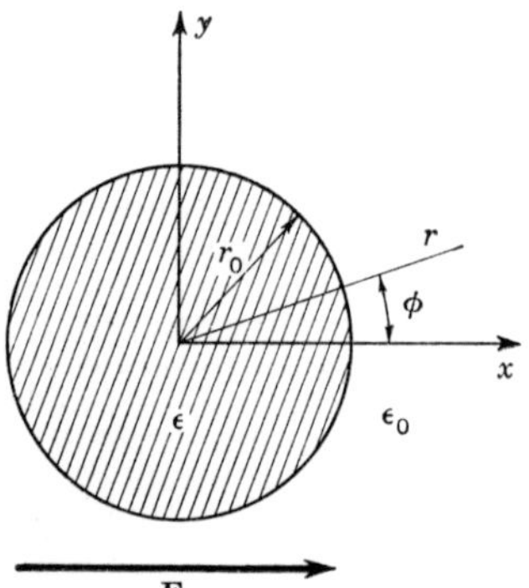

FIG. 4.4. A dielectric cylinder in a uniform applied field $\mathbf{E}_0$.

In cylindrical coordinates, $x = r \cos \phi$, and hence $\mathbf{E}_0$ may be considered as the field arising from an applied potential Φ_0 given by

$$\Phi_0 = -E_0 r \cos \phi \quad (4.39)$$

since $-\nabla\Phi_0 = \mathbf{E}_0$. Let Φ be the induced potential. Since Φ_0 varies with ϕ according to $\cos \phi$, the induced potential Φ will also. This may be seen by noting that the boundary conditions at $r = r_0$ must hold for all values of ϕ, and since $\cos \phi$ is orthogonal to $\cos n\phi$ and $\sin n\phi$ for $n \neq 1$, only the $n = 1$ term in the general solution (4.38) is coupled to the applied potential. The potential Φ must be finite at $r = 0$ and vanish as r approaches infinity. Hence a suitable form for Φ is

$$\Phi = \begin{cases} Ar \cos \phi & r \le r_0 \\ Br^{-1} \cos \phi & r \ge r_0 \end{cases}$$

At $r = r_0$, the total potential must be continuous across the boundary, so that

$$Ar_0 \cos \phi + \Phi_0(r_0) = Br_0^{-1} \cos \phi + \Phi_0(r_0)$$

or

$$B = r_0^2 A$$

Also at $r = r_0$ the radial component of the displacement flux density, that is, ϵE_r, must be continuous. Thus

$$-\epsilon \frac{\partial}{\partial r}(Ar \cos \phi - E_0 r \cos \phi) = -\epsilon_0 \frac{\partial}{\partial r}(Br^{-1} \cos \phi - E_0 r \cos \phi)$$

or

$$\epsilon(A - E_0) = -\epsilon_0 \left(\frac{B}{r_0^2} + E_0\right)$$

The solutions for A and B are now readily found and are

$$A = \frac{\epsilon - \epsilon_0}{\epsilon + \epsilon_0} E_0 \tag{4.40a}$$

$$B = r_0^2 A \tag{4.40b}$$

In the interior of the cylinder the total potential is

$$\Phi + \Phi_0 = -\frac{2\epsilon_0}{\epsilon + \epsilon_0} E_0 r \cos\phi \tag{4.41}$$

The field is still uniform but smaller in magnitude than the applied field E_0. This reduction in the internal field is produced by the depolarizing field set up by the equivalent dipole polarization charge on the surface of the cylinder. The total internal field is

$$E_i = \frac{2\epsilon_0}{\epsilon + \epsilon_0} E_0 \tag{4.42}$$

Outside the cylinder the induced field is $\mathbf{E}_e$, where

$$\mathbf{E}_e = \frac{\epsilon - \epsilon_0}{\epsilon + \epsilon_0} E_0 \left(\frac{r_0}{r}\right)^2 (\mathbf{a}_r \cos\phi + \mathbf{a}_\phi \sin\phi) \tag{4.43}$$

This field is identical with that produced by a line dipole located at the origin.

Solution When $k_z \neq 0$

When $k_z \neq 0$, the function $f(r)$ is a solution of Bessel's equation of order n for $\nu = n$:

$$\frac{1}{r}\frac{d}{dr} r \frac{df}{dr} + \left(\Gamma^2 - \frac{\nu^2}{r^2}\right) f = 0 \tag{4.44}$$

where we have chosen $k_z = j\Gamma$. Substitution of a general power series in r into this equation shows that the two independent solutions are

$$J_n(\Gamma r) = \sum_{m=0}^{\infty} \frac{(-1)^m (\Gamma r/2)^{n+2m}}{m!(n+m)!} \tag{4.45a}$$

$$\begin{aligned} Y_n(\Gamma r) = {} & \frac{2}{\pi}\left(\gamma + \ln\frac{\Gamma r}{2}\right) J_n(\Gamma r) \\ & - \frac{1}{\pi}\sum_{m=0}^{n-1} \frac{(n-m-1)!}{m!}\left(\frac{2}{\Gamma r}\right)^{n-2m} \\ & - \frac{1}{\pi}\sum_{m=0}^{\infty} \frac{(-1)^m (\Gamma r/2)^{n+2m}}{m!(n+m)!}\left(1 + \frac{1}{2} + \frac{1}{3} + \cdots + \frac{1}{m} + 1 \right. \\ & \left. + \frac{1}{2} + \frac{1}{3} + \cdots + \frac{1}{n+m}\right) \qquad \gamma = 0.5772 \end{aligned} \tag{4.45b}$$

The first series (4.45*a*) defines the Bessel function of the first kind and order n, while the second series defines the Bessel function of the second kind and order n. These functions are tabulated in many places. As seen from (4.45*b*), the function $Y_0(\Gamma r)$ has a logarithmic singularity at $r = 0$. For problems that include the origin, this function will not be

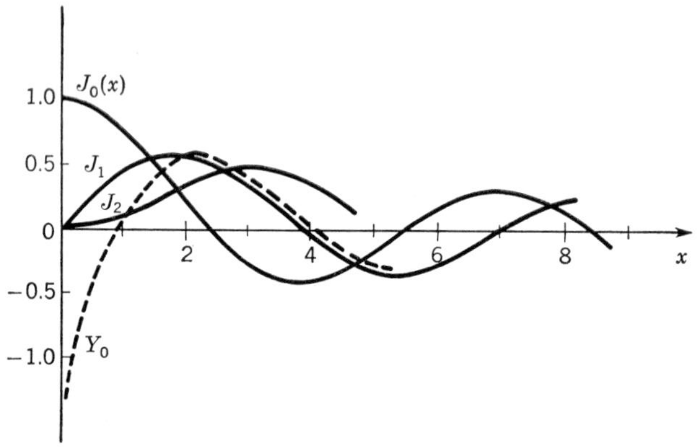

FIG. 4.5. A plot of a few of the lower-order Bessel functions.

part of the solution unless a line source is located at the origin. For $n > 0$, Y_n has a singularity of order r^{-n}. For large values of r, Bessel's functions reduce to damped sinusoids:

$$\lim_{r\to\infty} J_n(\Gamma r) = \sqrt{\frac{2}{\Gamma r \pi}} \cos\left(\Gamma r - \frac{\pi}{4} - \frac{n\pi}{2}\right) \tag{4.46a}$$

$$\lim_{x\to\infty} Y_n(\Gamma r) = \sqrt{\frac{2}{\Gamma r \pi}} \sin\left(\Gamma r - \frac{\pi}{4} - \frac{n\pi}{2}\right) \tag{4.46b}$$

A plot of a few of the lower-order Bessel functions is given in Fig. 4.5.

If k_z is real, then Γ is imaginary. In this case the two independent solutions to Bessel's equation are still given by the series (4.45*a*) and (4.45*b*). However, for convenience, new symbols have been adopted to represent Bessel functions of imaginary argument; that is, by definition,

$$I_n(x) = j^{-n}J_n(jx) = j^nJ_n(-jx) \tag{4.47a}$$

$$K_n(x) = \frac{\pi}{2} j^{n+1}[J_n(jx) + jY_n(jx)] \tag{4.47b}$$

The functions I_n and K_n are defined so that they are real when x is real. I_n is called the modified Bessel function of the first kind, while K_n is called the modified Bessel function of the second kind. In the definition of K_n, a linear combination of $J_n(jx)$ and $Y_n(jx)$ is chosen in order to make $K_n(x)$ a decaying exponential function for large values of x. When

x is very large, the asymptotic values of I_n and K_n are

$$I_n(x) \underset{x\to\infty}{=} \frac{e^x}{\sqrt{2\pi x}} \tag{4.48a}$$

$$K_n(x) \underset{x\to\infty}{=} \sqrt{\frac{\pi}{2x}}\, e^{-x} \tag{4.48b}$$

It should be noted that these functions do not ever equal zero and that only $K_n(x)$ is finite (vanishes) at infinity.

Some useful properties of the Bessel functions J_n and Y_n are given below. Although the formulas are written specifically for J_n, they apply without change if J_n is replaced by Y_n.

Differentiation Formulas

$$J_0'(\Gamma r) = \frac{dJ_0(\Gamma r)}{d(\Gamma r)} = -J_1(\Gamma r) \tag{4.49a}$$

$$xJ_n'(x) = nJ_n(x) - xJ_{n+1}(x) \tag{4.49b}$$

where x has been written for Γr.

Recurrence Formula

$$\frac{2n}{x} J_n(x) = J_{n+1}(x) + J_{n-1}(x) \tag{4.50}$$

If J_{n-1} and J_n are known, this formula permits J_{n+1} to be found.

Integrals

$$\int xJ_n(\alpha x)J_n(\beta x)\,dx = \frac{x}{\alpha^2-\beta^2}[\beta J_n(\alpha x)J_{n-1}(\beta x) - \alpha J_{n-1}(\alpha x)J_n(\beta x)] \qquad \alpha \neq \beta \tag{4.51a}$$

$$\int xJ_n^2(\alpha x)\,dx = \frac{x^2}{2}[J_n^2(\alpha x) - J_{n-1}(\alpha x)J_{n+1}(\alpha x)] = \frac{x^2}{2}\left[J_n'^2(\alpha x) + \left(1 - \frac{n^2}{\alpha^2x^2}\right)J_n^2(\alpha x)\right] \tag{4.51b}$$

Bessel functions have a useful orthogonality property that permits an arbitrary function $f(r)$ defined over an interval $0 \le r \le a$ to be expanded into a Fourier-type series. The function $f(r)$ must be at least piecewise continuous if the expansion is to be valid. Let Γ_{nm} $(m = 1, 2, 3, \ldots)$ be the sequence of values of Γ that makes $J_n(\Gamma a) = 0$; that is, $\Gamma_{nm}a$ is the mth root of $J_n(x) = 0$. For any two roots of $J_n(x) = 0$, say Γ_{nm} and Γ_{ns}, we have

$$J_n(\Gamma_{ns}r)\left[\frac{1}{r}\frac{d}{dr}r\frac{dJ_n(\Gamma_{nm}r)}{dr} + \left(\Gamma_{nm}^2 - \frac{n^2}{r^2}\right)J_n(\Gamma_{nm}r)\right] = 0$$

$$J_n(\Gamma_{nm}r)\left[\frac{1}{r}\frac{d}{dr}r\frac{dJ_n(\Gamma_{ns}r)}{dr} + \left(\Gamma_{ns}^2 - \frac{n^2}{r^2}\right)J_n(\Gamma_{ns}r)\right] = 0$$

Subtracting the two equations, multiplying by r, and integrating with respect to r over $0 \le r \le a$ gives

$$(\Gamma_{nm}^2 - \Gamma_{ns}^2) \int_0^a rJ_n(\Gamma_{nm}r)J_n(\Gamma_{ns}r)\,dr = \int_0^a \left[J_n(\Gamma_{nm}r)\frac{d}{dr}r\frac{dJ_n(\Gamma_{ns}r)}{dr} - J_n(\Gamma_{ns}r)\frac{d}{dr}r\frac{dJ_n(\Gamma_{nm}r)}{dr} \right] dr$$

Integrating the right-hand side by parts once gives

$$\begin{aligned}(\Gamma_{nm}^2 - \Gamma_{ns}^2) &\int_0^a rJ_n(\Gamma_{nm}r)J_n(\Gamma_{ns}r)\,dr \\ &= r\left[J_n(\Gamma_{nm}r)\frac{dJ_n(\Gamma_{ns}r)}{dr} - J_n(\Gamma_{ns}r)\frac{dJ_n(\Gamma_{nm}r)}{dr} \right]_0^a \\ &\quad - \int_0^a r\left[\frac{dJ_n(\Gamma_{ns}r)}{dr}\frac{dJ_n(\Gamma_{nm}r)}{dr} - \frac{dJ_n(\Gamma_{nm}r)}{dr}\frac{dJ_n(\Gamma_{ns}r)}{dr} \right] dr\end{aligned}$$

The integrand on the right-hand side vanishes. Likewise, the integrated terms vanish since $r = 0$ at the lower limit while at the upper limit $J_n(\Gamma_{nm}a) = J_n(\Gamma_{ns}a) = 0$. Hence we see that

$$\int_0^a rJ_n(\Gamma_{nm}r)J_n(\Gamma_{ns}r)\,dr = 0 \qquad m \ne s \tag{4.52}$$

From the nature of the proof it is clear that (4.52) is also true if Γ_{nm} and Γ_{ns} are chosen such that

$$\frac{dJ_n(\Gamma_{nm}r)}{dr} = \frac{dJ_n(\Gamma_{ns}r)}{dr} = 0 \qquad \text{at } r = a$$

or if Γ_{nm} makes $J_n(\Gamma_{nm}a) = 0$ and Γ_{ns} satisfies $J_n'(\Gamma_{ns}a) = 0$. These orthogonality properties are very similar to those for the sinusoidal functions $\sin nx$ and $\cos mx$ over the range $0 \le x \le 2\pi$. When $\Gamma_{nm} = \Gamma_{ns}$, the value of the integral is given by (4.51*b*). The example to be considered now will illustrate the use of the above orthogonality properties.

Example 4.4. Potential in a Cylindrical Region. Consider a cylinder of radius a and length d, as in Fig. 4.6. The end face at $z = d$ is kept at a constant potential V_0, while the remainder of the boundary is kept at zero potential. We wish to determine the potential field Φ within the cylinder.

The solution must be of the form $\Phi = f(r)g(\phi)h(z)$. However, in the present case, Φ is independent of the angle ϕ, and hence $g(\phi)$ is a constant. Since Φ is finite at $r = 0$, the solution for f is simply $J_0(\Gamma r)$. The unmodified Bessel function must be chosen here since we require a function of r that goes to zero at $r = a$, a property that I_0 and K_0 do not possess. The corresponding solution for $h(z)$ is hyperbolic and must be chosen as $\sinh \Gamma z$ in order to satisfy the boundary condition $\Phi = 0$ at $z = 0$.

Since $\Phi = 0$ at $r = a$, the allowed values for the separation constant Γ are the roots Γ_{0m} that make $J_0(\Gamma a) = 0$. The general solution for Φ is thus

$$\Phi = \sum_{m=1}^{\infty} A_m J_0(\Gamma_{0m} r) \sinh \Gamma_{0m} z \tag{4.53}$$

The amplitude constants A_m must be determined so that $\Phi = V_0$ at

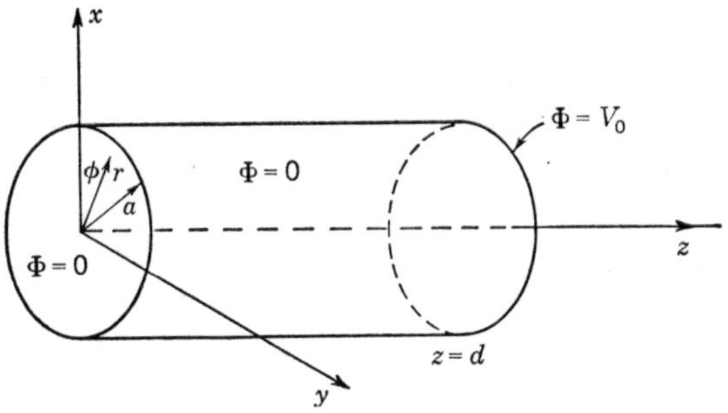

FIG. 4.6. A cylindrical boundary-value problem.

$z = d$, $0 \le r \le a$. Putting $z = d$, we have

$$V_0 = \sum_{m=1}^{\infty} A_m J_0(\Gamma_{0m} r) \sinh \Gamma_{0m} d$$

If we multiply both sides by $rJ_0(\Gamma_{0n} r)\, dr$ and integrate over $0 \le r \le a$, we obtain

$$V_0 \int_0^a rJ_0(\Gamma_{0n} r)\, dr = A_n \sinh \Gamma_{0n} d \int_0^a rJ_0^2(\Gamma_{0n} r)\, dr \tag{4.54}$$

by virtue of the orthogonality property (4.53). From (4.51b) we have

$$\int_0^a rJ_0^2(\Gamma_{0n} r)\, dr = \frac{r^2}{2} \left\{ \left[\frac{dJ_0(\Gamma_{0n} r)}{d(\Gamma_{0n} r)} \right]^2 + J_0^2(\Gamma_{0n} r) \right\}_0^a$$

$$= \frac{a^2}{2} J_0'^2(\Gamma_{0n} a) = \frac{a^2}{2} J_1^2(\Gamma_{0n} a) \tag{4.55}$$

by utilizing (4.49a). Employing the result

$$\int (\Gamma x)^{n+1} J_n(\Gamma x)\, d(\Gamma x) = (\Gamma x)^{n+1} J_{n+1}(\Gamma x) \tag{4.56}$$

gives

$$V_0 \int_0^a rJ_0(\Gamma_{0n} r)\, dr = \frac{V_0}{\Gamma_{0n}^2} (\Gamma_{0n} r) J_1(\Gamma_{0n} r) \Big|_0^a = \frac{V_0}{\Gamma_{0n}} a J_1(\Gamma_{0n} a) \tag{4.57}$$

Combining this result with (4.55) and (4.54) gives the solution for A_n:

$$A_n = \frac{2V_0}{\Gamma_{0n} a J_1(\Gamma_{0n} a) \sinh \Gamma_{0n} d} \tag{4.58}$$

Finally, by substituting into (4.53), we obtain the solution for Φ:

$$\Phi = 2V_0 \sum_{n=1}^{\infty} \frac{J_0(\Gamma_{0n}r) \sinh \Gamma_{0n}z}{\Gamma_{0n}aJ_1(\Gamma_{0n}a) \sinh \Gamma_{0n}d} \tag{4.59}$$

Tables for evaluating this series, i.e., for the roots Γ_{0n} and the values of the Bessel functions, may be found in "Tables of Functions" by Jahnke and Emde.

If we had specified that Φ be equal to zero at $z = 0$ and d, then we should be forced to choose for our functions $h(z)$ the form $\sin (m\pi z/d)$ $(m = 1, 2, \ldots)$. In this case the Bessel function to use must be either $I_0(m\pi r/d)$ or $K_0(m\pi r/d)$. Only I_0 is finite at $r = 0$ and is therefore the only allowed function. Our solution for Φ would now be of the form

$$\Phi = \sum_{m=1}^{\infty} A_m I_0\left(\frac{m\pi r}{d}\right) \sin \frac{m\pi z}{d}$$

At $r = a$ we could specify that Φ be a function of z, say $V(z)$, and by the usual Fourier series analysis determine the coefficients A_m. If Φ at $r = a$ has a ϕ dependence also, then the solution for Φ would be of the form

$$\Phi = \sum_{n=0}^{\infty} \sum_{m=1}^{\infty} C_{nm}(\cos n\phi + B_n \sin n\phi) I_n\left(\frac{m\pi r}{d}\right) \sin \frac{m\pi z}{d}$$

In this case a double Fourier series analysis has to be carried out in order to find the coefficients C_{nm} and B_n.

4.3. Spherical Coordinates

Problems such as that of a dielectric sphere placed in a uniform external field are best described in spherical coordinates r, θ, ϕ. With reference to Fig. 4.7, r is the radial coordinate, θ the polar angle, and ϕ the azimuth angle. In spherical coordinates the constant coordinate surfaces are spheres, cones, and planes.

In spherical coordinates Laplace's equation becomes

$$\nabla^2\Phi = \frac{1}{r^2}\frac{\partial}{\partial r}\left(r^2 \frac{\partial \Phi}{\partial r}\right) + \frac{1}{r^2 \sin \theta}\frac{\partial}{\partial \theta}\left(\sin \theta \frac{\partial \Phi}{\partial \theta}\right) + \frac{1}{r^2 \sin^2 \theta}\frac{\partial^2 \Phi}{\partial \phi^2} = 0 \tag{4.60}$$

As before, we assume a product solution of the form $f(r)g(\theta)h(\phi)$. Substituting into (4.60) and dividing by $fgh/(r^2 \sin^2 \theta)$ gives

$$\frac{\sin^2 \theta}{f}\frac{\partial}{\partial r}\left(r^2 \frac{\partial f}{\partial r}\right) + \frac{\sin \theta}{g}\frac{\partial}{\partial \theta}\left(\sin \theta \frac{\partial g}{\partial \theta}\right) + \frac{1}{h}\frac{\partial^2 h}{\partial \phi^2} = 0 \tag{4.61}$$

For this equation to be equal to zero for all values of r, θ, and ϕ it is necessary that

$$\frac{\partial^2 h}{\partial \phi^2} + n^2 h = 0 \tag{4.62}$$

where n^2 is a separation constant. The argument is analogous to that used previously; it is only necessary to note that the first two terms in

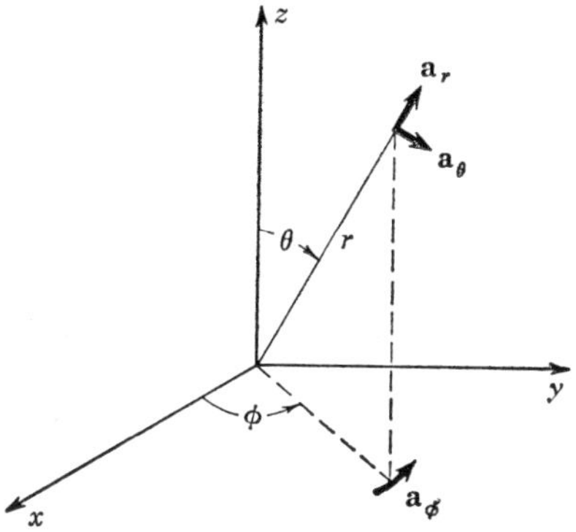

FIG. 4.7. Spherical coordinates.

(4.61) are functions of r and θ only and the last term a function of ϕ alone. For problems involving the whole range $0 \le \phi \le 2\pi$, the constant n must be an integer, so that h will be single-valued, i.e., so that $h(2\pi) = h(0)$. The solution to (4.62) is then

$$h(\phi) = C_1 \cos n\phi + C_2 \sin n\phi \tag{4.63}$$

Replacing $\frac{1}{h}\frac{\partial^2 h}{\partial \phi^2}$ by $-n^2$ in (4.61) and dividing through by $\sin^2\theta$ results in

$$\frac{1}{f}\frac{\partial}{\partial r}\left(r^2 \frac{\partial f}{\partial r}\right) + \frac{1}{g \sin\theta}\frac{\partial}{\partial \theta}\left(\sin\theta \frac{\partial g}{\partial \theta}\right) - \frac{n^2}{\sin^2\theta} = 0 \tag{4.64}$$

The first term is a function of r only, while the remaining terms are a function of θ only. For the sum to be equal to zero for all values of r and θ, it is necessary that each term be equal to a constant. Hence we may choose

$$\frac{\partial}{\partial r}\left(r^2 \frac{\partial f}{\partial r}\right) - m(m+1)f = 0 \tag{4.65}$$

where $m(m + 1)$ is the separation constant. The form $m(m + 1)$ is chosen for reasons that will be pointed out later. It is readily verified that the solutions to (4.65) are

$$f(r) = B_1 r^m + B_2 r^{-(m+1)} \tag{4.66}$$

From (4.64) and (4.65) we determine that $g(\theta)$ must satisfy the following differential equation:

$$\frac{d}{d\theta}\left(\sin\theta \frac{dg}{d\theta}\right) + \left[m(m+1)\sin\theta - \frac{n^2}{\sin\theta}\right] g = 0 \tag{4.67}$$

This is Legendre's equation. The standard form of this equation is obtained by making the substitution $\cos\theta = u$:

$$\frac{d}{d\theta} = \frac{d}{du}\frac{du}{d\theta} = -(1-u^2)^{1/2}\frac{d}{du}$$

The equation then becomes

$$\frac{d}{du}(1-u^2)\frac{dg(u)}{du} + \left[m(m+1) - \frac{n^2}{1-u^2}\right] g(u) = 0 \tag{4.68}$$

The solutions to this equation are the associated Legendre functions which we shall study briefly.

For the particular case $n = 0$, the equation becomes

$$\frac{d}{du}(1-u^2)\frac{dg}{du} + m(m+1)g = 0 \tag{4.69}$$

Unless the separation constant is chosen in the form $m(m+1)$ with $m = 0, 1, 2, \ldots$, all the solutions to (4.68) and (4.69) become infinite when either $u = 1$ or $u = -1$, that is, when $\theta = 0, \pi$. These solutions would not be suitable for physical problems that include the polar axis. As a differential equation of second degree, (4.69) has two independent solutions. These are called Legendre functions of the first and second kinds and are designated as $P_m{}^0(u)$ and $Q_m{}^0(u)$, respectively. When m is an integer, $P_m{}^0$ is a finite polynomial in u. However, $Q_m{}^0$ has a singularity at the poles $\theta = 0, \pi$. In the following we shall assume that the polar axis is part of the region of interest and that no singularity is to be expected there, so that Legendre functions of the second kind may be excluded.

When $n \neq 0$, the solutions to (4.68) that remain finite at the poles are associated Legendre polynomials that are designated by the symbol $P_m{}^n(u)$, where m and n are positive integers. The polynomials $P_m{}^n$ are readily obtained from the following generating function:

$$P_m{}^n(u) = \frac{(1-u^2)^{n/2}}{2^m m!}\frac{d^{n+m}(u^2-1)^m}{du^{n+m}} \tag{4.70}$$

Several of the Legendre and associated Legendre polynomials are given below; these may be easily confirmed through the use of (4.70).

$$P_0^0 = 1 \tag{4.71a}$$
$$P_1^0 = \frac{1}{2}\frac{d}{du}(u^2 - 1) = u = \cos\theta \tag{4.71b}$$
$$P_2^0 = \tfrac{3}{2}\cos^2\theta - \tfrac{1}{2} = \tfrac{3}{4}\cos 2\theta + \tfrac{1}{4} \tag{4.71c}$$
$$P_3^0 = \tfrac{1}{2}(5\cos^3\theta - 3\cos\theta) \tag{4.71d}$$
$$P_1^1 = \sin\theta \tag{4.71e}$$
$$P_2^1 = \tfrac{3}{2}\sin 2\theta \tag{4.71f}$$
$$P_m^n = 0 \qquad n > m \tag{4.71g}$$

From the differential equation the following orthogonality properties may be proved in a manner similar to that used for the Bessel functions:

$$\int_{-1}^{1} P_m^n P_l^n\,du = \int_0^{\pi} P_m^n P_l^n \sin\theta\,d\theta = 0 \qquad m \neq l \tag{4.72a}$$
$$\int_{-1}^{1} P_m^n P_m^l \frac{du}{1-u^2} = \int_0^{\pi} P_m^n P_m^l \frac{d\theta}{\sin\theta} = 0 \qquad n \neq l \tag{4.72b}$$

When $m = l$ or $n = l$ we obtain

$$\int_{-1}^{1} (P_m^n)^2\,du = \int_0^{\pi} [P_m^n(\cos\theta)]^2 \sin\theta\,d\theta = \frac{1}{n}\frac{(m+n)!}{(m-n)!} \tag{4.73}$$

By means of these formulas an arbitrary piecewise-continuous function $g(\theta)$ may be expanded into a Fourier-type series in terms of the polynomials P_m^n.

The general solution to Laplace's equation in spherical coordinates, subject to the assumptions already noted, is now seen to be

$$\Phi(r,\theta,\phi) = \sum_{m=0}^{\infty}\sum_{n=0}^{m} (A_n \cos n\phi + B_n \sin n\phi) \times (C_m r^m + D_m r^{-(m+1)}) P_m^n(\cos\theta) \tag{4.74}$$

The sum over n terminates at $n = m$ since P_m^n is zero for $n > m$. The coefficients A_n, B_n, C_m, and D_m are determined by the boundary conditions that Φ must satisfy.

Example 4.5. Potential Specified on the Surface of a Sphere. Let the surface of a sphere of radius R be kept at a potential

$$\Phi(R,\theta,\phi) = V_0 \sin\phi \sin\theta = V_0 P_1^1 \sin\phi \tag{4.75}$$

the latter since $P_1^1 = \sin\theta$. We require a solution for Φ in the interior of the sphere. Since Φ must remain finite at $r = 0$, the coefficent D_m in the general solution (4.74) must be zero. Equating (4.74) for $r = R$

to (4.75) gives

$$\sum_{m=0}^{\infty}\sum_{n=0}^{m}(A_n \cos n\phi + B_n \sin n\phi)C_m R^m P_m{}^n = V_0 P_1{}^1 \sin \phi \qquad (4.76)$$

If we multiply both sides by $\cos s\phi\, d\phi$ and integrate over $0 \le \phi \le 2\pi$, we find that $A_n = 0$ since we obtain

$$\sum_{m=0}^{\infty} A_s \pi C_m R^m P_m{}^s = V_0 P_1{}^1 \int_0^{2\pi} \sin \phi \cos s\phi\, d\phi = 0$$

by virtue of the orthogonal properties of the $\cos n\phi$ and $\sin n\phi$ functions. If we multiply (4.76) by $P_l{}^s \sin s\phi \sin \theta\, d\theta\, d\phi$, all terms on the left-hand side except $m = l$, $n = s$ integrate to zero between the limits $0 \le \phi \le 2\pi$ and $0 \le \theta \le \pi$. At the same time the integral of the right-hand side goes to zero unless $l = 1$, $s = 1$. Hence all B_n and C_m are zero except B_1 and C_1. For $l = s = 1$, we obtain

$$\begin{aligned}\tfrac{4}{3}B_1 C_1 R\pi &= V_0 \int_0^{2\pi}\int_0^{\pi} (P_1{}^1)^2 \sin^2 \phi \sin \theta\, d\theta\, d\phi \\ &= \tfrac{4}{3} V_0 \pi\end{aligned}$$

by using (4.73). Hence $B_1 C_1 = V_0/R$, and the solution for Φ is

$$\Phi = V_0 \frac{r}{R} P_1{}^1 \sin \phi = V_0 \frac{r}{R} \sin \theta \sin \phi \qquad (4.77)$$

Example 4.6. Dielectric Sphere in a Uniform Applied Field. Consider a dielectric sphere of permittivity ϵ, radius R, placed in a uniform external field $E_0\mathbf{a}_z$, as in Fig. 4.8. The applied field may be derived from a potential $\Phi_0 = -E_0 z = -E_0 r \cos \theta = -E_0 r P_1{}^0(\cos \theta)$.

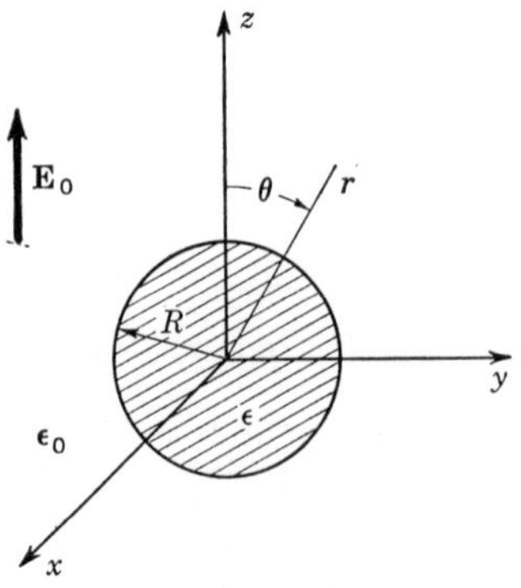

FIG. 4.8. A dielectric sphere in a uniform applied field.

Since the applied potential is independent of the angle ϕ, the induced potential Φ will also be independent of ϕ. Thus Φ must be of the form

$$\begin{aligned}\Phi &= \sum_{m=0}^{\infty} a_m r^m P_m{}^0(\cos \theta) \qquad & r \le R \\ &= \sum_{m=0}^{\infty} b_m r^{-(m+1)} P_m{}^0(\cos \theta) \qquad & r \ge R\end{aligned}$$

This potential is finite at $r = 0$ and vanishes at infinity. At $r = R$ the total potential must be continuous; hence

$$\begin{aligned}-E_0 R P_1{}^0(\cos \theta) &+ \sum_{m=0}^{\infty} a_m R^m P_m{}^0(\cos \theta) \\ &= -E_0 R P_1{}^0(\cos \theta) + \sum_{m=0}^{\infty} b_m R^{-(m+1)} P_m{}^0(\cos \theta) \qquad (4.78)\end{aligned}$$

Since the functions $P_m{}^0$ are orthogonal over the range $0 \le \theta \le \pi$, it is seen from (4.78) that

$$a_m = b_m R^{-(2m+1)} \tag{4.79}$$

An additional condition at $r = R$ is that the normal displacement D_r must be continuous. Hence

$$\epsilon \left[-E_0 P_1{}^0(\cos\theta) + \sum_{m=1}^{\infty} m a_m R^{m-1} P_m{}^0(\cos\theta) \right]$$
$$= \epsilon_0 \left[-E_0 P_1{}^0(\cos\theta) - \sum_{m=0}^{\infty} (m+1) b_m R^{-(m+2)} P_m{}^0(\cos\theta) \right] \tag{4.80}$$

Again since the functions $P_m{}^0$ are a set of orthogonal functions, the coefficients of each term $P_m{}^0$ must be equal. Thus we have

$$\epsilon(-E_0 + a_1) = \epsilon_0(-E_0 - 2b_1R^{-3}) \tag{4.81a}$$
$$\epsilon m a_m R^{m-1} = -\epsilon_0(m+1)b_m R^{-(m+2)} \qquad m > 1 \tag{4.81b}$$

Comparing (4.81*b*) with (4.79) shows that both of these equations can hold only if a_m and b_m equal zero for $m > 1$. For $m = 1$ we obtain, from (4.79) and (4.81*a*),

$$a_1 = \frac{\epsilon - \epsilon_0}{\epsilon + 2\epsilon_0} E_0 \tag{4.82a}$$
$$b_1 = \frac{\epsilon - \epsilon_0}{\epsilon + 2\epsilon_0} E_0 R^3 \tag{4.82b}$$

In (4.80) the coefficient of $P_0{}^0$ on the left-hand side is zero, and hence b_0 must equal zero. From (4.79) it is then seen that $a_0 = 0$ also.

The complete solution for the induced Φ is

$$\Phi = \frac{\epsilon - \epsilon_0}{\epsilon + 2\epsilon_0} E_0 r P_1{}^0 = \frac{\epsilon - \epsilon_0}{\epsilon + 2\epsilon_0} E_0 z \qquad r \le R \tag{4.83a}$$
$$\Phi = \frac{\epsilon - \epsilon_0}{\epsilon + 2\epsilon_0} E_0 \cos\theta \, \frac{R^3}{r^2} \qquad r \ge R \tag{4.83b}$$

Inside the sphere the induced potential gives rise to a uniform field $-(\epsilon - \epsilon_0)E_0/(\epsilon + 2\epsilon_0)$ directed in the z direction. The total potential within the sphere is consequently reduced from the free-space value. Outside the sphere the induced field is a dipole field, which may be considered due to a z-directed dipole of moment p given by

$$p = 4\pi R^3 \epsilon_0 \frac{\epsilon - \epsilon_0}{\epsilon + 2\epsilon_0} E_0 \tag{4.84}$$

and located at the origin. This result follows from the expression

$$\Phi = \frac{p}{4\pi\epsilon_0} \frac{\cos\theta}{r^2}$$

for the potential set up by a dipole of moment p.

Example 4.7. Potential from a Charged Disk. Figure 4.9 illustrates a circular disk of radius a, located in the xy plane and uniformly charged. The potential field Φ set up by the charged disk will have no variation with the azimuth angle ϕ because of the symmetry of the charge distribution about the polar axis. Furthermore, symmetry conditions also require that the potential field be symmetrical about the plane of the disk, that is, symmetrical about $\theta = \pi/2$, and hence $\Phi(\theta) = \Phi(\pi - \theta)$. The potential Φ is finite at the origin, vanishes at infinity, and is independent of ϕ and can therefore be represented by the following expansions:

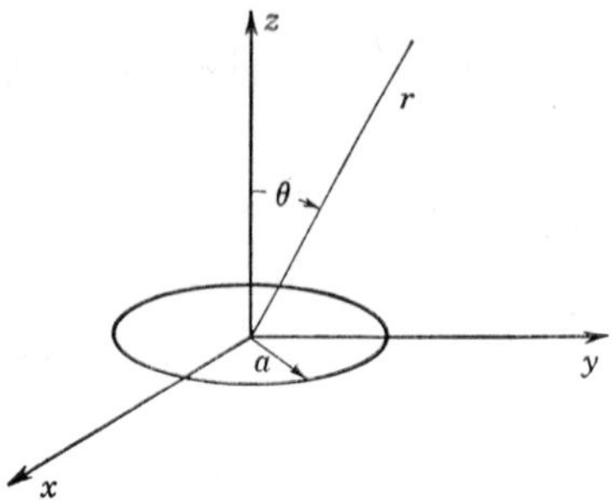

FIG. 4.9. A uniformly charged disk.

$$\Phi = \sum_{m=0}^{\infty} a_m r^m P_m{}^0(\cos\theta) \qquad r \le a,\ \theta < \frac{\pi}{2} \tag{4.85a}$$

$$\Phi = \sum_{m=0}^{\infty} a_m(-1)^m r^m P_m{}^0(\cos\theta) \qquad r \le a,\ \frac{\pi}{2} < \theta < \pi \tag{4.85b}$$

$$\Phi = \sum_{m=0}^{\infty} b_m r^{-(m+1)} P_m{}^0(\cos\theta) \qquad r \ge a,\ \theta < \frac{\pi}{2} \tag{4.85c}$$

$$\Phi = \sum_{m=0}^{\infty} b_m(-1)^m r^{-(m+1)} P_m{}^0(\cos\theta) \qquad r \ge a,\ \frac{\pi}{2} < \theta < \pi \tag{4.85d}$$

The polynomials $P_m{}^0$ for even values of m are even functions of θ about the plane $\theta = \pi/2$, while $P_m{}^0$ for odd values of m are odd functions of θ about the plane $\theta = \pi/2$. In order to satisfy the requirement of even symmetry of Φ about the plane $\theta = \pi/2$, the coefficients a_m and b_m for odd values of m must be replaced by $-a_m$ and $-b_m$ in the region $\pi/2 < \theta < \pi$. For this reason the factor $(-1)^m$ is included in (4.85b) and (4.85d). We shall see later on that all the coefficients b_m for odd values of m vanish. Along the positive z axis the above expansions reduce to the following:

$$\Phi = \sum_{m=0}^{\infty} a_m z^m \qquad z \le a \tag{4.86a}$$

$$\Phi = \sum_{m=0}^{\infty} b_m z^{-(m+1)} \qquad z \ge a \tag{4.86b}$$

since this corresponds to $\cos\theta = 1$ and $P_m{}^0(1) = 1$.

In Example 2.4 the potential along the axis of a uniformly charged disk was found by direct integration of the contribution from each element of charge. This potential was determined to be

$$\Phi = \frac{Q}{2\pi\epsilon_0 a^2}[(a^2 + z^2)^{1/2} - z] \qquad \text{for } z > 0$$

where Q is the total charge on the disk. We may use the binomial theorem to expand Φ into a power series in z and by comparing with (4.86) find the expansion coefficients a_m and b_m. In the region $0 < z < a$ we have

$$\begin{aligned}\Phi &= \frac{Q}{2\pi\epsilon_0 a^2}\left[-z + a\left(1 + \frac{z^2}{a^2}\right)^{1/2}\right] \\ &= \frac{Q}{2\pi\epsilon_0 a^2}\left[-z + a + \frac{a}{2}\left(\frac{z}{a}\right)^2 - \frac{a}{8}\left(\frac{z}{a}\right)^4 + \frac{a}{16}\left(\frac{z}{a}\right)^6 - \cdots\right] \qquad (4.87a)\end{aligned}$$

while for $z > a$ a similar expansion gives

$$\begin{aligned}\Phi &= \frac{Q}{2\pi\epsilon_0 a^2}\left[-z + z\left(1 + \frac{a^2}{z^2}\right)^{1/2}\right] \\ &= \frac{Q}{2\pi\epsilon_0 a^2}\left[\frac{z}{2}\left(\frac{a}{z}\right)^2 - \frac{z}{8}\left(\frac{a}{z}\right)^4 + \frac{z}{16}\left(\frac{a}{z}\right)^6 - \cdots\right] \qquad (4.87b)\end{aligned}$$

By comparing (4.87) with (4.86) we find, by equating coefficients of like powers of z, that

$$a_0 = \frac{Q}{2\pi\epsilon_0 a} \qquad a_1 = -\frac{Q}{2\pi\epsilon_0 a^2}$$

$$a_2 = \frac{Q}{4\pi\epsilon_0 a^3} \qquad a_3 = a_5 = a_7 = \cdots = 0$$

$$a_4 = -\frac{Q}{16\pi\epsilon_0 a^5}$$

$$\cdots\cdots\cdots$$

$$b_0 = \frac{Q}{4\pi\epsilon_0} \qquad b_2 = -\frac{Qa^2}{16\pi\epsilon_0} \qquad b_4 = \frac{Qa^4}{32\pi\epsilon_0}$$

$$b_1 = b_3 = b_5 = \cdots = 0$$

$$\cdots\cdots\cdots\cdots\cdots$$

As anticipated from symmetry conditions, all the coefficients b_m for m odd are zero. In the region $r > a$ the potential Φ must be a continuous function of θ with even symmetry about the plane $\theta = \pi/2$. This means that $\partial\Phi/\partial\theta$ is zero at $\theta = \pi/2$ $(r > a)$. Only the polynomials $P_m^0(\cos\theta)$ with m even satisfy these requirements. In the region $r < a$ the potential Φ must again be an even function of θ about the symmetry plane $\theta = \pi/2$. At the same time $\partial\Phi/(r\,\partial\theta)$ cannot vanish at $\theta = \pi/2$ since the normal derivative of Φ must equal $-\rho_s/2\epsilon_0$, where ρ_s is the surface density of charge on the disk, as we found in Example 2.4. Hence at least one odd polynomial $P_m^0(\cos\theta)$ has to be present in the expansion of Φ for the region

$r < a$. From the results obtained we see that this term is $a_1 r P_1{}^0(\cos\theta)$ for $\theta < \pi/2$ and $-a_1 r P_1{}^0(\cos\theta)$ for $\pi/2 < \theta < \pi$.

For the even polynomials $\partial P_m{}^0/\partial\theta$ equals zero at $\theta = \pi/2$. Consequently, for $r < a$,

$$\left.\frac{1}{r}\frac{\partial\Phi}{\partial\theta}\right|_{\theta=\pi/2} = a_1 \left.\frac{\partial \cos\theta}{\partial\theta}\right|_{\theta=\pi/2} = -a_1 = \frac{\rho_s}{2\epsilon_0} \tag{4.88}$$

The charge density ρ_s is equal to $Q/\pi a^2$, and thus our previous result $a_1 = -Q/2\pi\epsilon_0 a^2$ is verified.

It should be noted that the two expansions in (4.85) for $r > a$ and $r < a$ must be equal at $r = a$, but since the disk separates the region $r < a$ into two parts and the functions $P_m{}^0$ are not orthogonal over the range $0 \le \theta \le \pi/2$, the coefficients a_m cannot be found very readily in terms of the b_m.

For any axially symmetric charge distribution for which a power series expansion in z can be found for the potential along the axis, the above method may be used to determine the potential at all points outside the charge region. Since an expansion like (4.85) is a solution to Laplace's equation and gives the right value of potential along the polar axis, it gives a unique solution.

4.4. Solution of Two-dimensional Problems by Conformal Mapping

The theory of functions of a complex variable provides a powerful method for the solution of Laplace's equation in two dimensions. The viewpoint adopted here is one that considers an analytic function of a complex variable $Z = x + jy$ as generating a suitable orthogonal curvilinear coordinate system which is appropriate for the description of the problem being considered. It is assumed that the student has some familiarity with complex numbers and functions of a complex variable. However, in order to provide continuity, a brief review of the basic concepts that will be required is presented first.

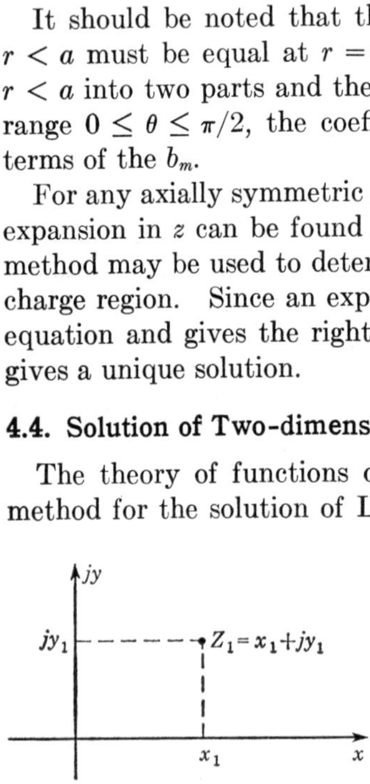

FIG. 4.10. The complex number $x_1 + jy_1$ in the complex Z plane.

The complex variable $Z = x + jy$ is the sum of the real variable x and the product of the imaginary number $j = \sqrt{-1}$ and the real variable y. The complex number $Z_1 = x_1 + jy_1$ is conveniently represented by the point with coordinates x_1, y_1 in the complex Z plane, as in Fig. 4.10. The sum of two complex numbers is the complex number obtained by adding the real parts and the imaginary parts; thus

$$(x_1 + jy_1) + (x_2 + jy_2) = (x_1 + x_2) + j(y_1 + y_2)$$

The product of two complex numbers is formed according to the usual rules of algebra and replacing j^2 by -1, j^3 by $-j$, j^4 by 1, etc. As an example,

$$(x_1 + jy_1)(x_2 + jy_2) = x_1x_2 + jx_1y_2 + jy_1x_2 + j^2y_1y_2 = (x_1x_2 - y_1y_2) + j(x_1y_2 + x_2y_1)$$

The complex conjugate of $x_1 + jy_1$ is obtained by replacing j by $-j$. An asterisk is used to denote this operation of taking the complex conjugate; thus

$$(x_1 + jy_1)^* = x_1 - jy_1$$

The quotient of two complex numbers is found by multiplying the numerator and denominator by the conjugate of the denominator; e.g.,

$$\frac{x_1 + jy_1}{x_2 + jy_2} = \frac{(x_1 + jy_1)(x_2 - jy_2)}{(x_2 + jy_2)(x_2 - jy_2)} = \frac{(x_1x_2 + y_1y_2) + j(x_2y_1 - x_1y_2)}{x_2^2 + y_2^2}$$

A function of the complex variable Z, say $W = F(Z)$, is called a complex function. An example is $W = Z^2$. The function W will be complex, with a real part u and an imaginary part jv, where u and v are obviously functions of x and y. For the above example,

$$W = u + jv = Z^2 = (x + jy)^2 = (x^2 - y^2) + 2jxy$$

and $u = x^2 - y^2$, $v = 2xy$. As the variable Z moves along some curve C in the complex Z plane, the variable $W = F(Z)$ will move along some

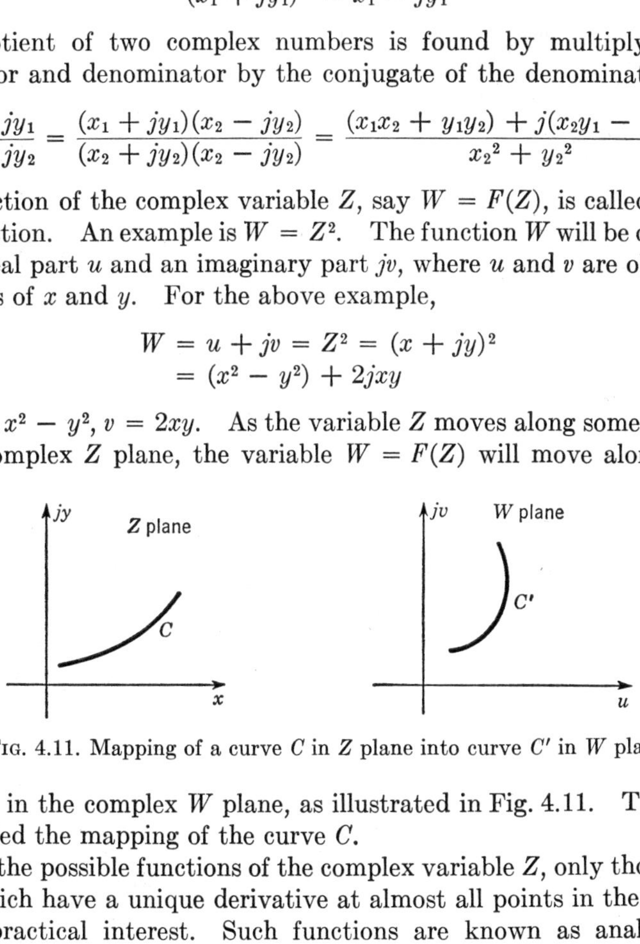

FIG. 4.11. Mapping of a curve C in Z plane into curve C' in W plane.

curve C' in the complex W plane, as illustrated in Fig. 4.11. The curve C' is called the mapping of the curve C.

Of all the possible functions of the complex variable Z, only those functions which have a unique derivative at almost all points in the Z plane are of practical interest. Such functions are known as analytic (or regular) functions. The derivative of $W = F(Z)$ is defined as

$$\frac{dW}{dZ} = \lim_{\Delta Z \to 0} \frac{\Delta W}{\Delta Z} = \lim_{\Delta z \to 0} \frac{W(Z + \Delta Z) - W(\Delta Z)}{\Delta Z} \tag{4.89}$$

If this limit exists and is independent of the direction along which ΔZ approaches zero in the complex Z plane, the function W is said to be analytic at that point. We may readily evaluate dW/dZ under the two conditions where $\Delta Z = \Delta x$ and $\Delta Z = j\,\Delta y$. For the first case we have

$$\frac{dW}{dZ} = \frac{dW}{dx} = \frac{\partial u}{\partial x} + j\frac{\partial v}{\partial x}$$

while in the second case we have

$$\frac{dW}{dZ} = \frac{dW}{j\,dy} = \frac{\partial u}{j\,\partial y} + \frac{\partial v}{\partial y} = -j\frac{\partial u}{\partial y} + \frac{\partial v}{\partial y}$$

If the derivative is to be unique, the two results must be equal, and hence

$$\frac{\partial u}{\partial x} + j\frac{\partial v}{\partial x} = -j\frac{\partial u}{\partial y} + \frac{\partial v}{\partial y}$$

or

$$\frac{\partial u}{\partial x} = \frac{\partial v}{\partial y} \qquad \frac{\partial v}{\partial x} = -\frac{\partial u}{\partial y} \tag{4.90}$$

This condition is seen to be a necessary condition, at least, in order for $W = F(Z)$ to be an analytic function. Equations (4.90) are known as the Cauchy-Riemann equations. It is possible to show that if u and v are continuous functions of x and y, the Cauchy-Riemann equations are both necessary and sufficient conditions for $F(Z)$ to be an analytic function. When the derivative exists it may be found by the same rules as are used for functions of a real variable. As examples we have

$$\frac{d\cos Z}{dZ} = -\sin Z$$

$$\frac{dZ^n}{dZ} = nZ^{n-1}$$

$$\frac{de^{2Z}}{dZ} = 2e^{2Z}$$

$$\frac{d\ln Z}{dZ} = \frac{1}{Z}$$

All the above functions are analytic, a result readily verified by showing that the Cauchy-Riemann equations are satisfied. Consider

$$W = \cos Z = \cos(x + jy) = \cos x \cosh y - j \sin x \sinh y$$

for which $u = \cos x \cosh y$, and $v = -\sin x \sinh y$. From (4.90) we obtain

$$\frac{\partial u}{\partial x} = -\sin x \cosh y = \frac{\partial v}{\partial y}$$

and

$$\frac{\partial u}{\partial y} = \cos x \sinh y = -\frac{\partial v}{\partial x}$$

and hence $\cos Z$ is an analytic function.

Coordinate Transformations

In order to appreciate the role played by complex functions in the solution of two-dimensional potential problems, we need to consider some fundamental properties of two-dimensional orthogonal curvilinear coordinates. As an example, if we require a solution for the potential between two infinitely long elliptic cylinders, as illustrated in Fig. 4.12, we would find it very difficult to obtain a suitable solution if we solved Laplace's

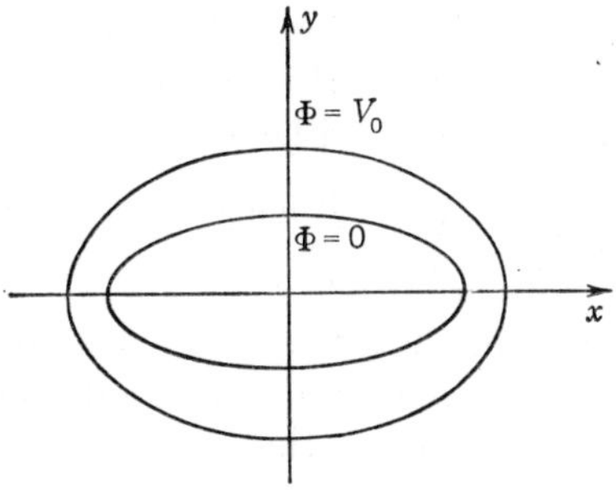

FIG. 4.12. Two concentric elliptic cylinders.

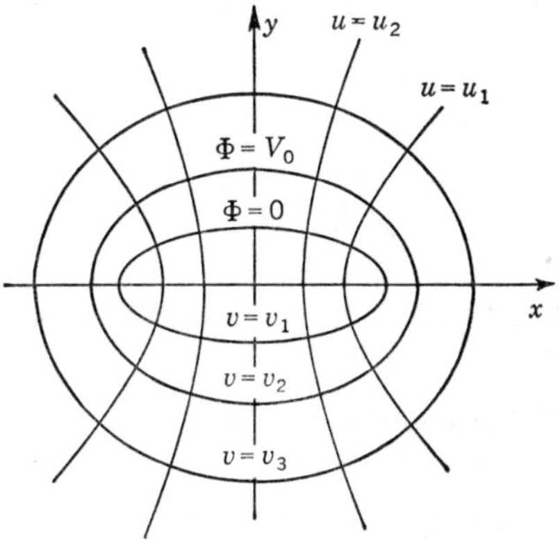

FIG. 4.13. Coordinates uv for elliptic cylinders.

equation in rectangular coordinates. Obviously, we should use elliptic coordinates, so that on each cylinder only one coordinate is variable. In Fig. 4.13 the appropriate coordinates u, v are illustrated. The v = constant curves are ellipses, and the u = constant curves are hyperbolas. Since the boundaries coincide with the v = constant coordinate curves, it follows that on the boundary of a cylinder the potential $\Phi(u,v) = \Phi(u,v_1)$ is a function of u only. For conducting cylinders the appropriate Φ must be independent of u in order that Φ be a constant on the cylinder.

In general, for the problem to be solved we should look for a suitable coordinate system in which the boundaries coincided with constant coordinate curves. Such a set of coordinates u, v are functions of x and y, so that

$$u = u(x,y)$$
$$v = v(x,y)$$

Instead of specifying a point by the coordinates x_1, y_1, we may equally well specify the point by the coordinates $u_1 = u(x_1,y_1)$ and $v_1(x_1,y_1)$. The curves u = constant and v = constant are called coordinate curves.

Along the u = constant curve only v varies, and vice versa. Along the coordinate curves we may construct two unit vectors $\mathbf{a}_u$ and $\mathbf{a}_v$ for the purpose of specifying a vector such as

$$\mathbf{A} = A_u\mathbf{a}_u + A_v\mathbf{a}_v$$

The unit vector $\mathbf{a}_u$ is tangent to the v = constant curves, and $\mathbf{a}_v$ is tangent to the u = constant curves.

When the coordinate curves intersect at right angles, as in Fig. 4.14, the uv coordinates are said to form an orthogonal curvilinear coordinate system. The function ∇u is a vector in the direction of the maximum rate of change of u, and hence

$$\nabla u = |\nabla u|\mathbf{a}_u$$

Similarly, $$\nabla v = |\nabla v|\mathbf{a}_v$$

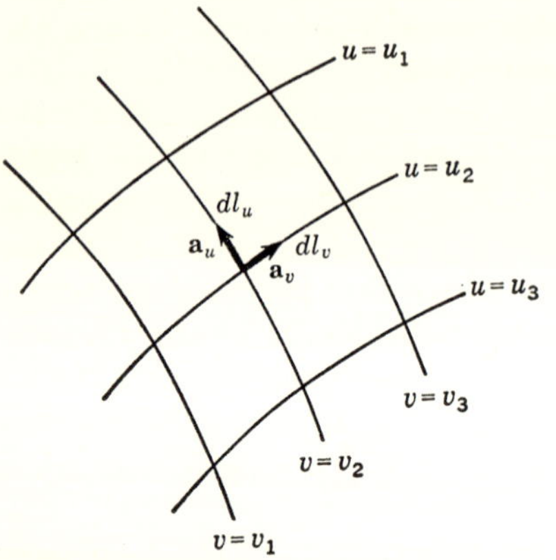

FIG. 4.14. An orthogonal curvilinear coordinate system.

If the coordinates form an orthogonal system, then $\mathbf{a}_u$ and $\mathbf{a}_v$ must be orthogonal everywhere. Hence the necessary and sufficient condition for u and v to form an orthogonal coordinate system is that

$$\nabla u \cdot \nabla v = \frac{\partial u}{\partial x}\frac{\partial v}{\partial x} + \frac{\partial u}{\partial y}\frac{\partial v}{\partial y} = 0 \tag{4.91}$$

The differentials du and dv are not in general measures of length along the coordinate curves. Thus the differentials of length dl_u and dl_v are given by

$$dl_u = h_u\, du \tag{4.92a}$$
$$dl_v = h_v\, dv \tag{4.92b}$$

where h_u and h_v are suitable scale factors. The directional derivative of u along the u coordinate curve is

$$\frac{du}{dl_u} = \nabla u \cdot \mathbf{a}_u = |\nabla u|$$

since ∇u and $\mathbf{a}_u$ are in the same direction. Comparison with (4.92a) shows that

$$dl_u = \frac{du}{|\nabla u|} = h_u\, du$$

and hence $$h_u = \left[\left(\frac{\partial u}{\partial x}\right)^2 + \left(\frac{\partial u}{\partial y}\right)^2\right]^{-1/2} \tag{4.93a}$$

Similarly, $$h_v = \left[\left(\frac{\partial v}{\partial x}\right)^2 + \left(\frac{\partial v}{\partial y}\right)^2\right]^{-1/2} \tag{4.93b}$$

Let us now consider the type of coordinate system generated by a function $W = u + jv = F(Z) = F(x + jy)$. We shall restrict F to be an analytic function, so that the Cauchy-Riemann equations (4.90) hold. The function F gives us two coordinate variables u and v. Using the Cauchy-Riemann equations we can verify that

$$\frac{\partial u}{\partial x}\frac{\partial v}{\partial x} + \frac{\partial u}{\partial y}\frac{\partial v}{\partial y} = -\frac{\partial u}{\partial x}\frac{\partial u}{\partial y} + \frac{\partial u}{\partial y}\frac{\partial u}{\partial x} = 0 \tag{4.94}$$

From this and the result of (4.91) we see that u and v form an orthogonal coordinate system. Thus the curves u = constant intersect the curves v = constant at right angles. A further consequence of the Cauchy-Riemann equations is that $h_u = h_v$. This follows since

$$h_v^{-2} = \left(\frac{\partial v}{\partial x}\right)^2 + \left(\frac{\partial v}{\partial y}\right)^2 = \left(\frac{\partial u}{\partial y}\right)^2 + \left(\frac{\partial u}{\partial x}\right)^2 = h_u^{-2}$$

Thus an analytic function $F(Z)$ generates an orthogonal curvilinear coordinate system in which the two scale factors h_u and h_v are equal.

By using the Cauchy-Riemann equations it is easy to show that both u and v satisfy the two-dimensional Laplace equation

$$\frac{\partial^2 u}{\partial x^2} + \frac{\partial^2 u}{\partial y^2} = 0$$

and similarly for v. For this reason u and v are called harmonic functions; i.e., any solution to Laplace's equation is called a harmonic function. Bessel functions and Legendre functions are often referred to as cylindrical and spherical harmonics, respectively, for the same reason.

In the uv coordinates Laplace's equation becomes

$$\frac{\partial}{\partial u}\frac{h_v}{h_u}\frac{\partial \Phi}{\partial u} + \frac{\partial}{\partial v}\frac{h_u}{h_v}\frac{\partial \Phi}{\partial v} = \frac{\partial^2 \Phi}{\partial u^2} + \frac{\partial^2 \Phi}{\partial v^2} = 0 \tag{4.95}$$

since $h_u = h_v$. We obtain the interesting result that $\Phi(x,y)$, which satisfies Laplace's equation

$$\frac{\partial^2 \Phi(x,y)}{\partial x^2} + \frac{\partial^2 \Phi(x,y)}{\partial y^2} = 0$$

also satisfies this equation when Φ is transformed to the uv system, i.e.,

$$\frac{\partial^2 \Phi(u,v)}{\partial u^2} + \frac{\partial^2 \Phi(u,v)}{\partial v^2} = 0$$

Thus in the uv coordinate system it is still only necessary to find a solution to Laplace's equation; furthermore, this will be very much simpler, assuming that an appropriate transformation has been found which makes u = constant or v = constant a boundary surface.

The usefulness of the coordinate transformation is often expressed by considering u and v as rectangular coordinates; that is, instead of plotting the u and v coordinate curves on the xy plane, we may distort the uv coordinates into a rectangular grid (W plane) along with the boundaries of the given problem. This means that we transform the curves representing the boundaries in the Z plane into new curves in the W plane. The representation in the W plane is called the conformal mapping of the boundary from the Z plane. The term conformal signifies the property that curves intersecting at right angles in the Z plane map into curves intersecting at right angles in the W plane, a property resulting from the orthogonality of the uv coordinate curves. In the W plane we may now solve for the potential Φ as a function of u and v by treating u and v as rectangular coordinates. The latter solution is possible because surfaces in the Z plane on which $\Phi =$ constant map into the W plane with the same value of constant potential. Similarly, a boundary condition such as $\partial\Phi/\partial n = 0$ is invariant under a conformal transformation. Consequently, the problem in the Z plane is replaced by one in the W plane with identical boundary conditions and the equivalent requirement that $\nabla^2\Phi(u,v) = 0$. But if an appropriate transformation has been found, the problem in the W plane is immeasurably simpler, since we make the boundaries lie along the rectangular $u =$ constant or $v =$ constant lines. Consequently, the solution $\Phi(u,v)$ can be found, and by an inverse transformation $\Phi(x,y)$ is then determined.

The previous remarks will be explained more fully in the course of solution of the following two problems. In the first we consider the case where a constant potential is assigned over the boundary surfaces. In the second the boundary condition $\nabla\Phi \cdot \mathbf{n} = \partial\Phi/\partial n = \rho_s/\epsilon_0$ is given. In this case the boundary value does not remain invariant under a conformal mapping since the scale factor h enters into the expression for $\nabla\Phi$ in the uv coordinate system. The correct technique to be applied is explained in this problem.

Example 4.8. Potential between Two Elliptic Cylinders. We wish to obtain a solution for the potential between two elliptic cylinders with the inner cylinder kept at zero potential and the outer cylinder kept at a potential V_0. The problem is illustrated in Fig. 4.13. In order to solve this problem we must find a complex function $W = F(Z)$ that will generate a uv coordinate system for which the elliptic-cylinder boundaries coincide with constant coordinate curves. There is no direct way in which the required function F may be found. We have to rely on our familiarity with the properties of various analytic functions in order to know which specific function is required. For the present problem the function $W = \cos^{-1} Z$ is a suitable one.

For the above function we have $u + jv = \cos^{-1}(x + jy)$ or

$$\cos(u + jv) = \cos u \cosh v - j \sin u \sinh v = x + jy$$

and hence

$$x = \cos u \cosh v \tag{4.96a}$$
$$y = -\sin u \sinh v \tag{4.96b}$$

Squaring both sides and adding leads to the result

$$\frac{x^2}{\cosh^2 v} + \frac{y^2}{\sinh^2 v} = \cos^2 u + \sin^2 u = 1 \tag{4.97}$$

If v is held constant, (4.97) is the equation of an ellipse. The family of

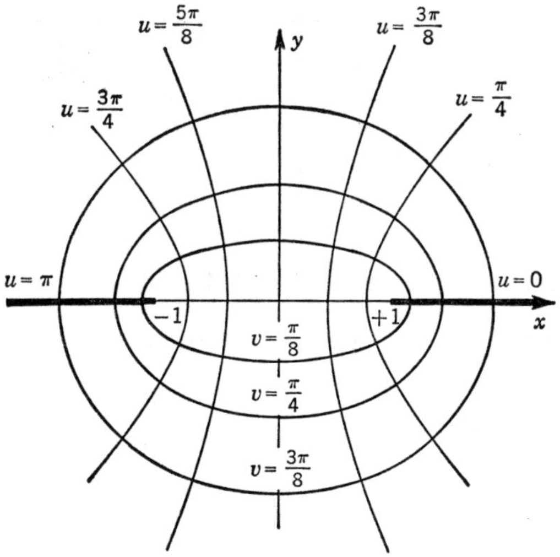

FIG. 4.15. Coordinate curves for inverse cosine function.

curves v = constant are confocal ellipses with foci at ± 1. From (4.96) we may obtain the following equation also by eliminating the variable v:

$$\frac{x^2}{\cos^2 u} - \frac{y^2}{\sin^2 u} = \cosh^2 v - \sinh^2 v = 1 \tag{4.98}$$

Thus the u = constant curves are a family of confocal hyperbolas which intersect the v = constant curves orthogonally. These coordinates are plotted in Fig. 4.15.

Let the boundaries coincide with the coordinate curves $v = \pi/8$ and $v = \pi/4$. Laplace's equation is

$$\frac{\partial^2 \Phi}{\partial u^2} + \frac{\partial^2 \Phi}{\partial v^2} = 0$$

A simple solution for Φ which is zero for $v = \pi/8$ and equals V_0 for $v = \pi/4$ and is independent of u is

$$\Phi = \frac{8}{\pi}\left(v - \frac{\pi}{8}\right) V_0 \tag{4.99}$$

This is the required solution for Φ since both the boundary conditions and Laplace's equation are satisfied, so that (4.99) is unique. In this

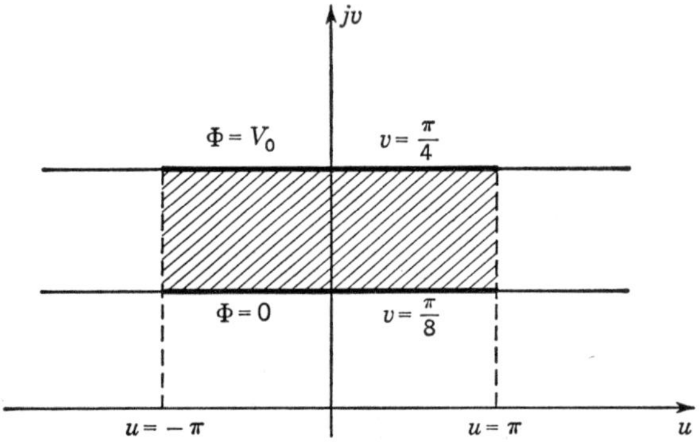

FIG. 4.16. Conformal mapping of Fig. 4.15 into W plane.

particular case Φ is a function of v only. If we solve (4.97) for v, we obtain

$$v = \sinh^{-1}\left[\frac{y^2 + x^2 - 1}{2} \pm \sqrt{\frac{(y^2 + x^2 - 1)^2}{4} + y^2}\right]^{1/2}$$

and this result may be substituted into (4.99) to obtain Φ as a function of x and y.

If we mapped the cylinders $v = \pi/8$ and $v = \pi/4$ into the W plane, our original problem becomes one of finding the potential between two infinite parallel planes separated by a distance $\pi/8$, as in Fig. 4.16. For this problem (4.99) is clearly the solution. As we move around the elliptic cylinder, the coordinate u varies from $-\pi$ to π. The mapping is a periodic one and repeats itself with a periodic 2π in u. The region between the two elliptic cylinders corresponds to the shaded area in Fig. 4.16.

If we are interested in the capacitance C per unit length between the two cylinders, this may be obtained from the equivalent problem in Fig. 4.16. For the latter case C is the capacitance of the finite portion of an infinite-parallel-plane capacitor and is, consequently,

$$C = \frac{2\pi\epsilon_0}{\pi/8} = 16\epsilon_0 \tag{4.100}$$

Note that in view of the infinite geometry there is no fringing field.

The capacitance C is invariant under a conformal mapping and is consequently the desired solution. This result may be proved as follows. The energy stored in the electric field between the two cylinders is given by $W_e = \frac{1}{2}CV_0^2$ per unit length. Now

$$W_e = \frac{\epsilon_0}{2}\int_S |\nabla\Phi|^2\,dx\,dy = \frac{\epsilon_0}{2}\int_{S'}\left[\left(\frac{1}{h_u}\frac{\partial\Phi}{\partial u}\right)^2 + \left(\frac{1}{h_v}\frac{\partial\Phi}{\partial v}\right)^2\right] h_u h_v\,du\,dv$$

$$= \frac{\epsilon_0}{2}\int_{S'}\left[\left(\frac{\partial\Phi}{\partial u}\right)^2\left(\frac{\partial\Phi}{\partial v}\right)^2\right] du\,dv \qquad (4.101)$$

since $h_u = h_v$. Hence we may evaluate W_e in the uv plane by treating u and v as rectangular coordinates, and we shall obtain the same result as we get by evaluating the integral in the xy plane. The area S' in the W plane maps into the area S in the Z plane.

Example 4.9. Potential from a Charged Infinitely Long Ribbon. Figure 4.17 illustrates a ribbon extending from -1 to 1 along the x axis. The ribbon may be thought of as a very thin conductor which is charged with a surface charge density

$$\rho = \rho_0 \frac{x}{(1 - x^2)^{\frac{1}{2}}} \qquad (4.102)$$

on both the upper and lower surface. The potential Φ set up by these sources may be determined by means of the inverse cosine function of Example 4.8. When $v = 0$, the elliptic cylinder degenerates into a straight line extending from -1 to 1 along the x axis, and hence $v = 0$ coincides with the surface of the ribbon. When $v = 0$,

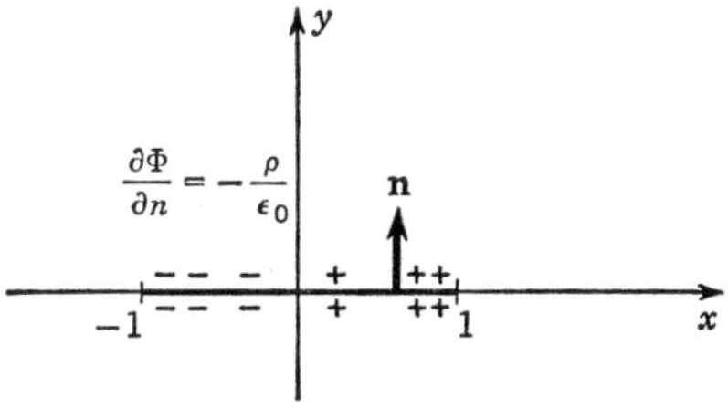

FIG. 4.17. A charged ribbon.

$$x = \cos u \cosh v = \cos u$$

and u therefore varies from 0 to $-\pi$ on the upper surface of the ribbon and from 0 to $+\pi$ on the lower surface. In uv coordinates the charge distribution ρ is given by

$$\rho = \rho_0 \frac{\cos u}{|\sin u|} \qquad (4.103)$$

If $\mathbf{n}$ is the outward normal to the ribbon surface, the boundary condition for Φ on this surface may be specified as

$$E_n = -\nabla\Phi\cdot\mathbf{n} = \frac{\rho}{\epsilon_0}$$

In the uv coordinate system $\mathbf{n} = \mathbf{a}_v$ and

$$\nabla\Phi = \frac{\mathbf{a}_u}{h}\frac{\partial\Phi}{\partial u} + \frac{\mathbf{a}_v}{h}\frac{\partial\Phi}{\partial v}$$

where $h = h_u = h_v$. Hence the boundary condition on Φ becomes

$$\frac{1}{h}\frac{\partial \Phi}{\partial v} = -\frac{\rho}{\epsilon_0} = -\frac{\rho_0}{\epsilon_0}\frac{\cos u}{|\sin u|} \tag{4.104}$$

on the upper surface.

The scale factor h is given by

$$h^{-2} = \left(\frac{\partial u}{\partial x}\right)^2 + \left(\frac{\partial u}{\partial y}\right)^2 = \left(\frac{\partial u}{\partial x}\right)^2 + \left(\frac{\partial v}{\partial x}\right)^2$$

since from the Cauchy-Riemann equations $\partial u/\partial y = -\partial v/\partial x$. If the derivative of $W = F(Z)$ with respect to Z is computed with $dZ = dx$, we obtain

$$\frac{dF}{dZ} = \frac{dW}{dZ} = \frac{\partial u}{\partial x} + j\frac{\partial v}{\partial x}$$

and

$$\left|\frac{dW}{dZ}\right|^2 = \left(\frac{\partial u}{\partial x}\right)^2 + \left(\frac{\partial v}{\partial x}\right)^2$$

Therefore the scale factor h is given by

$$h = h_u = h_v = \left|\frac{dW}{dZ}\right|^{-1} = \left|\frac{dZ}{dW}\right| \tag{4.105}$$

For the present problem

$$\begin{aligned} h &= \left|\frac{d \cos W}{dW}\right| = |\sin W| = |\sin u \cosh v + j \cosh u \sin v| \\ &= (\sin^2 u \cosh^2 v + \cosh^2 u \sin^2 v)^{1/2} \end{aligned} \tag{4.106}$$

On the ribbon where $v = 0$, we have

$$h = |\sin u|$$

Consequently, the boundary condition (4.104) becomes

$$\frac{\partial \Phi}{\partial v} = -\frac{\rho_0}{\epsilon_0}\cos u \tag{4.107}$$

Our problem may now be stated as follows. Obtain a solution to Laplace's equation

$$\frac{\partial^2 \Phi}{\partial u^2} + \frac{\partial^2 \Phi}{\partial v^2} = 0 \tag{4.108}$$

such that on the surface of the ribbon where $v = 0$ the boundary condition (4.107) is satisfied and also such that as $v \to \infty$, the potential Φ vanishes (note that x and y become infinite only when v does).

Equation (4.107) suggests that Φ will be of the following form: $f(v) \cos u$, where f is a function of v only. If we substitute this into (4.108), we find that f is a solution of

$$\frac{d^2f}{dv^2} - f = 0$$

Hence $f = Ae^{-v} + Be^{v}$, and

$$\Phi = (Ae^{-v} + Be^{v}) \cos u$$

Since Φ must vanish as v becomes infinite, the amplitude coefficient B must be zero. Imposing the boundary condition (4.107) now gives $A = \rho_0/\epsilon_0$; so the complete solution for Φ is

$$\Phi = \frac{\rho_0}{\epsilon_0} e^{-v} \cos u \tag{4.109}$$

The above potential function happens to be the induced potential that is set up if a conducting ribbon is placed in a uniform external field $E_0\mathbf{a}_x$. The applied field may be considered due to an applied potential

$$\Phi_0 = -E_0x = -E_0 \cos u \cosh v \tag{4.110}$$

The induced potential Φ_i must be such that it cancels the applied electric field along the surface of the ribbon and vanishes at infinity. It works out to be

$$\Phi_i = E_0e^{-v} \cos u \tag{4.111}$$

At $v = 0$, the total field E_u is

$$E_u = -\frac{1}{h}\frac{\partial(\Phi_0 + \Phi_i)}{\partial u} = \frac{1}{h}(-E_0 \sin u + E_0 \sin u) = 0$$

and vanishes as required. The charge induced on the ribbon is

$$\rho = -\epsilon_0 \frac{1}{h'}\frac{\partial\Phi}{\partial v}\bigg|_{v=0} = \frac{\epsilon_0 E_0 \cos u}{|\sin u|}$$

which equals that given by (4.103) if $\epsilon_0 E_0 = \rho_0$.

The induced potential may be written as

$$\begin{aligned}\Phi_i &= E_0(\cosh v - \sinh v) \cos u \\ &= E_0(x - \sinh v \cos u)\end{aligned}$$

by noting that $\cosh v - \sinh v = e^{-v}$. Solving (4.97) for $\sinh v$ and (4.98) for $\cos u$ gives for Φ_i the result

$$\begin{aligned}\Phi_i = E_0 \Bigg\{x - &\left[\frac{x^2 + y^2 - 1}{2} + \sqrt{\frac{(x^2 + y^2 - 1)^2}{4} + y^2}\right]^{1/2} \\ &\times \left[\frac{1 + x^2 + y^2}{2} - \sqrt{\frac{(1 + x^2 + y^2)^2}{4} - x^2}\right]^{1/2}\Bigg\}\end{aligned} \tag{4.112}$$

To obtain this solution by solving Laplace's equation in xy coordinates would be an extremely difficult task. The power of the conformal-mapping or complex-function technique is well illustrated by this example.

4.5. The Schwarz-Christoffel Transformation

The Schwarz-Christoffel transformation is a conformal transformation which will map the real axis in the Z plane into a general polygon in the

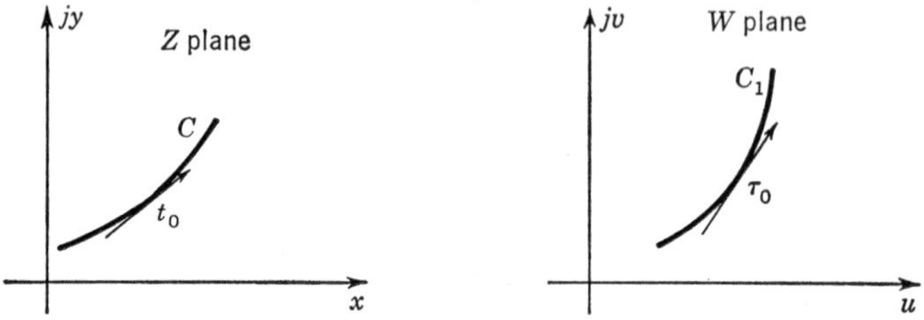

FIG. 4.18. Conformal mapping of C into C_1.

W plane, with the upper half of the Z plane mapping into the region interior to the polygon. To derive the basic transformation we consider first a curve C in the Z plane and its conformal mapping C_1 in the W plane, as illustrated in Fig. 4.18. Let the unit tangent to C in the Z plane be t_0 and the unit tangent to C_1 in the W plane be τ_0, where

$$t_0 = \lim_{\Delta Z \to 0} \frac{\Delta Z}{|\Delta Z|}$$

$$\tau_0 = \lim_{\Delta W \to 0} \frac{\Delta W}{|\Delta W|}$$

If the mapping function is $W = F(Z)$, we have

$$\frac{dW}{dZ} = F'(Z) = \lim_{\Delta Z \to 0} \frac{\Delta W}{\Delta Z} = \lim_{\Delta Z \to 0} \frac{\Delta W/|\Delta W|}{\Delta Z/|\Delta Z|} \left|\frac{\Delta W}{\Delta Z}\right| = \frac{\tau_0}{t_0} |F'(Z)|$$

and hence

$$\tau_0 = \frac{F'(Z)}{|F'(Z)|} t_0 \tag{4.113}$$

The angle that τ_0 makes with the u axis is given by

$$\angle\tau_0 = \angle t_0 + \angle F'(Z) \tag{4.114}$$

Consider next the following function:

$$F'(Z) = A(Z - x_1)^{-k_1}(Z - x_2)^{-k_2} \cdots (Z - x_N)^{-k_N}$$

$$= A \prod_{i=1}^{N} (Z - x_i)^{-k_i} \tag{4.115}$$

where A is an arbitrary constant, k_i is a real number, and $x_1 < x_2 < \cdots < x_N$. If the x axis is chosen as the curve C in the Z plane, the angle of the unit tangent to the mapping of C, that is, to C_1, in the W plane will be

$$\angle\tau_0 = \angle F' = \angle A - \sum_{i=1}^{N} k_i \angle(Z - x_i) \tag{4.116}$$

For $x < x_i$ we have $\angle(x - x_i) = \pi$, and for $x > x_i$ we have $\angle(x - x_i) = 0$. Thus as each point x_i is passed in the Z plane, the angle of τ_0 changes

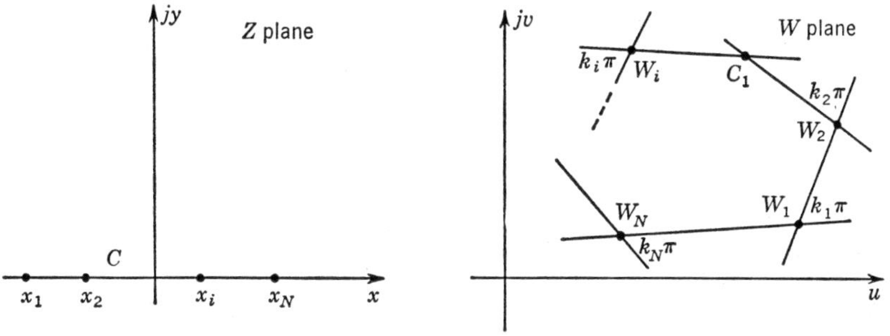

FIG. 4.19. The Schwarz-Christoffel transformation.

discontinuously by an amount $k_i\pi$ and a polygon such as illustrated in Fig. 4.19 is traced out. The exterior angles to the polygon are $k_i\pi$, and these angles must add up to 2π if the polygon is to be closed, and hence

$$\sum_{i=1}^{N} k_i = 2 \tag{4.117}$$

The constant A serves to rotate and magnify the figure in the W plane. The points x_i map into the points $W_i = F(x_i)$. The mapping function is given by

$$W = \int F'(Z)\,dZ + B = A \int \prod_{i=1}^{N} (Z - x_i)^{-k_i}\,dZ + B \tag{4.118}$$

where B is an arbitrary constant which serves to translate the figure in the W plane. The integration of (4.118) is usually not possible unless $|k_i| = 0, \frac{1}{2}, 1, \frac{3}{2}$, or 2. It should be noted that in this transformation there is a factor for each vertex corresponding to a finite value of x but no terms for the points $x = \pm\infty$.

In practice we normally wish to map a given polygon in the W plane into the x axis. This requires the inverse mapping function giving Z as a function of W. Generally, it is difficult to obtain this inverse transformation; so we usually proceed by trial and error to set up a transformation of the form (4.118) which will map the x axis into the given polygon. In this procedure we are aided by the condition that three of the points x_i may be arbitrarily chosen.

Example 4.10. Fringing Capacitance in a Parallel-plate Capacitor. As an example of the use of the Schwarz-Christoffel transformation, the

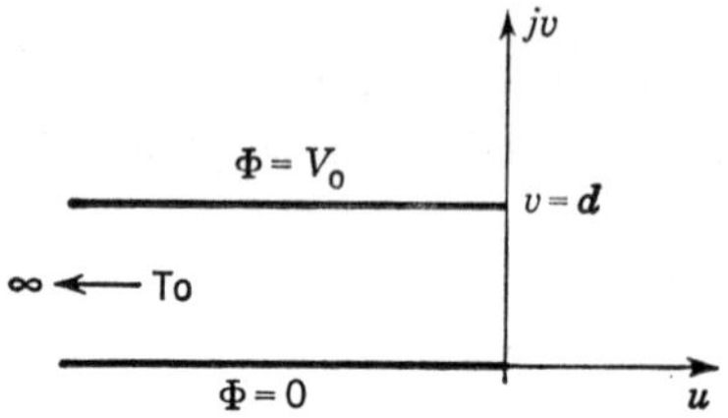

FIG. 4.20. A semi-infinite parallel-plate capacitor.

fringing capacitance C_f at the edge of a semi-infinite parallel-plate capacitor will be determined. The problem is illustrated in Fig. 4.20. In the complex W plane the capacitor boundaries are obtained by letting the points W_1 and W_3 tend to $-\infty$ in Fig. 4.21. The external angles at W_1, W_2, W_3, and W_4 in the limit become $-\pi$, $-\pi$, π, and $-\pi$, respectively.

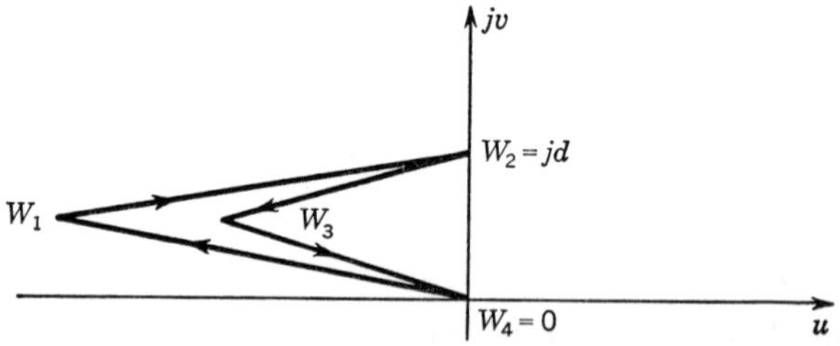

FIG. 4.21. Polygon corresponding to capacitor boundaries.

Let the points W_1, W_2, W_3, W_4 map into the points x_1, x_2, x_3, and x_4 in the Z plane. Our required transformation is thus

$$W(Z) = A_1 \int^Z (Z - x_1)(Z - x_2)(Z - x_3)^{-1}(Z - x_4)\,dZ + B$$
$$= A \int^Z \left(1 - \frac{Z}{x_1}\right)(Z - x_2)(Z - x_3)^{-1}(Z - x_4)\,dZ + B$$

where $A = -A_1x_1$, and A, B are complex constants.

To cover the complete x axis we now let x_1 tend to infinity and hence obtain

$$W(Z) = A \int^Z \frac{(Z - x_2)(Z - x_4)}{Z - x_3}\,dZ + B$$

Since we are free to choose three of the x_i's, let $x_2 = -1$, $x_3 = 0$, and $x_4 = 1$; thus

$$W = A \int^Z \frac{Z^2 - 1}{Z} dZ + B$$

which may be integrated to give

$$W = A\left(\frac{Z^2}{2} - \ln Z\right) + B \tag{4.119}$$

When $W = W_2 = jd$, $Z = -1$, and when $W = W_4 = 0$, $Z = 1$; so

$$jd = A(\tfrac{1}{2} - \ln e^{j\pi}) + B$$
$$0 = A(\tfrac{1}{2}) + B$$

where $e^{j\pi}$ is written for -1. Solving for A and B gives $A = -d/\pi$ and

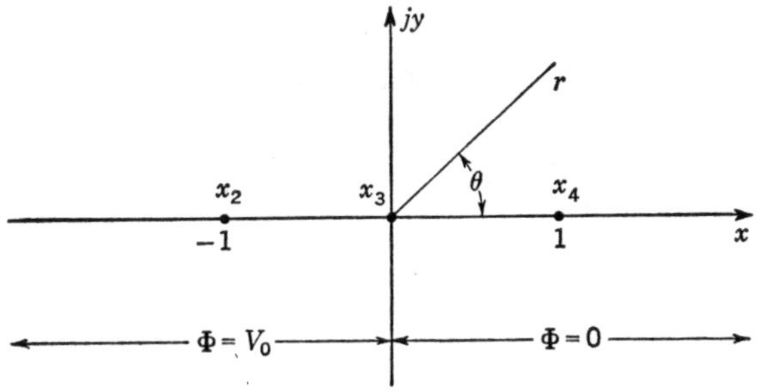

FIG. 4.22. Conformal mapping of capacitor boundaries into Z plane.

$B = d/2\pi$. Our final transformation is

$$W = \frac{d}{\pi}\left(\frac{1 - Z^2}{2} + \ln Z\right) \tag{4.120}$$

and is illustrated in Fig. 4.22.

In the xy plane the solution for Φ is clearly

$$\Phi = V_0 \frac{\theta}{\pi} = \frac{V_0}{\pi} \tan^{-1} \frac{y}{x} \tag{4.121}$$

The charge density on the plane $x < 0$ is given by

$$\rho_s = -\epsilon_0 \frac{\partial \Phi}{\partial n} = \epsilon_0 \frac{1}{r}\frac{\partial \Phi}{\partial \theta} = \frac{\epsilon_0 V_0}{|x|\pi} \tag{4.122}$$

The charge density on the plane $x > 0$ is the negative of (4.122). The total charge on each capacitor plate is, of course, infinite, since each plate is infinite. Let us therefore find the charge $Q(l)$ per unit length for a width l of the lower plate. When $W = -l$ on the inside of the lower plate, let $x = X_1$, where $0 < X_1 < 1$, and when $W = -l$ on the outside of the lower plate, let $x = X_2$, where $X_2 > 1$. For $Q(l)$ we have [$Q(l)$ is taken to be the magnitude of the charge]

$$Q(l) = \frac{\epsilon_0 V_0}{\pi} \int_{X_1}^{X_2} \frac{dx}{x} = \frac{\epsilon_0 V_0}{\pi} (\ln X_2 - \ln X_1)$$

The values of X_1 and X_2 may be found in terms of l from the transformation (4.120). We have

$$-l = \frac{d}{\pi}\left(\frac{1 - X^2}{2} + \ln X\right) = \frac{d}{\pi}\left(\frac{1 - X_1^2}{2} + \ln X_1\right)$$
$$= \frac{d}{\pi}\left(\frac{1 - X_2^2}{2} + \ln X_2\right) \tag{4.123}$$

If we choose $l \gg d$, we can obtain a good approximate solution to these transcendental equations. For X_1 we must have $0 < X_1 < 1$, and in particular for $l \gg d$, X_1 is very small. Hence $(1 - X_1^2)/2$ is negligible compared with $\ln X_1$ and $\ln X_1 \approx -\pi l/d$. For X_2 we must have $X_2 \gg 1$ when $l \gg d$, and we may then neglect the term $\frac{1}{2} + \ln X_2$ to get $X_2^2 \approx 2l\pi/d$ and $\ln X_2 = \frac{1}{2} \ln (2l\pi/d)$. The total charge $Q(l)$ is consequently given by

$$Q(l) = \frac{\epsilon_0 V_0}{\pi}\left(\frac{1}{2} \ln \frac{2l\pi}{d} + \frac{\pi l}{d}\right) \tag{4.124}$$

If there were no fringing field, the charge density on each plate would be constant and equal to $\epsilon_0 V_0/d$. For a width l, the total charge per unit length on the bottom plate would be

$$Q_0(l) = \frac{\epsilon_0 V_0 l}{d} \tag{4.125}$$

The additional charge $Q - Q_0$ gives rise to the fringing capacitance C_f. We therefore find that

$$C_f = \frac{Q - Q_0}{V_0} = \frac{\epsilon_0}{2\pi} \ln \frac{2l\pi}{d} \tag{4.126}$$

for a width l of the capacitor plate, per unit length.

As a practical application of this result consider a parallel-plate capacitor of width and length equal to b and with a spacing $d \ll b$. The parallel-plate capacitance is $\epsilon_0 b^2/d$. A first-order correction to this may be obtained by using (4.126) for the fringing capacitance per edge and choosing $l = b/2$. For the corrected value of capacitance we then obtain

$$C = \frac{\epsilon_0 b^2}{d} + \frac{2b\epsilon_0}{\pi} \ln \frac{\pi b}{d} \tag{4.127}$$

BIBLIOGRAPHY

Bowman, F.: "Introduction to Bessel Functions," Dover Publications, New York, 1958.

Churchill, R. V.: "Fourier Series and Boundary Value Problems," McGraw-Hill Book Company, Inc., New York, 1941.

———: "Introduction to Complex Variables and Applications," McGraw-Hill Book Company, Inc., New York, 1948.

Jahnke, E., and F. Emde: "Tables of Functions," Dover Publications, New York, 1945.

McLachlan, N. W.: "Bessel Functions for Engineers," 2d ed., Oxford University Press, New York, 1946.

MacRobert, T. M.: "Spherical Harmonics," 2d ed., Methuen & Co., Ltd., London, 1947.

Morse, P. M., and H. Feshbach: "Methods of Theoretical Physics," McGraw-Hill Book Company, Inc., New York, 1953.

Pipes, L. A.: "Applied Mathematics for Engineers and Physicists," 2d ed., McGraw-Hill Book Company, Inc., New York, 1958.

Smythe, W. R.: "Static and Dynamic Electricity," McGraw-Hill Book Company, Inc., New York, 1950.

Sommerfeld, A.: "Partial Differential Equations in Physics," Academic Press, Inc., New York, 1949.

Stratton, J. A.: "Electromagnetic Theory," McGraw-Hill Book Company, Inc., New York, 1941.

Weber, E.: "Electromagnetic Fields," John Wiley & Sons, Inc., New York, 1950.

CHAPTER 5

STATIONARY CURRENTS

In the previous chapters the characteristics of an electrical conductor were noted, namely, that it was a repository of free electronic charge which would readily move under the influence of an applied field. A particular consequence of this, that the electric field within a conductor must be zero in the presence of an electrostatic field, has already been described.

In this chapter we look into the conditions required for the production of steady current flow in conductors and for a description of the properties of current flow fields. We could take advantage of the existence of a large body of knowledge dealing with the flow of current in electric circuits that is available in electric circuit theory. We prefer, however, to describe current flow in terms of an appropriate electric field. The field-theory approach may then be related to the circuit approach.

In the region external to batteries, the field producing a current flow is the electrostatic field. Since the current density is linearly related to the electric field, an interesting duality between the current flow field and the displacement flux exists. This duality may often be made use of in the solution of current flow problems. In particular, we shall show that there exists a simple relationship between the capacitance and resistance between two electrodes.

Many current flow problems cannot be solved analytically. Therefore a discussion of flux-plotting techniques and experimental techniques, such as the electrolytic tank, is included in the latter part of this chapter.

5.1. Ohm's Law

In the conducting medium it is found experimentally that the current is related to the electric field by the following expression:

$$\mathbf{J} = \sigma \mathbf{E} \tag{5.1}$$

In this equation σ is the conductivity of the medium in mhos per meter and $\mathbf{J}$ is the current density in amperes per square meter. The past chapters have dealt with vector fields where, for conceptual reasons, we interpreted them in terms of flow (flux) fields. The current flow field, however, truly involves flow, and **J** represents the quantity of coulombs flowing across a unit cross-sectional area per second. As usual, to calculate total current flow across a surface, the following surface integral must be evaluated:

$$I = \int_S \mathbf{J} \cdot d\mathbf{S} \tag{5.2}$$

The phenomenon of conduction in a metal can be considered from an atomic viewpoint, in which case a fundamental understanding of the dependence of σ on atomic structure, impurities, and temperature can be developed. It will be sufficient for our purposes, however, to have in mind a very simple physical model. We may think of the conductor as composed of a lattice of fixed positive ions containing an electron gas free to move about. Ordinarily, these free electrons are in a state of random motion because of their thermal energy. The space-time-average charge density, however, is zero. Conduction arises from the drift of electrons because of the action of an applied electric field. Except for an initial transient, the electron velocity reaches a steady-state value when the accelerating force of the applied field is exactly balanced by the scattering effect of electron collisions with the lattice. These collisions may also be viewed as the mechanism whereby some of the energy of the electrons, hence of the field, is dissipated as heat. At equilibrium the current density at any point is simply the electron charge density at the point times its drift velocity. Thus it can be shown that the time-average drift velocity is†

$$\mathbf{v} = -\frac{e\mathbf{E}\lambda}{2mv_0} \tag{5.3}$$

where λ is the mean free path of the electrons, and v_0 the mean thermal velocity. Then, if the density of charge is N electrons per cubic meter,

$$\mathbf{J} = -Ne\frac{e\mathbf{E}\lambda}{2mv_0} = -\frac{Ne^2\lambda\mathbf{E}}{2mv_0} \tag{5.4}$$

This expression reveals the linear relation between current density and field and also relates the conductivity to the atomic quantities. Note that, by convention, positive current is associated with the flow of positive charge.

Equation (5.1) implies that conduction is both linear and isotropic.

† In obtaining this expression it is assumed that the scattering of electrons by the heavy-metal atoms occurs in a completely random manner so that the average velocity after collision is zero. Consequently, the average drift velocity is that which is acquired between collisions under the action of the electric field force, that is, $v = \frac{1}{2}at$, where $a = eE/m$ is the acceleration and $t = \lambda/v_0$ is the time between collisions. Ordinarily, the thermal velocity $v_0 \gg v$; hence the dependence of t on v_0 only.

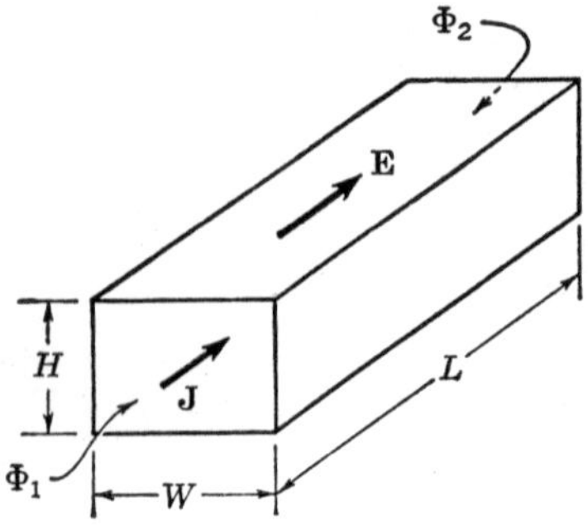

FIG. 5.1. Uniform current flow in a rectangular bar.

This is not always true. However, for metals, under a wide range of current densities it does apply. We shall assume that (5.1) correctly relates the current density and electric field in conductors.

Equation (5.1) is a point relationship and is true even if σ is a function of the coordinates. For a homogeneous body with uniform current, it is relatively easy to find the total current and to obtain a relationship between it and the applied field. Such a formulation would be desirable in a circuit analysis. A simple case is illustrated in Fig. 5.1, where a uniform axial applied field exists in a conductor of rectangular cross section. The total current that flows is

$$I = JWH = \sigma EWH \tag{5.5}$$

If we assume, for the moment, that over the extent of the conductor $\mathbf{E}$ is conservative, then $\mathbf{E} = -\nabla\Phi$, and the difference of potential between the ends of the conductor, V, is

$$V = \Phi_1 - \Phi_2 = EL \tag{5.6}$$

Combining (5.5) and (5.6) gives

$$I = \frac{\sigma WH}{L} V = GV \tag{5.7}$$

where

$$G = \frac{\sigma WH}{L}$$

is the total conductance of the parallelepiped.

It is more common to specify the resistance of the conductor. This is the reciprocal of the conductance and so may be written

$$R = \frac{L}{\sigma WH} = \frac{\rho L}{WH} \tag{5.8}$$

where ρ (not to be confused with charge density), the resistivity in ohm-meters, is the reciprocal of the conductivity; i.e.,

$$\rho = \frac{1}{\sigma} \tag{5.9}$$

From (5.6) and (5.7) we get the well-known statement of Ohm's law as applied to the macroscopic circuit:

$$V = IR \tag{5.10}$$

5.2. Nonconservative Fields—EMF

We should like to produce a steady current, and we inquire now into methods whereby this may be accomplished. As we know from (5.1), it will be necessary to start with an electric field. So far, however, we have considered only the production of an electrostatic field by stationary charges. Will this suffice?

Suppose we consider the electrostatic field set up within the parallel-plate capacitor of Fig. 5.2, into which we now insert the conductor of Fig. 5.1. At this instant the conductor finds itself in a uniform axial **E**

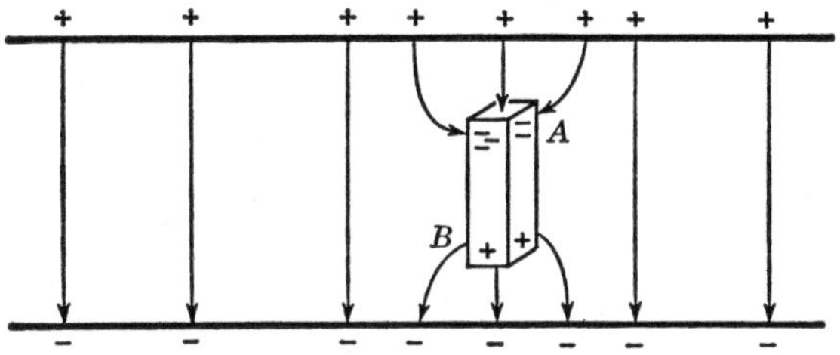

FIG. 5.2. A rectangular conducting bar placed in the electrostatic field between two charged plates.

field, and consequently a uniform current $I = \sigma WHE$ flows. But this current is nothing more than the movement of charge, and as time increases, it must be that negative charge accumulates at A while B becomes positively charged. These charges represent secondary sources of field, and it is not hard to see that they set up a field within the conductor that opposes the primary (capacitor) field. Actually, we have already considered this kind of problem and were led to the conclusion that the total field within the conductor would soon reach the equilibrium value of zero. In that case, though, the current would stop. It seems that an electrostatic field is not capable of setting up steady currents.

Further consideration explains why an electrostatic field alone cannot be the cause of steady currents. Consider an electron which is an element of a steady conduction stream. Since steady-state conditions exist, it must make a complete circuit and return to an arbitrary starting point, thence to repeat the circuit ad infinitum. In any such circuit the electron gives up energy to the conductors in the form of heat, as a consequence of the finite resistance of the conductors. The energy, however, comes ultimately from the field, since this is the basis for the current flow. But an electrostatic field is conservative; it is not capable of giving up energy indefinitely. As a matter of fact, if we assume the field to remain unchanged, as must be true where steady current exists, then an electron making a closed circuit in an electrostatic field gains no net energy from the field. Clearly, another source of field is required for the maintenance of steady currents, and this field must be nonconservative.

The action of a chemical battery may be interpreted from a field standpoint as producing such a nonconservative field $\mathbf{E}'$. In general, an electrostatic field will also be created by a battery as a result of the accumulation, at the battery terminals and elsewhere in the circuit, of stationary (capacitor) charge. Designating the latter field by $\mathbf{E}$, the total field is then

$$\mathbf{E}_t = \mathbf{E} + \mathbf{E}' \tag{5.11}$$

Equation (5.1) applies to the total field, so that

$$\mathbf{J} = \sigma(\mathbf{E} + \mathbf{E}') \tag{5.12}$$

This equation holds at all points, but it is important to note that $\mathbf{E}'$ may be zero at some points in the circuit; e.g., outside the battery $\mathbf{E}'$ is zero but $\mathbf{E}$ is not. If we integrate (5.11) over a closed circuit in which steady current flows and make use of (5.12) and the conservative nature of $\mathbf{E}$ (that is, $\oint_C \mathbf{E} \cdot d\mathbf{l} = 0$), then

$$\oint_C \mathbf{E}_t \cdot d\mathbf{l} = \oint_C \mathbf{E}' \cdot d\mathbf{l} = \mathcal{E} = \oint_C \frac{\mathbf{J} \cdot d\mathbf{l}}{\sigma} \tag{5.13}$$

where $\mathcal{E}$ is a measure of the strength of the nonconservative source. It is called the emf, an abbreviation of electromotive force. The current that flows depends on the conductivity, geometry, and the value of $\mathcal{E}$. If the current density is uniform over a constant cross section A, then we have

$$\mathcal{E} = I \oint \frac{dl}{\sigma A} = IR \tag{5.14}$$

In (5.14)

$$R = \oint \frac{dl}{\sigma A} \tag{5.15}$$

is the total circuit resistance and is a simple extension of the result given by (5.8).

Under open-circuit conditions (consider a battery with no circuit connections) an electrostatic field exists everywhere because of the accumulation of charge on the battery terminals. From a field standpoint the chemical action of the battery may be described by postulating a nonconservative field within the battery which just neutralizes the electrostatic field there. As viewed by a test charge, it is possible to acquire energy in moving from the positive to the negative terminal external to the battery, but by completing the circuit through the battery, where no field exists, this energy is not returned to the field (as would be the case in a purely electrostatic field). The test charge thus makes a complete circuit with a net accumulation of energy. With an actual circuit and real batteries, the accumulated energy is simultaneously dissipated as heat. The electron is capable of making repeated circuits, hence constituting a steady current.

With special arrangements, a nonconservative field can be set up so that the energy accumulated in a complete circuit by a unit of charge is available as kinetic energy. Repeated circuits continue to add energy to the charge, yielding the high-energy particles produced by devices such as the cyclotron, betatron, etc.

Making use of (5.11) under open-circuit conditions and integrating through the battery from terminals 1 to 2, we have

$$-\int_1^2 \mathbf{E} \cdot d\mathbf{l} = +\int_1^2 \mathbf{E}' \cdot d\mathbf{l} = \mathcal{E} \tag{5.16}$$

Thus the total emf just equals the open-circuit electrostatic voltage between the battery terminals. In general, points 1 and 2 may be arbitrarily chosen provided that the line integral just traverses the entire nonconservative field. Then the open-circuit electrostatic voltage between those points also equals the total emf of the source.

In the region external to that containing the nonconservative field, e.g., external to the battery, only an electrostatic field exists. Since the external region contains only a conservative field, it is possible to derive this field from the gradient of a scalar potential. This accounts for our ability to discuss d-c circuits in terms of potentials and potential difference, in spite of the nonconservative nature of the field as a whole. By describing the line integral of $\mathbf{E}_t$ through a battery from negative to positive terminal as a voltage rise (an increase in potential) equal to the emf of the battery, the multivalued nature of potential energy in a nonconservative field is avoided,† and one may now state the Kirchhoff loop equation

$$\text{Potential change over a closed loop} = 0 \tag{5.17a}$$

or

$$\text{emf} = \Sigma IR \text{ (over a closed loop)} \tag{5.17b}$$

5.3. Conservation of Charge

Since current consists of the flow of charge, a relationship between the two should be available. This is indeed the case. If we consider a volume V, then the net flow of current into this volume must be accom-

† As an analogy, consider the polar angle of a vector rotating in a counterclockwise direction. The angle increases from zero to 2π, thence from 2π to 4π, 4π to 6π, etc. But if we make a cut along $\theta = 0°$ and agree every time we cross it in a counterclockwise sense to subtract 2π, then we avoid the multivalued nature of the angle. The battery is analogous to the cut, and its strength is not -2π but whatever its emf might be.

panied by an increase of charge within V. We can express this as

$$-\int_S \mathbf{J} \cdot d\mathbf{S} = \frac{\partial}{\partial t} \int_V \rho \, dV \tag{5.18}$$

where the left-hand side of (5.18) gives the net inflow of current (coulombs per second) and the right-hand side represents the net rate of increase of total charge (coulombs per second). In (5.18) the surface S bounds the volume V, of course.

Using Gauss' theorem, (5.18) may be transformed to

$$\int_V \left(\nabla \cdot \mathbf{J} + \frac{\partial \rho}{\partial t} \right) dV = 0 \tag{5.19}$$

Since this must be true regardless of the choice of V, the integrand must itself be zero and we are led to the differential form of the law for conservation of charge.

$$\nabla \cdot \mathbf{J} + \frac{\partial \rho}{\partial t} = 0 \tag{5.20}$$

This equation is commonly referred to as the continuity equation also. Where steady currents are involved, then of course $\partial\rho/\partial t = 0$. In this case

$$\nabla \cdot \mathbf{J} = 0 \tag{5.21}$$

That is, for "stationary currents," the current density is solenoidal. We are restricting our attention in this chapter to steady currents and have already noted that such currents must form closed loops.

If (5.21) is applied to a volume that contains a junction of conductors in a network, then the second of the Kirchhoff equations results, namely,

$$\sum_i I_i = 0 \tag{5.22}$$

This equation states that the algebraic sum of the currents flowing into (or out of) a terminal is zero.

5.4. Relaxation Time

If a charge distribution is placed within a conducting body, the charge will move to the surface and distribute itself in such a way that zero field exists within and tangent to the conductor surface. The length of time required for this process to essentially take place is called the relaxation time. Whether this time is measured in millimicroseconds or in days is, of course, extremely important. A quantitative evaluation of this characteristic time is presented below.

Consider a homogeneous conducting region with a permittivity ϵ and a conductivity σ. From the divergence equation for $\mathbf{D}$ we have $\nabla \cdot \mathbf{E} = \rho/\epsilon$

when ϵ is constant. Within the conductor $\mathbf{J} = \sigma\mathbf{E}$, so that (5.20) can be written as

$$\nabla \cdot \mathbf{J} = \nabla \cdot \sigma\mathbf{E} = -\frac{\partial \rho}{\partial t} \tag{5.23}$$

From (5.23) and replacing $\nabla \cdot \mathbf{E}$ by ρ/ϵ, we obtain

$$\frac{\rho}{\epsilon} = -\frac{1}{\sigma}\frac{\partial \rho}{\partial t} \tag{5.24}$$

Thus

$$\frac{\sigma}{\epsilon}\int_0^t dt = -\int_{\rho_0}^{\rho} \frac{d\rho}{\rho}$$

and hence

$$\rho(x,y,z,t) = \rho_0(x,y,z)e^{-\sigma t/\epsilon} \tag{5.25}$$

where ρ_0 is the initial charge density when $t = 0$. The relaxation time τ is defined as

$$\tau = \frac{\epsilon}{\sigma} \tag{5.26}$$

and is the value of elapsed time required for the initial charge distribution to decay to $1/\epsilon$ of its initial value.

Table 5.1 gives the value of τ for several common materials. We note the extremely short duration for good conductors and the relatively large value for insulators. As a matter of fact, it is the relaxation time itself which truly measures what we choose to call a conductor or an insulator. When τ is extremely short compared with a measurement time, the material is considered as a "conductor"; however, if τ is very long compared with the duration of a measurement, we consider the material to behave like an "insulator." Note that our prior assumption of zero charge and field within a metallic conductor is amply justified by the numerical results in Table 5.1.

TABLE 5.1. RELAXATION TIMES FOR SOME COMMON MATERIALS

Material	*Relaxation time* τ
Copper	1.5×10^{-19} sec
Silver	1.3×10^{-19} sec
Sea water	2×10^{-10} sec
Distilled water	10^{-6} sec
Fused quartz	10 days

5.5. Resistance of Arbitrary Shaped Conductors

Equation (5.15) gives the total resistance of a uniform cylindrical conductor. For a homogeneous body of conductivity σ, but of an arbitrary shape, a more general formula is required. In order to derive it, we start with the conductor illustrated in Fig. 5.3, which is representative of a generalized shape.

If we visualize a battery connected to the ends of the body, then a current will flow and its density will be nonuniform. For simplicity we take the end surfaces A_1 and A_2 to be equipotentials; this could be assured, for example, by coating these surfaces with a perfect conductor.† Since,

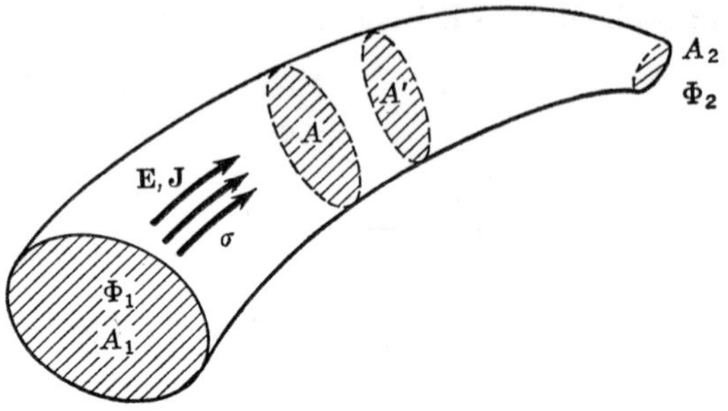

FIG. 5.3. An arbitrary conductor.

as we have already noted, the field external to the battery is conservative and can be derived from a scalar potential,

$$\int_C \mathbf{E} \cdot d\mathbf{l} = \Phi_1 - \Phi_2 \tag{5.27}$$

where C is any path starting at A_1 and terminating at A_2 and $\Phi_1 - \Phi_2$ is the difference of potential between the surfaces A_1 and A_2. If the battery and lead resistances are negligible, then it is also true that

$$\int_C \mathbf{E} \cdot d\mathbf{l} = \mathcal{E} \tag{5.28}$$

where $\mathcal{E}$ is the emf of the battery.

Consider any cross-sectional surface in the conductor such as A or A' in Fig. 5.3. Since the current is solenoidal, the same total current crosses surface A_1, A, A', and A_2. We can evaluate this current over any surface A as given by

$$I = \int_A \mathbf{J} \cdot d\mathbf{S} \tag{5.29}$$

Now $\mathbf{J} = \sigma\mathbf{E}$; so

$$I = \sigma \int_A \mathbf{E} \cdot d\mathbf{S} \tag{5.30}$$

By definition the resistance between the two faces A_1 and A_2 is

$$R = \frac{\Phi_1 - \Phi_2}{I} = \frac{\int_C \mathbf{E} \cdot d\mathbf{l}}{\sigma \int_A \mathbf{E} \cdot d\mathbf{S}} \tag{5.31}$$

† Coating the ends with a material whose conductivity was very much greater than that of the body would suffice.

Although the above formula is rather simple in concept and form, the integrals cannot be evaluated before a detailed solution for the field **E** (or current flow density **J**) has been obtained. For a general shaped conductor this is usually not feasible and one is forced to resort to approximate methods of analysis or experimental methods in order to determine the resistance R.

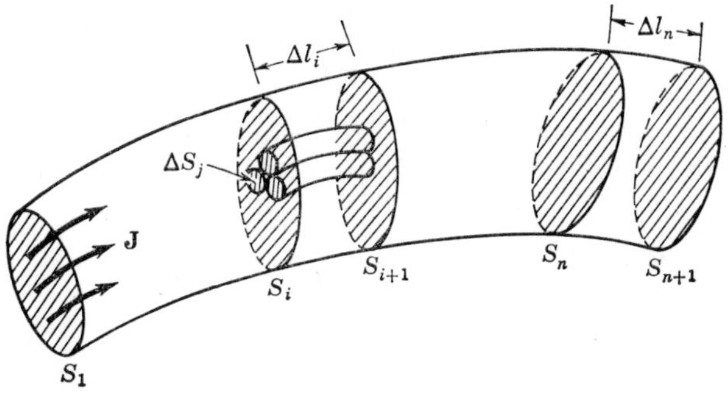

FIG. 5.4. Two equipotential surfaces within an arbitrary current-carrying conductor.

Another expression for R can be formulated that demonstrates the geometrical properties of the resistance more clearly. Again, it is necessary to know the field and current distribution everywhere within the conductor. In Fig. 5.4, a general conductor is illustrated and two equipotential cross-sectional surfaces S_i and S_{i+1} are indicated. Let the potential difference between these surfaces be designated by $\Delta\Phi_i$. The volume between S_i and S_{i+1} may be decomposed into a number of elementary flow tubes of length Δl_i and cross-sectional area ΔS_j. For each small tube the resistance r_j is given by

$$r_j = \frac{\Delta\Phi_i}{\Delta I_j} = \frac{E\,\Delta l_i}{\sigma E\,\Delta S_j} = \frac{\Delta l_i}{\sigma\,\Delta S_j} \tag{5.32}$$

The conductance of this flow tube is

$$g_j = r_j^{-1} = \frac{\sigma\,\Delta S_j}{\Delta l_i} \tag{5.33}$$

Since conductances in parallel add directly, the total conductance between the surfaces S_i and S_{i+1} is

$$\Delta G_i = \sum_j g_j = \sum_j \frac{\sigma\,\Delta S_j}{\Delta l_i} \tag{5.34a}$$

and the corresponding resistance is

$$\Delta R_i = \frac{1}{\displaystyle\sum_j \frac{\sigma\,\Delta S_j}{\Delta l_i}} \tag{5.34b}$$

In general, Δl_i will vary over the cross section since the spacing between the equipotential surfaces is not necessarily uniform. If we break up the whole conductor into n such sections, then (5.34*b*) is the resistance of the ith section. The total resistance is the series combination of all the ΔR_i and is given by

$$R = \sum_i \frac{1}{\sum_j \frac{\sigma \, \Delta S_j}{\Delta l_i}} \tag{5.35}$$

From (5.35) we can see how the resistance formula may be obtained by passing to the limit ΔS_j and Δl_i approaching zero. In order to obtain

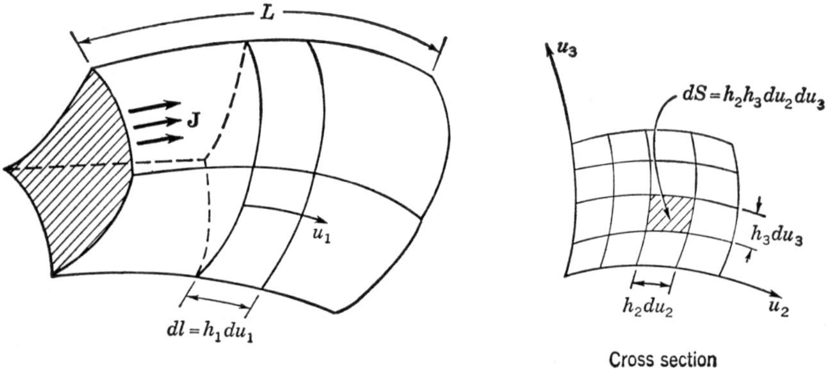

FIG. 5.5. Orthogonal curvilinear coordinates used to describe current flow in a conductor.

a meaningful formula, we have to digress for a moment and introduce a suitable set of curvilinear coordinates to express the variables in. Since $\sigma \mathbf{E} = \mathbf{J} = -\sigma \nabla \Phi$ and $\nabla \Phi$ is normal to the constant potential surfaces, we may introduce a curvilinear coordinate u_1 which increases in the direction parallel to the current flow lines and is normal to the constant potential surfaces. Distance dl along the flow lines is then given by $h_1 \, du_1$, where h_1 is a scale factor and will, in general, vary over the cross section of the conductor. Over the constant potential surface we shall assume that two additional orthogonal curvilinear coordinates u_2, u_3 can be introduced in order to measure the cross-sectional area of an elementary flow tube.† The cross-sectional area ΔS_j is now given by $\Delta S_j = h_2 h_3 \, \Delta u_2 \, \Delta u_3$, as in Fig. 5.5. The factors h_2 and h_3 are scale factors introduced so that $h_2 \, du_2$ and $h_3 \, du_3$ are differential lengths in the direction of increasing

† If u_2 and u_3 do not form an orthogonal system, the problem cannot, in general, be solved analytically anyway, so that the restriction is not a serious one from a practical standpoint.

u_2 and u_3, respectively. For (5.35) we may now write

$$R = \sum_i \frac{1}{\sum_j \frac{\sigma h_2 h_3 (\Delta u_2\, \Delta u_3)_j}{h_1 (\Delta u_1)_i}}$$
$$= \sum_i \frac{(\Delta u_1)_i}{\sum_j \frac{\sigma h_2 h_3}{h_1} (\Delta u_2\, \Delta u_3)_j}$$

since u_1, u_2, u_3 are independent variables because of their mutual orthogonality. In the limit we obtain

$$R = \int_0^L \frac{du_1}{\int_S \frac{\sigma h_2 h_3}{h_1}\, du_2\, du_3} \tag{5.36}$$

This equation is clearly a function of the geometry of the conductor only. In a later section we present a flux-mapping technique which is essentially a graphical procedure for evaluating the above expression for resistance. The following example will also help to clarify some of the concepts involved.

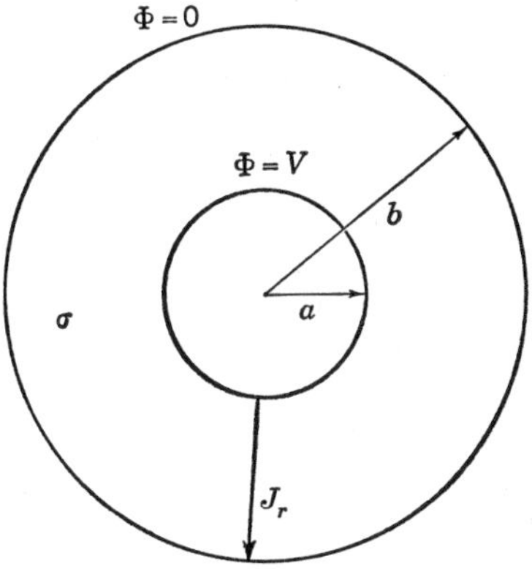

FIG. 5.6. A spherical resistor.

Example 5.1. Resistance of a Spherical Section. Consider two concentric spheres of radii a and b, as in Fig. 5.6. Let the inner sphere be kept at a potential V relative to the outer sphere, and let the medium between the spheres have a conductivity σ. From the work of previous chapters it is clear that the potential Φ is given by

$$\Phi = \frac{ab}{b-a} V \left(\frac{1}{r} - \frac{1}{b}\right) \tag{5.37}$$

since this makes $\Phi = V$ at $r = a$, $\Phi = 0$ at $r = b$, and $\nabla^2\Phi = 0$. The radial electric field E_r is given by $-\partial\Phi/\partial r$, and hence the current density is

$$J_r = \sigma E_r = \frac{ab}{b-a} V\sigma \frac{1}{r^2} \tag{5.38}$$

The total current is equal to $4\pi a^2$ times the current density at $r = a$ and is given by

$$I = \frac{4\pi ab}{b-a} \sigma V \tag{5.39}$$

The total resistance between the two shells is

$$R = \frac{V}{I} = \frac{b - a}{4\pi ab\sigma} \tag{5.40}$$

The above solution is a direct application of (5.31).

Let us now consider just a portion of this spherical resistor as obtained by lifting out a section contained within a cone of semiangle θ_0, as in Fig. 5.7*a*. The end surfaces $r = a, b$ are kept at a potential V and 0 as

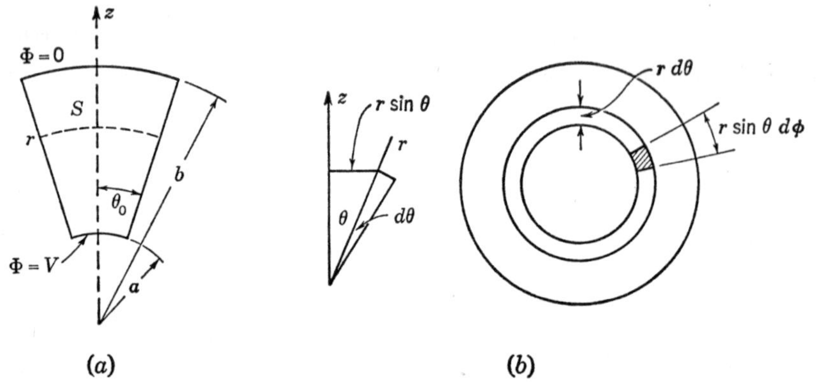

FIG. 5.7. (*a*) A conical resistor; (*b*) orthogonal curvilinear coordinates θ, ϕ on an equipotential surface.

before. Consequently, all surfaces r = constant are equipotential surfaces. On an equipotential surface the element of area dS may be described in terms of the spherical coordinates θ and ϕ, as in Fig. 5.7*b*. The separation between equipotential surfaces is simply dr. Our curvilinear coordinates u_1, u_2, u_3 and scale factors h_1, h_2, h_3 in the present case are

$$u_1 = r,\ h_1 = 1 \qquad u_2 = \theta,\ h_2 = r \qquad u_3 = \phi,\ h_3 = r \sin\theta$$

From (5.36) we obtain the following expression for the resistance:

$$R = \int_a^b \frac{dr}{\int_0^{2\pi}\int_0^{\theta_0} \sigma r^2 \sin\theta \, d\theta \, d\phi} = \int_a^b \frac{dr}{\sigma r^2 2\pi(1 - \cos\theta_0)}$$

$$= \frac{1}{2\pi\sigma(1 - \cos\theta_0)} \frac{b - a}{ab} \tag{5.41}$$

If $\theta_0 = \pi$, this result reduces to (5.40), as it should.

The result expressed by (5.41) could have been arrived at in another way also. Let the solid angle subtended by the end surface of the conical resistor be Ω. In the volume between the two spheres $4\pi/\Omega$, such resistors may be placed in parallel. The resistance of any one individual resistor is equal to $4\pi/\Omega$ times the combined total resistance, i.e., equal to $4\pi/\Omega$ times (5.40). For a surface such as that in Fig. 5.7*a*, the solid angle Ω is given by $2\pi(1 - \cos\theta_0)$, and hence (5.41) follows at once. For example, if $\theta_0 = \pi/2$, the resistor of Fig. 5.6 consists of two of the resistors of Fig. 5.7*a* in parallel. Therefore the resistance of a half-spherical section is twice the value given by (5.40). This result is verified at once from (5.41) by placing θ_0 equal to $\pi/2$.

In addition to the two formulas (5.31) and (5.36), the resistance of a circuit may be defined on an energy basis. Only the results are presented here; the derivation is given in Sec. 5.8. For the conductor as a whole, the power dissipated is given by I^2R. The power dissipation is given by one of the following volume integrals also, so that

$$I^2R = \int_V \mathbf{J} \cdot \mathbf{E}\, dV = \int_V \sigma\mathbf{E} \cdot \mathbf{E}\, dV = \frac{1}{\sigma}\int_V \mathbf{J} \cdot \mathbf{J}\, dV \tag{5.42}$$

Once the field has been found, the volume integral may be evaluated and R is then determined by the above equation. This latter method is often the easiest one to formulate.

5.6. Boundary Conditions and Refraction of Current Flow Lines

An examination of the flow of current across an interface between two media of different conductivity, σ_1 and σ_2, reveals that the flow lines are refracted. What happens is analogous to what has already been noted in electrostatics with respect to electric flux lines at a dielectric interface.

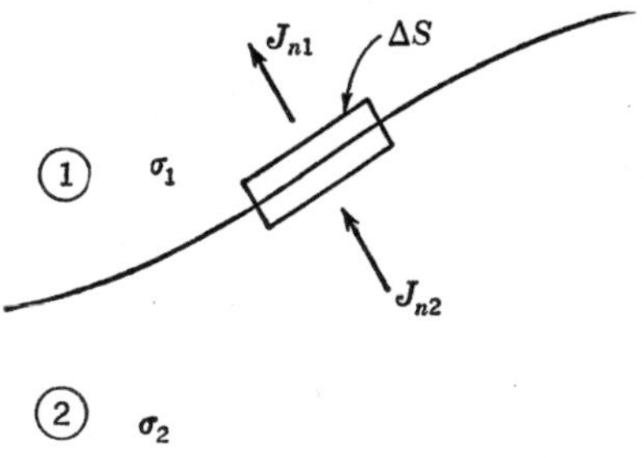

FIG. 5.8. Boundary between two conducting media.

Figure 5.8 shows the interface of medium 1 (conductivity σ_1) and medium 2 (conductivity σ_2). If a coin-shaped surface is considered which has a broad face in medium 1 and a broad face in medium 2, then over the volume V occupied by this surface

$$\int_V (\nabla \cdot \mathbf{J})\, dV = \oint_S \mathbf{J} \cdot d\mathbf{S} = 0 \tag{5.43}$$

Let the surface area of the coin faces be ΔS, and let the thickness of the coin approach zero. Then no contribution to the surface integral in (5.43) comes from the edges. The remainder of the integral can be written

$$(J_{n1} - J_{n2})\,\Delta S = 0$$

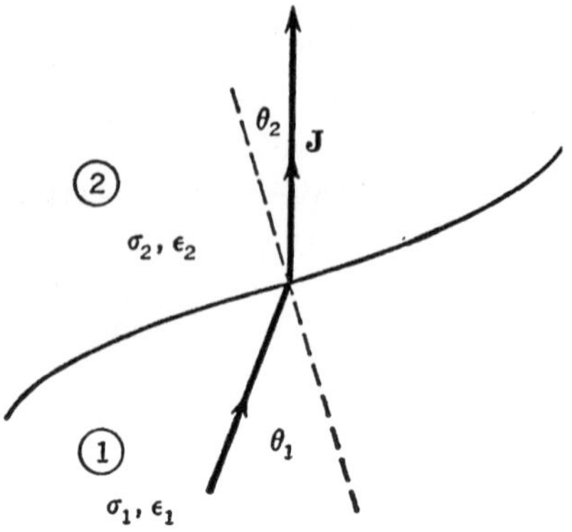

FIG. 5.9. Refraction of current flow lines.

The notation J_{n1} refers to the normal component of **J** at the interface in region 1, and similarly for J_{n2} in region 2. The element of surface ΔS is arbitrary; so we have

$$J_{n1} = J_{n2} \tag{5.44}$$

Since the electric field is conservative throughout the region in question (we assume this region to be outside the nonconservative source), it follows that $\nabla \times \mathbf{E} = 0$. As a consequence, the development in Sec. 3.3, which leads to continuity of tangential **E**, applies in this case also. This result can be stated in terms of the tangential current density as

$$\frac{J_{t1}}{\sigma_1} = \frac{J_{t2}}{\sigma_2} \tag{5.45}$$

By combining (5.44) and (5.45) and noting the definition of θ_1 and θ_2, as illustrated in Fig. 5.9, we may write

$$\tan\theta_1 = \frac{J_{t1}}{J_{n1}} \qquad \tan\theta_2 = \frac{J_{t2}}{J_{n2}} = \frac{\sigma_2 J_{t1}}{\sigma_1 J_{t2}}$$

Therefore

$$\tan\theta_2 = \frac{\sigma_2}{\sigma_1}\tan\theta_1 \tag{5.46}$$

If region 1 is a good conductor and region 2 an insulator, then $\sigma_1 \gg \sigma_2$, and the current leaves the surface in medium 2 at right angles. This corresponds to the requirement that the electric field be normal to the surface of a good conductor.

At the interface of lossy dielectrics, the above boundary condition which holds for the normal component of the current density is, in general, incompatible with the boundary conditions on the normal component of the displacement flux density **D**, for a dielectric material with finite conductivity, unless a layer of surface charge is assumed to exist on the boundary separating the two media. With reference to Fig. 5.9, let the permittivity of the two media be ϵ_1 and ϵ_2. From (5.44) we have $J_{n1} = \sigma_1 E_{n1} = J_{n2} = \sigma_2 E_{n2}$, or

$$\sigma_1 E_{n1} = \sigma_2 E_{n2} \tag{5.47}$$

If a surface layer of charge of density ρ_s exists on the boundary, the boundary condition on D_n gives

$$D_{n2} - D_{n1} = \rho_s$$

or

$$\epsilon_2 E_{n2} - \epsilon_1 E_{n1} = \rho_s \tag{5.48}$$

Only if $\epsilon_2/\epsilon_1 = \sigma_2/\sigma_1$ will ρ_s vanish. Combining (5.47) and (5.48) gives

$$\rho_s = \left(\epsilon_2 - \epsilon_1 \frac{\sigma_2}{\sigma_1}\right) E_{n2} = \left(\epsilon_2 \frac{\sigma_1}{\sigma_2} - \epsilon_1\right) E_{n1} \tag{5.49}$$

During the transient state while the current is building up to its final steady-state value, charge accumulates on the boundary. Once steady-state conditions have been reached, no further accumulation of charge takes place. If we introduce the relaxation time constants $\tau_1 = \epsilon_1/\sigma_1$, $\tau_2 = \epsilon_2/\sigma_2$ and replace σE_n by J_n, we have in place of (5.49)

$$\rho_s = (\tau_2 - \tau_1) J_n \tag{5.50}$$

Some of the practical implications of the above results are presented in the following example.

Example 5.2. Capacitor Filled with Lossy Dielectric Material. For simplicity consider a parallel-plate capacitor with spacing $2d$ and plate area A, as in Fig. 5.10. The region between the plates is filled with two

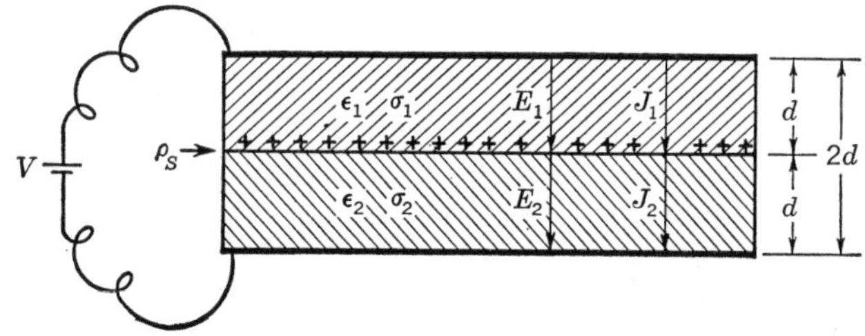

FIG. 5.10. Capacitor filled with lossy dielectric slabs.

lossy dielectric slabs of thickness d and with parameters ϵ_1, σ_1 and ϵ_2, σ_2. A potential V is applied across the plates. When steady-state conditions have been reached, the electric field between the plates must satisfy the following conditions:

$$E_1 d + E_2 d = V \tag{5.51}$$

$$J_1 = \sigma_1 E_1 = J_2 = \sigma_2 E_2 \tag{5.52}$$

$$\rho_s = \epsilon_2 E_2 - \epsilon_1 E_1 \tag{5.53}$$

and consequently

$$E_1 = \frac{\sigma_2}{\sigma_1 + \sigma_2} \frac{V}{d} \tag{5.54a}$$

$$E_2 = \frac{\sigma_1}{\sigma_1 + \sigma_2} \frac{V}{d} \tag{5.54b}$$

The surface charge density on the boundary separating the two dielectric media can now be found from (5.53) with the aid of (5.54) and is

$$\rho_s = \frac{\epsilon_2\sigma_1 - \epsilon_1\sigma_2}{\sigma_1 + \sigma_2}\frac{V}{d} \tag{5.55}$$

If now we turn our attention to the transient interval during which ρ_s increases to its steady-state value, then (5.52) no longer applies. This is because (5.52) is derived from the continuity equation under stationary conditions. For the time-varying case it is necessary to use (5.20). If this equation is applied to a coin-shaped surface centered at the interface using the, by now, familiar arguments, it is possible to establish the following general boundary conditions:

$$\underline{\frac{\partial \rho_s}{\partial t} = J_1 - J_2 = \sigma_1 E_1 - \sigma_2 E_2} \tag{5.56}$$

Note that if $\partial\rho_s/\partial t = 0$, then (5.56) reduces to (5.52). An expression for E_1 and E_2 in terms of ρ_s and V can be obtained from the simultaneous solution of (5.51) and (5.53). Substituting these values into (5.56) yields the following differential equation:

$$\frac{\partial \rho_s}{\partial t} = -\rho_s \frac{\sigma_1 + \sigma_2}{\epsilon_1 + \epsilon_2} + \frac{\epsilon_2\sigma_1 - \sigma_2\epsilon_1}{\epsilon_1 + \epsilon_2}\frac{V}{d} \tag{5.57}$$

The general solution to (5.57) is

$$\rho_s = A \exp\left(-\frac{\sigma_1 + \sigma_2}{\epsilon_1 + \epsilon_2}t\right) + \frac{\epsilon_2\sigma_1 - \sigma_2\epsilon_1}{\sigma_1 + \sigma_2}\frac{V}{d} \tag{5.58}$$

where A is an arbitrary constant. When $t \to \infty$, (5.58) correctly reduces to the steady-state value already found. The constant A is determined from the initial condition that $\rho_s = 0$ when $t = 0$. Consequently, we finally have

$$\rho_s = \frac{V}{d}\frac{\epsilon_2\sigma_1 - \sigma_2\epsilon_1}{\sigma_1 + \sigma_2}\left[1 - \exp\left(-\frac{\sigma_1 + \sigma_2}{\epsilon_1 + \epsilon_2}t\right)\right] \tag{5.59}$$

If the dielectric conductivity is very small, as is usual, then the time constant in (5.59) will be very large. Suppose that measurements are to be made which involve a lossy dielectric, under d-c conditions. If the duration of the experiment is short compared with the relaxation time $(\epsilon_1 + \epsilon_2)/(\sigma_1 + \sigma_2)$, then (5.59) ensures that the dielectric may be considered to be essentially perfect.

The problem discussed here can be analyzed by setting up an equivalent lumped-parameter circuit and proceeding to the analysis of this circuit by conventional techniques. While no new information can be expected by this procedure, it is of considerable interest since it relates field theory to circuit theory. Accordingly, we present and analyze a circuit structure in Fig. 5.11 which represents the lossy dielectric problem of Fig. 5.10. The conductances G_1, G_2 and capacitances C_1, C_2 are given by

$$R^{-1} = G_1 = \frac{\sigma_1 A}{d} \qquad R_2^{-1} = G_2 = \frac{\sigma_2 A}{d}$$

$$C_1 = \frac{\epsilon_1 A}{d} \qquad C_2 = \frac{\epsilon_2 A}{d}$$

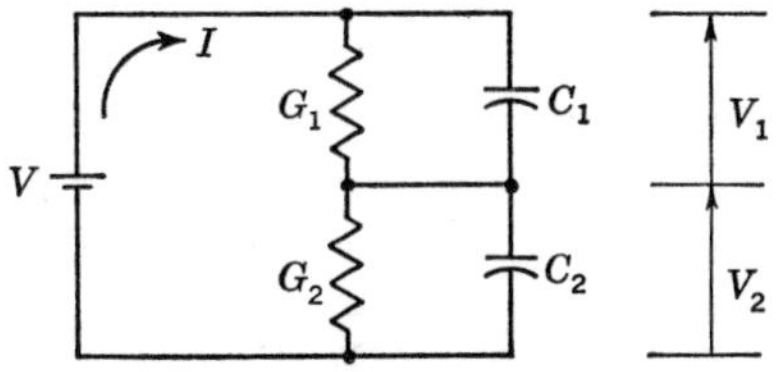

FIG. 5.11. Equivalent circuit of capacitor filled with lossy dielectric.

Under steady-state conditions it is clear that the division in voltage between the two halves of the capacitor is determined by the conductances only. From the equivalent circuit it is seen that for steady-state conditions

$$\frac{V_1}{V_2} = \frac{G_2}{G_1}$$

and
$$V_1 = \frac{G_2}{G_1 + G_2} V \qquad V_2 = \frac{G_1}{G_1 + G_2} V$$

But $E_1 d = V_1$, $E_2 d = V_2$, so that this is just the circuit equivalent of the field relations (5.54).

On the capacitor C_1 a total charge $Q_1 = C_1 V_1$ exists on the upper plate and a total charge $-Q_1$ on the lower plate. Similarly, on C_2 a total charge $Q_2 = C_2 V_2$ exists on the upper plate and a charge $-Q_2$ on the lower plate, under steady-state conditions. On the lower plate of C_1 and the upper plate of C_2, which together represent the boundary surface between the two dielectric slabs, the net total charge is $Q_2 - Q_1$ and is given by

$$Q_2 - Q_1 = C_2 V_2 - C_1 V_1 \tag{5.60}$$

This result is the same as that given by (5.53), since

$$A\rho_s = \frac{\epsilon_2 A}{d} E_2 d - \frac{\epsilon_1 A}{d} E_1 d = C_2 V_2 - C_1 V_1$$

Because of the finite conductivity, the structure of Fig. 5.10 does not behave as a pure capacitor; rather it is a parallel-series combination of resistance and capacitance, as illustrated in Fig. 5.11. The transient build-up of surface charge ρ_s may be found from a study of the transient behavior of the equivalent circuit. A transient analysis is readily carried out if it is assumed that the internal resistance of the battery is negligible (this assumption is really not valid, and its consequences will be pointed out later). Referring to the equivalent circuit, it is seen that the current flowing through the C_1, G_1 combination is

$$I = G_1(V - V_2) + C_1 \frac{d}{dt}(V - V_2)$$

The same current flows through the C_2, G_2 combination; so we have

$$G_1(V - V_2) - C_1 \frac{dV_2}{dt} = G_2 V_2 + C_2 \frac{dV_2}{dt}$$

since dV/dt is zero because V is constant. This equation may be written as

$$\tau \frac{dV_2}{dt} + V_2 = \frac{G_1}{G_1 + G_2} V \tag{5.61}$$

where
$$\tau = \frac{C_1 + C_2}{G_1 + G_2} = \frac{(C_1 + C_2)R_1R_2}{R_1 + R_2}$$

i.e., the product of the parallel combination of C_1, C_2 and R_1, R_2. The solution to (5.61) is

$$V_2 = \frac{G_1}{G_1 + G_2} V + Be^{-t/\tau} \tag{5.62}$$

where B is a constant to be determined. If we assume that the battery is connected at time $t = 0$, then at $t = 0$,

$$V_2 = \frac{C_1}{C_1 + C_2} V \qquad V_1 = V - V_2 = \frac{C_2}{C_1 + C_2} \tag{5.63}$$

However, this initial condition is not physically possible, since if we begin with zero charge on the capacitors, it implies an infinite current flow as soon as the battery is connected. If we actually had a battery with zero internal resistance, this could be accomplished since an uncharged capacitor behaves as a short circuit. An actual battery has finite internal resistance R_b, and the initial flow of current is finite. If, however, R_b is very small compared with R_1 and R_2, the current is initially limited only by R_b, C_1, and C_2, since the current flow into C_1 and C_2 will be much greater than the small amount of current flowing through R_1 and R_2. Thus the voltage across C_1 and C_2 builds up very rapidly. These remarks apply also to the field solution where the assumption that $\Phi = V$ when $t = 0$ implied a source with no internal losses. Using the idealized condition (5.63) in (5.62) gives

$$V_2 = \frac{G_1 V}{G_1 + G_2}\left(1 - \frac{\tau - \tau_1}{\tau} e^{-t/\tau}\right) \tag{5.64a}$$

$$V_1 = V - V_2 = \frac{G_2 V}{G_1 + G_2}\left(1 + \frac{G_1}{G_2}\frac{\tau - \tau_1}{\tau} e^{-t/\tau}\right) \tag{5.64b}$$

where $\tau_1 = C_1R_1$. As $t \to \infty$, these expressions clearly give the correct steady-state values of V_1 and V_2. From (5.60), the surface charge density ρ_s is found to be

$$\rho_s = \frac{Q_2 - Q_1}{A} = \frac{C_2}{A} V_2 - \frac{C_1}{A} V_1 = \frac{C_2G_1 - C_1G_2}{G_1 + G_2} \frac{V}{A} (1 - e^{-t/\tau})$$
$$= \frac{\epsilon_2\sigma_1 - \epsilon_1\sigma_2}{\sigma_1 + \sigma_2} \frac{V}{d} (1 - e^{-t/\tau}) \tag{5.65}$$

where $\tau = (\epsilon_2 + \epsilon_1)/(\sigma_1 + \sigma_2)$. This result checks with that found from the field point of view as given in (5.59).

As we have already noted, for time intervals that are short compared with τ, we can assume that ρ_s is negligible, and hence the boundary condition $D_{n2} = D_{n1}$ is a good approximation. This boundary condition could be assumed if the applied voltage were sinusoidal and the period much shorter than τ (high frequencies); that is, if the following inequality holds,

$$\frac{1}{2\pi f} = \frac{1}{\omega} \ll \frac{C_1 + C_2}{G_1 + G_2}$$

then the capacitor behaves essentially as if there were no losses. This inequality may be rewritten as

$$\frac{1}{\omega(C_1 + C_2)} \ll \frac{R_1R_2}{R_1 + R_2}$$

which from a circuit standpoint simply states that the resistances may be neglected if their parallel combination is much greater than the parallel capacitive reactance.

If we are interested in time intervals comparable with τ, that is, low frequencies, then the boundary condition $D_{n2} - D_{n1} = \rho_s$ must be used since the surface charge density will not be negligible. Since most dielectrics have a finite conductivity, these considerations are of importance and some care must be exercised in using the assumption of zero surface charge density, as is commonly done by many authors.

As a point of further interest, if the time constants are equal, that is, if $\tau = \tau_1$, then ρ_s is always equal to zero. An examination of (5.64) now shows that $V_1/V_2 = R_1/R_2$. Thus the circuit of Fig. 5.11 provides a frequency-independent voltage divider.

5.7. Duality between J and D

The current density **J** and displacement flux density **D** are both linearly related to the electric field **E** in many materials. A consequence of this property is the existence of dual relationships between **J** and **D**. In a region where nonconservative fields are absent, i.e., external to the battery, the following equations apply for linear, isotropic materials:

$$\begin{array}{cc} \underline{\textit{Conducting media}} & \underline{\textit{Dielectric media}} \\ \nabla \times \mathbf{E} = 0 & \nabla \times \mathbf{E} = 0 \\ \mathbf{J} = \sigma \mathbf{E} & \mathbf{D} = \epsilon \mathbf{E} \\ \nabla \cdot \mathbf{J} = 0 & \nabla \cdot \mathbf{D} = 0 \end{array} \tag{5.66}$$

In a homogeneous material where ϵ and σ are constant, we also have

$$\nabla \times \mathbf{J} = 0 \qquad \nabla \times \mathbf{D} = 0 \tag{5.67}$$

This latter property shows that both **J** and **D** may be derived from a scalar potential; hence

$$\mathbf{J} = -\nabla\Phi \qquad \mathbf{D} = -\nabla\Phi \tag{5.68}$$

and $\nabla^2\Phi = 0$ in both cases by virtue of the divergence relations given in (5.66). It should be noted that the divergence and curl of **J** and **D** can be zero only over part of space, or else **J** and **D** would vanish everywhere. In the present case we are limiting consideration to regions that are external to all sources, so that both the divergence and curl of **J** and **D** may be zero. The solution for **J** and **D** is uniquely determined by finding a scalar potential function Φ that satisfies Laplace's equation and any imposed boundary conditions.

An examination of the above equations shows that any solution for **J** can be transformed into a solution for **D**, and vice versa, by means of the following interchange of quantities:

$$\mathbf{J} \leftrightarrow \mathbf{D} \tag{5.69a}$$

$$\sigma \leftrightarrow \epsilon \tag{5.69b}$$

This means that if a solution to a boundary-value problem in electrostatics is known, it is also the solution to a corresponding problem in steady current flow. This procedure is valid only if the boundary conditions are equivalent in both cases. Where a boundary is an interface between different media, then

$$\begin{array}{cc} \textit{Conducting media} & \textit{Dielectric media} \\ J_{n1} = J_{n2} & D_{n1} = D_{n2} \\ \dfrac{J_{t1}}{\sigma_1} = \dfrac{J_{t2}}{\sigma_2} & \dfrac{D_{t1}}{\epsilon_1} = \dfrac{D_{t2}}{\epsilon_2} \end{array} \tag{5.70}$$

Note that (5.70) conforms to the duality relations expressed by (5.69), as we should expect.

Certain boundary-value problems involving steady current do not have a dual in electrostatics. This occurs when a region with $\sigma = 0$ is involved, i.e., a perfect insulator. Duality requires an electrostatic region with a relative dielectric constant of zero; however, it is not possible to achieve $\kappa < 1$. Thus consider, for example, the steady flow between parallel plates as illustrated in Fig. 5.12*a*. Since $\sigma_2 = 0$, the flow lines will be

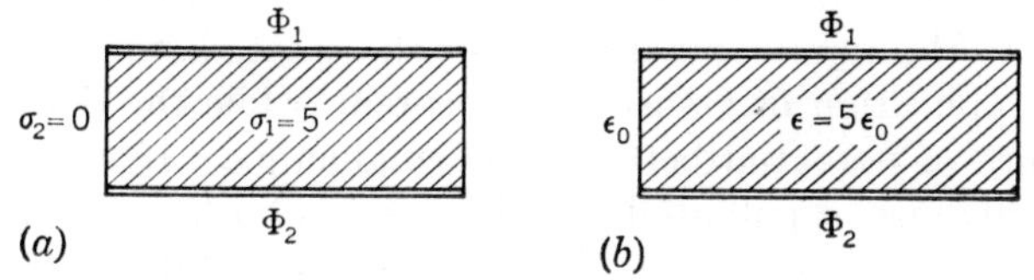

FIG. 5.12. Equivalent conductance and capacitance problems.

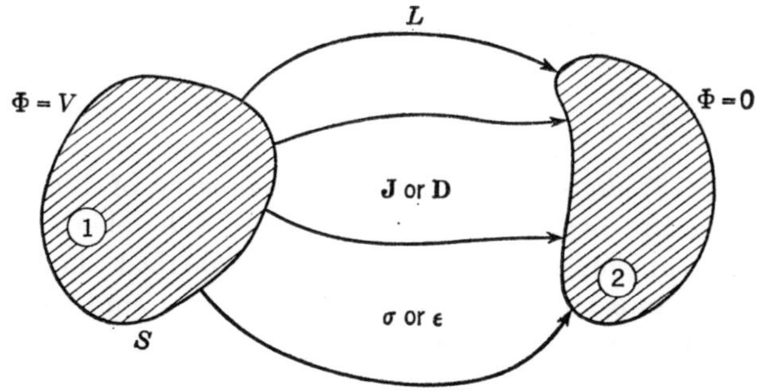

FIG. 5.13. Duality between conductance and capacitance for two arbitrary bodies.

uniform and directed normal to the parallel plates through the conducting medium of conductivity σ_1. In Fig. 5.12*b* a dielectric is placed between the parallel conducting plates. This is not quite the dual of Fig. 5.12*a* because it is impossible to provide a zero permittivity region surrounding the capacitor. As a consequence, fringing of the **D** lines occurs, such as did not happen in the case of current flow.

In Fig. 5.13 we show two arbitrarily shaped conducting bodies in a uniform, infinite medium. If the medium is a conductor, then the total resistance between the bodies is

$$R = \frac{\int_L \mathbf{E} \cdot d\mathbf{l}}{\oint_S \sigma \mathbf{E} \cdot d\mathbf{S}} \tag{5.71}$$

where L is any path from one body to the other and S is any surface enclosing either body. The numerator of (5.71) is the difference of potential between the bodies; the denominator evaluates the total current that flows between them.

If the conducting bodies are immersed in a dielectric, with a permittivity ϵ, then the capacitance C is given by

$$C = \frac{\int_S \epsilon \mathbf{E} \cdot d\mathbf{S}}{\int_L \mathbf{E} \cdot d\mathbf{l}} \tag{5.72}$$

where S and L are taken as before. That this evaluates the capacitance is clear, since the numerator determines the total charge on either conductor by application of Gauss' flux theorem, while the denominator is the difference of potential between the conductors. If we assume that the same difference of potential is maintained in both cases, then in view of the uniqueness theorem, $\mathbf{E}$ is the same in (5.71) and (5.72).

In comparing (5.71) and (5.72), one notes a dual relationship between $1/R = G$ and C. Provided that the conducting boundaries are identical and that everywhere σ is replaced by ϵ, then

$$RC = \frac{\epsilon}{\sigma} = \frac{C}{G} \tag{5.73}$$

Equation (5.73) does not hold for the geometry of Fig. 5.12 unless fringing effects are negligible. In the case of the spherical resistor in Fig. 5.6, the capacitance between the two spherical shells, when the intervening medium has a permittivity ϵ, may be found from (5.40) by using the relation (5.73). The result is

$$C = \frac{4\pi ab\epsilon}{b - a} \tag{5.74}$$

5.8. Joule's Law

The energy required to maintain a steady flow of current through an arbitrary conducting body of total resistance R (e.g., Fig. 5.3) can be found from basic principles. Let the total difference of potential across the body be V. Then the work done on a charge Q moving through this potential difference is $W = QV$. The rate at which work is expended by the field is then

$$\frac{dW}{dt} = P = V\frac{dQ}{dt} = VI \tag{5.75}$$

Or, since $V = IR$,

$$P = I^2R \tag{5.76}$$

This is known as Joule's law. As already noted, the work done by the field in moving the electrons through the conductor is in turn dissipated as heat as a consequence of electron-lattice interaction.

A formulation of Joule's law in differential form will also be useful.

Consider the differential volume element described in Fig. 5.14. The axial extent of the element is dl, and this is taken in the direction of current flow. If the electric field is $\mathbf{E}$, then the difference of potential across the ends of the element is

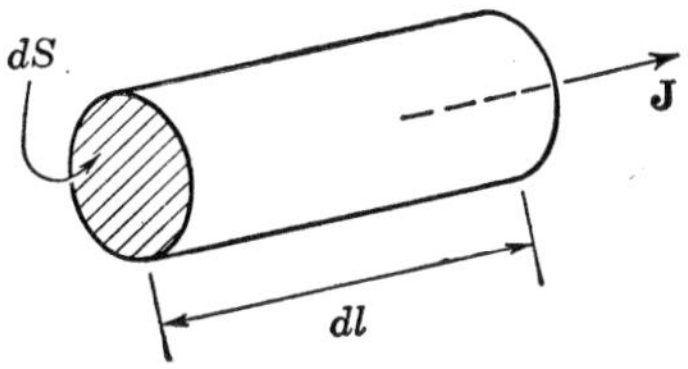

FIG. 5.14. A differential element of a resistor.

$$d\Phi = \mathbf{E} \cdot d\mathbf{l} = \frac{\mathbf{E} \cdot \mathbf{J}}{J} dl$$

The total current through the element, dI, equals $J\,dS$. According to (5.75), the total power dissipated in this element is

$$dP = dI\,d\Phi = \mathbf{E} \cdot \mathbf{J}\,dl\,dS$$

Consequently, the differential form of Joule's law is

$$\frac{dP}{dV} = U_J = \mathbf{E} \cdot \mathbf{J} \tag{5.77}$$

where dV is an element of volume $dS\,dl$. The electric field $\mathbf{E}$ therefore gives up $\mathbf{E} \cdot \mathbf{J}$ watts of power per unit volume to the steady electron flow $\mathbf{J}$. We have already noted that, since the current flows in a conductor, this power is converted into heat.

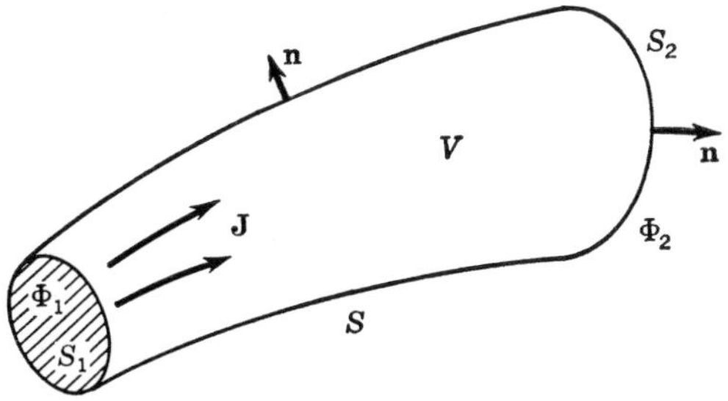

FIG. 5.15. An arbitrary resistor.

We are now in a position to derive the energy formula (5.42) for resistance. Figure 5.15 illustrates a conductor with end surfaces S_1, S_2 kept at constant potentials Φ_1 and Φ_2. The current density J flows normal to these end surfaces and parallel to the sides of the conductor. From (5.76) and (5.77) the total power (energy per second) dissipated in the resistor is

$$P = I^2R = \int_V \mathbf{E} \cdot \mathbf{J}\,dV = \int_V \frac{\mathbf{J} \cdot \mathbf{J}}{\sigma}\,dV = \int_V \sigma\mathbf{E} \cdot \mathbf{E}\,dV$$

and hence

$$R = \frac{\int_V \mathbf{E} \cdot \mathbf{J}\,dV}{I^2} \tag{5.78}$$

We may readily show that this definition of total resistance is identical with that given by (5.31). We note that $\mathbf{E} = -\nabla\Phi$ and

$$-(\nabla\Phi) \cdot \mathbf{J} = -\nabla \cdot (\Phi\mathbf{J})$$

since $\nabla \cdot \mathbf{J} = 0$. The volume integral in (5.78) becomes, upon application of the divergence theorem,

$$-\int_V \nabla \cdot (\Phi\mathbf{J})\, dV = -\oint_{S+S_1+S_2} \Phi\mathbf{J} \cdot \mathbf{n}\, dS$$

where $S + S_1 + S_2$ is the total surface of the resistor and $\mathbf{n}$ is the outward normal. Since $\mathbf{J} \cdot \mathbf{n} = 0$ except on the constant potential end surfaces, we obtain

$$I^2R = \Phi_1 \int_{S_1} \mathbf{J} \cdot (-\mathbf{n})\, dS - \Phi_2 \int_{S_2} \mathbf{J} \cdot \mathbf{n}\, dS = (\Phi_1 - \Phi_2)I$$

and hence $R = (\Phi_1 - \Phi_2)/I$, which is (5.31).

5.9. Convection Current

The flow of current in conductors, under the action of an electric field, that has been considered so far is called *conduction current.* In contrast, if an insulator carries charges and the entire body is in motion, an equivalent current also flows. An important special case is that of the movement of the charges themselves in a vacuum, under the influence, perhaps, of an electric field. In these cases the relation between current and electric field is no longer described by Ohm's law. The current developed by moving media depends on the mechanics of the moving particles; the motion of charges in an evacuated region containing an electric field can be described by the laws of mechanics with the inclusion of forces of electric origin. The term *convection current* is applied to describe this latter type of current.

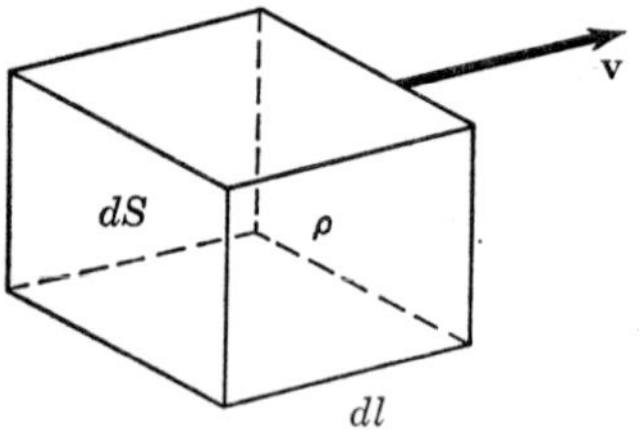

FIG. 5.16. A volume element of convection current.

Given a flow of convection current, then the electron stream is specified by a charge density ρ and a velocity $\mathbf{v}$ at any point. The current density at each point can be found by multiplying the charge density by the velocity; that is, if we consider a volume element with length dl in the direction of flow (see Fig. 5.16), then in a time interval $dt = dl/v$, all the charge ($\rho\, dS\, dl$) flows out of the region. The total current, by definition, is

$$dI = \frac{\rho\, dS\, dl}{dt} = \rho v\, dS$$

Hence the convection current density is given by

$$\mathbf{J} = \rho\mathbf{v} \tag{5.79}$$

The vector notation in (5.79) follows by inspection since **J** and **v** must obviously be in the same direction.

For convection currents in the presence of an electrostatic field, energy will be interchanged between the two. The development leading to (5.77) can be repeated for a conservative **E** field acting on a convection current, and the result is

$$U_J = \mathbf{E} \cdot \mathbf{J} = \rho\mathbf{E} \cdot \mathbf{v} \tag{5.80}$$

In this case the energy absorbed by the electron stream from the field, as given by (5.80), is not converted into heat, but into an increase in the kinetic energy of the particles. A further discussion of the interaction of charges and fields will be found in Chap. 12.

5.10. Flux Plotting

The solution of Laplace's equation forms the foundation not only for problems in electrostatics, but also for problems involving current flow fields. This is made clear by (5.68), which states that **J** can be derived from the gradient of a scalar potential Φ, where Φ is a solution to Laplace's equation. A full description of the **J** or **D** flow fields is consequently specified by the related function Φ. This may be of interest in itself or form the basis for further calculations such as for total resistance or capacitance.

The mathematical techniques for finding solutions of Laplace's equation are rather severely limited to certain geometry, e.g., boundaries that are spherical, circular, cylindrical, etc. For more arbitrary shapes other methods are required. This section is devoted to an explanation of an approximate graphical procedure known as the method of curvilinear squares, or simply flux plotting. In the next section the use of the electrolytic tank will be described. The technique of flux plotting is usually limited to two-dimensional problems.

By a two-dimensional problem, we refer to those cases where flow† is independent of one dimension; it can therefore be completely described in terms of the remaining two dimensions. Figure 5.17 shows the cross section of a solid, with surface S_1 at an equipotential Φ_1 and surface S_2 at an equipotential Φ_2. Flow takes place through the body, but in view of the axial uniformity, the direction of flow is confined to cross-sectional

† We have already noted the duality of **J** and **D** and that both may be derived from a scalar potential. When we refer to flow functions here, we do so in a general way; we refer to either **J** or **D**, or for that matter any vector function that can be derived as the gradient of a scalar potential.

planes. The uniformity further requires that the flux and potential distributions be the same in any transverse plane. For these reasons we can concentrate attention on finding a solution to the potential problem in such a typical plane.

The general technique involves making a guess as to the location of equipotential lines and flux lines. For reasons that we shall discuss in a

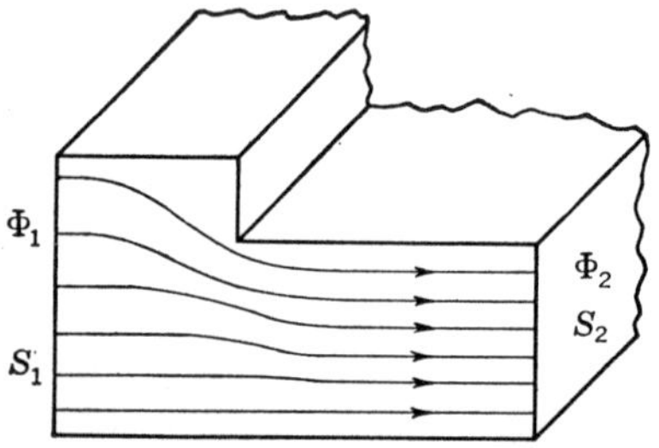

FIG. 5.17. A two-dimensional flow problem.

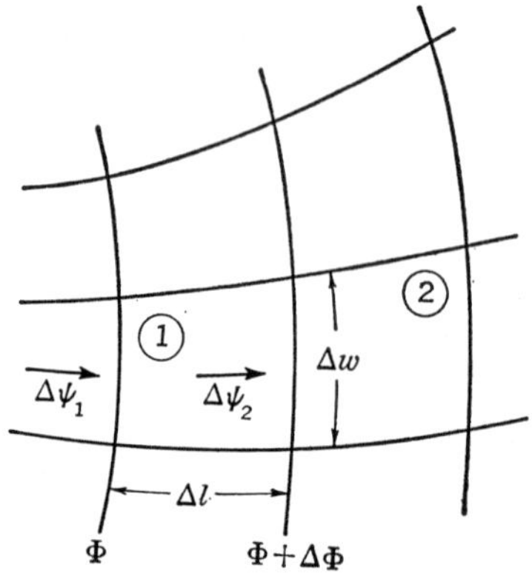

FIG. 5.18. A portion of a flux plot.

moment, equipotential lines are spaced at equal increments, with the magnitude of the spacing dependent on the coarseness or fineness of the desired plot. The flow lines will intersect these equipotentials orthogonally since the lines of flow are derived from the gradient of the potential. For example,

$$\mathbf{J} = -\sigma \nabla\Phi \tag{5.81}$$

We shall show that the flow lines should be spaced so that they form, as nearly as possible, curvilinear squares. Since certain boundary conditions are known, they serve to specify certain potential and flow lines at the outset. For example, in Fig. 5.17 the right- and left-hand edges must be equipotentials. Furthermore, if this represents a current flow problem with zero conductivity outside the body, then flow lines must be tangent to the remaining boundaries. Working from this point and by trial and error, the region can finally be covered with curvilinear squares. We proceed now to a justification of this method and an interpretation of the results.

For this purpose, consider a portion of a flux plot as just described. With respect to cell 1, in Fig. 5.18, since the width is Δw and the length in the direction of increasing potential is Δl, the field is approximately given by

$$E \approx -\frac{\Delta\Phi}{\Delta l} \tag{5.82}$$

From (5.81) the current density can be evaluated and is

$$J \approx -\sigma \frac{\Delta\Phi}{\Delta l} \tag{5.83}$$

The approximation improves as the size of the cell is made smaller. The total current in the tube of width Δw per unit length normal to the flux plot is then

$$\Delta I = \Delta\Psi_1 = -\sigma\,\Delta\Phi \frac{\Delta w}{\Delta l} \tag{5.84}$$

If the body were a dielectric and electric flux were being considered, then

$$\Delta\Psi_1 = -\epsilon\,\Delta\Phi \frac{\Delta w}{\Delta l} \tag{5.85}$$

In the above, Ψ represents total flow, either of current or of electric flux. For the latter case, since $\Delta\Psi_1 = D\,\Delta S$, the dimensions of Ψ are charge.

If the above procedure is followed at cell 2, then

$$\Delta\Psi_2 = -\sigma\,\Delta\Phi_2 \frac{\Delta w_2}{\Delta l_2} \tag{5.86}$$

Since the flux (either **J** or **D**) is solenoidal, $\Delta\Psi_2$ must equal $\Delta\Psi_1$ in order that the total flux within a tube be conserved. This can be accomplished by taking equal potential increments (itself a desirable procedure which simplifies the layout of the equipotential lines) and by making the aspect ratio $\Delta w/\Delta l$ a constant throughout the tube.

To facilitate interpretation of the entire flux plot, it is desirable that adjacent flux tubes represent equal quantities of flux. In this way the density of flow is graphically revealed by the spacing of flow lines. This will be accomplished by using the same potential increment $\Delta\Phi$ and aspect ratio $\Delta w/\Delta l$ everywhere throughout the plot. And since the eye can most readily gauge a square shape rather than a particular rectangular one, we choose $\Delta w/\Delta l$ as unity. This is the basis for the method of curvilinear squares just described. For a conducting medium, then,

$$\Delta\Psi = \Delta I = -\sigma\,\Delta\Phi \tag{5.87}$$

and in a dielectric medium

$$\Delta\Psi = -\epsilon\,\Delta\Phi \tag{5.88}$$

One objective in mapping a region may be to calculate the capacitance or resistance of a particular body. Let us see how this is done. Consider Fig. 5.19, which shows the plot of potential and flow fields between two cylindrical conductors of quite arbitrary space. The flux through any flux tube is given by (5.88). Then, if N_F is the total number of flux tubes, the total flux Ψ is

$$\Psi = N_F\,\Delta\Psi = -N_F\epsilon\,\Delta\Phi \tag{5.89}$$

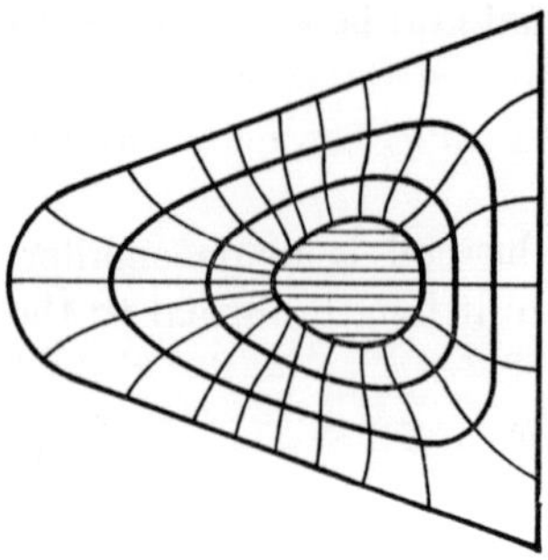

FIG. 5.19. Flux plot between two arbitrary conductors.

The total difference of potential, if N_P is the number of potential increments, is

$$V = N_P \, \Delta\Phi \tag{5.90}$$

The total charge per unit length on the conductors equals the total flux Ψ per unit length terminating thereon. Consequently, by definition, the capacitance C per unit length is

$$C = \frac{\Psi}{V} = \frac{\epsilon N_F}{N_P} \tag{5.91}$$

By duality, if the material between the conductors has a conductivity σ, the leakage conductance per unit length is

$$G = \frac{\sigma N_F}{N_P} \tag{5.92}$$

For the structure shown in Fig. 5.19, the number of potential divisions $N_P = 3$, while the number of flux divisions $N_F = 22$. Consequently, for an air dielectric,

$$C = \frac{1}{36\pi} \times 10^{-9} \times {}^{22}\!/_{3} = 65 \ \mu\mu\text{f/m}$$

A summary of the remarks concerning flux plotting, plus several additional suggestions on procedure, follows:

1. Examine the geometry to take advantage of any symmetry that may be present. For example, in Fig. 5.19, only the upper half need be plotted since mirror symmetry exists. If the cross section were elliptical, for example, only a quadrant would have to be considered.
2. Draw in the boundaries indicating those that are conducting and those that have zero conductivity.
3. Starting with known potentials and/or known flow lines, work out a rough sketch of the entire field, maintaining orthogonality between flow lines and equipotentials.
4. Refine the map to ensure that cells are curvilinear squares. If the rectangles remain very irregular, it may be desirable to cover the field with a finer net.
5. Several revisions may be necessary in order to achieve a satisfactory plot.

5.11. Electrolytic Tank

We have considered analytical solutions to Laplace's equation for conducting bodies in a homogeneous medium. We recall that solutions for potential and flow functions could be obtained only if the geometry were quite special. Under the restriction of two-dimensional variation, arbitrary shaped boundaries can be treated by means of the graphical method described in Sec. 5.10. A restricted class of such problems may also be handled analytically by means of conformal transformations, as was explained in Chap. 4.

This section is concerned with the determination of the potential field under completely arbitrary boundary conditions by means of an electrolytic analog, the electrolytic tank. Essentially, this involves setting up a model of the actual problem, using an electrolytic solution as the conducting medium and real electrodes of proper shape and positioning for the conducting boundaries. Potential and flow can now be measured using appropriate electrical instruments. Since the battery is external to the tank, the region in question contains a conservative electric field. As a consequence, a scalar potential that is a solution to Laplace's equation is, in fact, set up.

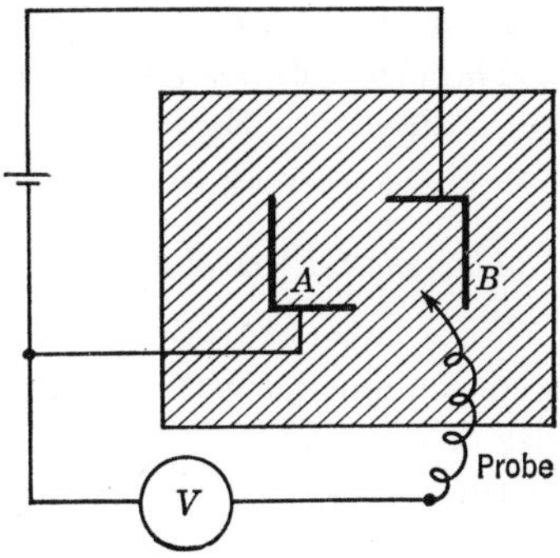

FIG. 5.20. The electrolytic tank.

Figure 5.20 illustrates a simple two-conductor problem where the shape of the field set up by electrodes A and B is desired; that is, A and B in Fig. 5.20 represent, to some scale, the actual electrodes in both shape and spacing. They are shown immersed in an electrolytic tank (shaded in the figure). A battery is connected across the electrodes, and this sets up a potential field within the electrolyte that is a solution of Laplace's equation and satisfies the appropriate boundary conditions on A and B.

The physical extent of the tank constitutes an additional boundary (the edge of the tank is characterized by requiring zero normal component of flux). Depending on the actual problem, this may represent only an approximation to actual conditions. For example, if the flow between the electrodes of Fig. 5.20 when immersed in a medium of infinite extent is desired, then the electrolytic-tank analog can be expected to be satisfactory only if the tank size is large compared with the over-all dimensions of the electrode system.

The potential field may be determined with the aid of a voltmeter. One lead is connected to an electrode (for example, A), and the free lead is used to probe the field in the electrolyte. The latter lead is insulated except for the tip. The locus of points for which the voltmeter reading is a constant establishes an equipotential surface. If the shapes of electrodes A and B do not vary in the direction normal to the page, then the potential will not vary with depth in the tank and a simpler two-dimensional problem exists. For such problems resistive coated paper may also be used in place of the electrolytic tank. In this case, electrodes can be painted on the paper with silver paint; the measurement procedure is essentially the same otherwise.

The electrolytic tank actually represents, to some scale, a current flow equivalent of the actual problem. If the original problem is, say, one in electrostatics, then both appropriate scale factors and duality conditions must be used to give the desired information. The actual problem may, itself, be a current flow problem, of course.

The electrolytic-tank technique serves to establish equipotential surfaces directly. By constructing a family of orthogonal trajectories, the flux paths may be found; i.e., these are the flux lines.

Double-sheet Electrolytic Tank

For a two-electrode problem current is confined to a finite region. As a consequence, electrodes can usually be scaled down in proportion to the size of the tank so that the medium appears infinite; that is, no serious disturbance of the current flow is caused by the limited extent of the tank.

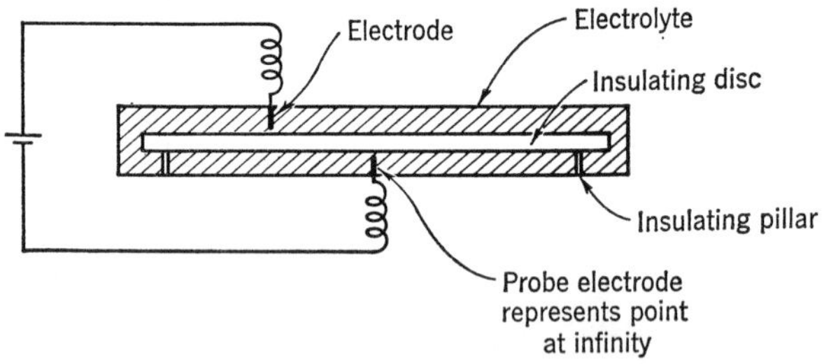

FIG. 5.21. A double-sheet electrolytic tank.

For those cases where a substantial current flow to infinity is involved, a double-sheet tank can be used. This tank, however, is applicable only to two-dimensional problems. The double-sheet tank is shown in Fig. 5.21. Let us consider the theory behind its construction, which will also serve to describe its operation.

Let us suppose the existence of an electrolytic tank of infinite extent. Electrodes could now be inserted in their proper geometric locations and a current flow set up. As before, a potential field is set up that satisfies Laplace's equation. For an arbitrary boundary at $r = R$, we could write the potential field for the region $r < R$ and for $r > R$ as follows:

$$\Phi = \Phi_1(r,\phi) \qquad r < R \tag{5.93a}$$

$$\Phi = \Phi_2(r,\phi) \qquad r > R \tag{5.93b}$$

A diagram of the coordinate system and of the location of these two regions is given in Fig. 5.22. Since Φ must be continuous, and because

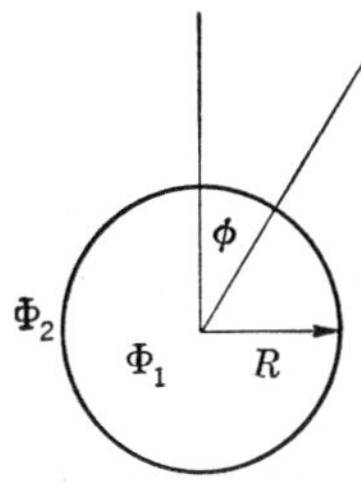

FIG. 5.22

$J_r = -\sigma\, \partial\Phi/\partial r$ is also continuous, it follows that

$$\Phi_1(R,\phi) = \Phi_2(R,\phi) \tag{5.94a}$$

$$\left.\frac{\partial\Phi_1}{\partial r}\right|_{r=R} = \left.\frac{\partial\Phi_2}{\partial r}\right|_{r=R} \tag{5.94b}$$

Assume now that all current sources lie in the region $r < R$, except possibly for a source or sink at infinity. Then

$$\nabla^2\Phi_2 = 0 \tag{5.95}$$

except possibly at infinity. Expanding (5.95) in cylindrical coordinates yields

$$\nabla^2\Phi_2 = \frac{1}{r}\frac{\partial}{\partial r}\left[r\,\frac{\partial\Phi_2(r,\phi)}{\partial r}\right] + \frac{1}{r^2}\frac{\partial^2\Phi_2}{\partial\phi^2} = 0$$

Let us introduce a new variable ρ such that

$$\rho = \frac{R^2}{r} \tag{5.96}$$

We desire to show that

$$\Psi(\rho,\phi) = \Phi_2\left(\frac{R^2}{\rho},\ \phi\right) \tag{5.97}$$

is a solution of Laplace's equation in the variable ρ; that is, we wish to verify that

$$\nabla^2\Psi = \frac{1}{\rho}\frac{\partial}{\partial\rho}\left(\rho\,\frac{\partial\Psi}{\partial\rho}\right) + \frac{1}{\rho^2}\frac{\partial^2\Psi}{\partial\phi^2} = 0 \tag{5.98}$$

From the relation between ρ and r we get

$$\frac{\partial}{\partial\rho} = \frac{\partial}{\partial r}\frac{\partial r}{\partial\rho} = -\frac{R^2}{\rho^2}\frac{\partial}{\partial r}$$

Replacing ρ in terms of r,

$$\frac{\partial}{\partial \rho} = -\frac{r^2}{R^2}\frac{\partial}{\partial r} \tag{5.99}$$

This expression in addition to (5.96) can now be used to convert (5.98) into a function of r. We get

$$\nabla^2\Psi(\rho,\phi) = \frac{r}{R^2}\left(-\frac{r^2}{R^2}\right)\frac{\partial}{\partial r}\left[\frac{R^2}{r}\left(-\frac{r^2}{R^2}\right)\frac{\partial}{\partial r}\Phi_2(r,\phi)\right] + \frac{r^2}{R^4}\frac{\partial^2\Phi_2(r,\phi)}{\partial\phi^2}$$
$$= \frac{r^4}{R^4}\left(\frac{1}{r}\frac{\partial}{\partial r}r\frac{\partial\Phi_2}{\partial r} + \frac{1}{r^2}\frac{\partial^2\Phi_2}{\partial\phi^2}\right) = 0$$

thereby proving the assertion expressed in (5.98).

If R is the radius of the double-sheet electrolytic tank, then we assert that $\Phi_1(r,\phi)$ represents the potential in the upper sheet while $\Psi(\rho,\phi)$ represents that in the lower sheet. (The radial variable in the lower sheet is taken to be ρ.) For we have noted that Φ_1 is the correct potential in the upper sheet (i.e., under conditions $r < R$) and have shown Ψ to be a solution of Laplace's equation. It is only required, now, to show that the necessary boundary conditions between Ψ and Φ_1 are satisfied at $r = R$.

From (5.94*a*) and (5.97) we establish that the potential is continuous, as it must be; that is,

$$\Phi_1(R,\phi) = \Phi_2(R,\phi) = \Psi(R,\phi) \tag{5.100}$$

The current density at $r = R$ must be radially outward in the upper sheet and radially inward in the lower sheet, the magnitudes being the same. Since $J_r = -\sigma\,\partial\Phi/\partial r$, this requires that

$$\left.\frac{\partial\Phi_1}{\partial r}\right|_{r=R} = -\left.\frac{\partial\Psi}{\partial\rho}\right|_{\rho=R} \tag{5.101}$$

If we start with (5.94*b*) and make use of (5.96) and (5.97), then we have

$$\frac{\partial\Psi(\rho,\phi)}{\partial\rho} = \frac{\partial}{\partial\rho}\Phi_2\left(\frac{R^2}{\rho},\phi\right) = -\frac{r^2}{R^2}\frac{\partial}{\partial r}\Phi_2(r,\phi)$$

When $\rho = R$, $r = R$, we get

$$\left.\frac{\partial\Psi}{\partial\rho}\right|_{\rho=R} = -\left.\frac{r^2}{R^2}\frac{\partial\Phi_2}{\partial r}\right|_{r=R} = -\left.\frac{\partial\Phi_2}{\partial r}\right|_{r=R} = -\left.\frac{\partial\Phi_1}{\partial r}\right|_{r=R}$$

hence proving (5.101). The proof that $\Psi(\rho,\phi)$ correctly gives the potential in the lower sheet is now complete.

A simple geometric connection exists between Ψ and Φ_2, that is, between potentials in the lower sheet and those in the infinite space for which $r > R$. This is revealed in (5.96). We note that an actual radial distance undergoes an inversion about the cylinder of radius R. In particular, the point at infinity ($r \to \infty$) goes into the origin in the lower sheet ($\rho \to 0$).

The double-sheet tank not only is useful for problems of field mapping, but also can be adapted to the solution of network problems. What follows is a very brief outline of this capability.

We shall agree that any impedance function can be defined, within an arbitrary constant, by the location of its poles and zeros; that is, we can write

$$Z(\lambda) = \frac{A(\lambda - \lambda_{Z1}) \cdots (\lambda - \lambda_{Zn})}{(\lambda - \lambda_{p1}) \cdots (\lambda - \lambda_{pm})} \tag{5.102}$$

where λ_{Zi} and λ_{pi} are the coordinates of the zeros and poles, respectively, in the complex plane. Taking the natural logarithm of (5.102) and writing the equation due to the real part of both sides gives

$$\ln |Z| = \sum_{i=1}^{n} \ln |\lambda - \lambda_{Zi}| - \sum_{i=1}^{m} \ln |\lambda - \lambda_{pi}| + \ln A \tag{5.103}$$

But this is equivalent in form to the potential set up by a system of line sources located at λ_{Zi} and λ_{pi}, the source at λ_{Zi} being positive, that at λ_{pi} negative. Such a problem can be simulated in the double-sheet tank since it is capable of representing the entire complex plane. It is only necessary to locate at λ_{Zi} a current input electrode adjusted to "unit amplitude," while at λ_{pi} an output electrode extracting unit current is provided. The potential at some arbitrary point λ is then a measure of the magnitude of the impedance at the corresponding complex value. A plot of the potential variation along the imaginary λ axis, that is, ω axis, gives the frequency dependence of $\ln |Z|$. A full discussion of the application of the double-sheet tank to network analysis, including the technique for determination of the phase of $Z(\lambda)$, is given in a paper by Boothroyd, Cherry, and Makar.†

† A. R. Boothroyd, E. C. Cherry, and R. Makar, An Electrolytic Tank for the Measurement of Steady State Response, Transient Response and Other Allied Properties of Networks, *J. IEE*, vol. 96, pt. 1, pp. 163–177, 1949.

CHAPTER 6

STATIC MAGNETIC FIELD IN VACUUM

Our knowledge of magnetism and magnetic phenomena is as old as science itself. According to the writings of the great Greek philosopher Aristotle, the attractive power of magnets was known by Thales of Miletus, whose life spanned the period 640?–546 B.C. It was not until the sixteenth century, however, that any significant experimental work on magnets was performed. During this century the English physician Gilbert studied the properties of magnets, realized that a magnetic field existed around the earth, and even magnetized an iron sphere and showed that the magnetic field around this sphere was similar to that around the earth. Several other workers also contributed to the knowledge of magnetism during this same period.

The eighteenth century was a period of considerable growth for the theory and understanding of electrostatics. It is therefore not surprising to find that in the eighteenth century the theory of magnetism developed along lines parallel to that of electrostatics. The basic law that evolved was the inverse-square law of attraction and repulsion between unlike and like magnetic poles. Indeed, it would have been difficult for the theory to develop along any other path since batteries for producing a steady current were nonexistent. With the development of the voltaic cell by Volta, it was not long before the magnetic effects of currents were discovered by Oersted in 1820. This was followed by the formulation, by Biot and Savart, of the law for the magnetic field from a long straight current-carrying wire. Further studies by Ampère led to the law of force between conductors carrying currents. In addition, Ampère's studies on the magnetic field from current-carrying loops led him to postulate that magnetism itself was due to circulating currents on an atomic scale. Thus the gap between the magnetic fields produced by currents and those produced by magnets was effectively closed.

Today it is expedient to base our entire theory of magnetism and static magnetic fields on the work of Biot, Savart, and Ampère. A formulation in terms of fields produced by currents or charges in motion is perfectly general and can account for all the known static magnetic effects. The magnetic effects of material bodies is accounted for by equivalent volume and surface currents. This is not to say that the early theory, based on concepts similar to those used in electrostatics, is of no value. On the contrary, it is often much simpler to use this alternative formulation when dealing with problems involving magnetized bodies (magnets) and the perturbing fields set up by permeable bodies placed in external magnetic fields. Throughout the next two chapters we shall have an opportunity to examine both theories. Our main efforts will be devoted to the study of the magnetic effects of currents, since this provides us with a general foundation for the understanding of all static magnetic phenomena. Ampère's law of force between two closed current-carrying conducting loops will be elevated to the position of the fundamental law or postulate from which we shall proceed.

6.1. Ampère's Law of Force

With reference to Fig. 6.1, let C_1 and C_2 be two very thin closed conducting loops (wires) in which steady currents I_1 and I_2 flow. The

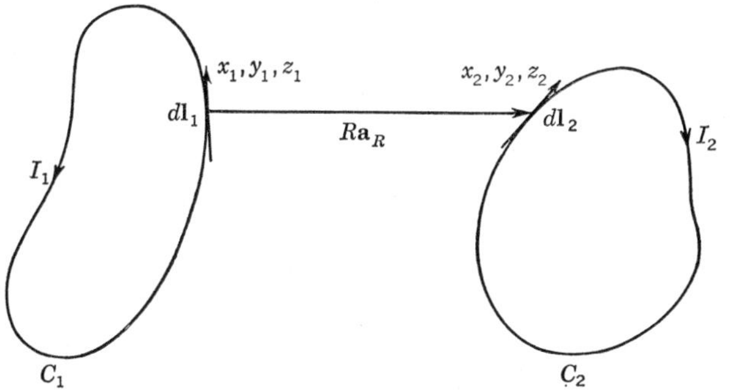

FIG. 6.1. Illustration of Ampère's law of force.

coordinates along the loop C_1 will be designated by x_1, y_1, z_1 and the vector arc length by $d\mathbf{l}_1$. Points along C_2 are designated by the variables x_2, y_2, z_2 and the vector arc length by $d\mathbf{l}_2$. The vector distance from $d\mathbf{l}_1$ to $d\mathbf{l}_2$ is $\mathbf{r}_2 - \mathbf{r}_1 = R\mathbf{a}_R$, where $\mathbf{a}_R$ is a unit vector directed from x_1, y_1, z_1 to x_2, y_2, z_2 and $R = [(x_2 - x_1)^2 + (y_2 - y_1)^2 + (z_2 - z_1)^2]^{1/2}$. From the work of Ampère it is found that the vector force $\mathbf{F}_{21}$ exerted on C_2 by C_1, as caused by the mutual interaction of the currents I_1 and I_2, may be expressed as

$$\mathbf{F}_{21} = \frac{\mu_0}{4\pi} \oint_{C_2} \oint_{C_1} \frac{I_2\, d\mathbf{l}_2 \times [I_1\, d\mathbf{l}_1 \times (\mathbf{r}_2 - \mathbf{r}_1)]}{|\mathbf{r}_2 - \mathbf{r}_1|^3} = \frac{\mu_0}{4\pi} \oint_{C_2} \oint_{C_1} \frac{I_2\, d\mathbf{l}_2 \times (I_1\, d\mathbf{l}_1 \times \mathbf{a}_R)}{R^2} \tag{6.1}$$

The force $\mathbf{F}_{21}$ is measured in newtons, the current in amperes, and all lengths in meters. The currents are assumed to be located in vacuum. The constant μ_0 arises because of the system of units (mks units) which we are using and is equal to $4\pi \times 10^{-7}$ henry per meter. Thus μ_0 has the dimensions of inductance per unit length. This constant is called the permeability of vacuum, and for practical purposes one may take the permeability of air equal to μ_0 also, with negligible error. An appreciation of the term permeability will have to be postponed until we take up the properties of magnetizable material bodies. It will suffice to note that permeability has much the same significance for magnetostatics that permittivity has for electrostatics.

Equation (6.1) reveals the inverse-square-law relationship. The differential element of force $d\mathbf{F}_{21}$ between $I_1\,d\mathbf{l}_1$ and $I_2\,d\mathbf{l}_2$ may be regarded as given by the integrand in (6.1) and is

$$d\mathbf{F}_{21} = \frac{\mu_0 I_1 I_2}{4\pi R^2}\,d\mathbf{l}_2 \times (d\mathbf{l}_1 \times \mathbf{a}_R) \tag{6.2a}$$

The triple-vector product may be expanded to give

$$d\mathbf{F}_{21} = \frac{\mu_0 I_1 I_2}{4\pi R^2}\,[(d\mathbf{l}_2 \cdot \mathbf{a}_R)\,d\mathbf{l}_1 - (d\mathbf{l}_2 \cdot d\mathbf{l}_1)\mathbf{a}_R] \tag{6.2b}$$

One should note that (6.2) does not correspond to a physically realizable condition since a steady-current element cannot be isolated. All steady currents must flow around continuous loops or paths since they have zero divergence.

A further difficulty with the relation (6.2a) or (6.2b) is that it is not symmetrical in $I_1\,d\mathbf{l}_1$ and $I_2\,d\mathbf{l}_2$. This superficially appears to contradict Newton's third law, which states that every action must have an equal and opposite reaction; i.e., the force exerted on $I_2\,d\mathbf{l}_2$ by $I_1\,d\mathbf{l}_1$ is not necessarily equal and opposite to the force exerted on $I_1\,d\mathbf{l}_1$ by $I_2\,d\mathbf{l}_2$. However, if the entire closed conductor, such as C_1 and C_2, is considered, no such difficulty arises and Newton's law is satisfied. Recalling that $\mathbf{a}_R/R^2 = -\nabla(1/R)$, we can replace the first term in (6.2b) with

$$\frac{-\mu_0 I_1 I_2\,d\mathbf{l}_1}{4\pi}\,\nabla\left(\frac{1}{R}\right) \cdot d\mathbf{l}_2$$

In an integration around C_2, this term vanishes, since $\nabla(1/R) \cdot d\mathbf{l}_2$ is a complete differential; that is, $\nabla(1/R) \cdot d\mathbf{l}_2$ is the directional derivative of $1/R$ along C_2 and is equal to

$$\frac{d(1/R)}{dl_2}\,dl_2 = d\left(\frac{1}{R}\right)$$

The integral of $d(1/R)$ is $1/R$, and since this is a single-valued function, it is equal to zero when evaluated around the closed contour C_2. For closed current loops, an equivalent form of Ampère's law of force may now be obtained by integrating (6.2b) and using the result that the first term vanishes; thus

$$\mathbf{F}_{21} = -\frac{\mu_0 I_1 I_2}{4\pi} \oint_{C_2} \oint_{C_1} \frac{(d\mathbf{l}_2 \cdot d\mathbf{l}_1)\mathbf{a}_R}{R^2} \tag{6.3}$$

This alternative relation is symmetrical with respect to loops 1 and 2 (that is, $\mathbf{F}_{21} = \mathbf{F}_{12}$) and therefore obeys Newton's third law.

Using the expansion of the integrand as given in (6.2b) shows that $d\mathbf{F}_{21}$ is a vector in the plane containing the vectors $d\mathbf{l}_1$ and $\mathbf{a}_R$ and in addition is perpendicular to $d\mathbf{l}_2$, as in Fig. 6.2a. When $d\mathbf{l}_2$ is perpendicular to $\mathbf{a}_R$, the force is entirely radial, as in Fig. 6.2b. When $d\mathbf{l}_2$ and $d\mathbf{l}_1$ are perpendicular, the force is directed parallel to $d\mathbf{l}_1$, since the component proportional to $d\mathbf{l}_2 \cdot d\mathbf{l}_1$ along $\mathbf{a}_R$ is zero. Finally, when $d\mathbf{l}_2$ is perpendicular to $\mathbf{a}_R$ and $d\mathbf{l}_2$ and $d\mathbf{l}_1$ are also mutually perpendicular, the force vanishes, as illustrated in Fig. 6.2c.

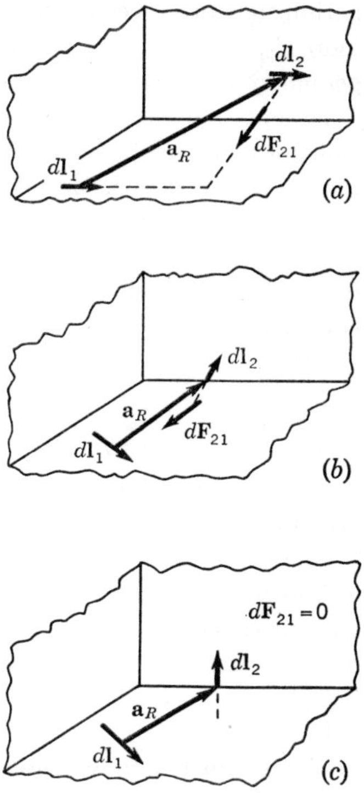

FIG. 6.2. Space relation between $d\mathbf{l}_1$, $d\mathbf{l}_2$, $\mathbf{a}_R$, and $d\mathbf{F}_{21}$.

6.2. The Magnetic Field B

In electrostatics the concept of an electric field was developed and found to be of great importance. This work stemmed from the definition of the electric field as the force acting on a unit charge. The field concept proves equally important in the magnetic case, and we find it possible to set up an analogous definition of the magnetic field **B**. In place of (6.1) we can write

$$\mathbf{F}_{21} = \oint_{C_2} I_2\, d\mathbf{l}_2 \times \mathbf{B}_{21} \tag{6.4}$$

where

$$\mathbf{B}_{21} = \frac{\mu_0}{4\pi} \oint_{C_1} \frac{I_1\, d\mathbf{l}_1 \times \mathbf{a}_R}{R^2} \tag{6.5}$$

Equation (6.1) may be thought of as evaluating the force between current-carrying conductors through an action-at-a-distance formulation. In contrast, (6.4) evaluates the force on a current loop in terms of the interaction of this current with the magnetic field **B**, which in turn is set up by the remaining current in the system. The current-field interaction that produces $\mathbf{F}_{21}$ in (6.4) takes place over the extent of the current loop C_2, while the magnetic field $\mathbf{B}_{21}$ depends only on the current and geometry of C_1 which sets up the field.

Except that the relations are vector ones, this work reiterates the field concept as developed in electrostatics. If we always assume orientation for maximum force, then $\mathbf{B}$ is the force per unit current element. In particular, (6.5) can be considered as a formulation for the magnetic field at any point in space which is independent of the existence of a test loop C_2 to detect the field. Furthermore, each element of current may be considered to contribute an amount

$$d\mathbf{B}(x,y,z) = \frac{\mu_0 I(x',y',z')\, d\mathbf{l}' \times (\mathbf{r} - \mathbf{r}')}{4\pi|\mathbf{r} - \mathbf{r}'|^3}$$

to the total field $\mathbf{B}(x,y,z)$. (The generalized notation here follows the definitions introduced in Sec. 1.17.) From this formula the field of an arbitrary current distribution can be found by superposition.

One of the advantages of the field formulation of (6.5) is that when $\mathbf{B}$ is known, this relation permits one to evaluate the force exerted on a current-carrying conductor placed in the field $\mathbf{B}$ without consideration of the system of currents which give rise to $\mathbf{B}$. Equation (6.5) is the law based on the experimental and theoretical work of Biot and Savart and is therefore usually called the Biot-Savart law. Since this law may also be extracted from Ampère's law of force, it is sometimes referred to as Ampère's law as well.

A charge q moving with a velocity v is equivalent to an element of current $I\, d\mathbf{l} = q\mathbf{v}$ and hence in the presence of a magnetic field experiences a force $\mathbf{F}$ given by

$$\mathbf{F} = q\mathbf{v} \times \mathbf{B} \tag{6.6}$$

This force is called the Lorentz force, and (6.6) is often taken as the defining equation for $\mathbf{B}$.

In practice, one does not always deal with currents flowing in thin conductors, and hence it is necessary to generalize the defining equation (6.5) for $\mathbf{B}$ so that it will apply for any arbitrary volume distribution of current. The steady-current flow field is divergenceless, and all flow lines form closed loops. Let us single out a short length dl of one current flow tube of cross-sectional area dS and compute its contribution to the total field $\mathbf{B}$. Let the current density in the current tube under consideration be $\mathbf{J}$, as in Fig. 6.3. We may associate the direction with the current density vector $\mathbf{J}$ rather than with the arc length dl, and hence the current-flow-tube element of length dl at (x',y',z') produces a partial field $d\mathbf{B}$ at (x,y,z) given by

$$d\mathbf{B} = \frac{\mu_0}{4\pi R^2}\mathbf{J}(x',y',z') \times \mathbf{a}_R\, dl\, dS$$

since the total current is $\mathbf{J}\,dS$. The total current contained in a volume V will therefore produce a field $\mathbf{B}$ given by

$$\mathbf{B}(x,y,z) = \frac{\mu_0}{4\pi}\int_V \frac{\mathbf{J}(x',y',z') \times \mathbf{a}_R}{R^2}\,dV' \tag{6.7}$$

where the integration is over the source coordinates x', y', z' and dV' is an element of volume $dS\,dl$. For a surface current $\mathbf{J}_s$ amperes per

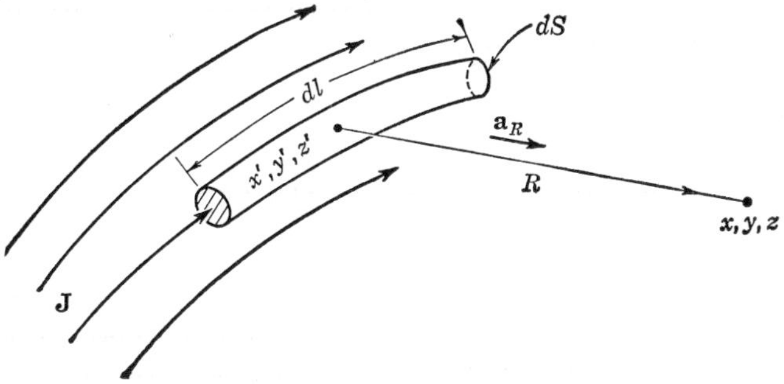

FIG. 6.3. A current flow tube.

meter flowing on a surface S, a similar derivation shows that the field produced is given by

$$\mathbf{B} = \frac{\mu_0}{4\pi}\int_S \frac{\mathbf{J}_s \times \mathbf{a}_R}{R^2}\,dS' \tag{6.8}$$

The unit for $\mathbf{B}$ is the weber per square meter, which is also equal to volt-seconds per square meter.

The equation for $\mathbf{B}$ is a vector equation, and its evaluation in practice is usually carried out by decomposing the integrand into components along three mutually perpendicular directions. If the current $\mathbf{J}$ is referred to a rectangular coordinate frame and has components J_x, J_y, and J_z, then

$$\mathbf{J} \times \mathbf{a}_R = \begin{vmatrix} \mathbf{a}_x & \mathbf{a}_y & \mathbf{a}_z \\ J_x & J_y & J_z \\ \dfrac{x-x'}{R} & \dfrac{y-y'}{R} & \dfrac{z-z'}{R} \end{vmatrix} = \mathbf{a}_x[J_y(z-z') - J_z(y-y')]\frac{1}{R}$$

$$+ \mathbf{a}_y[J_z(x-x') - J_x(z-z')]\frac{1}{R} + \mathbf{a}_z[J_x(y-y') - J_y(x-x')]\frac{1}{R}$$

The x component of magnetic field is thus given by

$$B_x(x,y,z) = \frac{\mu_0}{4\pi}\int_V \frac{(z-z')J_y(x',y',z') - (y-y')J_z(x',y',z')}{[(x-x')^2 + (y-y')^2 + (z-z')^2]^{3/2}}\,dx'\,dy'\,dz'$$

with similar expressions for B_y and B_z. Integrals of the above form are not always very easy to handle, and in practice it is convenient to compute an auxiliary potential function first from which **B** may subsequently be found by suitable differentiation. Such a procedure was used in electrostatics and found to be of considerable value as an intermediate step in finding the electric field. The next section will consider the potential function from which **B** may be obtained.

6.3. The Vector Potential

For convenience, the general equation defining the static magnetic field **B** is repeated here:

$$\mathbf{B}(x,y,z) = \frac{\mu_0}{4\pi} \int_V \frac{\mathbf{J}(x',y',z') \times \mathbf{a}_R}{R^2}\, dV' \tag{6.9}$$

If we replace $\mathbf{a}_R/R^2$ by $-\nabla(1/R)$, the integrand becomes $-\mathbf{J} \times \nabla(1/R)$. The vector differential operator ∇ affects only the variables x, y, z, and since $\mathbf{J}$ is a function of the source coordinates x', y', z' only, this latter relation may also be written as follows: $-\mathbf{J} \times \nabla(1/R) = \nabla \times (\mathbf{J}/R)$; that is, $\nabla \times (\mathbf{J}/R) = (1/R)\nabla \times \mathbf{J} - \mathbf{J} \times \nabla(1/R) = -\mathbf{J} \times \nabla(1/R)$, since $\nabla \times \mathbf{J} = 0$. Thus in place of (6.9) we may write

$$\mathbf{B}(x,y,z) = \frac{\mu_0}{4\pi} \int_V \nabla \times \frac{\mathbf{J}}{R}\, dV' = \nabla \times \frac{\mu_0}{4\pi} \int_V \frac{\mathbf{J}(x',y',z')}{R}\, dV' \tag{6.10}$$

The curl operation could be brought outside the integral sign since the integration is over the x', y', z' coordinates and the differentiation is with respect to the x, y, z coordinates. Equation (6.10) expresses the field **B** at the point (x,y,z) as the curl or circulation of a vector potential function given by the integral. From (6.10), the definition of the vector potential function, denoted by **A**, is

$$\mathbf{A}(x,y,z) = \frac{\mu_0}{4\pi} \int_V \frac{\mathbf{J}(x',y',z')}{R}\, dV' \tag{6.11}$$

The integral for **A** is a vector integral and must be evaluated by decomposing the integrand into components along the coordinate axis; e.g., the x component of **A** is given by $A_x = (\mu_0/4\pi) \int_V (J_x/R)\, dV'$. Note that the integral for each component of the vector potential **A** is of the same type as the integral for the scalar potential from a volume distribution of charge in electrostatics. Having computed **A**, the field **B** is obtained by taking the curl of **A**:

$$\mathbf{B} = \nabla \times \mathbf{A} \tag{6.12}$$

The integral for **A** is easier to evaluate than the original expression (6.9) for **B**, and since the curl operation is readily performed, the use of (6.11) as an intermediate step provides us with a simpler procedure for finding **B**.

Example 6.1. Field from an Infinite Wire Carrying a Current I. Consider an infinitely long straight wire in which a steady current I flows, as in Fig. 6.4. The magnetic field $\mathbf{B}$ will be determined at points which are much farther away from the wire than the diameter of the wire, so that we may assume that the wire is infinitely thin with negligible error. To form a closed loop we may imagine that the wire is closed by an infinitely large semicircular loop which does not contribute to the field in any finite region (as we could verify). According to the Biot-Savart law,

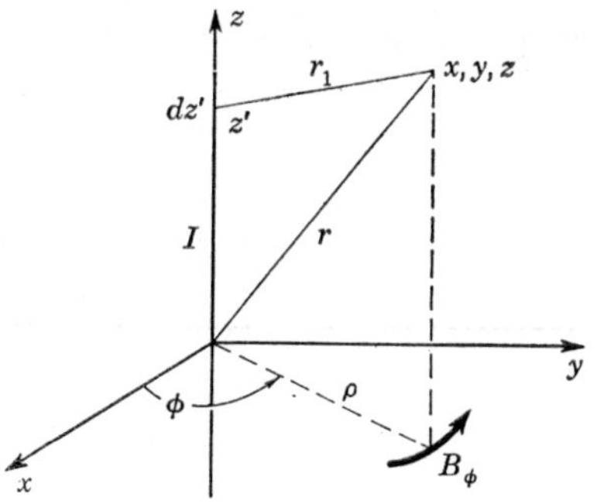

FIG. 6.4. An infinitely long wire with a current I.

$$\mathbf{B}(x,y,z) = \frac{\mu_0 I}{4\pi} \int_{-\infty}^{\infty} \frac{d\mathbf{l}' \times \mathbf{r}_1}{r_1^3}$$

The vector $\mathbf{r}_1$ is given by $\mathbf{r}_1 = \mathbf{a}_x x + \mathbf{a}_y y + \mathbf{a}_z(z - z')$, and $d\mathbf{l}' = \mathbf{a}_z\, dz'$, and hence

$$d\mathbf{l}' \times \mathbf{r}_1 = \begin{vmatrix} \mathbf{a}_x & \mathbf{a}_y & \mathbf{a}_z \\ 0 & 0 & dz' \\ x & y & z - z' \end{vmatrix} = -\mathbf{a}_x y\, dz' + \mathbf{a}_y x\, dz'$$

We may evaluate the x and y components of $\mathbf{B}$ separately. For later work it will be desirable to have an expression for the field contributed by a finite length L of wire so that the integral will be evaluated between $\pm L/2$ first. The x component is given by

$$B_x = \frac{-\mu_0 I y}{4\pi} \int_{-L/2}^{L/2} \frac{dz'}{[x^2 + y^2 + (z - z')^2]^{3/2}} \tag{6.13}$$

The integral may be evaluated by making the substitution

$$\tan \alpha = \frac{z - z'}{\rho}$$

where $\rho = (x^2 + y^2)^{1/2}$ and is the radial coordinate in a cylindrical coordinate system ρ, ϕ, z. The differential dz' becomes

$$-\rho d \tan \alpha = -\rho \sec^2 \alpha\, d\alpha$$

The term $x^2 + y^2 + (z - z')^2$ becomes $\rho^2(1 + \tan^2 \alpha) = \rho^2 \sec^2 \alpha$. When $z' = \pm L/2$, the corresponding values of the angle $\alpha = \alpha_{1,2}$ are given by $\tan \alpha_{1,2} = (z \mp L/2)/\rho$, or since $\tan^2 \alpha = \sec^2 \alpha - 1$, we get

$\cos \alpha_{1,2} = \rho/[\rho^2 + (z \mp L/2)^2]^{1/2}$ and

$$\sin \alpha_{1,2} = \frac{z \mp L/2}{[\rho^2 + (z \mp L/2)^2]^{1/2}}$$

Hence the component B_x is given by

$$B_x = \frac{\mu_0 I y}{4\pi\rho^2} \int_{\alpha_2}^{\alpha_1} \cos \alpha \, d\alpha = \frac{\mu_0 I y}{4\pi\rho^2} (\sin \alpha_1 - \sin \alpha_2)$$

Substituting $\sin^{-1} (z \mp L/2)/[\rho^2 + (z \mp L/2)^2]^{1/2}$ for α_1 and α_2 yields the final result

$$B_x = \frac{\mu_0 I y}{4\pi\rho^2} \left\{ \frac{z - L/2}{[\rho^2 + (z - L/2)^2]^{1/2}} - \frac{z + L/2}{[\rho^2 + (z + L/2)^2]^{1/2}} \right\} \tag{6.14}$$

The evaluation of B_y is similar and can be found from the expression for B_x by replacing y by $-x$. If we now note that the unit vector $\mathbf{a}_\phi$ is given by $\mathbf{a}_\phi = -\mathbf{a}_x \sin \phi + \mathbf{a}_y \cos \phi = (-y/\rho)\mathbf{a}_x + (x/\rho)\mathbf{a}_y$, we see that the two components of $\mathbf{B}$ combine to form a vector along the direction of the unit vector $\mathbf{a}_\phi$. The total field $\mathbf{B}$ is thus given, in this case, by

$$\begin{aligned} \mathbf{B} = B_x\mathbf{a}_x + B_y\mathbf{a}_y &= B_\phi \mathbf{a}_\phi \\ &= \frac{\mu_0 I}{4\pi\rho} \left\{ \frac{z + L/2}{[\rho^2 + (z + L/2)^2]^{1/2}} - \frac{z - L/2}{[\rho^2 + (z - L/2)^2]^{1/2}} \right\} \mathbf{a}_\phi \end{aligned} \tag{6.15}$$

For an infinitely long wire, L tends to infinity, and the limiting form of the expression for B_ϕ becomes

$$B_\phi = \frac{\mu_0 I}{2\pi\rho} \tag{6.16}$$

since for $L \gg \rho$ and $L \gg z$ the terms $(z + L/2)/[\rho^2 + (z + L/2)^2]^{1/2}$ and $(z - L/2)/[\rho^2 + (z - L/2)^2]^{1/2}$ approach 1 and -1, respectively. That $\mathbf{B}$ should have only a component B_ϕ could have been anticipated from the cylindrical symmetry of the problem.

Example 6.2. Field from a Conducting Ribbon with a Current I_s per Unit Width. Consider a thin conducting strip of width d, infinitely long and carrying a uniform current I_s amperes per meter, as in Fig. 6.5. The field from a strip of width dx' carrying a current $I_s\,dx'$ and located at $y = 0$, $x = x'$ is equivalent to that from a thin wire similarly located. From (6.16) this field is seen to have components dB_x and dB_y given by

$$dB_x = \frac{-\mu_0 I_s}{2\pi} \frac{y\,dx'}{y^2 + (x - x')^2}$$

$$dB_y = \frac{\mu_0 I_s}{2\pi} \frac{(x - x')\,dx'}{y^2 + (x - x')^2}$$

when dB_ϕ is decomposed into components along the x and y axis and x is replaced by $x - x'$ (new origin) throughout. The total field is found

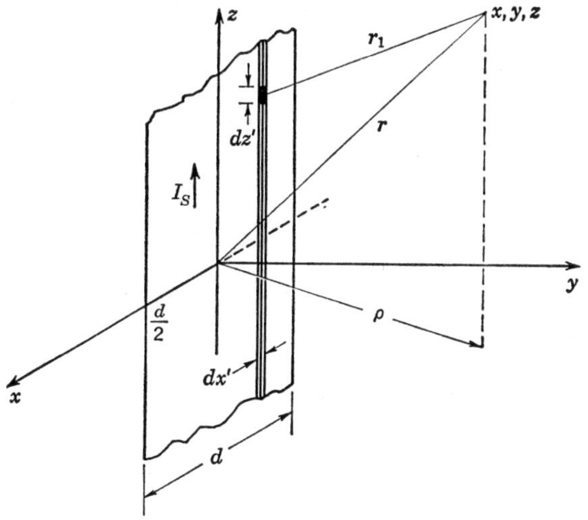

FIG. 6.5. Current I_s in an infinitely long strip.

by integrating over x' from $-d/2$ to $d/2$. The essential integrals to be evaluated are

$$\int_{-d/2}^{d/2} \frac{dx'}{y^2 + (x - x')^2} = -\frac{1}{y} \tan^{-1} \frac{x - x'}{y} \Bigg|_{-d/2}^{d/2}$$

$$\int_{-d/2}^{d/2} \frac{(x - x')\, dx'}{y^2 + (x - x')^2} = -\tfrac{1}{2} \ln \left[y^2 + (x - x')^2\right] \Bigg|_{-d/2}^{d/2}$$

Utilizing the above results, the components of the magnetic field become

$$B_x(x,y,z) = \frac{-\mu_0 I_s}{2\pi} \left(\tan^{-1} \frac{x + d/2}{y} - \tan^{-1} \frac{x - d/2}{y} \right) \qquad (6.17a)$$

$$B_y(x,y,z) = \frac{\mu_0 I_s}{4\pi} \ln \frac{y^2 + (x + d/2)^2}{y^2 + (x - d/2)^2} \qquad (6.17b)$$

Example 6.3. Force between Two Infinite Wires. Consider two thin infinite wires which are parallel and spaced at a distance d. The currents flowing in the wires are I_1 and I_2, as in Fig. 6.6. From (6.16) the magnetic field at C_2 due to C_1 has a ϕ component only and is given by

$$B_\phi = \frac{\mu_0 I_1}{2\pi d}$$

The force exerted on C_2 per unit length is given by Ampère's law of force (6.4) and is

$$\mathbf{F}_{21} = I_2\mathbf{a}_z \times \mathbf{a}_\phi B_\phi = \frac{-\mu_0 I_1 I_2}{2\pi d}\mathbf{a}_\rho \qquad \text{newtons/m} \tag{6.18}$$

When I_1 and I_2 are in the same direction, the two conductors experience

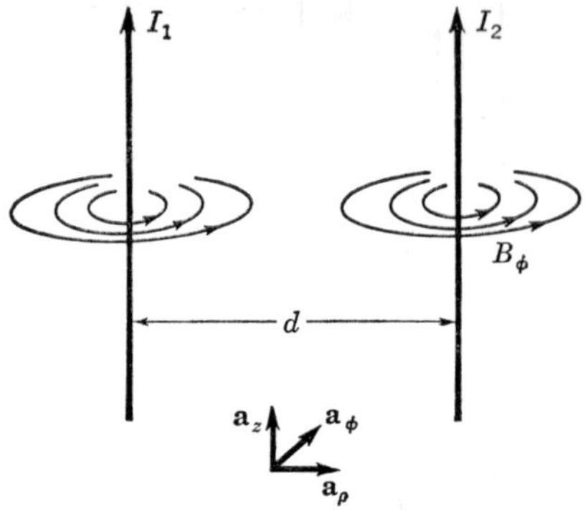

FIG. 6.6. Two parallel current-carrying wires.

an attractive force. When I_1 and I_2 are oppositely directed, the conductors repel each other.

Example 6.4. Field from a Circular Loop and Use of the Vector Potential. Consider a thin wire bent into a circular loop and carrying a current I. The radius of the loop is a, and it is located in the xy plane at the

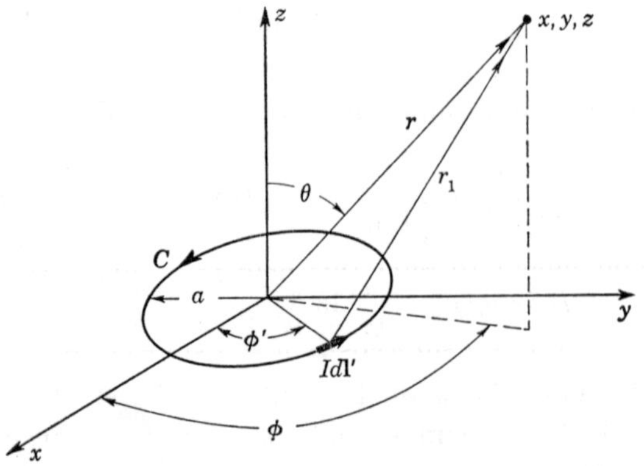

FIG. 6.7. A circular current loop.

origin, as in Fig. 6.7. We shall compute the field **B** at all points whose distance from the origin is much greater than the loop radius a by the direct method and by using the vector potential **A**. By the direct method **B** is given by the integral (6.5) as

$$\mathbf{B} = \frac{\mu_0 I}{4\pi} \oint_C \frac{d\mathbf{l}' \times \mathbf{r}_1}{r_1^3}$$

where in this case

$$d\mathbf{l}' = \mathbf{a}_\phi a\, d\phi' = (-\mathbf{a}_x \sin \phi' + \mathbf{a}_y \cos \phi')a\, d\phi'$$
$$\mathbf{r}_1 = \mathbf{a}_x(x - a \cos \phi') + \mathbf{a}_y(y - a \sin \phi') + \mathbf{a}_z z$$

Consequently,

$$d\mathbf{l}' \times \mathbf{r}_1 = [\mathbf{a}_x z \cos \phi' + \mathbf{a}_y z \sin \phi' - \mathbf{a}_z(y \sin \phi' + x \cos \phi' - a)]a\, d\phi'$$

The expression for r_1^3 is

$$\begin{aligned} r_1^3 &= [(x - a \cos \phi')^2 + (y - a \sin \phi')^2 + z^2]^{3/2} \\ &= (x^2 + y^2 + z^2 + a^2 - 2ax \cos \phi' - 2ay \sin \phi')^{3/2} \\ &\approx r^3 \left(1 - \frac{2ax}{r^2} \cos \phi' - \frac{2ay}{r^2} \sin \phi'\right)^{3/2} \end{aligned}$$

since $a^2 \ll r^2$. For r_1^{-3} we have approximately

$$r_1^{-3} \approx r^{-3} \left(1 + \frac{3ax}{r^2} \cos \phi' + \frac{3ay}{r^2} \sin \phi'\right)$$

upon using the binomial expansion and retaining only the leading terms.

We now obtain for the field **B** the expression

$$\begin{aligned} \mathbf{B} = \frac{\mu_0 I a}{4\pi r^3} \int_0^{2\pi} [\mathbf{a}_x z \cos \phi' + \mathbf{a}_y z \sin \phi' - \mathbf{a}_z(y \sin \phi' \\ + x \cos \phi' - a)] \left(1 + \frac{3ax}{r^2} \cos \phi' + \frac{3ay}{r^2} \sin \phi'\right) d\phi' \end{aligned}$$

The integration is straightforward, with most terms going to zero, and we are left with

$$\mathbf{B} = \frac{\mu_0 I \pi a^2}{4\pi r^3} \left[\mathbf{a}_x \frac{3xz}{r^2} + \mathbf{a}_y \frac{3yz}{r^2} - \mathbf{a}_z \left(\frac{3y^2}{r^2} + \frac{3x^2}{r^2} - 2\right)\right] \tag{6.19}$$

It will be convenient to refer this field to a spherical coordinate system r, θ, ϕ. For the purpose the following substitutions are required, namely, $z = r \cos \theta$, $x = r \sin \theta \cos \phi$, $y = r \sin \theta \sin \phi$. The component of **B** along the direction of the unit vector $\mathbf{a}_r$ is given by the projection of **B** on $\mathbf{a}_r$ and is $(\mathbf{B} \cdot \mathbf{a}_r)\mathbf{a}_r$; along the unit vector $\mathbf{a}_\theta$ it is $(\mathbf{B} \cdot \mathbf{a}_\theta)\mathbf{a}_\theta$, and similarly for the ϕ component. To evaluate these components note that

$$\begin{aligned} &\mathbf{a}_x \cdot \mathbf{a}_r = \sin \theta \cos \phi \qquad \mathbf{a}_y \cdot \mathbf{a}_r = \sin \theta \sin \phi \\ &\mathbf{a}_x \cdot \mathbf{a}_\theta = \cos \theta \cos \phi \qquad \mathbf{a}_y \cdot \mathbf{a}_\theta = \cos \theta \sin \phi \\ &\mathbf{a}_z \cdot \mathbf{a}_r = \cos \theta \qquad \mathbf{a}_z \cdot \mathbf{a}_\theta = -\sin \theta \\ &\mathbf{a}_x \cdot \mathbf{a}_\phi = -\sin \phi \qquad \mathbf{a}_y \cdot \mathbf{a}_\phi = \cos \phi \qquad \mathbf{a}_z \cdot \mathbf{a}_\phi = 0 \end{aligned}$$

Using the above relations in (6.19), the following expression for **B** is obtained:

$$\mathbf{B} = \frac{\mu_0 I \pi a^2}{4\pi r^3} (\mathbf{a}_r 2 \cos\theta + \mathbf{a}_\theta \sin\theta) \tag{6.20}$$

As may be anticipated from the symmetry about the z axis, **B** has no component along the direction of the unit vector $\mathbf{a}_\phi$.

For the second method we must evaluate the following integral for **A**:

$$\mathbf{A} = \frac{\mu_0 I}{4\pi} \oint_C \frac{d\mathbf{l}'}{r_1}$$

For $r^2 \gg a^2$ we have

$$r_1^{-1} \approx r^{-1}\left(1 + \frac{ax}{r^2}\cos\phi' + \frac{ay}{r^2}\sin\phi'\right)$$

by using the binomial expansion, as was done to obtain an approximate expression for r_1^{-3}. The integral for **A** becomes

$$\mathbf{A} = \frac{\mu_0 I a}{4\pi r} \int_0^{2\pi} (-\mathbf{a}_x \sin\phi' + \mathbf{a}_y \cos\phi')\left(1 + \frac{ax}{r^2}\cos\phi' + \frac{ay}{r^2}\sin\phi'\right) d\phi'$$

and integrates to give

$$\mathbf{A} = \frac{\mu_0 I \pi a^2}{4\pi r^3} (-\mathbf{a}_x y + \mathbf{a}_y x) \tag{6.21}$$

Referred to a spherical coordinate system, **A** is given by

$$\mathbf{A} = (\mathbf{A}\cdot\mathbf{a}_r)\mathbf{a}_r + (\mathbf{A}\cdot\mathbf{a}_\phi)\mathbf{a}_\phi + (\mathbf{A}\cdot\mathbf{a}_\theta)\mathbf{a}_\theta = \frac{\mu_0 I \pi a^2}{4\pi r^2}\mathbf{a}_\phi \sin\theta = A_\phi \mathbf{a}_\phi \tag{6.22}$$

and has only a component A_ϕ.

The magnetic field **B** is given by the curl of **A**. In spherical coordinates we have

$$\nabla \times \mathbf{A} = \mathbf{B} = \frac{1}{r^2 \sin\theta} \begin{vmatrix} \mathbf{a}_r & r\mathbf{a}_\theta & r\sin\theta\,\mathbf{a}_\phi \\ \frac{\partial}{\partial r} & \frac{\partial}{\partial \theta} & \frac{\partial}{\partial \phi} \\ 0 & 0 & r\sin\theta\, A_\phi \end{vmatrix} \tag{6.23}$$

where A_ϕ is given by (6.22). Expansion of this determinant gives

$$\mathbf{B} = \mathbf{a}_r \frac{1}{r\sin\theta}\frac{\partial}{\partial\theta}(\sin\theta A_\phi) - \mathbf{a}_\theta \frac{1}{r}\frac{\partial}{\partial r}(rA_\phi)$$

Substitution for A_ϕ from (6.22) will now yield the same expression as given by (6.20). The use of the vector potential in the present example leads to the end result in a simpler manner than the use of the defining relation (6.5) for **B**.

The previous results are restricted by the condition that $r \gg a$. If the field point is chosen to be along the axis of the loop (z axis), then a rigorous solution is easy to obtain. In the first method we note that since $x = y = 0$, we have

$$r_1^3 = (a^2 + z^2)^{3/2}$$
$$d\mathbf{l}' \times \mathbf{r}_1 = (\mathbf{a}_x z \cos \phi' + \mathbf{a}_y z \sin \phi' + a\mathbf{a}_z)a\, d\phi'$$

The expression for magnetic field

$$\mathbf{B} = \frac{\mu_0 I}{4\pi} \oint \frac{d\mathbf{l}' \times \mathbf{r}_1}{r_1^3} = \frac{\mu_0 I}{4\pi} \int_0^{2\pi} \frac{(\mathbf{a}_x z \cos \phi' + \mathbf{a}_y z \sin \phi' + a\mathbf{a}_z)a\, d\phi'}{(a^2 + z^2)^{3/2}}$$

then evaluates to

$$\mathbf{B} = \frac{\mu_0 I a^2}{2(a^2 + z^2)^{3/2}} \mathbf{a}_z$$

6.4. The Magnetic Dipole

The distant field **B** produced by a small current loop will be shown to be similar to the electric field from a small electric dipole. For this reason a small current loop is called a magnetic dipole. Its dipole moment **M** is defined as equal to the product of the area of the plane loop and the magnitude of the circulating current, and the vector direction of the moment is perpendicular to the plane of the loop and along the direction a right-hand screw would advance when rotated in the same sense as the current circulates around the loop. For a circular loop of radius a the magnitude of the dipole moment is $\pi a^2 I$, as in Fig. 6.8.

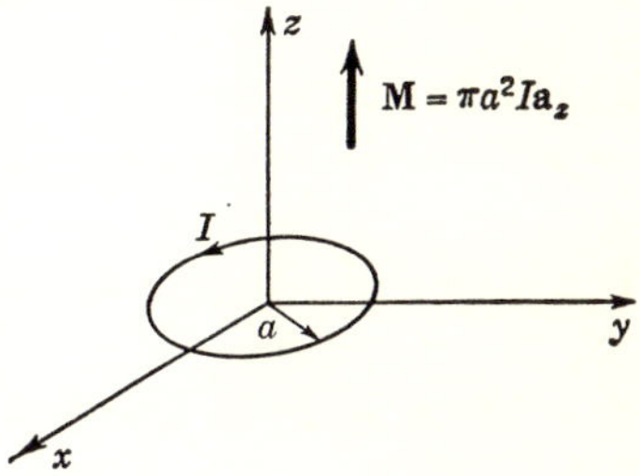

FIG. 6.8. The magnetic dipole.

From (6.21) in Example 6.4, we have $\mathbf{A} = (\mu_0 M/4\pi r^3)(-\mathbf{a}_x y + \mathbf{a}_y x)$ for the vector potential from a small circular current loop described as in Fig. 6.8. We now note that

$$\mathbf{a}_z \times \mathbf{r} = \mathbf{a}_z \times (x\mathbf{a}_x + y\mathbf{a}_y + z\mathbf{a}_z) = -\mathbf{a}_x y + \mathbf{a}_y x$$

and hence we may write

$$\mathbf{A} = \frac{\mu_0 M}{4\pi} \mathbf{a}_z \times \frac{\mathbf{r}}{r^3} = \frac{-\mu_0 M}{4\pi} \mathbf{a}_z \times \nabla\left(\frac{1}{r}\right) \tag{6.24}$$

since $\nabla(1/r) = -\mathbf{r}/r^3$. Now $\mathbf{M} = \mathbf{a}_z M$ is a constant; so we may also write in place of (6.24)

$$\mathbf{A} = \frac{\mu_0}{4\pi} \nabla \times \frac{\mathbf{M}}{r} \tag{6.25}$$

The steps involved in arriving at (6.25) are the same as those used to derive the integrand of (6.10).

The magnetic field $\mathbf{B}$ is given by

$$\mathbf{B} = \nabla \times \mathbf{A} = \frac{\mu_0}{4\pi} \nabla \times \left(\nabla \times \frac{\mathbf{M}}{r} \right)$$

This expression may be simplified by using the definition

$$\nabla \times \nabla \times = \nabla\nabla \cdot - \nabla^2$$

and the fact that $\nabla^2(1/r) = 0$ for $r \neq 0$. Remembering that $\mathbf{M}$ is a constant, we now get

$$\mathbf{B} = \frac{\mu_0}{4\pi} \left[\nabla \left(\nabla \cdot \frac{\mathbf{M}}{r} \right) - \nabla^2 \frac{\mathbf{M}}{r} \right] = \frac{\mu_0}{4\pi} \nabla \left(\nabla \cdot \frac{\mathbf{M}}{r} \right) = \frac{\mu_0}{4\pi} \nabla \left[\mathbf{M} \cdot \nabla \left(\frac{1}{r} \right) \right] \tag{6.26}$$

This last result is of the same form as the expression

$$\frac{1}{4\pi\epsilon_0} \nabla \left[\mathbf{p} \cdot \nabla \left(\frac{1}{r} \right) \right]$$

for the electric field from an electric dipole $\mathbf{p}$. Thus the electric and magnetic dipoles give rise to similar fields, as illustrated in Fig. 6.9. There is a fundamental difference, however, in that the electric lines of flux leave and terminate on charges while the magnetic lines of flux are continuous closed loops. In fact, we shall show in a later section that this property is always true for the magnetic field $\mathbf{B}$. This difference is not revealed by (6.26) and the equivalent electric dipole expression because these expressions are valid only for field points whose distance from the dipole is much greater than the extent of the dipole. Even for infinitesimal dipoles we may not use those expressions to reveal the behavior of the fields in the immediate vicinity of the sources.

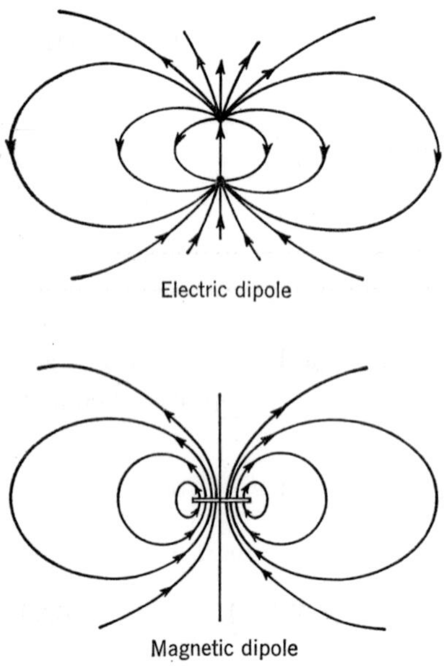

FIG. 6.9. Comparison of electric and magnetic dipoles.

For an arbitrary current loop the magnetic dipole moment is defined as equal to the product of the loop current I and the vector area $\mathbf{S}$ of the surface bounded by the loop; thus

$$\mathbf{M} = I\mathbf{S} \tag{6.27}$$

The positive direction of current flow is related to the positive direction of the surface by the usual right-hand-screw rule. This definition is illustrated in Fig. 6.10. It is not hard to show that the magnetic field from an arbitrary current loop of moment **M**, as defined in (6.27), is also given by (6.26), provided the distance to the field point is large

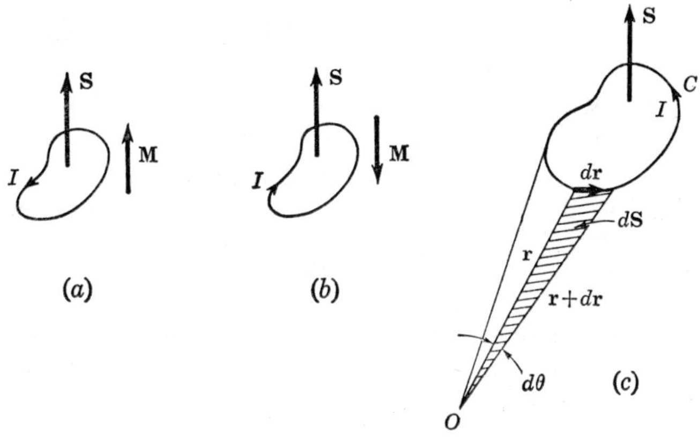

FIG. 6.10. General magnetic dipole.

compared with a characteristic linear dimension of the loop (see Prob. 6.16).

To facilitate the generalization of the magnetic dipole moment to a volume distribution of stationary currents, we first express the vector area **S** as an integral. Let **r** be a vector from some convenient origin to a point on the loop C. The arc length along the loop is $d\mathbf{r}$. The vector area of the infinitesimal triangle shown shaded in Fig. 6.10c is

$$\tfrac{1}{2}\mathbf{r} \times d\mathbf{r} = d\mathbf{S}$$

Note that this element of surface is in the positive direction as defined by the sense of the contour indicated in Fig. 6.10. The vector area of the cone with apex at O and subtended by the contour C is

$$\mathbf{S} = \oint_C d\mathbf{S} = \tfrac{1}{2} \oint_C \mathbf{r} \times d\mathbf{r} \tag{6.28}$$

This result is the same for any surface whose periphery is C, as we know from vector-analysis considerations. For example, if C is a plane curve, $|\mathbf{S}|$ is the plane area circumscribed. The magnetic dipole moment of an arbitrary shaped loop is therefore given by

$$\mathbf{M} = \frac{I}{2} \oint_C \mathbf{r} \times d\mathbf{r} \tag{6.29}$$

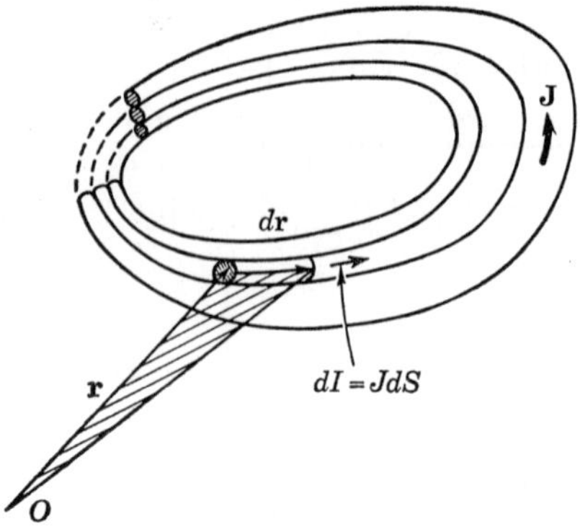

FIG. 6.11. A volume distribution of current separated into infinitesimal flow tubes.

The extension to a volume distribution of current is now a trivial one. Since the current is divergenceless, we may separate it into a large number of infinitesimal closed current flow tubes, as in Fig. 6.11. For any one flow tube the total current is $dI = J\,dS$, where dS is the cross-sectional area and J is the current density. Although dS may vary along the tube, the product $J\,dS$ is constant. A single flow tube contributes an amount

$$d\mathbf{M} = \frac{dI}{2} \oint \mathbf{r} \times d\mathbf{r} = \oint \frac{dS}{2} \mathbf{r} \times \mathbf{J}\, dr$$

to the total dipole moment, since we may associate the vector direction with $\mathbf{J}$ instead of with $d\mathbf{r}$. Now $dS\,dr$ is an element of volume dV, and summing over all current flow tubes (integration over dS), we get the general result

$$\mathbf{M} = \tfrac{1}{2} \int_V \mathbf{r} \times \mathbf{J}\, dV \tag{6.30}$$

Torque on a Magnetic Dipole

An electric dipole $\mathbf{p}$ placed in a uniform electrostatic field $\mathbf{E}$ experiences a torque $\mathbf{T} = \mathbf{p} \times \mathbf{E}$ but no translational force. A similar result holds for a magnetic dipole in a field $\mathbf{B}$ with the torque $\mathbf{T}$ given by

$$\mathbf{T} = \mathbf{M} \times \mathbf{B} \tag{6.31}$$

The torque is such that it tends to align the dipole with the field. The relation (6.31) is readily proved for a rectangular-loop dipole, as in Fig. 6.12.

Let the sides of the loop be L. The current in the loop is I, while $\mathbf{B}$ is chosen so that the plane defined by $\mathbf{B}$ and the surface normal is orthogonal to two sides of the loop, that is, C_2 and C_4 in Fig. 6.12*a*. $\mathbf{B}$ makes an angle θ with the surface normal. By Ampère's law of force (6.4), the force on the segments C_2 and C_4 is IBL and has the directions indicated in Fig. 6.12*b*, i.e., perpendicular to $\mathbf{B}$ and I. The forces on the segments C_1 and C_3 are equal, opposite, and directed along the axis of rotation and hence produce neither a torque nor a translational force. The product of the force F and the moment arm $L \sin \theta$ gives the torque as

$$T = IBL^2 \sin \theta = MB \sin \theta$$

or in vector form, $\mathbf{T} = \mathbf{M} \times \mathbf{B}$, since the magnitude of the dipole moment M is IL^2. If the direction of $\mathbf{B}$ is arbitrary, then it may be resolved into components of the above type and the resultant torque found by superposition. The analysis shows that the relation $\mathbf{T} = \mathbf{M} \times \mathbf{B}$ holds for arbitrary orientations of $\mathbf{B}$ relative to $\mathbf{M}$.

The generalization to an arbitrary current loop is obtained by considering the loop as made up

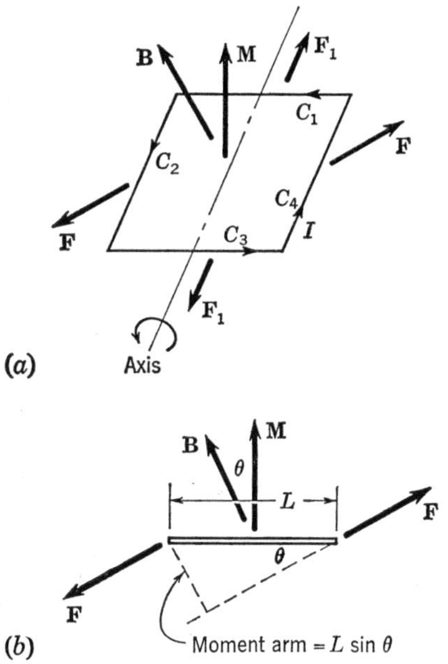

FIG. 6.12. Illustration of torque on a square-loop dipole.

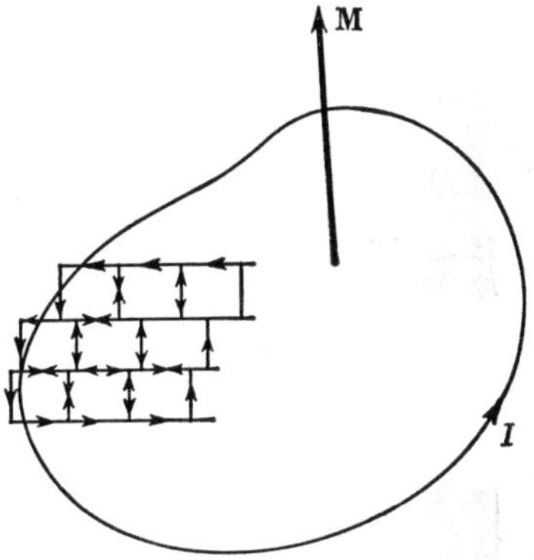

FIG. 6.13. Decomposition of dipole $\mathbf{M}$ into elementary square-loop dipoles $d\mathbf{M}$. (Currents along all interior boundaries cancel.)

of a large number of infinitesimal square loops, each with a dipole moment $d\mathbf{M}$, as in Fig. 6.13. The torque on each square-loop dipole is given by $d\mathbf{M} \times \mathbf{B} = I(d\mathbf{S} \times \mathbf{B})$, where I is the current magnitude and is constant. Integrating over all the infinitesimal dipoles gives the relation (6.31) if $\mathbf{B}$ is constant over the whole region. If $\mathbf{B}$ varies across the region occupied by the dipole, then (6.31) must be replaced by

$$\underline{\mathbf{T} = \int d\mathbf{M} \times \mathbf{B}} \tag{6.32}$$

Example 6.5. Torque on a D'Arsonval Movement. The D'Arsonval moving-coil instrument for measuring current consists of a rectangular coil, of n turns, which is free to rotate against the restoring torque of a hair spring. The coil is placed between the poles of a permanent magnet, which produces a field $\mathbf{B}$, as in Fig. 6.14, which we may assume to be uniform. The current I to be measured passes through the coil of cross-sectional area S. The magnetic dipole moment of the movement is nSI. The torque produced on the movement is then $BnSI \sin \theta = BnSI \cos \alpha$, where α is the angle of rotation from the zero current equilibrium position. Rotation stops when the restoring torque $k\alpha$ of the hair spring is equal to the torque produced by the field, i.e., when $k\alpha = BnSI \cos \alpha$. For small angles $\cos \alpha$ is approximately unity, and hence α is directly proportional to the current I. In practical instruments special shaped pole pieces

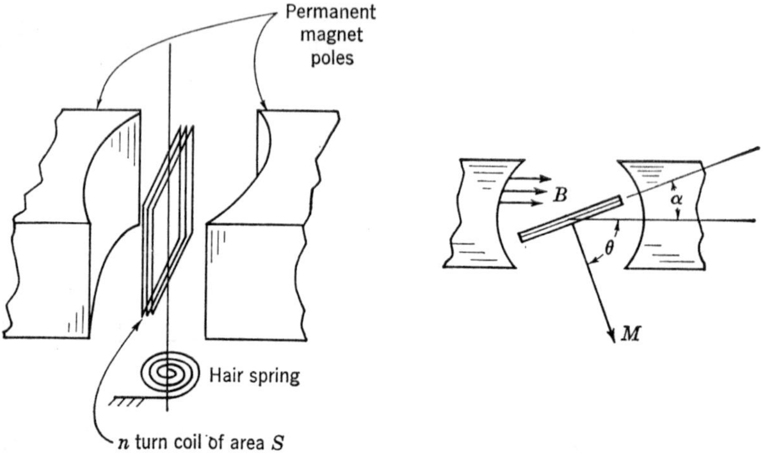

FIG. 6.14. The D'Arsonval moving-coil instrument.

are often used so as to produce a field **B** that will result in a linear scale over a larger range than that available with a uniform field.

6.5. Magnetic Flux and Divergence of B

We have seen that the magnetic field **B** can be derived from the curl of an auxiliary vector potential function **A**. This result leads at once to an important physical property for the field **B**. The divergence of the curl of any vector is identically equal to zero, and hence $\nabla \cdot \nabla \times \mathbf{A} = 0$, from which it follows that the divergence of **B** is also identically zero; i.e.,

$$\nabla \cdot \mathbf{B} = 0 \tag{6.33}$$

In the next chapter we shall show that the effects of material bodies can be accounted for by equivalent volume and surface magnetization currents. Thus even in the presence of material media it is possible to derive **B** from the curl of a vector potential, and hence the relation (6.33) is true in general. It now follows that the flux lines of **B** are always continuous and form closed loops. This property of **B** is the mathematical consequence of the formulation of Ampère's law, which in turn is based on experiments that do not reveal the existence of free magnetic poles, or "magnetic charge." All magnets have both a north and a south pole, and the field **B** is continuous through the magnet. For this reason the magnetostatic field **B** is fundamentally a different kind of field from the electrostatic field **E**. As discussed at several points earlier in this book, it is frequently advantageous to consider a vector field as representing the flow of something. The magnetic field **B** is often thought of as representing a magnetic flux density. Then the flux through an element of area

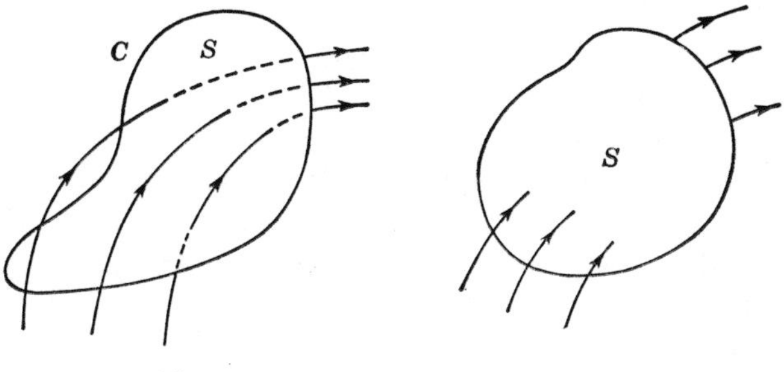

FIG. 6.15. Magnetic flux through a surface S. (a) Open surface; (b) closed surface.

$d\mathbf{S}$ is given by the dot product of **B** with $d\mathbf{S}$, $\mathbf{B} \cdot d\mathbf{S}$. The dot product selects the normal component of **B** through the surface $d\mathbf{S}$. For an arbitrary surface S bounded by a closed contour C, as in Fig. 6.15a, the total magnetic flux Ψ passing through the surface is given by

$$\Psi = \int_S \mathbf{B} \cdot d\mathbf{S} \tag{6.34}$$

The flux passing through the surface S bounded by the contour C is said to link the contour C and is commonly referred to as the "flux linkage."

For a closed surface S, as in Fig. 6.15b, just as much flux leaves the surface as enters because of the continuous nature of the flux lines. Hence the integral of $\mathbf{B} \cdot d\mathbf{S}$ over a closed surface is equal to zero. Mathematically, this result follows from (6.33) by an application of the divergence theorem. In the present case $\nabla \cdot \mathbf{B} = 0$; so we have

$$\int_V \nabla \cdot \mathbf{B}\, dV = \int_S \mathbf{B} \cdot d\mathbf{S} = 0 \tag{6.35}$$

The flux which links a contour C may be expressed in terms of the vector potential **A** also. Since $\mathbf{B} = \nabla \times \mathbf{A}$, we have

$$\Psi = \int_S \mathbf{B} \cdot d\mathbf{S} = \int_S \nabla \times \mathbf{A} \cdot d\mathbf{S}$$

The latter integral may be transformed to a contour integral by using

Stokes' law; thus

$$\Psi = \int_S \nabla \times \mathbf{A} \cdot d\mathbf{S} = \oint_C \mathbf{A} \cdot d\mathbf{l} \tag{6.36}$$

This latter integral is sometimes more convenient to evaluate than (6.34) is.

6.6. Ampère's Circuital Law

So far we have dealt only with integrals that give the field **B** or the vector potential **A**. What we need to do next is to obtain an equation for **B** that relates **B** to the current which exists at the point in space where **B** is being evaluated. A general vector field is a field which has both a divergence and a curl, neither of which is identically zero throughout all space. When the divergence and curl are both identically zero, the field vanishes everywhere. A field with a zero divergence but a nonzero curl is known as a pure rotational or solenoidal field. A field with a zero curl and a nonzero divergence is called an irrotational or lamellar field, of which the static electric field is a well-known example. We have seen that **B** is a solenoidal field with a zero divergence everywhere. Therefore the source **J** for the field **B** must be related to the curl of **B** in some manner.

To obtain the relation we are seeking we begin with (6.10) and take the curl of both sides to get

$$\nabla \times \mathbf{B}(x,y,z) = \nabla \times \nabla \times \frac{\mu_0}{4\pi} \int_V \frac{\mathbf{J}(x',y',z')}{R}\, dV' \tag{6.37}$$

The curl-curl operator may be expanded into the $\nabla\nabla \cdot - \nabla^2$ operator. We may also carry out the differentiation first and then the integration, because of the independence of the variables x, y, z and x', y', z'. Hence we have

$$\nabla \times \mathbf{B} = \frac{\mu_0}{4\pi} \int_V \left[\nabla\nabla \cdot \frac{\mathbf{J}(x',y',z')}{R} - \mathbf{J}(x',y',z')\nabla^2 \left(\frac{1}{R}\right) \right] dV' \tag{6.38}$$

We shall show later that the integral of the first term vanishes, so that

$$\nabla \times \mathbf{B} = -\frac{\mu_0}{4\pi} \int_V \mathbf{J}(x',y',z')\nabla^2 \left(\frac{1}{R}\right) dV' \tag{6.39}$$

By direction differentiation one readily finds that $\nabla^2(1/R) = 0$ for all finite values of R. Thus if the field point (x,y,z) is outside a finite source region, then R will never vanish, and it is clear that $\nabla \times \mathbf{B} = 0$. But if the field point is within the source region, then in the process of integration it is possible for R to be zero. This condition requires more careful attention since the integrand of (6.39) has a singularity at $R = 0$. Actually, as we shall now verify, the singularity of $\nabla^2(1/R)$ in (6.39) is integrable and yields a finite result. We proceed by noting that in the immediate neighborhood of the point $R = 0$ the current density function does not vary much from its value at the point $R = 0$, that is, at the point $x' = x$, $y' = y$, $z' = z$. Since the integrand is zero everywhere except at $R = 0$, we need only integrate (6.39) over a small sphere with center at (x,y,z), as in Fig. 6.16. We may take $\mathbf{J}(x',y',z')$ equal to $\mathbf{J}(x,y,z)$ throughout the volume of the sphere for the reason just given, and hence (6.39) becomes

$$\nabla \times \mathbf{B} = -\frac{\mu_0}{4\pi}\mathbf{J}(x,y,z)\int_{V_s}\nabla^2\left(\frac{1}{R}\right)dV' \qquad (6.40)$$

Now since $R = [(x - x')^2 + (y - y')^2 + (z - z')^2]^{1/2}$, we have the result $\nabla(1/R) = -\nabla'(1/R)$ and $\nabla^2(1/R) = \nabla'^2(1/R) = \nabla' \cdot \nabla'(1/R)$, where ∇' signifies differentiation with respect to x', y', and z'. In place of (6.40) we have

$$-\frac{\mu_0}{4\pi}\mathbf{J}(x,y,z)\int_{V_s}\nabla' \cdot \nabla'\left(\frac{1}{R}\right)dV'$$
$$= -\frac{\mu_0}{4\pi}\mathbf{J}(x,y,z)\oint_S \nabla'\left(\frac{1}{R}\right)\cdot d\mathbf{S}'$$

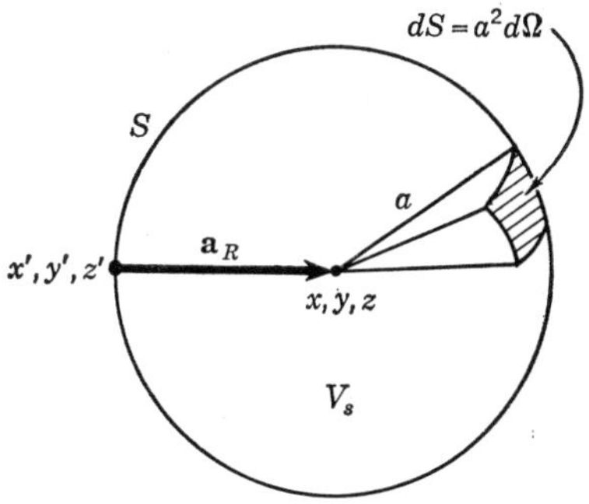

FIG. 6.16. Small sphere of radius a surrounding singularity point $x' = x$, $y' = y$, $z' = z$.

where we have also made use of the divergence theorem. The element of area in spherical coordinates is $\mathbf{n}R^2\,d\Omega$, where $\mathbf{n}$ is the unit outward normal and $d\Omega$ is an element of solid angle, that is,

$$d\Omega = \sin\theta\,d\phi\,d\theta$$

Also we have $\nabla'(1/R) = \mathbf{a}_R/R^2$, and hence

$$\nabla'\left(\frac{1}{R}\right)\cdot d\mathbf{S}' = \mathbf{n}\cdot\mathbf{a}_R\,d\Omega = -d\Omega$$

since $\mathbf{a}_R$ points toward the center of the sphere. Therefore (6.40) gives

$$\nabla \times \mathbf{B} = \frac{\mu_0}{4\pi}\mathbf{J}(x,y,z)\int_S d\Omega = \mu_0\mathbf{J}(x,y,z) \qquad (6.41)$$

since there are 4π steradians in the solid angle of a closed surface. Hence we see that the curl or rotation of **B** is equal to $\mu_0\mathbf{J}$.

Proof of $\int_V \nabla\nabla\cdot\frac{\mathbf{J}}{R}\,dV' = 0$

Having obtained the desired relation (6.41) between **B** and **J**, we must now return to (6.38) and show that the first integral vanishes as stated.

The integral in question may be rewritten as

$$\int_V \nabla\nabla' \cdot \frac{\mathbf{J}(x',y',z')}{R}\, dV' = \nabla \int_V \nabla \cdot \frac{\mathbf{J}}{R}\, dV'$$

where one of the differentiation operators has been brought outside the integral sign. To prove the desired result, rewrite the integrand as follows:

$$\nabla \cdot \frac{\mathbf{J}(x',y',z')}{R} = \mathbf{J} \cdot \nabla\left(\frac{1}{R}\right) = -\mathbf{J} \cdot \nabla'\left(\frac{1}{R}\right) = -\nabla' \cdot \frac{\mathbf{J}}{R}$$

since $\nabla' \cdot (\mathbf{J}/R) = (1/R)\nabla' \cdot \mathbf{J} + \mathbf{J} \cdot \nabla'(1/R)$ and $\nabla' \cdot \mathbf{J}$ is zero for stationary currents. The integral becomes

$$\nabla \int_V \nabla \cdot \frac{\mathbf{J}}{R}\, dV' = -\nabla \int_V \nabla' \cdot \frac{\mathbf{J}}{R}\, dV' = -\nabla \oint_S \frac{\mathbf{J}}{R} \cdot d\mathbf{S}' \qquad (6.42)$$

by using the divergence theorem, where S is a closed surface surrounding V.† Since $\mathbf{J}$ is a stationary current and confined to a finite region of space, we may choose S so large that all currents lie within and in particular so that $\mathbf{J} \cdot d\mathbf{S}'$ equals zero on the surface S. Hence the integral vanishes as stated.

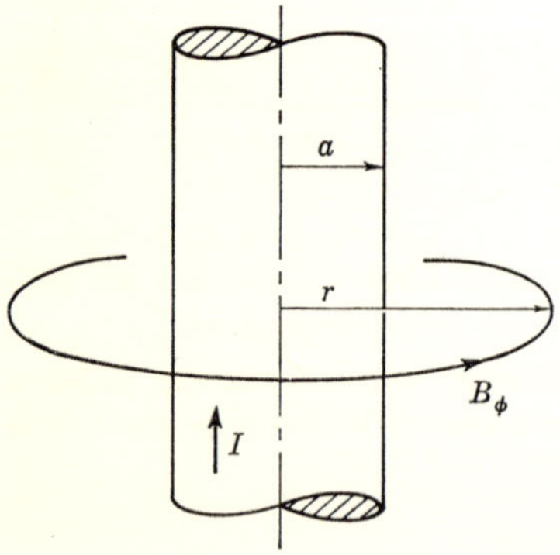

FIG. 6.17. An infinitely long coaxial line.

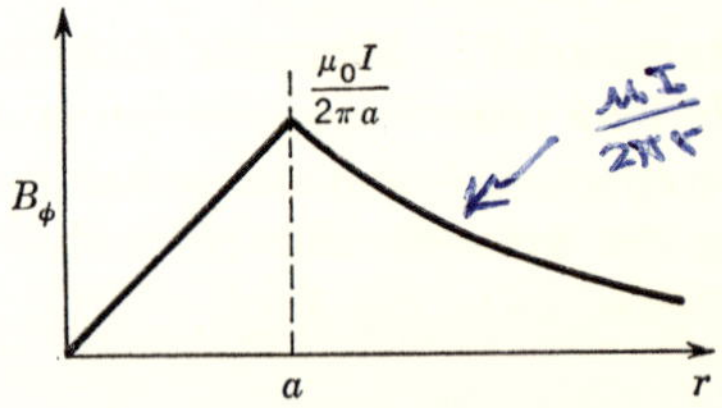

FIG. 6.18. Field B_ϕ as a function of radial distance from center of wire.

Equation (6.41) is Ampère's circuital law in differential form. By applying Stokes' law, an integral form of this law may be obtained. If we integrate $\nabla \times \mathbf{B}$ over a surface S bounded by a closed contour C and use Stokes' law, we get

$$\int_S \nabla \times \mathbf{B} \cdot d\mathbf{S} = \int_S \mu_0 \mathbf{J} \cdot d\mathbf{S} = \oint_C \mathbf{B} \cdot d\mathbf{l} \qquad (6.43)$$

This equation states that the line integral of $\mathbf{B} \cdot d\mathbf{l}$ around any closed contour C is equal to μ_0 times the total net current passing through the contour C. The law is particularly useful in solving magnetostatic prob-

† Note that the divergence theorem can be applied only to the second integral in (6.42), where the variables of the differential operator and of integration are the same.

lems having cylindrical symmetry, as the following examples will demonstrate.

Example 6.6. Field from an Infinite Wire of Finite Radius. Consider an infinite wire of radius a with total current I (Fig. 6.17). The current density J is equal to $I/\pi a^2$ and uniform over the cross section of the wire. From symmetry considerations the field $\mathbf{B}$ has only a component B_ϕ, which is a function of r only. Using Ampère's circuital law (6.43) and integrating around a circular contour of radius r gives

$$\oint B_\phi \, dl = \int_0^{2\pi} B_\phi r \, d\phi = \mu_0 \int_0^r \int_0^{2\pi} Jr \, d\phi \, dr$$

$$= \frac{\mu_0 I}{\pi a^2} \int_0^{2\pi} \int_0^r r \, d\phi \, dr \qquad r \leq a$$

or

$$B_\phi = \frac{\mu_0 I r}{2\pi a^2} \qquad r \leq a$$

For $r \geq a$, the total current enclosed is I; so

$$\int_0^{2\pi} B_\phi r \, d\phi = \mu_0 I$$

or

$$B_\phi = \frac{\mu_0 I}{2\pi r} \qquad a \leq r$$

A plot of the intensity of B_ϕ as a function of r is given in Fig. 6.18.

Example 6.7. Magnetic Field in a Coaxial Line. Consider an infinitely long coaxial line consisting of an inner conductor of radius a, an outer conductor of inner radius b, and thickness t. A current I flows along the inner conductor, and a return current $-I$ along the outer conductor, as in Fig. 6.19.

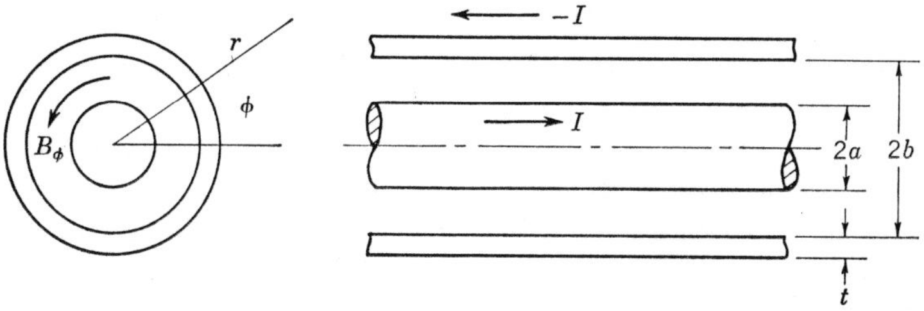

FIG. 6.19. An infinitely long coaxial line.

In the region $r \leq b$ the solution for B_ϕ is the same as in Example 6.6:

$$B_\phi = \frac{\mu_0 I r}{2\pi a^2} \qquad r \leq a$$

$$B_\phi = \frac{\mu_0 I}{2\pi r} \qquad a \leq r \leq b$$

In the region $b \leq r \leq b + t$ we have

$$\int_0^{2\pi} B_\phi r \, d\phi = \mu_0 I - \frac{\mu_0 I}{\pi[(b+t)^2 - b^2]} \int_b^r \int_0^{2\pi} r \, d\phi \, dr$$

or

$$2\pi r B_\phi = \mu_0 I - \frac{\mu_0 I (r^2 - b^2)}{(b+t)^2 - b^2}$$

since the current density in the outer conductor is $I/\pi[(b+t)^2 - b^2]$. Hence

$$B_\phi = \frac{\mu_0 I}{2\pi r} - \frac{\mu_0 I (r^2 - b^2)}{2\pi r[(b+t)^2 - b^2]} \qquad b \leq r \leq b + t$$

For $r \geq b + t$ the field B_ϕ is zero, since no net current is enclosed by the contour of integration. The above expression for B_ϕ is seen to vanish when r is placed equal to $b + t$. A plot of B_ϕ as a function of r is given in Fig. 6.20.

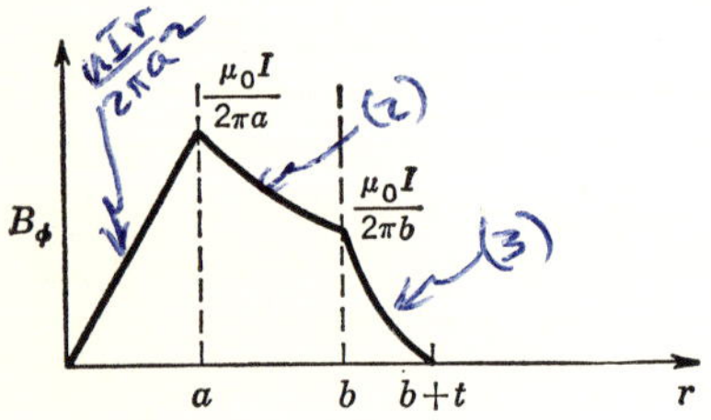

FIG. 6.20. The field B_ϕ as a function of r in a coaxial line.

A word in retrospect on the subject of vector fields. By definition, the vector field of a physical quantity is the totality of all points in a given region for which the direction and magnitude of the quantity are specified. Physically realizable fields are given by vector functions of position which are mathematically well behaved. To the list of vector fields given in Chap. 1, we may now add the fundamental field of electrostatics **E** and magnetostatics **B**.

The formal mathematical description of a vector field gives no insight into its physical properties. This means that an **E** or **B** field could also be thought of as representing a hydrodynamic velocity field, and vice versa. The hydrodynamic analogy was introduced in Chap. 1 in several places in order to develop some "feel" for the abstract vector calculus. We should like to follow this a bit further, but would like it clearly understood that only an analogy is being depicted. Other analogies could be formulated; indeed the reader may be satisfied with no analogy at all.

We plan to represent an arbitrary vector field by assigning a proportional value of velocity to each corresponding point in an infinite volume of incompressible fluid. Thus the condition that the fluid is at rest corresponds to a null field. There are two ways whereby we may bring about motion in the fluid and hence correspondingly set up a field. The simplest way is to "stir things up"; we could do this, for example, with a paddle wheel. But to allow for greatest generality we can think of the paddle wheel as infinitesimal since this permits us to synthesize a more arbitrary "stirrer" by means of an aggregate of paddle wheels (vortex source) whose direction and magnitude vary with position. Thus if a large paddle wheel is used to set the fluid in motion, i.e., create a vector field, this macroscopic source can be synthesized by an appropriate summation of infinitesimal vortex sources. Since the curl of the vector field is a measure of how effectively the source has stirred things up, it is an appropriate measure of the source strength.

In the case of the magnetostatic field we determined that

$$\underline{\nabla \times \mathbf{B} = \mu_0 \mathbf{J}}$$

which establishes $\mu_0\mathbf{J}$ as a vortex source. The existence of a current density in space may be thought of as causing a **B** field to exist by "stirring up" of the media.

A second way of causing fluid motion is to inject fluid or to remove fluid. If the total source plus sink is zero, then the net amount of fluid remains constant. Point sources refer to an idealization of sources where the fluid enters in a spherically symmetric uniform flow from a mathematical point. Similarly, a negative point source (or sink) removes fluid in an analogous pattern. While the over-all quantity of fluid is constant, there can be a net positive or negative flow in a limited region, and this net amount is proportional to the sum of sources within the region. The divergence of the vector function is a measure of the source strength at each point, where the source is considered as being distributed throughout a volume. In electrostatics the **E** field arises from divergence-producing types of sources and is given by the formula

$$\nabla \cdot \mathbf{E} = \frac{\rho}{\epsilon_0}$$

In the most general case the fluid may be set in motion, hence represent an arbitrary field, by a combination of sources such as described above. In this case the vector function has a nonzero value of both divergence and curl in at least some region of space. An example of such a field is the **D** field in electrostatics in the presence of dielectric materials, with permittivity that is a function of the coordinates. For in this case

$$\begin{aligned}
\nabla \cdot \mathbf{D} &= \rho \\
\nabla \times \mathbf{D} &= \nabla \times (\epsilon_0 \mathbf{E}) + \nabla \times \mathbf{P} \\
&= \nabla \times \mathbf{P}
\end{aligned}$$

The true charge causes an outflow of **D** from their positions, while the source $\nabla \times \mathbf{P}$ acts to stir up the **D** field.

6.7. Differential Equation for Vector Potential

In electrostatics we showed that the scalar potential Φ was related to the charge sources that produce it by means of the following equation:

$$\Phi = \frac{1}{4\pi\epsilon_0} \int \rho \, dV$$

We determined further that the scalar potential Φ satisfied the following partial differential equation (Poisson's):

$$\nabla^2\Phi = \frac{-\rho}{\epsilon_0}$$

so that the integral formulation could also be considered as a solution to Poisson's equation.

In magnetostatics we have found, analogously, that the vector potential $\mathbf{A}$ is related to the current sources through the following expression:

$$\underline{\mathbf{A} = \frac{\mu_0}{4\pi} \int \frac{\mathbf{J}\, dV}{R}}$$

If we consider the rectangular components of this equation, that is,

$$A_x = \frac{\mu_0}{4\pi} \int \frac{J_x \, dV}{R}$$
$$A_y = \frac{\mu_0}{4\pi} \int \frac{J_y \, dV}{R}$$
$$A_z = \frac{\mu_0}{4\pi} \int \frac{J_z \, dV}{R}$$

then each is a scalar equation of precisely the type dealt with in electrostatics. Then by analogy it is clear that each component must satisfy a Poisson equation with the corresponding current component as a source; that is,

$$\nabla^2 A_x = -\mu_0 J_x \qquad \nabla^2 A_y = -\mu_0 J_y \qquad \nabla^2 A_z = -\mu_0 J_z$$

If, now, each equation is multiplied by the corresponding unit vector and all three summed, we obtain the following vector Poisson equation:

$$\underline{\nabla^2 \mathbf{A} = -\mu_0 \mathbf{J}}$$

This result can be obtained in a more formal mathematical way. If we take the double curl of $\mathbf{A}$, which is equal to the curl of $\mathbf{B}$, and use (6.41) to replace $\nabla \times \mathbf{B}$ by $\mu_0\mathbf{J}$, we get $\nabla \times \nabla \times \mathbf{A} = \nabla \times \mathbf{B} = \mu_0\mathbf{J}$. Expanding the curl-curl operator gives

$$\nabla\nabla \cdot \mathbf{A} - \nabla^2\mathbf{A} = \mu_0\mathbf{J} \qquad (6.44)$$

Since $\mathbf{A}$ is given by $\mathbf{A} = (\mu_0/4\pi)\int(\mathbf{J}/R)\,dV'$, the divergence of $\mathbf{A}$ is given by $\nabla \cdot \mathbf{A} = (\mu_0/4\pi)\int\nabla \cdot (\mathbf{J}/R)\,dV'$. This latter integral occurred, and was shown to vanish, in the derivation of (6.41). It follows then that $\nabla \cdot \mathbf{A} = 0$ and (6.44) reduces to

$$\nabla^2\mathbf{A} = -\mu_0\mathbf{J} \tag{6.45}$$

which is an alternative proof that $\mathbf{A}$ is a solution of the vector Poisson equation.

We can also confirm that

$$\mathbf{A} = \frac{\mu_0}{4\pi}\int_V \frac{\mathbf{J}}{R}\,dV' \tag{6.46}$$

is a particular integral of (6.44). This can be established by using the singularity property of $\nabla^2(1/R)$, as was done to obtain (6.41). The details are left as a problem. To the particular solution (6.46) may be added any solution to the homogeneous equation $\nabla^2\mathbf{A} = 0$ as dictated by the boundary conditions which $\mathbf{A}$ must satisfy.

Gauge Transformation

In the earlier work $\mathbf{A}$ was defined by (6.11), as a consequence of which $\nabla \cdot \mathbf{A} = 0$ and $\nabla \times \mathbf{A} = \mathbf{B}$. If, alternatively, $\mathbf{A}$ was simply specified by the requirement that $\nabla \times \mathbf{A} = \mathbf{B}$, then there is a certain arbitrariness in the choice of $\mathbf{A}$. We could equally well use a new vector potential $\mathbf{A}'$ which differs from $\mathbf{A}$ by the addition of the gradient of a scalar function Φ, since $\nabla \times \nabla\Phi = 0$. Thus let

$$\mathbf{A}' = \mathbf{A} + \nabla\Phi \tag{6.47}$$

The field $\mathbf{B}$ is given by $\mathbf{B} = \nabla \times \mathbf{A}' = \nabla \times \mathbf{A} + \nabla \times \nabla\Phi = \nabla \times \mathbf{A}$ and is invariant to such a transformation to a new potential $\mathbf{A}'$. The transformation (6.47) is called a gauge transformation, and the invariant property of $\mathbf{B}$ is known as gauge invariance.

In view of the gauge invariance of $\mathbf{B}$ it is always possible to introduce a gauge transformation, as in (6.47), to make the new potential $\mathbf{A}'$ have zero divergence in the case when $\mathbf{A}$ does not have zero divergence. It is only necessary to choose Φ so that

$$\nabla \cdot \mathbf{A}' = 0 = \nabla \cdot \mathbf{A} + \nabla^2\Phi \tag{6.48}$$

Consequently, we can always work with a vector potential which has zero divergence if we wish, or on the other hand, if more convenient, we can choose a vector potential with nonzero divergence. In the general case $\mathbf{A}$ is then a solution of (6.44) rather than (6.45).

From another point of view we note that in defining the vector potential $\mathbf{A}$, only its curl has been specified; that is, $\nabla \times \mathbf{A} = \mathbf{B}$. From the Helmholtz theorem we understand that this does not completely specify the vector $\mathbf{A}$; in fact, the divergence of $\mathbf{A}$ is completely at our disposal. If we choose as a fundamental relation $\mathbf{A} = (\mu_0/4\pi)\int(\mathbf{J}/R)\,dV'$, then, as we have seen, we are inherently specifying $\nabla \cdot \mathbf{A} = 0$. This condition is usually a satisfactory one, but as pointed out above, it is not necessary. We could think of (6.45) as arising from a choice of integration constant equal to zero. This is equivalent to establishing an arbitrary reference potential, just as was discussed in the electrostatic case.

CHAPTER 7

MAGNETIC FIELD IN MATERIAL BODIES

According to the atomic theory, all matter is composed of a number of basic elementary particles of which the electron, proton, and neutron are perhaps the best known. These particles combine to form atoms, which we may, for our present purpose, think of as a number of neutrons and protons collected together into a central heavy core around which a number of electrons rotate along closed orbital paths. The electrons rotating in orbital paths are equivalent to circulating currents on an atomic scale. We may associate with each electron orbit a magnetic dipole moment in order to account for the external effects of the material body. The total dipole moment of an atom due to the orbital motion of electrons is called the orbital magnetic moment and is the vector sum of the moments of each orbit with its circulating electron. For a single electron with an electronic charge e rotating around a circular orbit of radius r and with a velocity v, the average circulating current is equal to the rate of movement of charge ev divided by the length $2\pi r$ of the orbit. Thus the magnetic dipole moment of the circulating electron is

$$\begin{aligned} m &= \text{circulating current} \times \text{area of loop} \\ &= \frac{ev}{2\pi r}\pi r^2 = \frac{evr}{2} \end{aligned} \tag{7.1}$$

In addition to the orbital magnetic moment, the electron itself is endowed with an intrinsic magnetic moment which plays an important role in the magnetic properties of material bodies. The physical structure of the electron that gives rise to its characteristic properties mass, charge, and mechanical and magnetic moment is not known, but this does not prevent us from discussing the external effects produced by these intrinsic properties. The presence of a mechanical and a magnetic moment has led physicists to think of the electron as having a spin, and its magnetic moment is therefore referred to as the spin magnetic moment.

The above discussion was introduced in order to provide some insight into the essential properties of matter which give rise to an interaction with externally applied magnetic fields. A more complete account of the physics of the magnetic properties of materials is given in a later section.

7.1. Magnetic Dipole Polarization per Unit Volume

Consider a material body of volume V and bounded by a surface S, as in Fig. 7.1*a*. In view of our earlier discussion we may think of this material body as consisting of a large number of small circulating current loops having a magnetic dipole moment $\mathbf{m}$. In the absence of any external applied magnetic field $\mathbf{B}_0$, the dipoles of individual atoms are randomly orientated and the body as a whole will have a zero net magnetic moment (this is not true for permanent magnets, which have a nonzero moment even in the absence of applied external fields). When we apply an external magnetic field $\mathbf{B}_0$, two things happen:

1. All atoms which have a nonzero magnetic moment tend to have their magnetic dipoles aligned with the applied field in accordance with the torque relations for magnetic dipoles, as discussed in Sec. 6.4.

2. Even if each atom has a zero moment, the presence of an external field distorts the electron orbits and thus creates or induces magnetic dipole moments.

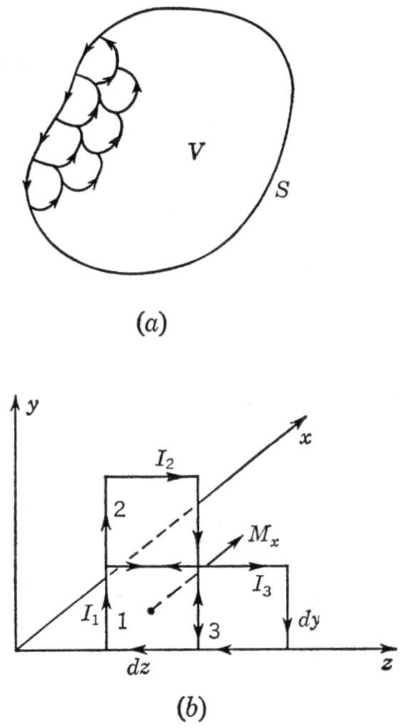

FIG. 7.1. A material body and the equivalent circulating atomic currents.

In many materials the individual atoms do not have an intrinsic magnetic dipole moment, so that only the second effect above takes place when an external field is applied. When induced dipoles are produced, they are always aligned in opposition to the inducing external field. This results in a net magnetic flux density in the interior of the material body, which is less than that of the external field itself. Such materials are referred to as diamagnetic, and the effects as diamagnetism. Some of the materials in this class are copper, bismuth, zinc, silver, lead, and mercury.

For materials with constituent atoms having nonzero dipole moments, both of the above effects are present when an external field is applied. However, the resultant dipole moment is always in the general direction of the applied field because the first effect overshadows that of the induced dipoles. Materials in this class are known as paramagnetic materials, and the effect as paramagnetism. In these materials the aligned dipoles contribute a net flux in the same direction as that of the applied field, and the resultant flux density in the interior of the material body is increased. Both diamagnetic and paramagnetic effects are usually very small. There does, however, exist a class of materials which exhibit very large paramagnetic effects, and these are referred to as ferromagnetic materials. Examples of ferromagnetic materials are iron, nickel, and cobalt. The properties of such materials will be considered in greater detail later.

When an external field is applied to a material body we may describe the external effects produced by the body by assigning to the material a volume distribution of magnetic dipoles. If $\mathbf{m}$ is the magnetic dipole moment of an individual atom or molecule and there are N such effective dipoles per unit volume, then the magnetic dipole polarization $\mathbf{M}$ per unit volume is given by

$$\mathbf{M} = N\mathbf{m} \qquad \text{amp/m}\dagger \tag{7.2}$$

Having assigned this dipole polarization to the material, we may evaluate the external effects of the material body without any further consideration as to the exact physical process responsible for the existence of such magnetic dipoles.

7.2. Equivalent Volume and Surface Polarization Currents

Using (6.25), the vector potential $\mathbf{A}$ at a point (x,y,z) due to an infinitesimal magnetic dipole $\mathbf{m}$ at (x',y',z') a distance R away is given by

$$\mathbf{A} = \frac{\mu_0}{4\pi} \nabla \times \frac{\mathbf{m}}{R}$$

The contribution to the vector potential from the magnetic polarization of a material body will be given by an integral over the body of the contribution from each elementary dipole element $\mathbf{M}\, dV'$. Thus we have

$$\mathbf{A}(x,y,z) = \frac{\mu_0}{4\pi} \int_V \nabla \times \frac{\mathbf{M}}{R}\, dV' \tag{7.3}$$

as illustrated in Fig. 7.2.

Since the physical basis of the magnetic dipole moment is the circulating atomic currents, it would appear that we should be able to describe the field produced by a magnetized body in terms of equivalent volume and surface currents. This is indeed the case, and by a suitable juggling of the expression (7.3), it can be put into a form that shows this explicitly.

† Because of the very limited size of the alphabet (both English and Greek) it is sometimes necessary that the same letter mean several things. In Chap. 6, the letter $\mathbf{M}$ designated the total dipole moment of an arbitrary current loop. For the remainder of the book, beginning in this chapter, $\mathbf{M}$ represents the dipole moment per unit volume. With a little care it should be possible to avoid confusing the two.

The desired result, which we shall prove, is that

$$\mathbf{A}(x,y,z) = \frac{\mu_0}{4\pi}\int_V \frac{\nabla' \times \mathbf{M}(x',y',z')}{R}\,dV' + \frac{\mu_0}{4\pi}\oint_S \frac{\mathbf{M} \times \mathbf{n}}{R}\,dS' \qquad (7.4)$$

where $\mathbf{n}$ is the unit outward normal to the surface S of the body, and ∇' indicates differentiation with respect to the source variables x', y', z'. If we compare (7.4) with the expression for the vector potential from a distribution of true currents, we are led to interpret the term $\nabla' \times \mathbf{M}$ in the volume integral as an equivalent volume density of polarization current $\mathbf{J}_m$ and similarly to interpret the term $\mathbf{M} \times \mathbf{n}$ in the surface integral as the equivalent surface polarization current $\mathbf{J}_{ms}$. Hence we may write

$$\nabla' \times \mathbf{M} = \mathbf{J}_m \qquad (7.5a)$$

$$\mathbf{M} \times \mathbf{n} = \mathbf{J}_{ms} \qquad (7.5b)$$

and (7.4) becomes

$$\mathbf{A} = \frac{\mu_0}{4\pi}\left(\int_V \frac{\mathbf{J}_m}{R}\,dV' + \oint_S \frac{\mathbf{J}_{ms}}{R}\,dS'\right) \qquad (7.5c)$$

Equations (7.5a) and (7.5b) are more than mathematical equivalents in that they represent net effective, though inaccessible, currents. These currents arise when adjacent molecular current loops do not quite cancel each other. This will take place if the magnetization is inhomogeneous, including, particularly, the discontinuity at the boundary of a magnetic material. Figure 7.1b illustrates three adjacent molecular current loops, which, for simplicity, are chosen to be square and to lie in the yz plane. They represent the x component of the net magnetic moment per unit volume for the cube with sides dx, dy, and dz. If $\mathbf{M}$ is the dipole moment per unit volume, then

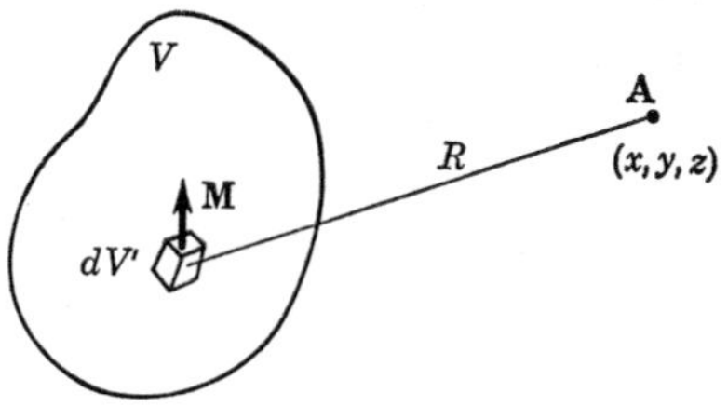

FIG. 7.2. Illustration of evaluation of vector potential from a volume distribution of magnetic dipoles.

$$\mathbf{m} = \mathbf{M}\,dx\,dy\,dz = I\,d\mathbf{S}$$

is the moment of a molecular loop of area $d\mathbf{S}$ with a net circulating current I. Applying this to loops 1, 2, and 3 in Fig. 7.1b leads to

$$I_1 = \frac{M_x\,dx\,dy\,dz}{dy\,dz} = M_x\,dx$$

$$I_2 = \left(M_x + \frac{\partial M_x}{\partial y}\,dy\right)dx = M_x\,dx + \frac{\partial M_x}{\partial y}\,dx\,dy$$

$$I_3 = \left(M_x + \frac{\partial M_x}{\partial z}\,dz\right)dx = M_x\,dx + \frac{\partial M_x}{\partial z}\,dx\,dz$$

A net current flows in the z direction if $I_2 \neq I_1$ and in the y direction if $I_1 \neq I_3$. The net current is given by

$$I_z = -\frac{\partial M_x}{\partial y}\,dx\,dy \qquad I_y = \frac{\partial M_x}{\partial z}\,dx\,dz$$

or in terms of a current density,

$$J_z = -\frac{\partial M_x}{\partial y} \qquad J_y = \frac{\partial M_x}{\partial z}$$

The remaining components follow from cyclic permutation; that is, we replace x by y, y by z, and z by x, and the final result is simply (7.5a). Consequently, that expression represents the physical picture presented here, of a net effective molecular current due to incomplete cancellation of adjacent current loops in the interior of a magnetizable body. A similar demonstration would show the physical basis for (7.5b).

If the magnetization of the body is uniform, **M** is constant throughout, the curl of **M** is zero, and the equivalent volume polarization current is zero. The surface polarization current vanishes only when **M** is perpendicular to the surface so that $\mathbf{M} \times \mathbf{n}$ is zero.

Derivation of Eq. (7.4)

From (7.3) we have

$$\mathbf{A}(x,y,z) = \frac{\mu_0}{4\pi}\int_V \nabla \times \frac{\mathbf{M}(x',y',z')}{R}\,dV' = \frac{-\mu_0}{4\pi}\int_V \mathbf{M}(x',y',z') \times \nabla\left(\frac{1}{R}\right)dV'$$

since $\nabla \equiv \mathbf{a}_x\,\partial/\partial x + \mathbf{a}_y\,\partial/\partial y + \mathbf{a}_z\,\partial/\partial z$ does not operate on $\mathbf{M}(x',y',z')$. Now

$$-\mathbf{M}(x',y',z') \times \nabla(1/R) = \mathbf{M}(x',y',z') \times \nabla'(1/R)$$

and

$$\nabla' \times \frac{\mathbf{M}(x',y',z')}{R} = \frac{1}{R}\nabla' \times \mathbf{M}(x',y',z') - \mathbf{M} \times \nabla'\left(\frac{1}{R}\right)$$

so that we obtain

$$\mathbf{A} = \frac{\mu_0}{4\pi}\int_V \frac{\nabla' \times \mathbf{M}}{R}\,dV' - \frac{\mu_0}{4\pi}\int_V \nabla' \times \frac{\mathbf{M}}{R}\,dV' \tag{7.6}$$

If we make use of the following vector identity,

$$-\frac{\mu_0}{4\pi}\int_V \nabla' \times \frac{\mathbf{M}}{R}\,dV' = \frac{\mu_0}{4\pi}\oint_S \frac{\mathbf{M} \times \mathbf{n}}{R}\,dS' \tag{7.7}$$

then (7.6) is converted into (7.4), which is the desired relation. Equation (7.7) can be thought of as a form of Stokes' theorem; a proof is given below.

Let **C** be a constant vector, and **F** a variable vector function. Then we have $\nabla' \cdot [\mathbf{F}(x',y',z') \times \mathbf{C}] = (\nabla' \times \mathbf{F}) \cdot \mathbf{C} - (\nabla' \times \mathbf{C}) \cdot \mathbf{F} = \mathbf{C} \cdot \nabla' \times \mathbf{F}$, since **C** is a constant vector and $\nabla' \times \mathbf{C} = 0$. Using this result and applying the divergence theorem permits us to establish the following vector relations, namely,

$$\int_V \nabla' \cdot (\mathbf{F} \times \mathbf{C})\,dV' = \oint_S \mathbf{F} \times \mathbf{C} \cdot \mathbf{n}\,dS' = \mathbf{C} \cdot \int_V \nabla' \times \mathbf{F}\,dV'$$

$$= -\mathbf{C} \cdot \oint_S \mathbf{F} \times \mathbf{n}\,dS' \tag{7.8}$$

where $\mathbf{F} \times \mathbf{C} \cdot \mathbf{n} = -\mathbf{C} \cdot \mathbf{F} \times \mathbf{n}$ has also been used. The constant vector $\mathbf{C}$ is arbitrary, so that the result above can hold only if $\int_V \nabla' \times \mathbf{F}\, dV' = -\oint_S \mathbf{F} \times \mathbf{n}\, dS'$. If we now let $\mathbf{F}$ be equal to $\mathbf{M}/R$, the relation (7.7) follows at once and the proof of the equivalence of (7.3) and (7.4) is completed.

7.3. The Magnetic Field Intensity H

When dealing with dielectric bodies in the presence of electrostatic fields, it was convenient to introduce the displacement vector $\mathbf{D}$ in order to eliminate the necessity of taking the electric dipole polarization $\mathbf{P}$ of the material into account explicitly. A similar procedure is used for the magnetic case.

Let us consider a material body of infinite extent in which a true current distribution $\mathbf{J}$ exists. This current gives rise to a partial magnetic field $\mathbf{B}_0$, which in turn gives rise to a magnetic polarization of the material. The secondary field $\mathbf{B}_i$ from the magnetic dipole polarization may be evaluated from the equivalent volume polarization current $\mathbf{J}_m$. The total magnetic field is thus given by

$$\mathbf{B} = \mathbf{B}_0 + \mathbf{B}_i = \nabla \times \mathbf{A}_0 + \nabla \times \mathbf{A}_i = \nabla \times \mathbf{A}$$

where $\mathbf{A}_0$ is the vector potential from the true current $\mathbf{J}$, and $\mathbf{A}_i$ arises from the polarization current $\mathbf{J}_m$. Introducing the current sources into the expression for $\mathbf{A}$ results in

$$\mathbf{B} = \nabla \times \frac{\mu_0}{4\pi} \int_V \frac{\mathbf{J} + \mathbf{J}_m}{R}\, dV' \tag{7.9}$$

In Sec. 6.6 it was shown that the curl of $\mathbf{B}$ was given by $\nabla \times \mathbf{B} = \mu_0 \mathbf{J}$ for currents located in vacuum. The derivation was performed by taking the curl of both sides of an expression similar to (7.9), replacing $\nabla \times \nabla \times$ by $\nabla\nabla \cdot - \nabla^2$ and then using the singularity property of the function $\nabla^2(1/R)$ to obtain the final result. If we take the curl of (7.9), we have a completely analogous development, the only difference being that the term $\mathbf{J} + \mathbf{J}_m$ appears instead of $\mathbf{J}$. Therefore (7.9) leads to

$$\nabla \times \mathbf{B} = \mu_0(\mathbf{J} + \mathbf{J}_m) \tag{7.10}$$

This result should not surprise us since the separation of current into the components $\mathbf{J}$ and $\mathbf{J}_m$ is, in a sense, an artifice. We have seen that $\mathbf{J}_m$ has all the physical characteristics of current flow, and if measurements were made on an atomic scale, there would be no basis for distinguishing $\mathbf{J}_m$ from $\mathbf{J}$. Consequently, (7.10) may also be thought of as evolving from (6.41) by generalizing the total current to include the sum of the true current $\mathbf{J}$ and magnetization current $\mathbf{J}_m = \nabla \times \mathbf{M}$. The viewpoint noted here is quite similar to that presented in electrostatics. In the absence of material bodies, the electric field $\mathbf{E}$ is related to true charge, just as magnetic field is related to true current. The presence of material bodies can be considered in terms of equivalent sources that are set up, sources which act in every way similarly to true sources. In electrostatics these are equivalent polarization charges ($\rho_p = -\nabla \cdot \mathbf{P}$, $\rho_{sp} = \mathbf{n} \cdot \mathbf{P}$), while in magnetostatics they are equivalent magnetization currents as given in (7.5*a*) and (7.5*b*).

Since the equivalent volume polarization current $\mathbf{J}_m$ is equal to the curl of the magnetic dipole polarization **M**, in place of (7.10) we also have $\nabla \times \mathbf{B} = \mu_0 \mathbf{J} + \mu_0 \nabla \times \mathbf{M}$, or

$$\nabla \times \left(\frac{\mathbf{B}}{\mu_0} - \mathbf{M}\right) = \mathbf{J} \tag{7.11}$$

The vector $\mathbf{B}/\mu_0 - \mathbf{M}$ on the left-hand side of (7.11) has as its source only the true current **J**. Therefore, to eliminate the necessity of dealing directly with the polarization **M**, a new vector **H** is introduced by the relation

$$\mathbf{H} = \frac{\mathbf{B}}{\mu_0} - \mathbf{M} \tag{7.12}$$

The vector function **H** is called the magnetic-field-intensity vector, and its (vortex) source is the true current distribution **J**; that is,

$$\nabla \times \mathbf{H} = \mathbf{J} \tag{7.13}$$

Note that $\nabla \cdot \mathbf{H} = -\nabla \cdot \mathbf{M}$, so that an irrotational component of **H** exists for which the sources correspond, mathematically, to an equivalent magnetic charge. The unit for **H** is amperes per meter.

For most materials (ferromagnetic materials excluded) it is found that the polarization **M** is directly proportional to the applied external field $\mathbf{B}_0$, and hence, because of the linear relation (7.12), **M** is then directly proportional to **H** also. Following the conventional procedure, we may therefore write

$$\mathbf{M} = \chi_m \mathbf{H} \tag{7.14}$$

where the dimensionless constant of proportionality χ_m is called the magnetic susceptibility of the material. It is a measure of how susceptible the material is to polarization by an applied field. For diamagnetic materials χ_m is negative, while for paramagnetic materials it is positive. A relation like (7.14) may be written for ferromagnetic materials also, but for these materials χ_m is a function of the field intensity **H** as well as the past magnetization history of the material.

If (7.14) is substituted into (7.12), we get

$$\mathbf{B} = \mu_0(1 + \chi_m)\mathbf{H} = \mu\mathbf{H} \tag{7.15}$$

where $\mu = \mu_0(1 + \chi_m)$ and is called the magnetic permeability of the material. In practice, the permeability μ is determined experimentally. When it is known, the effects of material bodies may be accounted for by using (7.13) to evaluate **H** and (7.15) to find the total net magnetic flux in the body. This process eliminates the necessity of taking the dipole polarization **M** into account explicitly.

In the interior of a material body both **B** and **M** vary rapidly in the region between individual atoms or molecules. However, in any practical procedure that might be used to measure **B** or **M**, the field is sampled in a space containing many atoms. Consequently, what is important are the space-average fields, where the average is taken over the domain of many atoms but which at the same time is small from the macroscopic viewpoint. Hence, in future work, when we speak of the polarization **M** and the total field **B**, such a space average is assumed. This concept is equivalent to the one introduced in discussing polarization in electrostatics.

For paramagnetic and diamagnetic materials μ differs from μ_0 by an amount which is entirely negligible for many practical situations; e.g., for bismuth, which is the strongest diamagnetic material known, $\mu = 0.99983\mu_0$, while for aluminum $\mu = 1.00002\mu_0$. For ferromagnetic materials μ is much greater than μ_0; for example, for silicon iron $\mu = 7{,}000\mu_0$; for permalloy 78 μ is about 10^5 greater than μ_0. Alloys with even larger values of μ exist. It must be kept in mind, however, that a unique value for μ does not exist for ferromagnetic materials because of strong nonlinearities. In ferromagnetic materials the relation between **B** and **H** is usually presented graphically in the form of a curve known as the hysteresis curve, or *B-H* curve. The nature of this relationship is discussed in the next section from an engineering viewpoint. The physical processes which give rise to the hysteresis curve are considered in Sec. 7.8.

7.4. *B-H* Curve

Consider a specimen of ferromagnetic material in the shape of a toroid, as in Fig. 7.3. The cross-section radius is r_0, and the mean toroid radius

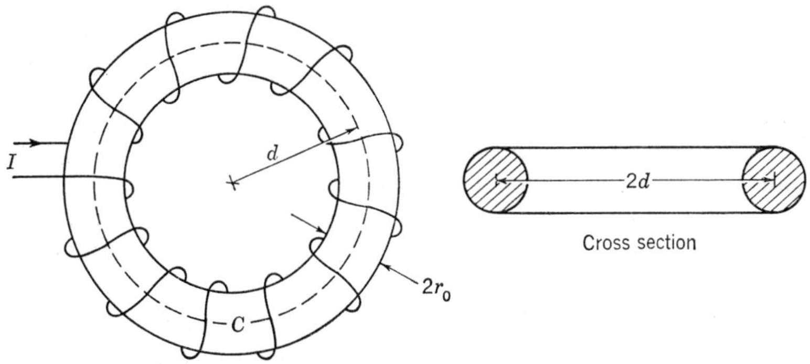

FIG. 7.3. A toroidal specimen of ferromagnetic material.

is d. The toroid is wound uniformly with N turns of wire through which an adjustable current I flows. The cross-sectional radius r_0 is assumed to be much smaller than the toroid radius d, so that the length of any closed circumferential path within the toroid may be taken as $2\pi d$.

The magnetic field intensity H in the interior is essentially uniform over the cross section and circumferentially directed when r_0 is small compared with d. The magnitude of H may be found from Ampère's circuital law. Starting with $\nabla \times \mathbf{H} = \mathbf{J}$, an application of Stokes' law gives

$$\int_S \nabla \times \mathbf{H} \cdot d\mathbf{S} = \oint_C \mathbf{H} \cdot d\mathbf{l} = \int_S \mathbf{J} \cdot d\mathbf{S}$$

If the line integral is taken along the path C in the interior of the toroid, as shown in Fig. 7.3, we get

$$\oint_C \mathbf{H} \cdot d\mathbf{l} = 2\pi dH = NI$$

since the total current linked by C, that is, cutting through a plane surface with boundary C, is NI. Hence the field intensity inside the toroid is $H = NI/2\pi d$ and may be varied by changing the current I.

If the ferromagnetic material is in an unmagnetized state (this condition can be arrived at by reversing the current I several times while at the same time gradually reducing the current magnitude to zero), it is found that as I, and hence H, is increased, the flux density B in the specimen increases from zero along a curve such as P_1P_2 in Fig. 7.4. If the current is now reduced from I_1 to $-I_1$ so as to change H from H_1 to $-H_1$, the flux density B will change according to the pattern illustrated by the curve $P_2P_3P_4P_5$. Increasing H from $-H_1$ to H_1 again causes the flux density to return to its value at the point P_2 along the curve $P_5P_6P_2$. The resultant closed curve $P_2P_3P_5P_6P_2$ is called a hysteresis curve. Further reversal of the field H between H_1 and $-H_1$ causes the same hysteresis loop to be retraced. The initial part of the curve from P_1 to P_2 is called the initial magnetization curve. When H is reduced from H_1, a value of $H = 0$ is reached for which the corresponding flux density B is nonzero (i.e., at P_3). The particular value $B = B_r$ at the point $H = 0$ is called the remanent flux density, i.e., the remaining flux density in the material. The value of H required to reduce B to zero is $-H_C$ and is called the coercive force and is the value of H at the point P_4 on the curve.

If H is increased from H_1 to H_2 and then reversed between the limits H_2 and $-H_2$, a new hysteresis loop with extremities P_7 and P_8 is traced out. When the magnitude of H increases beyond a certain range, the magnetization M does not increase any further because the material becomes magnetically saturated. In this condition all the magnetic dipoles are aligned with the field. In the saturated condition incremental changes in H cause changes in B according to their vacuum relations. For the hysteresis loop that brings the material into saturation, the value of remanent flux density is called the retentivity and the value of the field intensity H required to reduce B to zero is called the coercivity of the material. In the next chapter it will be shown that the area of the hysteresis loop is proportional to the energy dissipated in cycling the material around the hysteresis loop. This energy is dissipated as heat in the material.

The ratio of B to H is the slope of a line joining the origin to a point on the hysteresis curve and gives the permeability μ. Clearly, there is no

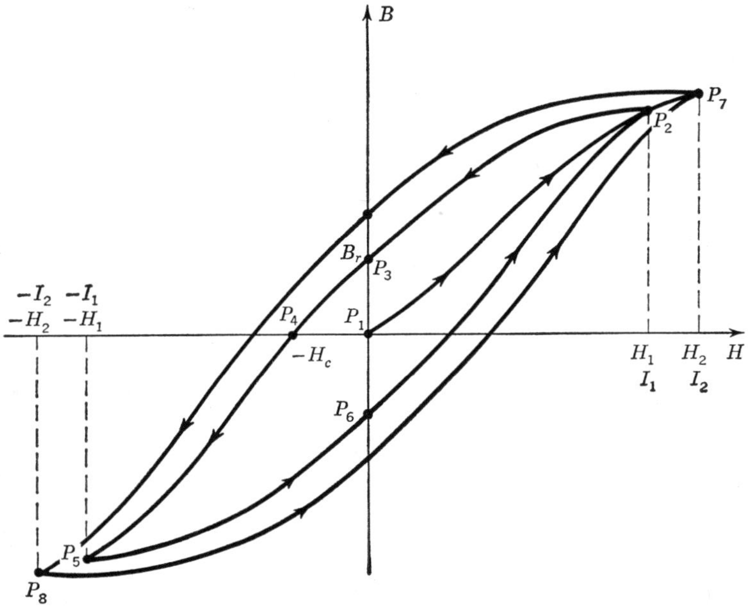

FIG. 7.4. A typical hysteresis (B-H) curve for a ferromagnetic material.

unique value of permeability. The tangent to the hysteresis curve at any particular point gives the differential permeability at that point. However, if H is varied between the limits $H_0 + \Delta H$ and $H_0 - \Delta H$, a new minor hysteresis loop centered about H_0 is traced out so that the differential permeability has little significance. The slope of the line joining the tips of the minor hysteresis loop is called the incremental permeability. It is the effective permeability for a small a-c current superposed on a d-c current.

7.5. Boundary Conditions for B and H

In the discussion leading to the introduction of the field intensity vector **H**, we considered an infinite medium and thereby avoided consideration of the equivalent surface polarization current $\mathbf{J}_{ms} = \mathbf{M} \times \mathbf{n}$. Consequently, the relationships that have been developed are applicable everywhere within or external to a finite body, but for points at the surface, either they must be modified to take account of the surface magnetization current, or a limiting procedure adopted that includes the discontinuity in **M**. Rather than deal with the surface currents explicitly, we shall demonstrate that they may be accounted for by specifying certain boundary conditions for the field vectors **B** and **H** at the boundary between two different media. This procedure duplicates that adopted in electrostatics where the effect of surface polarization charge on the **E** and **D** fields was considered in terms of certain boundary conditions.

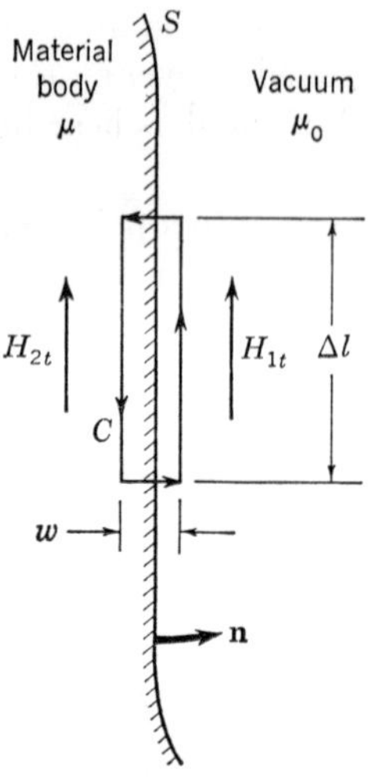

FIG. 7.5. Determination of boundary condition on the tangential component of **H**.

Consider a material body with a permeability μ, different from μ_0 and located in vacuum, as in Fig. 7.5. Let H_{1t} be the component of **H** which is tangent to the surface S on the vacuum side of the boundary, and let H_{2t} be the tangential component of **H** at the surface in the interior of the body. From (7.13) we have $\nabla \times \mathbf{H} = \mathbf{J}$. The integral form of this law is obtained by applying Stokes' theorem and leads to Ampère's circuital law for **H**; that is,

$$\int_{\Delta S} \nabla \times \mathbf{H} \cdot dS = \oint_C \mathbf{H} \cdot d\mathbf{l} = \int_{\Delta S} \mathbf{J} \cdot d\mathbf{S} \tag{7.16}$$

where C is the contour bounding the open surface ΔS. Let C be a small loop with sides parallel to the boundary surface S in Fig. 7.5 and an infinitesimal distance on either side. Applying (7.16) we have $\oint_C \mathbf{H} \cdot d\mathbf{l} = 0$, if, as we suppose, no true current is enclosed by the contour. For the small rectangular path illustrated, whose length is Δl, the line integral gives essentially $H_{1t}\,\Delta l - H_{2t}\,\Delta l = 0$, or

$$H_{1t} = H_{2t} \tag{7.17}$$

since the contribution from the ends is negligible; that is, the width w of the rectangular path can be made as small as we wish without affecting Δl. This relation tells us that the tangential magnetic field is continuous across a surface discontinuity separating a material body and vacuum. The same result is obviously also true at the boundary between two different material bodies.

Since $\mathbf{B} = \mu\mathbf{H}$, the corresponding boundary conditions for the tangential components of $\mathbf{B}$ are

$$\frac{B_{1t}}{\mu_0} = \frac{B_{2t}}{\mu} \tag{7.18}$$

Thus the tangential components of $\mathbf{B}$ are not continuous since μ and μ_0 are not equal. At first we might be somewhat puzzled over this result, but it can be explained quite readily. In fact, the tangential component of $\mathbf{B}$ is discontinuous because of the presence of the surface polarization current $\mathbf{J}_{ms} = \mathbf{M} \times \mathbf{n}$. This result may be demonstrated as follows. We have

$$\mathbf{B}_1 = \mu_0\mathbf{H}_1$$
$$\mathbf{B}_2 = \mu\mathbf{H}_2 = \mu_0(\mathbf{H}_2 + \mathbf{M})$$

and subtracting the tangential components of $\mathbf{B}$ gives

$$B_{2t} - B_{1t} = \mu_0(H_{2t} - H_{1t}) + \mu_0 M_t = \mu_0 M_t \tag{7.19}$$

since $H_{2t} = H_{1t}$ from (7.17). The polarization vector $\mathbf{M}$ may be written as the sum of a tangential component $\mathbf{M}_t$ and a normal component $\mathbf{M}_n$. The cross product of $\mathbf{M}$ with the surface normal $\mathbf{n}$ thus gives

$$\mathbf{M} \times \mathbf{n} = \mathbf{M}_t \times \mathbf{n} + \mathbf{M}_n \times \mathbf{n} = \mathbf{M}_t \times \mathbf{n} = \mathbf{J}_{ms}$$

The magnitude of $\mathbf{J}_{ms}$ is M_t since $\mathbf{M}_t$ and $\mathbf{n}$ are perpendicular. Thus the tangential components of $\mathbf{B}$ are discontinuous by an amount equal to μ_0 times the magnitude of the polarization current $\mathbf{J}_{ms}$. With reference to Fig. 7.6, let the surface polarization current flow into the paper. The field produced by this current sheet is denoted by $\mathbf{B}_s$, and the field produced by all other current sources as $\mathbf{B}_0$. The field $\mathbf{B}_0$ is continuous across the surface S, but the field $\mathbf{B}_s$ is oppositely directed on the two sides of the current sheet. It is for this reason that the tangential component of the total field $\mathbf{B}_0 + \mathbf{B}_s$ is discontinuous across S.

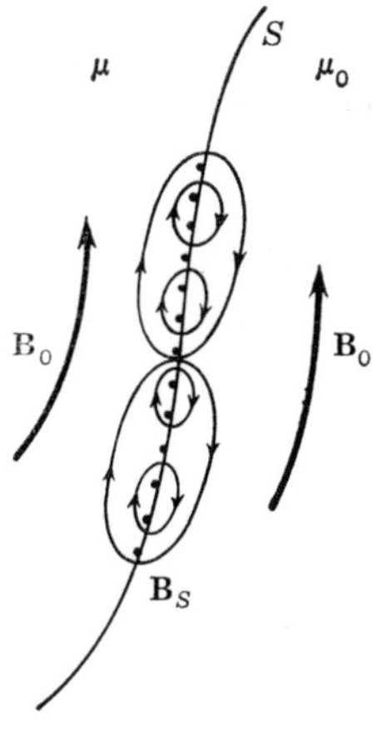

FIG. 7.6. Discontinuity in the tangential component of B as produced by the polarization surface current.

A further demonstration of the discontinuity in $\mathbf{B}$ may be had by applying (7.10) to the contour shown in Fig. 7.5. In this case the current term in the right-hand side of (7.10) is the total current, which includes polarization currents. Consequently, in place of the expression leading to (7.17), we get $B_{2t}\,\Delta l - B_{1t}\,\Delta l = \mu_0 J_{ms}\,\Delta l$, and this reduces to the result of (7.19). Note that since J_{ms} is a surface current, it does not disappear no matter how small the loop width w is allowed to get. The boundary conditions (7.17) and (7.18) take into account the effects of the surface polarization current, so that we do not need to include these explicitly when we are dealing with the interaction of material bodies with magnetic fields.

To determine the boundary conditions on the normal components of **B** and **H**, we note that since **B** is always equal to the curl of a vector potential **A**, the divergence of **B** is always zero; that is, $\nabla \cdot \mathbf{B} = 0$. An application of the divergence theorem gives

$$\int_V \nabla \cdot \mathbf{B}\, dV = \int_S \mathbf{B} \cdot d\mathbf{S} = 0$$

We now apply this result to the closed surface of a small coin-shaped box with end faces on adjacent sides of the surface S of the material body, as in Fig. 7.7. The height w of the box is assumed to be so small that the magnetic flux through the side is negligible. Since the total flux through the closed surface is zero, just as much flux leaves face 1 as enters in through face 2. Thus $B_{1n}\,\Delta S_0 = B_{2n}\,\Delta S_0$, or

$$B_{1n} = B_{2n} \qquad (7.20)$$

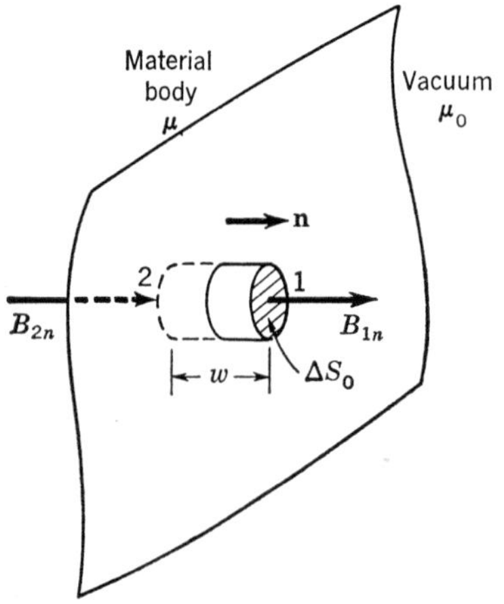

FIG. 7.7. Derivation of boundary conditions for the normal component of **B**.

The corresponding boundary condition on the normal component of **H** is obtained by using the relationship between **B** and **H** to get

$$\mu_0 H_{1n} = \mu H_{2n} \qquad (7.21)$$

It is seen that the normal component of **B** is continuous, but not so for the normal component of **H**. This result for the normal component of **H** is the consequence of the magnetic polarization of the material. We shall show later that we can think of the magnetic polarization as equivalent to a magnetic charge distribution in analogy with the similar situation in electrostatics. With this alternative viewpoint, the discontinuity in **H** arises because of an equivalent surface layer of fictitious magnetic charge.

Refraction of Magnetic Lines of Flux

At a boundary surface between two different materials with permeabilities μ_1 and μ_2, the boundary conditions on **B** and **H** are

$$B_{1n} = B_{2n} \tag{7.22a}$$
$$\mu_2 B_{1t} = \mu_1 B_{2t} \tag{7.22b}$$
$$H_{1t} = H_{2t} \tag{7.22c}$$
$$\mu_1 H_{1n} = \mu_2 H_{2n} \tag{7.22d}$$

These boundary equations are of a similar nature to those occurring in electrostatics as well as those occurring in the study of stationary current flow fields. For this reason one expects to have a refraction or bending of the flux lines associated with **B** at a surface separating two different materials.

Thus let us consider the geometry shown in Fig. 7.8, where the field **B** makes the angles θ_1 and θ_2 with the interface normal in mediums 1 and 2, respectively. In accord with (7.22) we have

$$B_1 \cos \theta_1 = B_2 \cos \theta_2$$
$$\mu_2 B_1 \sin \theta_1 = \mu_1 B_2 \sin \theta_2$$

It is possible to eliminate B_1 and B_2 by taking the quotient of the two equations, with the result that $\mu_2 \tan \theta_1 = \mu_1 \tan \theta_2$, or

$$\tan \theta_2 = \frac{\mu_2}{\mu_1} \tan \theta_1 \tag{7.23}$$

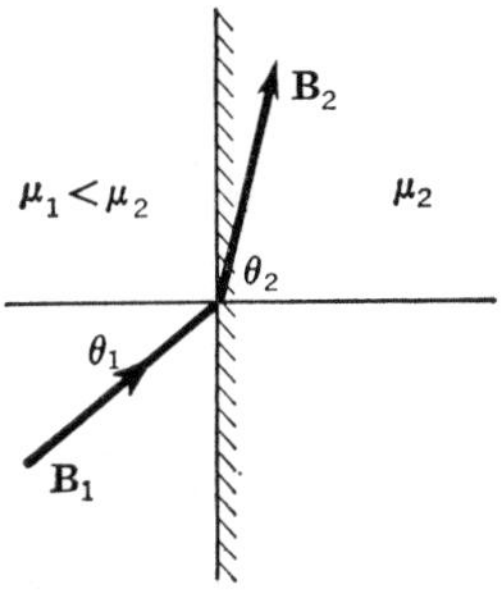

FIG. 7.8. Refraction of magnetic flux lines.

Equation (7.23) shows that the magnetic flux lines are bent away from the normal in the medium with the highest value of permeability. If medium 2 is a ferromagnetic material and medium 1 is vacuum or air, μ_2 is much greater than $\mu_1 = \mu_0$. In this case $\tan \theta_1$ is very small and the flux lines are for all practical purposes normal to the surface on the air side, provided θ_2 does not equal exactly 90°.

Discontinuity in **H** *at a Current Sheet*

In this section we shall examine the behavior of the magnetic field **H** in the vicinity of a thin sheet of conduction current. Such current sheets are of particular importance in the general boundary-value problem for time-varying magnetic fields. The surface of a closely wound solenoid also essentially constitutes a surface current sheet. Applying the right-hand rule, i.e., with the thumb of the right hand turned in the direction of current flow the fingers of the right hand are along the direction of the magnetic lines of flux, we see that the magnetic field is perpendicular to the current flow lines, as illustrated in Fig. 7.9. The line integral of **H** around the small contour C is equal to the total current passing through the contour C, that is, passing through the surface ΔS_0. If the enclosed current has a volume density **J**, we obtain

$$\oint_C \mathbf{H} \cdot d\mathbf{l} = \int_{\Delta S_0} \mathbf{J} \cdot d\mathbf{S}$$

Because the contour is of infinitesimal size and we assume $h \ll \Delta l$,

$$(H_{1t} - H_{2t})\,\Delta l = J\,\Delta S_0 \tag{7.24}$$

since in the contour integral the contribution from the ends is negligible. If we let the thickness h of the current sheet tend to zero and increase the

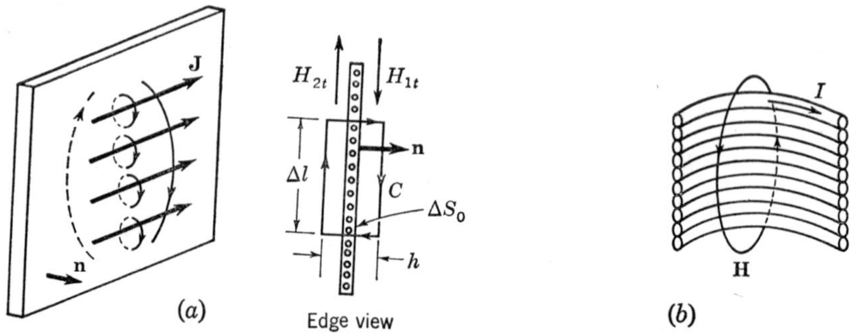

FIG. 7.9. Illustration of current sheets. (*a*) A uniform current sheet, (*b*) cross section of a solenoid.

volume density of current **J** so that the total enclosed current per unit length remains constant, i.e.,

$$\lim_{h\to 0} hJ = J_s$$

we obtain a true surface current sheet with a current J_s amperes per meter. In the limit, (7.24) gives

$$(H_{1t} - H_{2t})\,\Delta l = \lim_{h\to 0} hJ \frac{\Delta S_0}{h}$$
$$= J_s\,\Delta l$$

or

$$\underline{H_{1t} - H_{2t} = J_s} \tag{7.25}$$

since $\Delta S_0 = h\,\Delta l$. Thus the tangential component of **H** is discontinuous across a current sheet by an amount equal to the surface current density. This behavior is similar to that noted for the tangential component of **B** at a polarization current sheet. In vector form (7.25) may be written as

$$\underline{\mathbf{n} \times (\mathbf{H}_1 - \mathbf{H}_2) = \mathbf{J}_s} \tag{7.26}$$

since $\mathbf{n} \times \mathbf{H}$ selects the tangential component of **H**. In (7.26) **n** is the unit outward normal to the current sheet.

The normal component of **H** is continuous across the current sheet since the normal component of **B** is, and $\mathbf{H} = \mathbf{B}/\mu_0$ on both sides.

Example 7.1. Far Field from a Long Solenoid. Consider a long uniform solenoid having N turns per meter and carrying a current I, as in Fig. 7.10. The solenoid is of length L and has a radius a. We choose the coordinate system with the z axis coinciding with the axis of the solenoid.

We wish to determine the field at the point (x,y,z), which is at a much greater distance from the solenoid than the solenoid radius a. We may treat the solenoid as a stack of current loops of total height L. For a

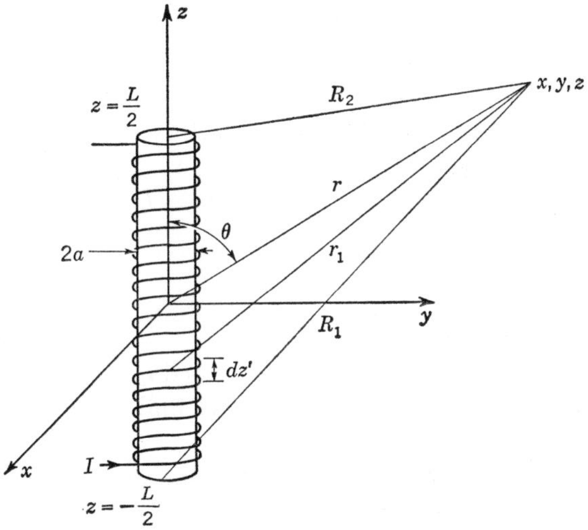

FIG. 7.10. A long uniform solenoid.

section of length dz' there are $N\,dz'$ turns, each with a current I, so that the dipole moment of such a section is

$$\mathbf{m} = NI\,dz'\,\pi a^2 \mathbf{a}_z \tag{7.27}$$

We could now proceed to compute the total vector potential set up by integrating over the length of the solenoid the contribution from each infinitesimal section. However, since we are looking for the field **B**, we shall use the relation (6.26) of Sec. 6.4 and find **B** directly. The field $d\mathbf{B}$ at a long distance from a dipole **m** is given by

$$d\mathbf{B} = \frac{\mu_0}{4\pi} \nabla \left[\mathbf{m} \cdot \nabla \left(\frac{1}{r_1} \right) \right]$$

where r_1 is the distance from the center of the dipole at z' to the field point (x,y,z). The total field B is given by

$$\mathbf{B} = \frac{\mu_0}{4\pi} \nabla \int_{-L/2}^{L/2} \mathbf{m} \cdot \nabla \left(\frac{1}{r_1} \right) dz' = \frac{\mu_0}{4\pi} \nabla \int_{-L/2}^{L/2} \nabla \left(\frac{1}{r_1} \right) \cdot \mathbf{a}_z NI\pi a^2\,dz' \tag{7.28}$$

Now $\nabla(1/r_1) = -\mathbf{r}_1/r_1^3$, where $\mathbf{r}_1 = \mathbf{a}_x x + \mathbf{a}_y y + \mathbf{a}_z(z - z')$. Hence (7.28) gives

$$\begin{aligned}\mathbf{B} &= \frac{-NIa^2\mu_0}{4} \nabla \int_{-L/2}^{L/2} \frac{(z - z')\,dz'}{[x^2 + y^2 + (z - z')^2]^{3/2}} \\ &= \frac{-NIa^2\mu_0}{4} \nabla[x^2 + y^2 + (z - z')^2]^{-1/2}\Big|_{-L/2}^{L/2} \\ &= \frac{-NIa^2\mu_0}{4}\left[\nabla\left(\frac{1}{r_1}\right)\right]_{-L/2}^{L/2} = \frac{-NIa^2\mu_0}{4}\nabla\left(\frac{1}{R_2} - \frac{1}{R_1}\right)\end{aligned}$$

where R_1 and R_2 are the distances from the ends of the solenoid to the field point, as in Fig. 7.10. We may approximate R_1 and R_2 as follows:

$$\begin{aligned}R_2^{-1} &= \left[x^2 + y^2 + \left(z - \frac{L}{2}\right)^2\right]^{-1/2} \\ &= \left[x^2 + y^2 + z^2 + \left(\frac{L}{2}\right)^2 - Lr\cos\theta\right]^{-1/2} \\ &\approx (r^2 - Lr\cos\theta)^{-1/2} \\ &\approx r^{-1}\left(1 + \frac{L}{2r}\cos\theta\right)\end{aligned}$$

where z is replaced by $r\cos\theta$, $(L/2)^2$ is dropped in comparison with r^2, and the binomial expansion used. We similarly find that

$$R_1^{-1} \approx r^{-1}\left(1 - \frac{L}{2r}\cos\theta\right)$$

and hence

$$\mathbf{B} = \frac{-\mu_0 NIa^2L}{4}\nabla\left(\frac{\cos\theta}{r^2}\right) = \frac{\mu_0}{4\pi}M\left(\mathbf{a}_r\frac{2\cos\theta}{r^3} + \mathbf{a}_\theta\frac{\sin\theta}{r^3}\right) \tag{7.29}$$

where $NI\pi a^2L$, the total dipole moment of the solenoid, is designated by M. The relation (7.29) is of the same form as that derived in Sec. 6.4 for a small elementary magnetic dipole. It is, of course, valid only at a distance r that is large compared with the length L.

Example 7.2. Far Field from a Long Cylindrical Bar Magnet. Figure 7.11 illustrates a cylindrical bar magnet of length L and radius a. The magnet is assumed to be permanently magnetized with a uniform magnetic dipole moment $\mathbf{M}_0$ per unit volume, where $\mathbf{M}_0$ is directed in the positive z direction.

The external field produced by the bar magnet may be computed in terms of the equivalent polarization currents. A cross section of the magnet is shown in Fig. 7.11b and illustrates the nature of the equivalent atomic circulating currents. In the interior the volume density of polarization current is given by $\mathbf{J}_m = \nabla \times \mathbf{M}_0$. If $\mathbf{M}_0$ is constant, $\mathbf{J}_m$ vanishes. This result is intuitively obvious from Fig. 7.11b, since with $\mathbf{M}_0$ constant, the circulating current loops all carry the same current, and hence all the interior currents effectively cancel. If $\mathbf{M}_0$ were not uniform throughout, there would be incomplete cancellation from one loop to the next and a residual volume current density would be left, as we have already discussed. Along the outer boundary there is no cancellation of the currents of each adjacent small loop. Thus all the individual current loops combine to produce a net surface polarization current flowing along the surface of the bar magnet. The value of this current has already been

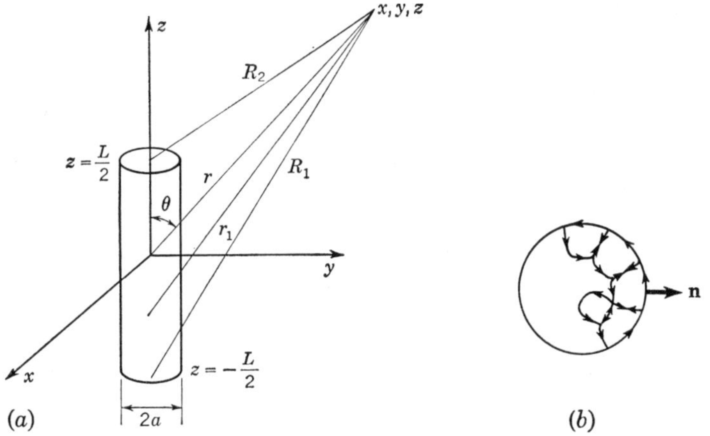

FIG. 7.11. (a) A cylinder bar magnet; (b) enlarged cross section illustrating equivalent circulating polarization currents.

determined to equal $\mathbf{M}_0 \times \mathbf{n}$. Since $\mathbf{M}_0 = \mathbf{a}_z M_0$ and $\mathbf{n} = \mathbf{a}_r$, the surface current is given by

$$\mathbf{J}_{ms} = \mathbf{M}_0 \times \mathbf{n} = M_0 \mathbf{a}_z \times \mathbf{a}_r = M_0 \mathbf{a}_\phi$$

and flows circumferentially around the cylindrical bar magnet. It is now apparent that the field from the bar magnet will be the same as that from an equivalent solenoid having an effective surface current of M_0 amperes per meter. A closely wound solenoid of N turns per meter carrying a current I would be equivalent to the bar magnet if $NI = M_0$. The total moment of the magnet is $\pi a^2 L M_0 = M$, and substitution of this into (7.29) gives the resultant field.

7.6. Scalar Potential for H

In the introduction to Chap. 6 it was pointed out that the early theory of magnetism developed along lines parallel to that of electrostatics. This alternative theory is useful for computing the fields produced by magnetized bodies, and it will therefore be worthwhile to examine it in some detail.

Let us consider a permanently magnetized body of volume V bounded by a surface S and located in a region of space where the true conduction current density $\mathbf{J}$ is zero. Since $\mathbf{J}$ is zero, we have $\nabla \times \mathbf{H} = \mathbf{J} = 0$ also. But this is just the condition that the vector field may be derived from the gradient of a scalar potential. Thus let

$$\mathbf{H} = -\nabla\Phi_m \tag{7.30}$$

where Φ_m is called the magnetic scalar potential. The negative sign in (7.30) is chosen only to make the analogy with electrostatics a closer one. A field $\mathbf{H}$ determined by (7.30) will have zero curl since $\nabla \times \nabla\Phi_m$ is identically zero.

In electrostatics the source for the scalar potential is the charge density ρ. In magnetostatics we do not have a physical magnetic charge (or unit magnetic pole), but we are nevertheless able to recast our expressions for the magnetic field produced by magnetic dipole polarization of a body into a form which gives the potential Φ_m in terms of equivalent magnetic charges.

According to (6.26) of Sec. 6.4, the field $\mathbf{B}$ from a single isolated magnetic dipole $\mathbf{m}$ is given by

$$\mathbf{B}(x,y,z) = \frac{\mu_0}{4\pi}\left\{\nabla\left[\nabla \cdot \frac{\mathbf{m}(x',y',z')}{R}\right] - \nabla^2\left(\frac{\mathbf{m}}{R}\right)\right\} \tag{7.31}$$

At all points except $R = 0$ we have $\nabla^2(1/R) = 0$, so that if we restrict attention to the fields external to a magnetized body, the second term is absent. If the body has a polarization $\mathbf{M}$ per unit volume, the field from the dipoles in a volume element dV' at (x',y',z') is given by

$$d\mathbf{H} = \frac{d\mathbf{B}}{\mu_0} = \frac{1}{4\pi}\nabla\left[\nabla \cdot \frac{\mathbf{M}(x',y',z')}{R}\right] dV'$$

The total field $\mathbf{H}$ is then

$$\mathbf{H} = \frac{1}{4\pi}\nabla\left(\int_V \nabla \cdot \frac{\mathbf{M}}{R}\, dV'\right)$$

$$= \frac{1}{4\pi}\nabla\left[\int_V \mathbf{M} \cdot \nabla\left(\frac{1}{R}\right) dV'\right]$$

If we compare this with (7.30), we see that the scalar potential Φ_m is given by

$$\Phi_m = -\frac{1}{4\pi}\int_V \mathbf{M} \cdot \nabla\left(\frac{1}{R}\right) dV' \tag{7.32}$$

Now

$$\nabla\left(\frac{1}{R}\right) = -\nabla'\left(\frac{1}{R}\right) \quad \text{and} \quad \nabla' \cdot \left(\frac{\mathbf{M}}{R}\right) = \mathbf{M} \cdot \nabla'\left(\frac{1}{R}\right) + \frac{1}{R}\nabla' \cdot \mathbf{M}$$

and hence $$-\mathbf{M} \cdot \nabla\left(\frac{1}{R}\right) = \mathbf{M} \cdot \nabla'\left(\frac{1}{R}\right) = \nabla' \cdot \frac{\mathbf{M}}{R} - \frac{1}{R}\nabla' \cdot \mathbf{M}$$

Substituting into (7.32) we get

$$\Phi_m = \frac{1}{4\pi}\left(\int_V \nabla' \cdot \frac{\mathbf{M}}{R}\,dV' - \int_V \frac{\nabla' \cdot \mathbf{M}}{R}\,dV'\right)$$
$$= \frac{1}{4\pi}\left(\oint_S \frac{\mathbf{M} \cdot \mathbf{n}}{R}\,dS' - \int_V \frac{\nabla' \cdot \mathbf{M}}{R}\,dV'\right) \quad (7.33)$$

after converting the first volume integral to a surface integral by means of the divergence theorem. We now compare (7.33) with the expression for the scalar potential Φ due to a volume and a surface distribution of charge in electrostatics and are thus led to interpret $\mathbf{M} \cdot \mathbf{n}$ as an equivalent surface density of magnetic charge ρ_{ms} and to interpret $-\nabla' \cdot \mathbf{M}$ as equivalent to a volume density of magnetic charge ρ_m. It must be stressed that this equivalence is a purely mathematical one and does not prove a physical existence of magnetic charge.

The result obtained in (7.33) can also be derived by analogy with the relationships found for the electrostatic field. We recall that the **E** field due only to a polarization source satisfies the following equations:

$$\nabla \times \mathbf{E} = 0 \qquad \nabla \cdot \mathbf{E} = -\frac{\nabla \cdot \mathbf{P}}{\epsilon_0} \quad (7.34a)$$

where **P** is the dipole moment per unit volume. Under these conditions it is also true, according to (2.97), that **E** is related to the sources that produce it by

$$\mathbf{E} = -\frac{1}{4\pi\epsilon_0}\nabla\left(\int_V \frac{\nabla' \cdot \mathbf{P}}{R}\,dV' - \oint_S \frac{\mathbf{P} \cdot \mathbf{n}}{R}\,dS'\right)$$

Now if the only source of magnetic field is a permanent magnetization **M**, the **H** field satisfies

$$\nabla \times \mathbf{H} = 0 \qquad \nabla \cdot \mathbf{H} = -\nabla \cdot \mathbf{M} \quad (7.34b)$$

If we compare (7.34*a*) with (7.34*b*), it is noted that the equations are intrinsically the same, it being necessary only to exchange **E** for **H** and $\mathbf{P}/\epsilon_0$ for **M**. Now the Helmholtz theorem states that a vector field is completely determined by its divergence and curl. Thus (7.34*a*) completely specifies **E**, as does (7.34*b*) specify **H**. Then since **E** also satisfies (2.97), it must be that the latter inevitably follows from (7.34*a*). Consequently, **H** must also satisfy such an equation, it being necessary only to replace **E** by **H** and $\mathbf{P}/\epsilon_0$ by **M**. If we carry out this substitution, the result obtained is precisely that given by (7.33).

The above is an excellent illustration of the power of the Helmholtz theorem. This theorem enables us to recognize basic common properties of vector fields independent of their individual physical properties. The results (7.33) and (2.97) are essentially embodied in the general vector formulation of (1.103).

The magnetic field intensity $\mathbf{H}$ is obtained by taking the negative gradient of (7.33) with respect to the coordinates x, y, z of the field point. The field $\mathbf{B}$ is given by $\mathbf{B} = \mu_0\mathbf{H}$. It is important to note that these results hold only for points exterior to the magnetized body since the term $-(\mu_0/4\pi)\nabla^2(\mathbf{m}/R)$ was dropped in (7.31). In the interior of the body Φ_m is still given by (7.32) and $\mathbf{H}$ by (7.30), as we shall show presently. However, the relation between $\mathbf{B}$ and $\mathbf{H}$ is

$$\mathbf{B} = \mu_0(\mathbf{H} + \mathbf{M}) \tag{7.35}$$

and this equation must be used to find $\mathbf{B}$ in the interior.

The above results for $\mathbf{B}$ and $\mathbf{H}$ in the interior of a magnetized body are readily proved. In the interior of the body we see from (7.31) that the field $\mathbf{B}$ is given by

$$\begin{aligned}\mathbf{B}(x,y,z) &= \frac{\mu_0}{4\pi}\nabla\int_V \nabla\cdot\frac{\mathbf{M}}{R}\,dV' - \frac{\mu_0}{4\pi}\int_V \nabla^2\left(\frac{\mathbf{M}}{R}\right)dV' \\ &= -\mu_0\,\nabla\Phi_m - \frac{\mu_0}{4\pi}\int_V \nabla^2\left(\frac{\mathbf{M}}{R}\right)dV'\end{aligned}$$

since the first term defines the magnetic scalar potential Φ_m. We have already noted that the remaining volume integral is zero for external field points since in this case $R \neq 0$ and $\nabla^2(1/R) = 0$. However, if the point (x,y,z) lies interior to V, then $R = 0$ is included in the integral. In this case it is necessary to consider the singularity property of $\nabla^2(1/R)$, as was done in the derivation of (6.41). With this result we have

$$-\frac{\mu_0}{4\pi}\int_V \nabla^2\left(\frac{\mathbf{M}}{R}\right)dV' = \begin{cases}\mu_0\mathbf{M}(x,y,z) & x,\ y,\ z \text{ in } V \\ 0 & x,\ y,\ z \text{ outside } V\end{cases}$$

Hence in the interior of the magnetized body the field $\mathbf{B}$ is given by $\mathbf{B} = -\mu_0\,\nabla\Phi_m + \mu_0\mathbf{M}$. Since $\mathbf{B}$ is also given by (7.35), it follows that (7.33) is a valid expression for the scalar magnetic potential Φ_m for $\mathbf{H}$ both interior and exterior to the magnetized body.

At this point it will be worthwhile to summarize briefly the major results that have been obtained thus far in this chapter. This is desirable in order to correlate the different approaches that may be used to treat the magnetic effects of material bodies. In the absence of any material bodies the magnetic field $\mathbf{B}$ is computed from the true current distribution $\mathbf{J}$, either directly or by means of the vector potential $\mathbf{A}$, according to the methods of Chap. 6. When we are dealing with the field produced by a permanently magnetized body in a region where $\mathbf{J}$ is zero, there are two equivalent approaches available.

For one, we represent the state of polarization of the body by an equivalent volume polarization current $\mathbf{J}_m = \nabla \times \mathbf{M}$ and an equivalent surface polarization current $\mathbf{J}_{ms} = \mathbf{M} \times \mathbf{n}$. From these current sources the field

$\mathbf{B}$ may be evaluated either directly or through the intermediate step of finding the vector potential $\mathbf{A}$ first. In the region surrounding the body the field $\mathbf{H}$ is given by $\mathbf{H} = \mathbf{B}/\mu_0$, while at all points interior to the body $\mathbf{H}$ is given by $\mu_0\mathbf{H} = \mathbf{B} - \mu_0\mathbf{M}$.

In the second approach the field intensity $\mathbf{H}$ is computed first. The magnetic polarization of the material is represented by an equivalent volume distribution of fictitious magnetic charge $\rho_m = -\nabla \cdot \mathbf{M}$, together with an equivalent surface charge $\rho_{ms} = \mathbf{M} \cdot \mathbf{n}$. The field $\mathbf{H}$ is given by $-\nabla\Phi_m$, and the magnetic scalar potential Φ_m is evaluated by methods analogous to those used in electrostatics. From the known value of $\mathbf{H}$, the field $\mathbf{B}$ may be found from the relations

$$\mathbf{B} = \mu_0\mathbf{H} \qquad \text{outside body}$$
$$\mathbf{B} = \mu_0(\mathbf{H} + \mathbf{M}) \qquad \text{interior to body}$$

The parameters μ and χ_m are unnecessary with a theory that takes explicit account of the magnetic polarization of the material.

When the problem involves both magnetizable material bodies and true conduction currents $\mathbf{J}$, the use of the magnetic scalar potential is usually avoided since $\mathbf{H}$ is no longer a conservative field and Φ_m becomes multivalued. An exception is the class of problems that are dual of the d-c electric circuit, where this approach is particularly useful. The best procedure is to obtain appropriate solutions for $\mathbf{B}$ in the regions internal and external to all material bodies and then match these solutions at the boundaries according to the boundary conditions on $\mathbf{B}$ and $\mathbf{H}$ which were presented in Sec. 7.5. For problems of this sort, if $\nabla \times \mathbf{M} = 0$, then it is preferable not to introduce the magnetic polarization of the material explicitly but to use the relation $\mathbf{B} = \mu\mathbf{H}$, together with the boundary conditions on $\mathbf{B}$ and $\mathbf{H}$ instead. The examples to be discussed now will clarify some of the above concepts.

Example 7.3. Use of Scalar Potential to Find Field from a Bar Magnet. The cylindrical magnet is of length L and radius a, as in Fig. 7.12. It is assumed to be uniformly magnetized with a magnetic dipole polarization $\mathbf{M}_0$ per unit volume.

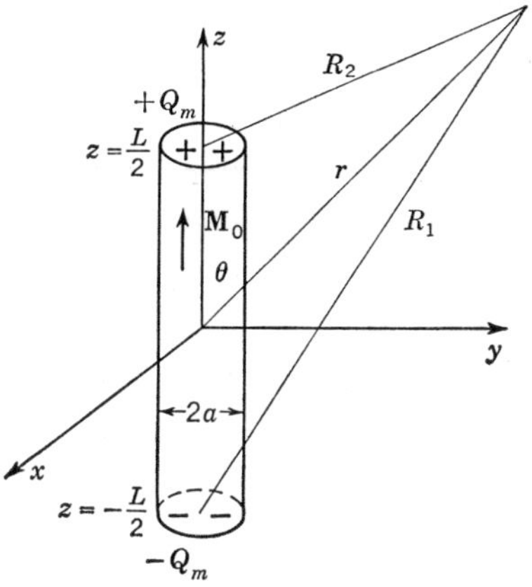

FIG. 7.12. A cylindrical bar magnet.

We shall compute $\mathbf{H}$ at all exterior points, remote from the magnet, by means of the scalar potential Φ_m. Since $\mathbf{M}_0$ is constant, $\nabla \cdot \mathbf{M}_0 = 0$ and there is no equivalent volume charge. On the two end faces at $z = \pm L/2$, we have an equivalent surface charge given by $\rho_{ms} = \mathbf{M}_0 \cdot \mathbf{n} = \mathbf{M}_0 \cdot \mathbf{a}_z$ or $\rho_{ms} = M_0$ at $z = L/2$ and $-M_0$ at $z = -L/2$. There are no surface charges on the sides since $\mathbf{M}_0$ is parallel to the sides. The total charge on each face is of magnitude $Q_m = \pi a^2 \rho_{ms} = \pi a^2 M_0$. For $r \gg a$ we may consider this charge to be lumped together at the center of each face. The scalar potential from two such charges is given by an expression similar to what one has in electrostatics and is

$$\Phi_m = \frac{Q_m}{4\pi}\left(\frac{1}{R_2} - \frac{1}{R_1}\right) = \frac{\pi a^2 M_0}{4\pi}\left(\frac{1}{R_2} - \frac{1}{R_1}\right) \tag{7.36}$$

The field $\mathbf{H}$ is thus given by

$$\mathbf{H} = \frac{\mathbf{B}}{\mu_0} = -\nabla\Phi_m = \frac{-\pi a^2 M_0}{4\pi}\nabla\left(\frac{1}{R_2} - \frac{1}{R_1}\right) \tag{7.37}$$

which is the same as that obtained in Example 7.2 by treating the magnet as an equivalent solenoid.

For $r \gg L$ the dipole relations obtained in Sec. 2.11 apply, and we obtain

$$\Phi_m = \frac{Q_m L \cos\theta}{4\pi r^2} = \frac{M_T \cos\theta}{4\pi r^2}$$

where $M_T = \pi a^2 L M_0$ is the total effective dipole moment of the magnetized rod.

Example 7.4. Field around a Cut Toroid. Figure 7.13*a* illustrates a highly permeable ($\mu \gg \mu_0$) toroid wound with N turns of wire carrying a current I. The mean radius of the toroid is d. The cross section of the toroid is circular, with a radius a much smaller than d.

The tangential field $\mathbf{H}$ is continuous across the boundary separating the toroid and the air region just outside but inside the helix winding. Therefore the flux density $\mathbf{B}$ in the interior is much greater than that outside the toroid since $\mathbf{B} = \mu\mathbf{H}$ inside and $\mu_0\mathbf{H}$ outside and we are assuming that μ is much greater than μ_0. Most of the flux lines are concentrated in the interior except for a small amount of leakage flux on the outside, as illustrated in Fig. 7.13*a*. If we apply Ampère's circuital law for $\mathbf{H}$ around a closed circular path in the interior of the toroid (path C in Fig. 7.13*a*), we obtain

$$\oint_C \mathbf{H} \cdot d\mathbf{l} = \int_S \mathbf{J} \cdot d\mathbf{S} = NI \tag{7.38}$$

since the total current cutting through the surface of a disk with boundary C, that is, total current linked by C, is NI. From symmetry considerations we conclude that $\mathbf{H}$ is not a function of the angle θ around the toroid and also that **H** is tangent to the curve C; so (7.38) gives

$$\oint_C \mathbf{H} \cdot d\mathbf{l} = H_\theta \int_0^{2\pi} r\, d\theta = 2\pi r H_\theta = NI$$

and

$$H_\theta = \frac{NI}{2\pi r} \approx \frac{NI}{2\pi d} \tag{7.39}$$

where r is the radius of the path C. Since $a \ll d$ we have $r \approx d$ for all closed paths within the toroid; so for a first approximation we may assume that H_θ is given by $NI/2\pi d$ at all points interior to the toroid. The flux density B_θ is given by μH_θ.

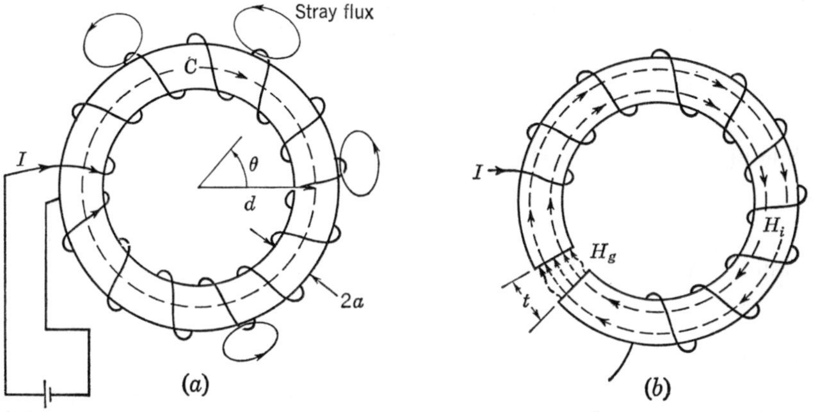

FIG. 7.13. (a) A toroid wound with N turns; (b) the same toroid with a small section removed (gap shown enlarged).

In the second situation illustrated in Fig. 7.13b, a small section of thickness t has been removed from the toroid. If $t \ll a$, the flux in the air gap is essentially uniform apart from a small amount of fringing or bulging of the field near the edge. The field B_θ, which is normal to the faces, must be continuous across the face, and hence B_θ has the same value in the air gap as in the interior of the toroid. The field H_θ is equal to B_θ/μ_0 in the gap and B_θ/μ in the toroid. If we let H_i be the interior field and H_g be the gap field, Ampère's circuital law gives

$$H_i(2\pi d - t) + H_g t = NI$$

and hence, substituting for H_i and H_g in terms of B_θ, we get

$$\frac{B_\theta}{\mu}(2\pi d - t) + \frac{B_\theta}{\mu_0} t = NI$$

Solving for B_θ gives

$$B_\theta = \frac{NI\mu\mu_0}{2\pi d\mu_0 + (\mu - \mu_0)t} \tag{7.40}$$

From (7.40) the fields H_g and H_i are readily found. When t is zero, the field H_i is given by $NI/2\pi d$, but when t is not equal to zero, H_i is reduced in value to

$$H_i = \frac{NI}{2\pi d + (\mu - \mu_0)t/\mu_0}$$

When the toroid is made of ferromagnetic material, the magnetic polarization will generally not be zero when the current I is reduced to zero. A determination of the exact values of B and H when $I = 0$ can be made only if the relationship of B to H is specified, as would be the case with an appropriate B-H curve. Let us consider how the desired result may be obtained. When I is zero, let the flux density in the toroid and air gap be B. In the air gap the magnetic field intensity is

$$H_g = \frac{B}{\mu_0}$$

while in the interior to the toroid

$$H_i = \frac{B}{\mu}$$

Applying Ampère's circuital law we get

$$H_i(2\pi d - t) + H_g t = 0$$

since $I = 0$, and hence

$$H_i = \frac{-H_g t}{2\pi d - t} \tag{7.41}$$

Equation (7.41) shows that inside the toroid H_i must be oppositely directed to the field H_g in the gap. However, in the gap H_g and B are in the same direction and B is continuous into the toroid. Therefore, in the toroid, B and H_i are oppositely directed, as illustrated in Fig. 7.14*a*.

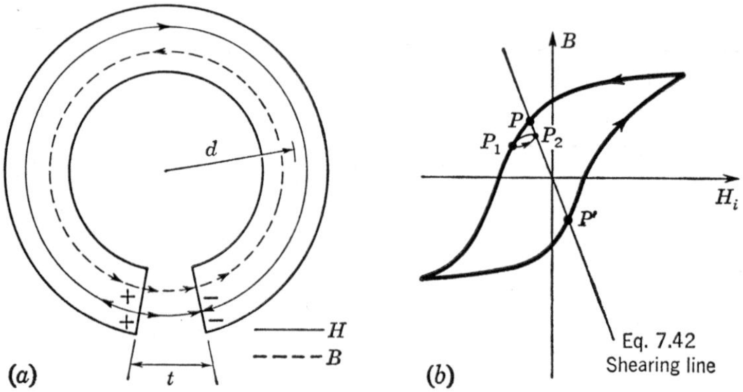

FIG. 7.14. The field in a magnetized cut toroid (gap shown enlarged).

Hence μ must be negative, a situation which is possible if the state of magnetization of the material corresponds to a point such as P on the B-H curve illustrated in Fig. 7.14*b*. If we substitute B/μ_0 for H_g in (7.41), we get $H_i = -Bt/\mu_0(2\pi d - t)$, or

$$B = \frac{-\mu_0 H_i(2\pi d - t)}{t} \tag{7.42}$$

This equation determines a relation between B and H_i. It is a straight line, called the shearing line, with negative slope, as shown plotted on the B-H curve in Fig. 7.14*b*. It intersects the B-H curve at the points P and P', which are the only two points that can simultaneously satisfy (7.42) and the relation between B and H_i given by the B-H curve. The flux density in the cut toroid is thus that given by the point P (or P').

The above behavior may be understood by recalling that the residual polarization $\mathbf{M}_0$ of the material when I is reduced to zero is equivalent to a magnetic surface charge density $\mathbf{M}_0 \cdot \mathbf{n}$ and $-\mathbf{M}_0 \cdot \mathbf{n}$ on the faces of the toroid at the gap. These charges result in a field $\mathbf{H}$, which is oppositely directed on the two sides of each face. From the point P on the B-H curve, B and H are determined. Thus, using the relation

$$B = \mu H = \mu_0(H + M_0)$$

we may calculate M_0 and μ. The value of M_0 turns out to be

$$M_0 = \frac{\mu - \mu_0}{\mu_0} H = \frac{B}{\mu_0} - H$$

This last result may be used to find the equivalent magnetic charge density on the end faces of the toroid.

An alternative interpretation of the results for the cut toroid involves consideration of the equivalent currents due to the residual magnetic polarization. Because of the assumed uniformity of $\mathbf{M}_0$, we have $\nabla \times \mathbf{M}_0 = 0$, so that $\mathbf{J}_m = 0$. However, $\mathbf{M}_0 \times \mathbf{n} \neq 0$ along the toroidal surface and thus leads to equivalent circulation currents along the meridians of the torus. These currents behave as a source for $\mathbf{B}$ just as do the true currents in the helical winding when $I \neq 0$. The residual $\mathbf{J}_{ms}$ accounts for the solenoidal nature of $\mathbf{B}$ and the persistence of flux when I is reduced to zero.

The above example illustrates several points which are of interest in the design of permanent magnets. For permanent magnets it is desirable to use materials that have a high retentivity and also a high coercivity. A magnet in the form of a "horseshoe" is magnetized by completing the magnetic circuit with an iron bar and winding the magnet with a coil through which a large steady current is passed. Upon removal of the coil and iron bar, the magnetization decreases in the manner indicated for the cut toroid because of the demagnetizing effect of the free poles, i.e., because of the equivalent magnetic charge on the end faces of the magnet.

A point such as P on the B-H curve in Fig. 7.14b is not a stable point since the application of a small magnetic field causes the magnetization to move along a new small hysteresis loop away from P. Since a magnet may at times be subjected to small stray fields, it is stabilized by applying a small negative field to bring the magnetization to the point P_1 and then removing this field to permit the magnetization to move along a new hysteresis loop up to the final point P_2, as in Fig. 7.14b. Application of a stray field now causes the magnetization to move along this new hysteresis loop. However, now upon removal of the stray field, the state of magnetization returns very nearly to the point P_2, whereas without the above stabilization technique the magnetization would not return to the initial value at P upon removal of the stray fields. In the absence of stray fields the magnetization will always lie on the shearing line; that is, P_2 lies on the shearing line.

7.7. The Magnetic Circuit

The solution to the general magnetostatic boundary-value problem involving conduction currents in the presence of ferromagnetic material bodies is extremely difficult to obtain. Fortunately, for many engineering applications involving ferromagnetic materials, good approximate solutions can be obtained by means of an analysis that parallels that used to analyze d-c circuits composed of series and parallel combinations of resistors. The ideas and limitations involved in this equivalent-circuit approach will be discussed below.

If we return to the cut-toroid problem of Example 7.4 and reexamine the method of solution presented, it will be seen to be similar to that which we would use for a simple d-c circuit of two resistors in series together with an applied voltage source. We consider the line integral of **H** around the circuit as being the magnetomotive force $\mathcal{H}$ which causes a total flux ψ to flow through the circuit. The magnitude of ψ is determined by $\mathcal{H}$ and a property of the circuit called the reluctance $\mathcal{R}$, where $\mathcal{R}$ is analogous to resistance.

According to (7.40), the flux density in the cut toroid is

$$B = \frac{NI\mu\mu_0}{2\pi d\mu_0 + (\mu - \mu_0)t}$$

The total flux ψ through the circuit is $\psi = AB = \pi a^2 B$, where A is the cross-sectional area of the toroid. The line integral of **H** around the toroid is equal to the ampere-turns NI and gives the magnetomotive force $\mathcal{H}$. The solution for ψ may be written as

$$\psi = BA = \frac{\mathcal{H}}{(2\pi d - t)/\mu A + t/\mu_0 A} \tag{7.43}$$

The first term in the denominator is similar to that for the d-c resistance of a conductor of length $2\pi d - t$, of cross-sectional area A, and having a specific conductivity μ. Thus this term is interpreted as the reluctance $\mathcal{R}_1$ of the section of toroid. For the same reasons the second term is interpreted as the reluctance $\mathcal{R}_2$ of the gap. We may now rewrite (7.43) as

$$\psi = \frac{\mathcal{H}}{\mathcal{R}_1 + \mathcal{R}_2} \tag{7.44}$$

where

$$\mathcal{R}_1 = \frac{2\pi d - t}{\mu A}$$

$$\mathcal{R}_2 = \frac{t}{\mu_0 A}$$

The close analogy between the d-c current circuit and the magnetic circuit may be seen from a comparison of the two following examples.

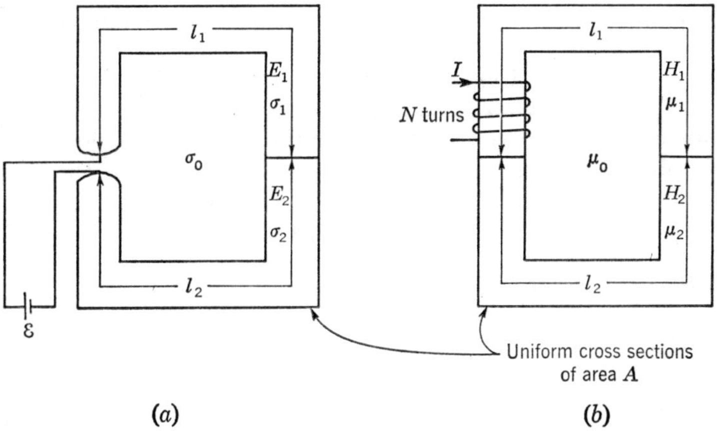

FIG. 7.15. (a) A d-c current circuit; (b) a similar magnetic circuit.

Figure 7.15*a* illustrates a d-c current circuit consisting of two sections of mean lengths l_1 and l_2 and having a uniform cross-sectional area A. The two sections have specific conductivities σ_1 and σ_2 and are submerged in a medium having a much lower conductivity σ_0. The two sections are connected in series with an applied electromotive force $\mathcal{E}$.

Since σ_0 is much smaller than σ_1 or σ_2, most of the current flow is confined to the highly conducting sections. The line integral of **E** around the circuit gives the applied electromotive force:

$$\oint_C \mathbf{E} \cdot d\mathbf{l} = \mathcal{E} = E_1 l_1 + E_2 l_2 \tag{7.45a}$$

The current density **J** is given by $\sigma\mathbf{E}$, and the total current is obtained by multiplying this quantity by the cross-sectional area A. Since the current is continuous, we get

$$I_1 = I_2 = I = \sigma_1 E_1 A = \sigma_2 E_2 A \tag{7.45b}$$

where the leakage current through σ_0 is neglected. Hence $E_2 = (\sigma_1/\sigma_2)E_1$, and from (7.45*a*)

$$E_1 = \frac{\mathcal{E}}{l_1 + (\sigma_1/\sigma_2)l_2} \tag{7.45c}$$

Thus

$$I = I_1 = \frac{\sigma_1 A \mathcal{E}}{l_1 + \sigma_1 l_2/\sigma_2} = \frac{\mathcal{E}}{l_1/A\sigma_1 + l_2/A\sigma_2} = \frac{\mathcal{E}}{R} \tag{7.45d}$$

where $R = l_1/A\sigma_1 + l_2/A\sigma_2$ and is the total resistance of the circuit.

Figure 7.15*b* illustrates a magnetic circuit which is similar to the above d-c circuit. Provided μ_1 and μ_2 are much larger than μ_0, most of the magnetic flux is confined to the highly permeable sections. The line integral of **H** around the circuit gives the applied magnetomotive force

$$\oint_C \mathbf{H} \cdot d\mathbf{l} = \mathcal{H} = NI = H_1 l_1 + H_2 l_2 \tag{7.46a}$$

The flux density **B** is given by $\mu\mathbf{H}$, while the total flux, analogous to current, is obtained by multiplying by the area A. Since the flux ψ is continuous,

$$\psi = \psi_1 = \psi_2 = \mu_1 H_1 A = \mu_2 H_2 A \tag{7.46b}$$

Hence $H_2 = (\mu_1/\mu_2)H_1$, and from (7.46*a*)

$$H_1 = \frac{\mathcal{H}}{l_1 + (\mu_1/\mu_2)l_2} \tag{7.46c}$$

Thus

$$\psi = \frac{\mathcal{H}}{l_1/A\mu_1 + l_2/A\mu_2} = \frac{\mathcal{H}}{\mathcal{R}} \tag{7.46d}$$

where $\mathcal{R} = l_1/\mu_1 A + l_2/\mu_2 A$ and is the reluctance of the circuit.

The validity of the solutions to the above two circuit problems depends on the accuracy of the assumption that the fields are uniform over the cross-sectional areas and that the current flow or flux is also uniform and confined to the sections of the circuit only, i.e., confined to the cross section of the two segments of length l_1 and l_2. In the case of the current circuit, this assumption is usually more nearly correct since the ratio

σ/σ_0 is generally very much larger than μ/μ_0 for the magnetic materials. In spite of these limitations, the concept of a magnetic circuit is of great utility in the engineering application of ferromagnetic materials. The ease with which it provides a solution for the magnetic flux ψ in a circuit makes it superior to other approximate methods of solution when the requirements on the accuracy of the solution are not too stringent.

Example 7.5. Iron-core Transformer. Figure 7.16*a* illustrates an iron-core transformer of uniform cross section and having a small air gap of thickness t in the center leg. Such air gaps are frequently used in practice in order to increase the reluctance of the magnetic circuit so as to prevent magnetic saturation of the material. The left leg is wound with N_1 turns of wire, and the right leg with N_2 turns. The problem is to compute the flux linkage for the coil with N_2 turns when a current I flows in the other coil.

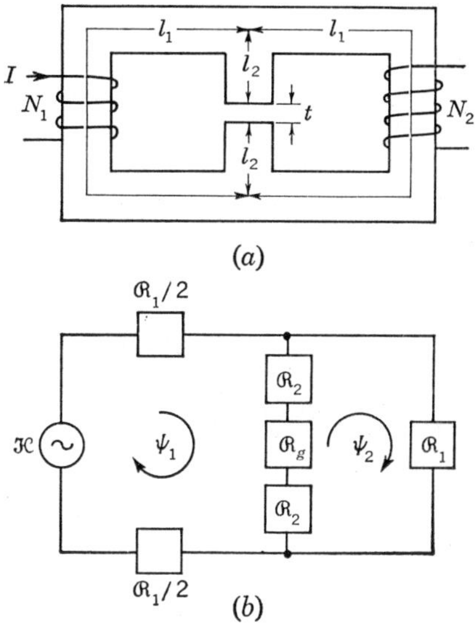

FIG. 7.16. An iron-core transformer and its equivalent magnetic circuit.

The equivalent magnetic circuit is illustrated in Fig. 7.16*b*. The magnetomotive force applied to the circuit is N_1I and acts in series with the reluctance $\mathcal{R}_1$ of the left side leg. This reluctance is shown arbitrarily split into two parts of magnitude $\mathcal{R}_1/2$ each. In series with $\mathcal{R}_1$ is the reluctance $\mathcal{R}_1$ of the right side leg in parallel with the reluctance of the center leg. The reluctances are given by the following expressions:

$$\mathcal{R}_1 = \frac{l_1}{\mu A}$$

$$\mathcal{R}_2 = \frac{l_2}{\mu A}$$

$$\mathcal{R}_g = \frac{t}{\mu_0 A}$$

Following the usual circuit-analysis approach, the following two equations may be written:

$$\mathcal{H} = \psi_1(\mathcal{R}_1 + 2\mathcal{R}_2 + \mathcal{R}_g) - \psi_2(2\mathcal{R}_2 + \mathcal{R}_g)$$
$$0 = \psi_1(2\mathcal{R}_2 + \mathcal{R}_g) - \psi_2(\mathcal{R}_1 + 2\mathcal{R}_2 + \mathcal{R}_g)$$

Solving for ψ_2 gives $\psi_2 = N_1I(2\mathcal{R}_2 + \mathcal{R}_g)/\mathcal{R}_1(\mathcal{R}_1 + 4\mathcal{R}_2 + 2\mathcal{R}_g)$. The flux linkage for the secondary coil is thus given by $N_2\psi_2$.

We should like to summarize in a more formal way the duality between electric and magnetic circuits that have been illustrated in the previous examples. As a starting point we note the similarity in the relations

$$\oint \mathbf{E} \cdot d\mathbf{l} = \mathcal{E} \qquad \oint \mathbf{H} \cdot d\mathbf{l} = NI = \mathcal{H}$$

In the study of stationary currents in Sec. 5.2 it was pointed out that external to the sources (i.e., battery, etc.) $\oint \mathbf{E} \cdot d\mathbf{l} = 0$, so that for the external region the electric field could be related to a scalar potential function. The potential concept must be handled carefully, however, since if a closed path is taken which includes the battery, the net change in potential is not zero but equals the emf of the source; that is, the potential is actually multivalued because the **E** field is not conservative. In electric-circuit theory this difficulty is avoided by assigning a potential rise of $\mathcal{E}$ to the integral of **E** through a source of emf. In this case, if $\Delta\Phi$ is the potential drop across a portion of a loop, then the requirement that $\oint \mathbf{E} \cdot d\mathbf{l} = \Sigma\mathcal{E}$ (the total emf) can be written

$$\sum_{\text{loop}} \Delta\Phi = \sum_{\text{loop}} \mathcal{E}$$

or

$$\sum_{\text{loop}} \Delta\phi + \sum_{\text{loop}} (-\mathcal{E}) = 0$$

The above relation is one of Kirchhoff's laws and symbolizes the statement that the sum of the potential drops taken around any closed loop equals zero.

In an analogous fashion the magnetic field **H** can be derived from a scalar magnetic potential provided the multivalued nature of the field is taken into account.† If we adopt a similar convention to that in electric circuits we can write

$$\sum_{\text{loop}} \mathcal{H} = \sum_{\text{loop}} \Delta\Phi_m$$

The magnitude of the mmf taken around a closed path equals the total number of ampere-turns linking that path. Often this quantity is con-

† We are assuming that the true current is not zero; otherwise $\nabla \times \mathbf{H} = 0$, in which case **H** can be derived from a scalar potential unambiguously. The multivalued nature of Φ_m in the presence of a current loop can be demonstrated directly, for it can be shown that if we were to require $\mathbf{H} = -\nabla\Phi_m$, then Φ_m must satisfy $\Delta\Phi_m = (I/4\pi)\,\Delta\Omega$; that is, in the presence of a current loop, the change in magnetic scalar potential as a consequence of a change in position of the field point equals the change in subtended solid angle times the current in the loop divided by 4π. For a path encircling the loop, $\Delta\Omega$ will change discontinuously by 4π at some point on this contour, hence producing the multivalued nature of Φ_m. The total change in potential for one complete loop equals I; that is, $\oint \mathbf{H} \cdot d\mathbf{l} = I$, as we already know.

centrated over a small region of the magnetic circuit, just as in the electric circuit the source of emf is usually highly localized. If this is not the case, as in a uniformly wound toroid, it is not possible to separate the sources from the circuit. It is as if in an electric circuit the battery were split up and continuously distributed through the circuit.

In addition to the above equations we also have the following similar relations:

$$\oint \mathbf{J} \cdot d\mathbf{S} = 0 \qquad \oint \mathbf{B} \cdot d\mathbf{S} = 0$$
$$\mathbf{J} = \sigma \mathbf{E} \qquad \mathbf{B} = \mu \mathbf{H}$$

In view of the fact that the three electric-circuit equations given above form the basis for all d-c circuit theory, magnetic-circuit theory follows immediately by duality. It is only necessary to replace **J** by **B**, **E** by **H**, $\mathcal{E}$ by $\mathcal{H}$, and σ by μ. For example, since the total current is obtained from

$$I = \int_S \mathbf{J} \cdot d\mathbf{S}$$

then by duality, total flux ψ is

$$\psi = \int_S \mathbf{B} \cdot d\mathbf{S}$$

As a further illustration we may obtain the reluctance of an arbitrary shaped magnetic material from (5.36). By duality this is clearly

$$\mathcal{R} = \int_0^L \frac{du_1}{\int_S (\mu h_2 h_3/h_1)\, du_2\, du_3}$$

It should be noted that the derivation in Sec. 5.5 tacitly assumed no current flow in the external medium, a relatively easy thing to achieve. In magnetic circuits the medium will invariably carry some leakage flux, so that true dual conditions cannot be provided, as we noted in an earlier example. However, with highly permeable materials, good results can be expected even though leakage is neglected.

7.8. Physical Properties of Magnetic Material†

As demonstrated earlier in this chapter, the magnetic properties of materials may be described in terms of a magnetic polarization or distribution of magnetic dipoles per unit volume. In this section we shall take a closer look at the properties of materials that give rise to a volume

† For a much more complete discussion, see, for example, A. J. Dekker, "Solid State Physics," 1957, or "Electrical Engineering Materials," 1959, Prentice-Hall, Inc., Englewood Cliffs, N.J.

density of magnetic dipoles. There are essentially three mechanisms, or properties, of atoms that give rise to a magnetic dipole moment:

1. The orbital motion of electrons around the nucleus is equivalent to a circulating current loop.

2. The spinning electron has an intrinsic magnetic dipole moment of magnitude $eh/4\pi w$, where e is the electron charge, w is the mass of the electron, and h is Planck's constant (6.62×10^{-34} joule-second). The magnitude of the spin magnetic moment of the electron is called a Bohr magneton.

3. The nucleus of an atom contains charged particles, and since the nucleus also has a spin, there is a magnetic dipole moment associated with the nucleus.

In most materials the dipole moment of the nucleus is negligible in comparison with the orbital and spin magnetic moments of the electron. The reason for this is the large mass of the nucleus in comparison with that of the electron (at least 10^3 larger). The angular momentum of the nuclear spin is about the same as that for the electron spin. Since the mass is so much greater, it follows that the angular velocity is much smaller, and hence the equivalent circulating current and dipole moment are also much smaller. It is the orbital magnetic dipole moment and, even more important, the electron-spin magnetic dipole moment that are largely responsible for the magnetic properties of materials.

Magnetic materials are classified according to the following scheme:

1. Diamagnetic Materials. These are materials which do not have a permanent magnetic dipole moment in the absence of an external applied magnetic field.

2. Paramagnetic Materials. These materials have a permanent magnetic dipole moment, but the interaction between neighboring dipoles is negligible, with the result that in the presence of an external field the flux density in the material is increased by only a small amount.

3. Ferromagnetic Materials. In these materials the dipoles interact strongly and all tend to line up parallel with the applied field so as to produce a large increase in the flux density in the material.

4. Antiferromagnetic Materials. In these materials the permanent dipoles tend to align themselves so that alternate dipoles are antiparallel to the applied field. The result is a cancellation of the effects of each dipole and a zero net increase in flux density in the material.

5. Ferrimagnetic Materials. In these materials magnetically polarized domains of unequal magnitude align with alternate domains parallel and antiparallel to the applied field. Usually a relatively large increase in flux density is produced since the strong dipoles align themselves with the field and the weak dipoles are aligned antiparallel to the applied field.

Diamagnetism

The nature of diamagnetism may be understood from the following example. Consider an atom with several electrons which rotate in orbits orientated so that the net magnetic dipole moment is zero. Instead of considering each orbit in detail, we consider a particular orbit for which the magnetic field has maximum effects. We shall then assume that the change in magnetic moment produced by the magnetic field is of the order of magnitude of the net change per orbit in the atom. The simplicity achieved outweighs the loss in quantitative precision. Accordingly, we specify our model to consist of an electron orbit whose plane is perpendicular to the externally applied field B_0, as in Fig. 7.17, and which is electrically balanced by a $+e$ nuclear charge. Let the radius of the orbit be r_0 and the angular velocity be ω_0 in the absence of the external field.

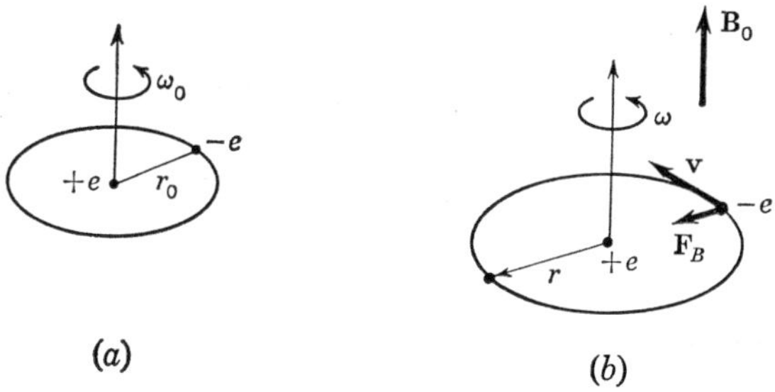

FIG. 7.17. Illustration of perturbation of electron orbit by an external field $\mathbf{B}_0$.

When B_0 is not present, the orbital angular velocity is determined by the condition that the outward centrifugal force $w\omega_0^2r_0$ equals the inward coulomb attraction force $e^2/4\pi\epsilon_0r_0^2$. Solving for ω_0 gives

$$\omega_0^2 = \frac{e^2}{4\pi\epsilon_0 w r_0^3} \tag{7.47}$$

In the presence of the external field B_0 there is an additional inward (Lorentz) force F_B whose magnitude is $evB_0 = er\omega B_0$. In place of the relation that led to (7.47) we must now have

$$\frac{e^2}{4\pi\epsilon_0 r^2} + er\omega B_0 = w\omega^2 r \tag{7.48}$$

To a first approximation the radius may be assumed to remain constant at the value r_0, and hence, using (7.47) in (7.48), we get

$$\omega^2 = \omega_0^2 + \frac{eB_0}{w}\,\omega \tag{7.49}$$

The field B_0 produces only a small perturbation in ω, so that

$$\omega^2 - \omega_0^2 = (\omega - \omega_0)(\omega + \omega_0) \approx 2\omega_0(\omega - \omega_0)$$

With this result, (7.49) may be written as

$$\omega - \omega_0 = \frac{eB_0}{2w} \tag{7.50}$$

In the absence of the field, the dipole moment of the rotating electron is $-ev_0\pi r_0^2/2\pi r_0 = -e\omega_0 r_0^2/2$, while in the presence of the field B_0 the total dipole moment is $-e\omega r_0^2/2$. Prior to the application of B_0 the contribution from all electronic orbits leads to a zero net dipole moment. The change due to B_0 is in a common direction for all orbits, hence resulting in a total net moment. The order of magnitude per orbit is the difference in dipole moments just evaluated and is

$$\begin{aligned} m_i &= \frac{-er_0^2(\omega - \omega_0)}{2} \\ &= \frac{-e^2r_0^2B_0}{4w} \\ &= \frac{-e^2r_0^2\mu_0 H_0}{4w} \end{aligned} \tag{7.51}$$

where $\mu_0 H_0 = B_0$. It should be noted that this induced moment is directed antiparallel to the applied field. If N is the effective number of such orbits per unit volume, the induced dipole polarization M per unit volume will be $M = -Ne^2r_0^2\mu_0 H_0/4w$, and hence the susceptibility χ_m is given by

$$\chi_m = \frac{-Ne^2r_0^2\mu_0}{4w} \tag{7.52}$$

Substitution of known typical values of N and r_0 shows that χ_m is of the order of -10^{-5}, a result in essential agreement with measured values. The small resultant value of χ_m justifies the assumption that the orbital radius remains essentially constant and that only a small change in the angular velocity ω_0 results upon application of an external field.

Although the above result was based on a classical physics approach, it still explains the basic process responsible for diamagnetism and may be verified by the methods of quantum mechanics.

Paramagnetism

Paramagnetism is due almost entirely to the spin magnetic dipole moment of the electron. Consider a material with N spinning electrons per unit volume. For paramagnetic materials the interaction between neighboring dipoles is negligible; so we may assume that the field acting on each dipole is just the applied field B_0. In the presence of the field B_0 some of the magnetic dipoles (each with a moment equal to one Bohr magneton) are aligned with the field while the remainder are aligned antiparallel to the field. The relative number in each position depends on the energy difference of the two positions and is essentially governed by Boltzmann's statistics. If N_p represents the number of dipoles per unit volume that are aligned parallel to the field and N_a is the number aligned antiparallel to the field, it is found that†

$$N_p = \frac{N}{1 + \exp\left(-ehB_0/2\pi wkT\right)}$$

$$N_a = \frac{N}{1 + \exp\left(ehB_0/2\pi wkT\right)}$$

where k is Boltzmann's constant and T is the absolute temperature. The net magnetic polarization is given by

$$M = \frac{N_p - N_a}{4\pi w} eh = \frac{Neh \tanh\left(ehB_0/4\pi wkT\right)}{4\pi w} \tag{7.53}$$

For normal temperatures and fields the argument of the hyperbolic tangent is small, so the approximation $\tanh x \approx x$ may be used and (7.53) gives

$$M = \frac{Ne^2h^2\mu_0 H_0}{(4\pi w)^2 kT} \tag{7.54}$$

Hence the susceptibility is given by

$$\chi_m = \frac{M}{H_0} = \frac{Ne^2h^2\mu_0}{(4\pi w)^2 kT} = \frac{C}{T} \tag{7.55}$$

where the constant C, defined by this equation, is called the Curie constant. Typical values of χ_m for paramagnetic materials are of the order of 10^{-3} in magnitude. As (7.55) shows, χ_m is inversely proportional to the absolute temperature T (Curie law). This is very reasonable, since at higher temperatures molecular activity opposes the effect of the applied field to orient the dipole moments. At the low temperatures the Curie law breaks down since the approximation used to obtain (7.55) is no longer valid.

Diamagnetic effects are also present in paramagnetic materials, but since the contribution to the susceptibility by the induced dipoles is of the order of -10^{-5}, it is completely masked by the paramagnetic effect.

Ferromagnetism

In ferromagnetic materials the effective field acting on each spin magnetic dipole is the vector sum of the applied field plus a strong inter-

† See Sec. 3.1, where an analogous derivation for electric dipoles is given.

action field arising from all the neighboring dipoles. This interaction field has been found experimentally to be much greater than the classically calculated magnetic field from the neighboring dipoles. These large "exchange forces" can be explained only by quantum mechanics and exist because of the wave nature of the electrons.

A theory for ferromagnetism was proposed by Weiss in 1907 and has been verified by quantum mechanics. The two basic postulates made by Weiss are: (1) There exists a strong interaction field from neighboring dipoles that tends to aid the alignment of the dipoles with the applied field. Thus the internal field acting on a dipole may be expressed as $B_i = \mu_0(H_0 + \alpha M)$, where α is called the internal field constant. (2) A ferromagnetic material consists of a number of domains with linear dimensions as large as or larger than 10^{-4} centimeter. In each domain all the spins are aligned in parallel but the direction of magnetization differs from one domain to the next. Also, each domain is spontaneously magnetized even in the absence of applied fields. However, because of the random orientation of the domains, the net flux density in the material is small.

By means of the domain theory a satisfactory explanation of the characteristic hysteresis curve for ferromagnetic materials can be given. Each domain can have its spins aligned along several possible directions or axes. The field strength required to produce magnetization along the various permissible axes differs, so that there are "easy" and "hard" directions of magnetization. In a ferromagnetic specimen which is originally unmagnetized, the application of an external field causes the following sequence of events to take place:

1. For weak applied fields those domains whose easy direction of magnetization is in the direction of the applied field, i.e., which are spontaneously magnetized in this direction, grow at the expense of the other domains. For small applied fields the domain wall movement is reversible. For large applied fields the wall movements are irreversible and a negative field must be applied to return the domain walls to their original positions. This irreversible wall movement gives rise to the hysteresis effect.

2. As the applied field is increased in strength, this domain growth continues until the whole specimen is essentially a single domain. At high field strengths some domain rotation takes place also. This process continues until the specimen becomes magnetically saturated. Figure 7.18 illustrates schematically this magnetization process.

Upon reducing the field the domain walls begin to move so as to produce more nearly equal sized domains again. When the applied field has been reduced to zero, a net magnetization remains, since many of the domains are still magnetized in the direction of the applied field. A microscopic examination of the magnetization curve shows that the process of magnetization is not a smooth one. This effect, known as the Barkhausen effect, is produced by random motion of the domain walls.

Curie-Weiss Law

If the expression $\mu_0(H_0 + \alpha M)$ for the internal field in a ferromagnetic material is substituted into (7.53), which gives the magnetization for

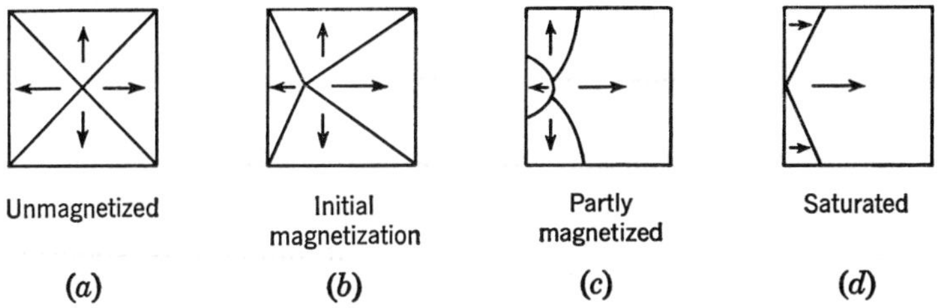

FIG. 7.18. Illustration of the growth of domains in the magnetization process.

paramagnetic materials, the equation for the magnetization in a ferromagnetic material is obtained; i.e.,

$$M = \frac{Neh}{4\pi w} \tanh \frac{eh\mu_0(H_0 + \alpha M)}{4\pi wkT} \tag{7.56}$$

This result is based on the paramagnetic model that electron spins are aligned either parallel or antiparallel to the applied field. For weak fields and high temperatures, (7.56) reduces to

$$M = N\left(\frac{eh}{4\pi w}\right)^2 \frac{\mu_0(H_0 + \alpha M)}{kT}$$

which may be solved for M to give

$$M = \frac{C}{T - \theta} \tag{7.57}$$

where $$C = \frac{N}{k}\left(\frac{eh}{4\pi w}\right)^2 \quad \text{and} \quad \theta = \alpha C$$

This law is known as the Curie-Weiss law, where C is the Curie constant and θ is the Curie temperature. For $T > \theta$, the behavior of a ferromagnetic material is similar to that of a paramagnetic material. Below the Curie temperature ($T < \theta$) the Curie-Weiss law is no longer applicable, since the hyperbolic tangent cannot be replaced by its argument in this range. For $T < \theta$ a finite value of the magnetization can exist even though the applied field H_0 is zero. This magnetization is called the spontaneous magnetization. With $H_0 = 0$ the magnetization is given by (7.56) as

$$M = \frac{Neh}{4\pi w} \tanh \frac{\mu_0 eh\alpha M}{4\pi wkT} \tag{7.58}$$

The value of M may be solved for as a function of the temperature T, and a curve of the form illustrated in Fig. 7.19 is obtained. At the Curie temperature $T = \theta$, the spontaneous magnetization vanishes. Above

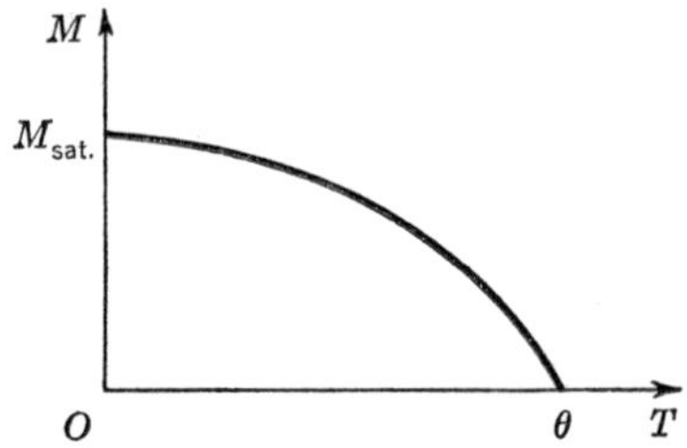

FIG. 7.19. Spontaneous magnetization as a function of temperature.

this temperature, (7.58) does not have a solution. At zero temperature the magnetization is equal to the saturation value $Neh/4\pi w$ corresponding to the condition where all the spin magnetic dipoles are aligned in the same direction. In a ferromagnetic medium it is actually each domain that is spontaneously magnetized in accord with (7.58). This spontaneous magnetization accounts for the residual magnetization in the absence of an external applied field.

CHAPTER 8

QUASI-STATIONARY MAGNETIC FIELD

In order to complete a discussion of magnetostatics, along lines analogous to that in electrostatics, it will be necessary to derive an expression for the stored magnetic energy. This will then make possible a full discussion of inductance (the counterpart of capacitance in electrostatics) and also an analysis of forces between current-carrying circuits. It turns out, however, that in order to determine the formula for stored magnetic energy due to time-stationary currents, it is necessary to know something about time-varying currents and time-varying magnetic fields. Consequently, this chapter starts out with a statement and discussion of Faraday's law of induction. Following this, inductance, energy, and force are considered, thereby completing the analysis of magnetostatics and preparing for the subject of general time-varying fields.

The procedure used to derive the field expression for electric stored energy involved evaluating the work done in assembling the charges that established the electric field. For the static magnetic field we might expect that a similar procedure could be used, except that in this case it would be necessary to evaluate the work done in assembling a system of current loops. This is true; however, the forces acting on the current loops multiplied by their respective displacements are not alone equal to the energy of assembly. In the process of moving the loops relative to each other, the magnetic flux linking each loop continually changes. It turns out that this changing flux results in an induced voltage in each loop and the battery must do work (or have work done on it) in order to keep the currents constant. This additional work must be taken into account in evaluating the net energy of assembly of the loops, which by the above definition equals the net stored magnetic energy.

Although we begin with a general formulation for time-varying magnetic effects, we shall be mostly concerned in this chapter with a quasi-static field. By a quasi-stationary magnetic field we mean a field that varies so slowly with time that all radiation effects are negligible. In a following chapter we shall discover that a system of conductors carrying sinusoidal currents must have dimensions of the order of a wavelength in order to radiate efficiently. For a frequency f, the wavelength in free space is given by $\lambda_0 = c/f$, where c is the velocity of light (3×10^8 meters per second). Thus, if the frequency is 100 kilocycles per second, $\lambda_0 = 3{,}000$ meters. A typical system of coils and loops used in the laboratory might have dimensions of around one meter; consequently, for frequencies less than 100 kilocycles, radiation effects would certainly be entirely negligible. The quasi-static application is thus widely applicable.

8.1. Faraday's Law

The discovery of electric induction by a changing magnetic field is credited to Michael Faraday. On Aug. 29, 1831, the classic experiment on induction was carried out. Faraday wound two separate coils on an iron ring and found that whenever the current in one coil was changed, an induced current would flow in the other coil. He also found that a similar induced current would be produced when a magnet was moved in the vicinity of the coil. At about the same time similar effects were being studied by Joseph Henry in America. However, Faraday was more fortunate in that he worked at the Royal Institution in London and his work was published and made known to the scientific world earlier than the work of Henry. As a consequence, the law of electric induction is known as Faraday's law.

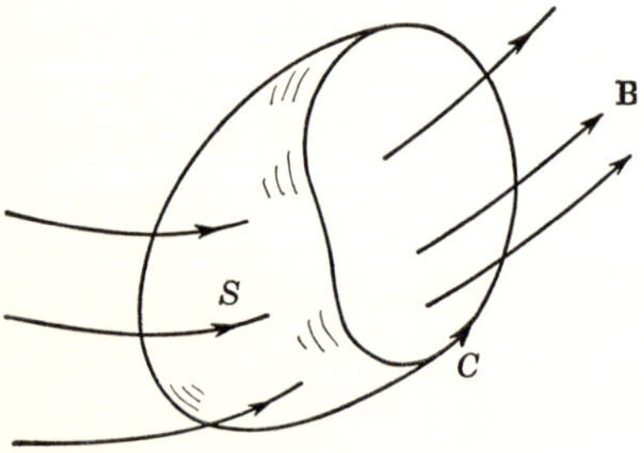

FIG. 8.1. Illustration of Faraday's law.

If we consider any closed stationary path in space which is linked by a changing magnetic field, it is found that the induced voltage around this path is equal to the negative time rate of change of the total magnetic flux through the closed path. Let C denote a closed path, as in Fig. 8.1. The induced voltage around this path is given by the line integral of the induced electric field around C and is

$$V_{ind} = \oint_C \mathbf{E} \cdot d\mathbf{l}$$

The magnetic flux through C is given by

$$\psi = \int_S \mathbf{B} \cdot d\mathbf{S}$$

where S is any surface with C as its boundary. Thus the mathematical statement of Faraday's law is

$$\oint_C \mathbf{E} \cdot d\mathbf{l} = -\frac{\partial}{\partial t} \int_S \mathbf{B} \cdot d\mathbf{S} + \oint_C \mathbf{v} \times \mathbf{B} \cdot d\mathbf{l} \qquad (8.1)$$

Basically, the law states that a changing magnetic field will induce an electric field. The induced electric field exists in space regardless of whether a conducting wire is present or not. When a conducting wire is present, a current will flow, and we refer to this current as an induced current. Faraday's law is the principle on which most electric generators operate. Note that the electric field set up by a changing magnetic field is nonconservative, as (8.1) clearly indicates. The changing magnetic field becomes a source for an electric field.

In addition to (8.1) there are several other equivalent statements of Faraday's law. Since **B** may be obtained from the curl of a vector potential **A**, we have

$$\oint_C \mathbf{E} \cdot d\mathbf{l} = -\frac{\partial}{\partial t}\int_S \nabla \times \mathbf{A} \cdot d\mathbf{S} = -\frac{\partial}{\partial t}\oint_C \mathbf{A} \cdot d\mathbf{l} \tag{8.2}$$

by using Stokes' law to convert the surface integral to a line integral. Equation (8.2) permits the induced voltage to be evaluated directly from the vector potential **A**.

The differential form of (8.1) is obtained by using Stokes' law to replace $\oint_C \mathbf{E} \cdot d\mathbf{l}$ by a surface integral, so that

$$\oint_C \mathbf{E} \cdot d\mathbf{l} = \int_S \nabla \times \mathbf{E} \cdot d\mathbf{S} = -\frac{\partial}{\partial t}\int_S \mathbf{B} \cdot d\mathbf{S}$$

or

$$\int_S \left(\nabla \times \mathbf{E} + \frac{\partial \mathbf{B}}{\partial t}\right) \cdot d\mathbf{S} = 0$$

Since S can be an arbitrary surface, the integrand must be equal to zero, and we obtain

$$\nabla \times \mathbf{E} = -\frac{\partial \mathbf{B}}{\partial t} \tag{8.3}$$

This result again shows that the electric field induced by **B** is not of the same nature as the electrostatic field for which the curl or rotation is zero. Our concept of the curl or rotation as being a measure of the line integral of the field around an infinitesimal contour per unit area makes (8.3) a natural consequence of (8.1).

Example 8.1. Induced Voltage in a Coil. Figure 8.2a illustrates a single-turn coil of wire of radius d. The coil is located in a uniform magnetic field $\mathbf{B} = \mathbf{B}_0 \sin \omega t$ and with the normal to the plane of the coil at an angle θ with respect to the lines of magnetic flux. The induced voltage measured between the two open ends of the coil is given by (8.1) as

$$V = -\frac{\partial}{\partial t}\int_S \mathbf{B} \cdot d\mathbf{S} = -\frac{\partial}{\partial t}(\pi d^2 B_0 \cos\theta \sin\omega t)$$
$$= -\omega\pi d^2 B_0 \cos\theta \cos\omega t$$

since the total magnetic flux linking the coil is $\pi d^2 B_0 \cos\theta \sin\omega t$. In Fig. 8.2*b* a coil with N turns is illustrated. To evaluate $\int_S \mathbf{B} \cdot d\mathbf{S}$, a surface must be constructed so that the coil forms the periphery and the total flux crossing the surface is evaluated. This surface resembles a spiral staircase. The net result is roughly equivalent to the notion that each turn is separately linked by the magnetic flux, a notion that is quite good for tightly wound coils. With this point of view, then, in each turn the induced voltage is given by the above expression. These voltages

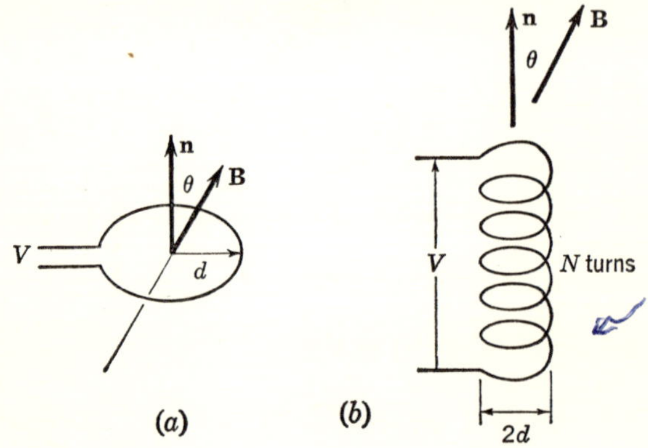

FIG. 8.2. Electric induction in a coil.

add in series, so that the total voltage across the complete coil is N times greater and hence given by

$$V = -N\omega\pi d^2 B_0 \cos\theta \cos\omega t$$

The induced voltage is proportional to the rate of change of the magnetic field, the number of turns, and the magnitude of the magnetic flux linking each turn.

8.2. Induced Electric Field Due to Motion

When conductors are moving through a static magnetic field, an induced voltage (we shall define this more precisely later) is produced in the conductor. This voltage is in addition to that calculated by (8.1). The magnitude of this voltage may be found from the Lorentz force equation. This states that a particle of charge q moving with a velocity $\mathbf{v}$ in a magnetic field $\mathbf{B}$ experiences a force $\mathbf{F}$ given by

$$\mathbf{F} = q\mathbf{v} \times \mathbf{B} \tag{8.4}$$

This force, known as the Lorentz force, is similar to the analogous relation $\mathbf{F} = I\,d\mathbf{l} \times \mathbf{B}$. (Note that $q\mathbf{v}$ can be interpreted as a current element.)

The force is seen to act in a direction perpendicular to both **v** and **B**. The interpretation of the Lorentz force gives rise to the concept that an observer moving through a static magnetic field sees, in addition to the magnetic field, an electric field also. A unit of charge moving with the observer appears to be stationary, and any force experienced by that charge is ascribed to the existence of an electrostatic field. But a force is experienced and is given by (8.4). Consequently, in the moving reference frame, this fact is interpreted as revealing the existence of an electric field **E** given by

$$\mathbf{E} = \frac{\mathbf{F}}{q} = \mathbf{v} \times \mathbf{B} \tag{8.5}$$

Equation (8.5) gives an alternative and more general method of evaluating the induced voltage in a moving conductor. This equation is the mathematical formulation of Faraday's second observation of induction by moving magnets.

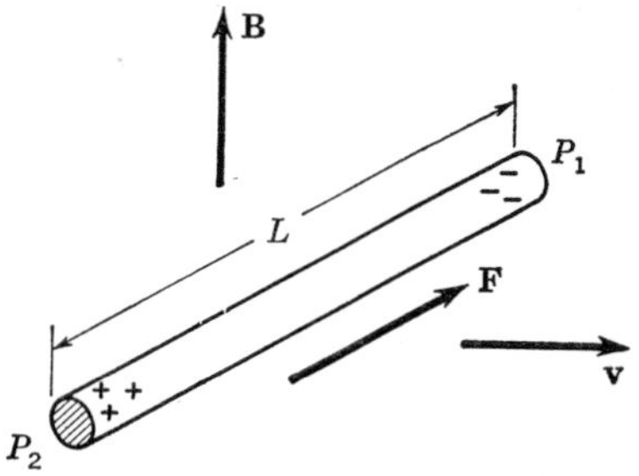

FIG. 8.3. Induced voltage in a moving conductor.

As an example, consider a conducting wire moving with a velocity **v** through a uniform field **B**, as in Fig. 8.3, where **B** is orthogonal to **v**. Each electron in the conductor experiences a force $F = -evB$, which tends to displace the electron along the wire in the direction indicated. As a result of this force electrons move toward the end marked P_1, leaving a net positive charge in the vicinity of the end marked P_2. When equilibrium has been reached, there is no further movement of the electrons along the wire, and this requires that there be no net force. What happens is that the displaced charges set up an electrostatic field which opposes the displacement of the charges due to the Lorentz force. When sufficient charge has been built up so that the electrostatic field produces a force equal and opposite to the Lorentz force, equilibrium is established. In this case $E = -vB$.

The induced voltage between the ends of the conductor is defined by

$$V = \int_{P_1}^{P_2} \mathbf{E} \cdot d\mathbf{l}$$

and in this example

$$V = vB \int_{P_1}^{P_2} dl = vBL$$

a result that is true when **v** and **B** are orthogonal. The induced voltage caused by motion of a conductor through a magnetic field is called motional emf (electromotive force).† The electrostatic field set up by the displaced charges may be observed in both the stationary frame of reference and the moving frame attached to the conductor.

Moving Conductor in a Time-varying Magnetic Field

When a closed conducting loop C, as in Fig. 8.4a, is moving with a constant velocity $\mathbf{v}$ through a nonuniform time-varying magnetic field $\mathbf{B}$, the

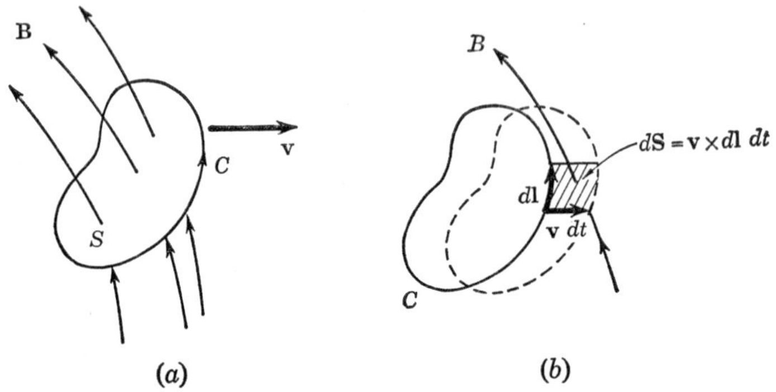

FIG. 8.4. Conductor C moving in a time-varying field $\mathbf{B}$.

induced voltage is given by

$$V_{ind} = -\int_S \frac{\partial \mathbf{B}}{\partial t} \cdot d\mathbf{S} + \oint_C \mathbf{v} \times \mathbf{B} \cdot d\mathbf{l} \tag{8.6}$$

In this expression the first term represents the contribution due to the time variation of $\mathbf{B}$ while the second term is the contribution representing the motional induced voltage.

The velocity $\mathbf{v}$ of different portions of the loop need not be the same, so that the loop C may be changing in shape as well as undergoing translation and rotation. However, in (8.6), the integral of $\partial \mathbf{B}/\partial t$ may be taken over the original surface S, since the contribution arising from an integration over the change ΔS in S is a second-order term.

The term $\oint_C \mathbf{v} \times \mathbf{B} \cdot d\mathbf{l}$ is the motional emf contribution. A further insight into the connection between this term and the changing-flux concept may be obtained as follows. With reference to Fig. 8.4b, it is clear that an element $d\mathbf{l}$ of C sweeps out an area $d\mathbf{S} = \mathbf{v} \times d\mathbf{l}\, dt$ in a time

† The field structure is similar to that described for an open-circuited battery. In the latter case chemical action sets up a nonconservative field within the battery (analogous to the Lorentz force field) and also an electrostatic field which pervades all space but cancels the nonconservative field within the battery (within the generator in the present case). Accordingly, $V = \text{emf} = vBL$ may be similarly viewed as an open-circuit voltage.

interval dt. The change in flux caused by the displacement of C is equal to the integral of **B** through the swept-out area, i.e., equal to

$$d\psi_1 = \oint_C \mathbf{B} \cdot (\mathbf{v} \times d\mathbf{l})\, dt \tag{8.7}$$

Hence, $-d\psi_1/dt = -\oint_C \mathbf{B} \cdot \mathbf{v} \times d\mathbf{l} = \oint_C \mathbf{v} \times \mathbf{B} \cdot d\mathbf{l}$, which is the usual form for the motional emf term. Consequently, we have shown that $V_{ind} = -d\psi/dt$, that is, equals the negative total time rate of change of flux linkage. Thus, a generalization of Faraday's law may be written

$$\oint_C \mathbf{E} \cdot d\mathbf{l} = -\frac{d}{dt}\int_S \mathbf{B} \cdot d\mathbf{S} \tag{8.8}$$

In the above case it was quite clear how the total change in flux linkage could be evaluated since a definite closed contour C was involved. In the case of a single conductor, as in Fig. 8.3, it is not clear how to evaluate a change in flux linkage since a definite closed contour is not involved. In a situation like this the use of the Lorentz force equation is the most straightforward.

Example 8.2. Motional EMF. Figure 8.5 illustrates a single-turn rectangular coil, with sides b and a, which is rotating with an angular

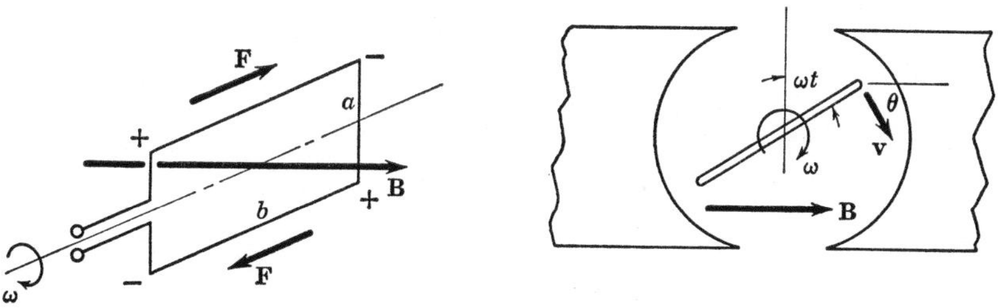

FIG. 8.5. Rotating coil in a magnetic field.

velocity ω about its axis. The coil is located between the pole pieces of a magnet which sets up a uniform magnetic field **B**. Since the magnetic field **B** does not vary with time, that is, $\partial\mathbf{B}/\partial t = 0$, the induced voltage is entirely of the motional type. We may calculate the induced voltage from the negative time rate of change of the total magnetic flux linking the coil. At any instant of time t the flux through the coil is

$$abB \cos \omega t$$

and hence the induced voltage is

$$V = -\frac{d}{dt}\int_S \mathbf{B} \cdot d\mathbf{S} = -\frac{d\psi}{dt} = \omega abB \sin \omega t$$

The above result may also be obtained by an application of the Lorentz force equation. The velocity of an electron along the sides of the coil is $v = a\omega/2$, and the sine of the angle between $\mathbf{v}$ and $\mathbf{B}$ is given by $\sin\theta = \sin\omega t$, as in Fig. 8.5. The force on an electron is then

$$F = e|\mathbf{v} \times \mathbf{B}| = evB\sin\theta = e\frac{a}{2}\omega B\sin\omega t$$

This is equivalent to the presence of an electric field E, where

$$E = \frac{F}{e} = \frac{a}{2}\omega B\sin\omega t$$

In each side arm of the coil the induced voltage is Eb. Consequently, the total voltage is just twice this amount; that is,

$$V = 2Eb = ab\omega B\sin\omega t$$

which is the same as that given earlier. The above result neglects the effect of the ends of the coil; however, there is no induced voltage in the ends since $\mathbf{F}$ is perpendicular to both $\mathbf{v}$ and $\mathbf{B}$. This analysis is seen to be equivalent to a formal evaluation of the motional emf term $\oint_C \mathbf{v} \times \mathbf{B} \cdot d\mathbf{l}$.

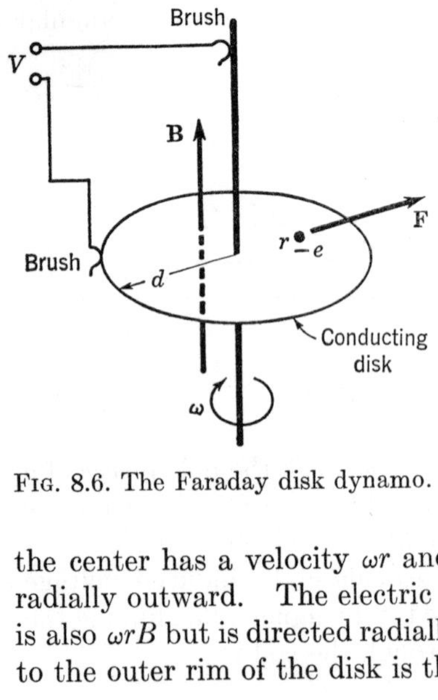

FIG. 8.6. The Faraday disk dynamo.

Example 8.3. Faraday Disk Dynamo. The Faraday disk dynamo is illustrated in Fig. 8.6. It consists of a circular conducting disk rotating in a uniform magnetic field B. Brushes make contact with the disk at the center and along the periphery. The problem is to determine if an induced voltage will be measured between the brushes. The answer is yes, and the magnitude of the voltage is readily found from the Lorentz force equation. An electron at a radial distance r from the center has a velocity ωr and hence experiences a force $e\omega rB$ directed radially outward. The electric field acting on the electron at equilibrium is also ωrB but is directed radially inward. The potential from the center to the outer rim of the disk is thus

$$V = \int_0^d E(r)\,dr = -\omega B\int_0^d r\,dr = -\frac{\omega Bd^2}{2} \tag{8.9}$$

The value computed by (8.9) is the open-circuit voltage of the Faraday disk dynamo and therefore also represents the emf of the generator.

8.3. Inductance

Consider a single current-carrying loop in which a constant current has been established. A magnetic field is set up which could be calculated from the given geometry of the loop and which is proportional to the current magnitude. If the current is caused to change, so will the magnetic field. But this means that the total flux linking the loop also changes and, by Faraday's law, a voltage is induced in the loop. If the problem is analyzed quantitatively, it will be discovered that the self-induced voltage always has such a polarity that tends to oppose the original change in current. For example, if the current begins to decrease, the induced voltage acts in a direction to offset this decrease.

If the problem involves two current loops, a somewhat more involved sequence of events takes place, but with the same qualitative outcome.

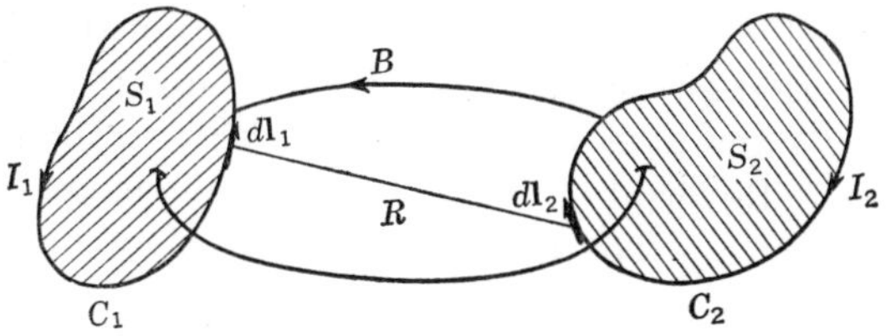

FIG. 8.7. Two circuits with magnetic coupling.

Thus Fig. 8.7 illustrates two circuits C_1 and C_2, with currents I_1 and I_2. The current I_1 produces a partial field $\mathbf{B}_1$, which causes a magnetic flux $\psi_{12} = \int_{S_2} \mathbf{B}_1 \cdot d\mathbf{S}$ to link C_2 and $\psi_{11} = \int_{S_1} \mathbf{B}_1 \cdot d\mathbf{S}$ to link C_1 (itself). Similarly, the partial field $\mathbf{B}_2$ due to I_2 is responsible for the flux

$$\psi_{21} = \int_{S_1} \mathbf{B}_2 \cdot d\mathbf{S}$$

linking C_1 and $\psi_{22} = \int_{S_2} \mathbf{B}_2 \cdot d\mathbf{S}$, which links itself. If now the current I_1 is allowed to change, this causes a corresponding variation in ψ_{11} and ψ_{12}. The latter effect results in an induced voltage in C_2, hence a change in I_2. This in turn causes ψ_{21} to be disturbed from its previous value, so that the net flux linking C_1 (that is, $\psi_{11} + \psi_{21}$) is altered. Again, if all possible cases are considered analytically, it turns out that the change in both ψ_{11} and ψ_{21} is always such that the induced voltage in C_1 is opposed to the original perturbation of I_1. The fact that the induced voltage always acts to oppose the change in current that produces the induced voltage is known as Lenz's law.

The property of a single circuit, such as C_1, that results in an induced voltage which opposes a change in the current flowing in the circuit is known as self-inductance. The similar effect of a changing current in one circuit producing an induced voltage in another circuit is known as mutual inductance. Inductance is analogous to inertia in a mechanical system. The symbol L is used for self-inductance, and M for mutual inductance. The symbol L, with appropriate subscripts, is also used for mutual inductance and is the notation we shall adopt. The unit for inductance is the henry, in honor of Joseph Henry, who contributed much to the early knowledge of magnetic fields and inductance.

There are several equivalent mathematical definitions of inductance. One definition is in terms of flux linkages. If ψ_{12} is the magnetic flux linking circuit C_2, due to a current I_1 flowing in circuit C_1, the mutual inductance L_{12} between circuits C_1 and C_2 is defined by

$$L_{12} = \frac{\text{flux linking } C_2 \text{ due to current in } C_1}{\text{current in } C_1} = \frac{\psi_{12}}{I_1} \tag{8.10}$$

The mutual inductance is considered to be positive if the flux ψ_{12} links C_2 in the same direction as the self-flux linkage ψ_{22} due to the field from the current I_2. If ψ_{12} and ψ_{22} are in opposite directions, the mutual inductance is negative. Reversal of either I_1 or I_2 will change the sign of the mutual inductance L_{12}. The self-inductance L_{11} of circuit C_1 is defined in a similar way; that is,

$$L_{11} = \frac{\text{flux linking } C_1 \text{ due to current in } C_1}{\text{current in } C_1} = \frac{\psi_{11}}{I_1} \tag{8.11}$$

The mutual inductance between C_1 and C_2 may be defined by

$$L_{21} = \frac{\psi_{21}}{I_2} \tag{8.12}$$

as well, where ψ_{21} is the flux linking C_1 due to a current I_2 in C_2. We shall show that $L_{12} = L_{21}$, so that (8.10) and (8.12) are equivalent.

Since C_1 and C_2 are two very thin current-carrying loops, it is a simple matter to formulate the expressions for the flux linkage. However, in the limit of zero cross section, (8.11) leads to an infinite value for L_{11}, although the magnetic energy associated with the field remains finite. The extension of the definition (8.11) to current loops or filaments of finite (large) cross section can be made and is done in a later section. The proper interpretation of ψ_{11} follows from a consideration of the magnetic energy associated with the circuit and is fully discussed later.

The above definition of inductance is satisfactory only for quasi-stationary magnetic fields where the current and the magnetic field have the same phase angle over the whole region of the circuit. At high frequencies the magnetic field does not have the same phase angle over the whole region of the circuit because of the finite time required to propagate the effects of a changing current and field through space. A more general definition in terms of the magnetic energy associated with a circuit will be given in the next section.

Neumann Formulas

Consider two very thin wires bent into two closed loops C_1 and C_2, as in Fig. 8.7. Let a current I_1 flow in C_1. Since the wire is assumed to be very thin, the value computed for $\mathbf{B}_1$ will not be much in error if the current is assumed concentrated in an infinitely thin filament along the center of the conductor, provided only field points external to the wire are considered. With this limitation in mind, the field $\mathbf{B}_1$ produced by I_1 is given by

$$\mathbf{B}_1 = \frac{\mu_0 I_1}{4\pi} \oint_{C_1} \frac{d\mathbf{l}_1 \times \mathbf{a}_R}{R^2} = \frac{\mu_0 I_1}{4\pi} \oint_{C_1} \left[\nabla\left(\frac{1}{R}\right)\right] \times d\mathbf{l}_1 \tag{8.13}$$

since $\nabla(1/R) = -\mathbf{a}_R/R^2$. The integration is over the source coordinates, while ∇ affects only the field coordinates; so we have

$$\nabla \times \frac{d\mathbf{l}_1}{R} = \left[\nabla\left(\frac{1}{R}\right)\right] \times d\mathbf{l}_1$$

since $d\mathbf{l}_1$ is a constant vector as far as ∇ is concerned. Then, in place of (8.13), we can write

$$\mathbf{B}_1 = \frac{\mu_0 I_1}{4\pi} \oint_{C_1} \nabla \times \frac{d\mathbf{l}_1}{R}$$

and hence the flux ψ_{12} linking circuit C_2 is

$$\begin{aligned}\psi_{12} &= \int_{S_2} \mathbf{B}_1 \cdot d\mathbf{S} = \frac{\mu_0 I_1}{4\pi} \int_{S_2} \oint_{C_1} \nabla \times \frac{d\mathbf{l}_1}{R} \cdot d\mathbf{S} \\ &= \frac{\mu_0 I_1}{4\pi} \oint_{C_1} \int_{S_2} \nabla \times \frac{d\mathbf{l}_1}{R} \cdot d\mathbf{S}\end{aligned}$$

upon changing the order of integration. By using Stokes' law the surface integral may be converted to a contour integral around C_2; so we get

$$\underline{\psi_{12} = \frac{\mu_0 I_1}{4\pi} \oint_{C_1} \oint_{C_2} \frac{d\mathbf{l}_1 \cdot d\mathbf{l}_2}{R}} \tag{8.14}$$

From the definition of mutual inductance stated in (8.10) we obtain Neumann's formula:

$$L_{12} = \frac{\psi_{12}}{I_1} = \frac{\mu_0}{4\pi} \oint_{C_1} \oint_{C_2} \frac{d\mathbf{l}_1 \cdot d\mathbf{l}_2}{R} \tag{8.15}$$

Since in (8.15) R is the distance between a point on C_1 to a point on C_2, the integral as a whole is symmetrical; that is, the subscripts 1 and 2 may be interchanged without changing the end result. This proves the reciprocity relation stated earlier:

$$L_{12} = L_{21} = \frac{\psi_{12}}{I_1} = \frac{\psi_{21}}{I_2} \tag{8.16}$$

Equation (8.15) may be derived in an alternative way by noting that $\mathbf{B}_1 = \nabla \times \mathbf{A}_1$, where the vector potential $\mathbf{A}_1$ is given by

$$\mathbf{A}_1 = \frac{\mu_0 I_1}{4\pi} \oint_{C_1} \frac{d\mathbf{l}_1}{R}$$

Thus

$$\psi_{12} = \int_{S_2} \mathbf{B}_1 \cdot d\mathbf{S} = \int_{S_2} \nabla \times \mathbf{A}_1 \cdot d\mathbf{S} = \oint_{C_2} \mathbf{A}_1 \cdot d\mathbf{l}_2$$

Using the expression for $\mathbf{A}_1$ and dividing by I_1 leads to the desired end result.

A formula similar to (8.15) may be written for the self-inductance also. However, it is not permissible to assume that the current is concentrated in a thin filament at the center since it is necessary to include values of **B** at the wire itself where the approximation breaks down. For an idealized infinitely thin wire, the analogous formula is

$$L_{11} = \frac{\mu_0}{4\pi} \oint_{C_1} \oint_{C_1} \frac{d\mathbf{l}_1' \cdot d\mathbf{l}_1}{R} \tag{8.17}$$

where $d\mathbf{l}_1$ and $d\mathbf{l}_1'$ are differential elements of length along C_1 and separated by a distance R. Since R can become zero, the integral is an improper one and leads to an infinite value of self-inductance, which is actually consistent with the assumption of infinitesimal wire diameter. To evaluate the self-inductance of a practical loop, the finite thickness of the conductor must be taken into account. A suitable procedure to be followed will be presented later, but first we shall consider some typical applications of (8.15) and also introduce the concept of internal inductance.

Example 8.4. Inductance of a Coaxial Line. Figure 8.8 illustrates a coaxial transmission line made of two thin-walled conducting cylinders

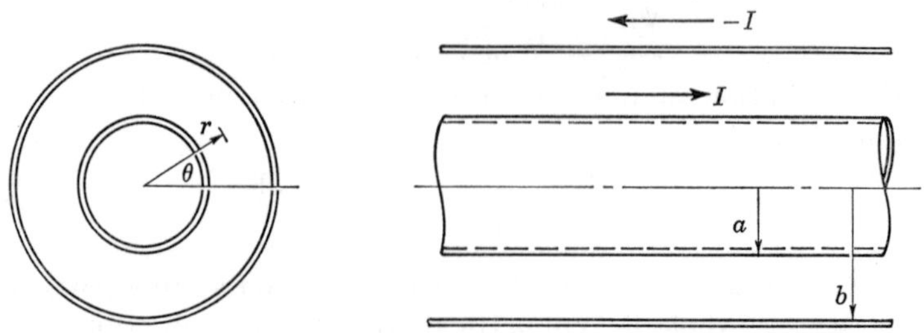

FIG. 8.8. A coaxial line made of two thin-walled cylinders.

with radii a and b. A current I flows along the inner cylinder, and a return current $-I$ along the outer cylinder. The inductance per unit length is to be evaluated.

It will be noted that this geometry does not correspond precisely to that of the thin-wire loops for which the definition of inductance has been formulated. At a later time a more fundamental definition of inductance will be given which allows generalization in terms of distributed current-carrying bodies. For the present we shall try to extend the definitions of (8.11) in a plausible way and with the understanding that future work will confirm its usefulness.

The field $\mathbf{B}$ is in the θ direction only and is given by $\mathbf{B} = (\mu_0 I/2\pi r)\mathbf{a}_\theta$. The total magnetic flux linking the inner conductor per unit length of line is

$$\psi = \frac{\mu_0 I}{2\pi}\int_a^b \frac{dr}{r} = \frac{\mu_0 I}{2\pi}\ln\frac{b}{a}$$

and hence the inductance per unit length of line is given by

$$L = \frac{\psi}{I} = \frac{\mu_0}{2\pi}\ln\frac{b}{a} \tag{8.18}$$

If the center conductor is solid, the above result is not valid, since the current I is distributed uniformly over the cross section of area πa^2. To treat this case the concept of partial flux linkages is required. The current flowing in the portion of the inner conductor between 0 and r is $I\pi r^2/\pi a^2 = Ir^2/a^2$. The field in the coaxial line is given by (see Example 6.7)

$$B = \frac{\mu_0 I r}{2\pi a^2} \qquad 0 < r < a$$

$$B = \frac{\mu_0 I}{2\pi r} \qquad a < r < b$$

Since the field is circularly symmetric, each element of current in the annular ring between r and $r + dr$ is linked by the same flux. The value of the magnetic flux linking this current is

$$\begin{aligned} d\psi' &= \int_r^b B\,dr = \frac{\mu_0 I}{2\pi a^2}\int_r^a r\,dr + \frac{\mu_0 I}{2\pi}\int_a^b \frac{dr}{r} \\ &= \frac{\mu_0 I}{4\pi a^2}(a^2 - r^2) + \frac{\mu_0 I}{2\pi}\ln\frac{b}{a} \end{aligned}$$

In the earlier calculation of inductance for the thin-walled inner conductor the flux linked the total current flow I. Within the solid conductor, however, we have flux which links only part of the current. Now since the flux $d\psi'$ does not link the entire current I, it seems plausible that we should reduce its contribution to the total flux linkage, for purposes of inductance calculation, by the ratio that the actual current linked bears to the total current. Since the current which is linked by $d\psi'$ is the current in an annular ring of area $2\pi r\,dr$, the reduction factor is $2\pi r\,dr/\pi a^2$, and the equivalent flux linkage $d\psi$ is given by

$$d\psi = \frac{2\pi r\,dr}{\pi a^2}\,d\psi'$$

At a later point a firmer basis for this procedure will be given. For the present example we have

$$d\psi = \frac{I2\pi r\,dr}{\pi a^2 I}\left[\frac{\mu_0 I}{4\pi a^2}(a^2 - r^2) + \frac{\mu_0 I}{2\pi}\ln\frac{b}{a}\right]$$

The total flux linkage is

$$\psi = \int_0^a d\psi = \frac{\mu_0 I}{\pi a^2}\left(\int_0^a \frac{a^2 r - r^3}{2a^2}\,dr + \ln\frac{b}{a}\int_0^a r\,dr\right)$$
$$= \frac{\mu_0 I}{8\pi} + \frac{\mu_0 I}{2\pi}\ln\frac{b}{a} \tag{8.19}$$

Hence the inductance per unit length is

$$L = \frac{\mu_0}{8\pi} + \frac{\mu_0}{2\pi}\ln\frac{b}{a} \tag{8.20}$$

The first term $\mu_0/8\pi$ is known as the internal inductance of the center conductor since this term arises from the flux linkages internal to the conductor. The second term is known as the external inductance since this corresponds to the external flux linkages.

Evaluation of Self-inductance

For an infinitely long single wire of circular cross section the internal inductance per unit length is obviously $\mu_0/8\pi$, since a single wire has a field internal to itself of the same form as the center conductor of the coaxial line we just considered. The external inductance per unit length is infinite, a result which may be obtained by letting b tend to infinity in (8.20). In practice, we do not have infinitely long wires; so this latter result is of no consequence. However, for a thin wire of total length l bent into an arbitrary loop, the magnetic field near the surface is very nearly the same as that for an infinitely long wire provided the radius of curvature of the loop is much greater than the conductor radius at all points. In other words, we may treat the wire locally as though it were part of an infinitely long wire. It follows that the internal inductance of any loop of mean length l is $\mu_0 l/8\pi$. This result is of great importance since it leads to a simple method of formulating an expression for the self-inductance of a circuit.

Consider a conductor of radius r_0 bent into a closed loop C_1, as in Fig. 8.9. Let the contour C_1 coincide with the interior edge of the conductor. The self-inductance of the circuit consists of the sum of the internal inductance and the external inductance. The external inductance arises from the flux linking the contour C_1. To evaluate this flux linkage we may assume that the current I is concentrated in an infinitely thin filament along the center C_0 of the conductor with negligible error. The problem is equivalent to that of evaluating the mutual inductance between the contours C_0 and C_1. Thus

$$L_e = \frac{\mu_0}{4\pi} \oint_{C_0} \oint_{C_1} \frac{d\mathbf{l}_0 \cdot d\mathbf{l}_1}{R}$$

The self-inductance L is thus given by

$$L = L_i + L_e = \frac{\mu_0 l}{8\pi} + \frac{\mu_0}{4\pi} \oint_{C_0} \oint_{C_1} \frac{d\mathbf{l}_0 \cdot d\mathbf{l}_1}{R} \tag{8.21}$$

Example 8.5. Self-inductance of a Circular Loop. Consider a conductor of radius r_0 bent into a circular loop of mean radius a, as in Fig. 8.10. The internal inductance of the loop is $\mu_0 2\pi a/8\pi = \mu_0 a/4$. The magnitudes of $d\mathbf{l}_0$ and $d\mathbf{l}_1$ are given by

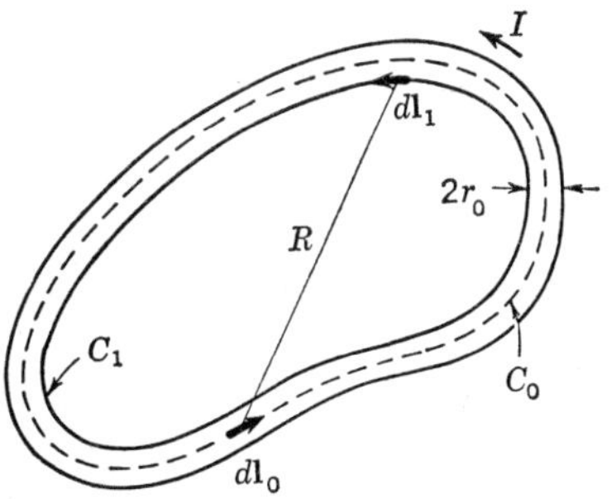

FIG. 8.9. A conductor of finite radius r_0 bent into a closed loop.

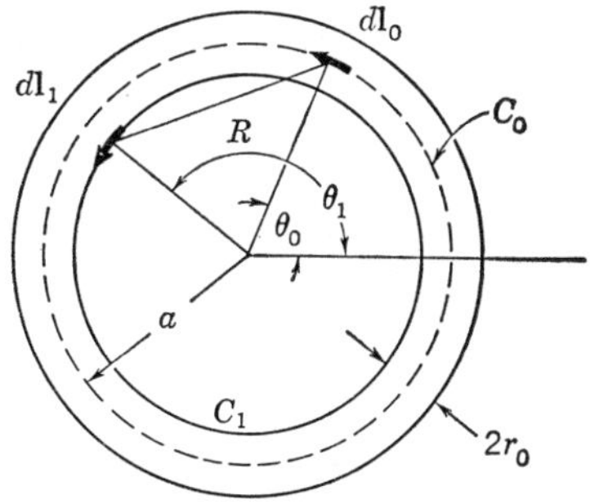

FIG. 8.10. A circular conducting loop.

$$|d\mathbf{l}_0| = a\, d\theta_0 \qquad |d\mathbf{l}_1| = (a - r_0)\, d\theta_1 \approx a\, d\theta_1$$

The angle between $d\mathbf{l}_0$ and $d\mathbf{l}_1$ is $\theta_1 - \theta_0$, and hence

$$d\mathbf{l}_0 \cdot d\mathbf{l}_1 = a^2 \cos(\theta_1 - \theta_0)\, d\theta_0\, d\theta_1$$

The distance R between the two elements of arc length is given by

$$R^2 = a^2 + (a - r_0)^2 - 2a(a - r_0)\cos(\theta_1 - \theta_0)$$

Hence the external inductance is given by

$$L_e = \frac{\mu_0 a^2}{4\pi} \int_0^{2\pi} \int_0^{2\pi} \frac{\cos(\theta_1 - \theta_0)\, d\theta_1\, d\theta_0}{[2a(a - r_0) + r_0^2 - 2a(a - r_0)\cos(\theta_1 - \theta_0)]^{1/2}}$$

If we integrate over θ_1 first, we may change variables and replace $\theta_1 - \theta_0$ by θ and $d\theta_1$ by $d\theta$; thus

$$L_e = \frac{\mu_0 a^2}{4\pi} \int_0^{2\pi} \int_0^{2\pi} \frac{\cos\theta\, d\theta\, d\theta_0}{[r_0^2 + 2a(a - r_0)(1 - \cos\theta)]^{1/2}}$$

It is not necessary to alter the limits of the integral because the origin for θ_1 is arbitrary in view of the circular symmetry. Since the result of the integration in θ is independent of θ_0, we may perform the integration over θ_0 at once, thereby obtaining a factor 2π. We now have

$$L_e = \frac{\mu_0 a^2}{2} \int_0^{2\pi} \frac{\cos\theta\, d\theta}{[r_0^2 + 2a(a - r_0)(1 - \cos\theta)]^{1/2}}$$

This expression can be evaluated in terms of elliptic integrals. Only the final result will be given here. It is found that

$$L_e = \mu_0 a \left[\left(\frac{2}{k} - k \right) K - \frac{2}{k} E \right] \tag{8.22}$$

where $k^2 = 4a(a - r_0)/(2a - r_0)^2$ and K and E are elliptic integrals given by

$$K = \int_0^{\pi/2} \frac{d\alpha}{(1 - k^2 \sin^2 \alpha)^{1/2}} \qquad E = \int_0^{\pi/2} (1 - k^2 \sin^2 \alpha)^{1/2}\, d\alpha$$

The above integrals are tabulated.† For $r_0 \ll a$, the result (8.22) reduces to

$$L_e = \mu_0 a \left(\ln \frac{8a}{r_0} - 2 \right) \tag{8.23}$$

Thus the self-inductance of a circular loop of mean radius a is

$$L = L_i + L_e = \mu_0 a \left(\ln \frac{8a}{r_0} - 1.75 \right) \qquad r_0 \ll a \tag{8.24}$$

8.4. Energy of a System of Current Loops

Consider two closed conducting loops C_1 and C_2, as in Fig. 8.7, with currents i_1 and i_2, which are initially zero. In the process of increasing i_1 and i_2 from zero to final values I_1 and I_2, work is done on the system. According to the field theory, this work results in stored energy in the magnetic field surrounding the conductors. To evaluate this quantity, let us maintain i_2 at zero while increasing i_1 from zero to its final value I_1

† E. Jahnke and F. Emde, "Tables of Functions," 4th ed., Dover Publications, New York, 1945.

first. When we change i_1 by an amount di_1 in a time interval dt, the magnetic field $\mathbf{B}_1$ due to i_1 changes at an average rate $d\mathbf{B}_1/dt$. Consequently, an induced voltage $\mathcal{E}_1 = -d\psi_{11}/dt$ is produced in C_1, and similarly, an induced voltage $\mathcal{E}_2 = -d\psi_{12}/dt$ is produced in C_2. Thus, in order to change i_1 an amount di_1 in a time interval dt, we must apply a voltage $-\mathcal{E}_1$ in the circuit C_1. At the same time we must apply a voltage $-\mathcal{E}_2$ in C_2 to maintain i_2 at zero. In the time interval dt the applied voltage $-\mathcal{E}_1$ does work of amount

$$dW_1 = -\mathcal{E}_1 i_1\,dt = i_1\,d\psi_{11} = L_{11}i_1\,di_1$$

since by definition $L_{11}i_1 = \psi_{11}$ and because L_{11} is constant $L_{11}\,di_1 = d\psi_{11}$. The applied voltage $-\mathcal{E}_2$ does zero work since i_2 is kept equal to zero. The total work done in increasing i_1 from zero to I_1 is thus

$$\underline{W_1 = \int_0^{I_1} L_{11}i_1\,di_1 = \tfrac{1}{2}L_{11}I_1^2} \tag{8.25}$$

This is the energy stored in the magnetic field surrounding a single circuit.

Next we keep I_1 constant and increase i_2 by an amount di_2 in a time interval dt. This results in an induced voltage

$$\mathcal{E}_2 = \frac{-d\psi_{22}}{dt} = -L_{22}\frac{di_2}{dt} \qquad \text{in } C_2$$

and

$$\mathcal{E}_1 = \frac{-d\psi_{12}}{dt} = -L_{12}\frac{di_2}{dt} \qquad \text{in } C_1$$

To maintain i_1 constant at its value I_1, we must apply a voltage $-\mathcal{E}_1$. In time dt this voltage does work (or has work done upon it, depending on whether $\mathcal{E}_1$ tends to increase or decrease i_1):

$$dW_{12} = -\mathcal{E}_1 I_1\,dt = I_1L_{12}\,di_2$$

Similarly, the voltage $-\mathcal{E}_2$ that must be applied to change i_2 by an amount di_2 does work of amount

$$dW_2 = -\mathcal{E}_2 i_2\,dt = L_{22}i_2\,di_2$$

The total work done in changing i_2 from zero to a final value I_2 is

$$\begin{aligned} W_{12} + W_{22} &= I_1L_{12}\int_0^{I_2} di_2 + L_{22}\int_0^{I_2} i_2\,di_2 \\ &= I_1I_2L_{12} + \tfrac{1}{2}I_2^2L_{22} \end{aligned} \tag{8.26}$$

The work done on the system is the sum of (8.25) and (8.26) and represents the energy W_m stored in the magnetic field. This energy is given by

$$\begin{aligned} W_m &= \tfrac{1}{2}L_{11}I_1^2 + L_{12}I_1I_2 + \tfrac{1}{2}L_{22}I_2^2 \\ &= \tfrac{1}{2}\sum_{i=1}^{2}\sum_{j=1}^{2} L_{ij}I_iI_j \end{aligned} \tag{8.27}$$

This result is easily generalized to a system of N loops; the result is given by (8.27) by increasing the summations in (8.27) for both i and j, up to N.

The magnetic energy in the field around a single current loop of finite cross section may be written in a form analogous to (8.27). We may divide the current-loop cross section into a large number of current filaments of cross-sectional area ΔS_i for the ith filament (see Fig. 8.11). If J_i is the current density in the ith filament, then $I_i = J_i\,\Delta S_i$ is the total current flow in this flow tube or filament. Let ψ_i be the flux linking the ith filament due to all the other current filaments in the current loop. This in turn is given by

$$\psi_i = \sum_{\substack{j=1 \\ j\neq i}}^{N} L_{ij}I_j = \sum_{\substack{j=1 \\ j\neq i}}^{N} L_{ij}J_j\,\Delta S_j$$

where N is the total number of current filaments or flow tubes making up the total current loop, L_{ij} is the mutual inductance between filaments i and j, and I_j is the current flowing in the jth filament. Since we have divided our original current loop into N current filaments, we have reduced the problem to one of a collection of N filamentary current loops and (8.27) may be applied to give

$$W_m = \tfrac{1}{2}\sum_{i=1}^{N} L_{ii}I_i^2 + \tfrac{1}{2}\sum_{i=1}^{N}\sum_{\substack{j=1 \\ i\neq j}}^{N} L_{ij}I_iI_j$$

$$= \tfrac{1}{2}\sum_{i=1}^{N} L_{ii}I_i^2 + \tfrac{1}{2}\sum_{i=1}^{N} \psi_i I_i \tag{8.28a}$$

Now, as demonstrated in Example 8.5, the self-inductance L_{ii} of a thin current filament of cross-sectional radius r_0 becomes infinite as $\ln r_0$. However, the total current in the filament decreases as r_0^2 as the cross-sectional area is made smaller, so that in the limit as r_0 goes to zero for each current filament, $L_{ii}I_i^2$ vanishes as $r_0^4 \ln r_0$. The number of current flux tubes N is inversely proportional to the cross-sectional area of the flux tube, that is, $N \propto r_0^{-2}$. Thus, for infinitely thin current filaments, the sum of the "self-energy" terms in (8.28a), i.e., the terms $L_{ii}I_i^2$, vanishes as $r_0^2 \ln r_0$. Each term in the double summation of (8.28a) is also of order r_0^4; however, the total number of such terms is $N^2 \propto r_0^{-4}$. Consequently, this summation may be expected to remain finite in the limit $r_0 \to 0$. Thus we are left with

$$W_m = \tfrac{1}{2}\sum_{i=1}^{N} \psi_i I_i \tag{8.28b}$$

This result expresses the energy of a single "thick" current loop in terms of the mutual energy between the current filaments that comprise the current loop. Equation (8.28*b*) will be used in the next section to establish a suitable definition for partial and total flux linkages.

Equation (8.27) gives an interpretation of the coefficients of inductance L_{ij} as the coefficients in the quadratic expression for the energy stored in the magnetic field. The terms L_{ij} $(i \neq j)$ may be either positive or negative, depending on the direction in which the mutual magnetic flux links the respective circuits. For two circuits with currents I_1 and I_2, we have

$$\underline{W_m = \tfrac{1}{2}I_1^2L_{11} + \tfrac{1}{2}I_2^2L_{22} \pm I_1I_2L_{12}}$$

which may be written as

$$W_m = \tfrac{1}{2}[(I_1\sqrt{L_{11}} - I_2\sqrt{L_{22}})^2 + I_1I_2(\sqrt{L_{11}L_{22}} \pm L_{12})]$$

The first term is always positive or zero. If we choose

$$I_1\sqrt{L_{11}} = I_2\sqrt{L_{22}}$$

so that the first term is zero, then since the energy stored in the field is always positive, we see that the mutual inductance L_{12} must satisfy the relation

$$\underline{L_{12} \leq \sqrt{L_{11}L_{22}}}$$

in order that the second term may also always be positive. The coefficient of coupling k is defined by

$$L_{12} = k\sqrt{L_{11}L_{22}} \tag{8.29}$$

and has a maximum value of unity when all the magnetic flux set up by the magnetic field of circuit 1 links circuit 2.

8.5. Energy as a Field Integral

In the preceding section the work done in setting up a system of current-carrying loops was evaluated in order to determine the energy stored in the magnetic field. As in electrostatics, it should be possible to express this energy in terms of the field alone. The analogy with the electrostatic field turns out to be a very close one, for we shall show that the energy in the magnetic field is given by the following integral:

$$\underline{W_m = \tfrac{1}{2}\int_V \mathbf{B}\cdot\mathbf{H}\,dV = \tfrac{1}{2}\mu\int_V \mathbf{H}\cdot\mathbf{H}\,dV} \tag{8.30}$$

where the integration is to be taken over the whole volume occupied by the field. The second expression in (8.30) is valid only if μ is a constant.

We shall prove the above result for the special case of a single conducting loop with finite thickness and carrying a current I, as in Fig. 8.11. Replacing $\mathbf{B}$ by $\nabla \times \mathbf{A}$ in (8.30), we obtain

$$W_m = \tfrac{1}{2} \int_V (\nabla \times \mathbf{A}) \cdot \mathbf{H}\, dV$$

Next we use the expansion $\nabla \cdot (\mathbf{A} \times \mathbf{H}) = (\nabla \times \mathbf{A}) \cdot \mathbf{H} - (\nabla \times \mathbf{H}) \cdot \mathbf{A}$ and replace $\nabla \times \mathbf{H}$ by $\mathbf{J}$, thereby obtaining

$$\begin{aligned} W_m &= \tfrac{1}{2} \int_V \mathbf{J} \cdot \mathbf{A}\, dV + \tfrac{1}{2} \int_V \nabla \cdot (\mathbf{A} \times \mathbf{H})\, dV \\ &= \tfrac{1}{2} \int_V \mathbf{J} \cdot \mathbf{A}\, dV + \tfrac{1}{2} \oint_S \mathbf{A} \times \mathbf{H} \cdot d\mathbf{S} \end{aligned}$$

where the divergence theorem has been used to convert the second volume integral to a surface integral over the closed surface S. If we choose S to

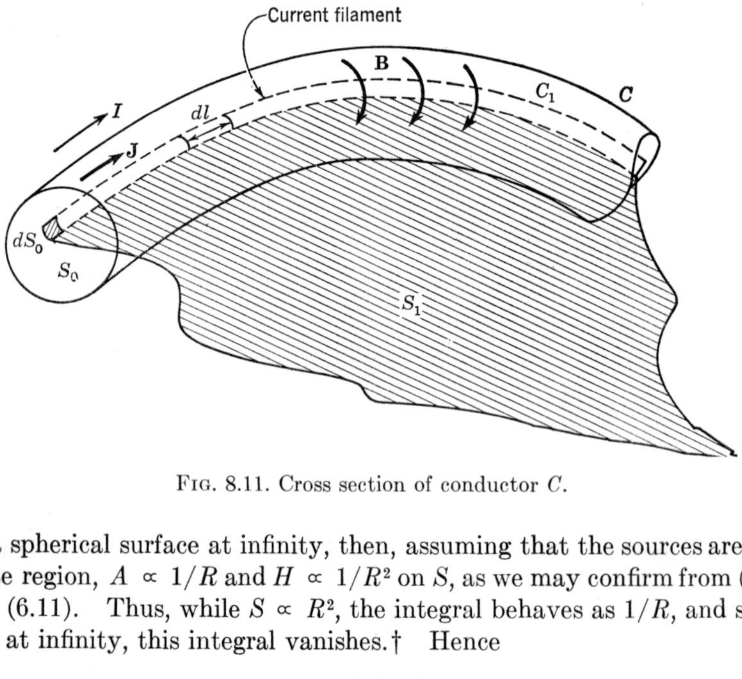

FIG. 8.11. Cross section of conductor C.

be a spherical surface at infinity, then, assuming that the sources are in a finite region, $A \propto 1/R$ and $H \propto 1/R^2$ on S, as we may confirm from (6.9) and (6.11). Thus, while $S \propto R^2$, the integral behaves as $1/R$, and since S is at infinity, this integral vanishes.† Hence

$$W_m = \tfrac{1}{2} \int_V \mathbf{J} \cdot \mathbf{A}\, dV \tag{8.31}$$

Now $\mathbf{J} = 0$ everywhere except along the circuit C, where $\mathbf{J}\, dV = J\, dS_0\, d\mathbf{l}$ and dS_0 is an element of area in the cross section of C, as in Fig. 8.11. We

† When we come to examine general time-varying fields, we shall discover that a radiation field can exist for which $H \propto 1/R$. Under these conditions the integral in question does not vanish but represents radiated energy.

may write (8.31) as follows:

$$W_m = \tfrac{1}{2} \int_{S_0} J \, dS_0 \oint_{C_1} \mathbf{A} \cdot d\mathbf{l}$$

where S_0 is the cross-sectional area of the conductor and C_1 is the contour of an elementary filament of current. If J is constant over the cross section S_0, we have $I = JS_0$, and we get

$$W_m = \tfrac{1}{2} I \int_{S_0} \left(\frac{dS_0}{S_0} \oint_{C_1} \mathbf{A} \cdot d\mathbf{l} \right) \tag{8.32}$$

Equation (8.32) is readily seen to be the integral form of (8.28*b*) since $\oint_{C_1} \mathbf{A} \cdot d\mathbf{l}$ is the flux that links the current filament $I \, dS_0/S_0$ and the integral over S_0 is merely the limit of the sum in (8.28*b*) as the number N of current filaments is made infinite; i.e., the cross section of each filament is made infinitesimally small. This result thus verifies the equivalence between the field integral (8.30) and the expression (8.28*b*) for the energy in the magnetic field surrounding a current loop.

The term $(dS_0/S_0) \oint_{C_1} \mathbf{A} \cdot d\mathbf{l}$ is called the partial flux linkage $d\psi$ because

$$\oint_{C_1} \mathbf{A} \cdot d\mathbf{l} = \int_{S_1} \nabla \times \mathbf{A} \cdot d\mathbf{S} = \int_{S_1} \mathbf{B} \cdot d\mathbf{S}$$

and is the flux linking the contour C_1, where S_1 is the surface bounded by C_1, as illustrated. It should be noted that the flux linking the contour C_1 is multiplied by the fraction of the total current that flows in the thin filament of cross-sectional area dS_0 to obtain the partial flux linkage. Completing the integration we have

$$W_m = \tfrac{1}{2} I \int_{S_0} d\psi = \tfrac{1}{2} I\psi = \tfrac{1}{2} LI^2 \tag{8.33}$$

This equation shows how the total flux linkage ψ of a single circuit must be defined in order that $\tfrac{1}{2}I\psi$ will give a correct result for the energy stored in the field. The alternative expression $W_m = \tfrac{1}{2}LI^2$ follows simply by defining L as equal to ψ/I, with ψ understood as the sum of all the partial flux linkages.

We now see that by a consideration of the energy stored in the magnetic field we are able to give a consistent and useful definition for the total flux linkage ψ. The resulting definition for the self-inductance L is thus based indirectly on energy considerations. We may, however, omit the intermediate step which introduced the flux linkage ψ and define L directly in terms of the magnetic energy stored in the field. Thus, consider a device with two terminals through which a current I enters and leaves. Let W_m be the energy stored in the magnetic field surrounding the device. Its inductance may now be defined as [Eq. (8.33)]

$$L = \frac{2W_m}{I^2} \tag{8.34}$$

This definition is often easier to apply in practice in order to evaluate L than the original definition in terms of flux linkages. A device of the type above is called an inductor, and its circuit applications are discussed at the end of Chap. 9.

It has thus been proved that (8.33) and (8.30) are equivalent expressions for the energy stored in the magnetic field. In the proof of this equivalence (8.30) was reduced to the form given by (8.32). The integrand in (8.32) was next identified as the partial flux linkage of the total current. This corresponds to the definition that was used in Example 8.4, where we chose

$$d\psi = \frac{dS_0}{S_0} \oint_{C_1} \mathbf{A} \cdot d\mathbf{l} \tag{8.35}$$

as the definition of the partial flux linkage. If the current density J is not constant over the cross section, the partial flux linkage must be taken as

$$d\psi = \frac{J\, dS_0}{I} \oint_{C_1} \mathbf{A} \cdot d\mathbf{l}$$

instead.

The above proof may be generalized to a system of N current loops as well, and hence (8.30) is a valid expression under all circumstances. At times it is convenient to think of the integrand $\mathbf{B} \cdot \mathbf{H}/2$ as the density of magnetic energy at a given point in space. However, it must be kept in mind that it is not possible to state where energy is located. Only the total energy associated with a given field has a physical meaning.

8.6. Forces as Derivatives of Coefficients of Inductance

The force between two separate current-carrying loops or circuits may be evaluated by means of Ampère's law of force. However, an alternative method that is much easier to apply in many cases may also be used. This alternative method consists essentially in evaluating the derivatives of mutual-inductance coefficients with respect to arbitrary virtual displacements of the circuits with respect to each other. When two circuits are displaced relative to each other, the mutual inductance, and hence the energy stored in the magnetic field, changes. The change in the magnetic energy is in turn related to the work done against the forces of the field in displacing the circuits.

Consider two circuits C_1 and C_2 with currents I_1 and I_2, as in Fig. 8.12. The force $\mathbf{F}$ exerted on C_2 by C_1 will be evaluated by finding the work done and the change in field energy when C_2 is displaced by an amount $d\mathbf{r}$. During the displacement the currents I_1 and I_2 will be kept constant. Initially, the flux ψ_{12} linking C_2 due to the current I_1 in C_1 is given by $\psi_{12} = L_{12}I_1$ by definition of L_{12}. The energy stored in the magnetic field is†

$$W_m = \tfrac{1}{2}I_1^2L_{11} + I_1I_2L_{12} + \tfrac{1}{2}I_2^2L_{22} \tag{8.36}$$

Consider that the displacement of C_2 by the amount $d\mathbf{r}$ occurs in a time

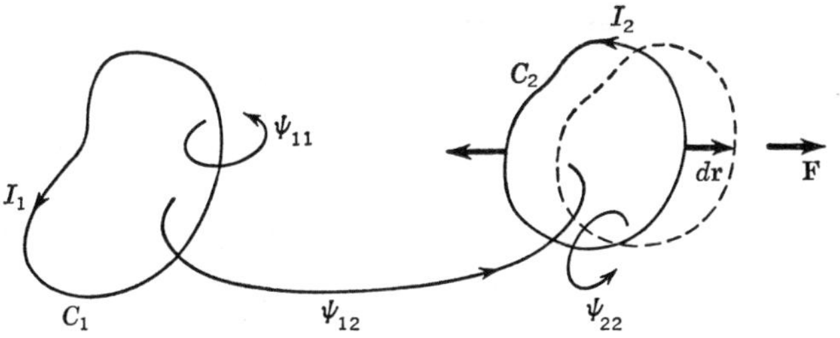

FIG. 8.12. Illustration of two circuits and their relative displacement (L_{12} negative).

interval dt. In this displacement the flux linking C_2 changes by an amount

$$d\psi_{12} = I_1\, dL_{12}$$

As a result of this change in flux linkage an induced voltage

$$\mathcal{E}_2 = -\frac{d\psi_{12}}{dt}$$

is produced in C_2. In order to keep I_2 constant we must apply a voltage $-\mathcal{E}_2$ in C_2. This voltage does work of amount

$$dW_{12} = -\mathcal{E}_2I_2\, dt = I_1I_2\, dL_{12}$$

† The interpretation of

$$W_m = \tfrac{1}{2}L_1I_1^2 + \tfrac{1}{2}L_2I_2^2 + L_{12}I_1I_2$$

may be made either in terms of infinitely thin circuit elements or in terms of finite cross-sectional current-carrying conductors. In the latter case the modified definition of self-inductance in terms of partial flux linkages must be used. For nonfilamentary conductors we define the mutual inductance in terms of the mutual energy and, by a derivation similar to that for self-inductance, are led to a generalized expression $L_{ij} = \psi_{ij}/I_i$, where ψ_{ij} is now the total partial flux linking the jth circuit due to I_i. Specifically, $\psi_{ij} = \int_{\text{cross section}} \psi_{0i}(dS_0/S_0)$, where ψ_{0i} is the flux due to i linking the current tube dS_0 of the jth circuit and S_0 is the total cross-sectional area of the jth circuit. From a practical standpoint the internal flux is often negligible, in which case $L_{ij} = \psi_{ij}/I_i$ and ψ_{ij} is the flux linking any mean current tube in the jth circuit.

in the time interval dt. Similarly, the flux linking C_1 changes by $d\psi_{12}$, and in order to keep I_1 constant, we must apply a voltage $-\mathcal{E}_1 = d\psi_{12}/dt$ in C_1. This voltage does work of amount $dW_{21} = -\mathcal{E}_1 I_1\,dt = I_1 I_2\,dL_{12}$ in the time dt. At the same time the energy stored in the magnetic field changes. This can be evaluated from (8.36) to be

$$dW_m = I_1 I_2\,dL_{12}$$

If we now assume that the force **F** due to C_1 on C_2 is in the direction of $d\mathbf{r}$, then we must apply a force $-\mathbf{F}$ along $d\mathbf{r}$ in order to displace C_2 relative to C_1. The work we shall do during the displacement is

$$dW = -\mathbf{F}\cdot d\mathbf{r}$$

In order to satisfy the law of energy conservation, the mechanical work done plus the work performed by the voltage sources in keeping I_1 and I_2 constant must be equal to the change in the field energy. Thus we get

$$-\mathbf{F}\cdot d\mathbf{r} + 2I_1 I_2\,dL_{12} = I_1 I_2\,dL_{12}$$

and hence the force exerted on C_2 in the direction $d\mathbf{r}$ is given by

$$F = I_1 I_2 \frac{dL_{12}}{dr} \tag{8.37}$$

The force between two circuits acts in the direction of increasing mutual inductance.

If we have N circuits and displace the jth circuit by an amount dr_j, we shall find in a similar way that the force F_j exerted on C_j by all the other circuits is given by

$$F_j = \sum_{\substack{n=1 \\ n\neq j}}^{N} I_j I_n \frac{dL_{jn}}{dr_j} \tag{8.38}$$

where F_j is the component of force along dr_j acting on the jth circuit. The result expressed by (8.38) is equivalent to

$$F_j = \frac{dW_m}{dr_j} \tag{8.39}$$

since

$$W_m = \tfrac{1}{2}\sum_{n=1}^{N}\sum_{s=1}^{N} I_n I_s L_{ns}$$

and we are assuming as a constraint that the currents be kept constant. When circuit j is displaced by an amount dr_j, it must be recalled that in differentiating W_m, the right-hand side is differentiated with respect to L_{nj} and L_{js} ($s = n$) and the factor $\frac{1}{2}$ is thus canceled

Usually one associates forces with the negative change in energy of a system; i.e., the system moves in such a manner that the energy stored in the system is decreased. The reason why the sign in (8.39) is positive is that, because of the changing flux linkages, the batteries in each circuit do work of twice the amount given by (8.39). Thus the batteries supply not only the increase in the field energy, but also an amount of energy equal to the work done by the field on the circuit during the displacement. The situation here is similar to the electrostatic one when a constant potential constraint is involved. Again it is important to note that the force exerted by the field is unique for a given system of loops with specified currents. The use of a constant-current constraint under a hypothetical displacement is only a matter of convenience; any other assumed constraint would lead to the same value for the force.

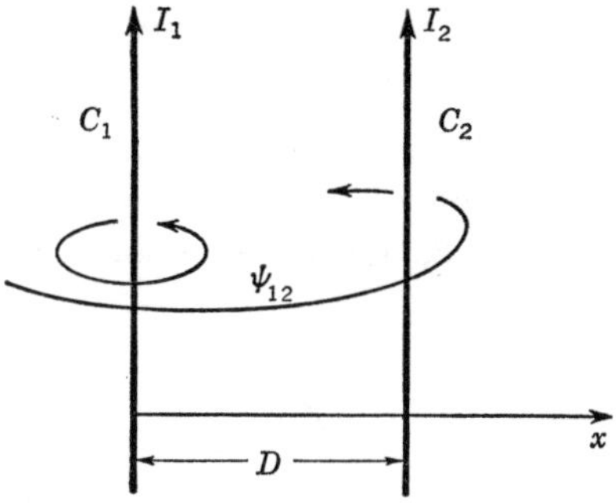

FIG. 8.13. Two infinite linear current-carrying conductors.

Example 8.6. Force on Two Parallel Wires. Consider two thin infinitely long and parallel conductors, as in Fig. 8.13. The conductors are separated by a distance D. The currents in the two conductors are I_1 and I_2. The flux linking C_2 due to the current I_1 in C_1 is

$$\psi_{12} = \frac{\mu_0 I_1}{2\pi} \int_D^\infty \frac{dx}{x} \qquad \text{per unit length}$$

The integral cannot be evaluated since it is not bounded at infinity. However, since we are going to differentiate it with respect to D, we do not need to evaluate it. From (8.37) the force per unit length exerted on C_2 by C_1 is

$$F = I_1 I_2 \left.\frac{dL_{12}}{dD}\right|_{I=\text{constant}} = I_2 \frac{d\psi_{12}}{dD} = -\frac{\mu_0 I_1 I_2}{2\pi D}$$

a result in accord with Ampère's law of force. The negative sign signifies that the force is an attractive one for currents I_1 and I_2, having the directions assumed in Fig. 8.13.

Example 8.7. Force between a Long Wire and a Rectangular Loop. Figure 8.14 illustrates a rectangular loop C_2 carrying a current I_2 and placed with its nearest side a distance D from an infinitely long conductor C_1 carrying a current I_1. With the assumed directions of current flow, the flux linking C_2 due to I_1 is oppositely directed to that due to I_2. Hence the mutual inductance is negative. The flux linking C_2 due to

I_1 is

$$\psi_{12} = \frac{\mu_0 I_1}{2\pi} a \int_D^{b+D} \frac{dx}{x} = \frac{\mu_0 I_1 a}{2\pi} \ln \frac{b+D}{D}$$

The mutual inductance L_{12} is thus

$$L_{12} = -\frac{\psi_{12}}{I_1} = -\frac{\mu_0 a}{2\pi} \ln \frac{b+D}{D}$$

The force exerted by the field on C_2 in the direction of increasing D is

$$F = I_1 I_2 \frac{dL_{12}}{dD}\bigg|_{I=\text{constant}} = \frac{\mu_0 ab I_1 I_2}{2\pi D(b+D)}$$

This result is readily verified by Ampère's law of force.

Example 8.8. Torque on a Rectangular Loop. The rectangular loop C_2 in Fig. 8.14 is rotated about its axis by an amount θ so that the resulting

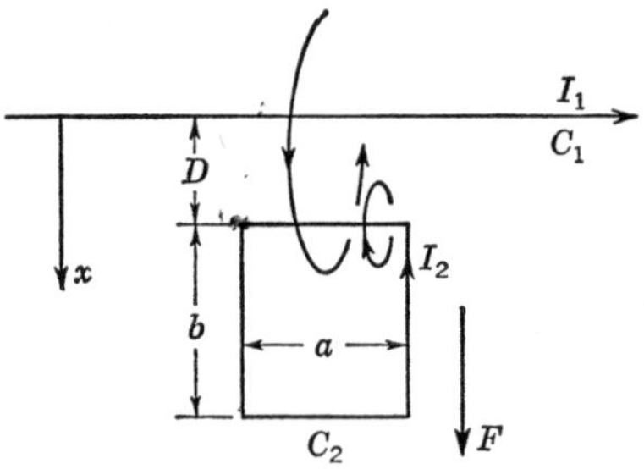

FIG. 8.14. Evaluation of force on a rectangular loop.

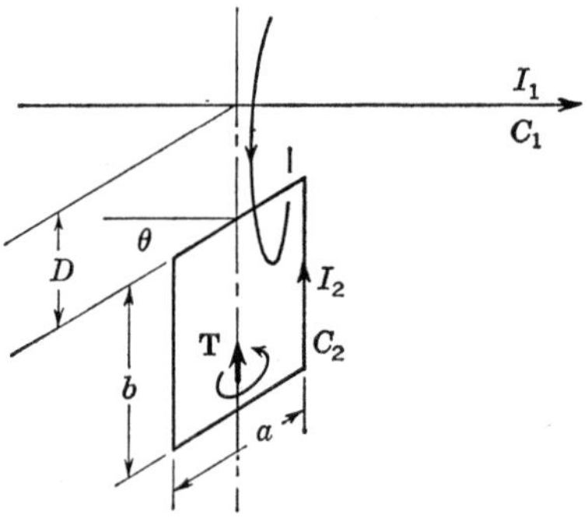

FIG. 8.15. Evaluation of torque on a rectangular loop.

configuration is given by Fig. 8.15. The torque exerted on the loop by the field is required. The flux linking the loop due to the field set up by I_1 differs from that of the previous example only by a factor $\cos\theta$. Hence we have in the present case

$$L_{12} = -\frac{\mu_0 a \cos\theta}{2\pi} \ln \frac{b+D}{D}$$

The torque exerted on the loop by the field, by a slight modification of (8.38), is

$$T = I_1 I_2 \frac{dL_{12}}{d\theta}\bigg|_{I=\text{constant}} = \frac{\mu_0 a \sin\theta}{2\pi} I_1 I_2 \ln \frac{b+D}{D}$$

8.7. Lifting Force of Magnets

An equation for the lifting force of a magnet may be obtained by means of an analysis similar to that used to obtain the force equation (8.39) in the previous section. Consider two U-shaped pieces of iron with uniform cross sections of area S and with mean lengths l. The two sections are separated by a small air gap of thickness x, as in Fig. 8.16. The upper section is wound with N turns of wire to produce an electromagnet. A current I flows in the coil. The permeability of the iron is assumed constant at the value μ. The case when μ varies is considered later. A solution of the magnetic circuit shows that the total flux in the cross section of the circuit is given by

$$\psi = \frac{NI}{2l/\mu S + 2x/\mu_0 S} \tag{8.40}$$

The inductance of the coil under the given conditions is

$$L = \frac{N\psi}{I} = \frac{SN^2}{2l/\mu + 2x/\mu_0} \tag{8.41}$$

since the flux ψ links N turns. If the lower U section is displaced by an amount dx in a time interval dt, then, assuming constant current, the flux ψ changes by an amount

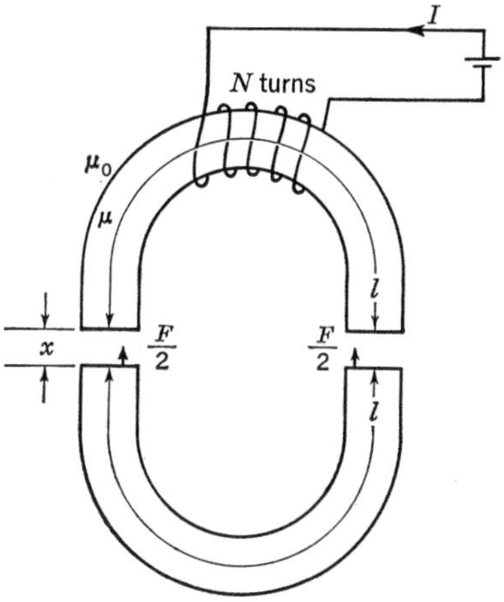

FIG. 8.16. An electromagnet.

$$d\psi = \left.\frac{d\psi}{dx}\right|_{I=\text{constant}} dx = -\frac{2NI\,dx}{\mu_0 S\left(\dfrac{2l}{\mu S} + \dfrac{2x}{\mu_0 S}\right)^2} = -\frac{\psi\,dx}{\mu_0\left(\dfrac{l}{\mu} + \dfrac{x}{\mu_0}\right)} \tag{8.42}$$

This results in an induced voltage $\mathcal{E}$ in the coil, where

$$\mathcal{E} = -N\left.\frac{d\psi}{dt}\right|_{I=\text{constant}} = -I\frac{dL}{dt}$$

In order to keep the current I constant, we must apply a voltage $-\mathcal{E}$ in series with the battery in the coil. This applied voltage does work of amount

$$dW_1 = -\mathcal{E}I\,dt = I^2\,dL$$

in the time interval dt. During the displacement, the energy in the magnetic field changes by an amount $dW_m = I^2\,dL/2$. If the field exerts a force F on the lower U section, we must apply a force $-F$ in order to increase the air gap by an amount dx. During the displacement we do work of amount $dW = -F\,dx$. Equating the work done on the system to the change in field energy, we get

$$dW + dW_1 \equiv -F\,dx + I^2\,dL = \tfrac{1}{2}I^2\,dL$$

and hence

$$F = \tfrac{1}{2}I^2\frac{dL}{dx} \tag{8.43}$$

Replacing IL by $N\psi$ and using (8.42), we obtain

$$\frac{F}{2} = -\frac{NI\psi}{4\mu_0 S(l/\mu S + x/\mu_0 S)} \tag{8.44}$$

for the force exerted on one pole or arm of the U section. In the air gap we have

$$B = \frac{\psi}{S}$$

$$SH = \frac{\psi}{\mu_0} = \frac{NI}{2\mu_0(l/\mu S + x/\mu_0 S)}$$

and hence (8.44) may also be written as

$$\frac{F}{2} = -\frac{BH}{2}S = -\frac{\mu_0 H^2}{2}S \tag{8.45}$$

and the negative sign indicates that the force is attractive under the given conditions. The interpretation of this result is that the magnetic field exerts an attractive force along the direction of the field with a magnitude or density given by

$$f = \frac{BH}{2} = \frac{\mu_0 H^2}{2} \qquad \text{newtons/sq m} \tag{8.46}$$

In (8.46) f is a force per unit area, the factor 2 having been absorbed by the two pole faces. With this interpretation it is clear that it does not matter if the magnet illustrated in Fig. 8.16 is an electromagnet or a permanent magnet. As long as the flux density in the air gap is the same, the lifting force produced by the field will be the same. This conclusion can be demonstrated to have very general application, and a discussion of this is reserved for the following section. First, however, we shall consider the effect of variable μ on the expression for the force exerted by the field.

We have seen that under a constant-current constraint the force exerted by the magnetic field could be expressed as [see Eq. (8.39)]

$$F_j = \frac{dW_m}{dr_j}$$

The derivation implicitly assumed that the permeability of all material bodies involved was constant. When the permeability is a function of the field $\mathbf{H}$, then it will be found that the above equation is not correct, although a very similar equation will be found to apply. For a ferromagnetic material the flux density $\mathbf{B}$ is a function of the field $\mathbf{H}$, and hence of the currents I that exist in the circuit. Thus the total flux in the magnetic circuit of Fig. 8.16 is a function of the current I as determined by the B-H curve for the iron involved. In Fig. 8.17 the total magnetic flux ψ is plotted as a function of the current I. If the current is changed by an amount dI in a time interval dt, an induced voltage $\mathcal{E} = -d\psi/dt$ is produced in the circuit. The battery consequently does work of amount $dW_1 = I\,d\psi$ in changing the current by the amount dI. The total work done in increasing ψ from zero to the final value ψ, that is, bringing the material from the point P_1 to P_2 in Fig. 8.17, is given by

$$W_1 = W_m = \int_0^\psi I\,d\psi \tag{8.47}$$

This is also equal to the energy stored in the magnetic field and is seen to be given by the area that is singly hatched in Fig. 8.17.

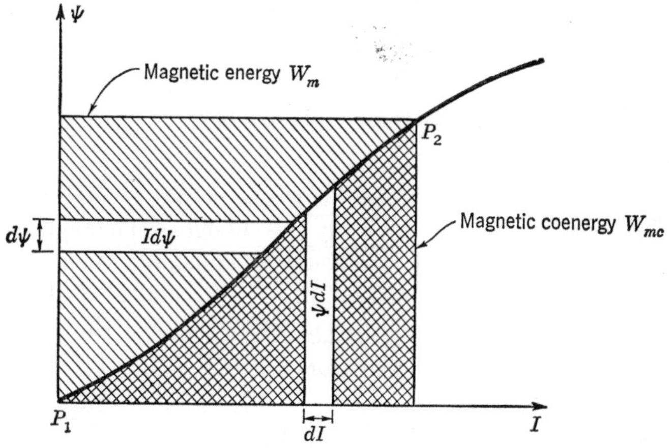

FIG. 8.17. Illustration of magnetic coenergy.

With the aid of (8.47) we shall now derive the expression for the force exerted by the magnetic field on the pole piece of the electromagnet illustrated in Fig. 8.16. If the field exerts a force F on the lower U-shaped section, then the external mechanical work done in a virtual displacement dx is $dW = -F\,dx$. If we assume a constant-current constraint as before, then, because of a change $d\psi$ in the flux through the circuit, the battery does work of amount $dW_1 = I\,d\psi$ in keeping the current constant during the displacement. The change in magnetic energy stored in the system may be found from (8.47) and is given by

$$dW_m = d\int_0^\psi I\,d\psi$$

Equating the external work done to the change in field energy we obtain

$$dW_1 + dW = dW_m$$

or

$$-F\,dx + I\,d\psi = d\int_0^\psi I\,d\psi$$

This expression may be simplified if we integrate the right-hand side by parts once to obtain

$$\int_0^{\psi} I\,d\psi = I\psi\Big|_0^{\psi} - \int_0^{I(\psi)} \psi\,dI$$

The latter integral, that is, $\int_0^{I(\psi)} \psi\,dI$, defines a quantity called the magnetic coenergy. With reference to Fig. 8.17 it is seen to be given by the double-crosshatched area. If we denote the magnetic coenergy by W_{mc}, then we have

$$d\int_0^{\psi} I\,d\psi = I\,d\psi - dW_{mc}$$

since $d(I\psi) = I\,d\psi$, because I is kept constant.

Our energy-balance equation now becomes

$$-F\,dx + I\,d\psi = I\,d\psi - dW_{mc}$$

or

$$F = \frac{dW_{mc}}{dx} = \frac{d(\text{coenergy})}{dx} \tag{8.48}$$

Thus, when μ is variable, the force due to the field is given by the change in the magnetic coenergy instead of by the change in the magnetic energy, as is the case for constant μ. When μ is constant, then the relation between I and ψ is a linear one and $W_m = W_{mc}$ and also $dW_m = dW_{mc}$. Since many practical problems for which we desire the forces acting involve iron, the concept of coenergy is a useful one in practice.†

8.8. Magnetic Stress Tensor

Ampère's law of force between two current elements,

$$\mathbf{F} = \mu_0 \frac{I_2\,d\mathbf{l}_2 \times (I_1\,d\mathbf{l}_1 \times \mathbf{a}_R)}{R^2}$$

is an action-at-a-distance law. It gives the force of one current element on another as though the force were acting at a distance from the current producing the force. When the field **B** is introduced, we have

$$\mathbf{F} = I_2\,d\mathbf{l}_2 \times \mathbf{B}$$

This alternative form explains the force on $I_2\,d\mathbf{l}_2$ by means of the interaction of the field **B**, set up by I_1, and the current I_2. In the preceding section we found that the force between an electromagnet and an iron bar could be expressed in terms of the field **B** alone. One of the aims of field theory is to express all observables such as forces and energy in terms of the field alone. Thus if we know the total field surrounding a body, the force exerted on the body is desired in terms of the field alone. This possibility was found to be true for the electrostatic field. In this section similar results for the magnetostatic field will be presented, but without proof.

The results of the previous section showed that along the lines of magnetic flux the

† An interesting discussion on the concept of coenergy from a thermodynamic viewpoint is given by O. K. Mawardi, On the Concept of Coenergy, *J. Franklin Inst.*, vol. 264, pp. 313–332, October, 1957.

field produced a force per unit area given by

$$f_t = \frac{\mu_0 H^2}{2} \tag{8.49}$$

This force per unit area is equivalent to a tension. We may therefore picture the magnetic lines of flux as elastic bands which are stretched, and hence are under tension. In addition to producing a tension along the lines of force, the magnetic field may produce a compressional force perpendicular to the lines of flux. The density of this compressional force turns out to be given by (8.49) also; i.e.,

$$f_c = \frac{\mu_0 H^2}{2} \tag{8.50}$$

We may consider (8.49) and (8.50) as the pressure that the field **H** exerts. In vector form the force acting on a body can be expressed by

$$\mathbf{F} = \oint_S \left[\mu_0 \mathbf{H}(\mathbf{n} \cdot \mathbf{H}) - \frac{\mu_0}{2} (\mathbf{H} \cdot \mathbf{H})\mathbf{n} \right] dS \tag{8.51}$$

where **n** is the unit outward normal to the surface S, and S is any closed surface surrounding the body. Since the force or pressure produced by the field has the dimensions of stress, (8.49) and (8.50) are called the components of the stress tensor. The components of the stress tensor acting on a surface element whose normal is **n** are given by the integrand in (8.51). This integrand is a vector force per unit area and was designated **T** in the analogous electrostatic case. The component of **T** along **n** is

$$\mathbf{n} \cdot \mathbf{T} = \mu_0(\mathbf{n} \cdot \mathbf{H})^2 - \frac{\mu_0}{2} (\mathbf{H} \cdot \mathbf{H})$$

since the component of **H** along **n** is $\mathbf{n} \cdot \mathbf{H}$. This force is a pressure force. The remaining term, $\mu_0(\mathbf{n} \cdot \mathbf{H})\mathbf{H}_t$, where $\mathbf{H}_t$ is the component of **H** tangential to the surface, represents a shearing force per unit area along the surface. As in the electrostatic case, the magnitude of the surface force density $|\mathbf{T}|$ is $\mu_0 H^2/2$, and analogously, **H** bisects the angle between **T** and **n**.

Example 8.9. Force on a Current Element. The stress tensor will be used to obtain an expression for the force exerted on a linear current element placed in a uniform field $\mathbf{B}_0$. Figure 8.18 illustrates a linear conductor carrying a current I and located in a field $\mathbf{B}_0$ directed perpendicular to it. To find the force acting on the conductor per unit length we shall evaluate (8.51) over the surface of a cylinder of unit length, radius r, and concentric with the conductor. The magnetic field that is required in (8.51) is in this case the sum of $\mathbf{B}_0 = \mu_0 \mathbf{H}_0$ and the field $\mathbf{B}_\phi = \mu_0 I/2\pi r$ due to the current I.

The field $\mathbf{B}_\phi$ will be decomposed into x and y components first in order to facilitate summation. We get

$$\mathbf{B}_\phi = \frac{\mu_0 I}{2\pi r} \mathbf{a}_\phi = \frac{\mu_0 I}{2\pi r} (-\mathbf{a}_x \sin \phi + \mathbf{a}_y \cos \phi)$$

and hence

$$\mathbf{B} = \mathbf{B}_\phi + \mathbf{B}_0 = -\frac{\mu_0 I}{2\pi r} \mathbf{a}_x \sin \phi + \mathbf{a}_y \left(\mathbf{B}_0 + \frac{\mu_0 I}{2\pi r} \cos \phi \right)$$

The unit normal $\mathbf{n}$ is equal to

$$\mathbf{n} = \mathbf{a}_r = \mathbf{a}_x \cos\phi + \mathbf{a}_y \sin\phi$$

and consequently the integrand in (8.51) becomes

$$\begin{aligned}\mu_0\mathbf{H}(\mathbf{n}\cdot\mathbf{H}) - \frac{\mu_0}{2}(\mathbf{H}\cdot\mathbf{H})\mathbf{n} &= \left[-\frac{\mu_0 I}{2\pi r}\mathbf{a}_x\sin\phi + \mathbf{a}_y\left(B_0 + \frac{\mu_0 I\cos\phi}{2\pi r}\right)\right] \\ &\quad\times\left(-\frac{I\sin\phi\cos\phi}{2\pi r} + H_0\sin\phi + \frac{I\sin\phi\cos\phi}{2\pi r}\right) \\ &\quad-\frac{1}{2}\left[\frac{\mu_0 I^2}{4(\pi r)^2}\sin^2\phi + \left(B_0 + \frac{\mu_0 I\cos\phi}{2\pi r}\right)\left(H_0 + \frac{I\cos\phi}{2\pi r}\right)\right] \\ &\quad\times(\mathbf{a}_x\cos\phi + \mathbf{a}_y\sin\phi)\end{aligned} \tag{8.52}$$

The element of area dS is $r\,d\phi$ for a unit length of cylinder. The expression (8.52) appears rather formidable to integrate but is in actual fact rather easy to handle. As a matter of fact, because of the orthogonality

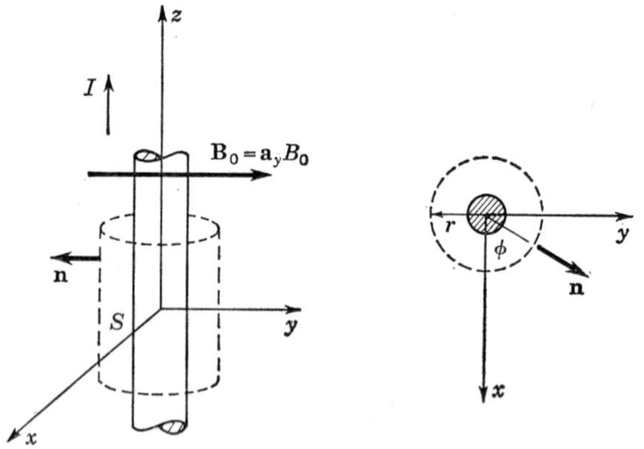

FIG. 8.18. A conductor in a field $\mathbf{B}_0$.

property of the trigonometric functions, all terms in (8.52), when integrated from zero to 2π, go to zero, except the terms in $\sin^2\phi$ or $\cos^2\phi$. Thus (8.52) simplifies to give

$$\mathbf{F} = -\frac{\mu_0 I H_0}{2\pi r}\mathbf{a}_x\int_0^{2\pi}\sin^2\phi\,d\phi - \frac{B_0 I}{2\pi r}\mathbf{a}_x\int_0^{2\pi}\cos^2\phi\,d\phi = -\mathbf{a}_x I B_0 \tag{8.53}$$

Over the ends of the cylinder $\mathbf{H}\cdot\mathbf{n} = 0$ and the integral of the term $-(\mu_0/2)(\mathbf{H}\cdot\mathbf{H})\mathbf{n}$ vanishes, since $\mathbf{n}$ is directed in the opposite sense on the two end faces. The result (8.53) is the same as given by Ampère's law of force:

$$\mathbf{F} = I\mathbf{a}_z \times \mathbf{a}_y B_0 = -\mathbf{a}_x I B_0$$

The use of the stress tensor in the present situation was considerably more cumbersome than the use of Ampère's law. However, our purpose in the present example was only to illustrate the application of (8.51).

8.9. Hysteresis Loss

In the previous chapter it was stated that the energy dissipated in cycling a ferromagnetic material around the hysteresis loop was proportional to the area of the hysteresis loop. The field concepts necessary to establish this result are now available; so a proof of this result will be presented here.

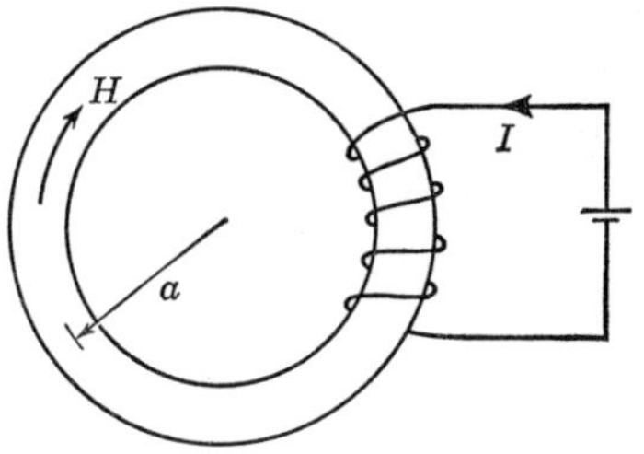

FIG. 8.19. A toroidal specimen of ferromagnetic material.

Consider a ferromagnetic specimen in the shape of a toroid, as in Fig. 8.19. The mean toroid radius is a, and its cross section is S. The cross section will be assumed to be sufficiently small so that the field in the specimen can be considered uniform. The toroid is wound with N turns, through which a current I flows.

Figure 8.20 illustrates the B-H, or hysteresis, loop for the material. Let the initial state of magnetization be represented by the point b. If we increase the current I by an amount dI, the field H increases by an amount dH and B by an amount dB. If this increase takes place in a

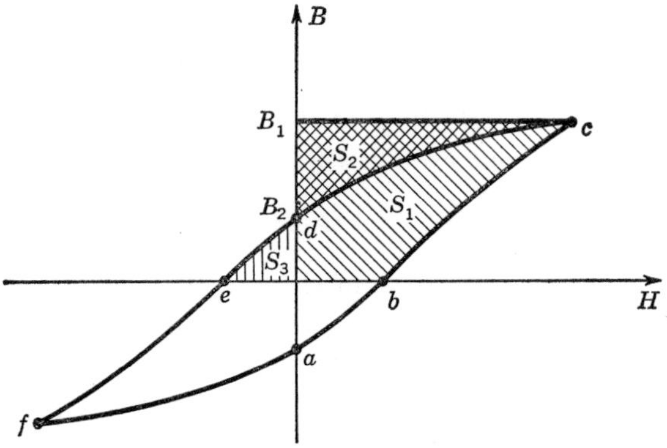

FIG. 8.20. B-H curve for ferromagnetic material.

time interval dt, the voltage induced in the toroid winding is

$$\mathcal{E} = -N\frac{d\psi}{dt} = -NS\frac{dB}{dt}$$

In order to increase I, the battery must do work against the induced voltage of amount

$$dW = -\mathcal{E}I\,dt = NSI\,dB \tag{8.54}$$

According to Ampère's circuital law, the field H is given by $H = NI/2\pi a$. Substituting into (8.54) gives

$$dW = 2\pi aSH\,dB = VH\,dB \tag{8.55}$$

where $V = 2\pi aS$ is the volume of the torus. In changing the field B up to B_1 along the path bc of the hysteresis loop, the work done by the battery is

$$W_1 = V\int_0^{B_1} H\,dB = V(S_1 + S_2) \tag{8.56}$$

where $S_1 + S_2$ is the area between the B axis and the portion bc of the B-H curve. When we decrease B from B_1 to B_2 along the path cd, the decreasing flux induces a voltage in the coil that tends to maintain the current I. Thus work is done against the battery. The amount of work done is

$$W_2 = V\int_{B_1}^{B_2} H\,dB = -VS_2 \tag{8.57}$$

where S_2 is the double-crosshatched area in Fig. 8.20. In reducing the field to zero along the path de, the battery again does work, since the direction of current flow is reversed but the induced voltage acts in the same direction as for (8.57). The work done is

$$W_3 = VS_3 \tag{8.58}$$

where S_3 is the area indicated in Fig. 8.20. To cycle the material from e to f and back up to b, the same amount of work is done as in bringing the material from b to e along the upper half of the loop. The total net work done by the battery in one cycle is therefore

$$W = 2V(W_1 + W_2 + W_3) = 2V(S_1 + S_2 - S_2 + S_3) = VS_t \tag{8.59}$$

where $S_t = 2(S_1 + S_3)$ is the area of the complete hysteresis loop.

Equation (8.59) shows that the area of the hysteresis loop represents the work done per unit volume in cycling a ferromagnetic material around the hysteresis loop once. This amount of energy is lost each cycle and is dissipated as heat in the material. The energy loss is caused by the work required to magnetize the material.

CHAPTER 9

TIME-DEPENDENT FIELDS

In all the previous chapters we have been primarily concerned with static fields, i.e., fields that are independent of time. In this chapter this restriction will be removed, and we shall then find that the electric and magnetic fields are intimately related to each other. An example of this interrelationship was provided by Faraday's law, which showed that a time-varying magnetic field induced an electric field. To complete the picture we shall show that a reciprocal effect, namely, that a time-varying electric field induces a magnetic field, also exists. This mutual support of each other, i.e., a magnetic field producing an electric field and an electric field producing a magnetic field, results in the phenomenon of wave propagation. The prediction of electromagnetic waves and the subsequent successful use of these waves in communication systems were an outstanding climax to the centuries of exploration and experimentation that preceded it.

9.1. Modification of Static Field Equations under Time-varying Conditions

Before presenting the general equations for the time-varying electromagnetic field we shall summarize the basic equations that govern the static electric and magnetic fields and the stationary current flow field. A number of equivalent choices are possible, but the following equations are chosen because they clearly show the irrotational property of the electrostatic field and the divergenceless property of the magnetostatic and stationary current flow fields.

For the electrostatic field we have

$$\nabla \times \mathbf{E} = 0 \tag{9.1a}$$

$$\nabla \cdot \mathbf{D} = \rho \tag{9.1b}$$

while for the magnetostatic field

$$\nabla \times \mathbf{H} = \mathbf{J} \tag{9.1c}$$

$$\nabla \cdot \mathbf{B} = 0 \tag{9.1d}$$

and for stationary currents

$$\nabla \cdot \mathbf{J} = 0 \tag{9.1e}$$

We already know that some of the above equations must be modified when the fields vary with time. In particular, Faraday's law of induction shows that (9.1*a*) must be replaced by

$$\nabla \times \mathbf{E} = -\frac{\partial \mathbf{B}}{\partial t} \tag{9.2}$$

when the field **B** varies with time. Also we know that when **J** and ρ vary with time, the continuity equation

$$\nabla \cdot \mathbf{J} = -\frac{\partial \rho}{\partial t} \tag{9.3}$$

must hold since current is a flow of charge, and hence the divergence of the current at a point must always equal the time rate of decrease of charge density at that point.

At this time we might very well ask whether there is any need to modify any of the other equations. The answer is, "Yes, there is," for we may easily show that the set of equations (9.1*b*), (9.1*c*), (9.1*d*), (9.2), and (9.3) do not form a self-consistent set. The divergence of the curl of any vector is identically zero, and hence from (9.1*c*) we obtain

$$\nabla \cdot \nabla \times \mathbf{H} = 0 = \nabla \cdot \mathbf{J} \tag{9.4}$$

This result is in contradiction with (9.3) when **J** varies with time. If the basic form of (9.1*c*) is to be retained under time-varying conditions, then the right-hand side must be solenoidal and reduce to **J** under time-stationary conditions. The necessary form of the right-hand side of (9.1*c*) can be deduced from (9.3) if we make use of (9.1*b*) in the expression for ρ. With this substitution we may write

$$\nabla \cdot \left(\mathbf{J} + \frac{\partial \mathbf{D}}{\partial t} \right) = 0 \tag{9.5}$$

The vector quantity in the parentheses of (9.5) is solenoidal and reduces to **J** if $\partial/\partial t = 0$. Consequently, the previous equations will become consistent if this quantity is substituted for the right-hand side of (9.1*c*); that is,

$$\nabla \times \mathbf{H} = \mathbf{J} + \frac{\partial \mathbf{D}}{\partial t} \tag{9.6}$$

The term $\partial \mathbf{D}/\partial t$ was originally introduced into the curl equation for **H** by Maxwell and is called the displacement current density because it has the dimensions of a current density. Although the way in which we introduced this term above in no way proves the correctness of (9.6), it has been found experimentally that the conclusions drawn from (9.6) are in accord with all known experimental facts; so there is no reason to doubt its validity. The sum of the terms on the right-hand side of (9.6) is in the form of a total current, which, as we are aware, is solenoidal.

As for the remaining equations, they are self-consistent and no experimental evidence for requiring any further modifications has been found. The above equations governing the time-varying electromagnetic field are known collectively as Maxwell's equations and will be discussed in greater detail in the next section.

9.2. Maxwell's Equations

From the previous section we have the following set of equations which govern the behavior of the time-varying electromagnetic field.

$$\nabla \times \mathbf{E} = -\frac{\partial \mathbf{B}}{\partial t} \tag{9.7a}$$

$$\nabla \cdot \mathbf{D} = \rho \tag{9.7b}$$

$$\nabla \times \mathbf{H} = \mathbf{J} + \frac{\partial \mathbf{D}}{\partial t} \tag{9.7c}$$

$$\nabla \cdot \mathbf{B} = 0 \tag{9.7d}$$

$$\nabla \cdot \mathbf{J} = -\frac{\partial \rho}{\partial t} \tag{9.7e}$$

In (9.7*c*) and (9.7*e*) the current $\mathbf{J}$ will consist, in general, of a conduction current $\sigma\mathbf{E}$, caused by the presence of an electric field $\mathbf{E}$ in a material with finite conductivity σ, and a convection current $\rho\mathbf{v}$, consisting of a free-charge distribution ρ flowing with a velocity $\mathbf{v}$. The convection current is of importance in many practical devices such as electron tubes, cathode-ray tubes, etc. In the majority of situations that we deal with in this book, however, the convection current is zero.

In place of the above equations, which are in derivative form, we may write an equivalent set of equations in integral form. Integration of (9.7*a*) over an open surface S and application of Stokes' law give

$$\oint_C \mathbf{E} \cdot d\mathbf{l} = -\frac{\partial}{\partial t} \int_S \mathbf{B} \cdot d\mathbf{S} \tag{9.8a}$$

where C is the boundary of S. For (9.7*b*) we integrate throughout a volume V and use the divergence theorem to obtain

$$\oint_S \mathbf{D} \cdot d\mathbf{S} = \int_V \rho \, dV \tag{9.8b}$$

Similarly, the remaining three equations give

$$\oint_C \mathbf{H} \cdot d\mathbf{l} = \frac{\partial}{\partial t} \int_S \mathbf{D} \cdot d\mathbf{S} + \int_S \mathbf{J} \cdot d\mathbf{S} \tag{9.8c}$$

$$\oint_S \mathbf{B} \cdot d\mathbf{S} = 0 \tag{9.8d}$$

$$\oint_S \mathbf{J} \cdot d\mathbf{S} = -\frac{\partial}{\partial t} \int_V \rho \, dV \tag{9.8e}$$

The integral form of Maxwell's equations is easier to interpret physically and is also useful in deriving the boundary conditions that the field vectors must satisfy. However, in the solution of a physical problem, the derivative form is invariably used.

The first equation in the set (9.8) is just Faraday's law of induction and states that the total voltage induced around a contour C is equal to the negative time rate of change of magnetic flux through this contour. Equation (9.8*b*) simply states that the total displacement flux through a closed surface S is equal to the enclosed charge (Gauss' law).

Equation (9.8*c*) is a generalization of Ampère's circuital law by the addition of the displacement current term. Without this term electromagnetic waves would not exist. It is not surprising that this term was not discovered experimentally, since it is only at radio frequencies that the displacement current becomes comparable to the conduction current in its effects. At the time of Maxwell, means for generating high-frequency currents and fields were virtually nonexistent and certainly, at

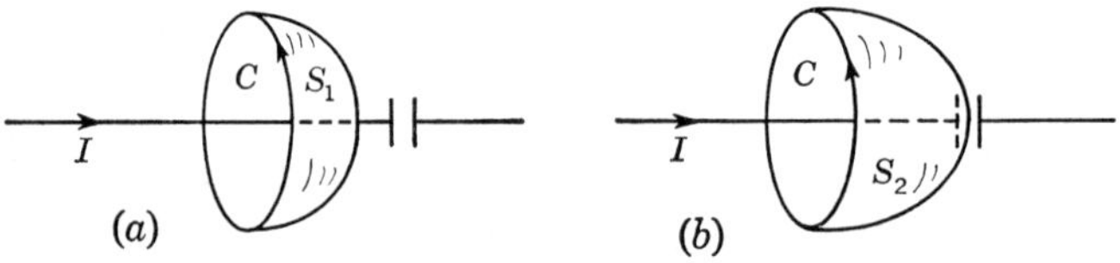

FIG. 9.1. Illustration of need for displacement current term in Ampère's circuital law.

best, very poorly understood. After introducing the displacement current Maxwell was able to show theoretically the existence of electromagnetic waves having a velocity of propagation equal to that of light. This prediction obviously led to the conclusion that light was electromagnetic in nature. It was only years later that the brilliant experimental work of Hertz, in generating electromagnetic waves by means of spark gaps and demonstrating that their properties were similar to that of light, verified the correctness of Maxwell's assumption.

As an example to illustrate the need for the displacement current, consider a parallel-plate capacitor connected to an a-c generator by means of two conducting wires. If we draw an arbitrary closed contour C through which the circuit passes and construct a surface S_1 that intersects the conductor, as in Fig. 9.1*a*, Ampère's circuital law gives

$$\oint_C \mathbf{H} \cdot d\mathbf{l} = \int_{S_1} \mathbf{J} \cdot d\mathbf{S} = I$$

where I is the total current flowing in the conductor. We could equally well draw our surface as a surface S_2 that passes between the plates of the capacitor, as in Fig. 9.1*b*. If we use the unmodified form of Ampère's circuital law, we should be led to the conclusion that the line integral of

H around C was equal to zero since no conduction current flows through the surface S_2. This would be an embarrassing situation, since the contour C is still the same contour as used in Fig. 9.1*a*. If we include the displacement current in Ampère's circuital law, we are able to resolve this difficulty. Let A be the area of the capacitor plate, and let d_0 be the separation. The capacitance C_0 is given by

$$C_0 = \frac{\epsilon_0 A}{d_0}$$

if fringing effects are neglected. When a current I is flowing into a capacitor, the voltage V across it is given by

$$I = C_0 \frac{dV}{dt}$$

But the voltage V is equal to Ed_0, where E is the electric field between the plates. Hence

$$I = C_0 d_0 \frac{dE}{dt} = \frac{C_0 d_0}{\epsilon_0 A} \frac{d\epsilon_0 AE}{dt} = \frac{d\epsilon_0 AE}{dt}$$

The latter term is the total displacement current flowing between the capacitor plates and is also given by

$$\frac{d}{dt}\left(\int_{S_2} \mathbf{D} \cdot d\mathbf{S}\right) = \frac{d\epsilon_0 AE}{dt} = I$$

Hence
$$\oint_C \mathbf{H} \cdot d\mathbf{l} = \frac{d}{dt}\left(\int_{S_2} \mathbf{D} \cdot d\mathbf{S}\right) = I$$

which is the same result as obtained by choosing the surface as S_1 in Fig. 9.1*a*. Of course, this example only confirms that consistent results are obtained through the inclusion of a displacement current term in Ampère's circuital law. It does not verify that the law can be extended to the time-varying case in the way stated. Only an appeal to experiment can confirm this, as it does.

9.3. Source-free Wave Equation

In a source-free region of free space, $\mathbf{J} = \rho = 0$, and Maxwell's equations reduce to the following:

$$\nabla \times \mathbf{E} = -\frac{\partial \mathbf{B}}{\partial t} = -\mu_0 \frac{\partial \mathbf{H}}{\partial t} \tag{9.9a}$$

$$\nabla \times \mathbf{H} = \frac{\partial \mathbf{D}}{\partial t} = \epsilon_0 \frac{\partial \mathbf{E}}{\partial t} \tag{9.9b}$$

$$\nabla \cdot \mathbf{D} = \epsilon_0 \nabla \cdot \mathbf{E} = 0 \tag{9.9c}$$

$$\nabla \cdot \mathbf{B} = \mu_0 \nabla \cdot \mathbf{H} = 0 \tag{9.9d}$$

If we eliminate either **E** or **H**, we obtain a three-dimensional wave equation for the remaining quantity. For example, take the curl of (9.9*a*) and use (9.9*b*) to eliminate $\nabla \times \mathbf{H}$ on the right-hand side and thus obtain

$$\nabla \times \nabla \times \mathbf{E} = -\mu_0 \frac{\partial}{\partial t} \nabla \times \mathbf{H} = -\mu_0\epsilon_0 \frac{\partial^2 \mathbf{E}}{\partial t^2}$$

We may expand $\nabla \times \nabla \times \mathbf{E}$ to get $\nabla\nabla \cdot \mathbf{E} - \nabla^2\mathbf{E}$, and since $\nabla \cdot \mathbf{E} = 0$ from (9.9*c*), we have

$$\nabla^2\mathbf{E} - \mu_0\epsilon_0 \frac{\partial^2 \mathbf{E}}{\partial t^2} = 0$$

or

$$\underline{\nabla^2\mathbf{E} - \frac{1}{c^2}\frac{\partial^2 \mathbf{E}}{\partial t^2} = 0} \tag{9.10}$$

where $c = (\mu_0\epsilon_0)^{-1/2}$. The parameter c has the dimensions of velocity and is numerically equal to 3×10^8 meters per second, i.e., the velocity of light in free space or vacuum. Equation (9.10) is the standard form of a three-dimensional vector wave equation. The field **H** satisfies the same equation, as may be readily shown by eliminating **E** from (9.9*b*). In practice, if we know **E**, we can obtain **H** by using (9.9*a*).

In order to examine the nature of (9.10) more closely, note that each rectangular component of **E** satisfies the scalar wave equation; e.g.,

$$\nabla^2 E_x - \frac{1}{c^2}\frac{\partial^2 E_x}{\partial t^2} = \frac{\partial^2 E_x}{\partial x^2} + \frac{\partial^2 E_x}{\partial y^2} + \frac{\partial^2 E_x}{\partial z^2} - \frac{1}{c^2}\frac{\partial^2 E_x}{\partial t^2} = 0$$

If we now assume that E_x is a function of the z coordinate and the time coordinate t only, a further simplification results. We obtain

$$\frac{\partial^2 E_x}{\partial z^2} - \frac{1}{c^2}\frac{\partial^2 E_x}{\partial t^2} = 0 \tag{9.11}$$

for which any function $f(z - ct)$ is a solution. This latter statement is readily verified. If we let $z - ct = u$, then

$$\frac{\partial f(z - ct)}{\partial z} = \frac{\partial f(u)}{\partial u}\frac{\partial u}{\partial z} = \frac{\partial f(u)}{\partial u} = f'$$

and

$$\frac{\partial^2 f(z - ct)}{\partial z^2} = f''$$

Similarly,

$$\frac{\partial f(z - ct)}{\partial t} = \frac{\partial f(u)}{\partial u}\frac{\partial u}{\partial t} = -cf'$$

and

$$\frac{\partial^2 f(z - ct)}{\partial t^2} = c^2 f''$$

Consequently, we obtain

$$\frac{\partial^2 f}{\partial z^2} - \frac{1}{c^2}\frac{\partial^2 f}{\partial t^2} = f'' - \frac{c^2}{c^2}f'' = 0$$

which verifies that $f(z - ct)$ is a solution of (9.11). A function such as $f(z - ct)$ represents a disturbance that propagates along the z axis with a velocity c. A typical plot of $f(z - ct)$ as a function of z for various values of t is given in Fig. 9.2 and clearly shows that the disturbance propagates in the z direction with a velocity c.

Another solution to (9.11) is any arbitrary function $f(z + ct)$. This solution is similar to the previous one, except that it represents a disturbance propagating in the negative z direction.

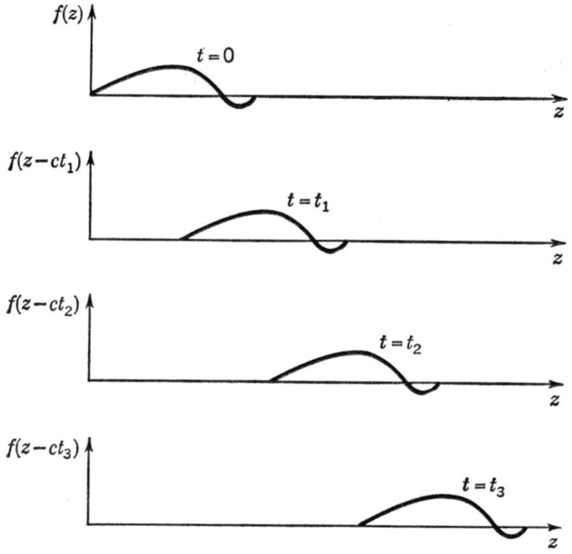

FIG. 9.2. Propagation of a disturbance $f(z)$.

In a homogeneous, isotropic, source-free material body with a permittivity ϵ and a permeability μ, a similar derivation shows that the wave equation satisfied by **E** is

$$\nabla^2\mathbf{E} - \mu\epsilon \frac{\partial^2\mathbf{E}}{\partial t^2} = 0 \tag{9.12}$$

This equation is similar to (9.10), with $\mu\epsilon$ replacing $\mu_0\epsilon_0$. The solution is also similar, with the exception that the velocity of propagation is now $v = (\mu\epsilon)^{-1/2}$ instead of c.

In a conducting body with parameters μ, ϵ, and σ, where σ is the conductivity, we expect to obtain a wave equation with a damping term present. The presence of an electric field **E** will cause a conduction current $\mathbf{J} = \sigma\mathbf{E}$ to flow, and this will result in a loss of energy because of joulean heat loss. In a conducting medium we must replace (9.9b) by

$$\nabla \times \mathbf{H} = \epsilon \frac{\partial \mathbf{E}}{\partial t} + \mathbf{J} = \epsilon \frac{\partial \mathbf{E}}{\partial t} + \sigma\mathbf{E} \tag{9.13}$$

The current that occurs here is not an impressed source current but rather the conduction current that flows as a result of the presence of the field **E**. The free charge in the conductor may be assumed to be zero, and hence $\nabla \cdot \mathbf{E}$ is still zero. Any free charge initially present decays to zero in an extremely short time interval since the relaxation time for a good conductor is very small. If we now use (9.13) to eliminate $\nabla \times \mathbf{H}$ in (9.9*a*), we find that

$$\nabla \times \nabla \times \mathbf{E} = \nabla\nabla \cdot \mathbf{E} - \nabla^2\mathbf{E} = -\nabla^2\mathbf{E} = -\mu\epsilon \frac{\partial^2\mathbf{E}}{\partial t^2} - \mu\sigma \frac{\partial \mathbf{E}}{\partial t}$$

or

$$\nabla^2\mathbf{E} - \mu\sigma \frac{\partial \mathbf{E}}{\partial t} - \mu\epsilon \frac{\partial^2\mathbf{E}}{\partial t^2} = 0 \tag{9.14}$$

As anticipated, we obtain a damping term $-\mu\sigma(\partial\mathbf{E}/\partial t)$ which is directly proportional to the conductivity σ. The presence of this term results in an exponential decay of the wave as it propagates away from the source. A fuller appreciation of (9.14) will be obtained in a later section when sinusoidal time-varying fields are analyzed.

9.4. Power Flow and Energy

Energy may be transported through space by means of electromagnetic waves. In addition, energy may be stored in the electromagnetic field, a result we could well anticipate in view of our earlier results in connection with energy storage in the static electric and magnetic fields. In this section relations will be derived that permit the evaluation of the energy stored in a given volume of space and the flow of energy in the electromagnetic field. In ordinary circuit theory, power flow is related to the product of voltage and current. For the electromagnetic field we shall find that the power flow across an element of area $d\mathbf{S}$ is given by $\mathbf{E} \times \mathbf{H} \cdot d\mathbf{S}$. In this expression **E** is analogous to voltage and has the dimensions of volts per meter, while **H** is analogous to current and has the dimensions of amperes per meter. Power flow is a vector quantity, and hence it is not surprising to find that it is given by a vector relation such as $\mathbf{E} \times \mathbf{H}$ watts per unit area.

To derive the relations that we wish to obtain, consider a volume V bounded by a closed surface S. Let the material inside S be isotropic, homogeneous, and characterized by electrical parameters μ, ϵ and conductivity σ. Consider the expression $\nabla \cdot \mathbf{E} \times \mathbf{H}$, which we may expand to obtain

$$\nabla \cdot \mathbf{E} \times \mathbf{H} = \mathbf{H} \cdot \nabla \times \mathbf{E} - \mathbf{E} \cdot \nabla \times \mathbf{H}$$

On the right-hand side we now replace $\nabla \times \mathbf{E}$ and $\nabla \times \mathbf{H}$ by $-\partial\mathbf{B}/\partial t$ and $\partial\mathbf{D}/\partial t + \sigma\mathbf{E}$, as obtained from Maxwell's equations, so that

$$\nabla \cdot \mathbf{E} \times \mathbf{H} = -\mathbf{H} \cdot \frac{\partial \mathbf{B}}{\partial t} - \mathbf{E} \cdot \frac{\partial \mathbf{D}}{\partial t} - \sigma\mathbf{E} \cdot \mathbf{E}$$

When μ and ϵ are constant we can write

$$\mathbf{H} \cdot \frac{\partial \mathbf{B}}{\partial t} = \mu \mathbf{H} \cdot \frac{\partial \mathbf{H}}{\partial t} = \frac{\mu}{2} \frac{\partial (\mathbf{H} \cdot \mathbf{H})}{\partial t} = \frac{\mu}{2} \frac{\partial H^2}{\partial t}$$

and similarly $\mathbf{E} \cdot \dfrac{\partial \mathbf{D}}{\partial t} = \dfrac{\epsilon}{2} \dfrac{\partial E^2}{\partial t}$

where H and E represent the magnitudes of $\mathbf{H}$ and $\mathbf{E}$. We now obtain the basic result

$$\nabla \cdot \mathbf{E} \times \mathbf{H} = -\frac{\partial}{\partial t}\left(\frac{\mu}{2} H^2 + \frac{\epsilon}{2} E^2\right) - \sigma E^2 \tag{9.15}$$

In this expression we interpret $(\mu/2)H^2$ and $(\epsilon/2)E^2$ as the density of energy stored in the magnetic and electric fields. This interpretation is carried over directly from the similar results that were derived earlier for the static fields. The term $-\sigma E^2$ is interpreted as the power loss per unit volume due to joulean heating brought about by the flow of conduction current $\sigma\mathbf{E}$. Equation (9.15) is thus understood to relate the divergence of power (which is a rate of flow of energy) from a unit element of volume to the sum of the time rate of decrease of the energy stored in the magnetic and electric fields per unit volume minus the power loss per unit volume.

A macroscopic form of (9.15) is obtained by integrating throughout the volume V and converting the volume integral of the divergence to a surface integral by means of the divergence theorem. We obtain

$$\oint_S \mathbf{E} \times \mathbf{H} \cdot d\mathbf{S} = -\frac{\partial}{\partial t} \int_V \left(\frac{\mu}{2} H^2 + \frac{\epsilon}{2} E^2\right) dV - \int_V \sigma E^2 \, dV \tag{9.16}$$

a result which states that the instantaneous flow of power across a closed surface S is equal to the time rate of decrease of the energy stored in the field in the interior of S minus the power loss due to joulean heating within the volume V. The vector

$$\mathbf{P} = \mathbf{E} \times \mathbf{H} \tag{9.17}$$

is called the Poynting vector and gives the instantaneous flow of power (both magnitude and direction) per unit area. The total power flow across a given surface is obtained by integrating the normal component of $\mathbf{E} \times \mathbf{H}$ over the surface in question. Energy flows in a direction that is perpendicular to both the $\mathbf{E}$ and $\mathbf{H}$ field vectors. While the interpretation of $\mathbf{E} \times \mathbf{H}$ as representing power density at a point is ordinarily a useful one, it should be noted that (9.16) states only that the total surface integral of $\mathbf{P}$ gives a net power flow across a closed surface.

Example 9.1. Plane Waves. Let us assume that the only electric field component present is E_x and that this component is a function of

z and t only, as in Sec. 9.3. If the field varies sinusoidally in time with a radian frequency ω, a possible solution to the wave equation is

$$E_x = f(z - ct) = E_0 \sin \frac{\omega}{c} (z - ct) = E_0 \sin \left(\frac{\omega}{c} z - \omega t \right) \quad (9.18a)$$

where E_0 is an amplitude constant. From the equation

$$\nabla \times \mathbf{E} = -\mu_0 \frac{\partial \mathbf{H}}{\partial t}$$

we obtain

$$-\mu_0 \frac{\partial \mathbf{H}}{\partial t} = \begin{vmatrix} \mathbf{a}_x & \mathbf{a}_y & \mathbf{a}_z \\ \dfrac{\partial}{\partial x} & \dfrac{\partial}{\partial y} & \dfrac{\partial}{\partial z} \\ E_x & 0 & 0 \end{vmatrix} = \mathbf{a}_y \frac{\partial E_x}{\partial z}$$

and hence
$$\frac{\partial \mathbf{H}}{\partial t} = -\frac{\omega E_0}{\mu_0 c} \mathbf{a}_y \cos \frac{\omega}{c} (z - ct)$$

or by integrating with respect to time,

$$H_y = \frac{E_0}{\mu_0 c} \sin \left(\frac{\omega}{c} z - \omega t \right) = E_0 \left(\frac{\epsilon_0}{\mu_0} \right)^{1/2} \sin \left(\frac{\omega}{c} z - \omega t \right) \quad (9.18b)$$

The parameter $(\epsilon_0/\mu_0)^{1/2}$ has the dimensions of admittance and is called the intrinsic admittance of free space. The reciprocal quantity $(\mu_0/\epsilon_0)^{1/2}$ is called the intrinsic impedance of free space and will be denoted by Z_0; that is,

$$Z_0 = \left(\frac{\mu_0}{\epsilon_0} \right)^{1/2} \quad (9.19)$$

Numerically, $Z_0 = 120\pi = 377$ ohms.

The particular solution of Maxwell's equations given by (9.18) is called a uniform-plane electromagnetic wave since both **E** and **H** lie in a plane (xy plane in this case) perpendicular to the direction of propagation (z direction in this example). The plane wave is uniform since neither E_x nor H_y varies with the transverse coordinates x and y. The space relationship between **E** and **H** is shown in Fig. 9.3.

The power flow is given by

$$\mathbf{P} = \mathbf{E} \times \mathbf{H} = E_x H_y \mathbf{a}_x \times \mathbf{a}_y = E_x H_y \mathbf{a}_z$$

and is seen to be in the direction of propagation. Substituting for E_x and H_y, we obtain

$$\mathbf{P} \cdot \mathbf{a}_z = \frac{E_0{}^2}{Z_0} \sin^2 \frac{\omega}{c} (z - \omega t)$$

as the instantaneous power flow across a unit area of the xy plane. The

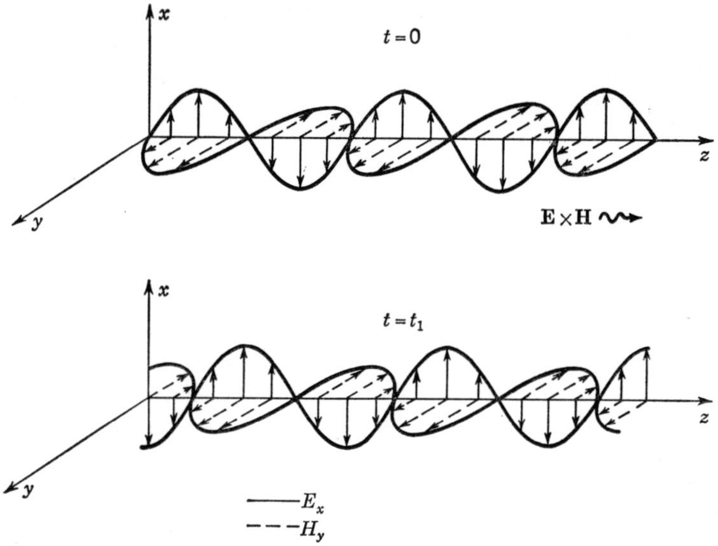

FIG. 9.3. Space relationship between **E** and **H** in a plane TEM wave.

time-average power flow per unit area is given by

$$P_{av} = \frac{1}{2}\frac{E_0^2}{Z_0}$$

The instantaneous energy stored in the electric field per unit volume is

$$\frac{\epsilon_0}{2} E_0^2 \int_0^1 \int_0^1 \int_0^1 \sin^2 \frac{\omega}{c}(z - ct)\, dx\, dy\, dz$$
$$= \frac{\epsilon_0}{2} E_0^2 \int_0^1 \tfrac{1}{2}\left[1 - \cos\frac{2\omega}{c}(z - ct)\right] dz$$
$$= \frac{\epsilon_0}{4} E_0^2 \left[z - \frac{c}{2\omega}\sin\frac{2\omega}{c}(z - ct)\right]_0^1$$

If we average over one period in time, we obtain

$$U_e = \frac{\epsilon_0 E_0^2}{4} \tag{9.20}$$

as the time-average energy stored in the electric field per unit volume. A similar derivation shows that the time-average energy stored in the magnetic field per unit volume is given by

$$U_m = \frac{\mu_0}{4}\left(\frac{E_0}{Z_0}\right)^2 = \frac{\epsilon_0}{4} E_0^2 = U_e \tag{9.21}$$

and is equal to the time-average energy stored in the electric field. Power is a rate of flow of energy, and hence if we multiply the total energy stored in the field per unit volume by the velocity of energy transport, we should obtain the expression for power flow. For the present example,

$$c(U_e + U_m) = \frac{c\epsilon_0}{2} E_0^2 = \frac{\epsilon_0}{2(\mu_0\epsilon_0)^{1/2}} E_0^2$$
$$= \frac{E_0^2}{2Z_0} = P_{\text{av}} \tag{9.22}$$

which checks with our earlier result. We are thus able to say that for a plane electromagnetic wave, energy is transported with the velocity c in free space; that is, the velocity of propagation of energy is the same as the phase velocity of the wave as given in (9.18*a*). Later we shall discover circumstances where the two velocities are not the same.

9.5. Sinusoidal Time-varying Fields

In practice, we generally deal with steady-state sinusoidal time-varying fields. Just as in circuit theory, it is convenient to introduce an abbreviated notation and to represent each field vector as a complex phasor. If the angular radian frequency is ω, we write, for the electric field,

$$\mathbf{E}'(x,y,z,t) = \text{Re}\,[\mathbf{E}(x,y,z)e^{j\omega t}] \tag{9.23}$$

where the prime is used to signify the real physical field. For brevity, we represent the electric field simply by the complex phasor $\mathbf{E}(x,y,z)$, where $\mathbf{E}(x,y,z)$ is a complex space vector and a function of x, y, z only. Logically, we should adopt a different notation for the phasor quantity, but the use of standard boldface type should provide very little confusion since we shall be dealing almost entirely with sinusoidal time-varying fields in the remainder of the book. Each space component of the phasor $\mathbf{E}$ is a complex quantity, for example, $E_x(x,y,z) = E_{xr}(x,y,z) + jE_{xi}(x,y,z)$, where E_{xr} is the real part and E_{xi} the imaginary part. Particular care must be used to avoid thinking of E_{xr} and E_{xi} as components of a space vector, as is sometimes done in circuit theory. The quantity $E_{xr} + jE_{xi}$ forms one component, the x component, of the complex phasor space vector $\mathbf{E}$. The physical field is always obtained by multiplying by $e^{j\omega t}$ and taking the real part (or the imaginary part).

When using complex phasor notation the time-average electric- and magnetic-energy densities are given by

$$U_e = \frac{\epsilon}{4}\mathbf{E}\cdot\mathbf{E}^* \tag{9.24a}$$

$$U_m = \frac{\mu}{4}\mathbf{H}\cdot\mathbf{H}^* \tag{9.24b}$$

where the asterisk denotes the complex conjugate phasor. The additional factor ½ arises because of averaging over one period in time. In (9.24) it is assumed that ϵ and μ are real; the case when ϵ and μ are complex will be considered later. The proof of (9.24) may be developed along lines similar to the following. If **E** has only an x component, then the physical field $E'_x(x,y,z,t)$ is given by

$$\begin{aligned} E'_x &= \text{Re}\,[E_{xr}(x,y,z) + jE_{xi}(x,y,z)]e^{j\omega t} \\ &= E_{xr}\cos\omega t - E_{xi}\sin\omega t \end{aligned}$$

where E_{xr} and E_{xi} are real. The time average of E'^2_x is obviously equal to $\frac{1}{2}(E^2_{xr} + E^2_{xi})$, a result which is seen to be equal to $\frac{1}{2}E_xE^*_x$ as well.

When we deal with steady-state sinusoidal time-varying fields, all time derivatives $\partial/\partial t$ may be replaced by $j\omega$. Thus Maxwell's equations reduce to the following form:

$$\nabla \times \mathbf{E} = -j\omega\mu\mathbf{H} \tag{9.25a}$$
$$\nabla \times \mathbf{H} = j\omega\epsilon\mathbf{E} + \mathbf{J} \tag{9.25b}$$
$$\nabla \cdot \mathbf{D} = \rho \tag{9.25c}$$
$$\nabla \cdot \mathbf{B} = 0 \tag{9.25d}$$
$$\nabla \cdot \mathbf{J} = -j\omega\rho \tag{9.25e}$$

When time-varying fields are applied to material bodies, the polarization vectors **P** and **M** vary with time at the same frequency as the applied fields. Because of damping forces which are always present to some extent, the polarization vectors **P** and **M** will usually lag behind the fields **E** and **H**. This means that, in general, ϵ and μ must be complex. The complex nature of ϵ and μ is a manifestation of power loss that will occur in the material because of the work that must be done in overcoming the frictional damping forces. As an example, let **E** be the complex phasor representing the field acting to polarize a dielectric material. The polarization **P** per unit volume is given by

$$\mathbf{P} = \epsilon_0 \alpha e^{-j\phi}\mathbf{E}$$

where α is a positive real constant and ϕ is the phase angle by which **P** lags **E**. We now see that the susceptibility is given by

$$\chi_e = \alpha e^{-j\phi}$$

and hence the dielectric permittivity ϵ is given by

$$\epsilon = \epsilon_0(1 + \chi_e) = \epsilon_0(1 + \alpha\cos\phi - j\alpha\sin\phi) = \epsilon' - j\epsilon'' \tag{9.26}$$

and is a complex quantity.

If in (9.25b) we let ϵ be complex and let **J** be a conduction current, we obtain

$$\nabla \times \mathbf{H} = j\omega\epsilon'\mathbf{E} + (\omega\epsilon'' + \sigma)\mathbf{E} \tag{9.27}$$

This result shows that the imaginary part of ϵ is equivalent to an increase in the conductivity of the medium. At high frequencies σ for a good dielectric is very small and most of the energy loss is caused by polarization damping forces that bring in the term ϵ''. For convenience when dealing with dielectric materials that have a finite conductivity σ, a single complex dielectric permittivity

$$\epsilon = \epsilon' - j\epsilon'' - \frac{j\sigma}{\omega}$$

is usually introduced so as to include the effect of both damping losses and conduction losses in a single term. The properties of the dielectric material are usually specified by giving its dielectric constant κ and its loss tangent $\tan \delta_l$. The value of ϵ is then

$$\epsilon = \kappa\epsilon_0(1 - j \tan \delta_l)$$

In this equation $\tan \delta_l$ includes the effects of finite conductivity as well as the effects of polarization damping forces.

Remarks similar to the above may be made about the permeability μ. When it is necessary to consider complex μ, we shall use the notation $\mu = \mu' - j\mu''$. In passive materials the imaginary parts of ϵ and μ are negative since these terms must correspond to a loss in the material. If the imaginary parts were positive, they would indicate a generation of energy by the material body, which is a violation of the condition that the material is passive in nature. The proof of these remarks will be given in Sec. 9.7.

9.6. Helmholtz's Equation

For sinusoidal time-varying fields the wave equation for waves in free space may be obtained from (9.10) simply by replacing $\partial^2/\partial t^2$ by $-\omega^2$; thus

$$\nabla^2\mathbf{E} + \omega^2\mu_0\epsilon_0\mathbf{E} = 0 \tag{9.28}$$

or

$$\nabla^2\mathbf{E} + k_0^2\mathbf{E} = 0 \tag{9.29}$$

where $k_0 = \omega(\mu_0\epsilon_0)^{1/2} = \omega/c$ is called the free-space wave number and $\mathbf{E}$ in (9.29) is now a complex phasor space vector that is independent of time. Equation (9.29) is commonly referred to as the vector Helmholtz equation. A simple application of this equation is given in the following example.

Example 9.2. Sinusoidal Time-varying Plane Wave. As a simple example of a solution to (9.29), consider the case where $\mathbf{E}$ is a function of z only and, furthermore, has only an x component. We now have

$$\frac{\partial^2 E_x}{\partial z^2} + k_0^2 E_x = 0$$

for which a general solution is

$$E_x(z) = A_1 e^{-jk_0 z} + A_2 e^{jk_0 z} \tag{9.30}$$

where A_1 and A_2 are amplitude constants. The physically real field $E'_x(z,t)$ is given by

$$\begin{aligned} E'_x(z,t) &= \text{Re}\,(A_1 e^{-jk_0 z + j\omega t} + A_2 e^{jk_0 z + j\omega t}) \\ &= A_1 \cos(k_0 z - \omega t) + A_2 \cos(k_0 z + \omega t) \end{aligned} \tag{9.31}$$

provided A_1 and A_2 are real. We thus see that $A_1 e^{-jk_0 z}$ represents a plane wave propagating in the positive z direction while $A_2 e^{+jk_0 z}$ represents a plane wave propagating in the negative z direction.

The distance a wave must propagate in order for its phase angle to change by an amount 2π is called the wavelength. In free space we shall denote the wavelength by the symbol λ_0. By definition we now have $k_0 \lambda_0 = 2\pi$, or

$$\lambda_0 = \frac{2\pi}{k_0} = \frac{2\pi}{\omega} c = \frac{c}{f} \tag{9.32a}$$

and also

$$k_0 = \frac{2\pi}{\lambda_0} \tag{9.32b}$$

The relationship between wavelength, velocity, and frequency obtained here is undoubtedly familiar to the reader from earlier courses in physics.

The magnetic field **H** corresponding to the electric field **E** given by (9.30) is readily found from (9.25*a*) and is

$$\mathbf{H} = \mathbf{a}_y (Y_0 A_1 e^{-jk_0 z} - Y_0 A_2 e^{jk_0 z}) \tag{9.33}$$

where $Y_0 = (\epsilon_0/\mu_0)^{1/2}$ and is the intrinsic admittance of free space. It is seen that the direction of **H** is reversed for the wave propagating in the negative z direction. This is a necessary requirement in order to obtain a reversal in the direction of power flow. Note that the electric and magnetic fields are in time phase but space quadrature.

The time-average power flow is given by one-half of the real part of the complex Poynting vector (details of the complex Poynting theorem are developed in the next section)

$$\mathbf{P} = \tfrac{1}{2} \mathbf{E} \times \mathbf{H}^* \tag{9.34}$$

If we consider the wave propagating in the positive z direction only ($A_2 = 0$), the power flow across a unit area in the xy plane is found to be

$$P_{\text{av}} = \tfrac{1}{2} Y_0 |A_1|^2 \tag{9.35}$$

Helmholtz's Equation in Dielectric and Conducting Media

In a dielectric medium with a permittivity ϵ (relative dielectric constant $\kappa = \epsilon/\epsilon_0$) and negligible losses, we have, in place of (9.29),

$$\nabla^2 \mathbf{E} + \omega^2 \mu_0 \epsilon \mathbf{E} = 0$$

or

$$\nabla^2 \mathbf{E} + k^2 \mathbf{E} = 0 \tag{9.36}$$

where $k = \kappa^{1/2} k_0 = (\kappa \omega^2 \mu_0 \epsilon_0)^{1/2}$. The solution to this equation is similar to that for (9.29) with k_0 replaced by k. The velocity of propagation is $v = \kappa^{-1/2} c$ instead of c. The parameter $\kappa^{1/2}$ is called the index of refraction and will be denoted by the symbol η. In a dielectric the wavelength of plane waves is λ, where

$$\lambda = \frac{2\pi}{k} = \frac{2\pi}{\eta k_0} = \frac{\lambda_0}{\eta} \tag{9.37}$$

and is less than the free-space wavelength.

In a medium with finite conductivity σ, the required Helmholtz equation is found by replacing $\partial/\partial t$ by $j\omega$ in the general time-varying wave equation (9.14). We obtain

$$\nabla^2 \mathbf{E} - j\omega\mu\sigma \mathbf{E} + \omega^2 \mu \epsilon \mathbf{E} = 0 \tag{9.38}$$

This may be rewritten as

$$\nabla^2 \mathbf{E} - j\omega\mu(j\omega\epsilon + \sigma)\mathbf{E} = 0$$

The term $j\omega\epsilon\mathbf{E}$ is the displacement current density, while $\sigma\mathbf{E}$ is the conduction current density. For metals, σ is of the order of 10^7 mhos per meter (for copper $\sigma = 5.8 \times 10^7$ mhos per meter) and ϵ is approximately equal to $\epsilon_0 = (36\pi)^{-1} \times 10^{-9}$ farad per meter. Consequently, $\sigma/\omega\epsilon \approx 10^{18}\omega^{-1}$, and hence for all frequencies up to the optical range we can neglect $\omega\epsilon$ in comparison with σ; for example, for $f = 10{,}000$ megacycles we have $\omega\epsilon/\sigma \approx 5 \times 10^{-8}$, so that $\omega\epsilon$ is certainly negligible in comparison with σ. This means that in metals the displacement current is entirely negligible compared with the conduction current.

In place of (9.38), we can now write

$$\nabla^2 \mathbf{E} - j\omega\mu\sigma \mathbf{E} = 0 \tag{9.39}$$

as the equation satisfied by **E** (and **H** also) in a metal. This equation is a diffusion equation and not a wave equation, since for the general time-varying case it would be of the form

$$\nabla^2 \mathbf{E}' - \mu\sigma \frac{\partial \mathbf{E}'}{\partial t} = 0 \tag{9.40}$$

In metals the fields may be thought of as diffusing into the material and will undergo both attenuation and phase retardation in the process. It is only when the displacement current term is predominant that we obtain true wave propagation. On the other hand, (9.39) is similar to the free-space Helmholtz equation, and consequently its solutions are also formally similar. In fact, any solution of (9.29) is a solution of (9.39) if we replace k_0 by $(-j\omega\mu\sigma)^{1/2} = (1 - j)(\omega\mu\sigma/2)^{1/2}$. For a plane wave depending on z only and having only an E_x component of electric field, a solution is

$$E_x = A \exp\left[-j\left(\frac{\omega\mu\sigma}{2}\right)^{1/2} z - \left(\frac{\omega\mu\sigma}{2}\right)^{1/2} z\right] \tag{9.41}$$

The rate of attenuation is $(\omega\mu\sigma/2)^{1/2}$ nepers per meter. The skin depth δ is defined as

$$\delta = \left(\frac{2}{\omega\mu\sigma}\right)^{1/2} \tag{9.42}$$

and is the distance the wave must propagate in order to decay by an amount e^{-1}. By using (9.42), we may rewrite (9.41) as

$$E_x = Ae^{-(1+j)z/\delta} \tag{9.43}$$

9.7. Complex Poynting Vector

The basic relation between power flow and energy storage in the sinusoidal time-varying electromagnetic field may be derived in a manner similar to that used in Sec. 9.4. The curl equations for **E** and **H*** are

$$\nabla \times \mathbf{E} = -j\omega\mu\mathbf{H} \qquad \nabla \times \mathbf{H}^* = -j\omega\epsilon^*\mathbf{E}^* + \sigma\mathbf{E}^*$$

where we have assumed that μ and ϵ may be complex. If we expand $\nabla \cdot \mathbf{E} \times \mathbf{H}^*$, we obtain

$$\begin{aligned}\nabla \cdot \mathbf{E} \times \mathbf{H}^* &= \mathbf{H}^* \cdot \nabla \times \mathbf{E} - \mathbf{E} \cdot \nabla \times \mathbf{H}^* \\ &= -j\omega\mu\mathbf{H} \cdot \mathbf{H}^* + j\omega\epsilon^*\mathbf{E} \cdot \mathbf{E}^* - \sigma\mathbf{E} \cdot \mathbf{E}^*\end{aligned} \tag{9.44}$$

after substituting for the curl of **E** and **H***. We now integrate (9.44) throughout a volume V bounded by a surface S and use the divergence theorem, with the result that

$$\oint_S \mathbf{E} \times \mathbf{H}^* \cdot d\mathbf{S} = -j\omega \int_V (\epsilon^*\mathbf{E} \cdot \mathbf{E}^* - \mu\mathbf{H} \cdot \mathbf{H}^*)\, dV + \int_V \sigma\mathbf{E} \cdot \mathbf{E}^*\, dV \tag{9.45}$$

by taking the vector area $d\mathbf{S}$ directed *into* the volume V. The time-average electric and magnetic energy stored in the field per unit volume is given by

$$U_e = \tfrac{1}{4} \operatorname{Re} \epsilon\mathbf{E} \cdot \mathbf{E}^* = \tfrac{1}{4}\epsilon'\mathbf{E} \cdot \mathbf{E}^* \tag{9.46a}$$

$$U_m = \tfrac{1}{4} \operatorname{Re} \mu\mathbf{H} \cdot \mathbf{H}^* = \tfrac{1}{4}\mu'\mathbf{H} \cdot \mathbf{H}^* \tag{9.46b}$$

The sum of the following terms will be shown to represent the time-average power loss per unit volume:

$$\tfrac{1}{2}\sigma \mathbf{E}\cdot\mathbf{E}^* + \frac{\omega}{2}\,\mathrm{Im}\,(\epsilon\mathbf{E}\cdot\mathbf{E}^* + \mu\mathbf{H}\cdot\mathbf{H}^*)$$

$$= \tfrac{1}{2}\sigma\mathbf{E}\cdot\mathbf{E}^* + \frac{\omega}{2}\,(\epsilon''\mathbf{E}\cdot\mathbf{E}^* + \mu''\mathbf{H}\cdot\mathbf{H}^*) \qquad (9.47)$$

In (9.46) ϵ', μ' are the real parts of ϵ and μ, while in (9.47) ϵ'', μ'' are the imaginary parts that represent loss due to polarization damping.

If we separate (9.45) into its real and imaginary parts we have

$$\tfrac{1}{2}\,\mathrm{Re}\oint_S \mathbf{E}\times\mathbf{H}^*\cdot d\mathbf{S} = \int_V \frac{\omega}{2}\,(\epsilon''\mathbf{E}\cdot\mathbf{E}^* + \mu''\mathbf{H}\cdot\mathbf{H}^*)\,dV + \int_V \frac{\sigma}{2}\,\mathbf{E}\cdot\mathbf{E}^*\,dV \qquad (9.48a)$$

and

$$\mathrm{Im}\oint_S \mathbf{E}\times\mathbf{H}\cdot d\mathbf{S} = 4\omega\int_V (U_m - U_e)\,dV \qquad (9.48b)$$

since $\epsilon = \epsilon' - j\epsilon''$ and $\mu = \mu' - j\mu''$. Equation (9.48*a*) states that the average power flow into the volume V is given by the integral of one-half the real part of the complex Poynting vector $\mathbf{E}\times\mathbf{H}^*$ over the surface S bounding V and that this is equal to the time-average power loss in V due to conduction current losses and polarization damping losses. The expressions for polarization damping losses given in (9.47) are identified by analogy with the expression for conduction current losses. For example, Maxwell's curl equation for $\mathbf{H}$ is

$$\nabla\times\mathbf{H} = j\omega\epsilon\mathbf{E} + \sigma\mathbf{E} = j\omega\epsilon'\mathbf{E} + (\omega\epsilon'' + \sigma)\mathbf{E}$$

and hence if $(\sigma/2)\mathbf{E}\cdot\mathbf{E}^*$ is the average power loss per unit volume due to joulean heating, then $(\omega\epsilon''/2)\mathbf{E}\cdot\mathbf{E}^*$ is the average power loss due to electric polarization damping forces. From (9.48*a*) it is readily seen that the imaginary parts of ϵ and μ must be negative, since these terms represent energy dissipation and not energy generation in passive materials.

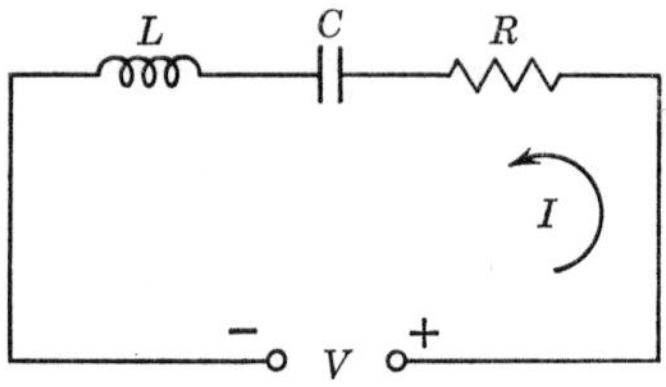

FIG. 9.4. A series *RLC* circuit.

Equation (9.48*b*) states that the integral of the imaginary part of the complex Poynting vector over the surface S bounding V is equal to 4ω times the difference in the average energy stored in the magnetic and electric field. This result is reminiscent of that obtained for low-frequency circuits, as is shown below.

Consider a simple *RLC* series circuit, as in Fig. 9.4. With an applied voltage V, a current $I = V/(R + j\omega L + 1/j\omega C)$ flows. For this circuit we have

$$\tfrac{1}{2}VI^* = \tfrac{1}{2}II^*Z_{\text{in}} = \tfrac{1}{2}II^*R + 2j\omega\left(\frac{II^*L}{4} - \frac{II^*}{4\omega^2C^2}C\right) \quad (9.49)$$

But $\frac{1}{2}II^*R$ is the average power loss in the resistor, $II^*L/4$ is the average energy stored in the inductor, and $II^*C/4(\omega C)^2$ is the average energy stored in the capacitor since $I/\omega C$ is the voltage across the capacitor. Thus (9.49) is the low-frequency circuit equivalent of (9.48). The result is not an unexpected one, since, after all, low-frequency circuit theory is based on Maxwell's equations. The relation between circuit theory and field theory is examined in greater detail in Sec. 9.10.

9.8. Boundary Conditions

In an infinite unbounded homogeneous medium the solutions to the field equations are relatively easy to obtain. In most practical situations, however, we require a solution for the fields in the presence of conducting bodies and boundaries separating material media with differing electrical parameters ϵ and μ. In order to obtain a solution, a knowledge of the boundary conditions to be applied to the field vectors is needed. The time-varying field satisfies boundary conditions similar to those obeyed by the static fields, as we show in the following analysis.

Boundary between Two Dielectric Media

Consider two dielectric media with electrical permittivity ϵ_1 and ϵ_2 and permeability μ_1 and μ_2 and having a common boundary, as in Fig. 9.5. Construct an infinitesimal "coin"-shaped box with end faces of area ΔS in adjacent media and the end surfaces parallel to the common boundary surface. Since $\nabla \cdot \mathbf{D} = 0$ in the present case, it follows by applying Gauss' law to the volume enclosed by the coin-shaped box that

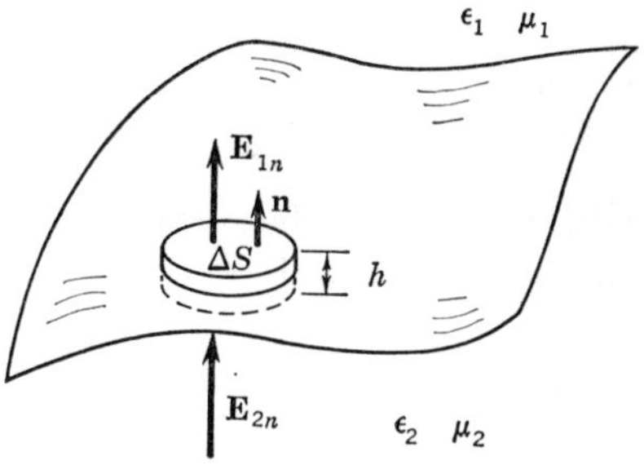

FIG. 9.5. Boundary between two dielectric media.

$$\int_V \nabla \cdot \mathbf{D}\, dV = 0 = \oint_S \mathbf{D} \cdot d\mathbf{S}$$

and hence

$$\lim_{h \to 0} \oint_S \mathbf{D} \cdot d\mathbf{S} = (D_{1n} - D_{2n})\,\Delta S = 0$$

where the subscript n denotes the component normal to the surface. The limit $h \to 0$ is taken to ensure that there will be no flux passing out through the sides of the box. We now have

$$D_{1n} = D_{2n} \qquad \text{or} \qquad \epsilon_1 E_{1n} = \epsilon_2 E_{2n}$$

which may be written in vector form as

$$\mathbf{n} \cdot \mathbf{D}_1 = \mathbf{n} \cdot \mathbf{D}_2 \tag{9.50}$$

where **n** is the unit normal to the boundary surface. This relation is the same as that derived for the static field.

If we apply Gauss' law to the field **B**, we obtain in a similar way the result

$$\mathbf{n} \cdot \mathbf{B}_1 = \mathbf{n} \cdot \mathbf{B}_2 \tag{9.51}$$

since $\nabla \cdot \mathbf{B} = 0$. Thus the normal components of **D** and **B** are continuous across a surface separating two dielectric media.

To derive the boundary relations that apply to tangential field components, consider a small contour C of length Δl with sides lying on

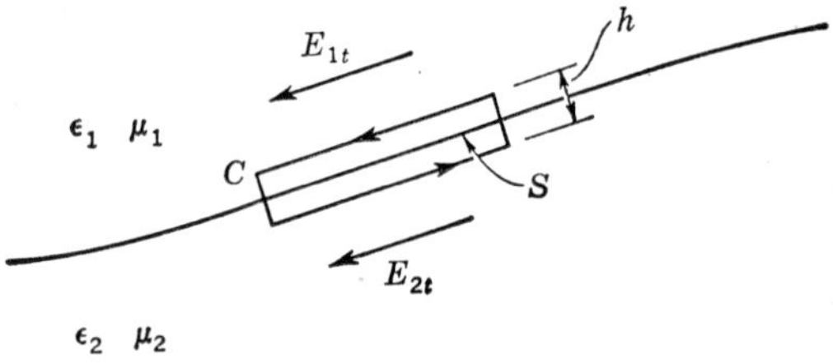

FIG. 9.6. Contour C for deriving boundary conditions for the tangential components of the field.

adjacent sides of the surface, as in Fig. 9.6. Application of Stokes' law to the equation $\nabla \times \mathbf{E} = -j\omega\mathbf{B}$ gives

$$\oint_C \mathbf{E} \cdot d\mathbf{l} = -j\omega \int_S \mathbf{B} \cdot d\mathbf{S}$$

Now
$$\lim_{h \to 0} \oint_C \mathbf{E} \cdot d\mathbf{l} = (E_{1t} - E_{2t})\,\Delta l$$

and
$$\lim_{h \to 0} - j\omega \int_S \mathbf{B} \cdot d\mathbf{S} = \lim_{h \to 0} - j\omega B_t h\,\Delta l = 0$$

so that we obtain

$$E_{1t} = E_{2t} \tag{9.52}$$

A similar application of Stokes' law to the equation $\nabla \times \mathbf{H} = j\omega\mathbf{D}$ gives

$$H_{1t} = H_{2t} \tag{9.53}$$

In (9.52) and (9.53) the subscript t denotes the component of the field that is tangential to the common surface separating the two media. These equations state that the tangential fields are continuous across a boundary between two different dielectric media.

Boundary of a Perfect Conductor

In the interior of a perfect conductor ($\sigma = \infty$), the electromagnetic time-varying field is zero. This may be seen from the expression (9.42) for the skin depth δ. As σ tends to infinity, $\delta = (2/\omega\mu\sigma)^{1/2}$ approaches zero. Thus the field decays infinitely fast (by an amount e^{-1} in a distance δ) and cannot penetrate into the conductor. Actually, we never have perfect conductors, but in most cases σ is so large that at high frequencies negligible error is made in assuming that the field in the interior of the conductor is zero. For example, for copper at a frequency of 1,000 megacycles, $\delta \approx 2 \times 10^{-3}$ millimeter; so we could consider the depth of penetration as zero without appreciable error. On the other hand, for a frequency of 1,000 cycles, we have $\delta \approx 2$ millimeters, which in many cases would not be negligible.

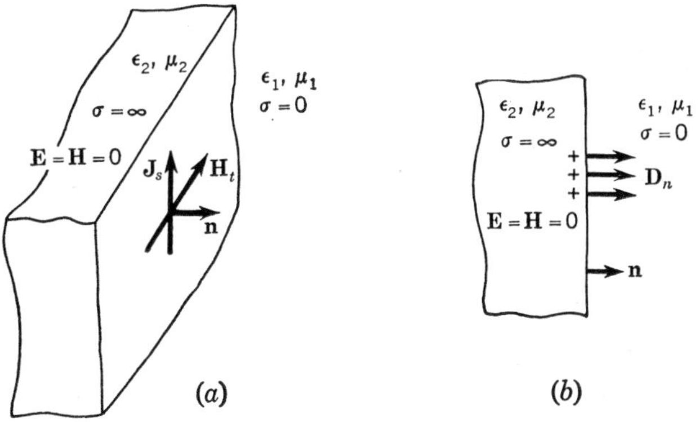

FIG. 9.7. Boundary conditions at a perfect conductor surface.

Because of the phenomenon described above, which is called "skin effect," as σ approaches infinity, the current flows in a narrower and narrower layer, until in the limit a true surface current exists on the surface (this problem is examined in detail in the next chapter). With reference to Fig. 9.7*a*, let $\mathbf{J}_s$ be the surface current density in amperes per meter. Since the displacement current in the conductor, as well as the field **H**, is zero, Ampère's circuital law shows that $\mathbf{H}_t$ is perpendicular to $\mathbf{J}_s$ and equal to $\mathbf{J}_s$ in magnitude; thus

$$H_t = J_s$$

or in vector notation,

$$\mathbf{n} \times \mathbf{H} = \mathbf{J}_s \tag{9.54}$$

Similarly, Gauss' law shows that

$$\mathbf{n} \cdot \mathbf{D} = \rho_s \tag{9.55}$$

where ρ_s is the surface charge density, as in Fig. 9.7*b*. While these results are rigorously true only for $\sigma \to \infty$, they are excellent approximations for practical conductors at high frequencies, where by high frequency we mean one that yields a value of skin depth that is small compared with all conductor dimensions.

Since the field in the interior of the conductor is zero and the tangential electric field E_t and normal magnetic field B_n are continuous across a boundary, it follows that

$$E_t = 0 \qquad \text{or} \qquad \mathbf{n} \times \mathbf{E} = 0 \tag{9.56}$$

$$\mathbf{n} \cdot \mathbf{B} = 0 \tag{9.57}$$

at the surface of a perfect conductor.

In any practical problem it is sufficient to ensure that the tangential components of the field satisfy the proper boundary conditions since this will automatically ensure that the normal components of **D** and **B** satisfy their respective boundary conditions. We may prove this statement as follows. Let $\nabla = \nabla_t + \nabla_n$, where ∇_t is the part of the del operator which represents differentiation with respect to the coordinates along the boundary surface separating two different media, and $\nabla_n = \mathbf{n}(\partial/\partial n)$ represents differentiation with respect to the coordinate normal to the boundary surface. The equation $\nabla \times \mathbf{E} = -j\omega\mathbf{B}$ separates into two parts,

$$\nabla_n \times \mathbf{E}_t + \nabla_t \times \mathbf{E}_n = -j\omega\mathbf{B}_t \tag{9.58a}$$

$$\nabla_t \times \mathbf{E}_t = -j\omega\mathbf{B}_n \tag{9.58b}$$

when the tangential and normal components are equated. This result is arrived at by noting that

$$\begin{aligned}\nabla \times \mathbf{E} &= \nabla_t \times (\mathbf{E}_t + \mathbf{E}_n) + \nabla_n \times (\mathbf{E}_t + \mathbf{E}_n) \\ &= \nabla_t \times \mathbf{E}_t + \nabla_n \times \mathbf{E}_t + \nabla_t \times \mathbf{E}_n\end{aligned}$$

since $\nabla_n \times \mathbf{E}_n = \partial(\mathbf{n} \times \mathbf{E}_n)/\partial n = 0$. The term $\nabla_t \times \mathbf{E}_t$ is a vector directed along the normal **n**, while $\nabla_n \times \mathbf{E}_t + \nabla_t \times \mathbf{E}_n$ is a vector in the boundary surface. If we make $\mathbf{E}_t$ continuous across the boundary surface, then the derivatives of $\mathbf{E}_t$ with respect to the coordinates along the boundary surface are also continuous. Therefore $\nabla_t \times \mathbf{E}_t$ is continuous, and likewise $\mathbf{B}_n$ must be continuous across the surface since

$$-j\omega\mathbf{B}_n = \nabla_t \times \mathbf{E}_t$$

For $\mathbf{D}_n$ and $\mathbf{H}_t$ we have the equation $\nabla_t \times \mathbf{H}_t = j\omega\mathbf{D}_n$, and a similar argument shows that $\mathbf{D}_n$ is continuous if $\mathbf{H}_t$ is continuous across the boundary. In the case when $\mathbf{H}_t$ is discontinuous across the boundary because of a surface current, $\mathbf{D}_n$ is also discontinuous. The discontinuity in $\mathbf{D}_n$ is equal to the surface charge density ρ_s. Furthermore, the surface current $\mathbf{J}_s$ and surface charge ρ_s satisfy a continuity equation

$$\nabla_t \cdot \mathbf{J}_s = -j\omega\rho_s \tag{9.59}$$

on the surface. It turns out that if $\mathbf{H}_t$ is made discontinuous by an amount equal to the surface current density, this automatically makes $\mathbf{D}_n$ discontinuous by an amount equal to the surface charge density.

The above results are of great importance in practice, since they make it necessary only to match the tangential field components at a discontinuity surface. This simplifies the analytical details of constructing a solution of Maxwell's field equations.

For the time-varying field a uniqueness theorem exists† which states that if a solution to the field equations has been found such that all boundary conditions are satisfied and also such that the fields have the proper behavior (singularity) at the position of the impressed sources, then this solution is unique. The proof may be constructed along lines similar to those employed in the proof of the uniqueness theorem for electrostatic boundary-value problems in Chap. 2. The details of the proof are not too important; so we omit them. The important fact is that such a theorem exists and thus guarantees the uniqueness of the solution once it has been found.

9.9. Scalar and Vector Potentials

The existence of an electromagnetic field implies a source of impressed currents and charges. If the impressed currents and charges are known, then the field may be determined by means of the equations to be derived in this section. We shall assume sinusoidal time variation, and hence all quantities we deal with are phasor quantities. For time-varying currents and charges the continuity equation $\nabla \cdot \mathbf{J} = -j\omega\rho$ serves to link the current $\mathbf{J}$ and the charge density ρ. As a consequence, we may not specify ρ and $\mathbf{J}$ independently.

The reader may readily verify that when $\mathbf{J}$ and ρ are not zero, the separation of Maxwell's equations into an equation for $\mathbf{E}$ alone or $\mathbf{H}$ alone gives

$$\nabla^2\mathbf{E} + k^2\mathbf{E} = j\omega\mu\mathbf{J} + \nabla\left(\frac{\rho}{\epsilon}\right) = j\omega\mu\mathbf{J} - \frac{1}{j\omega\epsilon}\nabla\nabla \cdot \mathbf{J} \tag{9.60}$$

$$\nabla^2\mathbf{H} + k^2\mathbf{H} = -\nabla \times \mathbf{J} \tag{9.61}$$

These equations are referred to as inhomogeneous Helmholtz equations. As seen, the impressed current density $\mathbf{J}$ enters into these equations in a relatively complicated way. For this reason we generally do not find the fields $\mathbf{E}$ and $\mathbf{H}$ directly, but rather first compute a scalar and a vector

† J. Stratton, "Electromagnetic Theory," sec. 9.2, McGraw-Hill Book Company, Inc., New York, 1941.

potential from which the fields may subsequently be found. The advantage of doing this is analogous to the similar procedure that was used for the static fields.

The field **B** always has zero divergence, and hence we may take

$$\mathbf{B} = \nabla \times \mathbf{A} \tag{9.62}$$

since $\nabla \cdot \nabla \times \mathbf{A}$ is identically zero. From Maxwell's equation

$$\nabla \times \mathbf{E} = -j\omega\mathbf{B}$$

so we can now write

$$\nabla \times \mathbf{E} = -j\omega\nabla \times \mathbf{A}$$

or

$$\nabla \times (\mathbf{E} + j\omega\mathbf{A}) = 0$$

The vector quantity $(\mathbf{E} + j\omega\mathbf{A})$ is irrotational; so it may be derived from the gradient of a scalar potential Φ; that is, the general integral of the above equation is

$$\mathbf{E} = -j\omega\mathbf{A} - \nabla\Phi \tag{9.63}$$

where Φ is, as yet, an arbitrary scalar function. If we now substitute (9.62) and (9.63) in the curl equation for **H**, we obtain

$$\nabla \times \mathbf{H} = \frac{1}{\mu}\nabla \times \nabla \times \mathbf{A} = j\omega\epsilon\mathbf{E} + \mathbf{J} = j\omega\epsilon(-\nabla\Phi - j\omega\mathbf{A}) + \mathbf{J}$$

Expanding $\nabla \times \nabla \times \mathbf{A}$ to give $\nabla\nabla \cdot \mathbf{A} - \nabla^2\mathbf{A}$, we get

$$\nabla\nabla \cdot \mathbf{A} - \nabla^2\mathbf{A} = -j\omega\epsilon\mu\,\nabla\Phi + k^2\mathbf{A} + \mu\mathbf{J} \tag{9.64}$$

According to the Helmholtz theorem, a vector function is completely specified by its divergence and curl. Since (9.62) gives only the curl of **A**, we are at liberty to specify the divergence of **A** in any way we choose. If we examine (9.64), it is clear that this equation simplifies considerably if we choose

$$\nabla \cdot \mathbf{A} = -j\omega\epsilon\mu\Phi \tag{9.65}$$

so that

$$\nabla\nabla \cdot \mathbf{A} = -j\omega\epsilon\mu\,\nabla\Phi$$

This particular choice is known as the Lorentz condition. Making use of (9.65) reduces (9.64) to

$$\nabla^2\mathbf{A} + k^2\mathbf{A} = -\mu\mathbf{J} \tag{9.66}$$

which is simpler than either (9.60) or (9.61).

Up to this point all Maxwell's equations except the equation $\nabla \cdot \mathbf{D} = \rho$ have been made use of and are therefore satisfied. To ensure that $\nabla \cdot \mathbf{D} = \rho$, we replace **E** by (9.63) and use the Lorentz condition (9.65) to obtain

$$\nabla \cdot \mathbf{E} = \nabla \cdot (-j\omega\mathbf{A} - \nabla\Phi) = -\nabla^2\Phi - k^2\Phi = \frac{\rho}{\epsilon}$$

or

$$\underline{\nabla^2\Phi + k^2\Phi = -\frac{\rho}{\epsilon}} \tag{9.67}$$

This equation determines Φ, and then $\nabla \cdot \mathbf{A}$ is obtained from (9.65). The divergence equation $\nabla \cdot \mathbf{D} = \rho$ is thus satisfied provided Φ is a solution of (9.67), and the divergence of $\mathbf{A}$ is determined from the Lorentz condition.

In practice, we do not need to solve for the scalar potential Φ. If we make use of the Lorentz condition, we can express both $\mathbf{B}$ and $\mathbf{E}$ in terms of the vector potential $\mathbf{A}$ alone. We have

$$\mathbf{B} = \nabla \times \mathbf{A} \tag{9.68a}$$

$$\mathbf{E} = -j\omega\mathbf{A} + (j\omega\epsilon\mu)^{-1}\,\nabla\nabla \cdot \mathbf{A} \tag{9.68b}$$

This result may seem rather strange at first, since normally we should expect to need both the scalar potential Φ and vector potential $\mathbf{A}$ in order to completely determine the field. The explanation lies in the fact that for time-varying sources the charge density ρ is determined by the current density $\mathbf{J}$ through the continuity equation. Thus specification of $\mathbf{J}$ alone is sufficient to completely determine all sources, and hence a solution for $\mathbf{A}$ in terms of the current density $\mathbf{J}$ contains all the necessary information to completely specify the time-varying field. In actual fact, the Lorentz condition is merely the continuity equation in disguise, as the following discussion shows.

If we take the Laplacian of (9.65), we find that

$$\nabla \cdot \nabla^2\mathbf{A} = -j\omega\epsilon\mu\nabla^2\Phi$$

since

$$\nabla^2(\nabla \cdot \mathbf{A}) = \nabla \cdot (\nabla^2\mathbf{A})$$

Replacing $\nabla^2\mathbf{A}$ from (9.66) and $\nabla^2\Phi$ from (9.67) gives

$$-k^2\nabla \cdot \mathbf{A} - \mu\nabla \cdot \mathbf{J} = -j\omega\epsilon\mu\left(-k^2\Phi - \frac{\rho}{\epsilon}\right)$$

or

$$-k^2(\nabla \cdot \mathbf{A} + j\omega\epsilon\mu\Phi) = \mu(\nabla \cdot \mathbf{J} + j\omega\rho) \tag{9.69}$$

The left-hand side vanishes by virtue of the Lorentz condition, and hence the right-hand side must also vanish. This means that

$$\nabla \cdot \mathbf{J} = -j\omega\rho$$

The Lorentz condition is seen to be a condition that ensures that the current and charge satisfy the continuity equation; that is, it provides a function Φ, corresponding to an $\mathbf{A}$ that satisfies (9.66), so that Φ is a solution of (9.67) for a source ρ that is related to $\mathbf{J}$ in (9.66) by the continuity equation. It is for this reason that Φ may be eliminated and the field determined in terms of the vector potential **A** alone.

When ω equals zero, the equations for **A** and Φ reduce to Poisson's equation, which is the appropriate result for static fields. In the static case ρ and **J** are no longer related, and hence Φ and **A** are now independent. Thus for static fields we require a scalar potential Φ to determine **E** and a vector potential **A** to determine **B**. In this respect the determination of the time-varying field is simpler.

The integration of (9.66) and (9.67) is very similar to the integration of Poisson's equation. If we let (x',y',z') be the source point and (x,y,z) be the field point, the solutions are

$$\mathbf{A}(x,y,z) = \frac{\mu}{4\pi}\int_V \frac{\mathbf{J}(x',y',z')}{R}\, e^{-jkR}\, dV' \tag{9.70a}$$

$$\Phi(x,y,z) = \frac{1}{4\pi\epsilon}\int_V \frac{\rho(x',y',z')}{R}\, e^{-jkR}\, dV' \tag{9.70b}$$

where $R = [(x - x')^2 + (y - y')^2 + (z - z')^2]^{1/2}$ and $k^2 = \omega^2\mu\epsilon$. These solutions represent waves propagating radially outward from the source point; that is, $e^{-jkR+j\omega t}$ is a radially outward propagating wave. As the wave propagates outward its amplitude falls off as $1/R$.

To verify the above solutions consider (9.70b) and note that

$$(\nabla^2 + k^2)\int_V \frac{\rho(x',y',z')}{4\pi\epsilon R}\, e^{-jkR}\, dV' = \frac{1}{4\pi\epsilon}\int_V \rho(x',y',z')(\nabla^2 + k^2)\frac{e^{-jkR}}{R}\, dV' \tag{9.71}$$

We may treat R as the radial coordinate in a spherical coordinate system. Since there is no θ or ϕ variation,

$$\nabla^2 \equiv \frac{1}{R^2}\frac{\partial}{\partial R} R^2 \frac{\partial}{\partial R}$$

By direct differentiation it is now found that

$$\nabla^2\left(\frac{e^{-jkR}}{R}\right) = -\frac{k^2}{R}\, e^{-jkR} \qquad R \neq 0$$

Hence the integrand in (9.71) vanishes at all points except $R = 0$, where it has a singularity. We now surround the singularity point $R = 0$, that is, the point (x,y,z), by a small sphere of radius δ and volume V_0. For all values of x', y', z' within this sphere we can replace $\rho(x',y',z')$ by $\rho(x,y,z)$ and e^{-jkR} by unity, provided we choose δ small enough. Note that the maximum value R can have is δ, so that e^{-jkR} can be made to approach unity with a vanishingly small error. The right-hand side of (9.71) thus reduces to

$$\frac{\rho(x,y,z)}{4\pi\epsilon}\int_{V_0}(\nabla^2 + k^2)\,\frac{1}{R}\,dV'$$

Now
$$\int_{V_0}\frac{k^2}{R}\,dV' = k^2\int_{V_0} R\sin\theta\,d\theta\,d\phi\,dR = 2\pi k^2\delta^2$$

and vanishes as δ tends to zero. We are therefore left with

$$\frac{\rho(x,y,z)}{4\pi\epsilon}\int_{V_0}\nabla^2\left(\frac{1}{R}\right)dV' = -\frac{\rho(x,y,z)}{\epsilon}$$

since
$$\int_{V_0}\nabla^2\left(\frac{1}{R}\right)dV' = -4\pi$$

as was demonstrated in Chap. 1, in connection with the integration of Poisson's equation. Therefore (9.70*b*) is a solution of (9.67).

Equation (9.66) for the vector potential **A** may be written as the sum of three scalar equations. For each component the above proof may be applied to show that (9.70*a*) is a solution. Application of the above solutions for the potentials will be made in Chap. 11, in connection with radiation from antennas.

Quasi-static Potentials

Let us assume that we have impressed sources located in free space where $k = k_0 = 2\pi/\lambda_0 = 2\pi f/c$. If we are interested in the fields in the immediate vicinity of the sources, and if the extent of the source region is small compared with a wavelength, then $k_0R = 2\pi R/\lambda_0$ is very small. We may now replace e^{-jk_0R} by unity, and the solutions for the potentials reduce to the static solutions

$$\mathbf{A} = \frac{\mu_0}{4\pi}\int_V \frac{\mathbf{J}(x',y',z')}{R}\,dV' \qquad (9.72a)$$

$$\Phi = \frac{1}{4\pi\epsilon}\int_V \frac{\rho(x',y',z')}{R}\,dV' \qquad (9.72b)$$

with the exception that both **J** and ρ have a time variation according to $e^{j\omega t}$. The fields derived from these potentials are called quasi-static fields, since the fields vary with time but the frequency is sufficiently low so that propagation effects are not important for the range of R of interest. In other words, for a region containing the sources that are small compared with the wavelength, the fields are quasi-static and similar in character to the static field.

For somewhat greater values of k_0R, the approximation

$$e^{-jk_0R} = 1 - jk_0R$$

may be used, and we obtain

$$\mathbf{A} = \frac{\mu_0}{4\pi} \int_V \mathbf{J} \frac{1 - jk_0R}{R} \, dV' \tag{9.73a}$$

$$\Phi = \frac{1}{4\pi\epsilon} \int_V \rho \frac{1 - jk_0R}{R} \, dV' \tag{9.73b}$$

The presence of the term $-jk_0R$ is an indication that propagation effects are becoming important and the contributions to the potentials from the various source elements no longer add in phase. Higher-order approximations are obtained if more terms in the expansion of e^{-jk_0R} are retained.

Retarded Potentials

If we replace k by $k_0 = \omega/c$ and restore the time function $e^{j\omega t}$, the solutions for the potentials may be written as

$$\mathbf{A} = \frac{\mu_0}{4\pi} \int_V \frac{\mathbf{J}}{R} e^{j\omega(t-R/c)} \, dV' \tag{9.74a}$$

$$\Phi = \frac{1}{4\pi\epsilon} \int_V \frac{\rho}{R} e^{j\omega(t-R/c)} \, dV' \tag{9.74b}$$

In this form the potentials are referred to as retarded potentials. The factor $e^{j\omega(t-R/c)}$ shows that at any point a distance R away from the source, the effects caused by changes in the source are not felt until a time interval R/c, the propagation time, has elapsed; that is, contributions to the potential at a point from current or charge sources must include the finite propagation time from each source element to the field point. This means that the potentials are related to source distributions in effect at an earlier time; i.e., they are retarded potentials.

The concept of a retarded potential, although introduced above for sinusoidal time-varying sources, is valid for arbitrary time variation as well. The retarded-potential concept is similar to the action-at-a-distance concept embodied in Coulomb's and Ampère's force laws as contrasted with the field concept.

9.10. Relation between Field Theory and Circuit Theory

Maxwell's equations provide a rigorous and detailed description of the electric and magnetic fields arising from arbitrary current sources in the presence of material bodies. Because of the complexities, rigorous solutions to time-varying electromagnetic problems are obtainable only under very special circumstances, usually where the geometry is particularly simple. In other cases simplifying approximations must be sought.

The reader may be familiar with the well-established techniques for discussing the properties of electrical networks. These are usually characterized by constant lumped-parameter elements such as resistors, capacitors, and inductors. Under steady-state conditions the properties of such networks may be established by setting up and solving a system of algebraic equations. The latter equations arise from an application of Kirchhoff's loop and node equations to the given network, with an assumed current-voltage relationship for each element.

It may seem very surprising that such a relatively simple procedure is available for circuit analysis. Viewed as an electromagnetic boundary-value problem, it is almost hopeless to find a field solution that satisfies the boundary conditions over the connecting leads, the coiled wires of the inductors, and the various shaped conductors that make up the variety of types of capacitors, and for which the primary source is, say, an electron stream within a high-vacuum tube. Circuit theory is obviously an approximation, and if we are to understand the nature and limitations of this technique, it is necessary to determine the assumptions that are required to deduce circuit theory from Maxwell's equations.

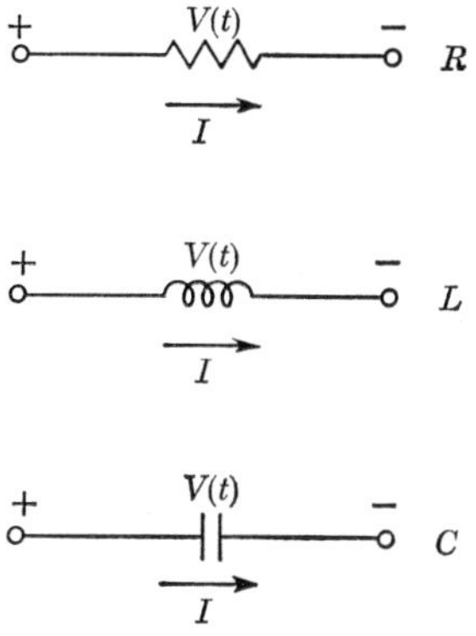

FIG. 9.8. Circuit elements.

As a start, let us briefly review the underlying structure of circuit theory. We assume the existence of four network parameters, the resistance R, the capacitance C, the inductance L, and mutual inductance M. The properties of these parameters are defined in terms of their voltage-current relationships as follows, where for simplicity the harmonic time variation $e^{j\omega t}$ is deleted:

$$V = RI \tag{9.75a}$$
$$V = j\omega LI \tag{9.75b}$$
$$V = \frac{I}{j\omega C} \tag{9.75c}$$

The above may also be specified in terms of the following inverse relations:

$$I = GV \tag{9.76a}$$
$$I = j\omega CV \tag{9.76b}$$
$$I = \frac{V}{j\omega L} \tag{9.76c}$$

where $G = 1/R$ is the conductance. Figure 9.8 illustrates schematically the three circuit elements described above. In each case the voltage V is considered to be the difference in potential between the terminals of the element. For an arbitrary network it is, furthermore, supposed that a unique potential may be assigned each circuit node. Consequently, the sum of the voltages taken around any closed path is zero, which is the statement of the Kirchhoff voltage law.

The current I is assumed to pass continuously through each element from one terminal to the other. Only conduction currents of this type are assumed to exist; consequently, their algebraic sum at any junction is zero, in order to conserve charge. This is the statement of Kirchhoff's current law.

When two coils are coupled so that some of the magnetic flux is common to both, a mutual-inductance term must be added to the circuit equations. Figure 9.9 illustrates schematically a circuit element which requires the use of a mutual inductance M. Since this is a four-terminal device, two equations must be written to describe its voltage-current characteristics. These equations are

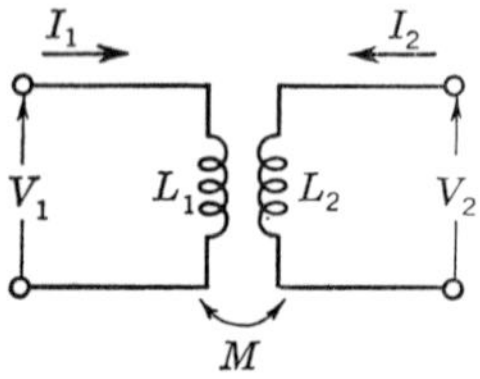

FIG. 9.9. Transformer-type circuit element.

$$V_1 = j\omega L_1 I_1 + j\omega M I_2 \tag{9.77a}$$
$$V_2 = j\omega M I_1 + j\omega L_2 I_2 \tag{9.77b}$$

In the absence of mutual coupling, (9.77) gives the voltage-current relations of the two separate inductors and is of the form (9.75*b*).

With these assumptions concerning the nature of V and I it is possible to establish Kirchhoff's laws. Using the latter and the properties of the circuit elements as described in (9.75) and (9.77), the entire steady-state theory of linear circuit analysis can be developed.† If we are to establish a justification of this theory, we must confirm that (9.75), (9.77), and unique relations for V and I can be derived from Maxwell's equations under suitable conditions.

In order to establish the desired result, we shall assume, first of all, that the maximum circuit dimensions are small compared with wavelength. If we consider any electronic device for which we intuitively feel that circuit theory should be applicable, we shall find the above assumption well justified. For example, an ordinary radio receiver has dimensions of much less than 1 meter, which is quite small compared with the smallest signal wavelength, which is around 200 meters. As a consequence of the above assumption, the fields will be quasi-static in nature; that is, when we expand e^{-jk_0R} we obtain the following expansions for the potentials:

$$\mathbf{A}(x,y,z) = \frac{\mu_0}{4\pi}\int_V \frac{\mathbf{J}(x',y',z')}{R}\,dV' - \frac{jk_0\mu_0}{4\pi}\int_V \mathbf{J}(x',y',z')\,dV' - \frac{k_0{}^2\mu_0}{4\pi}\int_V \mathbf{J}(x',y',z')R\,dV' + \cdots$$

† It is not difficult to extend this work to transient conditions as well, but it will be easier to emphasize the fundamentals if we assume steady-state harmonic time variations.

$$\Phi(x,y,z) = \frac{1}{4\pi\epsilon_0}\int_V \frac{\rho(x',y',z')}{R}\,dV' - \frac{jk_0}{4\pi\epsilon_0}\int_V \rho(x',y',z')\,dV' - \frac{k_0^2}{4\pi\epsilon_0}\int_V \rho(x',y',z')R\,dV' + \cdots$$

where $R = [(x - x')^2 + (y - y')^2 + (z - z')^2]^{1/2}$. The first term in the expansion for $\mathbf{A}$ and Φ is the same as for stationary sources, except that here both $\mathbf{J}$ and ρ are permitted to vary with time. These terms define the quasi-static potentials. The second term in each expansion integrates to a constant and drops out when the expressions $\nabla \times \mathbf{A}$ and $\nabla\Phi$ are formed. This term does contribute to the evaluation of the electric field through the quantity $-j\omega\mathbf{A}$, but the entire expression is proportional to k_0^2. This is the same order in k_0 as the third terms for Φ and $\mathbf{A}$ given above. These terms, along with the higher-order terms in k_0, are negligible for circuit elements that are small compared with the wavelength. Thus we may compute the instantaneous electric and magnetic fields from the charges and currents that exist at that instant as if the sources were time-stationary.

The characteristics of the R, L, and C elements in a circuit may be specified from the field-theory point of view on an energy basis. Thus we consider the ideal capacitor as a lossless element which stores electric energy. A practical capacitor would then be one for which the magnetic stored energy and the losses were negligible compared with the stored electric energy. The ideal inductor would, conversely, store only magnetic energy, and in the practical case it would be assumed that the stored electric energy and losses were negligible. The resistor, however, ideally dissipates energy; practically, some energy storage is unavoidable, but usually negligible. Let us consider each element in greater detail.

The Inductor

Figure 9.10 illustrates an inductor L. Physically, it usually consists of a solenoidal winding of good conducting wire. Because of its construction it sets up a magnetic field which tends to be localized in the region of the coil. This is easily confirmed for a solenoid whose length-to-diameter ratio is large. Because of the quasi-static nature of the field, the results of Chap. 6 apply and reveal (see Prob. 6.8) that the magnetic field is negligible, except within the solenoid, where it is essentially uniform. For moderate length-to-diameter ratios the magnetic field may still be assumed localized in the vicinity of the inductor. This characteristic is emphasized in Fig. 9.10 by crosshatching the assumed local region of the magnetic field.

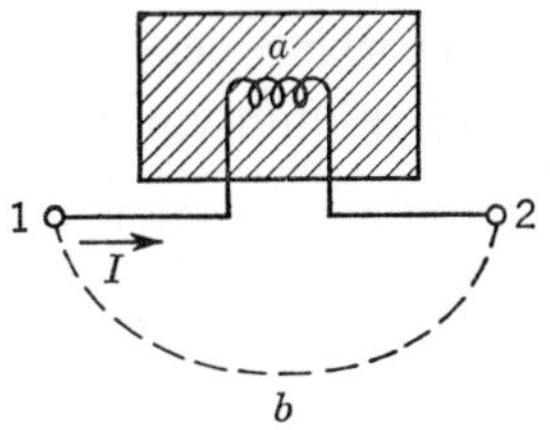

FIG. 9.10. An inductor.

If Faraday's law is applied to the closed contour $1a2b$ in Fig. 9.10, we obtain

$$\oint_{1a2b} \mathbf{E} \cdot d\mathbf{l} = \int_{1a2} \mathbf{E} \cdot d\mathbf{l} + \int_{2b1} \mathbf{E} \cdot d\mathbf{l} = -\frac{d}{dt} \int \mathbf{B} \cdot d\mathbf{S} \qquad (9.78)$$

where the path $1a2$ is along the coil wires, while $2b1$ is any return path in air. If the conductivity of the wire forming the inductor coil is sufficiently great, then within the wires $\mathbf{E} = \mathbf{J}/\sigma \approx 0$, so that (9.78) becomes

$$\int_{2b1} \mathbf{E} \cdot d\mathbf{l} = -\frac{d}{dt} \int \mathbf{B} \cdot d\mathbf{S} \qquad (9.79)$$

The assumption of quasi-static conditions means that we may use time-stationary concepts to evaluate $\int \mathbf{B} \cdot d\mathbf{S}$, which is the total flux linking the closed circuit indicated. By definition of inductance,

$$L = \frac{\int \mathbf{B} \cdot d\mathbf{S}}{I}$$

and consequently,

$$\int_{1b2} \mathbf{E} \cdot d\mathbf{l} = \frac{d}{dt}(LI) = j\omega LI \qquad (9.80)$$

after replacing d/dt by $j\omega$ and changing the sense of the path $2b1$ to $1b2$. We may note that the value of L depends somewhat on the external path $1b2$, but to the extent that most of the flux is localized as shown, moderate changes in $1b2$ will not particularly affect the net flux linkage. As a consequence, L may be thought of as a property of the inductor rather than the circuit; that is, we anticipate that the inductor will be connected to other circuit elements, and we assume that nearby resistors and capacitors contribute negligible flux in the evaluation of the right-hand side of (9.80). Furthermore, nearby coils are assumed also to provide negligible flux. When the latter condition is not fulfilled, then a mutual-inductance element is involved, for which a separate discussion is necessary. Thus whether as an individual element or an element in a circuit, the right-hand side of (9.80) depends on the self-inductance of the coil by itself and is essentially independent of the path $1b2$. This also means that $\int_{1b2} \mathbf{E} \cdot d\mathbf{l}$ is a unique quantity under steady-state conditions, and we may use it as a definition of difference in potential V across the inductor terminals; that is,

$$V = \int_{1b2} \mathbf{E} \cdot d\mathbf{l} = j\omega LI \qquad (9.81)$$

where the path $1b2$ is any path external to the inductor but in its general vicinity. This result confirms (9.75b).

For the case of mutual inductance, the procedure outlined above needs to be modified only slightly. In this case,

$$\int_{1b2} \mathbf{E} \cdot d\mathbf{l} = \frac{d}{dt} \int \mathbf{B} \cdot d\mathbf{S}$$

as before, except that **B** arises from currents in the coil itself and from currents in some other coil. Since quasi-static conditions are involved, we know from Sec. 8.3 that the total flux linkage $\int \mathbf{B} \cdot d\mathbf{S}$ is given by

$$\int \mathbf{B} \cdot d\mathbf{S} = LI_1 + MI_2 \tag{9.82}$$

where L is the self-inductance of the coil being considered and M is the mutual inductance to another coil carrying the current I_2. Consequently, we finally get

$$\int_{1b2} \mathbf{E} \cdot d\mathbf{l} = V = j\omega LI_1 + j\omega MI_2$$

Of course, the given coil can be magnetically coupled to more than one coil. The modification in that case should be fairly obvious from both the field and the circuit standpoint.

At this time it will be useful to consider how practical inductors depart from the above ideal inductor in their behavior. First of all, we know that the wire has a finite conductivity; so there will always be a potential drop across the coil resistance. At very low frequencies the current is uniform over the wire cross section, and hence if **J** is the current density, the electric field in the wire is $\mathbf{E} = \mathbf{J}/\sigma$. In (9.78) the first integral on the right now becomes

$$\int_{1a2} \mathbf{E} \cdot d\mathbf{l} = \int_{1a2} \frac{I}{S_0\sigma}\, d\mathbf{l} = \frac{lI}{\sigma S_0}$$

where l is the total length of wire, S_0 is its cross-sectional area, and the total current $I = S_0 J$. The low-frequency resistance is $R = l/\sigma S_0$, so that in place of (9.81), we have

$$V = j\omega LI + RI \tag{9.83}$$

In many applications $\omega L \gg R$, so that R may be neglected. At higher frequencies this approximation, however, becomes poorer, because the current now flows in a thin layer at the surface of the wire (skin effect), so that the resistance is much greater than the low-frequency value.

A second effect that exists in practical inductors is stray capacitance between turns. The association of this "stray capacitance" with the inductor is merely the recognition of the fact that the electric field and resultant accumulation of charge along the wire are not negligible. When there is a net accumulation of charge, the conduction current flowing in at one terminal does not equal that flowing at various other points in the inductor at each instant of time. The difference is equal to the rate of accumulation or depletion of charge between the two points under consideration; or in other words, the conduction current at the terminals of the inductor is not continuous through the coil winding at each instant of time. Only the total current, conduction plus displacement current, will be continuous. The displacement current, of course, accounts for the charging and discharging of the stray capacitance associated with the inductor.

The effect of stray capacitance is particularly noticeable at high frequencies since an increase in frequency is accompanied by an increase in the displacement current density $j\omega\mathbf{D}$. In fact, all practical inductors behave as capacitors at sufficiently high frequencies. In view of the above discussion, it is not surprising to find that most practical inductors must be characterized by all three ideal parameters, that is, inductance, resistance, and capacitance, particularly at the higher frequencies. An analysis for the equivalent circuit under these conditions is given in a subsequent section.

The Capacitor

Figure 9.11 is a schematic description of a capacitor. From a field standpoint the practical capacitor is characterized by the storage of electric energy, with a negligible accompanying magnetic energy or power loss. If we consider that a conduction current I flows in the leads, then from the continuity equation a charge $Q = I/j\omega$ accumulates on the plates. The electric field between the plates can be written in terms of the vector and scalar potentials as

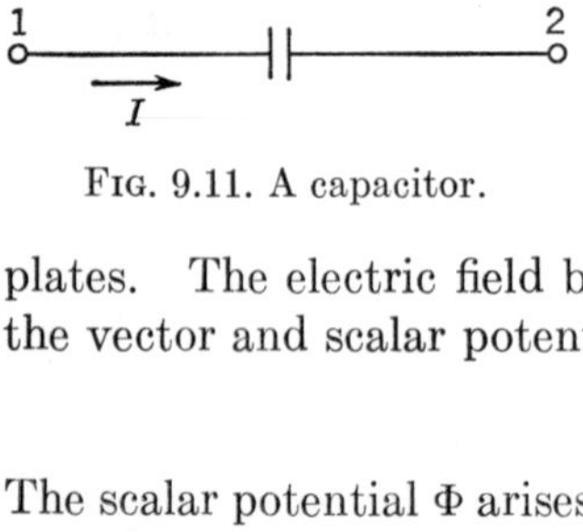

FIG. 9.11. A capacitor.

$$\mathbf{E} = -\nabla\Phi - j\omega\mathbf{A}$$

The scalar potential Φ arises essentially from charge stored on the capacitor plates, while the vector potential $\mathbf{A}$ is due mainly to current in the leads. For a given current magnitude, considering frequency as a variable, $j\omega\mathbf{A}$ is proportional to frequency while $\nabla\Phi$ is inversely proportional to frequency (that is, $\Phi \propto Q \propto 1/\omega$). Consequently, a low enough frequency exists for any capacitor so that $|\nabla\Phi| \gg |j\omega\mathbf{A}|$. The exact frequency below which this approximation is satisfactory can be specified only by a detailed consideration of the particular capacitor configuration involved.

When the vector potential contribution is negligible, the electric field may be derived from the negative gradient of Φ only. However, it is important to note that Φ may arise not only from the charge accumulated on the capacitor plates, but also from charge accumulations at other locations in the circuit. But this is nothing more than additional capacitive coupling between the capacitor under consideration and other nearby bodies and can be taken care of by introducing additional capacitors to describe the over-all circuit (see the discussion in Sec. 3.5 on multibody capacitors).

In order to arrive at the desired voltage-current relationship for the ideal capacitor, it is necessary to assume that the contribution to Φ from charges other than those on the capacitor plates is negligible in the region between the plates. Consequently, we may write, for the field between the capacitor plates,

$$\mathbf{E} = -\nabla\Phi$$

If we define the voltage drop across the capacitor terminals as

$$V = \int_1^2 \mathbf{E} \cdot d\mathbf{l} = \Phi_1 - \Phi_2 \tag{9.84}$$

where the path is through the leads and arbitrarily across the plate spacing, then the result depends only on the value of the scalar potential at each plate. (We again assume good conducting leads and neglect $\int \mathbf{E} \cdot d\mathbf{l}$ along them.) But the scalar potential Φ is derived from the charge on the capacitor according to the static formula, and so all the consequences of the work in Chap. 3 must hold. In particular, we may define the capacitance as $C = Q/(\Phi_1 - \Phi_2) = Q/V$, and in view of the continuity condition which requires $Q = I/j\omega$, we get

$$V = \frac{I}{j\omega C} \tag{9.85}$$

thus confirming (9.75*c*).

The Resistor

We consider, now, the properties of the resistor. As an ideal circuit element, it should set up negligible stored electric and magnetic energy, but is responsible for the dissipation of energy. We illustrate a resistor, schematically, in Fig. 9.12. At any point within the resistor we have

$$\mathbf{E} = \frac{\mathbf{J}}{\sigma}$$

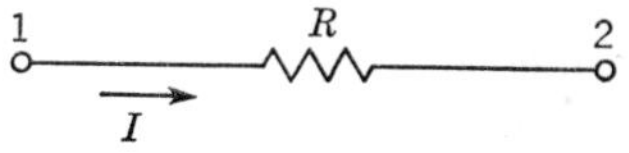

FIG. 9.12. A resistor.

If, for simplicity, we assume uniformity over the length of the resistor, then

$$E = \frac{I}{\sigma S}$$

where S is the effective cross-sectional area. The value of I is the total current and equals the terminal current provided that no substantial displacement current between the terminals exists. Since $\mathbf{E}$ is a constant, then

$$\int_1^2 \mathbf{E} \cdot d\mathbf{l} = \frac{Il}{\sigma S}$$

where l is the length of the resistor and the integration path is taken through the resistor. There is no ambiguity in this result; so we define it to be the potential drop V across the resistor; that is,

$$\int_1^2 \mathbf{E} \cdot d\mathbf{l} = V = IR \tag{9.86}$$

and $R = l/\sigma S$ is the usual definition. We note that (9.86) is in the form of (9.75a).

The Circuit

In order to connect the elements together to make a network, leads are necessary. From a field standpoint these should dissipate no energy nor store any energy. In this case we treat them like resistances with zero resistance. If the energy dissipated is not negligible, they will be treated like true resistances.

We have at this point established the validity of the assumed voltage-current relationships for the fundamental circuit elements, provided certain given conditions are met. We should like to illustrate the application of these ideas to a simple circuit which includes an applied emf and an R, L, and C in series. The circuit is illustrated in Fig. 9.13.

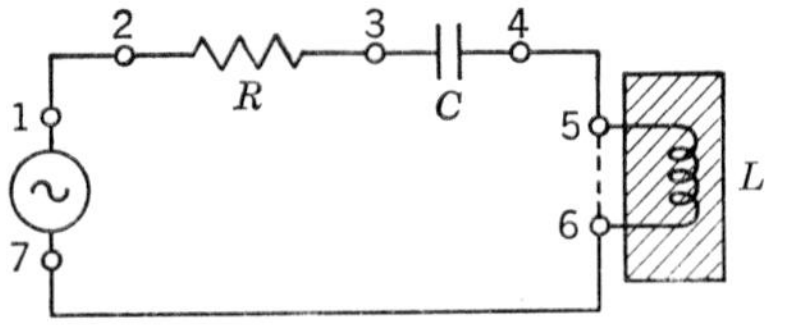

FIG. 9.13. RLC circuit with source of emf.

If we integrate the electric field around the circuit, but follow the dashed path 5-6 across the inductance, and integrate through the source of the emf, then

$$\oint \mathbf{E} \cdot d\mathbf{l} = \mathcal{E} - \frac{d\psi}{dt} \tag{9.87}$$

which is a generalization of Faraday's law when the path of integration includes a source of emf. Because the flux associated with the inductance is excluded from the surface bounded by the chosen contour,

$$\frac{d\psi}{dt} = j\omega\psi = 0$$

Thus

$$\mathcal{E} = \oint \mathbf{E} \cdot d\mathbf{l} = \int_1^2 \mathbf{E} \cdot d\mathbf{l} + \int_2^3 \mathbf{E} \cdot d\mathbf{l} + \int_3^4 \mathbf{E} \cdot d\mathbf{l} + \int_5^6 \mathbf{E} \cdot d\mathbf{l} + \int_6^7 \mathbf{E} \cdot d\mathbf{l} + \int_7^1 \mathbf{E} \cdot d\mathbf{l}$$

If the source of emf is, say, a rotating machine, then $\int_7^1 \mathbf{E} \cdot d\mathbf{l}$ taken through the high-conductivity armature winding is negligible, as is true for the high-conductivity leads in the case of $\int_1^2 \mathbf{E} \cdot d\mathbf{l}$ and $\int_6^7 \mathbf{E} \cdot d\mathbf{l}$. Consequently,

$$\mathcal{E} = IR + \frac{I}{j\omega C} + j\omega LI \tag{9.88}$$

which is the generalization of Kirchhoff's voltage law to include sources of emf.

In summary we may say that the usual circuit concepts hold provided:

1. Circuit dimensions are small compared with wavelength, so that quasi-static conditions prevail. This ensures, among other things, negligible radiated energy.
2. Inductors and capacitors dissipate negligible energy.
3. The magnetic field associated with resistances and capacitors is negligible.
4. The displacement current associated with all circuit elements, except that between capacitor plates, is negligible. Otherwise the entire terminal current may not flow through the element. Furthermore, the Kirchhoff current law can be violated if displacement current flows into or out of a node as a result of the physical arrangement of a network.

The expert in working with circuits is, of course, aware of the above limitations and often employs techniques to extend the realm of network analysis. For example, if condition 2 is not satisfied, we have seen that the lossy inductor may usually be satisfactorily represented by an ideal inductor in series with a resistor, while the lossy capacitor may be represented by a series or shunt resistor with an ideal capacitor. Condition 3 is recognized in the ultra-high-frequency range as requiring the use of short lead lengths to minimize lead inductance, as already noted. Finally, displacement current between windings of a coil is accounted for by the "stray capacitance," which can often be approximated by a circuit with an ideal inductor paralleled by an ideal capacitor. How well these approximate circuit techniques will represent the actual conditions cannot be completely decided in advance, since what is involved is the detailed account of construction, frequency, and dimensions of the actual device.

When the frequency is such that circuit dimensions become comparable to wavelength, field techniques begin to take over as it becomes more difficult to continue to separate electric and magnetic stored-energy regions. Even under these conditions it is still possible to specify an equivalent two-terminal lumped-parameter network that correctly describes the behavior of the physical device at any given frequency.

The following section provides an introduction to this general-equivalent-circuit analysis.

General Equivalent Circuits

Consider an arbitrary physical structure made up from conductors, dielectric material, and magnetizable material in general. Let the device have two terminals, as in Fig. 9.14. Furthermore, let a total current $Ie^{j\omega t}$ flow into the device at the upper terminal and flow out at the lower terminal. Also, let the voltage between the two terminals be $Ve^{j\omega t}$. If we could solve the boundary-value problem to determine the electric and magnetic fields around the structure, we could evaluate the energy W_e stored in the electric field, the energy W_m stored in the magnetic field, and the power loss P_l due both to energy dissipation within the structure and to power radiated. The integrals to be evaluated to obtain the time-average quantities are (note that $\epsilon = \epsilon' - j\epsilon''$, $\mu = \mu' - j\mu''$, and are assumed to be constant)

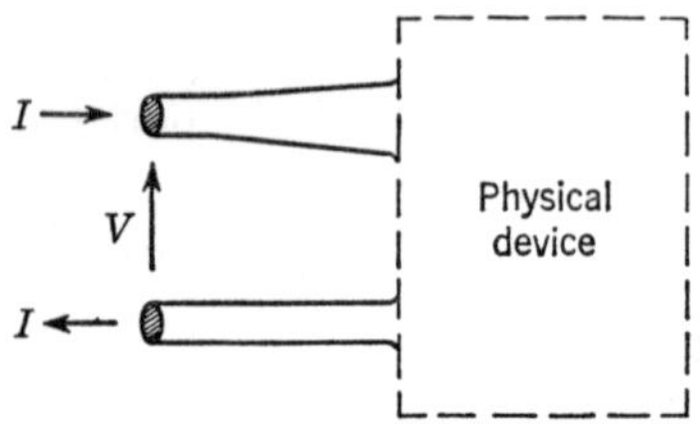

FIG. 9.14. Terminal voltage and current for a physical device.

$$W_e = \frac{\epsilon'}{4}\int_V \mathbf{E}\cdot\mathbf{E}^*\,dV \tag{9.89a}$$

$$W_m = \frac{\mu'}{4}\int_V \mathbf{H}\cdot\mathbf{H}^*\,dV \tag{9.89b}$$

$$P_l = \frac{\sigma}{2}\int_V \mathbf{E}\cdot\mathbf{E}^*\,dV + \frac{\omega}{2}\int_V (\epsilon''\mathbf{E}\cdot\mathbf{E}^* + \mu''\mathbf{H}\cdot\mathbf{H}^*)\,dV + \tfrac{1}{2}\oint_S \operatorname{Re}\mathbf{E}\times\mathbf{H}^*\cdot d\mathbf{S} \tag{9.89c}$$

In the expression for P_l, the integral of the real part of the Poynting vector $\mathbf{E}\times\mathbf{H}^*$ is to be taken over the surface of an infinite sphere surrounding the structure. This integral gives the total time-average power radiated. The terms involving ϵ'' and μ'' give the losses due to polarization damping forces present in the material.

We shall now define the capacitance, inductance, and resistance of the structure by means of the following relations:

$$C = \frac{II^*}{4\omega^2 W_e} \tag{9.90a}$$

$$L = \frac{4W_m}{II^*} \tag{9.90b}$$

$$R = \frac{2P_l}{II^*} \tag{9.90c}$$

In (9.90*a*) and (9.90*b*) the numerical factor is 4, since we are considering time-average quantities. The above definitions are consistent with those which arise under time-stationary conditions, and their justification will be discussed below.

The values of W_e, W_m, and P_l are unique under a given set of terminal conditions. If the voltage V and current I can also be specified uniquely, then the circuit parameters—capacitance C, inductance L, and resistance R—are uniquely defined by the above formulas. Under static conditions it has been shown in earlier chapters that the above definitions for R, L, and C are equivalent to the geometrical definitions. However, the above definitions are more general in that they recognize the fact that ideal circuit elements do not exist physically; e.g., a parallel-plate capacitor has some inductance and resistance associated with it, as we have noted earlier. In other words, by defining capacitance in terms of electric energy storage, account is taken of all portions of the physical structure that contribute to the capacitance of the over-all device, and similarly for the inductance and resistance. It might be noted that these definitions for R, L, and C are equally applicable to distributed circuits, lumped circuits, or a combination of both. Again we emphasize that it is necessary to be able to define unique terminal currents and voltages in order for these parameters to have unique values.

In order to establish a relationship between the terminal current and voltage, we shall make use of the complex Poynting vector theorem established in Sec. 9.7. At the same time conditions for the unique specification of the terminal voltage will be obtained. If the physical device has two conducting leads, the terminal current is clearly unique, it being simply the total conduction current flowing into the structure in one lead and out at the other lead.

Let us construct a closed surface, consisting of an infinite plane (xy plane for convenience) and the surface of a hemisphere at infinity, that completely encloses the physical device except for the terminal leads which protrude through the plane surface, as in Fig. 9.15. The electric and magnetic fields are uniquely determined by the scalar and vector potentials Φ, $\mathbf{A}$, where Φ and $\mathbf{A}$ are determined from the charge density ρ and current density $\mathbf{J}$ distributed throughout all space, both inside and outside S. The electric field $\mathbf{E}$ is given by $\mathbf{E} = -j\omega\mathbf{A} - \nabla\Phi$. In terms

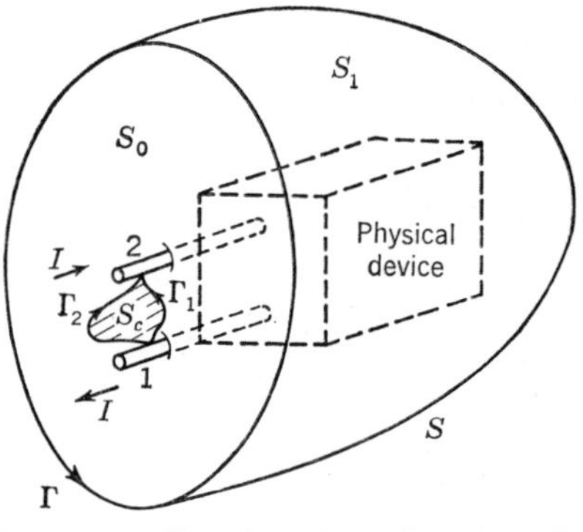

FIG. 9.15. Closed surface $S = S_0 + S_1$ surrounding a physical device.

of a line integral of **E**, the voltage between terminals 1 and 2 in Fig. 9.15 is

$$V = V_{12} = -\int_1^2 \mathbf{E}\cdot d\mathbf{l} = \int_1^2 \nabla\Phi\cdot d\mathbf{l} + j\omega\int_1^2 \mathbf{A}\cdot d\mathbf{l}$$
$$= \Phi_2 - \Phi_1 + j\omega\int_1^2 \mathbf{A}\cdot d\mathbf{l} \tag{9.91}$$

The integral of $\nabla\Phi$ from 1 to 2 may be taken over any arbitrary path without changing its value. However, this is not true for the line integral of **A**. We therefore see that the voltage V will be unique only if the contribution to its value from the vector potential **A** is zero. In general, this latter contribution is not zero, but as discussed earlier, it is negligible at sufficiently low frequencies. Physically, it is easy to see why there is a contribution from the vector potential **A** in general. If we integrate $\mathbf{A}\cdot d\mathbf{l}$ from 1 to 2 along the path Γ_1 and from 2 to 1 along the path Γ_2, we have

$$\int_{\Gamma_1} \mathbf{A}\cdot d\mathbf{l} + \int_{\Gamma_2} \mathbf{A}\cdot d\mathbf{l} = \oint_{\Gamma_1+\Gamma_2} \mathbf{A}\cdot d\mathbf{l} = \int_{S_c} \nabla\times\mathbf{A}\cdot d\mathbf{S} = \int_{S_c} \mathbf{B}\cdot d\mathbf{S} \tag{9.92}$$

where S_c is the area bounded by $\Gamma_1 + \Gamma_2$ and Stokes' law together with the relation $\mathbf{B} = \nabla\times\mathbf{A}$ has been used. The integral of **A** around the closed path gives the total magnetic flux through the area S_c. Since this flux induces an electric field when it changes with time, it follows that the condition for the vector potential not to contribute to the voltage V is that there be no magnetic lines of flux cutting through the boundary plane S_0 on which the line integral of **E** is taken. This requires that the vector potential **A** have no x or y (tangential) components on the surface S_0. When this is true a unique value for the terminal voltage V in terms of the line integral of **E** may be specified.

According to (9.48) the integral of the inward normal component of the complex Poynting vector over a closed surface S gives

$$\tfrac{1}{2}\oint_S \mathbf{E}\times\mathbf{H}^*\cdot d\mathbf{S} = -j\frac{\omega}{2}\int_V (\epsilon'\mathbf{E}\cdot\mathbf{E}^* - \mu'\mathbf{H}\cdot\mathbf{H}^*)\,dV$$
$$+ \tfrac{1}{2}\int_V (\omega\epsilon''\mathbf{E}\cdot\mathbf{E}^* + \omega\mu''\mathbf{H}\cdot\mathbf{H}^* + \sigma\mathbf{E}\cdot\mathbf{E}^*)\,dV$$
$$= 2j\omega(W_m - W_e) + P_{l1} \tag{9.93}$$

where W_m and W_e are the time-average magnetic and electric energy stored inside V, and P_{l1} is the time-average power loss inside V. We may split the integral over S into two parts:

$$\tfrac{1}{2}\oint_S \mathbf{E}\times\mathbf{H}^*\cdot d\mathbf{S} = \tfrac{1}{2}\int_{S_0} \mathbf{E}\times\mathbf{H}^*\cdot d\mathbf{S} + \tfrac{1}{2}\int_{S_1} \mathbf{E}\times\mathbf{H}^*\cdot d\mathbf{S}$$

where $-\tfrac{1}{2}\int_{S_1} \mathbf{E}\times\mathbf{H}^*\cdot d\mathbf{S}$ is the radiated power flowing out through the surface of the hemisphere. We may now rewrite (9.93) as

$$\frac{1}{2} \int_{S_0} \mathbf{E} \times \mathbf{H}^* \cdot d\mathbf{S} = 2j\omega(W_m - W_e) + P_l \tag{9.94}$$

where $P_l = P_{l1} - \frac{1}{2} \int_{S_1} \mathbf{E} \times \mathbf{H}^* \cdot d\mathbf{S}$.

We shall demonstrate shortly that under certain conditions $\mathbf{E}$ and $\mathbf{H}^*$ can be uniquely related to V and I^*, respectively, so that (9.94) becomes

$$\frac{1}{2} VI^* = 2j\omega(W_m - W_e) + P_l \tag{9.95}$$

Since I is linearly related to V, we may now define an input impedance Z_{in} for the physical device so that $V = IZ_{\text{in}}$. Hence

$$\frac{1}{2} Z_{\text{in}} II^* = 2j\omega(W_m - W_e) + P_l \tag{9.96}$$

From (9.90) we have

$$W_e = \frac{II^*}{4\omega^2 C}$$
$$W_m = \frac{L}{4} II^*$$
$$P_l = \frac{1}{2} RII^*$$

When we substitute the appropriate terms into (9.96) and solve for Z_{in}, we obtain

$$\begin{aligned} Z_{\text{in}} &= \frac{2}{II^*} \left[2j\omega \left(\frac{L}{4} II^* - \frac{II^*}{4\omega^2 C} \right) + \frac{1}{2} RII^* \right] \\ &= R + j\omega L - \frac{j}{\omega C} \end{aligned} \tag{9.97}$$

An analysis similar to the above may be carried out when R, L, and C are defined on a voltage basis as

$$R = \frac{VV^*}{2P_l} \tag{9.98a}$$

$$L = \frac{VV^*}{4\omega^2 W_m} \tag{9.98b}$$

$$C = \frac{4W_e}{VV^*} \tag{9.98c}$$

An admittance Y_{in} may be introduced so that $I = Y_{\text{in}} V$, and (9.95) then gives

$$\frac{1}{2} Y_{\text{in}} VV^* = -2j\omega(W_m - W_e) + P_l \tag{9.99}$$

This leads to the result

$$Y_{\text{in}} = \frac{1}{R} + \frac{1}{j\omega L} + j\omega C \tag{9.100}$$

and specifies an equally valid equivalent circuit for the physical device. However, it is important to note that the circuit parameters R, L, and C in (9.100) are not, in general, the same as those occurring in (9.97) since in the present case the normalization factor is VV^* in place of II^*. The equivalent circuit specified by (9.97) is a series RLC circuit, whereas (9.100) specifies an equivalent circuit consisting of R, L, and C in a parallel connection.

The use of the complex-Poynting-vector theorem and appropriate definitions for R, L, and C thus leads to the specification of an equivalent circuit for the physical device. It is important to note that since we have considered a general device, the parameters R, L, and C do not specify ideal elements, but rather only equivalent parameters for the device. They will therefore, in general, be functions of the applied frequency. As a matter of fact, this is also true of all practical circuit elements.

We must now return to an earlier point and demonstrate that under certain conditions

$$\tfrac{1}{2}VI^* = \tfrac{1}{2}\int_{S_0} \mathbf{E} \times \mathbf{H}^* \cdot d\mathbf{S}$$

Since we require a unique voltage V between terminals, we assume that $\mathbf{E} = -\nabla\Phi$; that is, the vector potential contribution $-j\omega\mathbf{A}$ is negligible. Consequently, we can write

$$\int_{S_0} \mathbf{E} \times \mathbf{H}^* \cdot d\mathbf{S} = -\int_{S_0} (\nabla\Phi) \times \mathbf{H}^* \cdot d\mathbf{S}$$

Now

$$\nabla \times (\Phi\mathbf{H}^*) = (\nabla\Phi) \times \mathbf{H}^* + \Phi\nabla \times \mathbf{H}^*$$

and hence

$$\int_{S_0} \mathbf{E} \times \mathbf{H}^* \cdot d\mathbf{S} = \int_{S_0} \Phi\nabla \times \mathbf{H}^* \cdot d\mathbf{S} - \int_{S_0} \nabla \times \Phi\mathbf{H}^* \cdot d\mathbf{S}$$

The last integral may be transformed to a contour integral around the boundary of S_0 by means of Stokes' law; thus

$$\int_{S_0} \nabla \times \Phi\mathbf{H}^* \cdot d\mathbf{S} = \oint_{\Gamma} \Phi\mathbf{H}^* \cdot d\mathbf{l}$$

Since Φ and $\mathbf{H}^*$ vanish at infinity to at least an order $1/R^2$ $(R \to \infty)$, while Γ increases only as R, this integral vanishes, and we are left with

$$\int_{S_0} \mathbf{E} \times \mathbf{H}^* \cdot d\mathbf{S} = \int_{S_0} \Phi(\mathbf{J}^* - j\omega\mathbf{D}^*) \cdot d\mathbf{S} \qquad (9.101)$$

upon replacing $\nabla \times \mathbf{H}^*$ by $\mathbf{J}^* - j\omega\mathbf{D}^*$ from Maxwell's equations. Provided the displacement current $j\omega\mathbf{D}^*$ can be neglected, (9.101) reduces simply to an integral over the cross section of the leads of the physical device. If one lead is at a potential $V/2$ and the other at a potential $-V/2$, we obtain

$$\int_{S_0} \mathbf{E} \times \mathbf{H}^* \cdot d\mathbf{S} = VI^* \tag{9.102}$$

which is the desired result.

From the above analysis it is seen that a unique voltage V can be specified only when a term multiplied by ω, that is, $j\omega\mathbf{A}$, can be neglected in comparison with the term $\nabla\Phi$. Likewise, the field $\mathbf{H}$ can be uniquely related to the terminal current I provided the displacement current $\omega\mathbf{D}$ (again a term multiplied by ω) can be neglected in comparison with the conduction current. This conclusion corresponds to that reached earlier when considering the individual elements. Again we remark that since the two terms to be neglected are multiplied by ω, it is clear that a frequency does exist such that below this frequency these terms are negligible.

When the terminal voltage and current can be specified uniquely, then it is possible to define equivalent-circuit parameters R, L, and C for the physical device so that $II^*/4\omega^2C$, $\frac{1}{4}LII^*$, and $\frac{1}{2}RII^*$ give unique values for the energy stored in the field and dissipated in or radiated from the device. The above discussion has been kept very general in order to make it applicable to any physical device with either lumped or distributed parameters. It enables us to understand more clearly the basis for and the limitations of the circuit concept in electromagnetic theory.

CHAPTER 10

PLANE WAVES, WAVEGUIDES, AND RESONATORS

We have now reached a point in the development of electromagnetic field theory where it is possible to consider a wide variety of important applications. It is, however, outside the scope of the present book to do much more than examine a small number of these. The topics to be considered in this chapter are those of plane waves in free space, reflection of plane waves from a dielectric interface and a conducting plane, the transmission line, rectangular and circular waveguides, and the cavity resonator. These particular topics are chosen because of their great importance at microwave frequencies (frequencies from about 1,000 megacycles up to and beyond 100,000 megacycles) in practical communication systems and also because the solutions are quite readily obtained and provide an elegant demonstration of the validity of Maxwell's field equations.

We shall be dealing entirely with steady-state sinusoidal fields with angular frequency ω. Thus all field vectors are represented by complex phasor vector quantities. Also, we shall assume that ϵ and μ for material bodies are real and constant, unless otherwise stated.

10.1. Classification of Wave Solutions

For most of the topics in this chapter, as well as for a large number of other problems of practical importance, it is possible to separate the solutions of Maxwell's equations in a source-free region into three basic types of fields. These three classifications are:

1. Transverse electromagnetic waves (TEM waves). The transverse electromagnetic wave is characterized by the condition that both the electric and magnetic field vectors lie in a plane perpendicular to the axis of propagation, i.e., have no components in the direction of propagation. The electric field for TEM waves may be derived from the transverse gradient (gradient in the plane transverse to the axis of propagation) of a scalar potential which satisfies the two-dimensional Laplace equation.

2. Transverse electric waves (TE, or *H*, waves). Transverse electric waves are characterized by having an electric field which is entirely in a plane transverse to the (assumed) direction of propagation. Only the magnetic field **H** has a component in the direction of propagation, and

hence this wave type is also known as an H wave. For TE waves it is possible to express all field components in terms of the axial-magnetic-field component.

3. Transverse magnetic waves (TM, or E, waves). Transverse magnetic waves are waves whose magnetic field vector is entirely in a plane transverse to the (assumed) axis of propagation. Only the electric field **E** has a component in the direction of propagation. For TM waves all field components may be expressed in terms of the axial electric field.

The above three types of solutions are sufficiently general so that any arbitrary field solution can be built up by superposing appropriate amounts of each wave type. The only basis for the above classification or division is that the solutions of many practical wave problems fall naturally into one or another of the above types. It is at times more convenient to choose other forms of solutions, but these are just suitable linear combinations of TE and TM wave types and so will not be considered here.

We shall apply the afore-mentioned classification of waves to a broad class of problems characterized by the fact that the geometry is uniform along a given direction; that is, if any material bodies are involved, they are assumed to be cylindrical, and we take the axis to be the z axis. Under these conditions the field solutions can vary axially only by a phase factor $e^{-j\beta z}$. The nature of the field variation in z depends on the propagation constant; if this is real, then unattenuated wave propagation exists. With this type of z dependence all derivatives with respect to z may be replaced by the factor $-j\beta$. The vector operator ∇ becomes

$$\nabla \equiv \mathbf{a}_x \frac{\partial}{\partial x} + \mathbf{a}_y \frac{\partial}{\partial y} - j\beta \mathbf{a}_z = \nabla_t - j\beta \mathbf{a}_z \tag{10.1}$$

where ∇_t signifies the transverse part (x and y part) of the ∇ operator.

It is convenient to separate out the z dependence and decompose all the fields into transverse and axial components. Thus we shall write

$$\begin{aligned} \mathbf{E}(x,y,z) &= \mathbf{E}_t(x,y,z) + \mathbf{E}_z(x,y,z) \\ &= \mathbf{e}(x,y)e^{-j\beta z} + \mathbf{e}_z(x,y)e^{-j\beta z} \end{aligned} \tag{10.2a}$$

$$\begin{aligned} \mathbf{H}(x,y,z) &= \mathbf{H}_t(x,y,z) + \mathbf{H}_z(x,y,z) \\ &= \mathbf{h}(x,y)e^{-j\beta z} + \mathbf{h}_z(x,y)e^{-j\beta z} \end{aligned} \tag{10.2b}$$

where **e**, **h** are transverse vector (x and y components only) functions of the transverse coordinates; $\mathbf{e}_z$, $\mathbf{h}_z$ are z-directed vector functions of x and y; and $\mathbf{E}_t$, $\mathbf{H}_t$ represent the transverse fields including the z dependence while $\mathbf{E}_z$, $\mathbf{H}_z$ represent the axial fields.

Maxwell's equations in a source-free region may now be written in the following form:

$$\nabla_t \times \mathbf{e} = -j\omega\mu \mathbf{h}_z \tag{10.3a}$$
$$\mathbf{a}_z \times \nabla_t e_z + j\beta \mathbf{a}_z \times \mathbf{e} = j\omega\mu \mathbf{h} \tag{10.3b}$$
$$\nabla_t \times \mathbf{h} = j\omega\epsilon \mathbf{e}_z \tag{10.4a}$$
$$\mathbf{a}_z \times \nabla_t h_z + j\beta \mathbf{a}_z \times \mathbf{h} = -j\omega\epsilon \mathbf{e} \tag{10.4b}$$
$$\nabla_t \cdot \mathbf{e} = j\beta e_z \tag{10.5a}$$
$$\nabla_t \cdot \mathbf{h} = j\beta h_z \tag{10.5b}$$

For example, we shall derive (10.3). The curl equation for **E** is

$$(\nabla_t - j\beta \mathbf{a}_z) \times (\mathbf{e} + \mathbf{e}_z)e^{-j\beta z} = -j\omega\mu(\mathbf{h} + \mathbf{h}_z)e^{-j\beta z}$$

Deleting the factor $e^{-j\beta z}$ and expanding give

$$\nabla_t \times \mathbf{e} + \nabla_t \times \mathbf{e}_z - j\beta \mathbf{a}_z \times \mathbf{e} - j\beta \mathbf{a}_z \times \mathbf{e}_z = -j\omega\mu(\mathbf{h} + \mathbf{h}_z) \tag{10.6}$$

Now $\mathbf{a}_z \times \mathbf{e}_z = 0$, $\nabla_t \times \mathbf{e}_z = \nabla_t \times \mathbf{a}_z e_z = -\mathbf{a}_z \times \nabla_t e_z$, and furthermore $\nabla_t \times \mathbf{e}$ is a z-directed vector function while $\mathbf{a}_z \times \nabla_t e_z$ and $\mathbf{a}_z \times \mathbf{e}$ are transverse vectors. Equating the transverse and axial parts of both sides of (10.6) now gives (10.3*a*) and (10.3*b*). In a similar way (10.4) follows from the curl equation for **H**. Equations (10.5*a*) and (10.5*b*) are the divergence equations $\nabla \cdot \epsilon \mathbf{E} = 0$, $\nabla \cdot \mu \mathbf{H} = 0$, with ϵ, μ considered as constant and with $\partial/\partial z$ replaced by $-j\beta$ and the z dependence deleted.

The remainder of this section is devoted to a derivation of the basic equations relating the field components for the three wave types. The following sections will make use of these results for constructing solutions to a variety of practical problems. A fuller appreciation of the properties of the various wave types will be obtained from a study of these examples.

TEM Waves

For TEM waves $\mathbf{e}_z = \mathbf{h}_z = 0$, and (10.3) reduces to the following:

$$\nabla_t \times \mathbf{e} = 0 \tag{10.7a}$$
$$\mathbf{h} = \frac{\beta}{\omega\mu} \mathbf{a}_z \times \mathbf{e} \tag{10.7b}$$

Equation (10.7*a*) is just the condition that **e** may be derived from the transverse gradient of a scalar potential $\Phi(x,y)$; thus

$$\mathbf{e}(x,y) = -\nabla_t \Phi(x,y) = -\mathbf{a}_x \frac{\partial \Phi}{\partial x} - \mathbf{a}_y \frac{\partial \Phi}{\partial y} \tag{10.8}$$

Since $\mathbf{e}_z = 0$, the divergence equation (10.5*a*) gives $\nabla_t \cdot \mathbf{e} = 0$, and hence

$$\nabla_t^2 \Phi = \frac{\partial^2 \Phi}{\partial x^2} + \frac{\partial^2 \Phi}{\partial y^2} = 0 \tag{10.9}$$

The relation (10.8) is physically understandable since, with $\mathbf{h}_z = 0$, the line integral of **e** around any closed contour C in the xy plane is zero, because there is no magnetic flux linking this contour. Thus there is associated with the electric field of a TEM wave a unique scalar potential (apart from a constant). Furthermore, since Φ is a solution of Laplace's equation in the transverse plane, the field $\mathbf{e}(x,y)$ has the same properties as a static electric field. This is a very interesting result in that, even though the fields may vary with time at a rate of thousands of megacycles per second, the field distribution in the transverse plane is a static field distribution.

The field $\mathbf{e}(x,y)e^{-j\beta z}$ is also a solution of the Helmholtz equation:

$$\nabla^2(\mathbf{e}e^{-j\beta z}) + k^2\mathbf{e}e^{-j\beta z} = 0$$

or

$$\nabla_t^2\mathbf{e} + (k^2 - \beta^2)\mathbf{e} = 0 \tag{10.10}$$

Expanding the relation $\nabla_t \times (\nabla_t \times \mathbf{e})$ gives

$$\nabla_t \times (\nabla_t \times \mathbf{e}) = \nabla_t\nabla_t \cdot \mathbf{e} - \nabla_t^2\mathbf{e} = 0$$

since $\nabla_t \times \mathbf{e} = 0$. The divergence of $\mathbf{e}$, that is, $\nabla_t \cdot \mathbf{e}$, is also zero, and hence $\nabla_t^2\mathbf{e} = 0$. Therefore (10.10) can be satisfied only if

$$\beta = \pm k = \pm\omega(\mu\epsilon)^{1/2} \tag{10.11}$$

From (10.7*b*) the magnetic field $\mathbf{h}$ is found to be given by

$$\mathbf{h} = \frac{k}{\omega\mu}\mathbf{a}_z \times \mathbf{e} = Y\mathbf{a}_z \times \mathbf{e} \tag{10.12}$$

where $Y = (\epsilon/\mu)^{1/2}$ is the intrinsic admittance of the medium.

The solution for TEM waves may be summarized as follows. First find a scalar potential Φ which satisfies the two-dimensional Laplace equation and any imposed boundary conditions. The electric and magnetic fields are then given by

$$\mathbf{E}_t = \mathbf{e}e^{-jkz} = -\nabla_t\Phi e^{-jkz} \tag{10.13a}$$

$$\mathbf{H}_t = Y\mathbf{a}_z \times \mathbf{e}e^{-jkz} \tag{10.13b}$$

For a wave propagating in the $-z$ direction, replace k by $-k$ and Y by $-Y$.

TE Waves

For transverse electric (TE) waves, $\mathbf{e}_z = 0$ but $\mathbf{h}_z \neq 0$. For these waves all field components may be expressed in terms of the axial magnetic field $\mathbf{h}_z$, as we shall presently establish.

The magnetic field $\mathbf{H} = (\mathbf{h} + \mathbf{h}_z)e^{-j\beta z}$ must be a solution of the Helmholtz equation $\nabla^2\mathbf{H} + k^2\mathbf{H} = 0$, and hence

$$\nabla_t^2 h_z + k_c^2 h_z = 0 \tag{10.14a}$$

$$\nabla_t^2\mathbf{h} + k_c^2\mathbf{h} = 0 \tag{10.14b}$$

where $k_c^2 = k^2 - \beta^2$. From (10.4*a*) $\nabla_t \times \mathbf{h} = 0$ since $\mathbf{e}_z = 0$, and hence

$$\nabla_t \times (\nabla_t \times \mathbf{h}) = \nabla_t \nabla_t \cdot \mathbf{h} - \nabla_t^2 \mathbf{h} = 0$$

Replacing $\nabla_t^2 \mathbf{h}$ from (10.14*b*) and using (10.5*b*) to replace $\nabla_t \cdot \mathbf{h}$ by $j\beta h_z$, we obtain

$$\nabla_t \nabla_t \cdot \mathbf{h} = j\beta \nabla_t h_z = \nabla_t^2 \mathbf{h} = -k_c^2 \mathbf{h}$$

or

$$\mathbf{h} = -\frac{j\beta}{k_c^2} \nabla_t h_z \tag{10.15}$$

This relation expresses the transverse-magnetic-field vector function in terms of h_z. The function $h_z(x,y)$ is seen to play the role of a scalar potential function from which $\mathbf{h}$ may be derived.

In order to express $\mathbf{e}$ in terms of $\mathbf{h}$, we take the vector product of (10.3*b*) by $\mathbf{a}_z$ to obtain

$$j\beta \mathbf{a}_z \times (\mathbf{a}_z \times \mathbf{e}) = j\beta[(\mathbf{a}_z \cdot \mathbf{e})\mathbf{a}_z - (\mathbf{a}_z \cdot \mathbf{a}_z)\mathbf{e}] = -j\beta \mathbf{e} = j\omega\mu \mathbf{a}_z \times \mathbf{h}$$

since $\mathbf{a}_z \cdot \mathbf{e} = 0$ and $\mathbf{e}_z = 0$. Replacing $\omega\mu$ by kZ, where $Z = (\mu/\epsilon)^{1/2}$, now gives

$$\mathbf{e} = -\frac{k}{\beta} Z \mathbf{a}_z \times \mathbf{h} \tag{10.16}$$

The factor kZ/β has the dimensions of an impedance and is called the wave impedance for TE or *H* waves. It will be designated by the symbol Z_h; that is,

$$Z_h = \frac{k}{\beta} Z \tag{10.17}$$

In component form (10.16) gives

$$\frac{e_x}{h_y} = -\frac{e_y}{h_x} = Z_h \tag{10.18}$$

Thus the ratio of the transverse electric field to the mutually perpendicular transverse magnetic field is equal to the wave impedance (apart from a minus sign in one case).

The solution for TE waves may be summarized as follows. First find a solution for h_z; that is, obtain a solution of

$$\nabla_t^2 h_z + k_c^2 h_z = 0$$

The parameter k_c is usually determined by boundary conditions that the field must satisfy. This will be elaborated on when specific examples are considered. Once k_c is determined, β may be obtained from the relation $\beta^2 = k^2 - k_c^2$. The vector function $\mathbf{h}$ is obtained from h_z by means of (10.15), and $\mathbf{e}$ is found from $\mathbf{h}$ by means of (10.16). The electric and magnetic fields are then given by

$$\mathbf{E} = \mathbf{e}e^{-j\beta z}$$
$$\mathbf{H} = (\mathbf{h} + \mathbf{h}_z)e^{-j\beta z}$$

For a wave propagating in the $-z$ direction, β is replaced by $-\beta$ in (10.15) and (10.16). The sign of $\mathbf{h}$ changes, but not the sign of $\mathbf{e}$. A reversal of the sign of either $\mathbf{h}$ or $\mathbf{e}$ is required in order to obtain a reversal in the direction of power flow (Poynting vector).

TM Waves

For TM waves $\mathbf{h}_z = 0$ but $\mathbf{e}_z \neq 0$. For this wave type all field components may be expressed in terms of e_z. The required equations may be derived in a manner similar to that used for TE waves but with the role of electric and magnetic fields interchanged. This possibility is a direct consequence of the symmetry of Maxwell's equations for $\mathbf{E}$ and $\mathbf{H}$ in a source-free region. In actual fact this symmetry forms the basis of the "principle of duality" for electromagnetic fields.

The duality principle states that if $\mathbf{E}_1$, $\mathbf{H}_1$ are solutions of the equations

$$\nabla \times \mathbf{E}_1 = -j\omega\mu\mathbf{H}_1 \qquad \nabla \times \mathbf{H}_1 = j\omega\epsilon\mathbf{E}_1$$
$$\nabla \cdot \mathbf{E}_1 = 0 \qquad \nabla \cdot \mathbf{H}_1 = 0$$

then a second field $\mathbf{E}_2$, $\mathbf{H}_2$, where

$$\mathbf{E}_2 = \pm Z\mathbf{H}_1 \tag{10.19a}$$
$$\mathbf{H}_2 = \mp Y\mathbf{E}_1 \tag{10.19b}$$

is also a solution. Substitution into Maxwell's equations verifies the result at once. For example,

$$\nabla \times \mathbf{E}_2 = \pm\nabla \times Z\mathbf{H}_1 = \pm j\omega\epsilon Z\mathbf{E}_1$$

But $Z\epsilon = Y\mu$ and $Y\mathbf{E}_1 = \mp\mathbf{H}_2$, and hence

$$\nabla \times \mathbf{E}_2 = \pm j\omega\mu(\mp\mathbf{H}_2) = -j\omega\mu\mathbf{H}_2$$

In a similar way it is readily verified that $\nabla \times \mathbf{H}_2 = j\omega\epsilon\mathbf{E}_2$, and hence $\mathbf{E}_2$, $\mathbf{H}_2$ as given by (10.19) is a solution if $\mathbf{E}_1$, $\mathbf{H}_1$ is a solution. This principle is very useful in practice for constructing solutions for TE waves from those for TM waves, and vice versa.

For TM waves we have, analogous to (10.14),

$$\nabla_t^2 e_z + k_c^2 e_z = 0 \tag{10.20a}$$
$$\nabla_t^2 \mathbf{e} + k_c^2 \mathbf{e} = 0 \tag{10.20b}$$

The equations analogous to (10.15) and (10.16) are obtained by using the duality principle, i.e., replacing $\mathbf{h}$, h_z, and $\mathbf{e}$ in these equations by $Y\mathbf{e}$, Ye_z, and $-Z\mathbf{h}$, respectively. Thus (10.15) becomes

$$Y\mathbf{e} = -\frac{j\beta}{k_c^2}\nabla_t Y e_z$$

or

$$\mathbf{e} = -\frac{j\beta}{k_c^2}\nabla_t e_z \qquad (10.21)$$

and (10.16) becomes

$$\mathbf{h} = \frac{k}{\beta} Y \mathbf{a}_z \times \mathbf{e} \qquad (10.22)$$

Equations (10.20), (10.21), and (10.22) are the required relations expressing the field components for TM waves in terms of the axial-electric-field function e_z. The relation (10.22) may be written in component form as

$$-\frac{h_x}{e_y} = \frac{h_y}{e_x} = \frac{k}{\beta} Y = Y_e = Z_e^{-1} \qquad (10.23)$$

where $Z_e = Z\beta/k$ is the wave impedance for TM waves. The combination of (10.23) with (10.17) shows that

$$Z_h Z_e = Z^2 \qquad (10.24)$$

a result which expresses the dual relationship between TE and TM waves.

The complete solution for TM waves is

$$\mathbf{E} = (\mathbf{e} + \mathbf{e}_z)e^{-j\beta z} \qquad (10.25a)$$

$$\mathbf{H} = \mathbf{h}e^{-j\beta z} \qquad (10.25b)$$

For a wave propagating in the $-z$ direction the sign of $\mathbf{e}_z$ and β is reversed. This changes the sign of $\mathbf{h}$ but leaves the sign of $\mathbf{e}$ unchanged (the sign of $\mathbf{e}_z$ is changed only to keep the sign of $\mathbf{e}$ unchanged).

10.2. Plane Waves

In Chap. 9 we considered the solution for a plane wave with components E_x, H_y and propagating in the z direction according to $e^{-jk_0 z}$. We should now like to reformulate the properties of a plane wave for an arbitrary direction of propagation, $\mathbf{n}$. We note that if $\mathbf{r} = x\mathbf{a}_x + y\mathbf{a}_y + z\mathbf{a}_z$ is the radius vector from the origin, then

$$\mathbf{n} \cdot \mathbf{r} = \text{constant} \qquad (10.26)$$

is the equation of a plane which is perpendicular to the unit vector $\mathbf{n}$, as in Fig. 10.1. Consequently, if we want to consider a wave propagating in a direction given by $\mathbf{n}$, then the appropriate propagation factor to use is $e^{-jk_0\mathbf{n}\cdot\mathbf{r}}$.

The mathematical formulation for the electric field $\mathbf{E}$ of a uniform plane wave propagating in a direction $\mathbf{n}$ can be written

$$\mathbf{E} = \mathbf{E}_0 e^{-jk_0\mathbf{n}\cdot\mathbf{r}} \tag{10.27}$$

where $\mathbf{E}_0$ is a constant vector. The restrictions on $\mathbf{E}_0$ can be found from the requirement that (10.27) be a solution of Maxwell's field equations in a

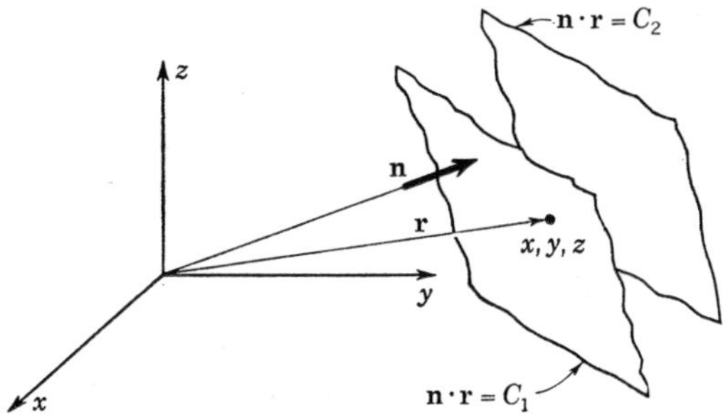

FIG. 10.1. Illustration of planes specified by $\mathbf{n}\cdot\mathbf{r}$ = constant.

source-free region of free space. The charge density is assumed to be zero; consequently, the divergence of $\mathbf{E}$ must be zero, and hence

$$\nabla\cdot\mathbf{E} = 0 = \nabla\cdot\mathbf{E}_0 e^{-jk_0\mathbf{n}\cdot\mathbf{r}} = \mathbf{E}_0\cdot\nabla e^{-jk_0\mathbf{n}\cdot\mathbf{r}} \tag{10.28}$$

since $\mathbf{E}_0$ is a constant vector. Now $\mathbf{n}\cdot\mathbf{r} = n_x x + n_y y + n_z z$, where n_x, n_y, n_z are the components of $\mathbf{n}$, and so

$$\mathbf{a}_x \frac{\partial}{\partial x} e^{-jk_0\mathbf{n}\cdot\mathbf{r}} = -jk_0 n_x \mathbf{a}_x e^{-jk_0\mathbf{n}\cdot\mathbf{r}}, \text{ etc.}$$

so that

$$\nabla e^{-jk_0\mathbf{n}\cdot\mathbf{r}} = -jk_0\mathbf{n}e^{-jk_0\mathbf{n}\cdot\mathbf{r}}$$

Therefore (10.28) gives

$$-jk_0(\mathbf{E}_0\cdot\mathbf{n})e^{-jk_0\mathbf{n}\cdot\mathbf{r}} = 0$$

or

$$\mathbf{n}\cdot\mathbf{E}_0 = 0 \tag{10.29}$$

Thus (10.27) is a possible solution only if $\mathbf{E}_0$ lies in a plane that is perpendicular to the direction of propagation specified by the unit vector $\mathbf{n}$.

The magnetic field may be found from the curl equation for $\mathbf{E}$ as follows:

$$-j\omega\mu_0\mathbf{H} = \nabla\times\mathbf{E} = \nabla\times\mathbf{E}_0 e^{-jk_0\mathbf{n}\cdot\mathbf{r}} = -\mathbf{E}_0\times\nabla e^{-jk_0\mathbf{n}\cdot\mathbf{r}}$$

or

$$\mathbf{H} = (\mathbf{E}_0\times\mathbf{n})\frac{-jk_0}{j\omega\mu_0}e^{-jk_0\mathbf{n}\cdot\mathbf{r}} = Y_0(\mathbf{n}\times\mathbf{E}_0)e^{-jk_0\mathbf{n}\cdot\mathbf{r}} \tag{10.30}$$

The magnetic field associated with the electric field given by (10.27) also lies in a plane transverse to the direction of propagation and furthermore is also perpendicular to $\mathbf{E}_0$, as in Fig. 10.2. Equations (10.27), (10.29), and (10.30) define a general plane transverse electromagnetic wave propagating in the direction **n**. The wave is called a plane wave since the constant phase surfaces given by $k_0\mathbf{n} \cdot \mathbf{r} =$ constant are planes.

If, for example, we wish to call the z axis the axis of propagation, then

$$\mathbf{E} = (\mathbf{E}_0 e^{-jk_0(xn_x+yn_y)})e^{-jk_0n_zz} \tag{10.31}$$

and $\beta_0 = k_0n_z$. Since $\mathbf{E}_0$ does not lie in the xy plane (excluding $n_x = n_y = 0$), the wave would not be classified as a TEM wave with respect to the z axis. Depending on the direction of $\mathbf{E}_0$, it could be a TE,

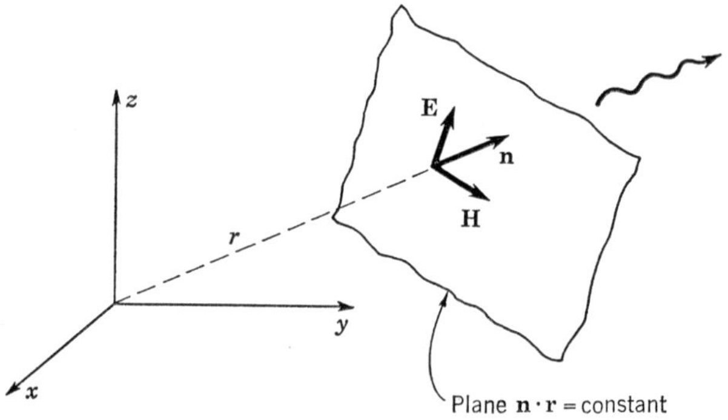

FIG. 10.2. Space relation between the field components and direction of propagation for a plane TEM wave.

TM, or a combination of a TE and a TM wave. In this respect the classification of a wave solution as a TEM, TE, or TM wave does depend on our choice of a preferred direction to be considered as the direction of propagation. In actual fact, the wave specified by (10.31) propagates in the direction **n**, and not in the z direction.

The classification into TE or TM categories is more meaningful for a wave of the type to be discussed now. Let us superpose on the solution (10.31) a similar wave solution with the direction of propagation given by

$$\mathbf{n}_1 = -\mathbf{a}_xn_x - \mathbf{a}_yn_y + \mathbf{a}_zn_z$$

Then

$$\mathbf{E} = \mathbf{E}_0(e^{-jk_0(xn_x+yn_y)} + e^{jk_0(xn_x+yn_y)})e^{-j\beta_0z} = 2\mathbf{E}_0 \cos [k_0(xn_x + yn_y)]e^{-j\beta_0z} \tag{10.32}$$

where $\beta_0 = k_0n_z$. This solution represents a wave propagating in the z direction only. In the transverse plane the solution is that of a standing wave. If $n_y = 0$, then **n** is a vector in the xz plane and it is not inconsistent to choose an $\mathbf{E}_0$ that lies in the y direction; that is, $\mathbf{E}_0 = E_0\mathbf{a}_y$.

The corresponding solution for the magnetic field is

$$-j\omega\mu_0\mathbf{H} = \nabla \times \mathbf{E} = -\mathbf{a}_y \times \nabla E$$
$$= 2E_0(j\beta_0\mathbf{a}_x \cos k_0n_xx - k_0n_x\mathbf{a}_z \sin k_0n_xx)e^{-j\beta_0z} \quad (10.33)$$

and is seen to have x and z components. This solution is clearly a TE wave, and under no circumstances could it be considered as a TEM wave since $H_z \neq 0$. In general, then, we are able to state that the combination of TEM waves propagating in different directions gives wave solutions of the TE and TM types. Nevertheless, it is convenient at times to classify obliquely propagating TEM waves as TE or TM waves also.

Reflection from a Dielectric Interface, Perpendicular Polarization

In this discussion we shall examine the problem of the reflection of an obliquely incident plane wave from a dielectric interface. With reference

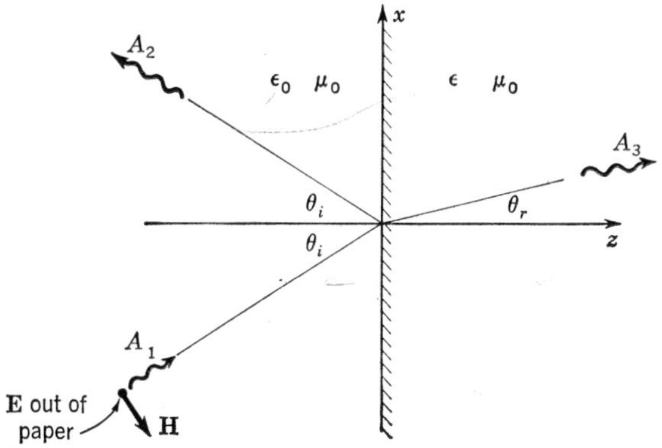

FIG. 10.3. Plane wave incident on a dielectric interface.

to Fig. 10.3, let the half space $z > 0$ be filled with a homogeneous, isotropic, lossless dielectric with a permittivity ϵ. The dielectric constant is $\kappa = \epsilon/\epsilon_0$, and the index of refraction is $\eta = \kappa^{1/2}$.

Without loss in generality we may choose the plane of incidence as the xz plane, and then

$$\mathbf{n} = \mathbf{a}_x \sin \theta_i + \mathbf{a}_z \cos \theta_i$$

where θ_i is the angle of incidence measured relative to the interface normal. Rather than consider an arbitrary polarized incident wave (i.e., we have yet to specify the orientation of the electric field), it is more convenient to treat two special cases separately. For one we choose a wave with the electric field in the y direction. This wave is called perpendicular-polarized since the electric field is perpendicular to the plane of incidence, where the latter is defined by the interface normal and the unit vector $\mathbf{n}$, that is, the xz plane. The corresponding magnetic field has both x and z components, and consequently the wave is a TE wave with respect to the z axis. In the other case the roles of the electric and magnetic fields are interchanged; i.e., the electric field lies in the xz plane while the magnetic field is directed along the y axis. This wave is a TM wave and is referred to as a parallel-polarized wave since the electric field is parallel to the plane of incidence. A superposition of these two cases gives the solution for an arbitrary-polarized incident wave. The two cases are treated separately because of the existence of certain basic differences, as we shall discover.

For the perpendicular-polarized incident wave let the electric field be

$$\mathbf{E}_i = \mathbf{a}_y A_1 e^{-jl_0 x - j\beta_0 z} \tag{10.34}$$

where $l_0 = n_x k_0 = k_0 \sin \theta_i$ and $\beta_0 = k_0 \cos \theta_i$. The corresponding magnetic field may be found from the curl of $\mathbf{E}_i$ and is, from (10.30),

$$\mathbf{H}_i = A_1(-Y_0 \cos \theta_i \mathbf{a}_x + Y_0 \sin \theta_i \mathbf{a}_z) e^{-jl_0 x - j\beta_0 z} \tag{10.35}$$

In order to satisfy the boundary conditions at the interface $z = 0$ when a plane wave is incident, it is necessary to assume that a part of it is reflected from the dielectric and a part of it is transmitted into the dielectric. At the interface the total tangential electric and magnetic fields must be equal on adjacent sides of the interface. This is possible only if all field components have the same variation with x on either side of the interface. Consequently, the form of the reflected and transmitted electric fields must be

$$\mathbf{E}_r = \mathbf{a}_y A_2 e^{-jl_0 x + j\beta_0 z} \tag{10.36}$$

$$\mathbf{E}_t = \mathbf{a}_y A_3 e^{-jlx - j\beta z} \tag{10.37}$$

where $l = k \sin \theta_r$, $\beta = k \cos \theta_r$, and θ_r is the angle of refraction, i.e., specifies the direction of propagation in the dielectric, as illustrated in Fig. 10.3. As noted above, l must equal l_0 in order to satisfy the boundary conditions for all values of x, and hence

$$k_0 \sin \theta_i = k \sin \theta_r$$

or

$$\sin \theta_i = \kappa^{1/2} \sin \theta_r = \eta \sin \theta_r \tag{10.38}$$

Equation (10.38) is the well-known Snell's law of refraction.

The magnetic fields for the reflected and transmitted waves are found from the corresponding electric fields using (10.30) and are

$$\mathbf{H}_r = A_2(Y_0 \cos \theta_i \mathbf{a}_x + Y_0 \sin \theta_i \mathbf{a}_z) e^{-jl_0 x + j\beta_0 z} \tag{10.39}$$

$$\mathbf{H}_t = A_3(-Y \cos \theta_r \mathbf{a}_x + Y \sin \theta_r \mathbf{a}_z) e^{-jl_0 x - j\beta z} \tag{10.40}$$

where $Y = (\epsilon/\mu_0)^{1/2} = \eta Y_0$. For the reflected wave the x component of magnetic field is reversed in sign corresponding to the use of the exponential function $e^{j\beta_0 z}$.

The amplitude coefficients A_2 and A_3 are determined by making the total tangential electric and magnetic field components continuous across the interface. The following two equations result from these two conditions:

$$\begin{aligned} A_1 + A_2 &= A_3 \\ (A_1 - A_2)Y_0 \cos\theta_i &= A_3 Y \cos\theta_r \end{aligned}$$

As noted in Sec. 9.8, matching the tangential fields at the boundary automatically ensures the proper behavior of the normal field components. The reflection coefficient ζ_1 is defined as the ratio of the reflected electric field amplitude A_2 to the incident electric field amplitude A_1. Similarly, the transmission coefficient T_1 is defined as the ratio of the transmitted electric field amplitude to the incident wave amplitude. We have

$$A_2 = \zeta_1 A_1 \qquad A_3 = T_1 A_1$$

and the boundary conditions become

$$1 + \zeta_1 = T_1 \tag{10.41a}$$

$$(1 - \zeta_1)Y_0 \cos\theta_i = T_1 Y \cos\theta_r = T_1 \eta Y_0 \cos\theta_r \tag{10.41b}$$

Solving these equations for ζ_1 and T_1 gives

$$\zeta_1 = \frac{\cos\theta_i - \eta\cos\theta_r}{\cos\theta_i + \eta\cos\theta_r} \tag{10.42a}$$

$$T_1 = 1 + \zeta_1 = \frac{2\cos\theta_i}{\cos\theta_i + \eta\cos\theta_r} \tag{10.42b}$$

These latter equations are called the Fresnel reflection and transmission equations for a perpendicular-polarized incident wave.

The student familiar with transmission-line circuit theory† will recognize the close analogy between the present problem and that of a junction of two transmission lines of different characteristic impedance. The transverse electric field is analogous to the voltage wave, while the transverse magnetic field is analogous to the current wave on a transmission line. The wave impedance is the counterpart of the characteristic impedance of a transmission line. The basis for the analogy is that the continuity at an interface of the total tangential fields $\mathbf{E}_t$ and $\mathbf{H}_t$ in the field analysis corresponds to the continuity at the transmission-line junction of the total V and I in the equivalent transmission-line analysis. Thus for the TE wave on the air side of the interface the wave impedance is

$$Z_{h0} = -\frac{E_y}{H_x} = Z_0 \sec\theta_i = Z_0 \frac{k_0}{\beta_0} \tag{10.43a}$$

† A discussion of the transmission line will be given in Sec. 10.4.

while for the dielectric region

$$Z_h = -\frac{E_y}{H_x} = Z \sec \theta_r = Z \frac{k}{\beta} \tag{10.43b}$$

where $$Z_0 = \left(\frac{\mu_0}{\epsilon_0}\right)^{1/2} \quad \text{and} \quad Z = \left(\frac{\mu_0}{\epsilon}\right)^{1/2}$$

The transmission-line circuit illustrated in Fig. 10.4 is formally equivalent to the problem being considered here. According to transmission-

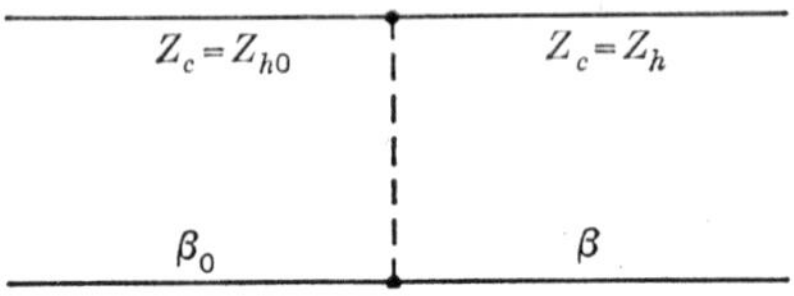

FIG 10.4. Transmission-line equivalent circuit for Fig. 10.3, perpendicular polarization.

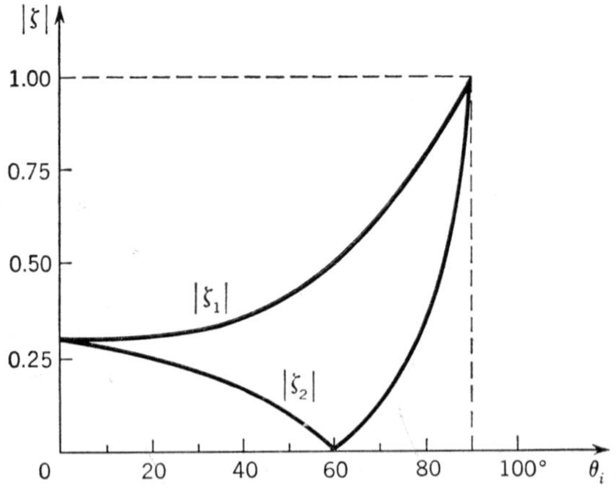

FIG. 10.5. Modulus of reflection coefficient for $\kappa = 3$.

line theory the reflection coefficient of the junction is

$$\zeta_1 = \frac{Z_h - Z_{h0}}{Z_h + Z_{h0}} = \frac{Z \sec \theta_r - Z_0 \sec \theta_i}{Z \sec \theta_r + Z_0 \sec \theta_i} = \frac{\cos \theta_i - \eta \cos \theta_r}{\cos \theta_i + \eta \cos \theta_r}$$

which is the same as (10.42a). Using Snell's law, we have

$$\eta \cos \theta_r = (\kappa - \sin^2 \theta_i)^{1/2}$$

and hence

$$\zeta_1 = -\frac{(\kappa - \sin^2 \theta_i)^{1/2} - \cos \theta_i}{(\kappa - \sin^2 \theta_i)^{1/2} + \cos \theta_i} \tag{10.44}$$

A plot of $|\zeta_1|$ as a function of θ_i for $\kappa = 3$ is given in Fig. 10.5. It is seen that $|\zeta_1|$ continually increases with increasing values of the angle θ_i.

Minimum reflection occurs at normal incidence, and the value of $|\zeta_1|$ for this condition depends on κ.

Reflection from a Dielectric Interface, Parallel Polarization

The solution for the case of a parallel-polarized incident wave is similar but with the role of electric and magnetic fields interchanged. The details are left as a problem. The reflection and transmission coefficients ζ_2, T_2 may be readily found from the equivalent transmission-line circuit

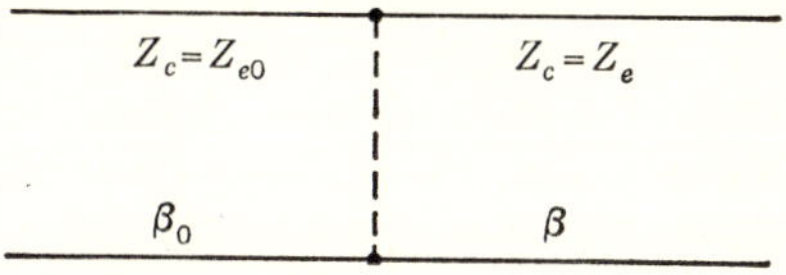

FIG. 10.6. Equivalent transmission-line circuit of dielectric interface, parallel polarization.

illustrated in Fig. 10.6. From (10.23) the wave impedances in the free-space and dielectric regions are

$$Z_{e0} = \frac{E_x}{H_y} = \frac{\beta_0}{k_0} Z_0 = Z_0 \cos \theta_i \tag{10.45a}$$

$$Z_e = \frac{\beta}{k} Z = \frac{Z_0}{\eta} \cos \theta_r \tag{10.45b}$$

Snell's law again holds; so $\kappa Z_e = Z_0(\kappa - \sin^2 \theta_i)^{1/2}$. Thus the Fresnel formulas for the reflection and transmission coefficients are

$$\zeta_2 = \frac{Z_e - Z_{e0}}{Z_e + Z_{e0}} = \frac{(\kappa - \sin^2 \theta_i)^{1/2} - \kappa \cos \theta_i}{(\kappa - \sin^2 \theta_i)^{1/2} + \kappa \cos \theta_i} \tag{10.46a}$$

$$T_2 = 1 + \zeta_2 = \frac{2(\kappa - \sin^2 \theta_i)^{1/2}}{(\kappa - \sin^2 \theta_i)^{1/2} + \kappa \cos \theta_i} \tag{10.46b}$$

An interesting property of ζ_2 is that for some particular value of θ_i it vanishes. From (10.46a), $\zeta_2 = 0$ when

$$\kappa - \sin^2 \theta_i = \kappa^2 \cos^2 \theta_i = \kappa^2 - \kappa^2 \sin^2 \theta_i$$

Denoting the solution for θ_i by θ_B, we have

$$\sin \theta_B = \left(\frac{\kappa}{\kappa + 1}\right)^{1/2} \tag{10.47}$$

This particular angle is called the Brewster angle. For a parallel-polarized wave incident at the angle θ_B, no reflection takes place and all the incident power is transmitted into the dielectric. A similar phenomenon does not occur for a perpendicular-polarized wave unless the dielectric medium has a permeability greater than unity. A plot of $|\zeta_2|$ is given in Fig. 10.5 for $\kappa = 3$. Up to an angle $\theta_i = \theta_B$ the reflection coefficient continually decreases. Beyond the angle θ_B the reflection coefficient increases rapidly up to a value of unity at grazing incidence when $\theta_i = 90°$.

The concept of a wave impedance is of great practical importance since it provides a formal analogy between wave problems and transmission-line problems. In transmission-line circuit problems the total line voltage and current are made continuous at the load termination. Similar boundary conditions are imposed on the tangential electric and magnetic fields at a discontinuity interface. For this reason the same formulas for reflection and transmission coefficients are applicable to wave problems. It is important to note, however, that the analogy holds only if the axis of propagation is chosen normal to the discontinuity interface. For obliquely incident TEM waves the use of a transmission-line equivalent circuit leads naturally to a classification of the incident wave as a TE or TM wave.

10.3. Reflection from a Conducting Plane

From an analysis of the problem of reflection of a plane wave from a conducting plane, the behavior of the electromagnetic field at the surface of a conductor may be deduced. We shall be able to show that the total current per unit width flowing in the conducting plane is essentially independent of the conductivity. As the conductivity is made to approach infinity, the current is squeezed into a narrower and narrower layer, until in the limit a true surface current is obtained. The conductor will be shown to be characterized as a boundary surface exhibiting a surface impedance

$$Z_m = R_m + jX_m$$

where $R_m = X_m = (\sigma\delta)^{-1}$ and δ is the skin depth. The power loss in the conductor is then readily shown to be given by

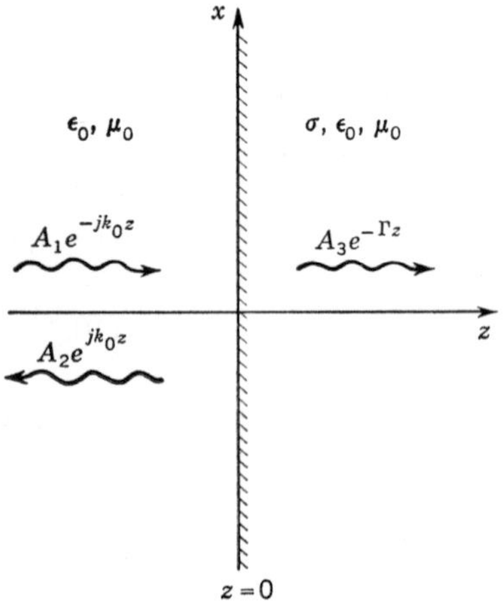

FIG. 10.7. Plane wave incident on a conducting plane.

$$P_l = \tfrac{1}{2}R_m\mathbf{J}_s \cdot \mathbf{J}_s^* \qquad \text{per unit area}$$

where $\mathbf{J}_s$ is the surface current density. Since $\mathbf{J}_s$ is also equal to $\mathbf{n} \times \mathbf{H}$, we have a very convenient method of evaluating the power loss in a conductor from a knowledge of the tangential magnetic field at the surface.

Let a plane TEM wave be incident on a conducting interface located at $z = 0$; that is, the half space $z \geq 0$ is filled with a conducting medium, as in Fig. 10.7. The incident field is chosen as follows:

$$\mathbf{E}_i = A_1\mathbf{a}_x e^{-jk_0z} \tag{10.48a}$$
$$\mathbf{H}_i = A_1 Y_0 \mathbf{a}_y e^{-jk_0z} \tag{10.48b}$$

At the interface there will be a reflected wave

$$\mathbf{E}_r = A_2\mathbf{a}_x e^{jk_0z} \tag{10.49a}$$
$$\mathbf{H}_r = -A_2 Y_0 \mathbf{a}_y e^{jk_0z} \tag{10.49b}$$

and a transmitted wave of the form

$$\mathbf{E}_t = A_3\mathbf{a}_x e^{-\Gamma z} \tag{10.50a}$$
$$\mathbf{H}_t = A_3 Y_m \mathbf{a}_y e^{-\Gamma z} \tag{10.50b}$$

where Γ and Y_m are yet to be determined. In a conducting medium the curl equation for $\mathbf{H}$ is $\nabla \times \mathbf{H} = j\omega\epsilon_0\mathbf{E} + \sigma\mathbf{E} \approx \sigma\mathbf{E}$, since the conduction current is much greater than the displacement current. If we rewrite this equation as $\nabla \times \mathbf{H} = j\omega(\sigma/j\omega)\mathbf{E}$, we see that $\sigma/j\omega$ may be considered as the permittivity in Maxwell's equations. Using this analogy, we may construct the solution for the plane wave in the conductor from the solution for the incident wave. Thus by analogy we have

$$\Gamma = j\omega\left(\frac{\mu_0\sigma}{j\omega}\right)^{1/2} = (j\omega\mu_0\sigma)^{1/2}$$

$$Z_m = Y_m^{-1} = \left(\frac{j\omega\mu_0}{\sigma}\right)^{1/2} = \frac{\Gamma}{\sigma}$$

since $\quad jk_0 = j\omega(\mu_0\epsilon_0)^{1/2} \quad$ and $\quad Z_0 = \left(\frac{\mu_0}{\epsilon_0}\right)^{1/2}$

The square root of j equals $(1 + j)/\sqrt{2}$, and hence

$$\Gamma = \frac{1+j}{\delta} \tag{10.51}$$

$$Z_m = \frac{1+j}{\sigma\delta} = R_m + jX_m \tag{10.52}$$

where δ is the skin depth and is given by

$$\delta = \left(\frac{2}{\omega\mu_0\sigma}\right)^{1/2} \tag{10.53}$$

It is seen that the conductor exhibits an impedance with equal resistive and inductive parts. Furthermore, the resistive part is just the d-c resistance of a sheet of metal 1 meter square and of thickness δ. (Actually, the resistance is independent of the area of the square plate.) Thus with reference to Fig. 10.8, the d-c resistance between the two faces 1 and 2 is given by

$$R_m = \frac{L}{L\delta\sigma} = \frac{1}{\delta\sigma} \quad \text{ohms/square}$$

Since the resistance is independent of the linear dimension L, it is called a surface resistance and is measured in ohms per square. The impedance Z_m is called the intrinsic impedance of the conductor. For the present case of normal incidence the ratio of the tangential electric and magnetic fields at the interface, called the surface impedance, equals the intrinsic impedance.

At the interface the tangential fields must be continuous; hence

$$A_1 + A_2 = A_3$$
$$(A_1 - A_2)Y_0 = A_3 Y_m$$

If we let $A_2 = \zeta A_1$, $A_3 = TA_1$, where ζ and T are the reflection and transmission coefficients, we have

$$1 + \zeta = T \tag{10.54a}$$
$$1 - \zeta = \frac{Z_0}{Z_m} \tag{10.54b}$$

Solving for ζ and T gives

$$\zeta = \frac{Z_m - Z_0}{Z_m + Z_0} \tag{10.55a}$$
$$T = \frac{2Z_m}{Z_m + Z_0} \tag{10.55b}$$

For any reasonably good conductor, Z_m is very small compared with Z_0. For example, for copper ($\sigma = 5.8 \times 10^7$ mhos per meter) at a frequency of 1,000 megacycles, $\delta = 2 \times 10^{-6}$ meter, $R_m = 0.0086$ ohm, while $Z_0 = 377$ ohms. For all practical purposes the field in front of the conductor ($z < 0$) is the same as would exist if σ were infinite, since ζ differs from -1 by a negligible amount. For the same reason the amount of power transmitted into the conductor is very small; that is, T is very small. We cannot, however, neglect the power transmitted into conducting surfaces in the case of transmission lines and waveguides since any loss that is present is important in determining the attenuation constant of a wave. While the attenuation due to conductor losses could be expected to be negligible for short laboratory connections, it would enter significantly in long transmission lines. A method for calculating the attenuation constant will be developed in the following discussion.

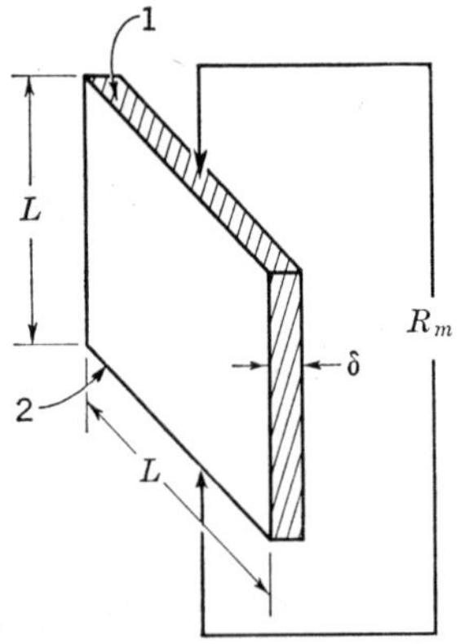

FIG. 10.8. Equivalent low-frequency resistance problem.

The current flowing in the conductor is $J_x = \sigma E_x = \sigma T A_1 e^{-\Gamma z}$. The total current per unit width of conductor is

$$J_s = \sigma T A_1 \int_0^\infty e^{-\Gamma z}\, dz = \frac{\sigma T A_1}{\Gamma} \qquad \text{amp/m}$$

Replacing Γ by $(1 + j)/\delta$ and noting that at the conductor surface $H_y = TA_1Y_m$, we get

$$J_s = \frac{\delta \sigma Z_m}{1 + j} H_y = H_y \tag{10.56}$$

since $Z_m = (1 + j)/\sigma\delta$. This result also can be obtained by applying $\oint_C \mathbf{H} \cdot d\mathbf{l} = \int_S \mathbf{J} \cdot d\mathbf{S}$ (displacement current being neglected) to a rectangular contour C whose long dimension runs from $z = 0$ to $z = +\infty$ and whose short dimension is a unit length parallel to the x axis. The details are left to the student.

If we now let σ tend to infinity, we find that $\delta \to 0$, $\zeta \to -1$, and $H_y \to 2H_i$. The total current J_s does not vanish since, from (10.56), it clearly approaches the value $2H_i$. However, it is squeezed into a narrower and narrower layer and in the limit becomes a true surface current measured in amperes per meter.

The power loss per unit area in the xy plane may be evaluated from the complex Poynting vector at the surface. We have

$$P_l = \tfrac{1}{2} \operatorname{Re} (E_x H_y^*) = \tfrac{1}{2} A_1 A_1^* T T^* \operatorname{Re} Y_m^* = \tfrac{1}{4} |TA_1|^2 \sigma \delta \tag{10.57}$$

We may also evaluate P_l by means of the following volume integral whose integrand expresses the joule heating loss per unit volume. We have

$$P_l = \tfrac{1}{2} \int_0^1 \int_0^1 \int_0^\infty \sigma E_x E_x^*\, dx\, dy\, dz = \frac{\sigma}{2} |A_1 T|^2 \int_0^\infty e^{-2z/\delta}\, dz$$
$$= \tfrac{1}{4} |A_1 T|^2 \sigma \delta \tag{10.58}$$

The two methods, of course, give the same results. This result can be put into another useful form if we replace $|A_1T|$ by $|J_s\Gamma/\sigma|$. We then obtain

$$P_l = \tfrac{1}{4} |J_s \Gamma|^2 \frac{\delta}{\sigma} = \tfrac{1}{2} |J_s|^2 R_m$$

In practice, the following approximate method is generally used to evaluate the power loss per unit area. The tangential magnetic field is first found using the assumption that σ is infinite. The surface current density is then determined from the boundary condition

$$\mathbf{J}_s = \mathbf{n} \times \mathbf{H}$$

where **n** is the outward normal to the conductor surface. Next it is recalled that the surface of a conductor exhibits a surface impedance Z_m, and hence the power loss per unit area is

$$P_l = \tfrac{1}{2}|H_t|^2 R_m \tag{10.59}$$

where H_t is the tangential magnetic field at the surface, evaluated for infinite σ. For the present problem $H_t = H_y = 2A_1Y_0$, and hence

$$P_l = 2|A_1|^2 Y_0^2 R_m$$

To compare this result with (10.57), note that

$$T \approx \frac{2Z_m}{Z_0} = \frac{2(1+j)}{\sigma\delta Z_0}$$

and hence (10.57) gives approximately

$$\tfrac{1}{4}|A_1|^2 \frac{4}{(\sigma\delta Z_0)^2} |1+j|^2 \sigma = 2|A_1|^2 Y_0^2 R_m$$

which is the result (10.59).

The approximation involved in (10.59) is that we take for H_t its value when σ is infinite. This is, however, a very good approximation, since $|Z_m| \ll Z_0$. Therefore in practice we are justified in determining the surface current density $\mathbf{J}_s$ by using the boundary condition $\mathbf{n} \times \mathbf{H} = \mathbf{J}_s$ and computing $\mathbf{H}$ as though the conductivity were infinite.

In the case of infinite conductivity the tangential electric field at the conductor surface is zero. For finite conductivity there has to be a finite value of tangential electric field in order to obtain a component of the Poynting vector directed into the conductor. The tangential electric field at the surface is given by $\mathbf{E}_t = \sigma\mathbf{J}_0$, where $\mathbf{J}_0$ is the current density at the surface. The student may verify that $\mathbf{E}_t$ is also given by

$$\mathbf{E}_t = \mathbf{J}_s Z_m \tag{10.60}$$

The above results were derived for a plane wave incident along the surface normal. For an obliquely incident wave the previous work can be utilized provided that (10.48) to (10.50) are interpreted as applying to the transverse fields and that the intrinsic admittances are replaced by the appropriate wave admittances. The wave impedance in the free-space region is, from (10.43*b*) and (10.45*a*),

$$Z_{h0} = Z_0 \sec\theta_i \qquad \text{perpendicular polarization}$$
$$Z_{e0} = Z_0 \cos\theta_i \qquad \text{parallel polarization}$$

The corresponding wave impedances in the conductor are

$$Z_h = Z_m \sec \theta_r$$
$$Z_e = Z_m \cos \theta_r$$

Since Snell's law must hold, we have

$$jk_0 \sin \theta_i = \Gamma \sin \theta_r = \frac{1+j}{\delta} \sin \theta_r$$

Now δ is very small compared with k_0^{-1}, and therefore $\sin \theta_r$ is very small and also complex. If $\sin \theta_r$ is very small, it follows that

$$\cos \theta_r = (1 - \sin^2 \theta_r)^{1/2}$$

is very nearly equal to unity. This shows that even for oblique incidence the conductor may be assumed to exhibit a surface impedance Z_m because Z_e and Z_h differ from Z_m by a negligible amount. Thus the procedure outlined earlier for the evaluation of the power loss in a conductor is valid for oblique incidence as well. The method breaks down only when the conductor is curved and has a radius of curvature not much greater than the skin depth. Conductors with such small radii of curvature are rarely encountered, except perhaps as small-diameter wires at the lower frequencies. In the majority of cases the use of the surface impedance Z_m and (10.59) is entirely valid for computing power loss at arbitrary conducting surfaces with arbitrary electromagnetic fields.

10.4. Transmission Lines

A transmission line consists of two or more uniform and parallel conductors. It is used to transmit high-frequency electromagnetic energy from a given source (generator) to a load, e.g., antenna. The cross sections of several typical transmission lines are illustrated in Fig. 10.9.

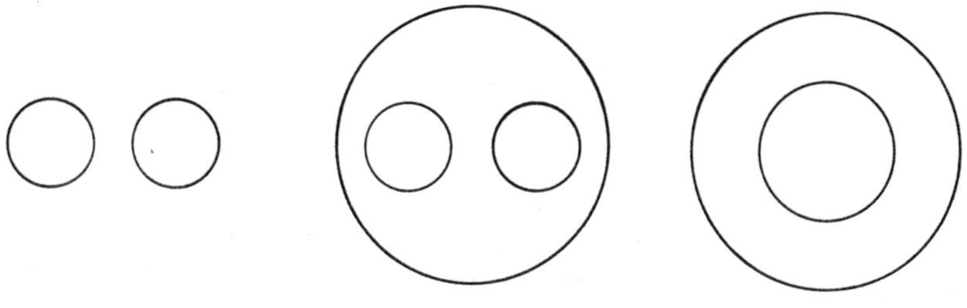

(a) (b) (c)

FIG. 10.9. Cross sections of typical transmission lines. (a) Two-wire line; (b) shielded two-wire line; (c) coaxial line.

The principal type of wave that may propagate along an ideal ($\sigma = \infty$) transmission line is a TEM wave. Thus the field surrounding the conductors of a transmission line is governed by the equations for TEM waves given in Sec. 10.1. With reference to Fig. 10.10, which illustrates a general two-conductor line, we can write

$$\mathbf{E} = \mathbf{E}_t = \mathbf{e}e^{-jkz} \tag{10.61a}$$
$$\mathbf{H} = \mathbf{H}_t = \mathbf{h}e^{-jkz} = Y\mathbf{a}_z \times \mathbf{e}e^{-jkz} \tag{10.61b}$$

where it is assumed that the medium surrounding the conductors has electrical parameters ϵ and μ_0. The field $\mathbf{e}$ is equal to the negative transverse

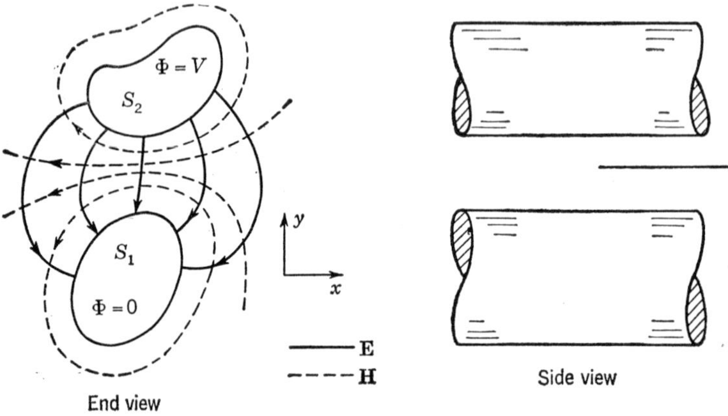

FIG. 10.10. A general two-conductor transmission line.

gradient of a scalar potential Φ, and Φ is a solution of the two-dimensional Laplace equation; that is,

$$\mathbf{e} = -\nabla_t \Phi \tag{10.62a}$$
$$\nabla_t^2 \Phi = \frac{\partial^2 \Phi}{\partial x^2} + \frac{\partial^2 \Phi}{\partial y^2} = 0 \tag{10.62b}$$

A nontrivial solution for Φ exists only if there is a potential difference V between the conductors. Thus associated with the electric field (10.61a) there is a unique voltage wave Ve^{-jkz}. The line integral

$$-\int_{S_1}^{S_2} \mathbf{e} \cdot d\mathbf{l} = \int_{S_1}^{S_2} d\Phi = V$$

is independent of the path by virtue of (10.62a), where the integration is taken from an arbitrary point on S_1 to an arbitrary point on S_2.

The boundary conditions on Φ, namely, that it equal a constant, say zero, on one conductor and V on the other, are independent of frequency. Since Φ must be a solution of Laplace's equation as well, then the uniqueness theorem requires that Φ be independent of frequency. In other words, the transverse field distribution of a transmission line is independent of frequency and, as a matter of fact, is precisely the distribution under static conditions. This is, of course, a general property of TEM waves, as was noted earlier.

From (10.61*b*) it is seen that the magnetic lines of flux are perpendicular to the electric lines of flux and hence must coincide with the constant-potential contours in the xy plane. The line integral of **h** around the conductor S_2 in Fig. 10.10 gives

$$\oint_{S_2} \mathbf{h} \cdot d\mathbf{l} = I \tag{10.63}$$

where I is the total z-directed current on S_2. The result follows from Ampère's circuital law since there is no z-directed displacement current but only a z-directed conduction current density $\mathbf{J}_s = \mathbf{n} \times \mathbf{h}$ on each conductor. On the conductor S_1 the total current flowing is $-I$. Thus associated with the magnetic field (10.61*b*) there is a unique current wave Ie^{-jkz}.

From the fact that the potential plot is independent of frequency, the direction and relative magnitude of the magnetic lines of flux must then be the same at all frequencies. Furthermore, since (10.63) holds at all frequencies, then if I remains the same, the absolute magnitude of **h** does not depend on the frequency as well. Thus the field distribution of **h** is nothing more than that under time-stationary conditions.

In view of the unique relationship between **e** and V on one hand and **h** and I on the other, it follows that the properties of a transmission line may be described in terms of the fields existing around the conductors or in terms of the associated voltage and current waves. In a field description the parameters of interest are the propagation constant k and the intrinsic impedance Z of the medium surrounding the conductors. In a voltage-current description the ratio V/I defines the characteristic impedance Z_c of the line. The characteristic impedance is the counterpart of the intrinsic impedance Z and in actual fact differs from Z by a factor which is a function of the line geometry only. The circuit parameters of a transmission line are discussed in the next section along with their method of derivation and will shed further light on the interrelationship between the field and circuit descriptions.

10.5. Transmission-line Parameters

In a circuit analysis of a transmission line it is postulated that the line can be characterized by the following distributed circuit parameters:

R = series resistance per meter
L = series inductance per meter
G = shunt conductance per meter
C = shunt capacitance per meter

By considering a differential section dz of line, as in Fig. 10.11, an equivalent circuit involving the above-mentioned parameters may be constructed as shown. The following circuit equations then arise from conventional circuit theory:

$$\mathcal{V}(z) - \left[\mathcal{V}(z) + \frac{d\mathcal{V}}{dz}\,dz\right] = -\frac{d\mathcal{V}}{dz}\,dz = (j\omega L + R)\,dz\,\mathcal{I} \qquad (10.64)$$

$$\mathcal{I}(z) - \left[\mathcal{I}(z) + \frac{d\mathcal{I}}{dz}\,dz\right] = -\frac{d\mathcal{I}}{dz}\,dz = (j\omega C + G)\,dz\,\mathcal{V} \qquad (10.65)$$

Equation (10.64) states that the decrease in voltage along a length dz of the line is equal to the voltage drop in the series impedance $(j\omega L + R)\,dz$, while (10.65) gives the decrease in current because of the shunt current

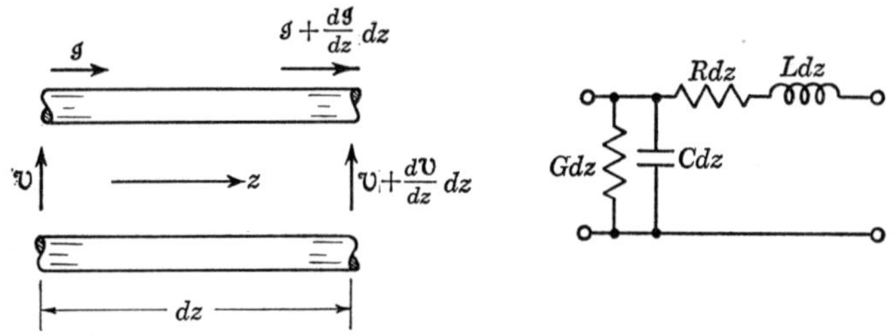

FIG. 10.11. A differential section of transmission line and its equivalent circuit.

flowing through the shunt admittance $(j\omega C + G)\,dz$. Differentiating (10.64) with respect to z and substituting into (10.65) give

$$\frac{d^2\mathcal{V}}{dz^2} - (j\omega L + R)(j\omega C + G)\mathcal{V} = 0 \qquad (10.66a)$$

Similarly, we may obtain

$$\frac{d^2\mathcal{I}}{dz^2} - (j\omega L + R)(j\omega C + G)\mathcal{I} = 0 \qquad (10.66b)$$

A solution to (10.66a) is

$$\mathcal{V} = Ve^{-\Gamma z} \qquad (10.67)$$

where V is an amplitude constant and

$$\Gamma = [(j\omega L + R)(j\omega C + G)]^{1/2} \qquad (10.68)$$

For any good line $\omega L \gg R$ and $\omega C \gg G$, so that (10.68) gives

$$\Gamma = j\beta + \alpha \approx j\omega\sqrt{LC} + \tfrac{1}{2}\left(R\sqrt{\frac{C}{L}} + G\sqrt{\frac{L}{C}}\right) \qquad (10.69)$$

From (10.64) the solution for $\mathcal{I}$ is

$$\mathcal{I} = \frac{\Gamma}{j\omega L + R} V e^{-\Gamma z} = V \left(\frac{j\omega C + G}{j\omega L + R} \right)^{1/2} e^{-\Gamma z} \approx V \sqrt{\frac{C}{L}} e^{-\Gamma z} \qquad (10.70)$$

for a line with small losses. The ratio $\mathcal{V}/\mathcal{I}$ defines the characteristic impedance Z_c; thus

$$Z_c = \sqrt{\frac{L}{C}} \qquad (10.71)$$

Lossless Line

In this section we shall establish the validity of the above approach as well as obtaining methods for the evaluation of L, C, R, and G. We consider a lossless line first with ϵ real and the conductivity σ of the conductors infinite. With reference to Fig. 10.10 and using the boundary condition $\epsilon \mathbf{n} \cdot \mathbf{e} = \rho_s$, where ρ_s is the surface charge density per meter on the conductor S_2, we have

$$Q = \epsilon \oint_{S_2} \mathbf{n} \cdot \mathbf{e} \, dl \qquad (10.72)$$

for the total charge per meter on S_2. Since the potential of S_2 is V, the capacitance per meter between S_2 and S_1 may be defined, as under static conditions, to be

$$C = \frac{Q}{V} \qquad (10.73)$$

At the surface S_2 the electric field has a normal component only while the magnetic field has only a tangential component. Furthermore, from (10.61*b*) we find that $|\mathbf{h}| = Y|\mathbf{e}| = Y\mathbf{n} \cdot \mathbf{e}$ on the surface of S_2. Hence the current flowing on S_2 is given by

$$I = \oint_{S_2} \mathbf{h} \cdot d\mathbf{l} = \oint_{S_2} Y\mathbf{n} \cdot \mathbf{e} \, dl = \frac{YQ}{\epsilon} \qquad (10.74)$$

Thus the characteristic impedance of the line is given by

$$Z_c = \frac{V}{I} = \frac{V}{Q} \epsilon Z = \frac{\epsilon Z}{C} \qquad (10.75)$$

Since C/ϵ is a function of the line geometry only, Z_c differs from Z only by a factor which is a function of the line geometry.

An alternative expression for Z_c may be derived from an energy definition of C and L, such as was introduced in electrostatics and magnetostatics. Thus the energy stored in the electric field per unit length of line is

$$W_e = \tfrac{1}{4} \epsilon \iint_{x,y} \mathbf{e} \cdot \mathbf{e} \, dx \, dy$$

and we define C by the following:

$$W_e = \tfrac{1}{4}CV^2 \tag{10.76}$$

The energy stored in the magnetic field per unit length of line is

$$\begin{aligned} W_m &= \tfrac{1}{4}\mu_0 \iint_{x,y} \mathbf{h} \cdot \mathbf{h}\, dx\, dy \\ &= \tfrac{1}{4}\mu_0 Y^2 \iint_{x,y} (\mathbf{a}_z \times \mathbf{e}) \cdot (\mathbf{a}_z \times \mathbf{e})\, dx\, dy \end{aligned}$$

But

$$\begin{aligned} (\mathbf{a}_z \times \mathbf{e}) \cdot (\mathbf{a}_z \times \mathbf{e}) &= \mathbf{a}_z \cdot [\mathbf{e} \times (\mathbf{a}_z \times \mathbf{e})] \\ &= \mathbf{a}_z \cdot [(\mathbf{e} \cdot \mathbf{e})\mathbf{a}_z - (\mathbf{e} \cdot \mathbf{a}_z)\mathbf{e}] = \mathbf{e} \cdot \mathbf{e} \end{aligned}$$

because $\mathbf{e} \cdot \mathbf{a}_z = 0$. Hence, since $Y^2 = \epsilon/\mu_0$,

$$W_m = \tfrac{1}{4}\epsilon \iint_{x,y} \mathbf{e} \cdot \mathbf{e}\, dx\, dy = W_e \tag{10.77}$$

The inductance L per meter may be defined from the energy relation $W_m = \tfrac{1}{4}LI^2$, and we see that

$$LI^2 = CV^2$$

or

$$\frac{V}{I} = \sqrt{\frac{L}{C}} \tag{10.78}$$

Consequently, we must have

$$\frac{\epsilon Z}{C} = \sqrt{\frac{L}{C}} = Z_c \tag{10.79}$$

by equating (10.75) and (10.78). From (10.79) we obtain the relation $\sqrt{LC} = \epsilon Z = \sqrt{\epsilon\mu_0}$. Thus for a line with no losses, $\alpha = 0$ and $\beta = \omega\sqrt{LC} = \omega\sqrt{\epsilon\mu_0} = k$. For the ideal line we can therefore conclude that the circuit approach is consistent with the rigorous analysis based on Maxwell's equations if L and C are defined as above. Since these definitions are in terms of time-stationary energy formulas and since we know that the electric and magnetic field distributions are precisely those under static conditions, then the static values of L and C are appropriate and correct at any frequency.

Line with Lossy Dielectric

Let us now consider the case where the dielectric surrounding the conductors is lossy. The dielectric loss may be due to a finite conductivity or polarization damping forces, or both. In all cases the effect of losses may be accounted for by a complex permittivity $\epsilon = \epsilon' - j\epsilon''$. The imaginary part ϵ'' is directly responsible for the loss. Substitution of a complex ϵ into the equations for TEM waves does not modify the form of these equations, so that a TEM wave solution is still possible. With ϵ complex, a shunt current of density $\mathbf{J} = \omega\epsilon''\mathbf{e}$ will flow from conductor S_2 to S_1 (note in the curl equation for $\mathbf{H}$ that $\omega\epsilon''$ is the equivalent conductivity). The total shunt conduction current I_s per meter is given by

$$I_s = \omega\epsilon'' \oint_{S_2} \mathbf{n} \cdot \mathbf{e}\, dl = \frac{\omega\epsilon'' Q}{\epsilon'} \tag{10.80}$$

since this is the total current flowing away from S_2. A shunt conductance G may be defined through the relation

$$I_s = VG \tag{10.81}$$

From (10.80) and the relation $C = Q/V$, we now have

$$G = \frac{I_s}{V} = \frac{\omega\epsilon''}{\epsilon'}\frac{Q}{V} = \frac{\omega\epsilon''}{\epsilon'} C \tag{10.82}$$

Thus the shunt conductance, since it is directly related to the capacitance C, depends on the geometry in precisely the same way as C, a result already established in Chap. 5. An alternative definition of G is in terms of the power loss in the dielectric per unit length of line. This definition is $P_{ld} = \frac{1}{2}V^2G = \frac{1}{2}\int_{xy \text{ plane}} \omega\epsilon''|\mathbf{e}|^2\, dS$ and is readily shown to be equivalent to that in (10.82) by a technique similar to that used in Sec. 5.8.

From the circuit equation for Γ, we have

$$\Gamma = [j\omega L(j\omega C + G)]^{1/2} \tag{10.83a}$$

while from the field equations

$$k = \omega(\mu_0\epsilon)^{1/2} = \omega[\mu_0(\epsilon' - j\epsilon'')]^{1/2} \tag{10.83b}$$

Since $LC = \mu_0\epsilon'$ and $G/\omega C = \epsilon''/\epsilon'$, (10.83a) becomes

$$\Gamma = j\omega(LC)^{1/2}\left(1 - \frac{jG}{\omega C}\right)^{1/2} = j\omega(\mu_0\epsilon')^{1/2}\left(1 - \frac{j\epsilon''}{\epsilon'}\right)^{1/2} = jk \tag{10.84}$$

Therefore again we find that the circuit analysis and field analysis are equivalent.

The General Lossy Line

When the conductors have a finite conductivity σ, a TEM wave solution is no longer possible. With a z-directed current along the conductors, there must be a z component of electric field $\mathbf{e}_z = Z_m\mathbf{J}$. The TEM wave is perturbed into a wave having at least a z component of electric field. However, for any good conductor, Z_m is so small that the solution is still essentially a TEM wave. We may find the power loss in the conductors by using the approximate technique outlined in Sec. 10.3.

The surface current density in the case where $\sigma = \infty$ is given by

$$\mathbf{J}_s = \mathbf{n} \times \mathbf{h}$$

The power loss in the conductors per meter for finite conductivity is thus approximated by

$$P_{lc} = \tfrac{1}{2} R_m \oint_{S_1+S_2} |\mathbf{n} \times \mathbf{h}|^2 \, dl$$

and this may be written as

$$P_{lc} = \tfrac{1}{2} R_m \oint_{S_1+S_2} |\mathbf{h}|^2 \, dl \tag{10.85}$$

since at the conductor surface $\mathbf{h}$ has a tangential component only. The series resistance R per meter may be defined by the relation

$$\tfrac{1}{2} R I^2 = P_{lc}$$

and hence

$$R = \frac{R_m}{I^2} \oint_{S_1+S_2} |\mathbf{h}|^2 \, dl = R_m \frac{\oint_{S_1+S_2} |\mathbf{h}|^2 \, dl}{\left(\oint_{S_2} |\mathbf{h}| \, dl\right)^2} \tag{10.86}$$

To compute the attenuation constant α, we note first that the power flow along the line will be of the form

$$P = P_0 e^{-2\alpha z}$$

where α is the attenuation constant for the electric and magnetic field waves and P_0 is the power flow at $z = 0$. The rate of decrease of P with z must equal the power loss per unit length arising from the dielectric and the conductors. Expressed mathematically,

$$-\frac{dP}{dz} = 2\alpha P = P_{lc} + P_{ld} \tag{10.87}$$

Since we define $\tfrac{1}{2} I^2 R = P_{lc}$ and $\tfrac{1}{2} V^2 G = P_{ld}$, then (10.87) can be expressed in terms of these line constants as well. This gives us

$$2\alpha P = \tfrac{1}{2} V^2 G + \tfrac{1}{2} I^2 R$$

For a low-loss line $P = \tfrac{1}{2} VI = \tfrac{1}{2} V^2 Y_c = \tfrac{1}{2} I^2 Z_c$, where

$$Z_c = Y_c^{-1} = \left(\frac{L}{C}\right)^{1/2}$$

Therefore (10.87) gives

$$\underline{\alpha = \tfrac{1}{2} \left[R \left(\frac{C}{L}\right)^{1/2} + G \left(\frac{L}{C}\right)^{1/2} \right]} \tag{10.88}$$

which is the same result as given by (10.69) for the voltage and current wave of a low-loss line, provided R and G are defined from the above energy relations.

As long as the losses are small, we can again justify the circuit approach to transmission lines by an analysis based on Maxwell's equations and the approximate method of evaluating power loss in a conductor. In the preceding analysis we neglected the small increase in the inductance of the line when σ is finite. This increase arises from a penetration of the magnetic field into the conductor. Since the effective depth of penetration is the skin depth, the internal inductance is very small compared with the external inductance and may usually be neglected.

Terminated Transmission Line

To complete the picture of the transmission line we shall consider a lossless line terminated in a load impedance Z_L at $z = 0$, as in Fig. 10.12.

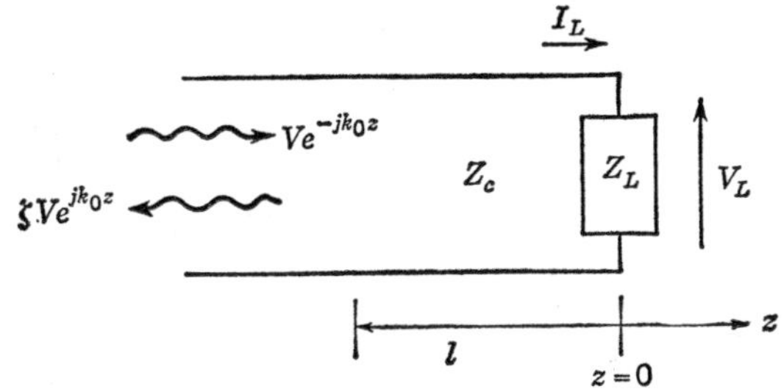

FIG. 10.12. A transmission line terminated in a load impedance.

Let a voltage wave Ve^{-jk_0z} be incident from the left. In general, a reflected wave ζVe^{jk_0z} will be produced by the load, where ζ is the reflection coefficient. The current waves associated with the incident and reflected voltage waves are $Y_cVe^{-jk_0z}$ and $-\zeta Y_cVe^{jk_0z}$, where Y_c is the characteristic admittance of the line. At $z = 0$ the total voltage across the load impedance Z_L is V_L, and the current flowing through the load is I_L, where

$$V_L = (1 + \zeta)V$$
$$I_L = Y_c(1 - \zeta)V$$

Since $V_L/I_L = Z_L$, we obtain

$$\frac{1+\zeta}{1-\zeta} = \frac{Z_L}{Z_c} \tag{10.89}$$

Solving for ζ gives

$$\zeta = \frac{Z_L - Z_c}{Z_L + Z_c} \tag{10.90}$$

In general, ζ is complex since Z_L may be complex. Only when $Z_L = Z_c$ will the reflection coefficient be zero. The maximum voltage amplitude on the line will be $(1 + |\zeta|)V$, while the minimum voltage amplitude will be $(1 - |\zeta|)V$. The ratio of the two is called the voltage standing-wave ratio, VSWR; that is,

$$\text{VSWR} = \frac{1 + |\zeta|}{1 - |\zeta|} \tag{10.91}$$

If we measure distance away from the load toward the generator by l, the total voltage and current at $z = -l$ may be expressed as

$$V(l) = Ve^{jk_0 l} + \zeta V e^{-jk_0 l} \tag{10.92a}$$

$$I(l) = Y_c V(e^{jk_0 l} - \zeta e^{-jk_0 l}) \tag{10.92b}$$

The ratio of $V(l)$ to $I(l)$ defines the input impedance, looking toward the load, at $z = -l$. Thus from (10.92),

$$Z_{\text{in}}(l) = Z_c \frac{e^{jk_0 l} + \zeta e^{-jk_0 l}}{e^{jk_0 l} - \zeta e^{-jk_0 l}}$$

Replacing $e^{jk_0 l}$ by $\cos k_0 l + j \sin k_0 l$ and using (10.90), this result is readily converted to the form

$$Z_{\text{in}} = Z_c \frac{Z_L + jZ_c \tan k_0 l}{Z_c + jZ_L \tan k_0 l} \tag{10.93}$$

This is the standard expression for the transformation of impedance along a transmission line. Two special cases of interest are the half-wavelength line and quarter-wavelength line. From (10.93) it is seen that

$$Z_{\text{in}}\left(\frac{\lambda_0}{2}\right) = Z_L \tag{10.94a}$$

$$Z_{\text{in}}\left(\frac{\lambda_0}{4}\right) = \frac{Z_c^2}{Z_L} \tag{10.94b}$$

A quarter-wavelength line inverts the impedance relative to the square of the characteristic impedance of the line.

10.6. Rectangular Waveguide

At the shorter wavelengths ($\lambda_0 < 10$ centimeters) the losses in a transmission line become quite large. At these wavelengths it is convenient to use instead a hollow uniform conducting tube for propagation of electromagnetic energy. These hollow conducting tubes constitute what are referred to as waveguides. The two most commonly used types have either a rectangular or a circular cross section. We shall discuss both types and shall note that the electrical properties of both are very similar. As a matter of fact, the general properties of all hollow cylindrical waveguides are the same, so that an understanding of the characteristics of, say, a rectangular guide automatically provides an understanding of other waveguide types as well.

In any hollow cylindrical waveguide a TEM wave solution is not possible. The reason involves the fact that the TEM electric field must be derivable as the transverse gradient of a scalar potential Φ that satisfies Laplace's equation. However, since Φ must also be constant on the trace of the conductors in the cross-sectional plane, the theory of harmonic functions demands that Φ be a constant when the conductor boundary is simply connected, as would arise with hollow cylindrical waveguides. In this event, a null **E** and **H** field results. The two basic types of waves that may propagate in a waveguide are the TE and TM waves (or modes).

It turns out that for a given wave or mode solution there exists a lower frequency, called the cutoff frequency, below which the mode will not propagate. Above the cutoff frequency the mode propagates and both the phase velocity v_p and the guide wavelength are greater than the corresponding quantities for plane waves in free space. If λ_c is the cutoff wavelength, corresponding to the cutoff frequency f_c, the guide wavelength λ_g is found to be given by

$$\lambda_g = \frac{\lambda_0}{[1 - (\lambda_0/\lambda_c)^2]^{1/2}} \tag{10.95}$$

and the phase velocity v_p is given by

$$v_p = \frac{\lambda_g}{\lambda_0} c \tag{10.96}$$

where λ_0 is the free-space wavelength and c is the velocity of light in free space. The velocity of energy propagation and the velocity with which a signal propagates (group velocity) are equal and given by

$$v_g = \frac{\lambda_0}{\lambda_g} c$$

or

$$v_g v_p = c^2 \tag{10.97}$$

Thus the group velocity is always less than c, as it must be since, according to the theory of relativity, energy or a signal cannot be propagated with a velocity exceeding c.

Equations (10.95) to (10.97) hold for any empty cylindrical waveguide. The only difference from one guide to the next is the specific value of the cutoff wavelength λ_c, which depends on the geometry of the guide cross section. We proceed now to a detailed derivation of the above results for both the rectangular and circular guides. For a more general treatment the reader is referred to one of the specialized texts in this field.

TE Waves

Figure 10.13 illustrates a rectangular waveguide of cross-sectional dimensions a and b. The conductivity of the walls will be assumed to be infinite, and the interior of the guide, free space. For the TE waves the equations derived in Sec. 10.1 are applicable. The fields are given by

$$\mathbf{H} = [\mathbf{h}(x,y) + \mathbf{h}_z(x,y)]e^{-j\beta z} \tag{10.98a}$$

$$\mathbf{E} = \mathbf{e}(x,y)e^{-j\beta z} \tag{10.98b}$$

$$\mathbf{h} = -\frac{j\beta}{k_c^2}\nabla_t h_z \tag{10.98c}$$

$$\mathbf{e} = -Z_h \mathbf{a}_z \times \mathbf{h} \tag{10.98d}$$

and h_z is a solution of

$$\nabla_t^2 h_z + k_c^2 h_z = 0 \tag{10.98e}$$

where $k_c^2 = k_0^2 - \beta^2$ and the wave impedance $Z_h = k_0 Z_0/\beta$. The first step is to find a solution for $h_z(x,y)$.

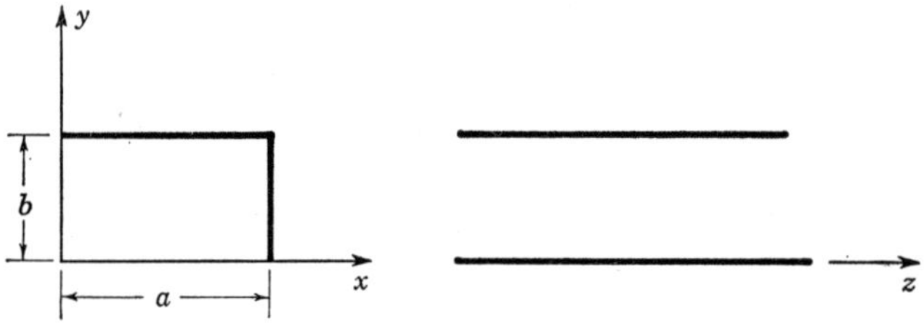

FIG. 10.13. The rectangular waveguide.

The standard method of solving a partial differential equation such as (10.98e) is the method of separation of variables. We assume that $h_z(x,y)$ can be expressed as a product of a function of x alone and a function of y alone; that is, $h_z(x,y) = f(x)g(y)$. Substitution of this type of solution into (10.98e) gives

$$\frac{\partial^2 h_z}{\partial x^2} + \frac{\partial^2 h_z}{\partial y^2} + k_c^2 h_z = g\frac{d^2 f}{dx^2} + f\frac{d^2 g}{dy^2} + k_c^2 fg = 0$$

Dividing by fg gives

$$\frac{1}{f}\frac{d^2 f}{dx^2} + \frac{1}{g}\frac{d^2 g}{dy^2} + k_c^2 = 0 \tag{10.99}$$

If we vary x, only the term $\frac{1}{f}\frac{d^2 f}{dx^2}$ in (10.99) can vary since the other terms are not functions of x. However, since (10.99) must hold for all values of x and y, it is necessary that the term involving f and also the term involving g be constant. Hence we can write

$$\frac{1}{f}\frac{d^2 f}{dx^2} = -k_x^2$$

or

$$\frac{d^2 f}{dx^2} + k_x^2 f = 0 \tag{10.100}$$

and

$$\frac{d^2 g}{dy^2} + k_y^2 g = 0 \tag{10.101}$$

where k_x^2 and k_y^2 are called separation constants. In order for (10.99) to hold, we must have

$$-k_x^2 - k_y^2 + k_c^2 = 0$$

or

$$k_c = (k_x^2 + k_y^2)^{1/2} \tag{10.102}$$

The separation-of-variables technique reduces a partial differential equation into two or more separate ordinary differential equations.

The solutions to (10.100) and (10.101) are (apart from an additional arbitrary amplitude constant)

$$f(x) = \cos k_x x + A \sin k_x x$$
$$g(y) = \cos k_y y + B \sin k_y y$$

and hence

$$h_z(x,y) = (\cos k_x x + A \sin k_x x)(\cos k_y y + B \sin k_y y)$$

The constants A, B, k_x, k_y will be determined by the boundary conditions that h_z must satisfy on the guide walls. Since the normal component of $\mathbf{h}$ is zero at a perfect conducting surface and $\mathbf{h}$ is proportional to $\nabla_t h_z$, it follows that

$$\frac{\partial h_z}{\partial x} = 0 \qquad x = 0, a$$
$$\frac{\partial h_z}{\partial y} = 0 \qquad y = 0, b$$

At $x = y = 0$, these conditions are satisfied if $A = B = 0$. At $x = a$ we must have $\sin k_x a = 0$, while at $y = b$, $\sin k_y b = 0$. Therefore the possible solutions for k_x and k_y are

$$k_x = \frac{n\pi}{a} \qquad n = 0, 1, 2, \ldots \tag{10.103a}$$

$$k_y = \frac{m\pi}{b} \qquad m = 0, 1, 2, \ldots \tag{10.103b}$$

Both n and m cannot be zero or the gradient of h_z will be zero. There is a double infinity of possible solutions. Each particular solution is called a mode and designated as a TE_{nm} or H_{nm} mode for the nmth solution. Thus for the nmth solution

$$h_{z,nm} = \cos \frac{n\pi x}{a} \cos \frac{m\pi y}{b} \tag{10.104}$$

From this solution for h_z all the other field components may be found by means of (10.98).

For the nmth solution the corresponding values of k_c and β will be written as $k_{c,nm}$ and β_{nm}. Similar subscripts will be used on the field components. From (10.102) and (10.103) we have

$$k_{c,nm}^2 = \left(\frac{n\pi}{a}\right)^2 + \left(\frac{m\pi}{b}\right)^2 \tag{10.105a}$$

$$\beta_{nm} = \left[k_0^2 - \left(\frac{n\pi}{a}\right)^2 - \left(\frac{m\pi}{b}\right)^2\right]^{1/2} \tag{10.105b}$$

For propagation in the z direction, β_{nm} must be real. This is possible only if

$$\frac{2\pi}{\lambda_0} = k_0 > \left[\left(\frac{n\pi}{a}\right)^2 + \left(\frac{m\pi}{b}\right)^2\right]^{1/2}$$

or

$$\lambda_0 < \frac{2}{[(n/a)^2 + (m/b)^2]^{1/2}} = \frac{2ab}{(n^2b^2 + m^2a^2)^{1/2}} \tag{10.106}$$

The particular value of λ_0 for which the left-hand side of (10.106) equals the right-hand side, i.e.,

$$\lambda_0 = \lambda_{c,nm} = \frac{2ab}{(n^2b^2 + m^2a^2)^{1/2}} \tag{10.107}$$

is called the cutoff wavelength for the *nm*th mode. For all values of $\lambda_0 < \lambda_{c,nm}$, β_{nm} is real, while for $\lambda_0 > \lambda_{c,nm}$, the propagation constant β_{nm} is imaginary. For the latter case the corresponding fields are exponentially damped in the z direction since $e^{-j\beta_{nm}z} = e^{-|\beta_{nm}|z}$. Such solutions are known as evanescent waves.

In practice, the waveguide dimensions a and b are chosen so that for the frequency band of interest only a single mode can propagate. Usually a is chosen as approximately equal to $2b$. If $a = 2b$, (10.107) gives

$$\lambda_{c,nm} = \frac{2a}{(n^2 + 4m^2)^{1/2}} \tag{10.108}$$

The largest cutoff wavelength occurs for the TE_{10} mode, that is, $n = 1$, $m = 0$. For this mode $\lambda_{c,10} = 2a$. The next modes to propagate are the TE_{20} and TE_{01} modes with $\lambda_{c,20} = \lambda_{c,01} = a$, followed by the TE_{11} mode with $\lambda_{c,11} = 2a/\sqrt{5}$. Provided we restrict the frequency to be in the range where $a < \lambda_0 < 2a$, only the TE_{10} mode will be able to propagate. This is the dominant mode of propagation in a rectangular guide (the first TM mode to propagate has $\lambda_c = \lambda_{c,11}$).

Multimode propagation is avoided in practice because each mode that could propagate has a different phase and group velocity and a different field configuration. The first difference means that the phase relation between the portions of the signal power carried by each mode continually varies along the guide and makes it difficult to extract all the energy from the guide at the receiving end. The second difference means that a different arrangement of coupling probes or loops must be used to excite each mode in the guide as well as to couple the energy out of the guide.

The TE_{10} Mode

For the dominant TE_{10} mode,

$$h_{z,10} = A \cos \frac{\pi x}{a}$$

where A is an amplitude constant. From (10.98) the fields are found to be

$$H_z = A \cos \frac{\pi x}{a} e^{-j\beta_{10}z}$$

$$H_x = j\beta_{10} \frac{a}{\pi} A \sin \frac{\pi x}{a} e^{-j\beta_{10}z}$$

$$E_y = -\frac{k_0}{\beta_{10}} Z_0 H_x$$

$$E_x = E_z = H_y = 0$$

where $k_{c,10} = 2\pi/\lambda_{c,10} = \pi/a$, and $\beta_{10} = [(2\pi/\lambda_0)^2 - (\pi/a)^2]^{1/2}$. The guide wavelength λ_g is defined as the distance the wave must propagate in order to undergo a phase change of 2π radians. Thus $\beta_{10}\lambda_g = 2\pi$, or

$$\lambda_g = \frac{2\pi}{\beta_{10}} = \frac{\lambda_0}{[1 - (\lambda_0/2a)^2]^{1/2}}$$

For this mode E_y is analogous to the voltage and $-H_x$ is analogous to the current on a transmission line. The wave impedance $Z_h = k_0 Z_0/\beta_{10}$ is the counterpart of the characteristic impedance. A sketch of the field configuration is given in Fig. 10.14. The density of lines is a measure of the relative amplitude or strength of the field.

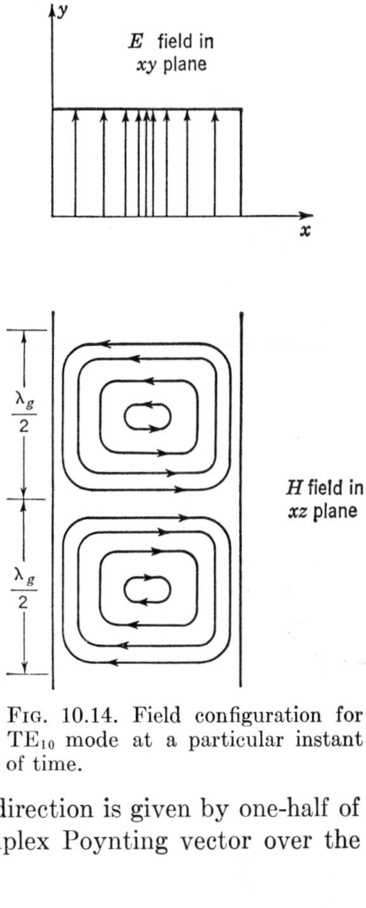

FIG. 10.14. Field configuration for TE_{10} mode at a particular instant of time.

The time-average power flow in the z direction is given by one-half of the real part of the integral of the complex Poynting vector over the guide cross section. We have

$$\begin{aligned} P &= \tfrac{1}{2} \text{Re} \int_0^b \int_0^a \mathbf{E} \times \mathbf{H}^* \cdot \mathbf{a}_z \, dx\, dy \\ &= \tfrac{1}{2} \text{Re} \int_0^b \int_0^a - E_y H_x^* \, dx\, dy \\ &= \frac{a^2}{2\pi^2} k_0 \beta_{10} Z_0 A A^* \int_0^b \int_0^a \sin^2 \frac{\pi x}{a} \, dx\, dy \\ &= \frac{a^3 b}{4\pi^2} k_0 \beta_{10} Z_0 A A^* \end{aligned} \tag{10.109}$$

The losses in the waveguide walls due to good but not perfect conductors may be found by the scheme previously used in discussing the transmission line. In this case the currents that flow in each of the four walls must be considered separately. Thus

$$\mathbf{J}_s\Big|_{x=0} = \mathbf{n} \times H_z\mathbf{a}_z\Big|_{x=0} = -Ae^{-j\beta_{10}z}\mathbf{a}_y$$

$$\mathbf{J}_s\Big|_{x=a} = -Ae^{-j\beta_{10}z}\mathbf{a}_y$$

$$\mathbf{J}_s\Big|_{y=0} = \mathbf{n} \times (H_z\mathbf{a}_z + H_x\mathbf{a}_x)\Big|_{y=0} = \left(\mathbf{a}_x \cos\frac{\pi x}{a} - \mathbf{a}_z j\beta_{10}\frac{a}{\pi}\sin\frac{\pi x}{a}\right)Ae^{-j\beta_{10}z}$$

$$\mathbf{J}_s\Big|_{y=b} = \left(-\mathbf{a}_x \cos\frac{\pi x}{a} + \mathbf{a}_z j\beta_{10}\frac{a}{\pi}\sin\frac{\pi x}{a}\right)Ae^{-j\beta_{10}z}$$

The losses per unit length are found by integrating $(R_m/2)\oint \mathbf{J}_s \cdot \mathbf{J}_s^* \, dl$ around the waveguide walls. In detail, we obtain

$$P_{lc} = R_m\left[\int_0^b AA^*\,dy + \int_0^a AA^*\left(\cos^2\frac{\pi x}{a} + \frac{\beta_{10}^2 a^2}{\pi^2}\sin^2\frac{\pi x}{a}\right)dx\right]$$
$$= AA^*R_m\left(b + \frac{a}{2} + \frac{\beta_{10}^2 a^3}{2\pi^2}\right)$$

The attenuation α may be found from the relation noted in (10.87); that is,

$$\alpha = \frac{1}{2}\frac{P_{lc}}{P} = \frac{R_m(b + \beta_{10}^2 a^3/2\pi^2 + a/2)}{(2a^3b/4\pi^2)k_0\beta_{10}Z_0}$$
$$= \frac{R_m k_0}{bZ_0\beta_{10}}\left[\frac{2b}{a}\left(\frac{\lambda_0}{2a}\right)^2 + 1\right] \tag{10.110}$$

The time-average electric energy stored in a unit length of guide is

$$W_e = \frac{\epsilon_0}{4}\int_0^b\int_0^a\int_0^1 E_yE_y^*\,dx\,dy\,dz$$
$$= \frac{\epsilon_0 a^2}{4\pi^2}k_0^2Z_0^2AA^*\int_0^b\int_0^a\int_0^1 \sin^2\frac{\pi x}{a}\,dx\,dy\,dz$$
$$= \frac{\epsilon_0 a^3 b}{8\pi^2}k_0^2Z_0^2AA^* \tag{10.111a}$$

The time-average magnetic energy stored in a unit length of guide is

$$W_m = \frac{\mu_0}{4}\int_0^b\int_0^a\int_0^1 (H_xH_x^* + H_zH_z^*)\,dx\,dy\,dz$$
$$= \frac{\mu_0}{4}b\int_0^a\left(\frac{a^2}{\pi^2}\beta_{10}^2AA^*\sin^2\frac{\pi x}{a} + AA^*\cos^2\frac{\pi x}{a}\right)dx$$
$$= \frac{\mu_0 a^3 b}{8\pi^2}AA^*\left(\beta_{10}^2 + \frac{\pi^2}{a^2}\right) = W_e \tag{10.111b}$$

since $\beta_{10}^2 + \pi^2/a^2 = k_0^2$ and $\mu_0 = Z_0^2\epsilon_0$. Since power flow is a rate of flow of energy, the velocity of energy transport v_g may be found by multiplying $W_e + W_m$ by v_g and equating the result to (10.109) for power flow. Thus

$$v_g = \frac{P}{2W_e} = \frac{\beta_{10}}{k_0\epsilon_0 Z_0} = \frac{\beta_{10}}{k_0}\frac{1}{(\mu_0\epsilon_0)^{1/2}} = \frac{\lambda_0}{\lambda_g}c \tag{10.112}$$

For the TE$_{10}$ mode we have for the z and t dependence

$$\exp\,(-j\beta_{10}z + j\omega t) = \exp j\omega\left[t - \left(\frac{\beta_{10}}{\omega}\right)z\right]$$

Thus the phase velocity v_p is given by $\omega/\beta_{10} = k_0c/\beta_{10}$, and hence $v_pv_g = c^2$, as stated earlier in (10.97).

We shall now show that the velocity with which a signal is propagated is equal to the velocity v_g. An amplitude-modulated signal

$$S = (1 + M\cos\omega_m t)\cos\omega t$$

may be expressed as

$$\begin{aligned} S &= \text{Re}\left(1 + \frac{M}{2}e^{j\omega_m t} + \frac{M}{2}e^{-j\omega_m t}\right)e^{j\omega t} \\ &= \text{Re}\left[e^{j\omega t} + \frac{M}{2}\left(e^{j(\omega+\omega_m)t} + e^{j(\omega-\omega_m)t}\right)\right] \end{aligned} \tag{10.113}$$

The term $M\cos\omega_m t$ is the signal being transmitted.

The modulation frequency ω_m is assumed to be very small compared with the carrier frequency ω. From (10.113) we see that we must consider the propagation of three components with frequencies ω, $\omega + \omega_m$, and $\omega - \omega_m$ along the guide. Since β_{10} is a function of frequency, each component will propagate with a different phase velocity. We may expand β_{10} in a Taylor series about the point ω to get

$$\beta_{10}(\omega + \Delta\omega) = \beta_{10}(\omega) + \beta_{10}'\,\Delta\omega + \cdots$$

where $\beta_{10}' = d\beta_{10}/d\omega$ at ω. Provided ω_m is small enough, we have

$$\beta_{10}(\omega \pm \omega_m) = \beta_{10}(\omega) \pm \omega_m\beta_{10}'$$

Thus as the signal S propagates along the guide, its z and t dependence will be

$$\begin{aligned} S &= \text{Re}\left[e^{j\omega t - j\beta_{10}(\omega)z} + \frac{M}{2}\left(e^{j(\omega+\omega_m)t}e^{-j(\beta_{10}+\omega_m\beta_{10}')z} + e^{j(\omega-\omega_m)t}e^{-j(\beta_{10}-\omega_m\beta_{10}')z}\right)\right] \\ &= \text{Re}\, e^{j\omega t - j\beta_{10}(\omega)z}\left[1 + \frac{M}{2}\left(e^{j\omega_m(t-\beta_{10}'z)} + e^{-j\omega_m(t-\beta_{10}'z)}\right)\right] \\ &= [1 + M\cos\omega_m(t - \beta_{10}'z)][\cos(\omega t - \beta_{10}z)] \end{aligned} \tag{10.114}$$

Thus the signal appears at z in an undistorted form but delayed in time by an amount $\tau = \beta'_{10}z$. The distance z traveled divided by the time delay τ defines the group velocity or signal velocity and is

$$v_g = \frac{1}{\beta'_{10}} = \frac{d\omega}{d\beta_{10}} = c\,\frac{dk_0}{d\beta_{10}}$$

since $ck_0 = \omega$. Now $\beta_{10}^2 + (\pi/a)^2 = k_0^2$; so $2k_0\,dk_0 = 2\beta_{10}\,d\beta_{10}$, and hence

$$v_g = \frac{\beta_{10}}{k_0}c \tag{10.115}$$

which is the same velocity as derived for the energy transport.

The above analysis is based on the assumption that ω_m is small enough so that only the first two terms in the Taylor series expansion are required. If more terms are required, it is then found that signal distortion takes place because of the phase dispersion between the various frequency components.

TE_{10} Mode as a Superposition of Plane Waves

A physical understanding of why the guide wavelength and phase velocity are greater than the corresponding quantities for plane waves may be obtained by decomposing the TE_{10} solution into two obliquely propagating plane waves. For the electric field E_y we may write

$$\begin{aligned} E_y &= -2jA_1 \sin \frac{\pi x}{a}\, e^{-j\beta_{10}z} \\ &= A_1(e^{-j\pi x/a} - e^{j\pi x/a})e^{-j\beta_{10}z} \end{aligned}$$

where $2jA_1 = jk_0Z_0Aa/\pi$. If we now write $\pi/a = k_0 \sin \theta_i$, $\beta_{10} = k_0 \cos \theta_i$, the relation $\beta_{10}^2 + (\pi/a)^2 = k_0^2$ is satisfied. The solution for E_y becomes

$$E_y = A_1(e^{-jk_0(x \sin \theta_i + z \cos \theta_i)} - e^{-jk_0(-x \sin \theta_i + z \cos \theta_i)})$$

which represents two plane waves propagating at angles θ_i and $-\theta_i$ relative to the z axis. One of these component waves is illustrated in Fig. 10.15. From this figure it is clear that when the plane wave has progressed a distance c in 1 second, the intersection of the phase front with the z axis has progressed a distance $c/\cos \theta_i = k_0c/\beta_{10} = v_p$. Similarly, it is clear that the spacing between adjacent wave crests in the z direction is greater by a factor $\sec \theta_i$ than the spacing in a direction normal to the phase front. Hence λ_g is greater than λ_0. Energy in a plane wave propagates in a direction normal to the phase front with a velocity c. The projection or component of this velocity along the z direction is $c \cos \theta_i = v_g$. It is because of the zigzag path the TEM waves follow as they reflect back and forth between the side walls of the guide, while progressing along the guide, that the properties of the guide as noted above arise. The cutoff condition, for example, corresponds to the case where $\theta_i \to \pi/2$. Under these conditions we obtain a picture of the wave

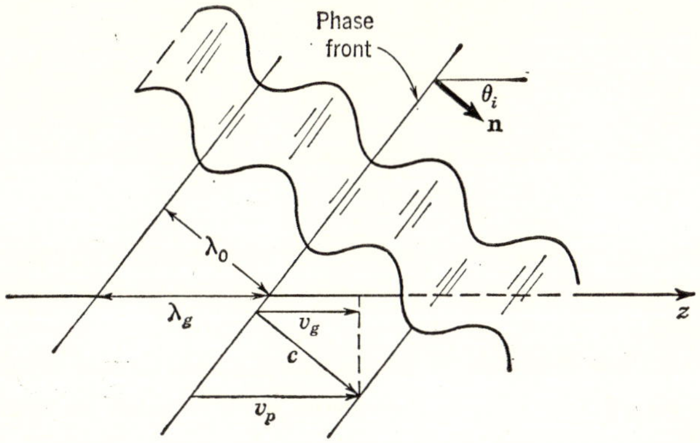

FIG. 10.15. Obliquely propagating plane wave.

propagating back and forth in the transverse plane with no component in the axial direction.

TM Waves

For TM waves all the field components may be expressed in terms of the axial electric field function $e_z(x,y)$. The solution for e_z is similar to that for h_z, with the exception that e_z must be of the form

$$e_z = \sin \frac{n\pi x}{a} \sin \frac{m\pi y}{b}$$

in order that it will vanish on the guide walls. A notable feature of this solution is that neither n nor m can be zero or e_z will vanish. As a result, the lowest-order TM mode is the TM_{11} mode. For $a = 2b$, this mode, as well as all the other TM_{nm} modes, will not propagate when $a < \lambda_0 < 2a$, hence confirming that the TE_{10} mode is dominant under those conditions. Apart from this difference the TM_{nm} modes are the duals of the TE_{nm} modes. The field components for TE_{nm} and TM_{nm} modes are listed in Table 10.1 along with other important information.

The attenuation constant α, measured in nepers per meter, for the TE and TM modes is given below. For the TE modes,

$$\alpha = \frac{2R_m}{bZ_0(1 - k_{c,nm}^2/k_0^2)^{1/2}} \left[\left(1 + \frac{b}{a}\right)\frac{k_{c,nm}^2}{k_0^2} + \frac{b}{a}\left(\frac{\epsilon_{0m}}{2} - \frac{k_{c,nm}^2}{k_0^2}\right)\frac{n^2ab + m^2a^2}{n^2b^2 + m^2a^2}\right] \qquad (10.116a)$$

TABLE 10.1. PROPERTIES OF MODES IN A RECTANGULAR WAVEGUIDE

	TE modes	TM modes
H_z	$\cos\frac{n\pi x}{a}\cos\frac{m\pi y}{b}e^{-j\beta_{nm}z}$	0
E_z	0	$\sin\frac{n\pi x}{a}\sin\frac{m\pi y}{b}e^{-j\beta_{nm}z}$
H_x	$\frac{j\beta_{nm}n\pi}{ak_{c,nm}^2}\sin\frac{n\pi x}{a}\cos\frac{m\pi y}{b}e^{-j\beta_{nm}z}$	$-\frac{E_y}{Z_{e,nm}}$
H_y	$\frac{j\beta_{nm}m\pi}{bk_{c,nm}^2}\cos\frac{n\pi x}{a}\sin\frac{m\pi y}{b}e^{-j\beta_{nm}z}$	$\frac{E_x}{Z_{e,nm}}$
E_x	$Z_{h,nm}H_y$	$-\frac{j\beta_{nm}n\pi}{ak_{c,nm}^2}\cos\frac{n\pi x}{a}\sin\frac{m\pi y}{b}e^{-j\beta_{nm}z}$
E_y	$-Z_{h,nm}H_x$	$-\frac{j\beta_{nm}m\pi}{bk_{c,nm}^2}\sin\frac{n\pi x}{a}\cos\frac{m\pi y}{b}e^{-j\beta_{nm}z}$
$Z_{h,nm}$	$Z_0\frac{k_0}{\beta_{nm}} = Z_0\left[1-\left(\frac{f_{c,nm}}{f}\right)^2\right]^{-1/2}$	
$Z_{e,nm}$	. . . .	$Z_0\frac{\beta_{nm}}{k_0} = Z_0\left[1-\left(\frac{f_{c,nm}}{f}\right)^2\right]^{1/2}$
$k_{c,nm}$	$\left[\left(\frac{n\pi}{a}\right)^2+\left(\frac{m\pi}{b}\right)^2\right]^{1/2}$	$\left[\left(\frac{n\pi}{a}\right)^2+\left(\frac{m\pi}{b}\right)^2\right]^{1/2}$
β_{nm}	$\left[k_0^2-\left(\frac{n\pi}{a}\right)^2-\left(\frac{m\pi}{b}\right)^2\right]^{1/2}$	$\left[k_0^2-\left(\frac{n\pi}{a}\right)^2-\left(\frac{m\pi}{b}\right)^2\right]^{1/2}$
$\lambda_{c,nm}$	$\frac{2ab}{[(nb)^2+(ma)^2]^{1/2}}$	$\frac{2ab}{[(nb)^2+(ma)^2]^{1/2}}$
$f_{c,nm}$	$(\lambda_{c,nm})^{-1}(\mu_0\epsilon_0)^{-1/2}$	$(\lambda_{c,nm})^{-1}(\mu_0\epsilon_0)^{-1/2}$

while for the TM modes,

$$\alpha = \frac{2R_m}{bZ_0(1-k_{c,nm}^2/k_0^2)^{1/2}}\frac{n^2b^3+m^2a^3}{n^2b^2a+m^2a^3} \qquad (10.116b)$$

where $R_m = (\omega\mu_0/2\sigma)^{1/2}$ and $\epsilon_{0m} = 1$ for $m = 0$ and $\epsilon_{0m} = 2$ for $m > 0$.

10.7. Circular Waveguides

The propagation of waves through a hollow waveguide of circular cross section can be readily considered on the basis of the general theory developed in Sec. 10.1. Following the scheme outlined in that section, we consider separately TEM, TM, and TE modes. In this case no TEM wave propagation is possible because the conducting-boundary contour in the transverse plane is simply connected. Let us then start with a consideration of TM waves.

For TM waves in cylindrical waveguides, all field components may be derived from the axial electric field e_z. This field, furthermore, satisfies the reduced Helmholtz equation

$$\nabla_t^2 e_z + k_c^2 e_z = 0$$

where

$$k_c^2 = k_0^2 - \beta^2$$

and $e^{-j\beta z}$ variation with z is assumed. In view of the circular cross-sectional geometry, as illustrated in Fig. 10.16, it is appropriate to expand

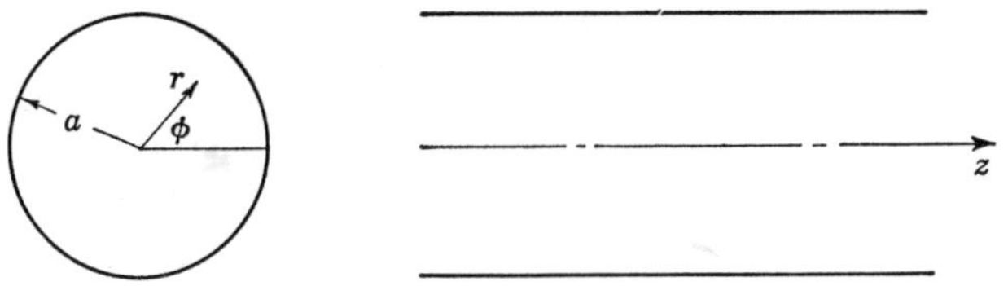

FIG. 10.16. The circular cylindrical waveguide.

the Laplacian in the above equation in circular cylindrical coordinates, so that we get

$$\frac{\partial^2 e_z}{\partial r^2} + \frac{1}{r}\frac{\partial e_z}{\partial r} + \frac{1}{r^2}\frac{\partial^2 e_z}{\partial \phi^2} = -k_c^2 e_z \tag{10.117}$$

We seek a solution of (10.117) by the method of separation of variables and therefore assume that we may express

$$e_z(r,\phi) = f(r)g(\phi) \tag{10.118}$$

Substituting (10.118) into (10.117) and then dividing by fg results in the following equation:

$$\frac{f''}{f} + \frac{1}{r}\frac{f'}{f} + \frac{1}{r^2}\frac{g''}{g} = -k_c^2 \tag{10.119}$$

where the primes represent derivatives with respect to the argument. If we multiply (10.119) by r^2 and rearrange, we can get

$$\frac{r^2 f''}{f} + \frac{rf'}{f} + k_c^2 r^2 = -\frac{g''}{g} \tag{10.120}$$

The left-hand side of (10.120) is a function only of r, while the right-hand side is a function only of ϕ. Since (10.120) must be identically correct for all r and ϕ, this could be true only if both sides were equal to the same constant ν^2. Consequently, we have been able to reduce the solution of (10.117) to the solution of the following two ordinary differential equations:

$$\frac{d^2f}{dr^2} + \frac{1}{r}\frac{df}{dr} + \left(k_c^2 - \frac{\nu^2}{r^2}\right)f = 0 \tag{10.121a}$$

$$\frac{d^2g}{d\phi^2} + \nu^2 g = 0 \tag{10.121b}$$

In the problem at hand the field must be periodic in ϕ with periodicity 2π. Consequently, it is necessary to choose $\nu = n$, an integer, and the solution to (10.121*b*) will be of the form

$$g(\phi) = A \cos n\phi + B \sin n\phi \tag{10.122a}$$

The differential equation given in (10.121*a*) may be recognized as Bessel's equation, the solution for which may be written

$$f(r) = CJ_n(k_c r) + DY_n(k_c r) \tag{10.122b}$$

The Bessel function of the second kind, $Y_n(r)$, has a singularity at the origin. Since such a singularity is inconsistent with the physical fields that are expected in the waveguide, we must choose $D = 0$. We finally get for e_z:

$$e_z = J_n(k_c r)(A \cos n\phi + B \sin n\phi) \tag{10.123}$$

Let us now consider the requirement that $e_z = 0$ when $r = a$; that is, the tangential electric field along the conducting boundary must vanish. A nontrivial solution is obtained only if certain values of k_c, the eigenvalues, are chosen. These values must be such that $J_n(k_c a) = 0$; that is, $k_c a$ must be a root of the nth-order Bessel function. An infinite number of roots exist, and we shall designate the mth root of the nth-order Bessel function as p_{nm}. This means that

$$J_n(p_{nm}) = 0$$

Values of p_{nm} for the first few modes are given in Table 10.2. We may

TABLE 10.2. VALUES OF p_{nm} FOR TM MODES

n	p_{n1}	p_{n2}	p_{n3}
0	2.405	5.520	8.654
1	3.832	7.016	10.174
2	5.135	8.417	11.620

now specify by the double-subscript notation $k_{c,nm}$ the doubly infinite set of eigenvalues for the TM modes of the circular cylindrical waveguide. Thus

$$k_{c,nm} = \frac{p_{nm}}{a} \tag{10.124}$$

Each choice of m and n specifies an eigenfunction solution or mode of the problem, which we designate TM_{nm}. The quantity n is related to the number of circumferential field variations, while m describes the number of radial variations of the field. The complete problem requires matching prescribed boundary conditions at some plane $z = z_1$ and $z = z_2$. In order to do this, a summation of different modes (including the TE type yet to be discussed) will be necessary, in general. The boundary conditions serve to specify the constants A and B for each mode.

The propagation constant for the nmth mode is given by

$$\beta_{mn} = \left[k_0^2 - \left(\frac{p_{nm}}{a}\right)^2\right]^{1/2} \tag{10.125}$$

and we note the same cutoff property that was found for the rectangular guide. Thus the eigenvalue $k_{c,nm}$ given by (10.124) is also the cutoff wave number for that mode. The cutoff wavelength is simply

$$\lambda_{c,nm} = \frac{2\pi a}{p_{nm}} \tag{10.126}$$

The remaining field components for the TM_{nm} wave are given in Table 10.3. They were derived from (10.123), with $k_{c,nm} = p_{nm}/a$, using the formulas developed in Sec. 10.1. The expression for the wave impedance is written in the form

$$Z_e = Z\left[1 - \left(\frac{f_c}{f}\right)^2\right]^{1/2}$$

which is similar to that given for the rectangular guide. Actually, we could show that this form is correct for any cylindrical waveguide; variations in cross-sectional shape affect the value of Z_e through the value of f_c. The lowest value of p_{nm} is the first root of the zero-order Bessel function for which $p_{01} = 2.405$. The cutoff wavelength, from (10.126), is then $2.61a$.

To explore the TE modes that exist in the circular cylindrical guide we must consider h_z as our potential function. It satisfies an analogous equation to (10.117); so we may write, immediately,

$$h_z = J_n(k_c r)(A' \cos n\phi + B' \sin n\phi) \tag{10.127}$$

The boundary conditions in this case require that $\partial h_z/\partial r = 0$ when $r = a$; i.e., the normal component of magnetic field at a conducting boundary vanishes. Let us designate by p'_{nm} the mth root of the following equation:

$$J'_n(x) \equiv \frac{dJ_n(x)}{dx} = 0$$

TABLE 10.3. PROPERTIES OF MODES IN A CIRCULAR CYLINDRICAL WAVEGUIDE

	TE modes	TM modes
H_z	$J_n\left(\frac{p'_{nm}r}{a}\right)e^{-j\beta_{nm}z}\begin{cases}\cos n\phi\\ \sin n\phi\end{cases}$	0
E_z	0	$J_n\left(\frac{p_{nm}r}{a}\right)e^{-j\beta_{nm}z}\begin{cases}\cos n\phi\\ \sin n\phi\end{cases}$
H_r	$-\frac{j\beta_{nm}p'_{nm}}{ak^2_{c,nm}}J'_n\left(\frac{p'_{nm}r}{a}\right)e^{-j\beta_{nm}z}\begin{cases}\cos n\phi\\ \sin n\phi\end{cases}$	$-\frac{E_\phi}{Z_{e,nm}}$
H_ϕ	$-\frac{jn\beta_{nm}}{rk^2_{c,nm}}J_n\left(\frac{p'_{nm}r}{a}\right)e^{-j\beta_{nm}z}\begin{cases}-\sin n\phi\\ \cos n\phi\end{cases}$	$\frac{E_r}{Z_{e,nm}}$
E_r	$Z_{h,nm}H_\phi$	$-\frac{j\beta_{nm}p_{nm}}{ak^2_{c,nm}}J'_n\left(\frac{p_{nm}r}{a}\right)e^{-j\beta_{nm}z}\begin{cases}\cos n\phi\\ \sin n\phi\end{cases}$
E_ϕ	$-Z_{h,nm}H_r$	$-\frac{jn\beta_{nm}}{rk^2_{c,nm}}J_n\left(\frac{p_{nm}r}{a}\right)e^{-j\beta_{nm}z}\begin{cases}-\sin n\phi\\ \cos n\phi\end{cases}$
β_{nm}	$\left[k_0^2-\left(\frac{p'_{nm}}{a}\right)^2\right]^{1/2}$	$\left[k_0^2-\left(\frac{p_{nm}}{a}\right)^2\right]^{1/2}$
$Z_{h,nm}$	$Z_0\left[1-\left(\frac{f_{c,nm}}{f}\right)^2\right]^{-1/2}=\frac{k_0}{\beta_{nm}}Z_0$	
$Z_{e,nm}$	. . . .	$Z_0\left[1-\left(\frac{f_{c,nm}}{f}\right)^2\right]^{1/2}=\frac{\beta_{nm}}{k_0}Z_0$
$k_{c,nm}$	$\frac{p'_{nm}}{a}$	$\frac{p_{nm}}{a}$
$\lambda_{c,nm}$	$\frac{2\pi a}{p'_{nm}}$	$\frac{2\pi a}{p_{nm}}$
$f_{c,nm}$	$(\lambda_{c,nm})^{-1}(\mu_0\epsilon_0)^{-1/2}$	$(\lambda_{c,nm})^{-1}(\mu_0\epsilon_0)^{-1/2}$

Then the eigenvalues for the problem are

$$k_{c,nm} = \frac{p'_{nm}}{a} \tag{10.128}$$

Values for p'_{nm} for the first few modes are given in Table 10.4. In other ways the results for the TE case follow analogously to the TM case already discussed. Table 10.3 summarizes the fields that exist for the TE_{nm} modes. The latter are obtained from (10.127), with $k_{c,nm} = p'_{nm}/a$, using the general equations developed in Sec. 10.1.

TABLE 10.4. VALUES OF p'_{nm} FOR TE MODES

n	p'_{n1}	p'_{n2}	p'_{n3}
0	3.832	7.016	10.174
1	1.841	5.331	8.536
2	3.054	6.706	9.970

The lowest value of p'_{nm} is p'_{11}, which equals 1.84 and for which the cutoff wavelength is $3.41a$. Consequently, the TE_{11} mode is the dominant mode in the circular cylindrical waveguide. A sketch of the field lines for this mode is given in Fig. 10.17. For $2.61a < \lambda_0 < 3.41a$, only the TE_{11} mode can propagate in a circular cylindrical waveguide.

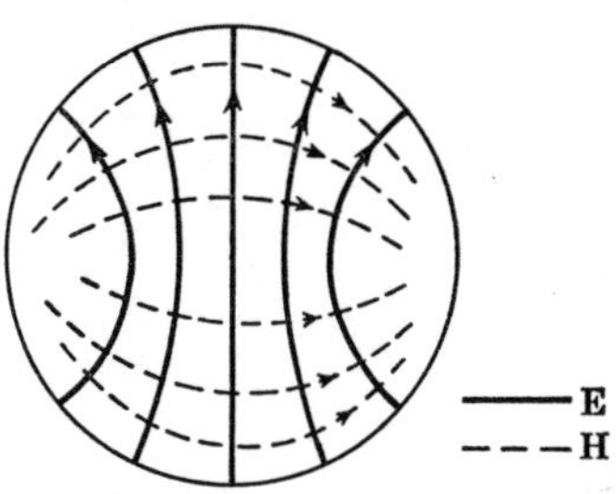

FIG. 10.17. TE_{11} field distribution in transverse plane.

The attenuation produced by imperfectly conducting walls may be calculated by means of the same technique that was used in the case of the transmission line and the rectangular waveguide. The results are, for TM modes,

$$\alpha = \frac{R_m}{aZ_0}\left[1 - \left(\frac{f_{c,nm}}{f}\right)^2\right]^{-1/2}$$

while for TE modes

$$\alpha = \frac{R_m}{aZ_0}\left[1 - \left(\frac{f_{c,nm}}{f}\right)^2\right]^{-1/2}\left[\left(\frac{f_{c,nm}}{f}\right)^2 + \frac{n^2}{p'^2_{nm} - n^2}\right]$$

where α is measured in nepers per meter, and $R_m = (\omega\mu_0/2\sigma)^{1/2} = 1/\sigma\delta$ and is the surface resistivity of the metallic walls. The attenuation constant for the TM waves as a function of increasing frequency decreases to a minimum at $f = \sqrt{3}\, f_{c,nm}$ and then increases indefinitely. The same general behavior occurs for the TE modes with the exception of the TE_{0m} waves. The latter are of very great interest because the corresponding attenuation constant decreases indefinitely with frequency and hence gives rise to the possibility of long-distance communication links.

10.8. Electromagnetic Cavities

At high frequencies the electromagnetic-cavity resonator replaces the lumped-parameter LC resonant circuit. Virtually any metallic enclosure, when properly excited, will function as an electromagnetic resonator; that is, for certain specific frequencies, electromagnetic field oscillations can be sustained within the enclosure with a very small expenditure of power. The only power that needs to be supplied is just that needed to compensate for the power loss in the cavity walls. Electromagnetic cavities are used as the resonant circuit in high-frequency tubes such as the klystron, for bandpass filters, and for wave meters to measure frequency, as well as for a number of other applications.

In this section we shall examine the basic properties of a rectangular cavity of the type illustrated in Fig. 10.18. Again, as was the case for

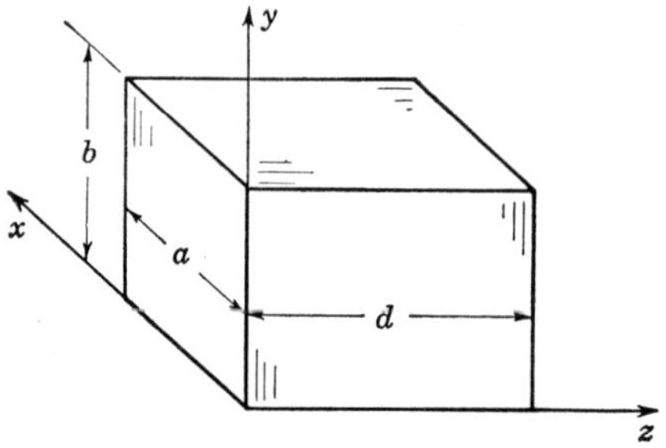

FIG. 10.18. A rectangular cavity resonator.

the rectangular guide, an understanding of the behavior of the rectangular cavity provides an understanding of other shapes of cavities also. The only essential difference between cavities of different shapes is the detailed structure of the interior fields, since this depends on the geometry or shape of the cavity.

The field solutions in a rectangular cavity are readily constructed from the corresponding solutions for the waveguide modes. For example, we may consider the cavity in Fig. 10.18 as being a part of a rectangular waveguide. Let a TE_{10} mode propagate in the positive z direction, and let a short-circuiting conducting plate be placed at $z = d$. Complete reflection will take place, and the electric field will be of the form

$$
\begin{aligned}
E_y &= A_1(e^{-j\beta_{10}z} - e^{-j\beta_{10}(2d-z)}) \sin \frac{\pi x}{a} \\
&= A_1 e^{-j\beta_{10}d} \sin \frac{\pi x}{a} (e^{-j\beta_{10}(z-d)} - e^{j\beta_{10}(z-d)}) \\
&= -2jA_1 e^{-j\beta_{10}d} \sin \frac{\pi x}{a} \sin \beta_{10}(z-d) \\
&= A \sin \frac{\pi x}{a} \sin \beta_{10}(z-d)
\end{aligned}
$$

where $A = -2jA_1e^{-j\beta_{10}d}$. At any plane where $\sin \beta_{10}(z - d)$ vanishes, we can place another conducting plate and thus obtain a rectangular enclosure. On the other hand, if the dimension d is given and the ends of the cavity are at $z = 0$ and $z = d$, we must have

$$\sin \beta_{10} d = 0$$

or

$$\beta_{10} = \frac{s\pi}{d} \qquad s = 1, 2, 3, \ldots$$

With β_{10}, k_x, and k_y all fixed by the cavity dimensions, it follows that only certain discrete values of k_0 will yield a possible solution. Since $\beta^2 = k_0^2 - k_x^2 - k_y^2$, we have

$$k_0 = \frac{2\pi f}{c} = \left[\left(\frac{s\pi}{d}\right)^2 + \left(\frac{n\pi}{a}\right)^2 + \left(\frac{m\pi}{b}\right)^2\right]^{1/2} \tag{10.129}$$

in general, or for the TE_{10} mode with a single sinusoidal variation along z,

$$f = c\left(\frac{1}{4d^2} + \frac{1}{4a^2}\right)^{1/2} \tag{10.130}$$

The mode of oscillation, whose frequency of oscillation is given by (10.130), is designated as the TE_{101} mode, since there is only a single standing-wave-pattern loop in the x and z directions and none in the y direction. For the higher-order TE_{n0s} modes there will be n loops along the x direction and s loops along the z direction. The corresponding resonant frequencies are given by (10.129) with $m = 0$.

In addition to the TE_{n0s} modes there are the TE_{nms} modes and their duals, the TM_{nms} modes. The TE and TM modes may be derived from the z component of the magnetic and electric field, respectively, by equations similar to those given in Sec. 10.1 (replace $-j\beta$ by $\partial/\partial z$). The solutions for the z components of the fields are readily found to be

$$H_z = \cos \frac{n\pi x}{a} \cos \frac{m\pi y}{b} \sin \frac{s\pi z}{d} \qquad \text{TE modes}$$

$$E_z = \sin \frac{n\pi x}{a} \sin \frac{m\pi y}{b} \cos \frac{s\pi z}{d} \qquad \text{TM modes}$$

and from these the remaining components may be found. Although there are an infinite number of discrete modes of oscillations in a rectangular cavity, we shall study only the TE_{101} mode in detail.

Since the TE_{101} mode is a TE_{10} waveguide standing-wave field, the only field components present are E_y, H_x, and H_z. For the electric field we have

$$E_y = A \sin \frac{\pi x}{a} \sin \frac{\pi z}{d} \tag{10.131a}$$

The magnetic field is readily determined from the curl equation

$$\nabla \times \mathbf{E} = -j\omega\mu_0\mathbf{H}$$

and is

$$H_x = \frac{-j}{\omega\mu_0}\frac{\partial E_y}{\partial z} = -\frac{j\pi A}{\omega\mu_0 d}\sin\frac{\pi x}{a}\cos\frac{\pi z}{d} \tag{10.131b}$$

$$H_z = \frac{j}{\omega\mu_0}\frac{\partial E_y}{\partial x} = \frac{j\pi A}{\omega\mu_0 a}\cos\frac{\pi x}{a}\sin\frac{\pi z}{d} \tag{10.131c}$$

The resonant frequency of the cavity for this mode of oscillation is, from (10.130),

$$\frac{\omega}{2\pi} = f = \frac{c}{2}(a^{-2} + d^{-2})^{1/2} \tag{10.132}$$

The electric and magnetic fields in a cavity are in phase quadrature, as is readily seen from (10.131), since a factor j multiplies the expressions for H_x and H_z.

The total time-average electric energy stored within the cavity is

$$\begin{aligned} W_e &= \frac{\epsilon_0}{4}\int_0^a\int_0^b\int_0^d AA^*\sin^2\frac{\pi x}{a}\sin^2\frac{\pi z}{d}\,dx\,dy\,dz \\ &= \frac{abd}{16}\epsilon_0 AA^* \end{aligned} \tag{10.133}$$

The total time-average magnetic energy stored in the cavity is

$$\begin{aligned} W_m &= \frac{\mu_0}{4}\int_0^a\int_0^b\int_0^d \frac{AA^*}{\omega^2\mu_0^2}\left(\frac{\pi^2}{d^2}\sin^2\frac{\pi x}{a}\cos^2\frac{\pi z}{d} + \frac{\pi^2}{a^2}\cos^2\frac{\pi x}{a}\sin^2\frac{\pi z}{d}\right)dx\,dy\,dz \\ &= \frac{AA^*}{\omega^2\mu_0}\frac{abd}{16}\left(\frac{\pi^2}{a^2} + \frac{\pi^2}{d^2}\right) = W_e \end{aligned} \tag{10.134}$$

since $k_0^2 = \omega^2\mu_0\epsilon_0 = (\pi/d)^2 + (\pi/a)^2$. Thus at the resonant frequency, $W_e = W_m$, a property similar to that for a resonant LC circuit at low frequencies.

The expressions given by (10.131) are the complex-phasor representations for the real physical field. The real-physical-field components, denoted by a prime, are obtained by multiplying by $e^{j\omega t}$ and taking the real part; thus

$$E_y' = A\sin\frac{\pi x}{a}\sin\frac{\pi z}{d}\cos\omega t \tag{10.135a}$$

$$H_x' = \frac{\pi A}{\omega\mu_0 d}\sin\frac{\pi x}{a}\cos\frac{\pi z}{d}\sin\omega t \tag{10.135b}$$

$$H_z' = -\frac{\pi A}{\omega\mu_0 a}\cos\frac{\pi x}{a}\sin\frac{\pi z}{d}\sin\omega t \tag{10.135c}$$

where ω is given by (10.132) and the amplitude constant A is assumed real.

In any practical cavity the walls have finite conductivity, and hence any mode of oscillation that has been excited, say by an impulse, must decay exponentially.† Thus the time behavior of the oscillations must be of the form $e^{-\alpha t}\cos\omega t$ and $e^{-\alpha t}\sin\omega t$, rather than the form given in (10.135), which is the steady-state solution for the ideal cavity. To determine the damping constant α, we must evaluate the power loss in the cavity walls. This may be done by assuming that the current flowing on the walls is the same as for the ideal cavity. We then have, from Sec. 10.3,

$$P_{lc} = \tfrac{1}{2}R_m \oint_{\text{walls}} \mathbf{H}_t \cdot \mathbf{H}_t^* \, dS \qquad (10.136)$$

where $\mathbf{H}_t$ is the tangential magnetic field at the cavity wall, and $R_m = 1/\sigma\delta$. The total time-average energy in the cavity is

$$W = W_e + W_m$$

and must decay with time as follows:

$$W = W_0 e^{-2\alpha t}$$

where W_0 is the time-average energy in the cavity at $t = 0$. The negative rate of change of W with time must equal the power loss in the walls, and hence

$$-\frac{dW}{dt} = 2\alpha W_0 e^{-2\alpha t} = 2\alpha W = P_{lc}$$

or

$$\alpha = \frac{P_{lc}}{2W} \qquad (10.137)$$

This relation permits the damping constant α to be determined.

The quality factor, or Q, of a resonant circuit may be defined as

$$Q = 2\pi \frac{\text{time-average energy stored}}{\text{energy loss per cycle of oscillation}} \qquad (10.138)$$

Since P_{lc} is the energy dissipated in the cavity walls per second, the energy loss in one cycle, or a time interval $\tau = 1/f$, is P_{lc}/f. From (10.137) and (10.138), we now see that

$$\alpha = \frac{P_{cl}/f}{2W/f} = \frac{f}{2}\frac{P_{lc}/f}{W} = \frac{f}{2}\frac{2\pi}{Q} = \frac{\omega}{2Q} \qquad (10.139)$$

† This type of behavior is characteristic of any low-loss oscillatory physical system where the power loss is directly proportional to the energy present at any instant of time. In this case $dW/dt = -kW$, so that $W = W_0 e^{-kt}$, where W_0 is the energy present at $t = 0$.

In order to determine the Q of the cavity for the TE_{101} mode, we must evaluate (10.136) for the power loss in the walls. On the end walls at $z = 0, d$, we have, from (10.131),

$$|H_t| = |H_x| = \frac{\pi A}{\omega\mu_0 d} \sin \frac{\pi x}{a}$$

and the power loss in these two walls is

$$\begin{aligned} P_{lc1} &= \frac{R_m}{2}\left(\frac{\pi A}{\omega\mu_0 d}\right)^2 2 \int_0^a \int_0^b \sin^2 \frac{\pi x}{a}\, dx\, dy \\ &= \frac{ab}{2} R_m \left(\frac{\pi A}{\omega\mu_0 d}\right)^2 \end{aligned}$$

where A is again assumed to be real. On the upper and lower walls at $y = 0, b$, we have

$$|H_t|^2 = |H_x|^2 + |H_z|^2 = \left(\frac{\pi A}{\omega\mu_0}\right)^2 \left(\frac{1}{d^2} \sin^2 \frac{\pi x}{a} \cos^2 \frac{\pi z}{d} + \frac{1}{a^2} \cos^2 \frac{\pi x}{a} \sin^2 \frac{\pi z}{d}\right)$$

and the power loss is

$$\begin{aligned} P_{lc2} = \frac{R_m}{2} \frac{\pi A}{\omega\mu_0} 2 \int_0^a \int_0^d \bigg(\frac{1}{d^2} \sin^2 \frac{\pi x}{a} \cos^2 \frac{\pi z}{d} \\ + \frac{1}{a^2} \cos^2 \frac{\pi x}{a} \sin^2 \frac{\pi z}{d}\bigg)\, dx\, dz = \frac{R_m}{4}\left(\frac{\pi A}{\omega\mu_0}\right)^2 \left(\frac{a}{d} + \frac{d}{a}\right) \end{aligned}$$

On the remaining two walls at $x = 0, a$,

$$|H_t| = |H_z| = \frac{\pi A}{\omega\mu_0} \sin \frac{\pi z}{d}$$

with a corresponding power loss

$$\begin{aligned} P_{lc3} &= \frac{R_m}{2}\left(\frac{\pi A}{\omega\mu_0 a}\right)^2 2 \int_0^b \int_0^d \sin^2 \frac{\pi z}{d}\, dy\, dz \\ &= \frac{bd}{2} R_m \left(\frac{\pi A}{\omega\mu_0 a}\right)^2 \end{aligned}$$

The total power loss is

$$\begin{aligned} P_{lc} = P_{lc1} + P_{lc2} + P_{lc3} &= \left(\frac{\pi A}{\omega\mu_0}\right)^2 R_m \left(\frac{ab}{2d^2} + \frac{d}{4a} + \frac{a}{4d} + \frac{bd}{2a^2}\right) \\ &= R_m \left(\frac{\pi A}{2\omega\mu_0 ad}\right)^2 (2a^3b + ad^3 + a^3d + 2bd^3) \end{aligned} \tag{10.140}$$

The total time-average energy stored in the cavity is, from (10.133) and (10.134),

$$W = 2W_e = \frac{abd}{8} \epsilon_0 A^2$$

Hence the Q of the cavity, for the TE_{101} mode, is given by

$$Q = \frac{\omega}{2\alpha} = \omega \frac{W}{P_{lc}} = \frac{\omega abd\epsilon_0}{8\pi^2 R_m} \frac{(2\omega\mu_0 ad)^2}{2a^3b + a^3d + ad^3 + 2bd^3}$$

$$= \frac{(k_0 ad)^3 b Z_0}{2\pi^2 R_m(2a^3b + 2d^3b + a^3d + d^3a)} \qquad (10.141)$$

where $k_0^2 = (\pi/a)^2 + (\pi/d)^2$.

As a typical example, consider a copper cavity ($\sigma = 5.8 \times 10^7$ mhos per meter) with dimensions $a = b = c = 3$ centimeters. The resonant frequency is found from (10.132) to be 7,070 megacycles per second. The surface resistance R_m is 0.022 ohm. The Q of the cavity is found to be 12,700, and α equals 1.74×10^6 nepers per second. A rather startling property of a cavity is its extremely high Q as compared with the Q of LC circuits at low frequencies, which is usually of the order of a few hundred only.

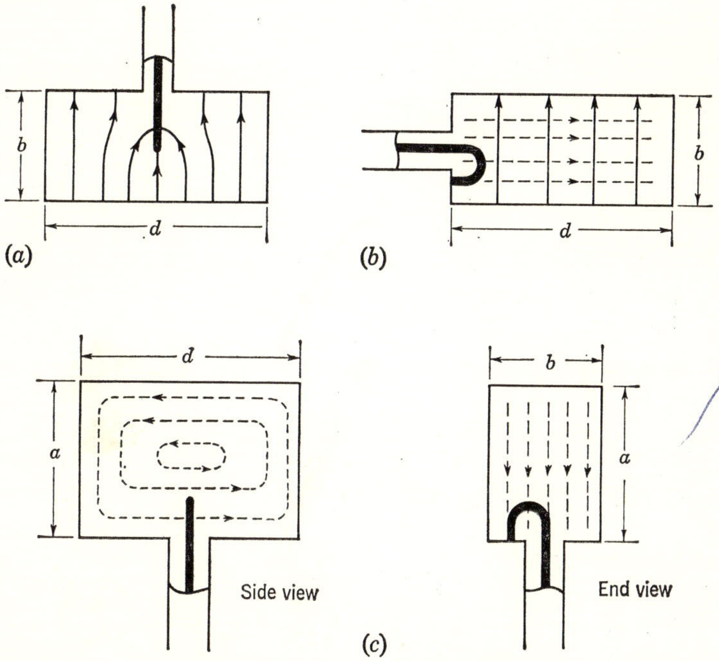

FIG. 10.19. Methods of exciting the TE_{101} mode from a coaxial line. (a) Probe coupling with E_y; (b) loop coupling with H_x; (c) loop coupling with H_z.

The oscillations in a cavity may be excited from a coaxial line by means of a small probe or loop antenna, as illustrated in Fig. 10.19. The probe couples to the electric field of the mode and is hence located in the center of the broad wall where E_y is a maximum. The loop antenna must be located at a point where the magnetic flux of the mode, through the loop, will be large. Similar probe and loop antennas may be used to excite the fields in a waveguide. The field configuration for the TE_{101} cavity mode is also illustrated in Fig. 10.19. Of course, the frequency of the incident wave in the coaxial line must be equal to the resonant frequency of the cavity if the mode is to be excited.

BIBLIOGRAPHY

Bronwell, A. B., and R. E. Beam: "Theory and Application of Microwaves," McGraw-Hill Book Company, Inc., New York, 1947.

Collin, R. E.: "Field Theory of Guided Waves," McGraw-Hill Book Company, Inc., New York, 1960.

Harrington, R. F.: "Time-harmonic Electromagnetic Fields," McGraw-Hill Book Company, Inc., New York, 1961.

Kraus, J. D.: "Electromagnetics," McGraw-Hill Book Company, Inc., New York, 1953.

Ramo, S., and J. R. Whinnery: "Fields and Waves in Modern Radio," 2d ed., John Wiley & Sons, Inc., New York, 1953.

Schelkunoff, S. A.: "Electromagnetic Waves," D. Van Nostrand Company, Inc., Princeton, N.J., 1943.

Southworth, G.: "Principles and Applications of Waveguide Transmission," D. Van Nostrand Company, Inc., Princeton, N.J., 1950.

CHAPTER 11

RADIATION AND ANTENNAS

We have noted that under time-varying conditions Maxwell's equations predict the radiation of electromagnetic energy from current sources. While such a phenomenon takes place at all frequencies, its relative magnitude is insignificant until the size of the source region is comparable to wavelength. In constructing circuits to operate at higher and higher frequencies, this means that a point is reached where radiation from the circuit will interfere with the desired circuit characteristics and the use of other techniques and devices, such as waveguides and resonators, is necessary. In this chapter, however, radiation is the desired end product. We shall, consequently, be interested in some of the characteristics of radiators, such as their efficiency and the resultant radiation patterns. We shall examine the transmitting properties of the dipole antenna and the array of dipoles and conclude with a discussion of the receiving antenna and reciprocity.

11.1. Radiation from a Linear Current Element

The simplest radiating structure is that of an infinitesimal current element. An understanding of the properties of such an antenna is of great use since, in principle at least, all radiating structures can be considered as a sum of small radiating elements. Furthermore, many practical antennas at low frequencies are very short compared with wavelength, and the results we obtain here will be sufficiently accurate to describe their behavior.

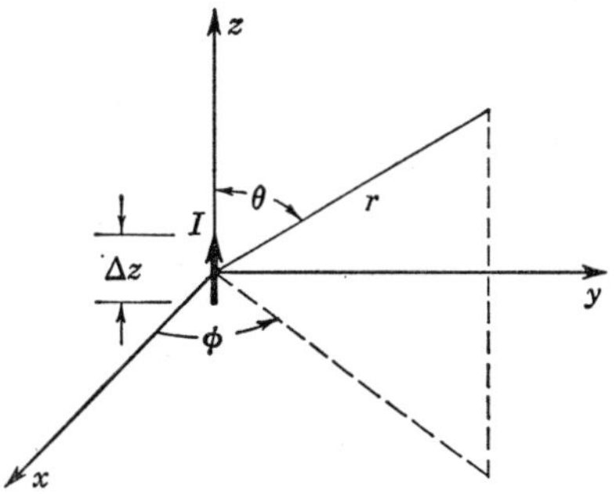

FIG. 11.1. An infinitesimal linear-current radiator.

Thus, consider a linear current element $I = I_0 e^{j\omega t}$ of length Δz, oriented in the z direction and located at the origin, as in Fig. 11.1. For convenience, we assume that I_0 is a real amplitude factor. The charge associated with this current element may be obtained by noting that the current flowing into the upper end must equal the time rate of increase of charge at the upper end. Thus $j\omega Q = I_0$ or $Q = -jI_0/\omega$ at the upper end and $-Q = jI_0/\omega$ at the lower end of the current element. The small linear current element may be viewed as two charges Q and $-Q$ oscillating back and forth.

From Sec. 9.9, the vector and scalar potentials from general volume distributions of current and charge are

$$\mathbf{A} = \frac{\mu_0}{4\pi} \int_V \frac{\mathbf{J}}{R} e^{-jk_0R}\, dV' \tag{11.1a}$$

$$\Phi = \frac{1}{4\pi\epsilon_0} \int_V \frac{\rho}{R} e^{-jk_0R}\, dV' \tag{11.1b}$$

In the present case we are dealing with a differential current element only, so that

$$\begin{aligned}\mathbf{A} &= \frac{\mu_0 I_0\, \Delta z}{4\pi r} \mathbf{a}_z e^{-jk_0 r} \\ &= \frac{\mu_0 I_0\, \Delta z}{4\pi r} (\mathbf{a}_r \cos\theta - \mathbf{a}_\theta \sin\theta) e^{-jk_0 r}\end{aligned} \tag{11.2}$$

The oscillating charge is equivalent to a small electric dipole of moment $Q\,\Delta z = -jI_0\,\Delta z/\omega$. (This accounts for the antenna being also referred to as an elementary dipole, or doublet.) The scalar potential Φ is readily seen to be given by

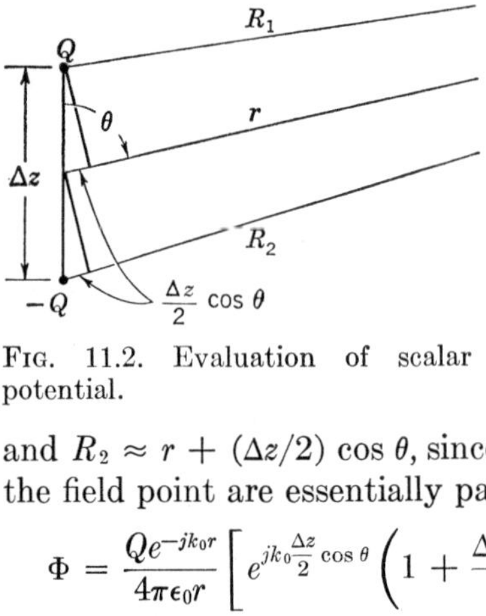

FIG. 11.2. Evaluation of scalar potential.

$$\Phi = \frac{Q}{4\pi\epsilon_0} \left(\frac{e^{-jk_0R_1}}{R_1} - \frac{e^{-jk_0R_2}}{R_2} \right)$$

where R_1 and R_2 are the distances specified in Fig. 11.2. From this figure it is seen that $R_1 \approx r - (\Delta z/2) \cos\theta$ and $R_2 \approx r + (\Delta z/2) \cos\theta$, since the paths from the ends of the dipole to the field point are essentially parallel. The expression for Φ becomes

$$\Phi = \frac{Qe^{-jk_0 r}}{4\pi\epsilon_0 r} \left[e^{jk_0\frac{\Delta z}{2}\cos\theta} \left(1 + \frac{\Delta z \cos\theta}{2r}\right) - e^{-jk_0\frac{\Delta z}{2}\cos\theta} \left(1 - \frac{\Delta z\cos\theta}{2r}\right) \right]$$

after replacing R_1^{-1} by

$$r^{-1}\left(1 - \frac{\Delta z \cos\theta}{2r}\right)^{-1} \approx r^{-1}\left(1 + \frac{\Delta z\cos\theta}{2r}\right)$$

and analogously for R_2^{-1}. Using the expansion $e^x \approx 1 + x$ for x small, we obtain

$$\Phi = \frac{Q\,\Delta z}{4\pi\epsilon_0} \left(\frac{\cos\theta}{r^2} + \frac{jk_0 \cos\theta}{r} \right) e^{-jk_0 r} \tag{11.3}$$

since $k_0\,(\Delta z/2)\cos\theta$ is small.

The electric and magnetic fields are given by (see Sec. 9.9)

$$\mathbf{H} = \mu_0^{-1}\nabla \times \mathbf{A} \tag{11.4a}$$

$$\mathbf{E} = -j\omega\mathbf{A} - \nabla\Phi \tag{11.4b}$$

$$\mathbf{E} = -j\omega\mathbf{A} + \frac{\nabla(\nabla\cdot\mathbf{A})}{j\omega\mu_0\epsilon_0} \tag{11.4c}$$

From (11.4*b*) we get, for **E**,

$$\mathbf{E} = -\frac{j\omega\mu_0 I_0\,\Delta z}{4\pi r}(\mathbf{a}_r\cos\theta - \mathbf{a}_\theta\sin\theta)e^{-jk_0r} + \frac{jI_0\,\Delta z}{4\pi\omega\epsilon_0}\nabla\left(\frac{\cos\theta}{r^2} + \frac{jk_0\cos\theta}{r}\right)e^{-jk_0r} \tag{11.5}$$

after replacing Q by $-jI_0/\omega$ in (11.3). We may readily show that the Lorentz condition $(j\omega\mu_0\epsilon_0)^{-1}\,\nabla\nabla\cdot\mathbf{A} = -\nabla\Phi$ is satisfied by **A** and Φ of (11.2) and (11.3). We have

$$\nabla\cdot\mathbf{A} = \frac{\mu_0 I_0\,\Delta z}{4\pi}\left[\frac{1}{r^2}\frac{\partial}{\partial r}\left(r^2\cos\theta\,\frac{e^{-jk_0r}}{r}\right) - \frac{1}{r\sin\theta}\frac{\partial}{\partial\theta}\left(\sin^2\theta\,\frac{e^{-jk_0r}}{r}\right)\right]$$

after expressing the divergence in spherical coordinates. Carrying out the differentiation gives

$$\nabla\cdot\mathbf{A} = -\frac{\mu_0 I_0\,\Delta z}{4\pi}\left(\frac{jk_0\cos\theta}{r} + \frac{\cos\theta}{r^2}\right)e^{-jk_0r} \tag{11.6}$$

We see from this expression that $(j\omega\mu_0\epsilon_0)^{-1}\,\nabla\nabla\cdot\mathbf{A}$ is equal to $-\nabla\Phi$, and hence we may compute the fields from the vector potential **A** alone, as in (11.4*c*). Again we point out that this possibility arises for time-varying fields since current and charge are not independent; i.e., they satisfy the continuity equation. In this specific case the choice of $j\omega Q = I_0$ (the continuity condition) is the necessary relationship for Φ and **A** to satisfy the Lorentz condition.

After carrying out the operations indicated in (11.4) we find that

$$\mathbf{H} = \frac{I_0\,\Delta z}{4\pi}\sin\theta\left(\frac{jk_0}{r} + \frac{1}{r^2}\right)e^{-jk_0r}\mathbf{a}_\phi \tag{11.7a}$$

$$\mathbf{E} = -\frac{I_0\,\Delta z}{2\pi}\frac{jZ_0}{k_0}\cos\theta\left(\frac{jk_0}{r^2} + \frac{1}{r^3}\right)e^{-jk_0r}\mathbf{a}_r - \frac{I_0\,\Delta z}{4\pi}\frac{jZ_0}{k_0}\sin\theta\left(\frac{-k_0^2}{r} + \frac{jk_0}{r^2} + \frac{1}{r^3}\right)e^{-jk_0r}\mathbf{a}_\theta \tag{11.7b}$$

The radial component of the complex Poynting vector is

$$\mathbf{E}\times\mathbf{H}^*\cdot\mathbf{a}_r = E_\theta H_\phi^* = \left(\frac{I_0\,\Delta z}{4\pi}\sin\theta\right)^2\frac{Z_0}{k_0}\left(\frac{k_0^3}{r^2} - \frac{j}{r^5}\right)$$

The integral of one-half of the real part of this expression over a sphere of radius r gives the total average power radiated into space by the current element. From the above we see that this radiated power is given by

$$P_r = \left(\frac{k_0 I_0 \,\Delta z}{4\pi}\right)^2 \frac{Z_0}{2} \int_0^{2\pi} \int_0^{\pi} \sin^3 \theta \, d\theta \, d\phi = \frac{(k_0 I_0 \,\Delta z)^2}{12\pi} Z_0 \tag{11.8}$$

The only part of the fields entering into this expression for the radiated power is the part consisting of the terms varying as r^{-1}, that is, the part

$$H_\phi = \frac{jk_0 I_0 \,\Delta z}{4\pi r} \sin \theta \, e^{-jk_0 r} \tag{11.9a}$$

$$E_\theta = \frac{jk_0 I_0 \,\Delta z \, Z_0}{4\pi r} \sin \theta \, e^{-jk_0 r} \tag{11.9b}$$

This part of the field is therefore called the far-zone, or radiation, field. For large values of r it is the only part of the total field which has a significant amplitude. Note in particular that E_r vanishes as r^{-2} for large r. The far-zone, or radiation, field is a spherical TEM wave since the constant-phase surfaces are spheres and E_θ, H_ϕ lie in a surface perpendicular to the direction of propagation (radial direction). From (11.9) it is seen that $E_\theta = Z_0 H_\phi$, which is the same relation that holds for plane TEM waves.

The part of the field varying as r^{-2} and r^{-3} is called the near-zone, or induction, field. It is similar in nature to the static fields surrounding a small linear-current element and an electric dipole. This induction field predominates in the region $r \ll \lambda_0$, where λ_0 is the wavelength. The induction field does not represent an outward flow of power, but instead gives rise to a storage of reactive energy in the vicinity of the radiating current element. This energy oscillates back and forth between the source and the region of space surrounding the source. The complex Poynting vector involving the near-zone-field components is a pure imaginary quantity.

The total power radiated by an antenna is conveniently expressed in terms of the power absorbed in an equivalent resistance called the radiation resistance. For the current element the radiation resistance R_0 is defined by the relation

$$\tfrac{1}{2} R_0 I_0^2 = P_r \tag{11.10}$$

From (11.8) we find that

$$R_0 = \frac{(k_0 \,\Delta z)^2}{6\pi} Z_0 = 80\pi^2 \left(\frac{\Delta z}{\lambda_0}\right)^2 \tag{11.11}$$

after replacing k_0 by $2\pi/\lambda_0$ and Z_0 by 120π ohms. As an example, if $\Delta z = \lambda_0/100$, we find that $R_0 = 0.079$ ohm. This example shows that for a current element which is 1 per cent of a wavelength long, the radiation resistance is very small. Appreciable power would be radiated only if the current amplitude I_0 were very large. A large current, on the other hand, would lead to large amounts of power dissipation in the conductor, and hence a very low efficiency. We can conclude from this analysis of the radiating properties of a short linear current element that current-carrying systems that have linear dimensions small compared with the wavelength radiate negligible power. An efficient radiator or antenna must have dimensions comparable to or greater than the wavelength.

A further property of the linear current radiator that is worthy of consideration is the directional property or relative amount of power radiated in different directions. The power density radiated in the direction specified by the polar angle θ and azimuth angle ϕ is

$$dP = \tfrac{1}{2}r^2 \operatorname{Re} (\mathbf{E} \times \mathbf{H}^* \cdot \mathbf{a}_r)$$
$$= \frac{(I_0 k_0 \, \Delta z)^2}{32\pi^2} Z_0 \sin^2 \theta \qquad \text{watts/unit solid angle} \tag{11.12}$$

The radiated power per unit solid angle is independent of the azimuth angle, as expected, because of the symmetry involved. As a function of θ, the power radiated per unit solid angle varies as $\sin^2 \theta$, and hence the radiation is most intense in the $\theta = \pi/2$ direction and zero in the direction $\theta = 0, \pi$. The directivity function $D(\theta,\phi)$ in the direction θ, ϕ is defined as the ratio of the power radiated per unit solid angle in the direction θ, ϕ divided by the total average power radiated per unit solid angle. From (11.8) the total power radiated is P_r. Since there are 4π steradians in a sphere, the average power radiated per unit solid angle is

$$\frac{P_r}{4\pi} = \frac{(I_0 k_0 \, \Delta z)^2 Z_0}{48\pi^2} \tag{11.13}$$

A fictitious isotropic radiator radiating a total power P_r uniformly in all directions would radiate an amount of power, per unit solid angle, given by (11.13). For the linear current element the directivity D is a function of θ only. Combining (11.12) and (11.13) shows that

$$D(\theta) = \tfrac{48}{32} \sin^2 \theta = 1.5 \sin^2 \theta \tag{11.14}$$

The maximum value of $D(\theta)$ is $D(\pi/2) = 1.5$, and this is commonly called the directivity of the radiator. The directivity is a measure of how effective the antenna is in concentrating the radiated power in a given direction.

The directivity function $D(\theta,\phi)$ defines a three-dimensional surface called the polar radiation pattern of the antenna. Figure 11.3 illustrates the polar radiation pattern for the short linear current radiator. In a plane ϕ = constant, the beamwidth between the half-power points is 90° [determined by solving the equation $D(\theta) = 0.5D(\pi/2)$ or $\sin^2 \theta = 0.5$].

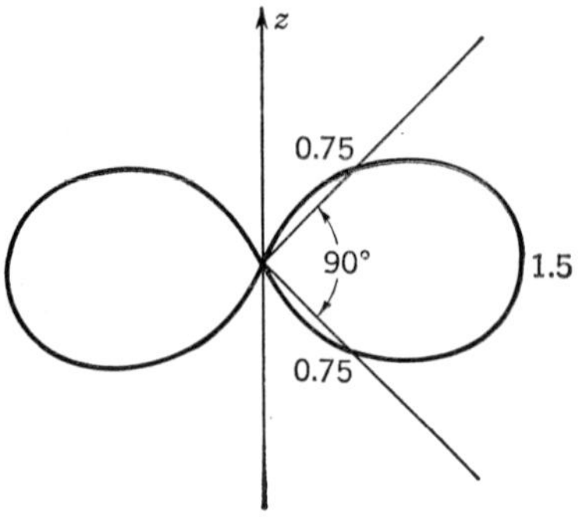

FIG. 11.3. Cross section of polar radiation pattern for infinitesimal current radiator (complete pattern is obtained by revolving cross section around polar axis).

The directivity function has been defined as the ratio of the power density in a given direction compared with the power density of a fictitious isotropic radiator with the same total radiated power. The antenna efficiency can be included if a function is defined as the ratio of power density in a given direction to the power density of an isotropic radiator with the same input power. The latter function is called the gain function $G(\theta,\phi)$, and it includes the losses in the antenna. The maximum value of G is often referred to as the "gain" of the antenna. The distinction between directivity and gain is not always carefully adhered to, in practice.

11.2. The Half-wave Antenna

The short linear current element considered above constitutes a mathematical ideal radiator. The results of the analysis, together with the principle of superposition, may be used to find the fields radiated by an antenna structure on which the current distribution is known. An antenna which is often used in practice is the half-wave dipole antenna illustrated in Fig. 11.4. This antenna consists of two thin linear con-

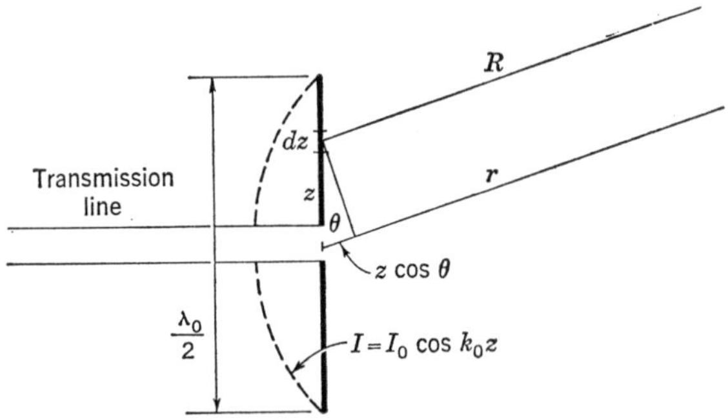

FIG. 11.4. Half-wave dipole antenna.

ductors each of length $\lambda_0/4$ and connected to a two-wire transmission line at the center. The radiation resistance of this antenna will be shown to be 73.13 ohms. This is a practical value of radiation resistance for which it is possible to obtain a good efficiency, i.e., a large amount of radiated power compared with the power loss in the conductors.

On a thin half-wave antenna it is found experimentally that the current along the antenna has a sinusoidal variation of the form

$$Ie^{j\omega t} = I_0 \cos k_0 z \, e^{j\omega t}$$

where I_0 is the current amplitude at the feeding point, assumed real for convenience. Each current element $I\,dz$ may be considered as a short linear current radiator and the total field obtained by summing up the fields radiated by each element. If we confine our attention to the far-zone or radiation field only, then, by using (11.9), we see that the current element $I_0 \cos k_0 z \, dz$ at z radiates a partial field

$$dE_\theta = \frac{jk_0 I_0 Z_0 \sin\theta}{4\pi R} \cos k_0 z \, e^{-jk_0 R}\, dz$$
$$dH_\phi = Y_0\, dE_\theta$$

From the law of cosines

$$R = (r^2 + z^2 - 2rz\cos\theta)^{1/2} = r\left[1 - \left(\frac{2z}{r}\right)\cos\theta + \frac{z^2}{r^2}\right]^{1/2}$$

The latter expression can be expanded in powers of z/r by the binomial theorem. Since $r \gg \lambda_0$ is assumed, we may retain the leading term only and get $R = r - z\cos\theta$. This result may be interpreted geometrically as equivalent to the assumption that the paths from each differential element to the distant field point are parallel. In the denominator of dE_θ we may replace R by r, but in the exponential we must use the more accurate expression $r - z\cos\theta$. The total radiated electric field is thus

$$\begin{aligned} E_\theta &= \frac{jk_0 I_0 Z_0 \sin\theta}{4\pi r} e^{-jk_0 r} \int_{-\lambda_0/4}^{\lambda_0/4} \cos k_0 z \, e^{jk_0 z\cos\theta}\, dz \\ &= \frac{jk_0 I_0 Z_0 \sin\theta}{2\pi r} e^{-jk_0 r} \int_0^{\lambda_0/4} \cos k_0 z \cos(k_0 z\cos\theta)\, dz \end{aligned}$$

after replacing the exponential by $\cos(k_0 z\cos\theta) + j\sin(k_0 z\cos\theta)$ and noting that the term involving the sine is an odd function and integrates to zero. The integration is readily performed by using the identity

$$\cos k_0 z \cos(k_0 z\cos\theta) = \tfrac{1}{2}\{\cos[k_0 z(1+\cos\theta)] + \cos[k_0 z(1-\cos\theta)]\}$$

The final result is

$$\begin{aligned} E_\theta &= \frac{jI_0 Z_0 \sin\theta}{4\pi r} e^{-jk_0 r}\left[\frac{\sin\frac{\pi}{2}(1+\cos\theta)}{1+\cos\theta} + \frac{\sin\frac{\pi}{2}(1-\cos\theta)}{1-\cos\theta}\right] \\ &= \frac{jI_0 Z_0}{2\pi r} e^{-jk_0 r} \frac{\cos\left(\frac{\pi}{2}\cos\theta\right)}{\sin\theta} \end{aligned} \qquad (11.15)$$

The total power radiated is obtained by integrating one-half of the real part of the complex Poynting vector $E_\theta H_\phi^* = Y_0|E_\theta|^2$ over a sphere of radius r. We have

$$P_r = \frac{I_0^2 Z_0}{8\pi^2} \int_0^{2\pi} \int_0^{\pi} \frac{\cos^2\left(\frac{\pi}{2}\cos\theta\right)}{\sin\theta}\, d\theta\, d\phi$$

$$= \frac{I_0^2 Z_0}{4\pi} \int_0^{\pi} \frac{\cos^2\left(\frac{\pi}{2}\cos\theta\right)}{\sin\theta}\, d\theta \qquad (11.16)$$

By an appropriate change of variables the integral is transformed to†

$$P_r = \frac{I_0^2 Z_0}{8\pi} \int_0^{2\pi} \frac{1 - \cos u}{u}\, du$$

The latter integral is

$$\int_0^{2\pi} \frac{1 - \cos u}{u}\, du = \ln 1.781 - \text{Ci}\,(2\pi) + \ln 2\pi$$

where Ci x is the cosine integral

$$\text{Ci}\, x = -\int_x^{\infty} \frac{\cos u}{u}\, du$$

and is tabulated. In particular, Ci $(2\pi) = -0.0226$. Thus we have

$$P_r = \frac{I_0^2 Z_0}{8\pi} [\ln 2\pi(1.781) - \text{Ci}\,(2\pi)]$$

$$= \frac{I_0^2 Z_0}{8\pi} (2.4151 + 0.0226) = 36.57 I_0^2 \qquad (11.17)$$

The current at the feeding point is I_0, and hence from the relation $\frac{1}{2} I_0^2 R_0 = P_r$, the radiation resistance is found to be 73.13 ohms.

The near-zone field for the half-wave dipole does not contribute to the radiated power. In actual fact, the near-zone field represents a storage of reactive energy in the immediate space surrounding the antenna. This reactive energy gives rise to a reactive term in the input impedance presented by the antenna to the transmission-line feeder. By choosing a proper antenna length, the average electric and magnetic energy stored in the near-zone field can be made equal and the input reactive term will vanish. This is equivalent to adjusting a tuned circuit to a resonant condition. For a thin half-wave dipole antenna this resonant length is found to be a few per cent shorter than a half wavelength.

† J. Stratton, "Electromagnetic Theory," sec. 8.7, McGraw-Hill Book Company, Inc., New York, 1941.

The directivity of the half-wave dipole is given by

$$D(\theta) = \frac{60}{36.57}\left[\frac{\cos\left(\frac{\pi}{2}\cos\theta\right)}{\sin\theta}\right]^2 \tag{11.18}$$

The maximum value of D is 1.64, which is only slightly larger than that for the short linear current radiator. A plot of the radiation pattern is given in Fig. 11.5. This pattern is similar to that of the short current radiator except that the half-power beamwidth is 78° instead of 90°.

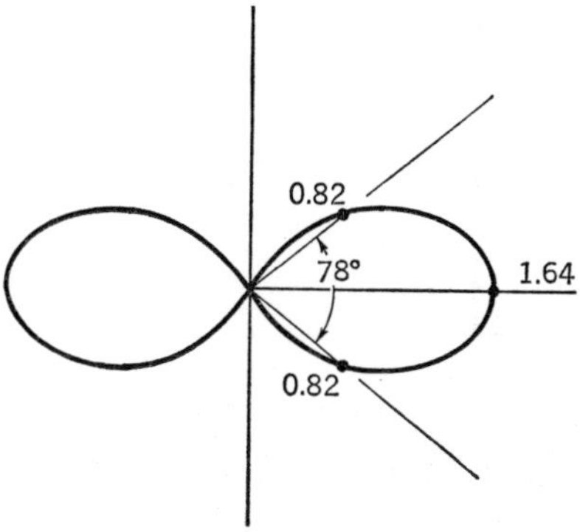

FIG. 11.5. Radiation pattern of a half-wave dipole antenna.

11.3. Introduction to Arrays

In examining the radiation pattern of an elementary dipole, we note that very little directivity is achieved. Maximum radiation takes place at right angles to the axis; however, this falls off relatively slowly as the polar angle decreases toward zero, and furthermore the pattern is uniform with respect to azimuth. Although the half-wave dipole achieves a somewhat greater concentration of energy in the direction normal to the axis, its pattern does not differ substantially from that of the elementary dipole. We should find that other-length resonant-wire antennas, while producing more complicated patterns, do not result in highly directive patterns. One way in which greater control of the radiation pattern may be achieved is by the use of an array of dipole (or other) antennas. Such arrays are capable of producing directional patterns or special characteristics of other sorts.

If we visualize the total primary current source as made up of differential radiators, then the resultant pattern is the superposition of the field contributions from each elementary source. This means that a highly directive antenna will result if the amplitude and phase of each element can be suitably chosen so that cancellation of the fields in all but the desired direction is essentially achieved. The resonant-wire antenna does not permit sufficient flexibility of assignment of phase and amplitude since all elements are in the same phase while the amplitude variation is sinusoidal. The desired freedom can be achieved by arranging together a number of separately driven antennas whose spacing and excitation are at our disposal. In general, it is found that desired results can be achieved through the use of identical elements equally spaced, and this case only will be considered. As a consequence of using identical array elements, it turns out that the total pattern can be formulated as the product of the pattern of the element times the pattern of the array, as if each element were an isotropic radiator. In this way the characteristics of an array can be discussed independently of the characteristics of its elements.

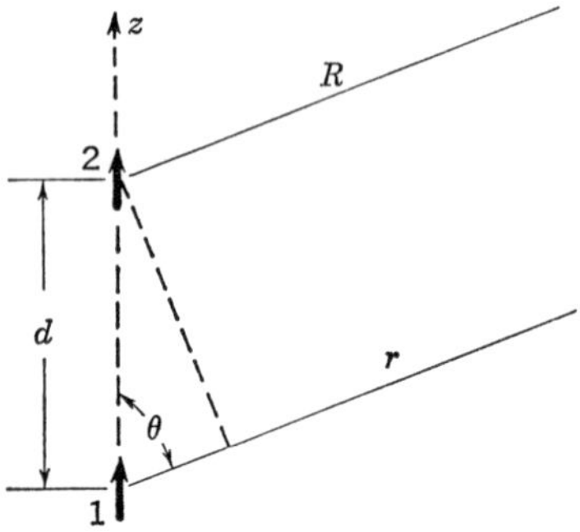

FIG. 11.6. Array of two collinear infinitesimal current elements.

As a very simple example we consider two identical infinitesimal linear current elements which are collinear and along the z axis with a spacing d. The geometric arrangement is illustrated in Fig. 11.6. Antenna 1 is chosen at the origin, and it produces an electric field at a far-zone point P that, according to (11.9b), equals

$$E_{\theta 1} = \frac{jk_0 Z_0 I_0 \,\Delta z}{4\pi r} \sin\theta \, e^{-jk_0 r} \tag{11.19}$$

The field due to antenna 2 is similar except that the distance to P is R instead of r. Using the law of cosines we have

$$R = (r^2 + d^2 - 2rd\cos\theta)^{1/2} = r\left(1 - \frac{2d\cos\theta}{r} + \frac{d^2}{r^2}\right)^{1/2}$$

The latter may be expanded by means of the binomial theorem in powers of d/r. The statement that P is in the far zone of the array requires that r be sufficiently greater than d so that for phase calculations we can take the two leading terms $R = r - d\cos\theta$, with little error.† This procedure is completely equivalent to that followed in the analysis of the half-wave dipole and similarly permits the geometrical interpretation that the paths from each array element to the field point may be considered to be parallel. It is sufficient to take $R = r$ when only magnitudes are concerned. With these "far-zone approximations," the field due to antenna 2 is

$$E_{\theta 2} = \frac{jk_0 Z_0 I_0 \,\Delta z}{4\pi r} \sin\theta \, e^{-jk_0(r - d\cos\theta)} \tag{11.20}$$

and the total field is

$$E_\theta = \frac{jk_0 Z_0 I_0 \,\Delta z}{4\pi} \sin\theta \, \frac{e^{-jk_0 r}}{r} (1 + e^{jk_0 d\cos\theta}) \tag{11.21}$$

† The criterion is usually specified as

$$r > \frac{2d^2}{\lambda_0}$$

although the numerical coefficient is sometimes taken as unity or 4.

If instead of infinitesimal current elements we had assumed arbitrary antennas 1 and 2, which were identical geometrically and were excited by identical currents but were displaced from each other a distance d along the polar axis, then the field of antenna 1 could be written

$$\mathbf{E}_1 = \mathbf{f}(\theta,\phi) \frac{e^{-jk_0r}}{r} \tag{11.22a}$$

while that of antenna 2 would be

$$\mathbf{E}_2 = \mathbf{f}(\theta,\phi) \frac{e^{-jk_0r}}{r} e^{jk_0d\cos\theta} \tag{11.22b}$$

Equation (11.22a) expresses the fact that any antenna, including arrays, may be considered as made up of a large number of elementary sources, where, provided the field point is in the far zone, the resultant field due to each element contains a common factor e^{-jk_0r}/r just as in the previous example. Apart from this factor we have a complex-phasor-vector summation, $\mathbf{f}(\theta,\phi)$, which in general is a function of the direction (θ,ϕ) to the field point. Antenna 2 being identical with antenna 1 results in a superposition of partial fields from elements that correspond to those of antenna 1, except for a displacement d along the z axis, hence resulting in an additional common phase factor $e^{jk_0d\cos\theta}$ for the current elements of antenna 2. The total field due to the two-element array may be written

$$\mathbf{E} = \mathbf{f}(\theta,\phi) \frac{e^{-jk_0r}}{r} (1 + e^{jk_0d\cos\theta}) \tag{11.23}$$

Note that the specific example leading to (11.21) conforms to this general result.

For N radiators spaced a distance d apart along the z axis and equally excited, the previous result can be readily generalized to give

$$\mathbf{E} = \mathbf{f}(\theta,\phi) \frac{e^{-jk_0r}}{r} \left[1 + \sum_{n=1}^{N-1} e^{jk_0nd\cos\theta}\right] \tag{11.24}$$

and it is now necessary that $r \gg Nd$ for the far-zone approximation to hold. The form of (11.24) displays the total field as the product of the pattern of the element antenna and what we call the array factor. The latter, in this case, is

$$1 + \sum_{n=1}^{N-1} e^{jk_0nd\cos\theta}$$

and may also be thought of as the pattern of an array of isotropic radiators excited in the same way as the actual antennas. More generally, if the nth antenna has a relative amplitude C_n and phase $e^{j\alpha_n}$, then the array factor A becomes

$$A = 1 + \sum_{n=1}^{N-1} C_n e^{j(k_0 n d \cos\theta + \alpha_n)} \tag{11.25}$$

Normally, only the absolute value of the field is required. The array factor in (11.23), for example, becomes

$$|A| = |1 + e^{jk_0 d \cos\theta}| = \left| \cos \frac{k_0 d \cos\theta}{2} \right| \tag{11.26}$$

Example 11.1. A Two-element Array. Let us calculate the pattern due to two infinitesimal dipoles whose axes are "horizontal" but which are spaced "vertically," as shown in Fig. 11.7. They are identical, and the magnitude of their excitation is the same, but a relative phase shift of $e^{-jk_0 d}$ is imposed on antenna 2, as noted in Fig. 11.7. Since this geometry no longer has the axial symmetry of earlier cases, a pattern that is a function of ϕ in addition to θ must be expected.

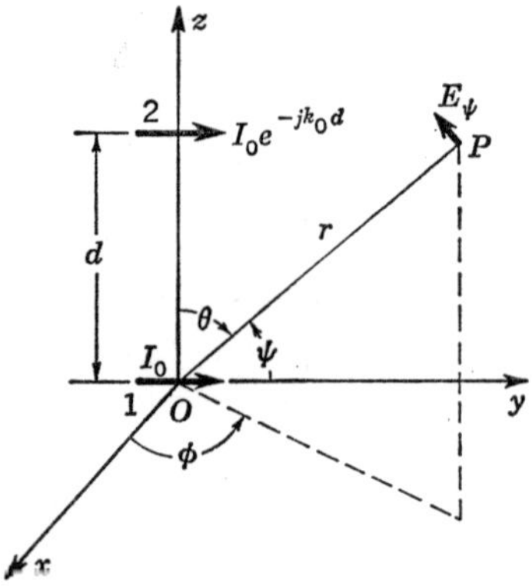

FIG. 11.7. Vertical array of two horizontal elementary dipoles.

The electric field of antenna 1 at point P is found from (11.9b) to be

$$E_{\psi 1} = \frac{jk_0 Z_0 I_0 \,\Delta z}{4\pi r} \sin\psi \, e^{-jk_0 r} \tag{11.27a}$$

where ψ is measured from the axis of the current element and is the polar angle relative to the dipole axis. The direction of E_ψ is normal to r in the OP-OY plane, as illustrated in Fig. 11 7. For the field of element 2 we get

$$E_{\psi 2} = \frac{jk_0 Z_0 I_0 \,\Delta z}{4\pi r} \sin\psi \, e^{jk_0 d\cos\theta} e^{-jk_0 d} e^{-jk_0 r} \tag{11.27b}$$

where the factor $e^{-jk_0 d}$ is due to the relative phase of excitation of the second element. The total field E_ψ is then

$$E_\psi = \frac{jk_0 Z_0 I_0 \,\Delta z}{4\pi r} e^{-jk_0 r} \sin\psi \, [1 + e^{jk_0 d(\cos\theta - 1)}] \tag{11.28}$$

and the quantity in the bracket is readily identified as the array factor. The absolute value of this array factor is found to be

$$|A| = |1 + e^{jk_0 d(\cos\theta - 1)}| = 2 \left| \cos \frac{k_0 d(\cos\theta - 1)}{2} \right| \tag{11.29}$$

Equation (11.28) can be put into a more useful form by expressing ψ in terms of ϕ and θ. From the geometry of Fig. 11.7 the unit vector in the r direction, $\mathbf{a}_r$, is

$$\mathbf{a}_r = \mathbf{a}_x \sin\theta\cos\phi + \mathbf{a}_y \sin\theta\sin\phi + \mathbf{a}_z\cos\theta$$

so that
$$\sin\psi = |\mathbf{a}_r \times \mathbf{a}_y| = (1 - \sin^2\phi\sin^2\theta)^{1/2}$$

Furthermore,
$$\mathbf{E}_\psi = E_\theta\mathbf{a}_\theta + E_\phi\mathbf{a}_\phi$$

where
$$E_\theta = \frac{-E_\psi\cos\theta\sin\phi}{\sqrt{1-\sin^2\phi\sin^2\theta}}$$

$$E_\phi = \frac{-E_\psi\cos\phi}{\sqrt{1-\sin^2\phi\sin^2\theta}}$$

since
$$\begin{aligned}\mathbf{a}_\psi &= -\mathbf{a}_y\csc\psi + \mathbf{a}_r\cot\psi \\ &= -(\mathbf{a}_\theta\cos\theta\sin\phi + \mathbf{a}_\phi\cos\phi)\csc\psi + \mathbf{a}_r(\cot\psi + \sin\theta\sin\phi)\end{aligned}$$

We may now write the total field $\mathbf{E}$ as

$$\mathbf{E} = \frac{-jk_0Z_0I_0\,\Delta z\, e^{-jk_0r}}{2\pi r}\cos\frac{k_0d(\cos\theta-1)}{2}\,e^{jk_0d(\cos\theta-1)/2} \times (\mathbf{a}_\theta\cos\theta\sin\phi + \mathbf{a}_\phi\cos\phi) \tag{11.30}$$

The full three-dimensional pattern of the antenna is given by (11.30). However, instead of treating the pattern as a whole, it is (usually) sufficient to describe the antenna pattern in the principal coordinate planes. We obtain, then, the following:

yz-plane pattern ($\phi = \pi/2$)

$$|\mathbf{E}| = E_\theta = \frac{k_0Z_0I_0\,\Delta z}{2\pi r}|\cos\theta|\left|\cos\frac{k_0d(\cos\theta-1)}{2}\right| \tag{11.31}$$

xz-plane pattern ($\phi = 0$)

$$|\mathbf{E}| = E_\phi = \frac{k_0Z_0I_0\,\Delta z}{2\pi r}\left|\cos\frac{k_0d(\cos\theta-1)}{2}\right| \tag{11.32}$$

xy-plane pattern ($\theta = \pi/2$)

$$|\mathbf{E}| = E_\phi = \frac{k_0Z_0I_0\,\Delta z}{2\pi r}|\cos\phi|\left|\cos\frac{k_0d}{2}\right| \tag{11.33}$$

For the case where $d = \lambda_0/4$, hence $k_0d = \pi/2$, the pattern in the xz plane is proportional to $\cos[(\pi/4)(\cos\theta - 1)]$, while the pattern in the yz plane is obtained by multiplying the former pattern by $\cos\theta$. The results are plotted in Fig. 11.8. For the case where $d = 3\lambda_0/4$, the results are plotted in Fig. 11.9. Note that the pattern in the xy plane is independent of d and is simply a sinusoid in ϕ.

The patterns of Figs. 11.8 and 11.9 show a maximum of radiation in the direction of the positive z axis. This could have been foreseen from the nature of the excitation. We note that antenna 2 is excited with a lagging phase that corresponds precisely to the phase delay of a wave

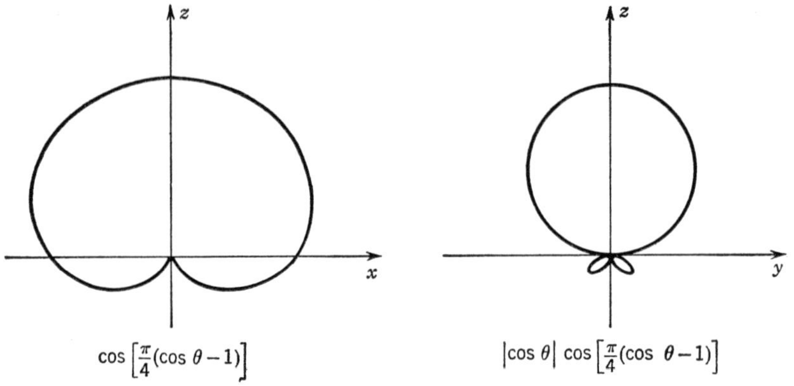

FIG. 11.8. Normalized patterns in xz and yz planes for vertically stacked horizontal dipoles separated by $\lambda_0/4$.

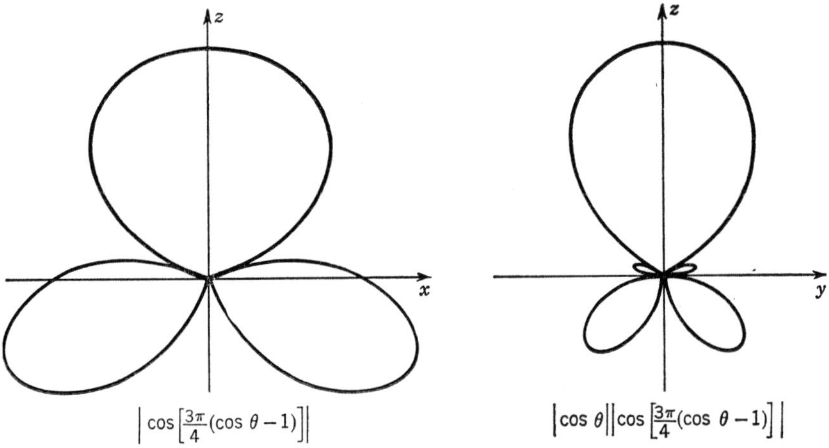

FIG. 11.9. Normalized patterns in xz and yz planes for vertically stacked horizontal dipoles separated by $3\lambda_0/4$.

leaving antenna 1 and propagating in the positive z direction. Consequently, the partial fields contributed by antennas 1 and 2 arrive in phase and therefore add together in the $+z$ direction. For other directions there will usually be only partial addition. If the number of elements is increased and a progressive phase delay given each successive element, then the partial fields of each element can be made to add in the $+z$ direction. However, when a large number of elements are involved, then usually, in other than the "forward" direction, the phase of each contribution will tend to cause cancellation of the field from each element, with a net result that the field strength in these directions is relatively small. An array of the type just described is called an end-fire array since the maximum occurs along the line of the array. Let us consider long arrays of this type analytically.

11.4. Linear Arrays

We consider now a linear array of N elements equally spaced a distance d apart. If we choose the line of the array to be the polar axis, then the geometry is as illustrated in Fig. 11.10 and the array factor is given by (11.25). The latter can be easily rederived if the array of Fig. 11.10 is considered to be composed of isotropic elements where the amplitude and phase of the nth element, relative to the reference element at the origin, are $C_n e^{j\alpha_n}$.

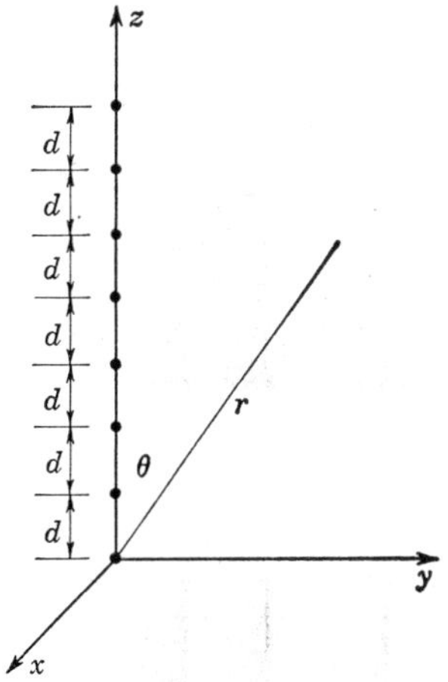

FIG. 11.10. Linear array of many elements along z axis.

A case of considerable interest will be considered where the amplitude of excitation of each element is the same and where the relative phase shift of the excitation is of the form

$$\alpha_n = -nk_0 d \cos \theta_0 \tag{11.34}$$

where θ_0 is a constant. In this form, by choosing $\theta_0 = 0$, the end-fire case previously discussed results. If $\theta_0 = \pi/2$, then all α_n's are zero and each element is excited in phase. In this case we see by inspection that the maximum radiation occurs in a direction normal to the line of the array, since in this direction the partial contribution from each element adds directly. An array for which this condition holds is designated a broadside array. By considering the excitation phase according to (11.34), we see that the special cases of end-fire and broadside arrays are included.

Putting (11.34) into (11.25) and setting $C_n = 1$, we obtain

$$|A| = |1 + e^{jk_0 d(\cos\theta - \cos\theta_0)} + e^{j2k_0 d(\cos\theta - \cos\theta_0)} + \cdots + e^{j(N-1)k_0 d(\cos\theta - \cos\theta_0)}| \tag{11.35}$$

for an array of N elements. Since (11.35) is a geometric progression with a ratio $e^{jk_0 d(\cos\theta - \cos\theta_0)}$, the sum can be expressed as

$$|A| = \left| \frac{e^{jNk_0 d(\cos\theta - \cos\theta_0)} - 1}{e^{jk_0 d(\cos\theta - \cos\theta_0)} - 1} \right| = \left| \frac{\sin \dfrac{Nk_0 d(\cos\theta - \cos\theta_0)}{2}}{\sin \dfrac{k_0 d(\cos\theta - \cos\theta_0)}{2}} \right| \tag{11.36}$$

This pattern, considered as a function of $x = k_0d(\cos\theta - \cos\theta_0)$, is of the form

$$|A| = f(x) = \left| \frac{\sin \frac{Nx}{2}}{\sin \frac{x}{2}} \right| \qquad x = k_0 d(\cos\theta - \cos\theta_0) \tag{11.37}$$

A typical curve of $f(x)$ for large N is shown in Fig. 11.11. Note that for $x = 0$, $f(x)$ approaches the value N. Furthermore, the nulls will occur

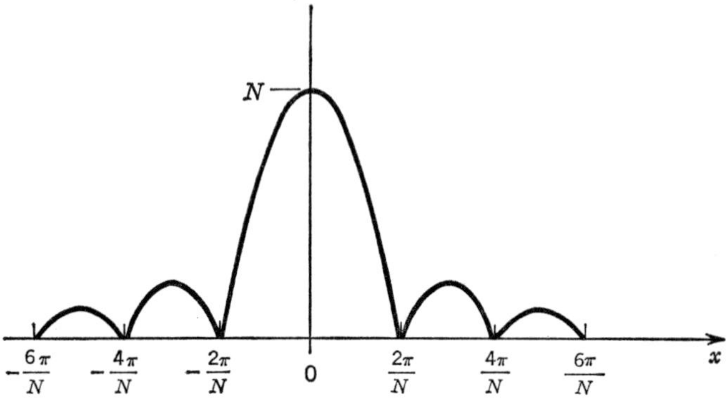

FIG. 11.11. Universal pattern for a linear array with uniform amplitude and progressive phase shift.

for $Nx/2 = \pi, 2\pi, 3\pi, \ldots, (N-1)\pi$ (larger values correspond to pattern repetition). Also if N is large, the subsidiary maxima correspond approximately to maxima of the numerator; that is, $x = 3\pi/N, 5\pi/N, \ldots$. At these points the magnitude of $f(x)$ is given by

$$f\left(\frac{3\pi}{N}\right) \approx \frac{2N}{3\pi} \qquad f\left(\frac{5\pi}{N}\right) \approx \frac{2N}{5\pi} \qquad f\left(\frac{7\pi}{N}\right) \approx \frac{2N}{7\pi} \qquad \cdots \tag{11.38}$$

a result that depends on N being sufficiently large so that $\sin(3\pi/N) \approx 3\pi/N$, etc. Under these conditions we see that the sidelobe level, i.e., the ratio of the peak amplitude of the main lobe to the subsidiary-lobe peak amplitude, is independent of N, for large N. Its value is $3\pi/2$, or about 13.5 decibels.

Since the maximum value of $f(x)$ occurs for $x = 0$, then in the actual pattern the main beam is in the direction $\theta = \theta_0$. We note, then, that the direction of the main beam can be shifted by altering the progressive phase shift. The position of the first nulls of the main beam is important since it characterizes the beam width. We shall designate the "upper" null by the value of $\theta = \theta_0 + \theta^+$ and the "lower" null by $\theta = \theta_0 - \theta^-$. Then θ^+ satisfies the equation

$$k_0 d[\cos(\theta_0 + \theta^+) - \cos\theta_0] = \frac{2\pi}{N} \tag{11.39a}$$

while θ^- is given by

$$k_0d[\cos(\theta_0 - \theta^-) - \cos\theta_0] = -\frac{2\pi}{N} \tag{11.39b}$$

If N is very large, it seems likely (we shall confirm this later) that θ^+ and θ^- will be very small. In this case we can approximate $\cos\theta$ by $(1 - \theta^2/2)$ and $\sin\theta$ by θ, hence obtaining

$$\cos(\theta_0 + \theta^+) - \cos\theta_0 \approx -\theta^+ \sin\theta_0 - \frac{(\theta^+)^2}{2}\cos\theta_0 \tag{11.40a}$$

$$\cos(\theta_0 - \theta^-) - \cos\theta_0 \approx \theta^- \sin\theta_0 - \frac{(\theta^-)^2}{2}\cos\theta_0 \tag{11.40b}$$

Provided that θ_0 is sufficiently greater than $\theta^\pm$, the quadratic term in (11.40) can be dropped. With the resultant expression, (11.39) can be solved for θ^+ and θ^-, yielding

$$\theta^+ = \theta^- = \frac{2\pi}{N}\frac{1}{k_0d\sin\theta_0} = \frac{\lambda_0}{Nd\sin\theta_0} \tag{11.41}$$

The assumption that θ^+ and θ^- are small is seen to be justified provided Nd/λ_0 is large and θ_0 not too close to zero, as we have already required. The beamwidth Δ may be defined by the total angle between nulls, a quantity which is simply $2\theta^+$. Assuming that N is large, then the total array length $L = (N - 1)d \approx Nd$, and

$$\Delta = \frac{2\lambda_0}{L\sin\theta_0} \tag{11.42}$$

For a given array length in wavelengths, the minimum beamwidth occurs for the broadside array, where $\theta_0 = \pi/2$, in which case

$$\Delta_{bs} = \frac{2\lambda_0}{L} \tag{11.43}$$

For the end-fire case $\theta_0 = 0$, and (11.41) does not hold. However, we can return to (11.40) and utilize the quadratic term (the linear terms go out) to establish $\theta^+ = \theta^- = \sqrt{2\lambda_0/Nd}$. Consequently,

$$\Delta_{ef} = 2\sqrt{\frac{2\lambda_0}{L}} \tag{11.44}$$

for large N.

11.5. Two-dimensional Arrays

The linear arrays discussed above produce patterns which are axially symmetric. If, for example, a highly directive pencil beam is desired, then the linear array by itself cannot be used to produce such a pattern; that is, we can construct an array that is long compared with wavelength and thereby achieve a narrow beam but the array pattern will be the same in each longitudinal plane. Further shaping is required in this case. We recall, however, that the array pattern must be multiplied by the directivity of the element of the array to obtain the over-all pattern. The element can itself be an array with its elements spaced at right angles to the original array, as suggested in Fig. 11.12a. In this case the array element can produce a narrow beam symmetric about its own axis. The product of the two patterns, in this case, yields a maximum only over a

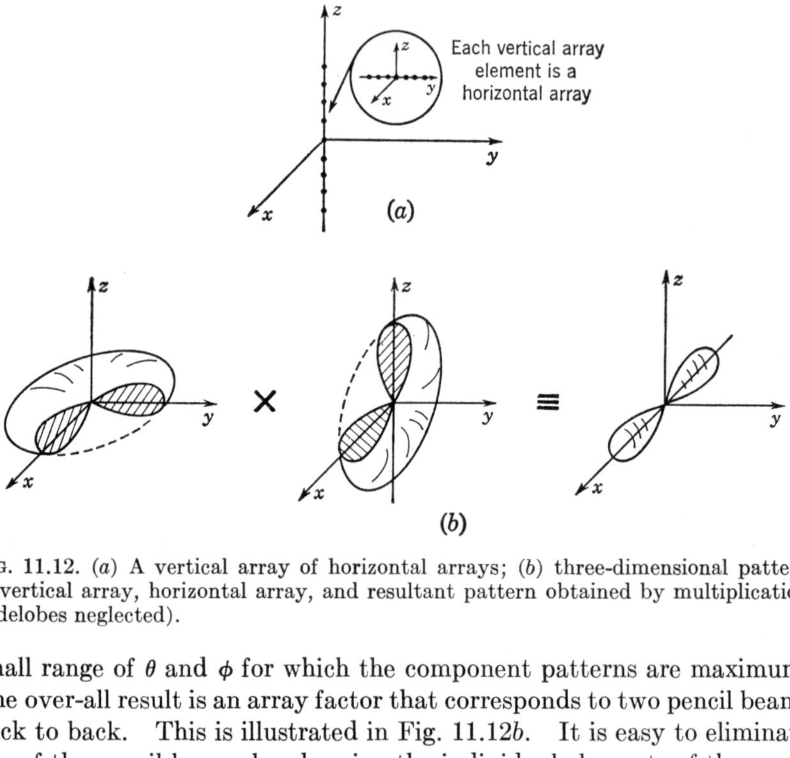

FIG. 11.12. (a) A vertical array of horizontal arrays; (b) three-dimensional pattern of vertical array, horizontal array, and resultant pattern obtained by multiplication (sidelobes neglected).

small range of θ and ϕ for which the component patterns are maximum. The over-all result is an array factor that corresponds to two pencil beams back to back. This is illustrated in Fig. 11.12b. It is easy to eliminate one of the pencil beams by choosing the individual elements of the array to have a null in the direction of one of the two beams.

We note in the example discussed above that the final physical arrangement is that of a two-dimensional array. The pattern capabilities of such an array are much more versatile than those of the linear array. For a class of problems the pattern of a two-dimensional array, such as the one discussed above, can be reduced to that of determining the pattern of two linear arrays. Let us formulate this more specifically.

Figure 11.13 illustrates a two-dimensional array with uniform horizontal and vertical spacing h and v, where h and v are not necessarily equal. The total number of horizontal rows is N, and the total number of vertical columns is M. We designate by the double subscript mn the element in the mth column and nth row. Then if the excitation of the mnth element can be written in the form

$$e^{j(m\alpha+n\beta)}$$

where α and β are constants, we have a uniform (progressive) phase shift from one element to the next along either the horizontal or vertical direction. Furthermore, we are considering the case where the amplitude of excitation is constant, the so-called uniform array. Under these conditions it is possible to think of the array as a linear array of vertical elements, the latter being linear arrays of horizontal elements, or vice versa. In either case the array factor A is the product of the array factor of the horizontal array A_h and the vertical array A_v. This result can be substantiated analytically, and we turn now to this task.

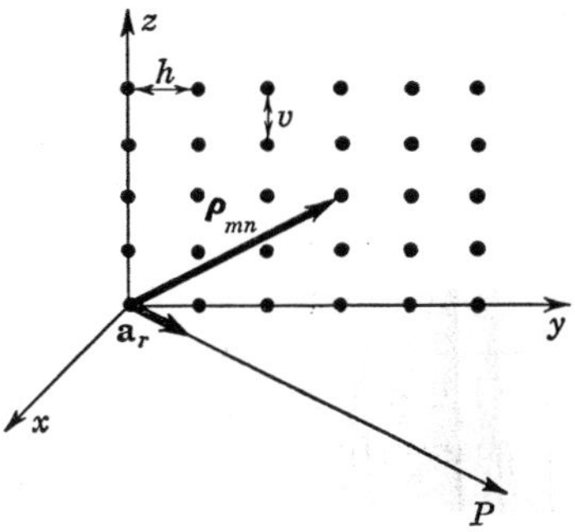

FIG. 11.13. Two-dimensional array.

Let the direction to the field point $P(\theta,\phi)$ be given by the unit vector $\mathbf{a}_r$, where

$$\mathbf{a}_r = \mathbf{a}_x \sin\theta\cos\phi + \mathbf{a}_y \sin\theta\sin\phi + \mathbf{a}_z\cos\theta$$

The mnth element can be described by a vector $\boldsymbol{\varrho}_{mn}$ from the origin to its location; that is,

$$\boldsymbol{\varrho}_{mn} = mh\mathbf{a}_y + nv\mathbf{a}_z$$

In accordance with the far-zone assumptions, the contribution from the mnth element arrives at P with a phase, relative to that from the reference element at the origin, which arises owing to the difference in the respective path lengths, where the latter is simply the projection of $\boldsymbol{\varrho}_{mn}$ on $\mathbf{a}_r$. The total pattern or array factor is thus

$$A = \sum_{n=0}^{N}\sum_{m=0}^{M} e^{-jk_0\boldsymbol{\varrho}_{mn}\cdot\mathbf{a}_r}e^{j(m\alpha+n\beta)} \tag{11.45}$$

Expanding the dot product and combining terms allow us to write

$$A = \sum_{m=0}^{M} e^{-jm(k_0h\sin\theta\sin\phi-\alpha)}\sum_{n=0}^{N} e^{-jn(k_0v\cos\theta-\beta)} \tag{11.46}$$

and it is clear that the total array factor is the product of the separate linear array factors; that is,

$$A = A_hA_v \tag{11.47}$$

where
$$A_h = \sum_{m=0}^{M} e^{-jm(k_0h \sin\theta \sin\phi - \alpha)} \tag{11.48a}$$
$$A_v = \sum_{n=0}^{N} e^{-jn(k_0v \cos\theta - \beta)} \tag{11.48b}$$

If the electric field of an array element is $\mathbf{f}(\theta,\phi)e^{-jk_0r}/r$, then the over-all field produced by the array is

$$\mathbf{E} = A_hA_v\mathbf{f}(\theta,\phi)\frac{e^{-jk_0r}}{r} \tag{11.49}$$

One of the conclusions reached concerning the linear array was that the sidelobe level, for a large number of elements, is a constant. For the two-dimensional array of the type considered, this same conclusion must hold, since the total array factor is simply the product of the individual linear array factors. In some cases it is of importance to adjust the sidelobe level below the values achieved by the class of arrays described here. In this instance the restriction to uniform amplitude excitation, such as has been assumed, must be removed. Even greater generality may be desired where both amplitude and phase of each element are independently specified. A fairly elaborate theory for optimizing array patterns with respect to certain desired parameters exists, and a means for synthesizing such arrays is available. A number of references are given at the end of the chapter for the reader who wishes to pursue this topic further.

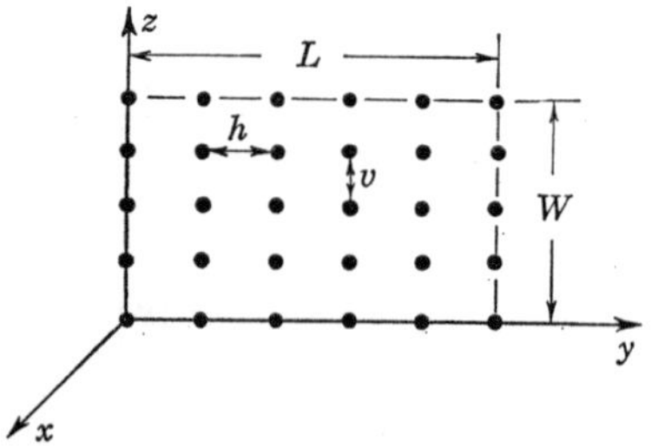

FIG. 11.14. A two-dimensional array of length L and width W.

11.6. Continuous Distributions

Figure 11.14 again illustrates a two-dimensional array where the over-all length is L and the width is W. In terms of numbers of elements and spacing, $L \approx hM$ and $W \approx vN$. If all elements are excited in phase, then a pencil beam in the directions ($\theta = \pi/2$, $\phi = 0, \pi$) is produced. Considered in the xz plane ($\phi = 0$), we see from (11.47) and (11.48) that the pattern is that of the vertical array (the horizontal array factor is a constant in this case), and hence the beam width is simply $2\lambda_0/W$. In the xy plane ($\theta = \pi/2$), Eqs. (11.48) reveal that the pattern is due to the horizontal array factor alone, so that the beam width is $2\lambda_0/L$. This result does not depend on the actual spacing of the elements so long as the total number is large. One can increase the number of elements and decrease their spacing, until in the limiting case a continuous distribution is obtained.

A broad class of radiating antennas, particularly useful in the microwave region, are horn-type radiators, paraboloidal reflectors, and lenses, some of which are illustrated in Fig. 11.15. In each case the energy must pass through a physically distinct aperture. It is possible to show that the field in the aperture behaves like an equivalent source. This fact is related to Huygens' principle, which states that each element of a wavefront may be considered as a secondary source. One may therefore use the results just obtained for the uniform array to predict the beam width

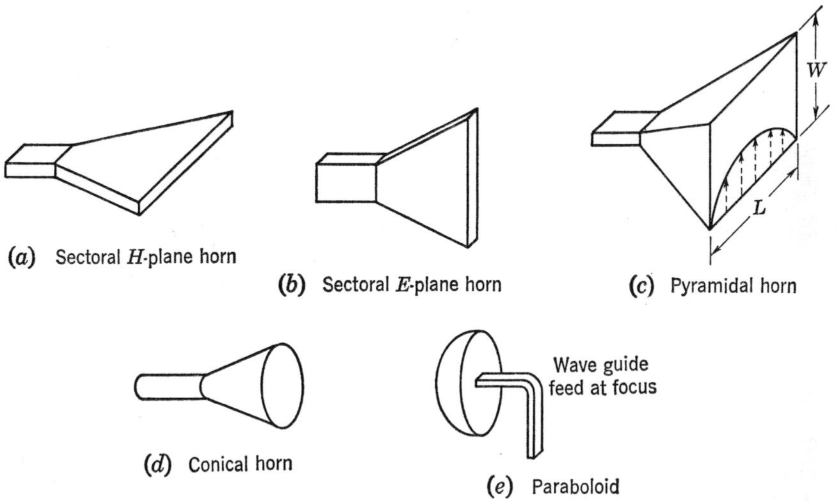

FIG. 11.15. Aperture-type antennas.

due to an aperture-type antenna for the case of uniform-aperture field intensity.

In Table 11.1 the theoretical results for several types of radiators are given. In addition to the case of uniform "illumination," tapered intensities and waveguide distributions are also shown. For the TE_{10} aperture distribution, the equivalent vertical array is uniform but the horizontal array is tapered sinusoidally. Results are also given for circular apertures, and although the numerical factor cannot be checked from the array theory, the behavior as a function of size follows the expected form. The results in Table 11.1 assume that the aperture dimensions are at least several wavelengths.

11.7. Network Formulation for Transmitting-Receiving System

So far we have discussed the antenna under transmitting conditions only. In this and the following section we shall show that the behavior of the receiving antenna can be determined from its characteristics when

TABLE 11.1. BEAM WIDTH FOR SEVERAL APERTURE-TYPE ANTENNAS
Dimensions large compared with wavelength

Type	Field distribution across aperture	Width of major lobe between nulls	
		xy plane	xz plane
Rectangular paraboloid	Uniform	$\frac{115°}{L/\lambda_0}$	$\frac{115°}{W/\lambda_0}$
Rectangular paraboloid	Sinusoidal	$\frac{172°}{L/\lambda_0}$	$\frac{172°}{W/\lambda_0}$
Circular paraboloid (diameter D)	Uniform	$\frac{140°}{D/\lambda_0}$	$\frac{140°}{D/\lambda_0}$
Pyramidal horn †	TE_{10}	$\frac{172°}{L/\lambda_0}$	$\frac{115°}{W/\lambda_0}$
Conical horn † (diameter D)	TE_{11}	$\frac{194°}{D/\lambda_0}$	$\frac{140°}{D/\lambda_0}$

† Assumes small flare angle.

transmitting. To assist in this analysis we first derive the Lorentz reciprocity theorem.

Let $\mathbf{E}_a$, $\mathbf{H}_a$ and $\mathbf{E}_b$, $\mathbf{H}_b$ represent solutions to Maxwell's equations, in a source-free region of space, which arise from different sources outside the region under consideration. Then we may form

$$\nabla \cdot (\mathbf{E}_a \times \mathbf{H}_b) = \mathbf{H}_b \cdot \nabla \times \mathbf{E}_a - \mathbf{E}_a \cdot \nabla \times \mathbf{H}_b$$

and if we utilize the fact that the fields satisfy the homogeneous Maxwell equations, this may be written

$$\nabla \cdot (\mathbf{E}_a \times \mathbf{H}_b) = -j\omega\mu\mathbf{H}_a \cdot \mathbf{H}_b - j\omega\epsilon\mathbf{E}_a \cdot \mathbf{E}_b$$

By interchanging the subscripts a and b, we also have

$$\nabla \cdot (\mathbf{E}_b \times \mathbf{H}_a) = -j\omega\mu\mathbf{H}_a \cdot \mathbf{H}_b - j\omega\epsilon\mathbf{E}_a \cdot \mathbf{E}_b$$

Consequently,
$$\nabla \cdot (\mathbf{E}_a \times \mathbf{H}_b) - \nabla \cdot (\mathbf{E}_b \times \mathbf{H}_a) = 0 \tag{11.50}$$

If (11.50) is integrated throughout the given source-free region and use is made of the divergence theorem, we have

$$\oint_S (\mathbf{E}_a \times \mathbf{H}_b - \mathbf{E}_b \times \mathbf{H}_a) \cdot d\mathbf{S} = 0 \tag{11.51}$$

which is the desired form of the reciprocity theorem.

Figure 11.16 illustrates two antennas and their associated equipment, and it is understood that either antenna may be transmitting with the other receiving. For simplicity in the analysis, the feed line to each antenna is assumed to be a coaxial line, and it is further assumed that the electromagnetic sources (the vacuum tubes) may be entirely enclosed by a conducting surface that forms an extension to the outer conductor of the coaxial line. The dipole antenna shown represents any of the wide variety of actual practical antennas that might be used. Also, of course,

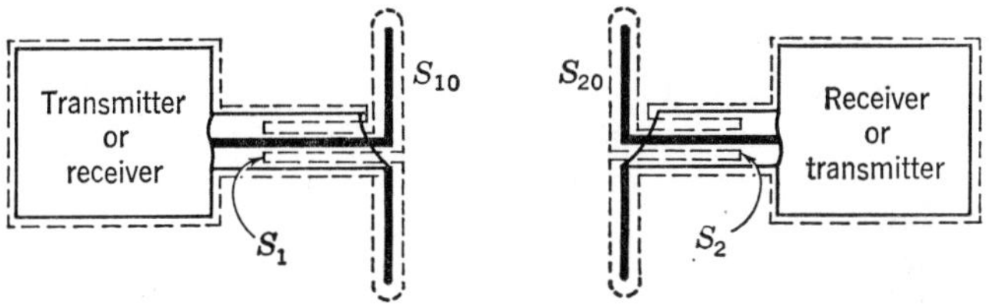

FIG. 11.16. Transmitting and receiving antennas (arbitrarily located).

the relative locations of the transmitter and receiver are completely arbitrary.

At antenna 1 we specify input conditions at an arbitrary transverse reference surface S_1 in the coaxial transmission line. The only restriction on the location of S_1 is that it must be sufficiently far from the antenna or other discontinuity so that only the dominant TEM mode exists. Similar considerations apply to the specification of reference surface S_2. At each of these surfaces both voltage and current are defined in terms of conventional transmission-line concepts. If we consider the current I_1 that flows at S_1, then because of the linearity of Maxwell's equations, it must be linearly related to the voltage at S_1 and at S_2. (In terms of the fields, the magnetic field at S_1, which is proportional to the current, is linearly related to the electric field at S_1 and S_2.) We may thus write

$$I_1 = Y_{11}V_1 + Y_{12}V_2 \tag{11.52a}$$

Similarly,

$$I_2 = Y_{21}V_1 + Y_{22}V_2 \tag{11.52b}$$

Equations (11.52) are in the same form as that for a standard four-terminal network composed of linear bilateral lumped elements. The similarity should not be surprising since in both cases the linearity of Maxwell's equations leads to the formulation. The fact that coupling in the antenna problem involves radiation fields whereas in circuit analysis only induction fields are involved is only a matter of detail. That I_1 is uniquely related to V_1 and V_2 has not been shown, but is a consequence of the general uniqueness theorem in electromagnetic theory. Equations

(11.52) can be solved simultaneously for V_1 and V_2, leading to an alternative form

$$V_1 = I_1Z_{11} + I_2Z_{12} \tag{11.53a}$$
$$V_2 = I_1Z_{21} + I_2Z_{22} \tag{11.53b}$$

Just as in network theory, we can show that $Y_{12} = Y_{21}$, that is, reciprocity. For this proof we choose a volume of space bounded by a fairly complicated surface which consists of the surface at infinity and the doubly connected surfaces of the two antenna installations, as illustrated by dashed lines in Fig. 11.16, that is, $S_1 + S_{10}$ and $S_2 + S_{20}$. The surface S_{10} (and also S_{20}) is taken along the antenna surface and runs along the outer conductor of the coaxial line and includes the conducting surface that excludes the energy sources, used to generate the high-frequency currents, from the volume being considered. This surface then connects with S_1 (S_2) along the inner surface of the outer conductor and the outer surface of the inner conductor of the coaxial line. Surfaces S_1 and S_2 are the transverse planar surfaces in the coaxial line that has already been described. Let $\mathbf{E}_a$, $\mathbf{H}_a$ be the fields set up by a source inside the surface $S_1 + S_{10}$, while $\mathbf{E}_b$, $\mathbf{H}_b$ are the fields set up by a source inside the surface $S_2 + S_{20}$. If we apply (11.51) to the region bounded by the aforementioned surfaces, then since $\mathbf{E} \times d\mathbf{S}$ is zero along the conducting surfaces S_{10} and S_{20}, these integrals vanish. Furthermore, the contribution f.om the surface at infinity can be set equal to zero if we assume a vanishingly small amount of dissipation in the medium, so that $\mathbf{E}$ and $\mathbf{H}$ decrease slightly faster than $1/R$.† As a result the following relationship is arrived at, namely,

$$\int_{S_1} (\mathbf{E}_a \times \mathbf{H}_b - \mathbf{E}_b \times \mathbf{H}_a) \cdot d\mathbf{S} + \int_{S_2} (\mathbf{E}_a \times \mathbf{H}_b - \mathbf{E}_b \times \mathbf{H}_a) \cdot d\mathbf{S} = 0 \tag{11.54}$$

since $S_1 + S_2$ is the only portion of the total surface for which the integral does not vanish. Now on either S_1 or S_2 we have, for a coaxial line, that

$$\mathbf{H} = \mathbf{a}_\phi H_\phi = \frac{I}{2\pi r}\mathbf{a}_\phi \tag{11.55a}$$
$$\mathbf{E} = \mathbf{a}_r E_r \tag{11.55b}$$

and

$$V = \int_{ri}^{ro} E_r\,dr \tag{11.55c}$$

† Actually, if the nature of the radiation fields from finite sources is considered, then in absence of dissipation,

$$\oint_{S_\infty} \mathbf{E}_a \times \mathbf{H}_b \cdot d\mathbf{S} = \oint_{S_\infty} \mathbf{E}_b \times \mathbf{H}_a \cdot d\mathbf{S}$$

and the result of (11.54) is also obtained.

This means, for example, that

$$\int_{S_1} (\mathbf{E}_a \times \mathbf{H}_b) \cdot d\mathbf{S} = 2\pi \int_{ri}^{ro} E_{ra} H_{\phi b} r \, dr = I_{1b} \int_{ri}^{ro} E_{ra} \, dr = I_{1b} V_{1a} \qquad (11.56)$$

The double subscripts identify the reference plane and the source condition, that is, I_{1b} is the current at S_1 caused by the field radiated by the source inside $S_2 + S_{20}$, while V_{1a} is the voltage at S_1 as produced by the field from the source inside $S_1 + S_{10}$. Carrying out the integration of (11.54) as in the example above leads to

$$V_{1a} I_{1b} + V_{2a} I_{2b} - V_{1b} I_{1a} - V_{2b} I_{2a} = 0 \qquad (11.57)$$

When only the field $\mathbf{E}_a$, $\mathbf{H}_a$ is present, (11.52) gives

$$I_{1a} = Y_{11} V_{1a} + Y_{12} V_{2a}$$
$$I_{2a} = Y_{21} V_{1a} + Y_{22} V_{2a}$$

while if $\mathbf{E}_b$, $\mathbf{H}_b$ is the only field present,

$$I_{1b} = Y_{11} V_{1b} + Y_{12} V_{2b}$$

and

$$I_{2b} = Y_{21} V_{1b} + Y_{22} V_{2b}$$

Using these results in (11.57) yields

$$(V_{1a} V_{2b} - V_{2a} V_{1b})(Y_{12} - Y_{21}) = 0 \qquad (11.58)$$

Since the a and b conditions are completely arbitrary, (11.58) can be satisfied, in general, only if

$$Y_{12} = Y_{21} \qquad (11.59)$$

as we wished to show.

The formulation of (11.56) and the reciprocity of (11.59) were facilitated by choosing the transmission lines to be coaxial and the reference planes at a point such that only TEM waves need be considered. However, these results will also apply to waveguide feed systems for each mode separately, provided appropriate definitions of voltage and current are made. A detailed discussion of this situation may be found in Collin and other references, given at the end of the chapter.

11.8. Antenna Equivalent Circuits

Ordinarily, the separation of transmitting and receiving antennas is very great, so that if antenna 1 is transmitting, (11.53a) may be written as

$$V_1 = I_1 Z_{11} \qquad (11.60)$$

and Z_{11} is the internal impedance of the antenna at the chosen reference plane S_1; that is, for large separation, the coupling, represented by Z_{12}, can be neglected, reflecting the fact that the transmitting antenna will hardly be affected by a very distant receiving antenna. The equivalent circuit of antenna 1 is shown in Fig. 11.17, where $\mathcal{E}_1$ is the effective emf of the source and Z_{i1} its internal impedance. Under matched conditions $Z_{i1} = Z_{11} = R_{11}$ and $V_1 = \mathcal{E}_1/2$.

When antenna 1 is operating under receiving conditions, the equivalent circuit that follows from (11.53*a*) is represented as shown in Fig. 11.18.

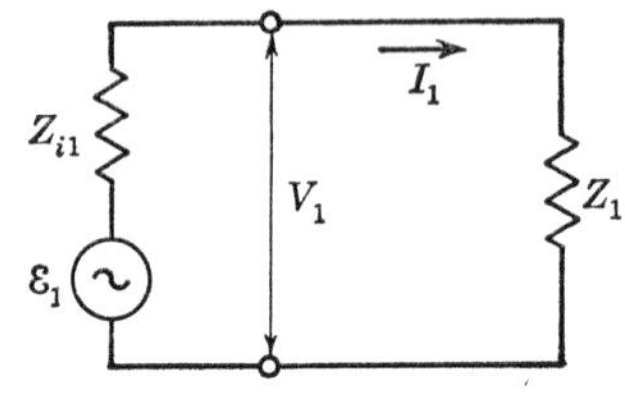

FIG. 11.17. Equivalent circuit for transmitting antenna.

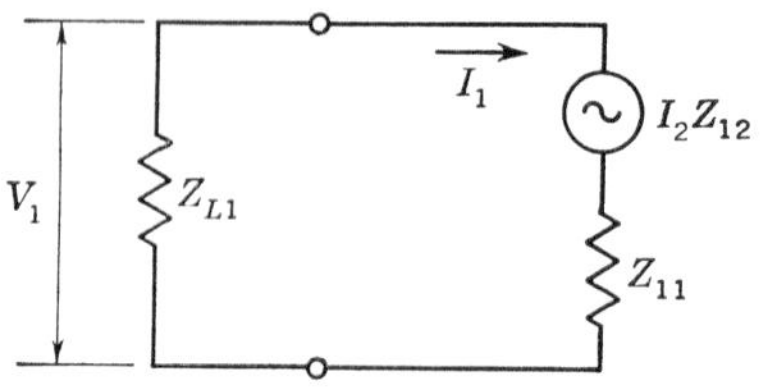

FIG. 11.18. Equivalent circuit for receiving antenna.

Here Z_{L1} is the impedance of the load. In this case we may no longer ignore Z_{12} even though it is very small, since it represents the equivalent source of energy, and in its absence no current flows in the receiving circuit. We can interpret $-I_2Z_{12}$ as the effective emf due to transmitter 2, and Z_{11} as the internal impedance of this equivalent source. In general, I_2 is a function of Z_{L1}, but since it is assumed that the antennas are weakly coupled, we can consider I_2Z_{12} as a true voltage source independent of load conditions at the receiver. Note that the impedance of an antenna as measured when transmitting is the same as its effective internal impedance when acting as a receiver and equals Z_{11} when the antennas are sufficiently separated to justify the circuit of Fig. 11.17. The received voltage V_1 can be readily calculated from the circuit of Fig. 11.18:

$$V_1 = \frac{I_2Z_{12}}{Z_{L1} + Z_{11}} Z_{L1} \tag{11.61}$$

Under matched conditions $Z_{L1} = Z_{11} = R_{11}$, and then

$$V_1 = \frac{I_2Z_{12}}{2} \tag{11.62}$$

In (11.62) the received signal depends on the current I_2 and on the parameter Z_{12}, which in turn is a function of the type of antennas and their location. But the actual mechanism of reception involves, usually, a plane wave incident on antenna 1 from a particular direction and with a certain polarization. We should be able to develop a formula that yields the received signal, given only the properties of antenna 1 and the incident wave, thereby removing specific reference to the source of the wave, i.e., antenna 2. The following discussion is devoted to this objective.

11.9. Receiving Antennas

In the last section the operation of a receiving antenna was described in terms of direct coupling to a transmitting source. A description of this interaction was formulated in terms of a four-terminal network which is capable of transforming the voltage and current at the input to the transmitting antenna into voltage and current conditions in the receiving-antenna feed line. Another point of view is possible where we consider the antenna as an element that transforms the incident electromagnetic wave into voltage and current in its transmission line. In place of the direct interaction of transmitting and receiving antennas we now view the transmitting antenna as setting up a field which the receiving antenna in turn transforms into the received signal. Since the latter method of analysis requires only a description of the incident wave on a receiving antenna, the source of that wave is unimportant.

In this section we shall determine how the voltage at the receiving-antenna reference plane S_1 is related to an incident field $\mathbf{E}_i$ whose magnitude, direction, and polarization are known. In the following discussion we shall assume that the transmitter is sufficiently far from the receiver so that the incident field may be considered to be a plane wave. Such conditions usually are obtained in practice. Since the actual source of $\mathbf{E}_i$ is irrelevant, we may suppose it to arise from an equivalent elementary current source; that is, we can always define a dipole of length Δl and current I_e such that if properly oriented and located, a field $\mathbf{E}_i$ is produced at the receiving antenna. In order to determine the desired relation between $\mathbf{E}_i$ and the received voltage, we first need to develop a slightly modified form of the reciprocity theorem.

Let us begin with Maxwell's equations, which we apply to a given region containing fixed material bodies and impressed source $\mathbf{J}_1$ or $\mathbf{J}_2$. Then

$$\nabla \times \mathbf{E}_1 = -j\omega\mu\mathbf{H}_1 \tag{11.63a}$$
$$\nabla \times \mathbf{H}_1 = j\omega\epsilon\mathbf{E}_1 + \mathbf{J}_1 \tag{11.63b}$$
$$\nabla \times \mathbf{E}_2 = -j\omega\mu\mathbf{H}_2 \tag{11.64a}$$
$$\nabla \times \mathbf{H}_2 = j\omega\epsilon\mathbf{E}_2 + \mathbf{J}_2 \tag{11.64b}$$

If we dot-multiply (11.63a) by $\mathbf{H}_2$ and (11.64b) by $\mathbf{E}_1$ and subtract (11.64b) from (11.63a), we obtain

$$\mathbf{H}_2 \cdot \nabla \times \mathbf{E}_1 - \mathbf{E}_1 \cdot \nabla \times \mathbf{H}_2 = -j\omega\mu\mathbf{H}_1 \cdot \mathbf{H}_2 - j\omega\epsilon\mathbf{E}_1 \cdot \mathbf{E}_2 - \mathbf{E}_1 \cdot \mathbf{J}_2 \tag{11.65}$$

Similarly, by dot-multiplying (11.63b) by $\mathbf{E}_2$ and (11.64a) by $\mathbf{H}_1$ and then subtracting, we obtain

$$\mathbf{E}_2 \cdot \nabla \times \mathbf{H}_1 - \mathbf{H}_1 \cdot \nabla \times \mathbf{E}_2 = j\omega\epsilon\mathbf{E}_1 \cdot \mathbf{E}_2 + j\omega\mu\mathbf{H}_1 \cdot \mathbf{H}_2 + \mathbf{E}_2 \cdot \mathbf{J}_1 \tag{11.66}$$

The left side of (11.65) can be identified as $\nabla \cdot (\mathbf{E}_1 \times \mathbf{H}_2)$, while the left side of (11.66) is $\nabla \cdot (\mathbf{H}_1 \times \mathbf{E}_2)$. If (11.65) and (11.66) are integrated over a given volume V and the divergence theorem applied to the left-hand sides and the results added, we obtain

$$\int_V (\mathbf{E}_2 \cdot \mathbf{J}_1 - \mathbf{E}_1 \cdot \mathbf{J}_2)\, dV = \oint_S (\mathbf{E}_1 \times \mathbf{H}_2 - \mathbf{E}_2 \times \mathbf{H}_1) \cdot d\mathbf{S} \quad (11.67)$$

If we now let the volume under consideration be all of space, then S is

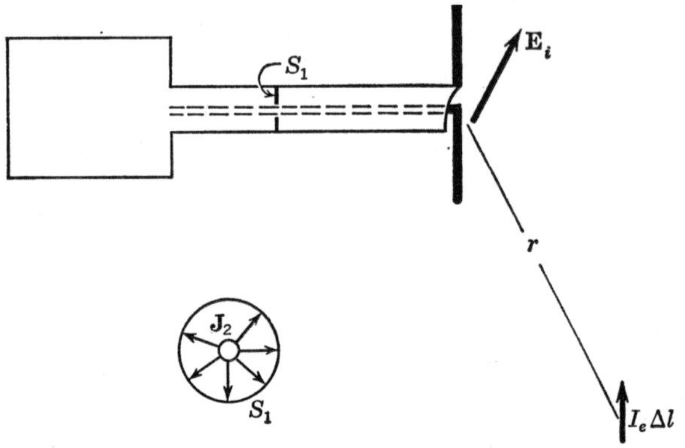

FIG. 11.19. Current source $\mathbf{J}_2$ on surface S_1.

the surface at infinity. In this case the surface integral vanishes, as noted earlier. Consequently

$$\int_V \mathbf{E}_2 \cdot \mathbf{J}_1\, dV = \int_V \mathbf{E}_1 \cdot \mathbf{J}_2\, dV \quad (11.68)$$

Note that (11.68) involves impressed currents only since induced currents are included by making ϵ complex.

Figure 11.19 illustrates the problem at hand where the antenna represents an arbitrary receiving antenna whose output voltage and current are referred to plane S_1 in the (assumed) coaxial feed line. The incident field $\mathbf{E}_i$ is taken as arising from the equivalent current element $I_e\, \Delta l$. The current element represents the source $\mathbf{J}_1$ in utilizing (11.68). For the source $\mathbf{J}_2$ in (11.68) we take a radial current source over the interconductor space on S_1. We choose the surface current density of this source to be $\mathbf{J}_s$, where

$$\mathbf{J}_s = \frac{I_0}{2\pi r}\, \mathbf{a}_r$$

This choice is a completely arbitrary one and has been made only to facilitate the use of (11.68) to arrive at an expression for the received signal.

The current source will cause a discontinuity in H_ϕ, but because of the chosen functional form, this boundary condition can be satisfied by a TEM mode alone. Consequently, the current source may be thought of as launching TEM waves propagating toward the antenna and toward the receiving load. We shall assume matched conditions so that there are no reflected waves and, further, assume that the load impedance and the radiation resistance R_{11} are both equal to the characteristic impedance of the coaxial line. Note that the terminating impedance designated as the receiving load impedance above may also be considered as the internal impedance of the current generator that maintains the source $\mathbf{J}_2$. The field radiated by $\mathbf{J}_s$ will be designated $\mathbf{E}_t$. With this notation the left side of (11.68) becomes

$$E_t I_e \, \Delta l \cos \alpha \tag{11.69}$$

since $\mathbf{J}_1$ is zero at all points in the volume except over the differential current element where $J_1 \, dV = I_e \, \Delta l$. The angle α in (11.69) is the angle between the direction of $\mathbf{E}_t$ and the polar axis of the dipole. In terms of the geometry given in Fig. 11.20 we have $\cos \alpha = \cos \psi \sin \theta$,

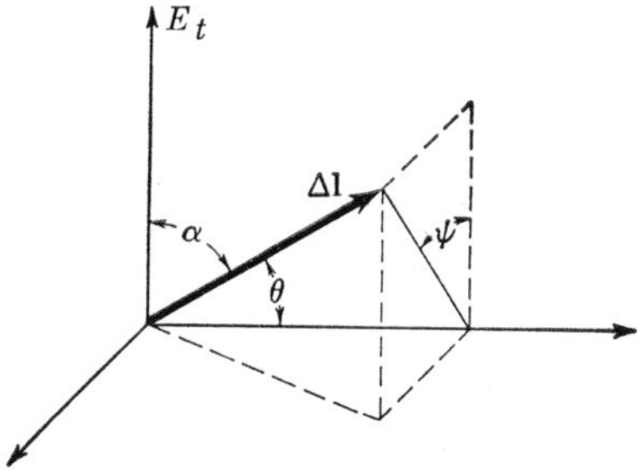

FIG. 11.20. Dipole geometry.

where ψ is the angle between the plane of polarization (that formed by $\mathbf{E}_t$ and the direction of the wave normal) and the plane of the dipole, and θ is the polar angle referred to the dipole axis. The entire expression (11.68) reduces to

$$E_t I_e \, \Delta l \cos \psi \sin \theta = \int_{S_1} \frac{E_r I_0}{2\pi r} \, dS = I_0 \int_{r_1}^{r_2} E_r \, dr = I_0 V_r \tag{11.70}$$

where E_r is the TEM received field at S_1 due to the incident field $\mathbf{E}_i$, while V_r is the received signal voltage.

Equation (11.70) can be put into a more useful form if the following energy relationships are incorporated. When the source is $\mathbf{J}_1$ (the elementary current radiator), then the radiated power is

$$P_r = \tfrac{1}{2} I_e^2 R_{rad} = \tfrac{1}{2} I_e^2 80\pi^2 \left(\frac{\Delta l}{\lambda_0}\right)^2 \tag{11.71}$$

from (11.11). But if $\mathbf{E}_i$ is the field set up at the receiving antenna, then

$$\frac{E_i^2}{2Z_0} = \frac{P_r G_e}{4\pi r^2} \tag{11.72}$$

where G_e is the gain of the current element radiator in the direction of the receiving antenna, that is, $G_e = 1.5 \sin^2 \theta$, and r is the separation between antennas. Consequently,

$$E_i = \frac{60\pi I_e \, \Delta l \sin \theta}{r\lambda_0} \tag{11.73}$$

When $\mathbf{J}_s$ is active as the source (condition 2), then it equals the discontinuity in H_ϕ across S_1. From the assumption that the antenna and load are matched to the transmission line, symmetry requires that the input current to the antenna be $I_0/2$. Consequently, the radiated power is

$$P_t = \frac{1}{2}\left(\frac{I_0}{2}\right)^2 R_{11} \tag{11.74}$$

The field set up by the antenna at the location of the current element is then such that

$$\frac{1}{2}\frac{E_t^2}{Z_0} = \frac{P_t G}{4\pi r^2} \tag{11.75}$$

where $G = G(\phi,\theta)$ is the gain of the antenna in the direction of the current element. Solving for E_t from (11.74) and (11.75) gives

$$E_t = \frac{(30GR_{11})^{1/2}}{2r} I_0 \tag{11.76}$$

Finally, by substituting (11.76) and (11.73) into (11.70), we obtain the desired result:

$$V_r = \frac{E_t I_e \, \Delta l}{I_0} \cos \psi \sin \theta = \frac{\lambda_0}{2\pi}\left(\frac{R_{11}G}{120}\right)^{1/2} E_i \cos \psi \tag{11.77}$$

When the receiving antenna is oriented for maximum reception,† $\cos \psi = 1$. In this case,

$$V_r = \frac{1}{2}\frac{\lambda_0 E_i}{\pi}\left(\frac{R_{11}G}{120}\right)^{1/2} \tag{11.78}$$

This result is perfectly general, except for the assumptions of matched conditions, linear polarization, and orientation for maximum signal reception. It relates the received voltage to the incident wave. The fact that

† This supposes that the antenna on transmission produces a linearly polarized wave, as has been tacitly assumed. If this is not the case, the problem must be reformulated in terms of the sum of two linearly polarized waves of appropriate relative magnitude and phase.

we chose to think of the incident wave as arising from a dipole source is completely irrelevant to the relationship described by (11.78). Note also that (11.78) is independent of the choice of the assumed source $\mathbf{J}_s$. In fact, (11.78) can be derived without introducing the source $\mathbf{J}_s$. This alternative approach is left as a problem.

From the above result we see that for an incident wave of a given magnitude the maximum received signal depends only on the direction from which the wave comes. This functional dependence is described by the gain $G(\phi,\theta)$ of the receiving antenna in the direction of the incident wave. We may now understand why receiving antennas are not treated as such in the general literature. This is because the characteristics of the transmitting antenna, namely, R_{11} and $G(\phi,\theta)$, are precisely those needed to analyze the same antenna on reception. The transmitting pattern is, as noted, identical with the receiving pattern.

For the matched system the total power absorbed in the load is given by

$$P_{abs} = \frac{V_r^2}{2R_{11}} = \frac{1}{8}\frac{\lambda_0^2}{\pi^2}\frac{E_i^2}{120}G(\phi,\theta)$$

The absorption cross section of an antenna is the effective area of interception of an incident plane wave; that is,

$$P_{abs} = \tfrac{1}{2}A\,\frac{E^2}{120\pi}$$

where A is the absorption cross section. For any matched antenna we have

$$A(\theta,\phi) = \frac{G(\theta,\phi)\lambda_0^2}{4\pi} \tag{11.79}$$

In the treatment of antennas in this chapter we tacitly assumed that both the transmitting and receiving antennas were in free space. In a practical communications system additional (parasitic) sources must be considered. For example, the effect of the ground must be considered if substantial energy is directed earthward. This may often be disposed of by assuming perfect conductivity and using the method of images. Greater accuracy can be obtained if the earth is considered flat and homogeneous but a complex dielectric constant used to describe its properties. The formula of Chap. 10 for reflection from a dielectric interface may then be used.

For certain ranges of frequency, propagation effects such as those due to the ionosphere or the atmosphere must be considered. Furthermore, diffraction effects due to the earth itself or to features on the earth may have to be taken into account. For a consideration of these problems the student is referred to the suggested references below.

BIBLIOGRAPHY

Antennas

Brown, J.: "Microwave Lenses," Methuen & Co., Ltd., London, 1953.

Jasik, H.: "Antenna Engineering Handbook," McGraw-Hill Book Company, Inc., New York, 1961.

Jordan, E. C.: "Electromagnetic Waves and Radiating Systems," Prentice-Hall, Inc., Englewood Cliffs, N.J., 1950.

Kraus, J. D.: "Antennas," McGraw-Hill Book Company, Inc., New York, 1950.

Schelkunoff, S. A.: "Advanced Antenna Theory," John Wiley & Sons, Inc., New York, 1952.

——— and H. T. Friis: "Antennas: Theory and Practice," John Wiley & Sons, Inc., New York, 1952.

Silver, S.: "Microwave Antenna Theory and Design," McGraw-Hill Book Company, Inc., New York, 1949.

Wait, J. R.: "Electromagnetic Radiation from Cylindrical Structures," Pergamon Press, New York, 1959.

Arrays

Dolph, C. L.: A Current Distribution for Broadside Arrays Which Optimizes the Relationship between Beam Width and Side-lobe Level, *Proc. IRE*, vol. 34, no. 6, p. 335, 1946.

Ruze, J.: Physical Limitations on Antennas, *MIT Research Lab. Electronics Tech. Rept.* 248, 1952.

Schelkunoff, S. A.: A Mathematical Theory of Linear Arrays, *Bell System Tech. J.*, vol. 22, p. 80, 1943.

Taylor, T. T.: Design of Line Source Antennas for Narrow Beamwidth and Low Side-lobes, *IRE Trans.*, vol. AP-3, p. 16, 1955.

Propagation

Bremmer, H.: "Terrestrial Radio Waves," Elsevier Publishing Company, Amsterdam, 1949.

Jordan, E. C.: "Electromagnetic Waves and Radiating Systems," Prentice-Hall, Inc., Englewood Cliffs, N.J., 1950.

Kerr, D. E. (ed.): "Propagation of Short Radio Waves," vol. 13, MIT Radiation Laboratory Series, McGraw-Hill Book Company, Inc., New York, 1951.

Reed, H. R., and C. M. Russell: "Ultra High Frequency Propagation," John Wiley & Sons, Inc., New York, 1953.

Microwave circuits

Collin, R. E.: "Field Theory of Guided Waves," McGraw-Hill Book Company, Inc., New York, 1960.

Montgomery, C. G., R. H. Dicke, and E. M. Purcell: "Principles of Microwave Circuits," vol. 8, MIT Radiation Laboratory Series, McGraw-Hill Book Company, Inc., New York, 1948.

CHAPTER 12

INTERACTION OF CHARGED PARTICLES WITH FIELDS

In this chapter we shall consider a class of applications which arise from the interaction of fields and moving charge. In the simplest case we consider the movement of charged particles in static fields, where it is assumed that the perturbation of the existing fields by the particles themselves can be neglected. The problem is essentially a ballistic one, and it will be necessary to make use of the electromagnetic-force laws in conjunction with the laws of mechanics. Applications to electron-tube devices such as the cathode-ray tube and electron lenses will be given.

We then proceed to a consideration of the electron stream under the influence of time-varying electric fields for conditions where magnetic forces may be neglected. However, the modifying effect of the charges themselves on the electric field will be included in this analysis. With the theory developed, it will be possible to consider applications to space-charge control vacuum tubes with parallel-plane geometry.

A more thorough account of the interaction of fields and electron streams follows. The theory will be developed for cylindrical geometry and will serve to illustrate the existence, under appropriate conditions, of space-charge waves. It will be possible to explain the operation of high-frequency velocity-modulated-type tubes such as the klystron and traveling-wave amplifier on the basis of this work.

Next discussed is propagation in gyrotropic media, with specific application to propagation through the ionosphere and ferrites. The chapter concludes with an introduction to magnetohydrodynamics. As in the earlier sections, the student will have an opportunity to study applications to systems that must be described by both electromagnetic and mechanical laws.

12.1. Charged Particles in Static Fields—Fundamental Equations

In order to determine the trajectory of a charged particle moving in a static electromagnetic field, we shall need to know the force that acts owing to the field, as well as the appropriate laws of mechanics. The pertinent equations are

$$\mathbf{F} = q(\mathbf{E} + \mathbf{v} \times \mathbf{B}) \tag{12.1}$$

$$\mathbf{F} = \frac{d}{dt}(m\mathbf{v}) \tag{12.2}$$

Equation (12.2) is a generalization of Newton's law which allows for a relativistic correction through taking

$$m = \frac{m_0}{(1 - v^2/c^2)^{1/2}} \tag{12.3}$$

where m_0 is the rest mass, v the particle velocity, and c the velocity of light. For $v \ll c$ the mass is essentially constant and (12.2) assumes the classical form

$$\mathbf{F} = m\frac{d\mathbf{v}}{dt} \tag{12.4}$$

Consider a charged particle q on which a force, given by (12.1), acts. The work done by the field in moving the particle from some point 1 to a point 2 can be expressed as

$$W_{12} = \int_1^2 \mathbf{F} \cdot d\mathbf{l} = \int_1^2 q(\mathbf{E} + \mathbf{v} \times \mathbf{B}) \cdot d\mathbf{l} \tag{12.5}$$

This can also be written in terms of the mechanical quantities given in (12.4). In this case,

$$W_{12} = \int_1^2 \mathbf{F} \cdot d\mathbf{l} = m\int_1^2 \frac{d\mathbf{v}}{dt} \cdot d\mathbf{l} \tag{12.6}$$

Since velocity and displacement are both functions of time, (12.6) can be written

$$W_{12} = m\int_1^2 d\mathbf{v} \cdot \frac{d\mathbf{l}}{dt} = m\int_1^2 \mathbf{v} \cdot d\mathbf{v} = \frac{m}{2}\int_1^2 d(\mathbf{v} \cdot \mathbf{v}) = \frac{m}{2}(v_2^2 - v_1^2) \tag{12.7}$$

thus obtaining the familiar result that the work done by the field equals the net increase in kinetic energy of the particle.

For static fields **E** may be derived from the gradient of a scalar potential Φ, so that (12.5) simplifies to

$$W_{12} = -q\int_1^2 \nabla\Phi \cdot d\mathbf{l} + q\int_1^2 \mathbf{v} \times \mathbf{B} \cdot d\mathbf{l} \tag{12.8}$$

In the triple scalar product of (12.8) the dot and cross may be interchanged. But then

$$q\int_1^2 \mathbf{v} \times \mathbf{B} \cdot d\mathbf{l} = q\int_1^2 (d\mathbf{l} \times \mathbf{v}) \cdot \mathbf{B} = 0 \tag{12.9}$$

the integral going to zero since $d\mathbf{l} = \mathbf{v}\,dt$. That the Lorentz force contributes nothing to the total energy of the particle is understandable, since $\mathbf{v} \times \mathbf{B}$ is always directed at right angles to the instantaneous displacement. The presence of the Lorentz force will, however, affect the direction along which the particle moves. The first term in (12.8) can be integrated, so that combined with (12.7) there results

$$\frac{m}{2}(v_2^2 - v_1^2) = -q(\Phi_2 - \Phi_1) \tag{12.10}$$

For an electron with charge $q = -e$, starting from rest and at zero potential ($v_1 = \Phi_1 = 0$), it acquires a velocity v in accelerating to a potential V, which from (12.10) is given by

$$v = \left(\frac{2eV}{m}\right)^{1/2} \tag{12.11}$$

Putting in the electronic charge of $e = 1.6 \times 10^{-19}$ coulomb and mass of 0.91×10^{-30} kilogram, (12.11) becomes

$$v = 5.93 \times 10^5 V^{1/2} \qquad \text{m/sec} \tag{12.12}$$

For V equal to 1 volt, the energy acquired by an electron is 1.6×10^{-19} joule. This is often a convenient unit of energy and is called an electron volt.

For $V = 5{,}000$ volts, $v/c = 0.14$ from (12.12), and the error in neglecting the relativistic correction to the electron's mass is approximately 1 per cent. In the following work, where relativistic effects are neglected, we may assume this to be justified when the maximum accelerating potential is less than about 5,000 volts.

12.2. Motion of Charged Particles in an Electrostatic Field

A charged particle in a uniform electric field is subjected to a constant force and consequently follows a trajectory that, by analogy with a

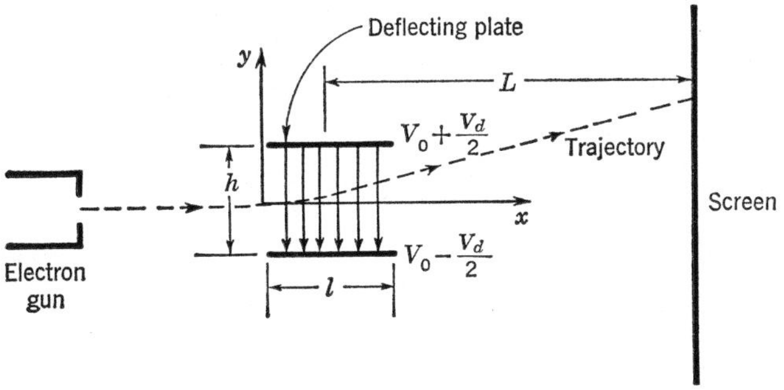

FIG. 12.1. Cathode-ray tube with electrostatic deflection.

particle in a gravitational field, is parabolic. The cathode-ray tube with electrostatic deflection furnishes a practical application for illustrative purposes.

The main elements of the cathode-ray tube are illustrated in Fig. 12.1.

The gun furnishes a supply of electrons which are collimated by that structure into a narrow beam. The deflecting plates and the region between the plates and screen are kept at a high accelerating potential of V_0. A deflection potential is superposed at the deflecting plates and consists of a voltage of magnitude $V_d/2$ at the upper plate and $-V_d/2$ at the lower plate. Between the deflection plates, therefore, an electric field V_d/h exists, where h is the plate separation. We assume that the modification of the electric field due to the electron stream itself can be neglected (this assumes relatively low values of beam current) and also that relativistic effects can be ignored. The velocity of an electron at the left-hand edge of the deflection plates is

$$v_x = \left(\frac{2eV_0}{m}\right)^{1/2} \tag{12.13}$$

and this velocity component is unaffected by the field within the deflecting-plate region (fringing fields being neglected). For the y component of velocity, (12.1) and (12.4) give

$$\frac{dv_y}{dt} = \frac{eV_d}{mh}$$

or

$$v_y = \frac{eV_d t}{mh} \tag{12.14}$$

The trajectory of the electron path is given in parametric form by

$$x = \left(\frac{2eV_0}{m}\right)^{1/2} t \tag{12.15}$$

$$y = \frac{eV_d}{mh}\frac{t^2}{2} \tag{12.16}$$

if (12.13) and (12.14) are integrated with respect to time. If the parameter t is eliminated, the parabolic nature of the trajectory is revealed in the more familiar form; i.e.,

$$y = \frac{1}{4h}\frac{V_d}{V_0}x^2 \tag{12.17}$$

The time required for the electron to pass through the plate region is simply the length of the plates divided by the x component of velocity. If we denote this time interval by t_d, we have

$$t_d = \frac{l}{v_x} \tag{12.18}$$

Consequently, the exit velocity has the components

$$v_x = \left(\frac{2eV_0}{m}\right)^{1/2} \tag{12.19a}$$

$$v_y = \frac{eV_d l}{mhv_x} \tag{12.19b}$$

Since the potential in the region between the deflection plates and screen is constant, i.e., field-free, the electron continues with the velocities given by (12.19) until it hits the screen. Its path is therefore linear, with a deflection angle θ given by

$$\theta = \tan^{-1}\frac{v_y}{v_x} = \tan^{-1}\frac{eV_d l}{mhv_x^2} = \tan^{-1}\frac{V_d l}{2hV_0} \tag{12.20}$$

The geometry is redrawn in Fig. 12.2. If the linear path of the electron as it leaves the plate region is projected backward, then it is easy to show

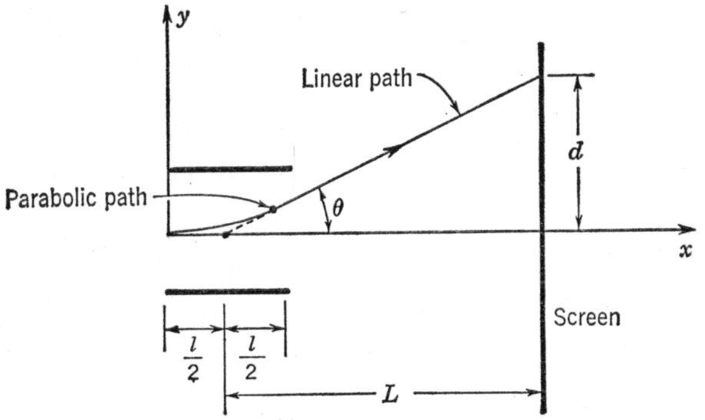

FIG. 12.2. Deflection in an electrostatic-type oscilloscope.

that it passes through the point $(l/2, 0)$. If L is the distance from the center of the deflection-plate region to the screen, the deflection d of the electron beam is given by

$$d = L \tan\theta = \frac{lL}{2hV_0} V_d \tag{12.21}$$

We note a very satisfactory proportionality between the deflecting voltage V_d and the resultant beam deflection d. Although this result was obtained under static conditions, it may be assumed that if V_d is a time-varying signal, then (12.21) will still be true provided the time variations are not too rapid. The significant criterion, in this connection, is the relative magnitude of the electron transit time through the region of the deflecting plates as compared with the time over which significant deflection-voltage changes occur. When the latter ratio is small, (12.21) is satisfactory as a quasi-static solution. Under harmonic conditions the requirement would be that the transit time be short compared with a period of the applied deflection-plate voltage.

The motion of electrons in nonuniform fields is of considerable interest in the field of electron optics. Consider, as in Fig. 12.3, an electron with velocity $\mathbf{v}_1$ at an equipotential surface Φ_1. The electron leaves a closely adjoining equipotential surface Φ_2 with velocity $\mathbf{v}_2$. The change in velocity is given approximately by

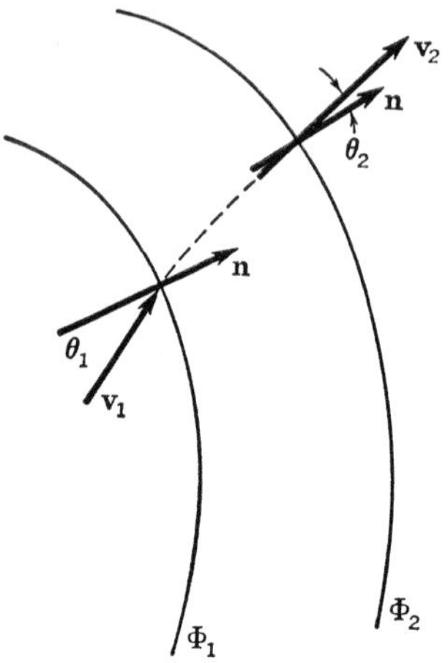

FIG. 12.3. Snell's law for electron optics.

$$\Delta\mathbf{v} = \frac{\mathbf{F}\,\Delta t}{m} = \nabla\Phi \frac{e}{m} \Delta t \tag{12.22}$$

with the positive sign arising since electrons are being considered. Since $\Delta\mathbf{v}$ is in the direction of the gradient, the component of velocity parallel to the equipotential surface is unperturbed. Consequently, with reference to Fig. 12.3, the following equation can be obtained:

$$v_1 \sin\theta_1 = v_2 \sin\theta_2 \tag{12.23}$$

Making use of (12.11) transforms (12.23) into

$$\frac{\sin\theta_1}{\sin\theta_2} = \left(\frac{\Phi_2}{\Phi_1}\right)^{1/2} \tag{12.24}$$

This equation is the analogy of Snell's law in optics. Results from work in optics may thus be applied in electron optics if the principle of duality is used to convert the index of refraction in light optics into an equivalent square root of potential in electron optics. This result can actually be established on a more general and fundamental basis. In optics Fermat's law requires the following integral to be stationary (usually a minimum):

$$\delta \int_1^2 \eta \, dl = 0 \tag{12.25}$$

where η is the index of refraction, and δ signifies a variation in the integrated path length. For an electron in an electrostatic field, the principle of least action in mechanics requires that

$$\delta \int_1^2 v \, dl = C\delta \int_1^2 \Phi^{1/2} \, dl = 0 \tag{12.26}$$

Again the duality between η and $\Phi^{1/2}$ is apparent.

A direct application of the above results is the double-layer lens illustrated in Fig. 12.4. The interior is maintained at a potential Φ_2, while the outer region is kept at a potential Φ_1. To accomplish this and still allow the passage of electrons through the lens, the material used for the lens is wire mesh. If the object side has a radius of R_1 and the image

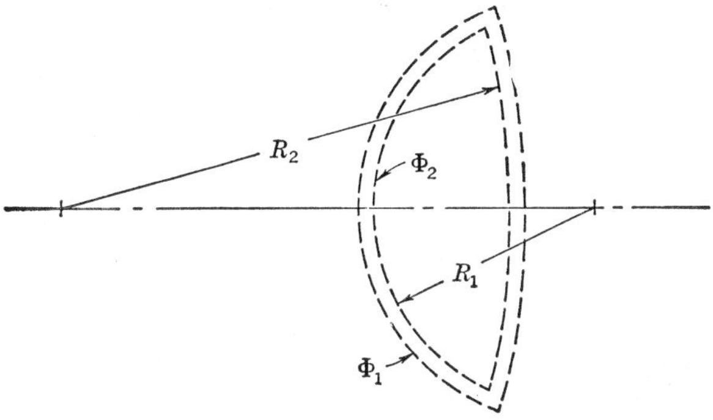

FIG. 12.4. Double-layer lens.

side a radius of R_2, then the thin-lens formula of optics can be used if η is replaced by $\Phi^{1/2}$. The result is

$$\frac{1}{f} = \left(\frac{\Phi_2^{1/2}}{\Phi_1^{1/2}} - 1\right)\left(\frac{1}{R_1} + \frac{1}{R_2}\right) \tag{12.27}$$

where f is the focal length of the lens.

The double-layer lens is actually a rather crude approximation since in practice there is considerable difficulty in providing the required potential distribution and unimpeded transmission of electrons. The type of electron lens used in practice makes use of gradual variations in potential and does not correspond to the lens of optics, which depends on abrupt changes in the index of refraction between regions with no variations. The counterpart of the latter class of electron lens would be an optical lens with continuously varying index of refraction. A complete discussion of electron lenses is outside the scope of the present work. The interested student is referred to several references in the field of electron optics, given at the end of this chapter, where the subject is developed more completely.

12.3. Motion of Charged Particles in a Magnetostatic Field

We shall investigate the motion of an electron in a uniform static magnetic field by considering the operation of the magnetic focusing coil illustrated in Fig. 12.5. It will be assumed that the coil produces an essentially uniform axial magnetic field. Electrons emerging from the gun will depart from being truly axial in some statistical way; thus injection velocities are described by such possibilities as are labeled A, B, C in

Fig. 12.5. The Lorentz force on the electron is

$$\mathbf{F} = -e\mathbf{v} \times \mathbf{B}$$

and since **B** is directed axially, a force results whose magnitude is

$$F = ev_rB_z \tag{12.28}$$

where v_r is the component of initial velocity in a direction transverse to

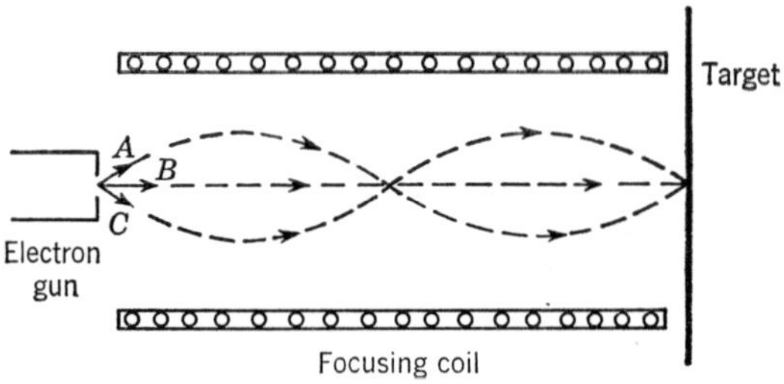

FIG. 12.5. Magnetic focusing coil.

the axis. Since the direction of this force is and remains normal to the transverse component of velocity, the magnitude of v_r does not change, and hence F in (12.28) is constant. Consequently, the particle executes a circular motion in the transverse plane. At the same time the axial velocity v_z is unaffected by the magnetic field, so that the electron's trajectory is a circular helix.

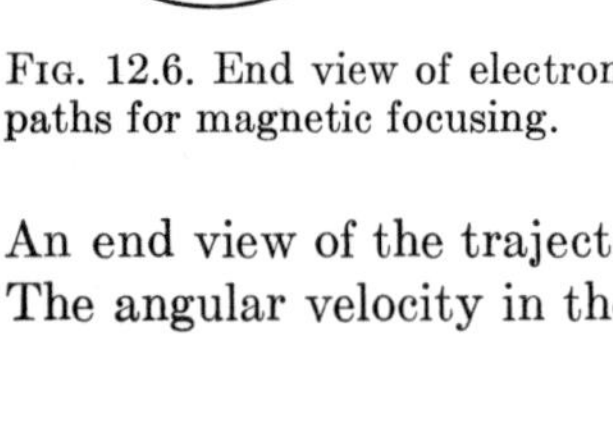

FIG. 12.6. End view of electron paths for magnetic focusing.

The force given by (12.28) is a centripetal force and under equilibrium conditions must equal the mass of the electron times the centripetal acceleration. Consequently,

$$\frac{mv_r^2}{R} = v_reB_z \tag{12.29}$$

where R is the radius of the projected circle. An end view of the trajectories for paths A and C is shown in Fig. 12.6. The angular velocity in the projected circular path is

$$\omega = \frac{v_r}{R} = \frac{eB_z}{m} \tag{12.30}$$

and the period is, consequently,

$$T = \frac{2\pi}{\omega} = \frac{2\pi m}{eB_z} = \frac{3.57 \times 10^{-11}}{B_z} \quad \text{sec} \tag{12.31}$$

Note that T does not depend on the magnitude of the radial velocity, although the radius does depend on v_r, since

$$R = \frac{v_r m}{eB_z} \tag{12.32}$$

If the electrons emerge from the gun with $v_r \ll v_z$, then the transit time

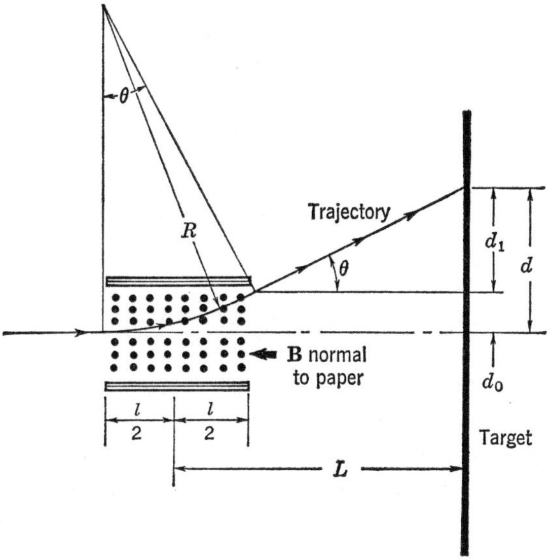

FIG. 12.7. Magnetic deflection system.

from gun to target, $t_d = l/v_z$, is essentially the same for all electrons. If B_z is chosen so that $t_d = nT$, that is,

$$B_z = \frac{3.57 \times 10^{-11}}{t_d} n \tag{12.33}$$

where n is an integer, then all electrons will have executed an integral number of revolutions by the time they reach the target, thus achieving a point focus. If the finiteness of the gun aperture is considered, then a virtual image of it is formed at the target plane. However, the focusing system is not perfect when the variation in initial velocity and mutual repulsive forces of the electrons is considered.

A static magnetic field can also be used to achieve electron-beam deflection in a cathode-ray tube. As illustrated in Fig. 12.7, the magnetic field is approximately uniform over the length of the deflection coil and is directed normal to the plane of the paper. Since the electrons enter at right angles to the field, they follow a circular path with radius

$$R = \frac{mv}{eB} = \frac{1}{B}\left(\frac{2m\Phi}{e}\right)^{1/2} \tag{12.34}$$

In (12.34) Φ is the accelerating potential which exists throughout the region depicted in Fig. 12.7. Referring to the geometry in Fig. 12.7, we see that

$$d_0 = R - (R^2 - l^2)^{1/2} \tag{12.35}$$

$$d_1 = \left(L - \frac{l}{2}\right)\tan\theta = \frac{l(L - l/2)}{(R^2 - l^2)^{1/2}} \tag{12.36}$$

$$d = d_0 + d_1 \tag{12.37}$$

For $R \gg l$ (small angles of deflection) the approximation

$$(R^2 - l^2)^{1/2} \approx R\left[1 - \frac{1}{2}\left(\frac{l}{R}\right)^2\right]$$

can be used, with the result that

$$d = \frac{lL}{R} = \frac{lLB}{\Phi^{1/2}}\left(\frac{e}{2m}\right)^{1/2} \tag{12.38}$$

As in the discussion of electrostatic deflection under quasi-static conditions, (12.38) will correctly describe the deflection for a relatively slowly varying field B also. Note that the deflection d is directly proportional to the deflecting field B.

12.4. Motion of Charged Particles in a Static Electric and Magnetic Field

In the previous types of deflection systems the electron strikes the screen obliquely. In low-velocity scanning devices such as the orthicon and vidicon, it is desirable that the electron be directed orthogonally to the screen. We shall examine a system that accomplishes this since it furnishes an example of motion of an electron in a combined static **E** and **B** field at right angles to each other.

Consider electrons emitted at the origin with zero initial velocity into a region of uniform electric and magnetic fields. The field **E** is in the negative z direction, and **B** is in the positive y direction. Equation (12.1) can be written in rectangular coordinates to give, for the present problem,

$$m\frac{dv_x}{dt} = ev_zB_y \tag{12.39a}$$

$$m\frac{dv_y}{dt} = 0 \tag{12.39b}$$

$$m\frac{dv_z}{dt} = eE_z - ev_xB_y \tag{12.39c}$$

Note that as written E_z is positive for **E** directed in the negative z direction.

Differentiating (12.39c) with respect to time and substituting from (12.39a) result in the following differential equation for v_z:

$$\frac{d^2v_z}{dt^2} + \left(\frac{eB_y}{m}\right)^2 v_z = 0 \tag{12.40}$$

Equation (12.40), when solved subject to the initial conditions that

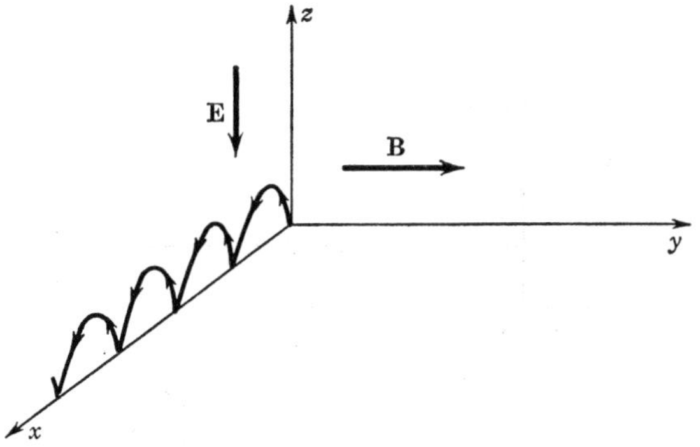

FIG. 12.8. Electron motion in a crossed **E** and **B** field.

$v_x = v_z = 0$, $dv_z/dt = eE_z/m$ at $t = 0$, results in

$$v_z = \frac{E_z}{B_y} \sin \omega_r t \tag{12.41}$$

where $\omega_r = eB_y/m$ and is called the cyclotron frequency. An expression for v_x can be obtained from (12.39a) and (12.41) which gives

$$v_x = \frac{eB_y}{m} \int v_z \, dt = \frac{E_z}{B_y} (1 - \cos \omega_r t) \tag{12.42}$$

where the initial condition $v_x = 0$ at $t = 0$ has been used to evaluate the constant of integration.

The electron's trajectory is in the xz plane, since v_y is initially zero, and from (12.39b) it remains zero. The parametric expressions for the trajectory can now be found. We have

$$x = \int_0^t v_x \, dt = \frac{mE_z}{eB_y^2} \left(\frac{eB_y}{m} t - \sin \omega_r t\right) \tag{12.43a}$$

$$z = \int_0^t v_z \, dt = \frac{mE_z}{eB_y^2} (1 - \cos \omega_r t) \tag{12.43b}$$

These are parametric equations for a cycloid. An illustration of the trajectory is given in Fig. 12.8. There will be zero displacement in

z whenever $\omega_r t = 2n\pi$. Under these conditions it is also true that $v_z = v_x = 0$. Thus when $\omega_r t = 2n\pi$,

$$z = v_x = v_y = v_z = 0 \tag{12.44a}$$

$$x = \frac{2n\pi E_z}{\omega_r B_y} \tag{12.44b}$$

where n is any integer.

On the basis of the previous analysis it is clear that an initial velocity v_y would be undisturbed. Considering such a case, there is a superposed

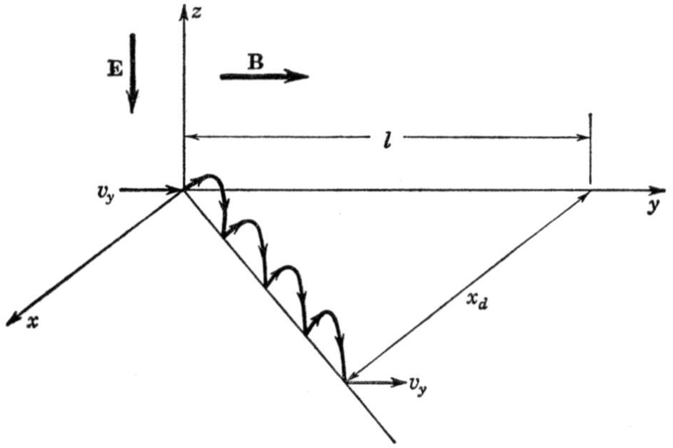

FIG. 12.9

displacement in the y direction in addition to that in the x and z directions, as already determined. The displacement in y is simply given by

$$y = v_y t \tag{12.45}$$

and the final path followed by the electron is that illustrated in Fig. 12.9. The electric field which acts in the z direction, that is, E_z, is arranged to be zero outside the region $0 < y < l$, as in Fig. 12.9; the transit time through the region is $t_d = l/v_y$. If we make $\omega_r t_d = 2n\pi$, then the electron reaches the plane $y = l$ at the moment when $v_x = v_z = z = 0$, so that it emerges with a velocity v_y in the y direction and with no net displacement in z. It will thus hit a screen in a plane parallel to the xz plane at right angles. The effect of the crossed fields is to obtain a deflection x_d given by

$$x_d = \frac{2n\pi E_z}{\omega_r B_y} \tag{12.46}$$

The parameters can be more clearly related by solving for B_y from

$$\omega_r = \frac{eB_y}{m} = \frac{2n\pi}{l} v_y$$

whereupon

$$B_y = \frac{2n\pi}{l}\frac{m}{e}\left(\frac{2e\Phi_0}{m}\right)^{1/2} \tag{12.47}$$

with Φ_0 being the accelerating potential. If the deflection voltage is Φ_d and is applied to parallel plates with a spacing h, then

$$x_d = \frac{l^2}{4n\pi h}\frac{\Phi_d}{\Phi_0} \tag{12.48}$$

Note that the axial magnetic field can extend into the region $y > l$ and will then behave as a focusing field.

The electron motion in the crossed electric and magnetic fields in Fig. 12.8 is also of importance in the operation of the traveling-wave magnetron. For this application the initial conditions are chosen, so that when $t = 0$, we have $x = y = z = v_y = v_z = 0$, $v_x = E_z/B_y$. In this case, since $v_x \neq 0$ for $t = 0$, the initial conditions that led to (12.41) are changed. We now have, instead,

$$v_z = \left(\frac{E_z}{B_y} - v_x\bigg|_{t=0}\right)\sin \omega_r t \tag{12.49}$$

which reduces to $v_z = 0$. Equation (12.42) is also modified and becomes

$$v_x = \text{constant} = \frac{E_z}{B_y} \tag{12.50}$$

The initial conditions are consequently those for which the trajectory degenerates into a straight line along the x axis.

12.5. Fields in the Presence of Space Charge

In the previous sections we described the motion of charges in static fields under the assumption that the charges themselves do not affect the field. In other words, we assumed the charge density of the electron stream to be sufficiently small so that charge interaction was negligible. In this section we shall remove this restriction. Attention will be confined, however, to infinite-plane geometry, so that only one-dimensional current flow will be considered. For a wide class of problems this restriction will not be too severe.

Figure 12.10 illustrates the flow of current in the negative x direction, that is, flow of electrons in the positive x direction. The interaction with an axial electric field $E_x(x,t)$ is desired. The current is $i(x,t)$, which, along with the electric field, has no variation with y or z. The total current at any plane x is made up of a convection current $i(x,t)$ and a displacement current $-\partial D_x/\partial t$. If the total current is designated $i_T(x,t)$, then

$$i_T(x,t) = i(x,t) - \epsilon_0 \frac{\partial E_x(x,t)}{\partial t} \tag{12.51}$$

Since $i = -\rho v$, this can be written as

$$i_T(x,t) = -\rho v - \epsilon_0 \frac{\partial E_x}{\partial t} \tag{12.52}$$

The currents i_T and i are the currents that flow across a unit cross-sectional area, and ρ is negative for electrons. The minus sign in (12.51) and (12.52) arises because the positive direction of current flow is taken to be in the negative x direction.

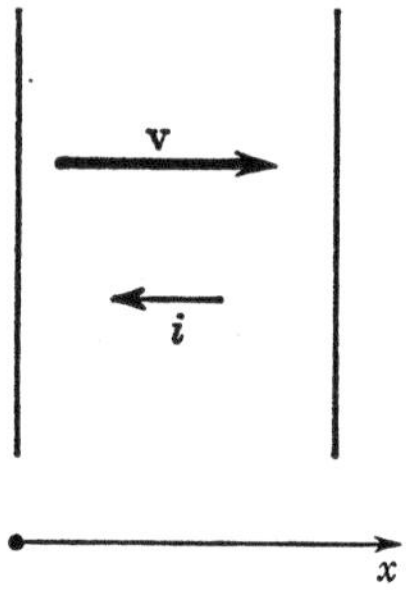

FIG. 12.10. Parallel-plane current flow.

The reason for the choice of positive direction of current flow is that when the equivalent impedance between the two electrodes is evaluated, with the right-side electrode positive relative to the left-side one, the usual circuit definition of impedance may be applied without introducing a minus sign. The total conduction current that would flow in an external circuit connected to the two electrodes in Fig. 12.10 is also equal to that given by (12.52). This is because the total current is solenoidal in nature, as demonstrated in Sec. 9.2. Consequently, the conduction current external to the electrode space must equal the sum of the convection and displacement currents flowing across any plane in the interelectrode region. The parallel-plate-capacitor problem discussed in Sec. 9.2 is an example of this type of phenomenon when the convection current is zero.

From the divergence equation for the electric field, we have

$$\nabla \cdot \mathbf{D} = \rho = \epsilon_0 \frac{\partial E_x}{\partial x} \tag{12.53}$$

If we consider an electron in the current stream as it moves in the x direction, we may replace v by dx/dt in (12.52), and by substituting from (12.53), we obtain

$$i_T = -\epsilon_0 \left(\frac{\partial E_x}{\partial x} \frac{dx}{dt} + \frac{\partial E_x}{\partial t} \right) \tag{12.54}$$

But the total change in E_x with time, as seen by an observer moving with the electron stream, is simply

$$\frac{dE_x}{dt} = \frac{\partial E_x}{\partial x} \frac{dx}{dt} + \frac{\partial E_x}{\partial t}$$

Consequently,

$$i_T(t) = -\epsilon_0 \frac{dE_x(t)}{dt} \tag{12.55}$$

Note that in this formulation i_T and E_x are explicit functions of time only, since we are following a particle in the electron stream.

Let us now consider the application of Newton's law to this particular particle in the electron stream. From (12.1) and (12.4) we get

$$m\frac{d^2x}{dt^2} = -eE_x \tag{12.56}$$

The Lorentz force term may be neglected in comparison with the electric force if, as we assume, $v \ll c$ for the electron stream.†

A brief review of the previous steps may be helpful at this point. In determining the effect of the electron stream on the electric field, Poisson's equation has been used. In doing so, the "granular" nature of the electron stream has been ignored and an average value of charge density ρ has been employed instead. This procedure recognizes that we are actually interested only in statistically average effects. The resultant electric field is also, then, a statistical average field and allows us to compute the dynamic effect on a representative electron by means of (12.56).

Differentiating both sides of (12.56) with respect to time and inserting (12.55) gives the Benham-Müller-Llewellyn equation

$$\frac{d^3x}{dt^3} = \frac{e}{m\epsilon_0} i_T(t) \tag{12.57}$$

Equation (12.57) is a dynamic equation of motion for the moving electrons in the stream. In this equation x is the coordinate of the particular electron under consideration.

Example 12.1. Parallel-plane Diode under D-C Conditions. Let us consider the parallel-plane diode illustrated in Fig. 12.11, with a d-c voltage V_0 applied between the cathode K and plate P. Since static conditions prevail, (12.10) allows writing

$$\tfrac{1}{2}mv^2(x) = e\Phi(x) \tag{12.58}$$

where we have taken the cathode potential and energy of the emitted electrons as zero. Actually, of course, there is always a finite energy of emission. From Poisson's equation

$$\frac{d^2\Phi}{dx^2} = -\frac{\rho(x)}{\epsilon_0} \tag{12.59}$$

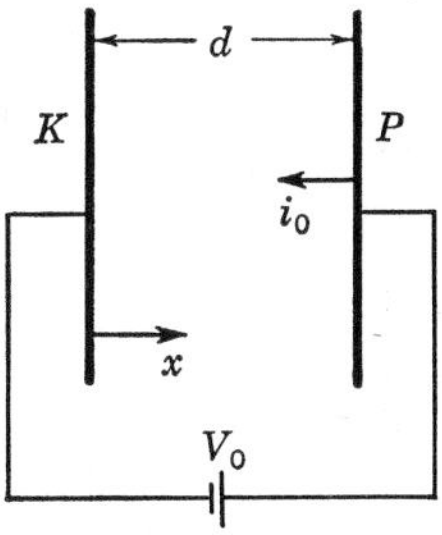

FIG. 12.11. Parallel-plane diode.

The convection current flow across a unit area in the negative x direction is

$$i_0 = -\rho(x)v(x)$$

and is the same for any x since charge must be conserved.

† This will be discussed in a little more detail in Sec. 12.10.

We shall assume full space-charge operation. Physically, this means that the supply of electrons is very much greater than can be collected at the plate for the given plate voltage. What happens then is that electrons accumulate near the cathode, so that further emission is prevented. The cloud of electrons near the cathode constitutes a space charge, and it will depress the potential gradient to the extent that equilibrium is achieved. For full space-charge operation this requires that the field at the cathode, hence the acceleration, be essentially zero. Actually, for the conditions stated, it would be slightly greater than zero; otherwise no electrons would leave the cathode plane. We may summarize the initial conditions for full space-charge-limited operation as follows:

$$E_x = \frac{d^2x}{dt^2} = \frac{dx}{dt} = 0 \qquad \text{when } x = 0$$

If we replace $\rho(x)$ by $-i_0/v(x)$ in (12.59) and make use of (12.58), the following differential equation results:

$$\frac{d^2\Phi}{dx^2} = \frac{i_0}{\epsilon_0 v(x)} = \frac{i_0}{\epsilon_0(2e\Phi/m)^{1/2}} \tag{12.60}$$

Equation (12.60) can be integrated if both sides are first multiplied by $d\Phi/dx$, for then

$$\frac{d}{dx}\left(\frac{d\Phi}{dx}\right)^2 = \frac{2i_0}{\epsilon_0}\left(\frac{m}{2e}\right)^{1/2}\Phi^{-1/2}\frac{d\Phi}{dx} \tag{12.61}$$

and

$$\left(\frac{d\Phi}{dx}\right)^2 = \frac{4i_0}{\epsilon_0}\left(\frac{m}{2e}\right)^{1/2}\Phi^{1/2} + C_1 \tag{12.62}$$

The integration constant C_1 equals zero, since $\Phi = d\Phi/dx = 0$ at $x = 0$. Extracting the square root of (12.62) and integrating again give

$$\tfrac{4}{3}\Phi^{3/4} = \left(\frac{4i_0}{\epsilon_0}\right)^{1/2}\left(\frac{m}{2e}\right)^{1/4} x \tag{12.63}$$

where the constant of integration is again zero since $\Phi = 0$ at $x = 0$. Solving for Φ gives

$$\Phi = V_0\left(\frac{x}{d}\right)^{4/3} \tag{12.64}$$

where the condition $\Phi = V_0$ at $x = d$ has been imposed.

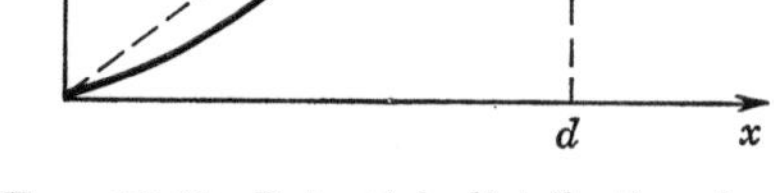

FIG. 12.12. Potential distribution in a space-charge-limited parallel-plane diode.

The potential distribution described by (12.64) is plotted in Fig. 12.12. In the absence of space charge the potential would be a linear function of x, as shown by the dashed line. The effect of space charge on the potential distribution is clearly illustrated by this example.

If $x = d$, then $\Phi = V_0$ is substituted into (12.63) and i_0 can be related to V_0 as

$$i_0 = \tfrac{4}{9}\epsilon_0 \left(\frac{2e}{m}\right)^{1/2} \frac{V_0^{3/2}}{d^2} \tag{12.65}$$

The current is related to the three-halves power of the plate-cathode voltage. This expression, along with (12.64), is known as the Child-Langmuir relation.

Example 12.2. Parallel-plane Diode with an Applied Signal. We shall illustrate the application of (12.57) to the parallel-plane diode in Fig. 12.13, where the plate-cathode voltage consists of a d-c term V_0 on which a harmonic signal voltage $V_1 e^{j\omega t}$ is superposed. We assume small-signal conditions; that is, $V_1 \ll V_0$. Electron emission will also be assumed to be space-charge-limited.

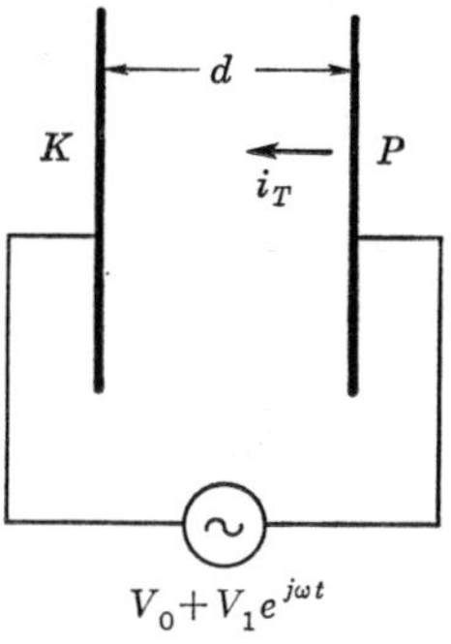

FIG. 12.13. Parallel-plane diode, time-varying case.

Under the conditions stated, $i_T(t)$ is a time-varying function, as yet unknown. Because of the small-signal conditions it can be assumed to be of the form

$$i_T = i_0 + i_1 e^{j\omega t}$$

Under large-signal conditions harmonic terms such as $i_2 e^{2j\omega t}$, $i_3 e^{3j\omega t}$, etc., would not be negligible. Let t_0 be the time of departure of a reference electron or charge element from the cathode. Then at any time t the acceleration, velocity, and position of the charge can be found by successive integration of (12.57). If we let $t - t_0 = \tau$ and $m\epsilon_0/e = \alpha$ and utilize the initial conditions due to full space-charge operation, that is, $d^2x/dt^2 = dx/dt = x = 0$ when $t = t_0$, then

$$\alpha \frac{d^3x}{dt^3} = i_0 + i_1 e^{j\omega t} \tag{12.66a}$$

$$\alpha \frac{d^2x}{dt^2} = i_0\tau + \frac{i_1}{j\omega} e^{j\omega t}(1 - e^{-j\omega\tau}) \tag{12.66b}$$

$$\alpha \frac{dx}{dt} = \frac{i_0\tau^2}{2} - \frac{i_1}{\omega^2} e^{j\omega t}[1 - (1 + j\omega\tau)e^{-j\omega\tau}] \tag{12.66c}$$

$$\alpha x = \frac{i_0\tau^3}{6} - \frac{i_1}{j\omega^3} e^{j\omega t}\left[1 - \left(1 + j\omega\tau - \frac{\omega^2\tau^2}{2}\right)e^{-j\omega\tau}\right] \tag{12.66d}$$

The plate-cathode voltage can be related to the electric field by evaluating the line integral

$$V_{PK}(t) = -\int_0^d E_x\, dx \tag{12.67}$$

at a given instant of time t. Since $E_x(t)$ can be expressed in terms of the current i_T, we shall find, by evaluating this integral at the specific instant of time t, that we can relate the a-c part of the voltage $V_{PK}(t)$ to the a-c part of the current $i_T(t)$ by an impedance relation of the form

$$V_{PK}\Big|_{a\text{-}c} = [R(\omega) + jX(\omega)]i_1$$

But since we have $V_{PK}\Big|_{a\text{-}c} = V_1$, we can now obtain explicit expressions for the a-c diode load impedance $R + jX$. It is important to note that since V_{PK} and i_T are functions of time, they can be related to each other only at the same instant of time t.

The relationship expressed by (12.67) can be interpreted in a quasi-static sense since, in practice, magnetic and propagation effects can be neglected; that is, rigorously,

$$\begin{aligned} V_{PK} &= -\int_0^d \left(-\nabla\Phi - \frac{\partial \mathbf{A}}{\partial t}\right) \cdot d\mathbf{l} \\ &= \Phi_d(t) - \Phi_0(t) + \frac{\partial}{\partial t}\int_0^d \mathbf{A} \cdot d\mathbf{l} \end{aligned}$$

If $d \ll \lambda$, propagation effects are negligible and $\Phi(t)$ is a quasi-static potential. Furthermore, the vector potential contribution will be negligible since $d \ll \lambda$ and $v \ll c$, as discussed in Sec. 9.10. For finite plates we must expect that a one-dimensional-flow description will fail near the boundaries. But if the area of the plates is large compared with d^2, or if by other means edge effects are minimized, then we may proceed as if the plates were infinite in extent.

Equation (12.67) can be integrated if we recognize that for the particular charge element under consideration its position is a function of $t - t_0 = \tau$; that is, $x = f(\tau)$. This permits us to make the following change in variables:

$$x = f(\tau) \qquad dx = \frac{\partial x}{\partial \tau}\Big|_{t=\text{constant}} d\tau$$

and to use the limits $\tau = 0$, $T(t)$ when $x = 0, d$. Equation (12.67) now becomes

$$V_{PK}(t) = -\int_0^{T(t)} E_x(\tau) \frac{\partial x}{\partial \tau}\Big|_{t=\text{constant}} d\tau \tag{12.68}$$

The integration over τ in (12.68) is performed for a constant value of t as in (12.67). The upper limit $T(t)$ is the transit time for an electron arriving at the plate ($x = d$) at the instant t and is clearly a function of t. We may obtain $E_x(\tau)$ from the relation

$$E_x(\tau) = -\frac{m}{e}\frac{d^2x}{dt^2}$$

while $(\partial x/\partial \tau)_{t=\text{constant}}$ is obtained by differentiating (12.66d) with respect to τ, holding t constant. The results are

$$E_x(\tau) = -\frac{1}{\epsilon_0}\left[i_0\tau + \frac{i_1 e^{j\omega t}}{j\omega}(1 - e^{-j\omega t})\right] \tag{12.69}$$

$$\left.\frac{\partial x}{\partial \tau}\right|_{t=\text{constant}} = \frac{\tau^2}{2\alpha}(i_0 + i_1 e^{j\omega(t-\tau)}) \tag{12.70}$$

Substituting (12.69) into (12.68) and integrating give

$$V_{PK}(t) = \frac{1}{2\alpha\epsilon_0}\left\{\frac{i_0^2 T^4}{4} + \frac{i_0 i_1 e^{j\omega t}}{\omega^4}\left[-\frac{j\omega^3 T^3}{3} - e^{-j\omega T}(-j\omega^3 T^3 - 2\omega^2 T^2 + 4j\omega T + 4) + 4\right]\right\} \tag{12.71}$$

where the small term involving i_1^2 has been dropped. Since the applied voltage consists of a d-c term plus a small a-c term, it can be expected that the transit time for various charge elements will vary periodically about some average value T_0. Consequently, $T(t)$ can be expanded in a power series in $e^{j\omega t}$; however, consistent with our assumption of small a-c signals, we shall not include terms beyond the linear ones; that is, we shall put

$$T = T_0 + T_1 e^{j\omega t} \qquad T_1 \ll T_0$$

The exponential term $e^{-j\omega T} = \exp[-j\omega(T_0 + T_1 e^{j\omega t})]$ is approximately equal to $e^{-j\omega T_0}(1 - j\omega T_1 e^{j\omega t})$, and the constant part of this is simply $e^{-j\omega T_0}$. Hence, if we are to retain only linear terms in (12.71), then T must be replaced by T_0 in the expression in brackets since the coefficient already involves the linear quantity i_1. However, a linear term involving T_1 is obtained from the first term in braces. For that term

$$\frac{i_0^2 T^4}{4} = \frac{i_0^2}{4}(T_0 + T_1 e^{j\omega t})^4 \approx \frac{i_0^2 T_0^4}{4} + i_0^2 T_0^3 T_1 e^{j\omega t}$$

Thus (12.71) is approximated by

$$V_{PK}(t) = \frac{1}{2\alpha\epsilon_0}\left\{\frac{i_0^2 T_0^4}{4} + i_0^2 T_0^3 T_1 e^{j\omega t} + \frac{i_0 i_1}{\omega^4} e^{j\omega t}\left[-j\frac{\omega^3 T_0^3}{3} - e^{-j\omega T_0}(-j\omega^3 T_0^3 - 2\omega^2 T_0^2 + 4j\omega T_0 + 4) + 4\right]\right\} \tag{12.72}$$

To complete the problem, it will be necessary to obtain an expression for T_0 and also for T_1. These can be obtained from (12.66d) and the requirement that when $\tau = T$, $x = d$. Retaining linear terms only, we obtain

$$\alpha d = \frac{i_0 T_0^3}{6} + \frac{3 i_0 T_0^2 T_1 e^{j\omega t}}{6} - \frac{i_1 e^{j\omega t}}{j\omega^3}\left[1 - \left(1 + j\omega T_0 - \frac{\omega^2 T_0^2}{2}\right)e^{-j\omega T_0}\right] \tag{12.73}$$

From (12.73) it follows that

$$\alpha d = \frac{i_0 T_0^3}{6}$$

or

$$T_0 = \left(\frac{6m\epsilon_0 d}{i_0 e}\right)^{1/3} \tag{12.74}$$

and

$$T_1 = -\frac{2i_1 j}{T_0^2 i_0 \omega^3}\left[1 - \left(1 + j\omega T_0 - \frac{\omega^2 T_0^2}{2}\right)e^{-j\omega T_0}\right] \tag{12.75}$$

We can now equate the d-c and a-c terms on both sides of (12.72), recalling that $V_{PK}(t) = V_0 + V_1 e^{j\omega t}$ and making use of (12.74) and (12.75). After algebraic reduction the results are

$$V_0 = \frac{1}{8}\left(\frac{i_0}{\epsilon_0}\right)^{2/3}\left(\frac{m}{e}\right)^{1/3}(6d)^{4/3}$$

or

$$i_0 = \tfrac{4}{9}\epsilon_0\left(\frac{2e}{m}\right)^{1/2}\frac{V_0^{3/2}}{d^2} \tag{12.76}$$

and

$$V_1 = i_1 Z(\omega) = i_1[R(\omega) + jX(\omega)] \tag{12.77}$$

where

$$R(\omega) = \frac{12}{g_1}\,\frac{2(1 - \cos\theta) - \theta\sin\theta}{\theta^4} \tag{12.78a}$$

$$X(\omega) = \frac{12}{g_1}\,\frac{2\sin\theta - \theta(1 + \cos\theta) - \theta^3/6}{\theta^4} \tag{12.78b}$$

In (12.78) the substitutions $\theta = \omega T_0$ and $g_1 = \tfrac{3}{2}i_0/V_0$ have been made.

If there were no time-varying voltage the prevailing d-c conditions would be completely described by (12.76), which is the Child-Langmuir relation already obtained in Example 12.1. With the inclusion of the a-c plate-cathode signal, the diode presents an effective load impedance $Z(\omega)$ that depends on the transit angle θ. The variation of the diode load with θ is given in (12.78) and plotted in Fig. 12.14. The variation in loading results from the interchange of energy between the electron stream and the field as a consequence of the variation in transit time, hence of interaction time. Of particular interest is the condition $2\pi < \theta < 3\pi$, since the diode load resistance is then negative, and this suggests the possibility of extracting energy from the electron stream to obtain oscillations. This can actually be done, but the efficiency turns out to be very low.†

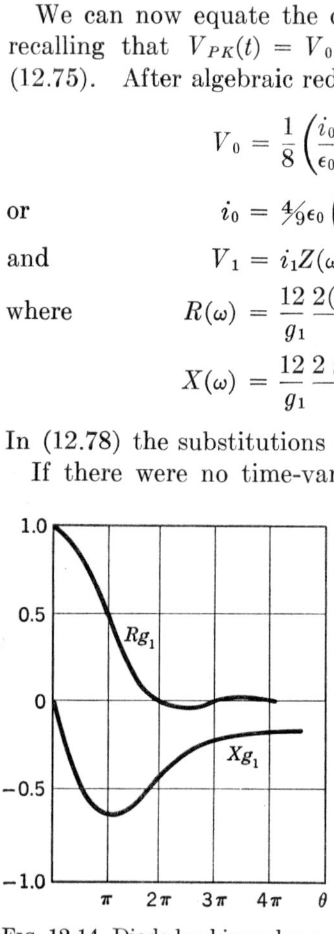

FIG. 12.14. Diode load impedance as a function of transit angle.

† F. B. Llewellyn and A. E. Bowen, The Production of Ultra-high Frequency Oscillations by Means of Diodes, *Bell System Tech. J.*, vol. 18, pp. 280–291, April, 1939.

12.6. Llewellyn-Peterson Equations

Based on a small-signal analysis where $i_T(t)$ is replaced by $i_0 + i_1e^{j\omega t}$, $i_1 \ll i_0$, (12.57) can be successively integrated to give acceleration, velocity, and displacement of an arbitrary charge element. A general expression can be written by specifying the initial velocity and acceleration arbitrarily. Following an analysis such as performed in Example 12.2, an expression can be obtained for the a-c voltage V_1 between arbitrary planes at x_1 and x_2, as a function of i_1, v_1, and the initial convection current q_1. (The convection current is denoted by the symbol q in this section so as not to confuse it with the total current.) The net result is summarized in a set of equations known as the Llewellyn-Peterson equations. In terms of conditions at planes 1 and 2, as illustrated in Fig. 12.15, the L-P equations are

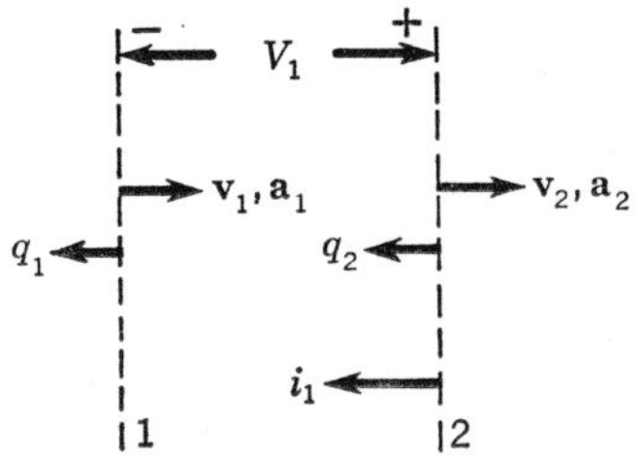

FIG. 12.15. Parallel-plane two-electrode system.

$$
\begin{aligned}
V_1 &= Ai_1 + Bq_1 + Cv_1 \\
q_2 &= Di_1 + Eq_1 + Fv_1 \\
v_2 &= Gi_1 + Hq_1 + Iv_1
\end{aligned}
\tag{12.79}
$$

The values of the L-P coefficients are determined by the d-c transit time, d-c potentials, and the space-charge factor. The coefficients are given in the paper by Llewellyn and Peterson.†

Equations (12.79) can also be rewritten in the following form:

$$
\begin{aligned}
i_1 &= a_{11}V_1 + a_{12}q_1 + a_{13}v_1 \\
q_2 &= a_{21}V_1 + a_{22}q_1 + a_{23}v_1 \\
v_2 &= a_{31}V_1 + a_{32}q_1 + a_{33}v_1
\end{aligned}
\tag{12.80}
$$

The coefficients a_{ij} appearing in (12.80) are given in Table 12.1 for complete space charge ($\zeta = 1$) and for negligible space charge ($\zeta \ll 1$). In these equations velocity and electron flow are from plane 1 to plane 2, while current flow is taken from plane 2 to plane 1.‡

The purpose of developing the L-P equations is that they permit solving a broad class of problems involving parallel-plane multielectrode tubes. This is possible because the equations may be applied successively from the first to last electrode pair, where the initial conditions in an electrode pair arise from the output conditions of the previous pair.

† F. B. Llewellyn and L. C. Peterson, Vacuum-tube Networks, *Proc. IRE*, vol. 32, pp. 144–166, March, 1944.

‡ The coefficients in Table 12.1 are believed to correct several minor errors that occur in Beck and Kleen (see references at the end of this chapter).

The following example will serve as an illustration. Another equivalent example involving the klystron will be found in Sec. 12.8.

We wish to analyze a conventional but parallel-plane triode, as illustrated in Fig. 12.16. In order to simplify the analysis, the grid will be assumed to be always negative and to effectively shield the grid-plate region from the grid-cathode region (essentially infinite amplification μ). Then the cathode-grid region can be analyzed as if it were an independently functioning diode. Assuming space-charge-limited operation, we have $q_1 = v_1 = 0$, and hence from (12.80),

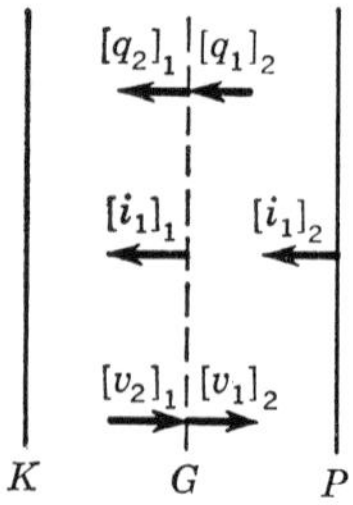

FIG. 12.16. Planar triode-tube structure.

$$(i_1)_1 = (a_{11})_1 V_{gK} \tag{12.81}$$

The subscript 1 on the parentheses refers to the cathode-grid space, the subscript 2 to the grid-plate region, and V_{gK} is the a-c grid-cathode signal voltage. If (12.81) is interpreted from a circuit viewpoint, then it specifies an equivalent admittance

$$Y_{11} = \frac{(i_1)_1}{V_{gK}} = (a_{11})_1 \tag{12.82}$$

connected between the grid and cathode. In view of the assumed zero coupling from the grid-plate region into the grid-cathode region, the latter behaves as an isolated diode. The coefficient $(a_{11})_1$ for full space charge should then equal the reciprocal of impedance, as computed in Example 12.2 and expressed by (12.78). The reader can readily verify this result from Table 12.1.

The grid-plate region can also be analyzed as a separately operating diode. However, its input convection current and electron velocity equal the output values of the grid-cathode region. The assumption of negative grid conditions ensures that the grid will not collect current, so that the above conditions will hold. We may summarize the boundary conditions at the input to the grid-plate region as follows (see Fig. 12.16):

$$(q_2)_1 = (q_1)_2 = (a_{21})_1 V_{gK} \tag{12.83}$$
$$(v_2)_1 = (v_1)_2 = (a_{31})_1 V_{gK} \tag{12.84}$$

Consequently, the current flowing to the plate can be expressed as

$$\begin{aligned}(i_1)_2 &= (a_{11})_2 V_{pg} + (a_{12})_2 (q_1)_2 + (a_{13})_2 (v_1)_2 \\ &= (a_{11})_2 V_{pg} + (a_{12})_2 (a_{21})_1 V_{gK} + (a_{13})_2 (a_{31})_1 V_{gK}\end{aligned} \tag{12.85}$$

Equation (12.85) can be interpreted from a circuit standpoint in terms of

TABLE 12.1. COEFFICIENTS FOR LLEWELLYN-PETERSON EQUATIONS

	$\zeta = 1$	$\zeta \ll 1$
a_{11}	$\frac{3e}{m}\frac{i_0}{(v_{01}+v_{02})^2}f_6^{-1}$	$\frac{j\omega\epsilon_0}{d}+\frac{2e}{m}\frac{i_0}{(v_{01}+v_{02})^2}f_5$
a_{12}	$\frac{v_{01}}{v_{01}+v_{02}}\frac{f_4}{f_6}$	$\frac{v_{01}f_2+v_{02}f_3}{v_{01}+v_{02}}$
a_{13}	$\frac{3i_0}{v_{01}+v_{02}}\frac{f_3}{f_6}$	$\frac{i_0 j\theta f_3}{v_{01}+v_{02}}$
a_{21}	$\frac{3e}{m}\frac{i_0}{v_{02}(v_{01}+v_{02})}\frac{f_3}{f_6}$	$\frac{e}{m}\frac{j\theta f_3 i_0}{v_{02}(v_{01}+v_{02})}$
a_{22}	$\frac{v_{01}}{v_{02}}\left(-e^{-j\theta}+\frac{f_3f_4}{f_6}\right)$	$e^{-j\theta}$
a_{23}	$\frac{i_0}{v_{02}}\left(j\theta e^{-j\theta}+\frac{3f_3^2}{f_6}\right)$	$\frac{i_0}{v_{02}}j\theta e^{-j\theta}$
a_{31}	$\frac{e}{m}(v_{01}+v_{02})^{-1}\frac{f_4}{f_6}$	$\frac{e}{m}\frac{f_2+v_{01}f_3/v_{02}}{v_{01}+v_{02}}$
a_{32}	$\frac{v_{01}}{3i_0}\frac{f_4^2}{f_6}$	$\frac{e}{m}\left(\frac{j\omega\epsilon_0 v_{02}}{d}\right)^{-1}[-e^{-j\theta}+(f_2v_{02}+f_3v_{01})(f_2v_{01}+f_3v_{02}) \times (v_{01}+v_{02})^{-2}]$
a_{33}	$-e^{-j\theta}+\frac{f_3f_4}{f_6}$	$\frac{v_{01}}{v_{02}}e^{-j\theta}$

v_{01} = d-c velocity at plane 1 $\qquad v_{02}$ = d-c velocity at plane 2

$\theta = \omega T_0$ = d-c transit angle $\qquad d$ = electrode separation

i_0 = d-c current $\qquad T_0 = \frac{2d}{v_{01}+v_{02}}$ for $\zeta = 0$

$$T_0 = \frac{3d}{v_{01}+v_{02}} \quad \text{for } \zeta = 1 \text{ [see also Eq. (12.74)]}$$

$$f_1 = \frac{1-e^{-j\theta}}{j\theta} \qquad f_2 = \frac{2(j\theta - 1 + e^{-j\theta})}{(j\theta)^2} \qquad f_3 = 2f_1 - f_2$$

$$f_4 = \frac{3(f_2 - f_3)}{j\theta} \qquad f_5 = \tfrac{1}{3}j\theta f_4 \qquad f_6 = \frac{2(1-f_4)}{j\theta}$$

For θ very small,

$$f_1 = 1 - \frac{j\theta}{2} \qquad f_2 = 1 - \frac{j\theta}{3} \qquad f_3 = 1 - \frac{j2\theta}{3}$$

$$f_4 = 1 - \frac{j\theta}{2} \qquad f_5 = \frac{j\theta}{3} \qquad f_6 = 1 - \frac{j3\theta}{10}$$

a self-admittance Y_{22} and a mutual admittance Y_{12} such that

$$(i_1)_2 = Y_{22}V_{pg} + Y_{12}V_{gK} \tag{12.86}$$

where

$$Y_{22} = (a_{11})_2 \tag{12.87a}$$

$$Y_{12} = (a_{12})_2(a_{21})_1 + (a_{13})_2(a_{31})_1 \tag{12.87b}$$

The mutual-admittance term behaves, actually, as a current source of strength $Y_{12}V_{gK}$. An equivalent circuit embodying Eqs. (12.82) and (12.86) is given in Fig. 12.17.

The results in the previous work are based on assumptions which include infinite μ, negative grid, and neglect of initial velocity of emission.

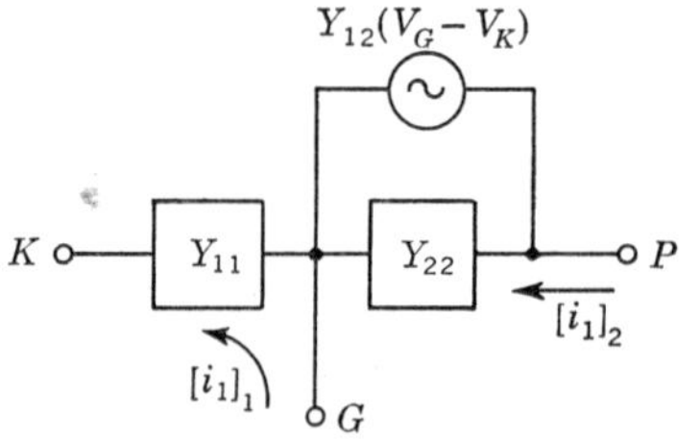

FIG. 12.17. Equivalent circuit for triode.

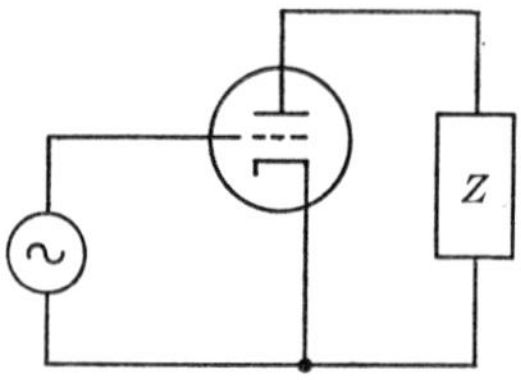

FIG. 12.18

Suitable modifications in the theory which allow partial or complete removal of these limitations can be found in papers by Llewellyn. One can, however, obtain a great deal of insight into the transit-time effects in practical tubes even if the simplifying assumptions are retained.

A discussion of the equivalent circuit of Fig. 12.17 permits us to point out the essential characteristics of space-charge control tubes. The function of the control grid is to modulate the electron flow into the grid-plate region. As a consequence an a-c signal is developed in the plate load which corresponds to the signal impressed on the grid. A relatively large amount of power can be controlled in the plate circuit with the expenditure of very little power in the grid circuit.

For a conventional grounded-cathode connection, as illustrated in Fig. 12.18, Y_{11} gives the load presented to the signal source. We usually seek conditions that will make this very small. On the other hand, we try also to maintain the transconductance Y_{12} at as large a value as possible since this means a large signal developed in the plate load. With the aid of the L-P equations it is possible to formulate an expression for power gain as a function of transit time, tube voltages, and spacings. Examples and details are left for the student with the assistance of problems.

The distinguishing feature of space-charge control tubes such as the above-illustrated triode, as well as conventional tetrodes and pentodes, etc., is that the space-charge-limited current is caused to vary as a result of varying potentials on the grid adjacent to the cathode. In contrast, another class of tubes operate with an initially steady current at the grid nearest the cathode. Such tubes operate by varying the velocity of the electrons in the beam and are called velocity-modulated tubes, the klystron being a typical example. For small signals and short drift space the L-P equations are adequate for analyzing the behavior of the klystron. However, the space-charge-wave theory discussed in the next section is required to explain the operation of the more common long-drift-space tube. We defer to that section a discussion of the klystron by means of the L-P equations since it will then be possible to contrast the analysis with that based on space-charge-wave theory.

12.7. Space-charge Waves

The Benham-Müller-Llewellyn equation (12.57) made possible an analysis of the properties of an electron stream in a quasi-static field which included the effects of space charge. A wide class of modern electron tubes depend for their operation on a continuous interaction of electron stream and electromagnetic fields over distances which are comparable to wavelength. To describe this interaction we shall find most satisfactory an essentially rigorous theory based on Maxwell's equations. This is the space-charge-wave theory first developed by Hahn and Ramo.†

We are going to consider the properties of an essentially uniform electron stream flowing through a circular conducting tube which it completely fills, as in Fig. 12.19. We shall see that such a configuration can support the propagation of electromagnetic waves, one of which has a phase velocity slightly greater and one slightly less than the velocity of the stream of electrons. We can think of the beam as behaving in the manner of an elastic medium in supporting these "space-charge waves."

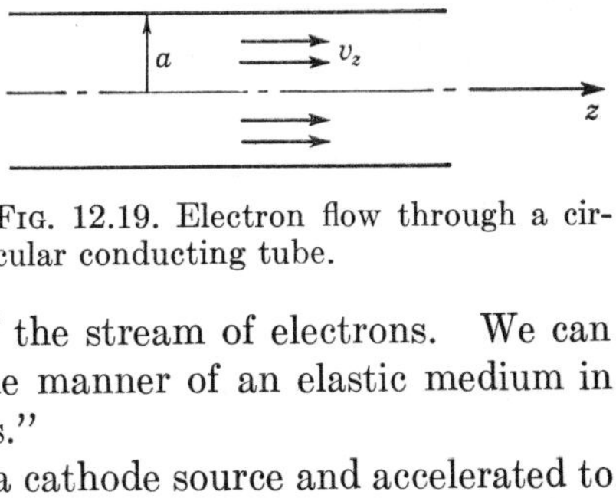

FIG. 12.19. Electron flow through a circular conducting tube.

The electron stream is produced at a cathode source and accelerated to some value of d-c potential. We consider its properties at the point where its d-c velocity is held constant; i.e., the beam is caused to flow through a d-c field-free region. We shall leave for later a consideration of the practical means for setting up the space-charge waves and the discussion of their usefulness. For the moment we wish only to show that their existence is confirmed by Maxwell's equations and to determine some of their properties.

† W. C. Hahn, Small Signal Theory of Velocity Modulated Electron Beams, *Gen. Elec. Rev.*, vol. 42, pp. 258–270, June, 1939. S. Ramo, Space Charge and Field Waves in an Electron Beam, *Phys. Rev.*, vol. 56, pp. 276–283, August, 1939.

Our procedure will be to assume the existence of space-charge waves; then we shall show that, subject to certain conditions, such waves are consistent with Maxwell's equations. One of the main restrictions is that we consider small signals so that only linear terms need be retained, and all a-c quantities will be of the form $e^{j\omega t - \gamma z}$. This function is appropriate for representing a harmonically time-varying quantity propagating in the $+z$ direction. Thus the electron charge density will be

$$\rho^- = -(\rho_0 + \rho_1 e^{j\omega t - \gamma z}) \tag{12.88}$$

where for convenience ρ_0 and ρ_1 are taken as positive quantities.

Now it is impossible to completely evacuate the space through which the electrons flow. Some production of positive ions is unavoidable, and because of their weight and the fact, as we shall see, that they lie in an almost field-free space, they remain for long periods of time compared with the electrons streaming by. They tend, therefore, to offset the d-c space charge of the electron beam. Very often, in solving a complicated physical problem, certain idealizations are made in order to simplify the analysis. This makes it possible to obtain a rigorous solution to an ideal problem, which is then an approximation to the nonideal physical problem. Such a procedure is well justified, particularly where a small loss in quantitative accuracy is balanced by the ability to substitute a simple model. In the problem at hand it will be very convenient if we can ignore the d-c space-charge forces. This would be the case, actually, if a positive-ion density equal to the electron density existed. We shall take this to be the case; our resultant solution will then be exact only for such a case. So far as a-c space-charge effects are concerned, we shall deal with these rigorously.

We then have, for the positive-ion density ρ^+,

$$\rho^+ = \rho_0 \tag{12.89}$$

Note that no time-varying part has been included in (12.89). This is a consequence of the large relative mass of the positive ions, so that the a-c current is due principally to electron motion only. As a further simplification, we assume the existence of an extremely large axial magnetic field. Such a field constrains the electron motion to the axial direction only, and consequently electron velocity may be written in the one-dimensional form

$$v_z = v_0 + v_1 e^{j\omega t - \gamma z} \tag{12.90}$$

where z is the axial coordinate.

The total convection current density can be computed from (12.90) and (12.88) as follows:

$$\begin{aligned} J_z = \rho^- v_z &= -(\rho_0 + \rho_1 e^{j\omega t - \gamma z})(v_0 + v_1 e^{j\omega t - \gamma z}) \\ &\approx -\rho_0 v_0 - (\rho_1 v_0 + \rho_0 v_1) e^{j\omega t - \gamma z} \end{aligned} \tag{12.91}$$

when second-order a-c terms are neglected. The last result may be expressed as

$$J_z = J_0 + J_1 e^{j\omega t - \gamma z} \tag{12.92a}$$

where

$$J_0 = -\rho_0 v_0 \tag{12.92b}$$

$$J_1 = -(\rho_1 v_0 + \rho_0 v_1) \tag{12.92c}$$

The d-c quantities v_0, ρ_0, and J_0 are assumed constant over the cross section of the beam. However, the a-c quantities v_1, ρ_1, and J_1 will vary with the transverse coordinates in addition to having the z dependence $e^{-\gamma z}$.

The a-c convection current may be considered to give rise to an electromagnetic field. Since the current flow is in the z direction only, it will be most convenient to determine the vector potential due to the a-c current first, since the vector potential will have only a single (that is, z) component. The fields may then be readily found from the vector potential. From (9.66) the axial component of the vector potential satisfies the following inhomogeneous wave equation:

$$\nabla^2 A_z + k_0^2 A_z = -\mu_0 J_1 e^{j\omega t - \gamma z} \tag{12.93}$$

Since we are assuming that all a-c field quantities may be expressed in the form of a function of the transverse coordinates times the factor $e^{j\omega t - \gamma z}$, we can write

$$A_z = A_1 e^{j\omega t - \gamma z} \tag{12.94}$$

and (12.93) becomes

$$\nabla_t^2 A_1 + (\gamma^2 + k_0^2) A_1 = -\mu_0 J_1 \tag{12.95}$$

where ∇_t^2 signifies the transverse Laplacian operator.

The electric field that arises can be calculated from the vector potential by using (9.68*b*). We are particularly interested in the axial component, and this is given by

$$E_z = \frac{1}{j\omega\epsilon_0\mu_0} \frac{\partial^2 A_z}{\partial z^2} - j\omega A_z \tag{12.96}$$

If we let

$$E_z = E_1 e^{j\omega t - \gamma z} \tag{12.97}$$

and make use of (12.94), we obtain for (12.96) the result

$$E_1 = \frac{\gamma^2 + k_0^2}{j\omega\epsilon_0\mu_0} A_1 \tag{12.98}$$

In order to solve (12.95), we shall work out an expression for J_1 in terms of A_1. This is possible, since A_1 determines a field E_1, which in turn can be related to the convection current through the dynamical equations. Let us consider the details. We first note that since this is a one-dimensional-current-flow problem the continuity equation gives

$$\frac{\partial J_z}{\partial z} = -\frac{\partial \rho}{\partial t}$$

From (12.92a) and (12.88),

$$-\gamma J_1 e^{j\omega t - \gamma z} = j\omega\rho_1 e^{j\omega t - \gamma z}$$

whereupon

$$J_1 = -\frac{j\omega\rho_1}{\gamma} \tag{12.99}$$

Let us now substitute (12.99) into (12.92c) to obtain

$$\frac{j\omega\rho_1}{\gamma} = \rho_1 v_0 + \rho_0 v_1$$

Solving for ρ_1 results in

$$\rho_1 = \frac{\rho_0 v_1 \gamma}{j\omega - \gamma v_0} \tag{12.100}$$

Finally, we make use of the force equation

$$-\frac{e}{m} E_z = \frac{dv_z}{dt} = \frac{\partial v_z}{\partial t} + \frac{\partial v_z}{\partial z}\frac{dz}{dt} = \frac{\partial v_z}{\partial t} + v_z \frac{\partial v_z}{\partial z}$$

Substituting derived values gives

$$-\frac{e}{m} E_1 e^{j\omega t - \gamma z} = j\omega v_1 e^{j\omega t - \gamma z} + (v_0 + v_1 e^{j\omega t - \gamma z})(-\gamma v_1 e^{j\omega t - \gamma z})$$

Neglecting higher-order terms, we obtain

$$-\frac{e}{m} E_1 = j\omega v_1 - \gamma v_0 v_1$$

and hence

$$v_1 = -\frac{eE_1}{m(j\omega - \gamma v_0)} \tag{12.101}$$

It is now possible to express J_1 in terms of A_1. The end result is obtained by the successive substitution of (12.98) into (12.101) into (12.100) into (12.99) and is

$$J_1 = \frac{\omega_0^2(\gamma^2 + k_0^2)}{\mu_0(j\omega - \gamma v_0)^2} A_1 \tag{12.102}$$

where the "plasma-electron-resonance" frequency, defined by

$$\omega_0^2 = \frac{e\rho_0}{m\epsilon_0} \tag{12.103}$$

has been introduced. We can now simplify (12.95) so that it becomes

$$\nabla_t^2 A_1 = -(\gamma^2 + k_0^2)\left[1 + \frac{\omega_0^2}{(j\omega - \gamma v_0)^2}\right] A_1 \tag{12.104}$$

At this point it is necessary to solve (12.104) subject to imposed boundary conditions. Since we are considering a perfectly conducting circular cylinder completely filled by the beam, A_1 must give rise to an electric field whose tangential components vanish on the walls. Since the axial electric field is linearly related to A_1, this means that A_1 is zero on the boundaries. Thus we are faced with a boundary-value problem which is completely analogous to that for TM waves in a circular waveguide. If we call

$$q^2 = (\gamma^2 + k_0^2)\left[1 + \frac{\omega_0^2}{(j\omega - \gamma v_0)^2}\right] \tag{12.105}$$

then

$$\nabla_t^2 A_1 = -q^2 A_1 \tag{12.106}$$

We shall assume axial symmetry ($\partial/\partial\phi = 0$), so that (12.106) results in Bessel's equation in terms of the radial coordinate r, namely,

$$\frac{d^2A_1}{dr^2} + \frac{1}{r}\frac{dA_1}{dr} + q^2 A_1 = 0 \tag{12.107}$$

The solution to this equation can be written as

$$A_1 = CJ_0(qr) \tag{12.108}$$

where J_0 is the Bessel function of the first kind and zero order (this should not be confused with the d-c current density). Since $A_1 = 0$ for $r = a$, where a is the radius of the cylindrical tube,

$$q = q_m = \frac{p_{0m}}{a} \tag{12.109}$$

In (12.109) p_{0m} is the mth root of the equation $J_0(x) = 0$. Corresponding to each eigenvalue q_m, (12.105), which is a quartic in γ, yields four values of γ; that is, having obtained a value for q_m from (12.109), we now determine corresponding values γ_m for γ from the equation

$$(\gamma_m^2 + k_0^2)\left[1 + \frac{\omega_0^2}{(j\omega - \gamma_m v_0)^2}\right] = q_m^2 \tag{12.110}$$

For typical operating conditions of electron beams at microwave frequencies, $\omega \gg \omega_0$ and the term $\omega_0^2/(j\omega - \gamma_m v_0)^2$ is negligible to a first approximation. In this case, (12.110) reduces to what is obtained in the analysis of a guide filled with air. Two solutions for γ thus result from (12.110) that correspond to slightly modified field waves, one propagating in the positive z direction, the other in the negative z direction. Using a perturbation method (see Prob. 12.13) the propagation constants are determined to be

$$\gamma_m^2 = \Gamma_m^2 - \frac{q_m^2\omega_0^2}{(j\omega - \Gamma_m v_0)^2} \tag{12.111}$$

where Γ_m is the propagation constant for the waveguide in the absence of the electron beam.

Of much greater interest are the remaining two propagation constants. They turn out to have values that correspond to phase velocities slightly greater and slightly less than the beam velocity v_0. These are the space-charge waves, and their importance lies in the fact that since they have a phase velocity approximately equal to the beam velocity, extensive beam-field interaction is facilitated. For the space-charge waves we may assume that

$$\gamma = \frac{j\omega}{v_0(1 \pm \delta)} \tag{12.112}$$

where, as we shall verify later, $\delta \ll 1$. We are assuming that the phase velocities differ by equal amounts from the d-c beam velocity v_0. Since $\delta \ll 1$, (12.112) is approximately equal to

$$\gamma = \frac{j\omega}{v_0}(1 \mp \delta) \tag{12.113}$$

If we substitute (12.113) into (12.110), we can determine δ. Now

$$\frac{\omega_0^2}{(j\omega - \gamma v_0)^2} = \frac{\omega_0^2}{[j\omega - (j\omega/v_0)v_0(1 \mp \delta)]^2} = -\frac{\omega_0^2}{\omega^2\delta^2}$$

and consequently,

$$1 - \frac{\omega_0^2}{\omega^2\delta^2} = \frac{q_m^2}{\gamma^2 + k_0^2} = \frac{q_m^2}{k_0^2 - \omega^2/v_0^2}$$

In the last term on the right-hand side above, γ^2 has been replaced by $-\omega^2/v_0^2$, since in this instance inclusion of δ involves a negligible correction. If, further, we note that $k_0^2 = \omega^2/c^2 \ll \omega^2/v_0^2$, since we are neglecting relativistic effects here and hence tacitly assuming $v_0^2 \ll c^2$, then

$$\frac{\omega_0^2}{\omega^2\delta^2} = 1 + \frac{q_m^2}{\omega^2/v_0^2}$$

and

$$\delta^2 = \frac{\omega_0^2}{\omega^2 + v_0^2 q_m^2} \tag{12.114}$$

But $(\omega_0/\omega)^2 \ll 1$; consequently, $\delta^2 \ll 1$, thereby justifying our original assumption and confirming the existence of space-charge waves. The latter propagate without attenuation and at velocities slightly greater and slightly less than the beam velocity, as we see by examining (12.112) and (12.114).

If instead of a circular beam we had chosen to consider an infinite beam with no transverse variations, then (12.105) reduces to

$$(\gamma^2 + k_0^2)\left[1 + \frac{\omega_0^2}{(j\omega - \gamma v_0)^2}\right] = 0 \tag{12.115}$$

Two solutions for γ then correspond to free-space propagation,

$$\gamma = \pm jk_0 \tag{12.116}$$

and two additional solutions arise from setting

$$1 + \frac{\omega_0^2}{(j\omega - \gamma v_0)^2} = 0 \tag{12.117}$$

The latter leads to space-charge waves with propagation constants

$$\gamma = j\frac{\omega \pm \omega_0}{v_0} \tag{12.118}$$

The free-space waves with propagation constants $\pm jk_0$ are transverse electromagnetic waves with no z component of electric or magnetic field, as is easily verified by (12.98). For this reason no interaction with the electron beam takes place for these waves. Note that the assumption of a very large d-c axial magnetic field prevents any electron motion in the transverse direction, and hence no interaction with transverse-electric-field components will occur.

For the circular beam, combining (12.114) and (12.113), we have

$$\gamma = j\left\{\frac{\omega}{v_0} \pm \frac{\omega_0}{v_0}\frac{1}{[1 + (p_{0m}v_0/\omega a)^2]^{1/2}}\right\} \tag{12.119}$$

If (12.118) is compared with (12.119), then the effect of the transverse variations in the latter may be thought of as reducing the plasma frequency ω_0. When we wish to talk rather generally about space-charge waves, it will be convenient to write the propagation-constant equation as

$$\gamma = j\left(\frac{\omega}{v_0} \pm \frac{\omega_q}{v_0}\right) \tag{12.120}$$

where ω_q is the "effective plasma frequency." This, in turn, may be expressed as

$$\omega_q = F\omega_0 \tag{12.121}$$

where F is called the "space-charge reduction factor" and depends on the geometry of the beam and the cylindrical boundary. For the beam completely filling a circular cylindrical tube, as described in (12.119), and for the dominant mode, the space-charge reduction factor is

$$F_1 = \left[1 + \left(\frac{2.405v_0}{\omega a}\right)^2\right]^{-1/2} \tag{12.122}$$

For a more general case where the beam is cylindrical and of diameter $2b$ within a conducting tube of diameter $2a$, as illustrated in Fig. 12.20, the value of F may be computed and is plotted for the dominant mode in Fig. 12.21. Higher-order modes would result in a smaller value of F.

In the next two sections applications of space-charge-wave theory to microwave tubes will be discussed. The following section considers the

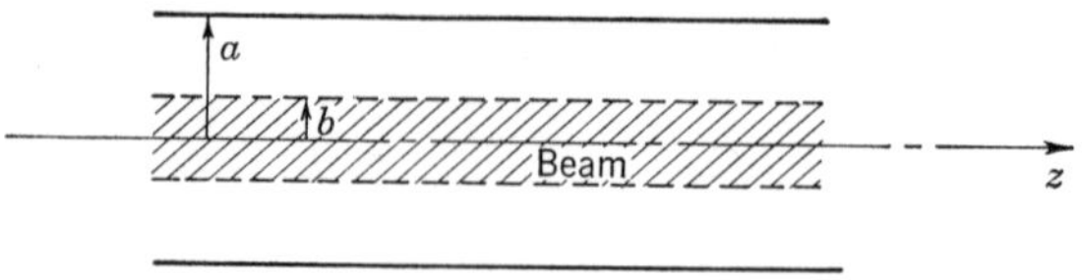

FIG. 12.20. Cylindrical beam in a conducting circular waveguide.

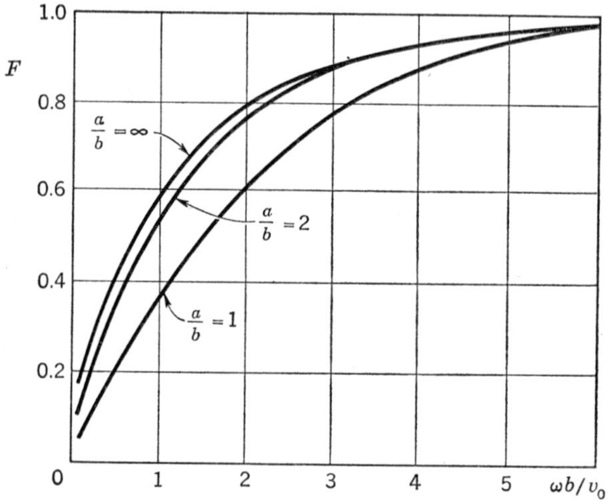

FIG. 12.21. Space-charge reduction factor for the cylindrical beam in a circular waveguide (dominant mode).

klystron; then follows an analysis of the traveling-wave amplifier. We defer until the end a summary of the space-charge-wave theory.

12.8. The Klystron

The klystron is a tube designed to operate as an amplifier or oscillator at frequencies where transit times are sufficiently large so that conventional tubes fail. It operates on the principle of velocity modulation of the electron stream rather than density modulation. Let us consider briefly a qualitative description of its operation.

The essential mechanism of the klystron is illustrated in Fig. 12.22. An electron gun is the source of the electron beam, which is collimated to flow in the axial direction. Beyond the gun structure the remaining region of the tube is maintained at a uniform d-c potential so that the velocity of the electrons entering the *buncher*-cavity grids is constant. The buncher grids form part of a reentrant cavity. We assume that this cavity is externally excited in a mode that develops an axial electric field across the buncher grids. The electron beam, consequently, in traversing the buncher will be accelerated or decelerated, depending on the portion of the r-f cycle in which transit occurs. Averaged over a cycle no net energy is transferred to the electron stream, since what is lost by the cavity field during acceleration of the electrons is regained during deceleration, neglecting electron-beam loading of the cavity.

The electron stream emerges from the buncher still uniform, but with a varying velocity. Consider as a reference the electrons that leave the

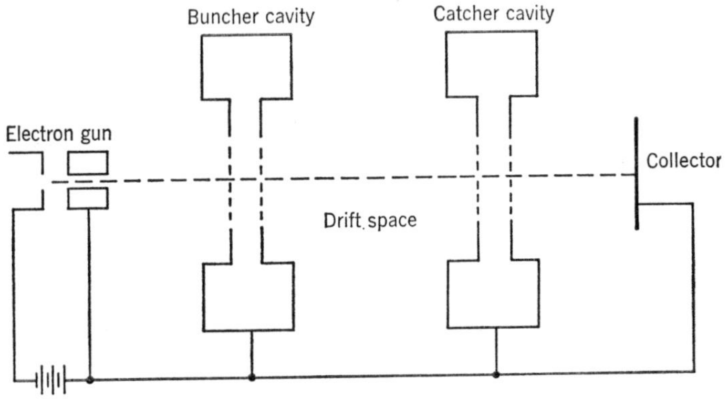

FIG. 12.22. Schematic diagram of klystron.

buncher when the field is changing from decelerating to accelerating. Electrons that left the buncher earlier than the reference were then decelerated and emerge with smaller velocity. Electrons leaving after the reference have an increased velocity. The faster but later electrons tend to overtake the earlier but slower electrons. With proper design it is possible to cause the electrons to bunch together when they reach the end of the *drift space*. At this point the electrons enter the *catcher* cavity.

At the catcher grids an interchange of energy between the electron stream and an assumed r-f field is possible. If the r-f field has the proper phase, the electron bunches will be decelerated in transit through the catcher, thus giving up energy to the r-f field in the cavity. Since the number of electrons not in the bunches is relatively small, there is a net transfer of energy to the r-f field, hence a net power gain.

If we assume small signals at both the buncher and catcher grids and if we assume that the drift space is not long, then an analysis is possible, based on the use of the Llewellyn-Peterson equations. The problem is reduced to parallel-plane geometry, as in Fig. 12.23. Since the electron density is low, we take the space-charge factor to be very small compared with unity. For the input gap (buncher-grid region), we have

$$\begin{aligned}(V_{a\text{-}c})_1 &= (V_1)_1 \\ (q_1)_1 &= 0 \\ (v_1)_1 &= 0\end{aligned} \tag{12.123}$$

and this leads to [see Eqs. (12.80)]

$$(i_1)_1 = (a_{11})_1(V_1)_1 \tag{12.124}$$

The subscript on the parentheses refers to the region designated in Fig. 12.23, while the inner subscripts 1 and 2 refer to input and output of the

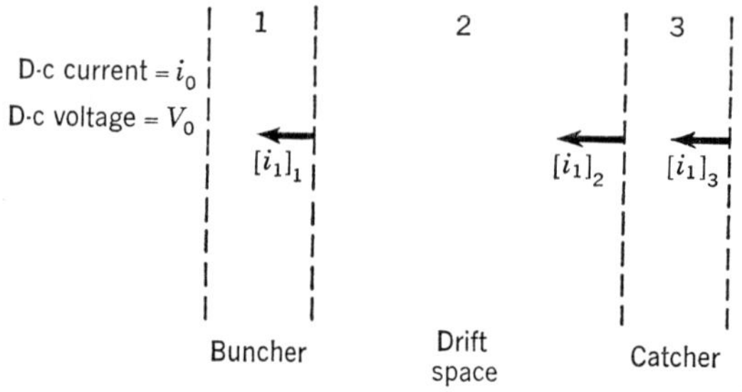

FIG. 12.23. Parallel-plane klystron.

respective region with the exception of the a-c current i_1, which is that flowing to the left from the output plane of each respective region, and V_1, which is the a-c voltage across the gap. For the buncher grids we also have

$$(v_2)_1 = (a_{31})_1(V_1)_1 = (v_1)_2 \tag{12.125}$$
$$(q_2)_1 = (a_{21})_1(V_1)_1 = (q_1)_2 \tag{12.126}$$

For gap 2, the drift space, we have

$$(V_1)_2 = 0 \tag{12.127}$$
$$(v_2)_2 = (a_{32})_2(a_{21})_1(V_1)_1 + (a_{33})_2(a_{31})_1(V_1)_1 \tag{12.128}$$
$$(q_2)_2 = (a_{22})_2(a_{21})_1(V_1)_1 + (a_{23})_2(a_{31})_1(V_1)_1 \tag{12.129}$$

Finally, for gap 3, the catcher-grid region, we have

$$\begin{aligned}(i_1)_3 = (a_{11})_3(V_1)_3 &+ (a_{12})_3(a_{22})_2(a_{21})_1(V_1)_1 + (a_{13})_3(a_{32})_2(a_{21})_1(V_1)_1 \\ &+ (a_{13})_3(a_{33})_2(a_{31})_1(V_1)_1 + (a_{12})_3(a_{23})_2(a_{31})_1(V_1)_1\end{aligned} \tag{12.130}$$

By inspection of the previous set of equations, the main characteristics of the klystron can be described. From (12.124), one obtains the self-admittance of the buncher gap. Designating this by Y_{11}, we find that

$$Y_{11} = (a_{11})_1 = \frac{j\omega\epsilon_0}{d} + \frac{g_B}{2}\left[\frac{2(1-\cos\theta_1) - \theta_1 \sin\theta_1}{\theta_1^2} + j\,\frac{2\sin\theta_1 - \theta_1(1+\cos\theta_1)}{\theta_1^2}\right] \tag{12.131}$$

where $g_B = i_0/V_0$, and the expression for $(a_{11})_1$ is that given in Table 12.1 with $\zeta \ll 1$. The first term in (12.131) is a cold, capacity loading per unit area of the buncher-grid region, where d is the grid spacing. The

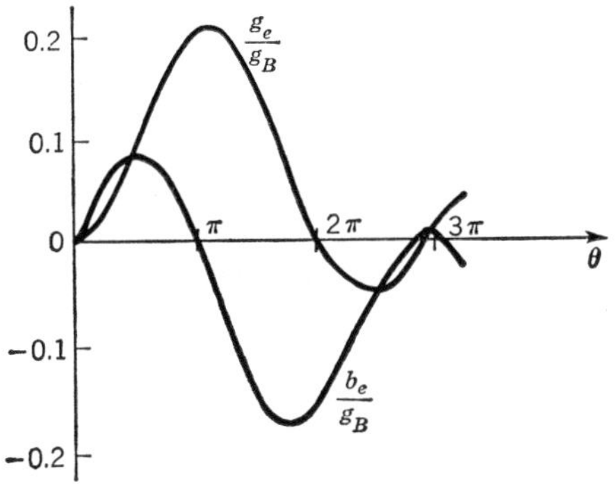

FIG. 12.24. Plot of real and imaginary parts of the electron-beam admittance of a parallel-plane gap as a function of transit angle and d-c admittance $g_B = i_0/V_0$.

remaining terms express the electron-beam admittance. A plot of the real and imaginary parts of the electron-beam admittance $g_e + jb_e$, as a function of the transit angle, is given in Fig. 12.24. Zero loading occurs when $\theta = 0$, the loading in general increasing with increasing values of θ. This explains why a cavity design is required with short electron transit distances. Since a high Q is also required, hence a high ratio of volume to surface area, the reentrant type of cavity evolved.

From (12.130) we note that the self-admittance of the output gap, Y_{33}, is of the same form as that for the input gap; that is,

$$Y_{33} = (a_{11})_3 \tag{12.132}$$

Consequently, Fig. 12.24 is also applicable to the output gap. For the same reasons, the transit angle θ_3 is made as small as possible to reduce the electronic loading of the output cavity.

The principal measure of klystron performance can be found from the ratio of total current in the output to a-c input voltage. If we call this ratio the transadmittance Y_m, then from (12.130) we find that

$$Y_m = (a_{12})_3(a_{22})_2(a_{21})_1 + (a_{13})_3(a_{32})_2(a_{21})_1 + (a_{13})_3(a_{33})_2(a_{31})_1 + (a_{12})_3(a_{23})_2(a_{31})_1 \tag{12.133}$$

This can be simplified if we assume that θ_1 and θ_3 are very small. In this case we find that Y_m is given approximately by

$$Y_m \approx (a_{12})_3(a_{23})_2(a_{31})_1 \qquad (12.134)$$

Since the Llewellyn-Peterson equations are valid for small signals only, it is necessary to assume that the drift space is short in the present treatment. This is to avoid substantial bunching of the electron beam and

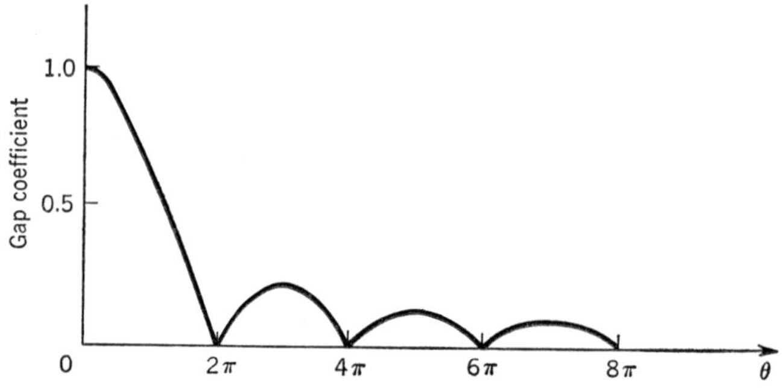

FIG. 12.25. Gap coefficients A and B as a function of their respective transit angle.

hence large signals. The reader can readily verify that [note that for the klystron $v_{01} = v_{02} = v_0 = (2eV_0/m)^{1/2}$ in all three regions]

$$\begin{aligned} |(a_{12})_3| &= \left|\frac{\sin(\theta_3/2)}{\theta_3/2}\right| = A \\ \frac{mv_0}{e}|(a_{31})_1| &= \left|\frac{\sin(\theta_1/2)}{\theta_1/2}\right| = B \\ \frac{e}{mv_0}(a_{23})_2 &= j\frac{i_0}{2V_0}\theta_2 e^{-j\theta_2} \end{aligned} \qquad (12.135)$$

by using Table 12.1 and taking $\zeta \ll 1$. Thus we find that

$$|Y_m| = \frac{i_0\theta_2}{2V_0}AB \qquad (12.136)$$

where V_0 is the d-c accelerating voltage impressed between the cathode and the cavity grids and collector, and i_0 is the average or d-c current that flows. The factors A and B above are called the "gap coefficients" for the input and output gaps. As can be seen by reference to Fig. 12.25, short transit time in the input and output gap is required if the transadmittance is not to be seriously degraded.

Equation (12.136) suggests that the transadmittance increases indefinitely as θ_2 increases. We have, however, imposed a restriction that θ_2 be small. There are actually two reasons for this. First, as already mentioned, this ensures small-signal conditions so that the Llewellyn-

Peterson equations may be applied. But even if small signals were assured, the L-P equations may be applied only to short drift spaces since they neglect a-c space-charge debunching. On the other hand, the latter is included in space-charge-wave theory. Let us therefore apply this theory to the klystron and contrast the solution it yields with that just obtained.

The drift space of the klystron is now viewed as a region in which space-charge waves are propagating with propagation constants

$$\gamma_{1s} = j\frac{\omega + \omega_q}{v_0} \tag{12.137}$$

and

$$\gamma_{1f} = j\frac{\omega - \omega_q}{v_0} \tag{12.138}$$

as obtained from (12.120). The effective plasma frequency ω_q depends on the geometry and on the constants of the beam. Now if the transit angle in the buncher is very small, the output a-c current is still zero, while a velocity modulation of the beam has been produced. The action of the buncher may thus be viewed as setting up a slow γ_{1s} and a fast γ_{1f} space-charge wave such that they combine to yield a net zero current at the buncher and a net velocity v_1. We shall now show how these conditions may be achieved.

Let i_{1f} be the a-c current amplitude of the fast space-charge wave and i_{1s} be that of the slow wave. Then the total current for a unit area of the beam is

$$i_1 = i_{1f}e^{j[\omega t - (\omega - \omega_q)z/v_0]} + i_{1s}e^{j[\omega t - (\omega + \omega_q)z/v_0]} \tag{12.139}$$

The positive direction of flow for i_1, i_{1f}, and i_{1s} is from right to left. The beam velocity can be found from the current by combining (12.99) and (12.100). The required relations are (note the change in sign since the positive direction of current flow in Sec. 12.7 is in the $+z$ direction)

$$i_{1f} = j\frac{\omega\rho_0 v_{1f}}{j\omega - \gamma_{1f}v_0} = \frac{\omega\rho_0}{\omega_q}v_{1f} \tag{12.140}$$

$$i_{1s} = j\frac{\omega\rho_0 v_{1s}}{j\omega - \gamma_{1s}v_0} = -\frac{\omega\rho_0}{\omega_q}v_{1s} \tag{12.141}$$

The total beam velocity from the sum of the fast- and slow-wave components is

$$v_1 = \frac{\omega_q}{\omega\rho_0}\left(i_{1f}e^{j[\omega t - (\omega - \omega_q)z/v_0]} - i_{1s}e^{j[\omega t - (\omega + \omega_q)z/v_0]}\right) \tag{12.142}$$

The requirement of zero a-c current at the buncher $z = 0$ is satisfied if we take

$$i_{1f} = -i_{1s} \tag{12.143}$$

If we designate by $(v_1)_1$ the a-c velocity at the buncher, then

$$v_1 = (v_1)_1 \cos \frac{\omega_q z}{v_0} e^{j(\omega t - \omega z/v_0)} \tag{12.144}$$

and

$$i_1 = j i_0 \frac{(v_1)_1}{v_0} \frac{\omega}{\omega_q} \sin \frac{\omega_q z}{v_0} e^{j(\omega t - \omega z/v_0)} \tag{12.145}$$

where ρ_0 has been replaced by i_0/v_0 (again a change in sign).

The results expressed by (12.144) and (12.145) arise from the principal space-charge mode. Depending on the boundary conditions at the buncher, higher-order modes may be launched as well. In this case the above equations should, at least, indicate the main effects.

As before, we calculate the transadmittance as a measure of merit of the klystron. If the signal applied at the buncher is $V_1 \sin \omega t$ and the d-c accelerating voltage is V_0, then from (12.11),

$$v = \left[\frac{2e}{m}(V_0 + V_1 e^{j\omega t})\right]^{1/2} \approx \left(\frac{2e}{m} V_0\right)^{1/2} \left(1 + \frac{V_1}{2V_0} e^{j\omega t}\right) \tag{12.146}$$

We may evaluate v_0 and $(v_1)_1$ from this equation and thus obtain

$$v_0 = \left(\frac{2eV_0}{m}\right)^{1/2} \tag{12.147a}$$

$$(v_1)_1 = \frac{v_0 V_1}{2V_0} \tag{12.147b}$$

Substituting these results into (12.145) and taking the length of the drift space as z_0 permit us to solve for the transadmittance; thus

$$Y_m = \frac{i_1(z_0)}{V_1} = j \frac{i_0}{2V_0} \frac{\omega}{\omega_q} \sin \frac{\omega_q z_0}{v_0} e^{j(\omega t - \omega z_0/v_0)} \tag{12.148}$$

For a small drift space, $(\omega_q z_0/v_0)$ will be small and the sine term in (12.148) can be replaced by its argument. With this restriction, (12.148) becomes

$$Y_m \approx j \frac{i_0}{2V_0} \frac{\omega z_0}{v_0} e^{j(\omega t - \omega z_0/v_0)}$$

But the drift-space transit angle θ is defined as $\omega z_0/v_0$; hence

$$Y_m \approx j \frac{i_0}{2V_0} \theta e^{j(\omega t - \theta)} \tag{12.149}$$

which is the same result as obtained by using the Llewellyn-Peterson equations. It is now clear that a linear dependence of Y_m on transit angle, as determined by the L-P equations, does not continue to hold as the drift space increases, but rather assumes a sinusoidal variation. The transadmittance therefore does not increase indefinitely with drift-space length but reaches an optimum value when

$$z_0 = \frac{\pi}{2} \frac{v_0}{\omega_q} \tag{12.150}$$

The ratio $2\pi v_0/\omega_q$ is referred to as the space-charge wavelength λ_{sc}, so that the maximum gain condition occurs for a drift-space length of $\lambda_{sc}/4$.

While the result given in (12.148) is more accurate than that available from the L-P analysis, it is still a small-signal theory. A satisfactory large-signal analysis that fully accounts for space charge is not available. However, a ballistic treatment is available which makes possible an approximate description of large-signal conditions. It indicates that optimum performance occurs for a specific amount of bunching, an amount that is well beyond linear theory. This is the theory of Webster, and an account can be found in Kleen, Beck, or Hutter (see references at the end of this chapter).

12.9. Traveling-wave-tube Amplifier

In the analysis of Sec. 12.7 we established that space-charge waves can be set up on electron beams and that these waves propagate at velocities

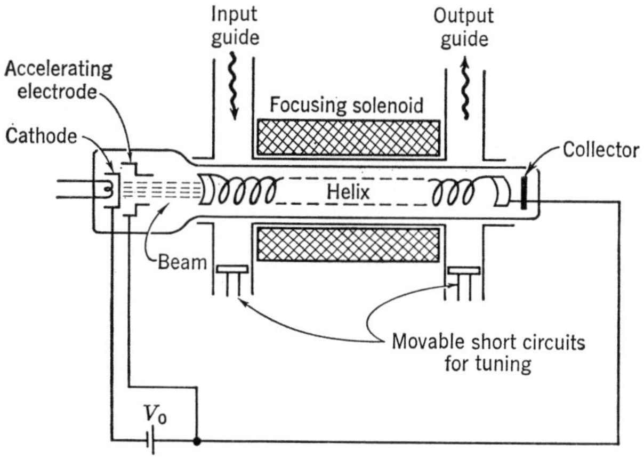

FIG. 12.26. Traveling-wave amplifier.

slightly greater or slightly less than the beam velocity. If the cylindrical conducting boundary is modified so that an electromagnetic field wave could propagate at the beam velocity, then in the presence of the beam an interaction between beam and field can take place such that the field acquires energy from the beam. We shall find that under proper conditions this energy exchange does take place, with consequent amplification of the r-f wave. Such a device, as described, is the traveling-wave amplifier.

Figure 12.26 is a sketch of the main parts of the traveling-wave tube

(TWT). An electron gun furnishes a collimated electron beam which travels through the drift space to a collector. An electromagnetic wave is coupled onto the helix at the input, and this tends to propagate along the helix wire, hence with an approximate phase velocity v_p given by

$$v_p \approx c \sin \psi$$

where ψ is the pitch angle of the helix. As a consequence, the helix furnishes a method of obtaining a "slow wave," that is, a wave with a

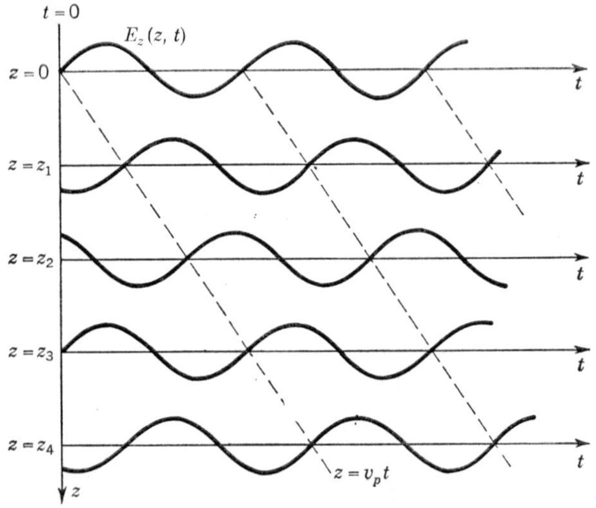

FIG. 12.27. Axial electric field in traveling-wave tube.

velocity considerably less than the velocity of light. We shall see that conditions for amplification require a slow wave at slightly less than beam velocity. The amplified r-f signal is coupled out at the output guide. The entire length of the drift space is placed in an axial d-c magnetic field for focusing purposes.

Before proceeding to a rigorous consideration of the problem, we shall give a qualitative ballistic picture of what takes place in the TWT and why amplification may be expected. We shall first of all suppose that an axial electric field exists and propagates along the tube with a slow velocity (approximately beam velocity) v_p, so that

$$E_z = E_0 \cos \omega \left(t - \frac{z}{v_p} \right)$$

A sketch of the field E_z as a function of time for several values of z is given in Fig. 12.27. At $z = 0$ those electrons that enter the helix region when

E_z is positive have a force $-eE_z$ exerted on them and are therefore slowed down. Electrons entering when E_z is negative are accelerated. In other words, a velocity modulation of the beam is produced. Electrons which are retarded are continually slowed down until they have slipped sufficiently far back through the electromagnetic wave to be in a region where E_z is zero. The accelerated electrons continually advance through the wave until they get into a region where E_z is positive and a retarding force is exerted on them. They are now slowed down and move back relative

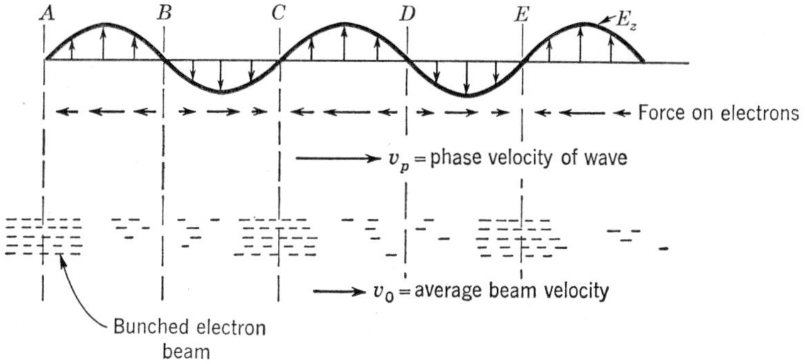

FIG. 12.28. Illustration of formation of electron bunches.

to the wave into a position where E_z is zero. The result of this mechanism is to produce a density modulation of the beam; i.e., electron bunches are formed.

Figure 12.28 illustrates qualitatively the bunching process. Those electrons that have bunched in positions where the field is changing from accelerating to decelerating, such as A, C, and E, are in a stable position, since if they advance relative to the wave, the wave exerts a retarding force, while if they slip back relative to the wave, an accelerating force is exerted on them. When the electrons are being accelerated, the field is giving energy to the beam, while the beam gives up energy to the wave during the decelerating process. In order to get a net gain for the wave, the phase velocity v_p is chosen slightly smaller than the beam velocity v_0. Thus the bunched beam as a whole moves up through the wave as the two progress along the helix. As a result, a continuous decelerating force is exerted on the beam by the wave. This continues until the initial average beam velocity v_0 has been reduced to v_p and the beam and wave are in synchronism. Some of the kinetic energy of the beam thus becomes converted into an increased amplitude of the wave, and amplification results.

In order to obtain a rigorous solution for the traveling-wave-tube fields we shall set up solutions to Maxwell's equations for the region internal

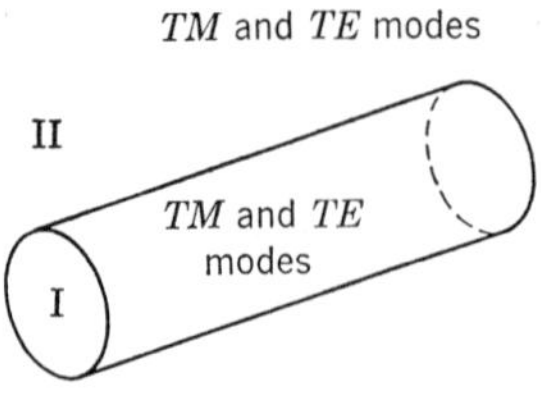

FIG. 12.29. Mode structure in a traveling-wave tube.

and external to the helix, and then impose suitable boundary conditions over the helix surface. The boundary conditions will permit setting up a determinantal equation for the permissible propagation constants. The general scheme is indicated diagrammatically in Fig. 12.29. Both TE and TM modes will be required in order to satisfy all the boundary conditions, and this fact is noted in Fig. 12.29. The characteristics of the electron stream are assumed to be similar to those discussed in Sec. 12.7, namely:

1. Current is in the axial z direction only.
2. The ratio of a-c to d-c quantities is small.
3. The d-c space charge is neutralized by positive ions.
4. All a-c quantities are of the form $e^{j\omega t - \gamma z}$.

We furthermore assume that the beam completely fills region 1 and is bounded by an ideal helix. The drift space is taken to be, essentially, infinite. Finally, azimuthal symmetry is assumed so that $\partial/\partial\phi = 0$.

The TM waves for the internal region can be found from an axial vector potential A_z, where

$$A_z = A_1 e^{j(\omega t - \gamma z)} \tag{12.151}$$

But A_z arises in precisely the same way as it did in the analysis of space-charge waves; that is, a review of the above conditions will show that Eqs. (12.88) to (12.104) may be rewritten for the present problem. Accordingly, we have, from (12.104), that the vector potential satisfies

$$\nabla_t^2 A_1 = -(\gamma^2 + k_0^2)\left[1 + \frac{\omega_0^2}{(j\omega - \gamma v_0)^2}\right] A_1$$

where ω_0 is the plasma frequency as determined by

$$\omega_0^2 = \frac{e\rho_0}{m\epsilon_0}$$

and v_0 is the d-c beam velocity. The a-c current of the electron beam enters the equation by means of the term involving the plasma frequency. Let us designate

$$p^2 = -(\gamma^2 + k_0^2) \tag{12.152a}$$

and

$$g^2 = p^2\left[1 + \frac{\omega_0^2}{(j\omega - \gamma v_0)^2}\right] \tag{12.152b}$$

Note that since $k_0^2 = \omega^2/c^2$ and $\gamma^2 \approx -\omega^2/v_0^2$, we may expect p^2 and g^2 to be predominantly real. In cylindrical coordinates with $\partial/\partial\phi = 0$, the wave equation for A_1 becomes

$$\frac{d^2A_1}{dr^2} + \frac{1}{r}\frac{dA_1}{dr} - g^2 A_1 = 0 \tag{12.153}$$

We recognize (12.153) as Bessel's equation of zero order in the argument (jgr). Since we cannot admit a singularity at the axis, only the Bessel function of the first kind is involved, and we have

$$A_1 = CJ_0(jgr) \equiv CI_0(gr) \tag{12.154}$$

The function $I_0(x)$ is the modified Bessel function of the first kind, and $I_0(x)$ equals $J_0(jx)$. As can be verified by substituting the variable jx into the Bessel function series given in (4.45a), $I_0(x)$ is real for real x. Its asymptotic expansion for large argument is

$$I_0(x) \sim \sqrt{\frac{1}{2\pi x}}\, e^x \tag{12.155}$$

A second solution to (12.153) is $K_0(gr)$, where K_0 is the modified Bessel function of the second kind. Its asymptotic form for large x is

$$K_0(x) \sim \sqrt{\frac{\pi}{2x}}\, e^{-x} \tag{12.156}$$

Some additional details are given in Sec. 4.2. We may also note that

$$I_1(x) = \frac{d}{dx} I_0(x) \qquad K_1(x) = -\frac{d}{dx} K_0(x)$$

and that for large argument the asymptotic form for the first-order functions is the same as that for the corresponding zero-order functions.

The electric and magnetic fields for the TM mode for region I are now found from the vector potential A_1. Applying (9.68) to (12.154) results in the following fields (the subscript 1 refers to region I).

For TM waves:

$$\left.\begin{aligned} E_z &= -\frac{p^2}{j\omega\epsilon_0\mu_0} B_1 I_0(gr)e^{-\gamma z} \\ E_r &= -\frac{\gamma g}{j\omega\epsilon_0\mu_0} B_1 I_1(gr)e^{-\gamma z} \\ \mu_0 H_\phi &= -gB_1 I_1(gr)e^{-\gamma z} \end{aligned}\right\} \quad r < a \tag{12.157}$$

For TE waves in region I and for TE and TM waves in region II we have homogeneous equations; that is, for the latter cases there are no source functions. For $r > a$ we may consider (E_z,H_z) as potential functions which must satisfy

$$\nabla_t^2 \begin{pmatrix} E_z \\ H_z \end{pmatrix} = -(\gamma^2 + k_0^2) \begin{pmatrix} E_z \\ H_z \end{pmatrix} = p^2 \begin{pmatrix} E_z \\ H_z \end{pmatrix} \tag{12.158}$$

In cylindrical coordinates, and using $\partial/\partial\phi = 0$, the equation for E_z (and an identical one for H_z) becomes

$$\frac{d^2E_z}{dr^2} + \frac{1}{r}\frac{dE_z}{dr} - p^2E_z = 0 \tag{12.159}$$

The solution is in the form of modified Bessel functions. Since I_0 increases indefinitely with r, as is revealed by (12.155), solutions of (12.159) must be expressed in terms of K_0. Thus TM waves in region II can be derived from

$$E_z = B_2 K_0(pr) e^{-\gamma z} \tag{12.160}$$

while TE waves arise from

$$H_z = C_2 K_0(pr) e^{-\gamma z} \tag{12.161}$$

For the TE waves in region I we must choose as a solution of (12.159)

$$H_z = C_1 I_0(pr) e^{-\gamma z} \tag{12.162}$$

since $K_0(pr)$ becomes infinite for $r \to 0$.

We are now in a position to compute TE and TM waves in region II and TE waves in region I. These are obtained in the usual way from the axial electric and magnetic fields given in Eqs. (12.160) to (12.162) (see Sec. 10.1). If we include TM waves in region I, for completeness, then we have

For TM waves:

$$\left.\begin{aligned} E_z &= -\frac{p^2}{j\omega\epsilon_0\mu_0} B_1 I_0(gr) e^{-\gamma z} \\ E_r &= -\frac{\gamma g}{j\omega\epsilon_0\mu_0} B_1 I_1(gr) e^{-\gamma z} \\ \mu_0 H_\phi &= -g B_1 I_1(gr) e^{-\gamma z} \end{aligned}\right\} \quad r < a$$
$$\left.\begin{aligned} E_z &= B_2 K_0(pr) e^{-\gamma z} \\ E_r &= -\frac{\gamma}{p} B_2 K_1(pr) e^{-\gamma z} \\ H_\phi &= -\frac{j\omega\epsilon_0}{p} B_2 K_1(pr) e^{-\gamma z} \end{aligned}\right\} \quad r > a \tag{12.163}$$

For TE waves:

$$\left.\begin{aligned} H_z &= C_1 I_0(pr) e^{-\gamma z} \\ H_r &= \frac{\gamma}{p} C_1 I_1(pr) e^{-\gamma z} \\ E_\phi &= -\frac{j\omega\mu_0}{p} C_1 I_1(pr) e^{-\gamma z} \end{aligned}\right\} \quad r < a$$
$$\left.\begin{aligned} H_z &= C_2 K_0(pr) e^{-\gamma z} \\ H_r &= -\frac{\gamma}{p} C_2 K_1(pr) e^{-\gamma z} \\ E_\phi &= \frac{j\omega\mu_0}{p} C_2 K_1(pr) e^{-\gamma z} \end{aligned}\right\} \quad r > a \tag{12.164}$$

We must now consider the boundary conditions that exist over the surface of the helix. The helix is often formed from a conducting tape wound with a uniform pitch. The spacing between turns may then be small compared with the width of the tape. Our problem will be simplified if we idealize the helix. We shall suppose it to consist of an infinitely thin conducting cylinder with infinitely thin helical slits. The effect of the latter may be described in terms of a cylindrical sheath that has perfect conductivity along the slits and zero conductivity at right angles.

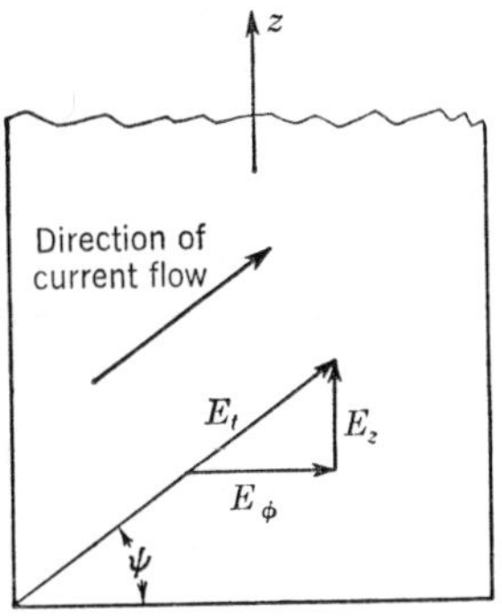

FIG. 12.30. Developed helical sheath.

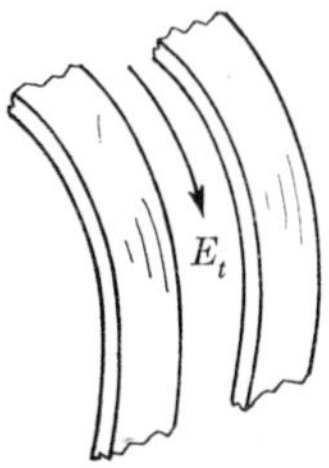

FIG. 12.31. Two adjacent turns of helix.

Figure 12.30 is the developed cylindrical surface showing the direction of the helical cuts and hence that of current flow. The angle Ψ is the pitch angle of the helix.

Figure 12.31 is a magnified view of two adjacent turns of the helix. If E_t is the electric field tangential to the turns, then $E_t = 0$. This condition must be satisfied by both the internal and external fields. In cylindrical coordinates we have (see Fig. 12.30 for geometry)

$$\text{Region I } (r = a) \qquad E_{\phi_1} \cos \Psi + E_{z_1} \sin \Psi = 0 \tag{12.165a}$$

$$\text{Region II } (r = a) \qquad E_{\phi_2} \cos \Psi + E_{z_2} \sin \Psi = 0 \tag{12.165b}$$

The component of **E**, in the helical surface, that is normal to the cuts must be continuous from region I to II through the helix. This leads to the condition

$$E_{z_1} \cos \Psi - E_{\phi_1} \sin \Psi = E_{z_2} \cos \Psi - E_{\phi_2} \sin \Psi \qquad r = a \tag{12.165c}$$

Finally, the component of **H** in the helical surface that is parallel to the cuts must be continuous from region I to II. This is the consequence of the fact that no current flows normal to the cuts. This condition yields

$$H_{\phi_1} \cos \Psi + H_{z_1} \sin \Psi = H_{\phi_2} \cos \Psi + H_{z_2} \sin \Psi \qquad r = a \tag{12.165d}$$

Equations (12.163) and (12.164) completely describe the fields inside and outside the helix. There are four constants to be determined, and these can be found by substitution into the boundary conditions given by (12.165). The result is a system of four homogeneous simultaneous equations in four unknowns. This can have a nontrivial solution only if the determinant is zero. The resultant equation is known as the determinantal equation, or characteristic equation. The student may confirm that the algebra leads to

$$\frac{-gI_1(ga)}{I_0(ga)} = \frac{pK_1(pa)}{K_0(pa)} - \frac{p^3 \tan^2 \Psi}{k_0^2}\left[\frac{I_0(pa)}{I_1(pa)} + \frac{K_0(pa)}{K_1(pa)}\right] \tag{12.166}$$

Let us first examine (12.166) for the condition that an electron stream is absent. In this case it is necessary to modify (12.166) by setting $g = p$. If we furthermore speculate on the occurrence of a slow wave, in which case $|\gamma^2| \gg k_0^2$ and $p^2 \approx -\gamma^2$, then if a is on the order of a free-space wavelength, $pa \gg 1$. But in this event we can show that, asymptotically for large pa,

$$\frac{I_1(pa)}{I_0(pa)} \rightarrow 1 \qquad \frac{K_1(pa)}{K_0(pa)} \rightarrow 1 \tag{12.167}$$

With this approximation (12.166) becomes

$$-g = -p = p - 2\frac{p^3}{k_0^2}\tan^2 \Psi$$

or

$$p^2 = -(\gamma^2 + k_0^2) = k_0^2 \cot^2 \Psi \tag{12.168}$$

This equation is quite well justified provided that $pa \geq 4$. Solving for γ gives

$$\gamma = \frac{j\omega}{c \sin \Psi} \tag{12.169}$$

which is the slow wave as predicted. The phase velocity, from (12.169), is

$$v_p = c \sin \Psi$$

The wave may be considered to propagate along the helical conductor with a velocity c. It then progresses along the axial direction at a velocity $c \sin \Psi$ only.

Let us return now to the case where an electron stream is present. Let p_0 be the value of p in the absence of the electrons; that is,

$$p_0 = k_0 \cot \Psi$$

Then if we continue to assume $pa \gg 1$, and in addition that $ga \gg 1$, the substitution of (12.167) into (12.166) leads to

$$-g = p - \frac{2p^3}{k_0^2}\tan^2 \Psi = p\left(1 - \frac{2p^2}{p_0^2}\right) \tag{12.170}$$

If we equate (12.170) to the value of g given in (12.152*b*), there results

$$-\sqrt{1+\frac{\omega_0^2}{(j\omega-\gamma v_0)^2}}=1-\frac{2p^2}{p_0^2}$$

which simplifies to

$$1+\frac{\omega_0^2}{(j\omega-\gamma v_0)^2}=\left(1-\frac{2p^2}{p_0^2}\right)^2\approx\left(1-\frac{2\gamma^2}{\gamma_0^2}\right)^2 \qquad (12.171)$$

where $\gamma_0 = j\omega/v_0$.

We shall seek solutions for γ in (12.171) by assuming that it deviates only slightly from γ_0, that is, that the phase velocity is approximately the electron velocity. If δ is a quantity small compared with unity, then

$$\gamma=\frac{j\omega}{v_0}(1+\delta) \qquad (12.172)$$

Putting (12.172) into (12.171) gives

$$1+\frac{\omega_0^2}{(-j\omega\delta)^2}=[1-2(1+\delta)^2]^2$$

If we retain only the leading term in δ (this is the cubic term), then

$$\delta^3=-\frac{\omega_0^2}{8\omega^2}$$

or

$$\delta=C(-1)^{1/3} \qquad (12.173)$$

where $C = \frac{1}{2}(\omega_0/\omega)^{2/3}$. Thus the deviation δ arises from the three roots of -1. The three values of δ are

$$\delta_1=-C \qquad (12.174a)$$

$$\delta_2=C\left(\tfrac{1}{2}-j\frac{\sqrt{3}}{2}\right) \qquad (12.174b)$$

$$\delta_3=C\left(\tfrac{1}{2}+j\frac{\sqrt{3}}{2}\right) \qquad (12.174c)$$

We have already noted that $\omega_0 \ll \omega$ in practice, so that $C \ll 1$, and the assumption that $\delta \ll 1$ is justified. This, furthermore, justifies $ga \gg 1$ if $pa \gg 1$.

Let us consider the physical interpretation of the three values of propagation constant that we have just found. Substituting δ_1 into (12.172) reveals the presence of a propagating wave with a phase velocity slightly greater than that of the stream. The value of δ_2 results in a propagation constant where the phase velocity is slightly less than that of the electron beam and where, in addition, an attenuation of the wave results. In the value of δ_3, a propagation constant results where the phase velocity (equal to that produced by δ_2) is somewhat less than that of the stream and where an exponential increase in amplitudes results. This is the growing-wave solution, and if all waves are present at the tube input, this wave will predominate at the output; that is, the three roots for δ give rise to propagation factors

$$\exp\left[-j\omega\frac{(1-C)z}{v_0}\right] \qquad \exp\left(-j\omega\frac{1+C/2}{v_0}z - \frac{\sqrt{3}\,\omega C}{2v_0}z\right)$$

and

$$\exp\left(-j\omega\frac{1+C/2}{v_0}z + \frac{\sqrt{3}\,\omega C}{2v_0}z\right)$$

of which the last is clearly an exponentially increasing wave in the $+z$ direction. The amplification per unit length of tube is given by the growth constant α, where

$$\alpha = \frac{0.866C\omega}{v_0} \qquad \text{nepers/m} \tag{12.175}$$

As illustrated in Fig. 12.26, the electromagnetic wave to be amplified is coupled from the waveguide to the helix by a suitable adapter. While a cavity provides suitable control of coupling, it is frequency-sensitive and restricts the bandwidth from what might otherwise be attained. The traveling-wave tube by itself is inherently capable of operation over a very wide band of frequencies.

Conditions at $z = 0$ are that the a-c convection current and a-c velocity are zero. We assume the presence of the three possible modes and note that they must satisfy the condition

$$i_1 + i_2 + i_3 = 0$$
$$v_1 + v_2 + v_3 = 0$$

Using the relationship between current and velocity and also the values of the corresponding propagation constants leads to the results

$$\frac{\dot{i}_1}{\dot{i}_2} = e^{j2\pi/3}$$

$$\frac{\dot{i}_3}{\dot{i}_2} = e^{-j2\pi/3}$$

Consequently, all waves at $z = 0$ have equal magnitude, but they are 120° apart in phase. As already noted, only one wave will be amplified, and this wave will eventually predominate. Consequently, only one-ninth of the input r-f power is useful. The over-all power gain may then be written

$$\text{Gain} = \tfrac{1}{9}e^{2\pi\sqrt{3}CN} \tag{12.176}$$

where

$$N = \frac{\text{frequency} \times \text{length}}{\text{beam velocity}}$$

The results obtained in this section depend on the property of the helix that it be capable of propagating an electromagnetic wave with a velocity comparable to that of the electron stream, i.e., a velocity much less than the velocity of light. In addition to the helix, other slow-wave structures are available with this property. Examples of such structures are discussed in the books by Beck, Kleen, and Watkins listed in the references at the end of this chapter.

The interaction of the electron stream and electromagnetic wave is designed, in the traveling-wave amplifier, so that energy is given up by the stream to the field. We have commented on this earlier, where we sought to explain in a qualitative way the amplifying property of the traveling-wave tube. Now it is also possible to set up an appropriate structure such that the traveling electromagnetic wave gives up energy to the electron stream. It will be necessary that the wave travel at approximately the same velocity as the stream for interaction to occur. Under these conditions continuous acceleration of the electrons will occur and very high energies may be achieved. This device is the linear accelerator, and details of its operation may be found in the book by Slater.

12.10. Electromagnetic-wave Propagation in Gyrotropic Media

There are two classes of media that are referred to as gyrotropic media. The one is an ionized gas or plasma in the presence of a d-c magnetic field, and the other is a ferrite medium, also in the presence of a d-c magnetic field. In the first medium the application of an electric field causes the ionized particles to gyrate in circular orbits about the d-c magnetic lines of flux in accordance with the laws of motion discussed in Sec. 12.4. In a ferrite material the spinning electrons are forced to gyrate about the d-c magnetic field lines because of the torque acting on the dipole moment when a magnetic field is applied. Both media involve gyrating particles and hence are referred to as "gyrotropic." Both media will produce "Faraday rotation" of the plane of polarization of a plane transverse electromagnetic wave propagating in a direction along the d-c magnetic lines of flux. Furthermore, this Faraday rotation is a nonreciprocal effect, and both media may be used to construct nonreciprocal devices. The ionosphere in the presence of the earth's magnetic field is a familiar example of a gyrotropic medium. The propagation of electromagnetic waves through the ionosphere is, of course, of great importance in practical communication systems.

In this section we shall examine wave propagation in both plasmas and ferrites. It will be found that the natural modes of propagation in both media are circular-polarized plane waves. For this reason it will be expedient to consider the mathematical formulation for circular-polarized waves first.

Circular-polarized Waves

A plane TEM wave propagating in the z direction may be described by the equations

$$\mathbf{E} = \mathbf{a}_x E_1 e^{j\omega t - jk_0 z}$$
$$\mathbf{H} = \mathbf{a}_y Y_0 E_1 e^{j\omega t - jk_0 z}$$

where $Y_0 = (\epsilon_0/\mu_0)^{1/2}$ and E_1 is an amplitude constant, assumed real for simplicity. The above wave is said to be linearly polarized, and the plane of polarization is the plane containing the **E** vector and wave normal, the xz plane above.

If we superpose on the above wave another linearly polarized wave propagating in the same direction and of equal amplitude but whose corresponding **E** and **H** fields are in space quadrature and 90° apart in time phase relative to the first wave, then the resultant wave is said to be circular-polarized. Such a wave would be described mathematically as

$$\mathbf{E} = (\mathbf{a}_x E_1 + j\mathbf{a}_y E_1) e^{j\omega t - jk_0 z} \tag{12.177a}$$
$$\mathbf{H} = (\mathbf{a}_y Y_0 E_1 - j\mathbf{a}_x Y_0 E_1) e^{j\omega t - jk_0 z} \tag{12.177b}$$

The real physical electric field corresponding to (12.177*a*) is the real part of the above:

$$\mathbf{E} = \mathbf{a}_x E_1 \cos(\omega t - k_0 z) - \mathbf{a}_y E_1 \sin(\omega t - k_0 z) \tag{12.178}$$

The magnitude of the resultant electric field is

$$E = (E_x^2 + E_y^2)^{1/2} \tag{12.179a}$$

where

$$E_x = E_1 \cos(\omega t - k_0 z) \tag{12.179b}$$
$$E_y = -E_1 \sin(\omega t - k_0 z) \tag{12.179c}$$

If we square both sides of (12.179*b*) and (12.179*c*) and add, we obtain

$$\begin{aligned} E_x^2 + E_y^2 &= E_1^2[\cos^2(\omega t - k_0 z) + \sin^2(\omega t - k_0 z)] \\ &= E_1^2 \end{aligned}$$

If this is compared with (12.179*a*), it is seen that **E** is constant in magnitude. The trace of the tip of the electric vector in the xy plane must then be a circle. The vector rotates at a uniform rate of ω radians per second, as is confirmed by (12.179*b*) and (12.179*c*). It is for this reason that the wave is called a circular-polarized wave.

In Fig. 12.32 the resultant **E** vector is plotted for several values of ωt in the $z = 0$ plane [the equations used are (12.179*b*) and (12.179*c*)]. For this particular case the **E** vector rotates in a counterclockwise direction when looking in the direction of propagation. This particular wave is therefore said to be negative-circular-polarized. If the y component of electric field lags the x component in time by 90° instead of leading by 90°, the **E** vector will rotate in a clockwise direction. This results in a positive-circular-polarized wave.

In general, if we superpose two linearly polarized waves of unequal amplitude whose planes of polarization do not coincide and for which a

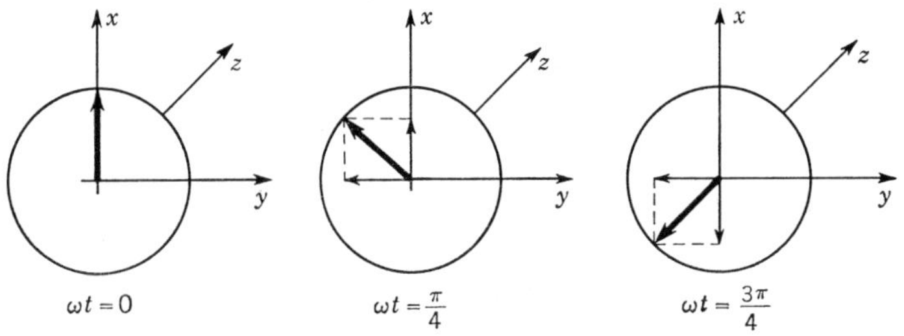

FIG. 12.32. Trace of **E** vector in xy plane for a negative-circular-polarized wave.

relative phase shift exists, we obtain an elliptical-polarized wave. The tip of the **E** vector in this case traces out an ellipse in the transverse plane.

Wave Propagation in the Ionosphere

The constant bombardment of the upper layers of the atmosphere by ultraviolet energy results in the ionization of the gas molecules. Equal numbers of electrons and positive ions are present. However, since the positive ions are so much heavier than the electrons, only the electrons interact appreciably with an electromagnetic wave. For the discussion here it will be assumed that there are N electrons present per unit volume and that the gas is so rarefied that interaction of electrons with each other may be neglected.

Consider an ionized gas of the type described above with a d-c magnetic field B_0 applied in the z direction. The force acting on the electron of charge $-e$ and mass w is

$$F = -e[\mathbf{E} + \mathbf{v} \times (\mathbf{B} + \mathbf{a}_z B_0)]$$

where **E**, **B** are the a-c fields and **v** is the electron velocity. We are using the symbol w for the electron mass here since we wish to reserve the symbol m for the magnetic dipole moment of the spinning electron later. In a plane wave $\mu_0|\mathbf{H}| = |\mathbf{B}| = \mu_0 Y_0|\mathbf{E}| = (\mu_0\epsilon_0)^{1/2}|\mathbf{E}| = |\mathbf{E}|c^{-1}$, where c is the velocity of light. Thus the Lorentz force $-e\mathbf{v} \times \mathbf{B}$, with a magnitude $e|\mathbf{E}|v/c$, is smaller by a factor v/c compared with the force $-e\mathbf{E}$ due to the electric field. Since we shall assume $v \ll c$, which is the usual case in practice, the force due to the time-varying part of the magnetic field may be neglected.

In order to study the behavior of the electrons in the presence of an electromagnetic field, the analysis is simplified if we assume that the electrons are gyrating about the d-c field B_0 in a circular orbit and in synchronism with the **E** vector of a circular-polarized wave. With reference to Fig. 12.33, let **E** be a positive-circular-polarized wave propagating in the z direction. In the xy plane let the electron rotate with an angular velocity ω in synchronism with **E**. The force $\mathbf{F}_e$ due to the electric field

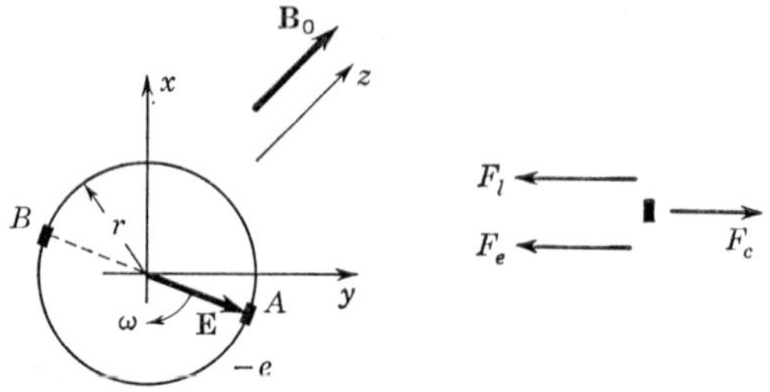

FIG. 12.33. A gyrating electron in the presence of a positive-circular-polarized electric field.

is radially inward and equal in magnitude to eE_1, where $E_1 = |\mathbf{E}|$. The Lorentz force $-e\mathbf{v} \times B_0\mathbf{a}_z$ is also directed inward with a magnitude $F = evB_0$. The sum of these two forces must equal the outward centrifugal force $F_c = wv^2/r$ if the postulated electron orbit is to be a stable one. Thus

$$eE_1 + evB_0 = \frac{wv^2}{r}$$

or

$$eE_1 + e\omega rB_0 = w\omega^2 r \tag{12.180}$$

Solving for the radius r of the orbit gives

$$r = \frac{eE_1}{w\omega^2 - e\omega B_0} = \frac{(e/w)E_1}{\omega^2 - \omega\omega_g} \tag{12.181}$$

where the gyrofrequency ω_g is given by

$$\omega_g = \frac{eB_0}{w} \tag{12.182}$$

If $\omega < \omega_g$, then the electron's position is diametrically across from that assumed (position B in Fig. 12.33), so that the force due to the electric field is radially outward and the expression for r is still positive.

The circulating electron constitutes a rotating electric dipole of moment re, 180° out of phase with the electric field and rotating in synchronism with it. Thus for N electrons per unit volume the dipole polarization P_+ per unit volume associated with the positive-circular-polarized electric field is $P_+ = -Nre$ and from (12.181) leads to

$$\mathbf{P}_+ = -\frac{N(e^2/w)\mathbf{E}_1}{\omega^2 - \omega\omega_g} \tag{12.183}$$

As is clear from the model used, P_+ and E_1 are parallel, and hence (12.183) is written in vector form. The circular-polarized displacement field $\mathbf{D}_1$ is

$$\mathbf{D}_1 = \epsilon_0\mathbf{E}_1 + \mathbf{P}_+ = \epsilon_+\mathbf{E}_1 = \left[\epsilon_0 - \frac{N(e^2/w)}{\omega^2 - \omega\omega_g}\right]\mathbf{E}_1 \tag{12.184}$$

This equation defines the effective permittivity ϵ_+ for the positive-circular-polarized wave. It is given by

$$\epsilon_+ = \epsilon_0\left[1 - \frac{N(e^2/\epsilon_0 w)}{\omega^2 - \omega\omega_g}\right] \tag{12.185}$$

Since the electromagnetic field is a solution of Maxwell's equations and represents a plane wave, its propagation constant k_+ is given by

$$k_+ = \omega\sqrt{\mu_0\epsilon_+} \tag{12.186a}$$

and its phase velocity is

$$v_+ = c\sqrt{\frac{\epsilon_0}{\epsilon_+}} \tag{12.186b}$$

For ω smaller than the gyrofrequency ω_g, $\epsilon_+ > \epsilon_0$, while for $\omega > \omega_g$, the permittivity is less than ϵ_0 and the phase velocity exceeds that of waves in free space. When $\omega = \omega_g$, (12.185) predicts an infinite value for ϵ_+. In practice, this does not occur because of electron collisions which introduce a damping term.

The above solutions for ϵ_+ and k_+ may be arrived at in a more formal way by treating the rotating electron as a current element $-Ne\mathbf{v}$. The curl equation for $\mathbf{H}$ may then be written

$$\nabla \times \mathbf{H} = j\omega\epsilon_0\mathbf{E} + \mathbf{J} = j\omega\epsilon_0\mathbf{E} - Ne\mathbf{v} \tag{12.187}$$

Now it is assumed that we have a positive-circular-polarized wave

$$\mathbf{E} = (\mathbf{a}_x E_1 - j\mathbf{a}_y E_1)e^{-jk_+z} \tag{12.188}$$

The corresponding magnetic field is found from the relation

$$\nabla \times \mathbf{E} = -j\omega\mu_0\mathbf{H}$$

Hence

$$\nabla \times \mathbf{H} = \frac{j}{\omega\mu_0}\nabla \times \nabla \times \mathbf{E} = -\frac{j}{\omega\mu_0}\nabla^2\mathbf{E}$$

since $\nabla \times \nabla \times \mathbf{E} = \nabla\nabla \cdot \mathbf{E} - \nabla^2\mathbf{E}$ and $\nabla \cdot \mathbf{E} = 0$ because there is no x or y variation. Upon resolving $\mathbf{v}$ in rectangular coordinates, (12.187) may be rewritten as

$$-\frac{j}{\omega\mu_0}\nabla^2\mathbf{E} = j\omega\epsilon_0\mathbf{E} - Nev(j\mathbf{a}_x + \mathbf{a}_y) \tag{12.189}$$

since $\mathbf{v} = \text{Re}\,[v(j\mathbf{a}_x + \mathbf{a}_y)e^{j(\omega t - k_+ z)}]$, that is, $\mathbf{v}$ is formulated to be orthogonal to $\mathbf{E}$. There is no x or y variation, so $\nabla^2\mathbf{E} = -k_+^2\mathbf{E}$. In component form, (12.189) gives

$$\frac{jk_+^2 E_1}{\omega\mu_0} = j\omega\epsilon_0 E_1 - jNev$$

for the x component and a similar result for the y component. This equation is satisfied if

$$k_+^2 = \omega^2\mu_0\epsilon_0 - \frac{Ne\omega\mu_0 v}{E_1}$$

Replacing v by ωr and using (12.181) to eliminate r yield the solution (12.186a) for k_+. The permittivity ϵ_+ may then be defined by the relation $k_+^2 = \omega^2\mu_0\epsilon_+$ and is obviously given by (12.185).

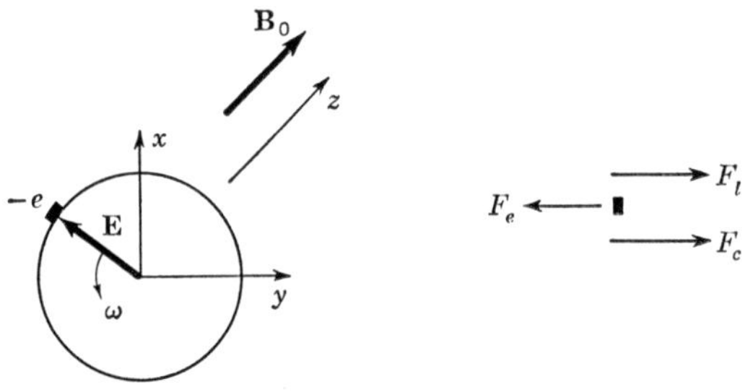

FIG. 12.34. Gyrating electron in the presence of a negative-circular-polarized wave.

An analysis similar to the above will now be carried out for a negative-circular-polarized wave. With reference to Fig. 12.34, let $\mathbf{E}$ be a negative-circular-polarized wave. The $\mathbf{E}$ vector, of magnitude E_1, rotates in a counterclockwise sense in this case. The electron must also gyrate in the same direction in order to remain in equilibrium with the force exerted by the fields. The electric field produces a force eE_1 directed inward. The Lorentz force in this case is, however, directed radially outward. Corresponding to (12.180) we have

$$eE_1 - evB_0 = \frac{wv^2}{r} = w\omega^2 r$$

Solving for r gives

$$r = \frac{(e/w)E_1}{\omega^2 + \omega\omega_g} \tag{12.190}$$

If we introduce the dipole moment $P_- = -Ner$ per unit volume, associated with the negative-circular-polarized wave, it is found that the permittivity ϵ_- is given by

$$\epsilon_- = \epsilon_0\left[1 - \frac{N(e^2/\epsilon_0 w)}{\omega^2 + \omega\omega_g}\right] \tag{12.191}$$

The corresponding propagation constant k_- is

$$k_- = \omega \sqrt{\mu_0 \epsilon_-} \tag{12.192}$$

As before, this solution may be verified by means of Maxwell's equations and considering $-Ne\mathbf{v}$ as a current density $\mathbf{J}$.

When the d-c field $B_0 = 0$, then

$$\epsilon_+ = \epsilon_- = \epsilon_0 \left(1 - \frac{Ne^2}{\omega^2 \epsilon_0 w}\right)$$

This is actually the condition that describes the first-order solution for "sky-wave" radio propagation in the ionosphere; although $B_0 \neq 0$, for the frequencies involved (3 to 25 megacycles per second) $\omega \gg \omega_g$, where $\omega_g \approx 1.4$ megacycles per second, and the above equation is a good approximation to (12.185) and (12.191). An exploration of the mechanism of this propagation involves a detailed description of the charged ionospheric layers which exist at great heights (50 to 500 kilometers) above the earth's surface, where N increases with height to a maximum value in each layer, and then falls off. As a consequence we may think of the ionosphere as consisting of a continuum of layers of varying dielectric-constant material, and the analysis is then a successive application of Snell's law of refraction. The portion of the ionosphere corresponding to increasing N, hence decreasing refractive index, causes an obliquely incident ray to be bent away from the normal. As a result, the wave may be bent back to a point on the earth's surface. If this occurs for an incident ray with a large angle of incidence, and considering the height of the ionosphere, the refracted energy will return to the earth at a considerable distance from the transmitter. It is also possible to obtain several successive reflections from the ground and ionosphere.

For a frequency $\omega < \omega_c = (Ne^2/\epsilon_0 w)^{1/2}$, the effective permittivity becomes negative. In this case the propagation constants k_+ and k_- are imaginary. If N_{max} represents the maximum charge density overhead, then a wave radiated directly upward will not be able to penetrate the ionosphere if $\omega < (N_{max}e^2/\epsilon_0 w)^{1/2} = \omega_c$. By measuring the maximum frequency that yields a reflected wave, we obtain ω_c, and hence the maximum electron density. Typical values for the critical frequency $f_c = \omega_c/2\pi$ range from 3 to 8 megacycles per second, depending on season, time of day, and sunspot activity. It should be noted that even though a frequency of, say, 15 megacycles per second would penetrate the ionosphere if transmitted vertically upward, an oblique transmission may be bent sufficiently to be returned to the earth.

The above results apply to linearly polarized waves as well, a conclusion that will be readily apparent after study of the following section on Faraday rotation. In this case the electrons execute an essentially linear vibration.

Faraday Rotation

In this section we shall show that a gyrotropic medium causes the plane of polarization of a linearly polarized wave to rotate as the wave propagates through the medium in a direction parallel to the d-c magnetic field. The essential steps in the analysis are to decompose the linearly polarized wave into the sum of a positive- and negative-circular-polarized wave. These component waves will suffer a relative change in phase in propagating a distance d in the medium. Upon recombination, it is found that the resultant wave is a linearly polarized wave but with its plane of polarization rotated.

At $z = 0$, let the electric field of a linearly polarized wave be

$$\mathbf{E} = \mathbf{a}_x E_1 e^{j\omega t} \qquad z = 0$$

We may write this wave in the following form also:

$$\begin{aligned}\mathbf{E} &= \frac{E_1}{2}(\mathbf{a}_x - j\mathbf{a}_y)e^{j\omega t} + \frac{E_1}{2}(\mathbf{a}_x + j\mathbf{a}_y)e^{j\omega t} \\ &= \mathbf{E}_+ + \mathbf{E}_- \end{aligned} \tag{12.193}$$

where the positive- and negative-circular-polarized components are

$$\mathbf{E}_+ = \frac{E_1}{2}(\mathbf{a}_x - j\mathbf{a}_y)e^{j\omega t}$$

$$\mathbf{E}_- = \frac{E_1}{2}(\mathbf{a}_x + j\mathbf{a}_y)e^{j\omega t}$$

At $z = d$ these waves become

$$\mathbf{E}_+ = \frac{E_1}{2}(\mathbf{a}_x - j\mathbf{a}_y)e^{j\omega t - jk_+ d}$$

$$\mathbf{E}_- = \frac{E_1}{2}(\mathbf{a}_x + j\mathbf{a}_y)e^{j\omega t - jk_- d}$$

When we add these two waves together again we obtain

$$\mathbf{E} = \frac{E_1}{2}\mathbf{a}_x(e^{-jk_+ d} + e^{-jk_- d})e^{j\omega t} - j\frac{E_1}{2}\mathbf{a}_y(e^{-jk_+ d} - e^{-jk_- d})e^{j\omega t}$$

By taking out a factor $e^{-j(k_+ + k_-)d/2}$, we obtain

$$\mathbf{E} = E_1 e^{-j(k_+ + k_-)d/2} e^{j\omega t}\left[\mathbf{a}_x \cos\frac{(k_+ - k_-)d}{2} - \mathbf{a}_y \sin\frac{(k_+ - k_-)d}{2}\right] \tag{12.194}$$

This is a linearly polarized wave, and the plane of polarization makes an angle

$$\begin{aligned}\theta &= \tan^{-1}\frac{E_y}{E_x} = -\tan^{-1}\tan\frac{(k_+ - k_-)d}{2} \\ &= (k_- - k_+)\frac{d}{2}\end{aligned} \tag{12.195}$$

with respect to the x axis. Consequently, the plane of polarization rotates by an amount $(k_- - k_+)/2$ per unit distance. The wave propagates with an effective phase constant $(k_+ + k_-)/2$ and undergoes rotation at the same time.

Below the gyrofrequency ω_g, we have $k_+ > k_-$ and the plane of polarization rotates in a counterclockwise sense about the d-c magnetic field when looking in the direction of the applied field. If a similar analysis is

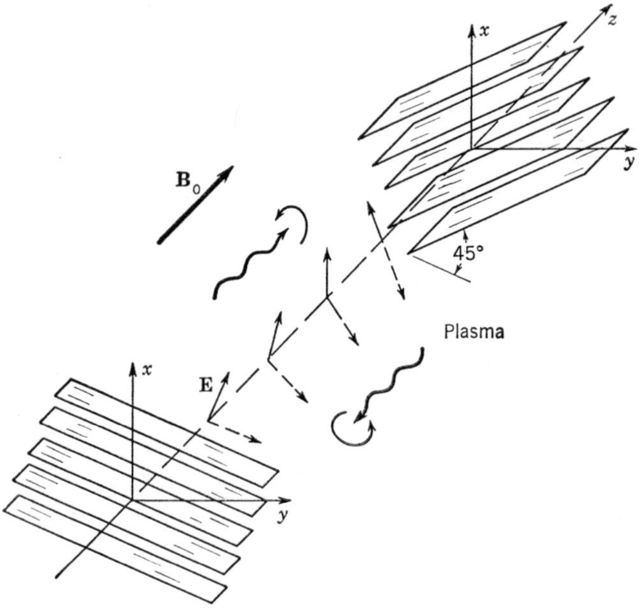

FIG. 12.35. A nonreciprocal device.

carried out for a plane wave propagating in the $-z$ direction, it is found that for $k_+ > k_-$, the plane of polarization continues to rotate in a counterclockwise sense about the d-c magnetic field; that is, if we examine the negative traveling wave, we note that it is propagating in opposition to the d-c magnetic field. If the derivations are consulted, it will be found that k_+ and k_- are interchanged from the positive traveling wave. This involves one change in sign. The second sign change is related simply to the reversal in direction of propagation, hence in description of positive θ, and the two effects cancel each other. It is this nonreciprocal property of the Faraday rotation that permits nonreciprocal microwave devices to be constructed.

As a simple example of a nonreciprocal device, consider that illustrated in Fig. 12.35. This consists of a set of parallel plates at the left end and a similar set of plates at the right end, but rotated by 45° about the axis relative to the first set of plates. The region between the two sets of plates contains an ionized gas (plasma) with an applied d-c field B_0 sufficient to produce 45° of Faraday rotation in a counterclockwise sense. The plates themselves are spaced by less than a half wavelength, so that a plane wave with the **E** vector parallel to the plates cannot propagate through (similar to a waveguide beyond cutoff), but a plane wave with the **E** vector perpendicular to the plates is essentially unperturbed by the plates.

Let a plane wave with the **E** vector perpendicular to the plates be incident from the left. This wave is transmitted through the plates and will have its plane of polarization rotated by 45° in propagating through the plasma region. It therefore arrives at the right-side set of plates with a polarization that permits it to be freely transmitted into the region to the right.

A plane wave incident from the right with the **E** vector perpendicular to the plates is transmitted freely into the plasma-filled space. Upon propagation through the plasma, the plane of polarization is rotated by 45° in a counterclockwise sense about B_0. Consequently, the wave arrives at the left-side set of plates with the **E** vector parallel with the plates. It is therefore reflected and does not propagate into the region to the left. This device thus permits free transmission from left to right but no transmission in the reverse direction. It is clearly a nonreciprocal device.

Propagation in Ferrite Media

Ordinary ferromagnetic materials are of little use at high frequencies because of their large conductivities and hence excessive eddy-current losses. However, a class of materials known as ferrites have been developed and are usable at frequencies up to and including microwave frequencies. A ferrite is a ceramiclike material with a specific resistivity 10^6 or more greater than that of metals. It is a chemical compound of iron, oxygen, zinc, and a small content of such metals as nickel or manganese. The relative permeability may be as large as several thousand, and the relative dielectric constant usually lies in the range 5 to 25. The magnetic properties of these materials are due principally to the magnetic dipole moment m associated with the spinning electron.

The electron has an angular momentum **P** equal to $h/4\pi$, or 0.527×10^{-34} joule-meter, where h is Planck's constant. The magnetic moment m is designated as one Bohr magneton, where $m = eh/2w = 9.27 \times 10^{-24}$ ampere–square meter. The angular moment vector **P** and magnetic dipole moment **m** are antiparallel, in view of the fact that the electron charge is negative. The ratio $\gamma = m/P$ is called the gyromagnetic ratio.

A spinning electron located in a d-c magnetic field $\mathbf{B}_0$ has a torque

$$\mathbf{T} = \mathbf{m} \times \mathbf{B}_0$$

exerted on it. This torque causes the dipole axis to precess about the d-c magnetic field, as in Fig. 12.36. The equation of motion for the angular momentum $\mathbf{P}$ is

$$\mathbf{T} = \frac{d\mathbf{P}}{dt} = \mathbf{m} \times \mathbf{B}_0 = \boldsymbol{\omega}_0 \times \mathbf{P} = -\gamma^{-1}\boldsymbol{\omega}_0 \times \mathbf{m} \qquad (12.196)$$

In Fig. 12.36 the torque $\mathbf{T}$ appears to give rotation in the wrong direction, but this is only because $\mathbf{m}$ and $\mathbf{P}$ are antiparallel. If ϕ is the precession angle, (12.196) gives

$$mB_0 \sin \phi = \gamma^{-1}\omega_0 m \sin \phi$$

or

$$\omega_0 = \gamma B_0 \qquad (12.197)$$

The free-precession frequency ω_0 is independent of the angle ϕ.

A ferrite material may be considered as consisting of N effective spinning electrons with a moment $M = Nm$ per unit volume. In (12.196) the field $\mathbf{B}_0$ is the net internal magnetic field acting on a single dipole. This field consists, in general, of the external applied field, a demagnetizing field dependent on sample shape and size, and local molecular interaction fields. The local interaction field $\mathbf{B}_i$ may be shown to be approximately proportional to $\mathbf{M}$. As such, it does not contribute to the torque, since $\mathbf{M} \times \mathbf{B}_i = C\mathbf{M} \times \mathbf{M} = 0$. The field acting to produce a torque is thus only μ_0 times the internal $\mathbf{H}$ field. The macroscopic behavior of a ferrite may be determined by considering it to consist of dipoles of moment $\mathbf{M}$ per unit volume, whose behavior is the same as that of a single spinning electron.

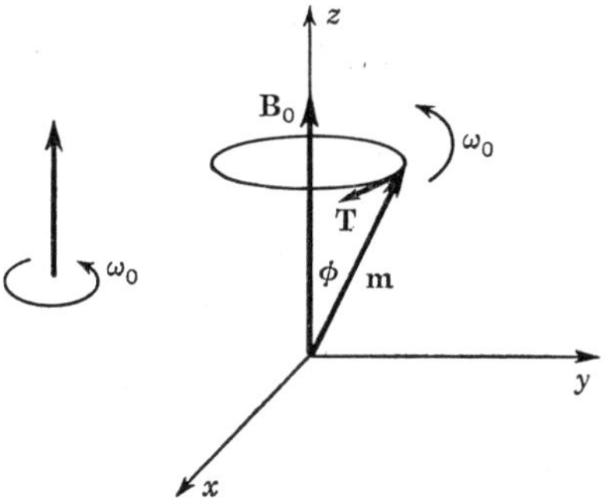

FIG. 12.36. Free precession of a spinning electron.

If in addition to the d-c field $\mathbf{B}_0$ we apply a positive-circular-polarized a-c magnetic field $\mathbf{B}_1^+ = (\mathbf{a}_x - j\mathbf{a}_y)B_1^+ = \mu_0 H_1^+(\mathbf{a}_x - j\mathbf{a}_y)$, the resultant field $\mathbf{B}_t$ rotates about the z axis at an angular frequency ω, where ω is the radian frequency of the applied a-c field. The dipole moment $\mathbf{M}$ must also precess about $\mathbf{B}_0$ at this same rate in order to maintain equilibrium. The precession angle ϕ must be greater than the angle θ made by the field vector $\mathbf{B}_t$ and the z axis if a torque that will cause clockwise rotation is to be produced, as illustrated in Fig. 12.37a.

The equation of motion (12.196), but with $\mathbf{M} \times \mathbf{B}_t$ replacing the expression for $\mathbf{T}$, leads to

$$\begin{aligned}\omega M \sin\phi &= MB_t\gamma \sin(\phi - \theta) \\ &= MB_t\gamma(\sin\phi\cos\theta - \cos\phi\sin\theta)\end{aligned}$$

Replacing $B_t \sin\theta$ by B_1^+ and $B_t \cos\theta$ by B_0, we may solve for $\tan\phi$ to obtain

$$\tan\phi = \frac{\gamma B_1^+}{\omega_0 - \omega} \tag{12.198}$$

where $\omega_0 = \gamma B_0$.

The component of $\mathbf{M}$ in the xy plane that rotates in synchronism with the positive-circular-polarized field B_1^+ is $M_+ = M\sin\phi = M_0 \tan\phi$,

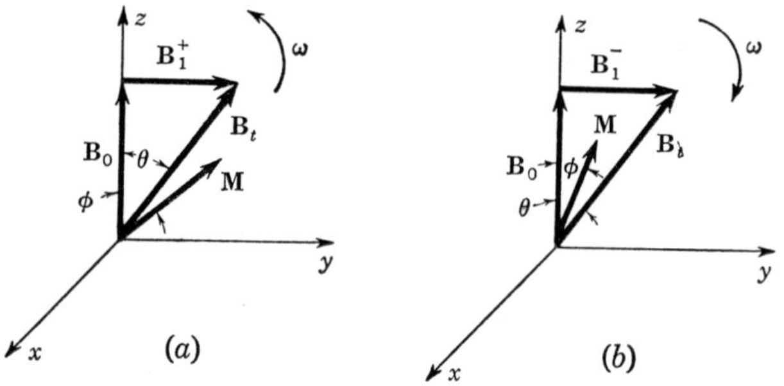

FIG. 12.37. Forced precession of magnetic dipole **M**.

where $M_0 = M\cos\phi$ is a static magnetization per unit volume in the z direction. When r-f quantities are small compared with the corresponding d-c quantities, then M_0 is essentially the d-c magnetic moment in the absence of signal. Using (12.198), we obtain

$$M_+ = \frac{\gamma M_0 B_1^+}{\omega_0 - \omega} = \frac{\gamma M_0 \mu_0 H_1^+}{\omega_0 - \omega} \tag{12.199}$$

Now $B_1^+ = \mu_0 H_1^+$ and is only a partial internal a-c magnetic field. The total a-c circular-polarized magnetic field, according to (7.12), is given by

$$B_+ = \mu_0(H_1^+ + M_+) = \mu_0\left(1 + \frac{\gamma\mu_0 M_0}{\omega_0 - \omega}\right)H_1^+ = \mu_+ H_1^+ \tag{12.200}$$

The effective permeability μ_+ for the circular-polarized wave is given by

$$\mu_+ = \mu_0\left(1 + \frac{\gamma\mu_0 M_0}{\omega_0 - \omega}\right) \tag{12.201}$$

The corresponding propagation constant for a positive-circular-polarized plane wave, in the direction of B_0, is

$$k_+ = \omega\sqrt{\epsilon\mu_+} \tag{12.202}$$

where ϵ is the permittivity of the ferrite.

If we apply a negative-circular-polarized field

$$\mathbf{B}_1^- = B_1^-(\mathbf{a}_x + j\mathbf{a}_y) = \mu_0 H_1^-(\mathbf{a}_x + j\mathbf{a}_y)$$

the precession angle ϕ must be less than θ, as in Fig. 12.37*b*, in order to obtain a torque that will give counterclockwise precession. The equation of motion now gives

$$-MB_t\gamma \sin(\theta - \phi) = -\omega M \sin \phi$$

Expanding and solving for $\tan \phi$ give

$$\tan \phi = \frac{\gamma B_1^-}{\omega_0 + \omega} = \frac{\gamma \mu_0 H_1^-}{\omega_0 + \omega} \tag{12.203}$$

The negative-circular-polarized magnetic polarization M_- is

$$M_- = M_0 \tan \phi = \frac{\gamma \mu_0 M_0 H_1^-}{\omega_0 + \omega} \tag{12.204}$$

The resultant total a-c magnetic field B_- is given by

$$B_- = \mu_0(H_1^- + M_-) = \mu_- H_1^- \tag{12.205}$$

where

$$\mu_- = \mu_0\left(1 + \frac{\gamma \mu_0 M_0}{\omega_0 + \omega}\right) \tag{12.206}$$

The corresponding propagation constant is thus

$$k_- = \omega \sqrt{\epsilon \mu_-} \tag{12.207}$$

Since the propagation constants for the positive- and negative-circular-polarized waves differ, it follows that Faraday rotation of the plane of polarization of a plane wave will take place. The analysis is similar to that for the plasma and, for this reason, will not be repeated.

If we add the two fields $H_1(\mathbf{a}_x - j\mathbf{a}_y)$ and $H_1(\mathbf{a}_x + j\mathbf{a}_y)$, we can describe conditions arising from a linearly polarized field $2H_1\mathbf{a}_x$. The corresponding magnetic field $\mathbf{B}$ may be obtained from (12.200) and (12.205) and is

$$\begin{aligned}
\mathbf{B} &= B_+(\mathbf{a}_x - j\mathbf{a}_y) + B_-(\mathbf{a}_x + j\mathbf{a}_y) \\
&= (B_+ + B_-)\mathbf{a}_x + j(B_- - B_+)a_y \\
&= 2\mu_0 H_1\left(1 + \frac{\omega_0 \gamma \mu_0 M_0}{\omega_0^2 - \omega^2}\right)\mathbf{a}_x - 2j\mu_0 H_1 \frac{\omega \gamma \mu_0 M_0}{\omega_0^2 - \omega^2}\mathbf{a}_y
\end{aligned} \tag{12.208}$$

By subtracting the two fields $(\mathbf{a}_x + j\mathbf{a}_y)H_1$ and $(\mathbf{a}_x - j\mathbf{a}_y)H_1$ we can investigate a y-directed linearly polarized wave where $\mathbf{H} = 2j\mathbf{a}_y H_1$. The corresponding magnetic field $\mathbf{B}$ is

$$\begin{aligned}
\mathbf{B} &= B_-(\mathbf{a}_x + j\mathbf{a}_y) - B_+(\mathbf{a}_x - j\mathbf{a}_y) \\
&= -2\mu_0 H_1 \frac{\omega \gamma \mu_0 M_0}{\omega_0^2 - \omega^2}\mathbf{a}_x + 2j\mu_0 H_1\left(1 + \frac{\omega_0 \gamma \mu_0 M_0}{\omega_0^2 - \omega^2}\right)\mathbf{a}_y
\end{aligned} \tag{12.209}$$

We may relate the components of the **B** field to those of the **H** field in matrix form as follows:

$$\begin{bmatrix} B_x \\ B_y \\ B_z \end{bmatrix} = \begin{bmatrix} \mu_0\left(1 + \dfrac{\omega_0\gamma\mu_0 M_0}{\omega_0^2 - \omega^2}\right) & -j\mu_0 \dfrac{\omega\gamma\mu_0 M_0}{\omega_0^2 - \omega^2} & 0 \\ j\mu_0 \dfrac{\omega\gamma\mu_0 M_0}{\omega_0^2 - \omega^2} & \mu_0\left(1 + \dfrac{\omega_0\gamma\mu_0 M_0}{\omega_0^2 - \omega^2}\right) & 0 \\ 0 & 0 & \mu_0 \end{bmatrix} \begin{bmatrix} H_x \\ H_y \\ H_z \end{bmatrix} \qquad (12.210)$$

since $H_x = 2H_1$ and $H_y = 2jH_1$. In our case H_z and the a-c field B_z are zero. These terms are included in (12.210) only for completeness.

A ferrite is seen to be characterized by an unsymmetrical matrix (tensor) permeability, and this accounts for its nonreciprocal properties. As long as **B** is related to **H** by (12.210), Maxwell's equations may be used to obtain the solutions for waves in the medium. In particular, Maxwell's equations give the previous results for k_+ and k_- for circular-polarized waves. The tensor permittivity for the plasma discussed earlier may also be constructed in a manner similar to that used to obtain (12.210).

12.11. Magnetohydrodynamics

The subject of magnetohydrodynamics deals with the coupling or interaction between a moving conducting fluid and the magnetic field. Under suitable conditions wave propagation can occur. Electrons moving with the fluid through a magnetic field are subjected to a Lorentz force $-e\mathbf{v} \times \mathbf{B}$. This force, equivalent to an electric field $\mathbf{v} \times \mathbf{B}$, results in a current flow of density $\mathbf{J} = \sigma\mathbf{v} \times \mathbf{B}$, where σ is the conductivity of the fluid. The presence of the current $\mathbf{J}$ results in a body force $\mathbf{J} \times \mathbf{B}$ in addition to the mechanical forces. This magnetic body force will in turn influence the motion of the fluid. The fundamental relations involved in this class of problems will be developed in this section.

In order to have a significant interaction between the moving fluid and a magnetic field, it is necessary for the conductivity σ of the fluid to be quite large. In fact, σ must be so large that in the fluid the displacement current is negligible in comparison with the conduction current. In view of this, Maxwell's equations may be taken as

$$\nabla \times \mathbf{H} = \mathbf{J} \qquad (12.211a)$$

$$\nabla \times \mathbf{E} = -\mu \frac{\partial \mathbf{H}}{\partial t} \qquad (12.211b)$$

$$\nabla \cdot \mathbf{B} = \nabla \cdot \mathbf{H} = 0 \qquad (12.211c)$$

$$\nabla \cdot \mathbf{D} = \nabla \cdot \mathbf{E} = 0 \qquad (12.211d)$$

$$\mathbf{J} = \sigma(\mathbf{E} + \mathbf{v} \times \mathbf{B}) \qquad (12.211e)$$

Equation (12.211e) recognizes that the field acting to produce a current flow in a moving volume element of fluid is the sum of the field **E** (measured in the stationary system) and the Lorentz field $\mathbf{v} \times \mathbf{B}$ which arises from the motion of the fluid through a field **B**.

Consider a fluid in which a scalar pressure field p and a velocity field **v** exist. The pressure field gives rise to a force $-\nabla p$ acting on each volume element. If we let Ψ be a scalar potential which gives the potential energy of a volume element located in the gravitational field, then the force due to gravity may be written as $-m_v \nabla \Psi$, where m_v is the density of the fluid. In addition to the above mechanical body forces, there is a frictional force due to the viscosity of the fluid. This force may be expressed as $\eta m_v \nabla^2 \mathbf{v}$, where η is the kinematic viscosity. To these body forces must be added the magnetic force $\mathbf{J} \times \mathbf{B}$ to obtain the total body force. The acceleration of a volume element of fluid is equal to the sum of the body forces acting, so that we have

$$m_v \frac{d\mathbf{v}}{dt} = -\nabla p + m_v \eta \nabla^2 \mathbf{v} - m_v \nabla \Psi + \mathbf{J} \times \mathbf{B} \qquad (12.212)$$

where $d/dt = \partial/\partial t + \mathbf{v} \cdot \nabla$, that is, the total time derivative. In addition, we have the continuity equation for the fluid; i.e.,

$$\nabla \cdot m_v \mathbf{v} = -\frac{\partial m_v}{\partial t} \qquad (12.213)$$

For an incompressible fluid, m_v is constant, and from (12.213), it then follows that

$$\nabla \cdot \mathbf{v} = 0$$

The above equations are the basic equations that govern the motion of the fluid. We wish to examine the fluid behavior when the velocity is small. In this case **B** consists of a uniform static field $\mathbf{B}_0 = \mu \mathbf{H}_0$ and a small a-c field $\mu \mathbf{h}$. In the basic equations we shall now be able to neglect terms involving h^2, v^2, and vh. The basic aim of the analysis to follow is to obtain an equation for **v** alone and **h** alone.

First we shall find two equations relating the small quantities **v** and **h**. From (12.211e) we have

$$\mathbf{E} = \frac{\mathbf{J} - \sigma \mathbf{v} \times \mathbf{B}}{\sigma}$$

which may be substituted into (12.211b) to yield

$$\frac{\partial \mathbf{H}}{\partial t} = \frac{\partial \mathbf{h}}{\partial t} = -\frac{1}{\mu\sigma} [\nabla \times (\mathbf{J} - \sigma \mathbf{v} \times \mu \mathbf{H}_0 - \sigma \mathbf{v} \times \mu \mathbf{h})]$$

Replacing **J** by $\nabla \times \mathbf{H} = \nabla \times \mathbf{h}$ from (12.211a) and neglecting second-order terms lead to

$$\frac{\partial \mathbf{h}}{\partial t} = -\frac{1}{\mu\sigma} \nabla \times (\nabla \times \mathbf{h}) + \nabla \times (\mathbf{v} \times \mathbf{H}_0)$$

The first term on the right may be expanded to give

$$\nabla \times \nabla \times \mathbf{h} = \nabla\nabla \cdot \mathbf{h} - \nabla^2\mathbf{h} = -\nabla^2\mathbf{h}$$

since $\nabla \cdot \mathbf{h} = 0$. The second term expands to

$$\begin{aligned}\nabla \times (\mathbf{v} \times \mathbf{H}_0) = \mathbf{v}(\nabla \cdot \mathbf{H}_0) - \mathbf{H}_0(\nabla \cdot \mathbf{v}) - (\mathbf{v} \cdot \nabla)\mathbf{H}_0 + (\mathbf{H}_0 \cdot \nabla)\mathbf{v} \\ = (\mathbf{H}_0 \cdot \nabla)\mathbf{v}\end{aligned}$$

since $\mathbf{H}_0$ is constant and we are assuming an incompressible fluid for which $\nabla \cdot \mathbf{v} = 0$. Collecting our results, we finally obtain one of the desired equations relating $\mathbf{h}$ and $\mathbf{v}$; that is,

$$\frac{\partial \mathbf{h}}{\partial t} = \frac{1}{\mu\sigma} \nabla^2\mathbf{h} + (\mathbf{H}_0 \cdot \nabla)\mathbf{v} \tag{12.214a}$$

Replacing $\mathbf{J}$ by $\nabla \times \mathbf{h}$ in (12.212) gives

$$\begin{aligned}m_v \frac{d\mathbf{v}}{dt} &= -\nabla(p + m_v\Psi) + m_v\eta \nabla^2\mathbf{v} + (\nabla \times \mathbf{h}) \times (\mu\mathbf{H}_0 + \mu\mathbf{h}) \\ &\approx -\nabla(p + m_v\Psi) + m_v\eta \nabla^2\mathbf{v} - \mu\mathbf{H}_0 \times (\nabla \times \mathbf{h})\end{aligned}$$

by dropping the second-order term $(\nabla \times \mathbf{h}) \times \mathbf{h}$. Now if we use the vector identity (1.122) and the fact that $(\nabla \times \mathbf{H}_0) = 0$, we have

$$\begin{aligned}\mathbf{H}_0 \times (\nabla \times \mathbf{h}) &= \nabla(\mathbf{H}_0 \cdot \mathbf{h}) - (\mathbf{h} \cdot \nabla)\mathbf{H}_0 - (\mathbf{H}_0 \cdot \nabla)\mathbf{h} \\ &= \nabla(\mathbf{H}_0 \cdot \mathbf{h}) - (\mathbf{H}_0 \cdot \nabla)\mathbf{h}\end{aligned}$$

Also $d\mathbf{v}/dt = \partial\mathbf{v}/\partial t + (\mathbf{v} \cdot \nabla)\mathbf{v} \approx \partial\mathbf{v}/\partial t$ to a first approximation. Using these results, the second equation relating $\mathbf{h}$ and $\mathbf{v}$ becomes

$$m_v \frac{\partial \mathbf{v}}{\partial t} = -\nabla(p + m_v\Psi + \mu\mathbf{H}_0 \cdot \mathbf{h}) + \mu(\mathbf{H}_0 \cdot \nabla)\mathbf{h} + \eta m_v \nabla^2\mathbf{v} \tag{12.214b}$$

The gradient term in (12.214*b*) may be shown to vanish as follows. First it is assumed that the disturbance $\mathbf{v}$ and $\mathbf{h}$ is confined to a finite volume V_0 of the fluid. The potential $p + m_v\Psi + \mu\mathbf{H}_0 \cdot \mathbf{h}$ will be a continuous function across the surface S that bounds V_0, with $\mathbf{h}$ going to zero outside S. Since $\mathbf{v}$ is also zero outside S, the potential $p + m_v\Psi$ must be constant outside in order for the body force $-\nabla(p + m_v\Psi)$ to equal zero. The net body force must be zero if the fluid is to be in a static ($\mathbf{v} = 0$) equilibrium condition. The divergence of (12.214*b*) shows that

$$\nabla^2(p + m_v\Psi + \mu\mathbf{H}_0 \cdot \mathbf{h}) = 0$$

because $\nabla \cdot \mathbf{v} = \nabla \cdot \mathbf{h} = 0$. Since $p + m_v\Psi + \mu\mathbf{H}_0 \cdot \mathbf{h}$ satisfies Laplace's equation everywhere and is constant on a boundary S, we can show that $p + m_v\Psi + \mu\mathbf{H}_0 \cdot \mathbf{h}$ must be constant throughout the whole volume V_0 bounded by S. Let $\Phi = p + m_v\Psi + \mu\mathbf{H}_0 \cdot \mathbf{h}$ and apply the divergence theorem to the function $\Phi\,\nabla\Phi$. We obtain

$$\int_{V_0} \nabla \cdot \Phi\,\nabla\Phi\,dV = \int_{V_0} (|\nabla\Phi|^2 + \Phi\,\nabla^2\Phi)\,dV$$
$$= \oint_S \Phi \frac{\partial\Phi}{\partial n}\,dS = 0$$

since Φ is constant on S; that is, if Φ equals a constant, say Φ_0, on S, the integral becomes $\Phi_0 \oint_S \nabla\Phi \cdot d\mathbf{S} = \Phi_0 \int_V \nabla^2\Phi\,dV = 0$ by using the divergence theorem again and the condition that Φ satisfies Laplace's equation. Now $\nabla^2\Phi = 0$, so it follows that

$$\int_{V_0} |\nabla\Phi|^2\,dV = 0$$

and this is possible only if Φ is constant throughout V_0, which is the result that we wished to prove.

With the application of the above result we obtain for our basic equations

$$\frac{\partial\mathbf{h}}{\partial t} = \frac{1}{\mu\sigma}\nabla^2\mathbf{h} + (\mathbf{H}_0 \cdot \nabla)\mathbf{v} \tag{12.215a}$$

$$\frac{\partial\mathbf{v}}{\partial t} = \eta\,\nabla^2\mathbf{v} + \frac{\mu}{m_v}(\mathbf{H}_0 \cdot \nabla)\mathbf{h} \tag{12.215b}$$

Solutions to the above equations that correspond to propagating waves are possible. The terms involving $\nabla^2\mathbf{h}$ and $\nabla^2\mathbf{v}$ are damping terms and vanish if the viscosity η goes to zero and σ becomes infinite (perfectly conducting fluids). A particular type of wave solution to (12.215) is considered below.

Alfvén Waves

We shall consider only the case where $\eta = 0$ and σ is infinite. Equations (12.215) then become

$$\frac{\partial\mathbf{h}}{\partial t} = (\mathbf{H}_0 \cdot \nabla)\mathbf{v} \tag{12.216a}$$

$$\frac{\partial\mathbf{v}}{\partial t} = \frac{\mu}{m_v}(\mathbf{H}_0 \cdot \nabla)\mathbf{h} \tag{12.216b}$$

We may assume that $\mathbf{H}_0$ is in the z direction, whereby (12.216) may be replaced by

$$\frac{\partial\mathbf{h}}{\partial t} = H_0\frac{\partial\mathbf{v}}{\partial z} \tag{12.217a}$$

$$\frac{\partial\mathbf{v}}{\partial t} = \frac{\mu}{m_v}H_0\frac{\partial\mathbf{h}}{\partial z} \tag{12.217b}$$

By eliminating **v** from the first and **h** from the second, we obtain

$$\frac{\partial^2 \mathbf{h}}{\partial t^2} = \frac{\mu}{m_v} H_0{}^2 \frac{\partial^2 \mathbf{h}}{\partial z^2} \tag{21.218a}$$

$$\frac{\partial^2 \mathbf{v}}{\partial t^2} = \frac{\mu}{m_v} H_0{}^2 \frac{\partial^2 \mathbf{v}}{\partial z^2} \tag{12.218b}$$

These are one-dimensional wave equations and have solutions that are called Alfvén waves (after Alfvén, who first demonstrated their existence). A typical solution for sinusoidal time dependence is

$$\mathbf{h} = \mathbf{A}_1 e^{j\omega(t-z/u)} + \mathbf{A}_2 e^{j\omega(t+z/u)} \tag{12.219}$$

where $\mathbf{A}_1$ and $\mathbf{A}_2$ are constant vector amplitudes and the phase velocity u is given by

$$u = H_0 \sqrt{\frac{\mu}{m_v}} \tag{12.220}$$

A similar solution exists for **v**. It should be noted that these waves are also called magnetohydrodynamic waves. These waves have been observed experimentally in mercury. With a field of 1,000 gauss, the phase velocity is about 75 centimeters per second.

The general field of magnetohydrodynamics is of importance in connection with the physical theories of sunspots, the stars, the behavior of heavily ionized gases, supersonic ionized shock waves, and the direct generation of electricity by passing hot plasmas through d-c magnetic fields. The field of study is a relatively new one, and considerable research activity is at present taking place.

BIBLIOGRAPHY

Electron optics

Cosslett, V. E.: "Introduction to Electron Optics," Oxford University Press, New York, 1946.

Maloff, I. G., and D. W. Epstein: "Electron Optics in Television," McGraw-Hill Book Company, Inc., New York, 1938.

Pierce, J. R.: "Theory and Design of Electron Beams," D. Van Nostrand Company, Inc., Princeton, N.J., 1950.

Spangenberg, K. R.: "Vacuum Tubes," McGraw-Hill Book Company, Inc., New York, 1948.

Zworykin, V. K., et al.: "Electron Optics and the Electron Microscope," John Wiley & Sons, Inc., New York, 1946.

Microwave tubes

Beck, A. H. W.: "Thermionic Valves," Cambridge University Press, New York, 1953.

———: "Space Charge Waves," Pergamon Press, Inc., New York, 1958.

Hutter, R. G.: "Beam and Wave Electronics in Microwave Tubes," D. Van Nostrand Company, Inc., Princeton, N.J., 1960.

Kleen, W. J.: "Electronics of Microwave Tubes," Academic Press, Inc., New York, 1958.

Pierce, J. R.: "Travelling Wave Tubes," D. Van Nostrand Company, Inc., Princeton, N.J., 1950.

Slater, J. C.: "Microwave Electronics," D. Van Nostrand Company, Inc., Princeton, N.J., 1950.

Periodic structures

Brillouin, L.: "Propagation in Periodic Structures," Dover Publications, New York, 1953.

Collin, R.: "Field Theory of Guided Waves," McGraw-Hill Book Company, Inc., New York, 1960.

Watkins, D. A.: "Topics in Electromagnetic Theory," John Wiley & Sons, Inc., New York, 1958.

Ferrites

Soohoo, R. F.: "Theory and Application of Ferrites," Prentice-Hall, Inc., Englewood Cliffs, N.J., 1960.

Magnetohydrodynamics

Cowling, T. G.: "Magnetohydrodynamics," Interscience Publishers, Inc., New York, 1957.

TABLE OF NUMERICAL CONSTANTS

Conductivity of Common Materials	
Silver	6.17×10^7 mhos/meter
Copper	5.80×10^7 mhos/meter
Aluminum	3.54×10^7 mhos/meter
Brass	1.57×10^7 mhos/meter
Steel	$0.5\text{–}1.0 \times 10^7$ mhos/meter
Sea water	3–5 mhos/meter
Physical Constants	
Mass of proton	1.67×10^{-27} kg
Mass of electron	9.107×10^{-31} kg
Charge of electron	1.602×10^{-19} coulomb
(Charge of electron)/(mass of electron)	1.76×10^{11} coulombs/kg
Boltzmann's constant	1.380×10^{-23} joule/°K
Free-space Properties	
Permeability μ_0	$4\pi \times 10^{-7}$ henry/meter
Permittivity ϵ_0	$8.854 \times 10^{-12} \approx \frac{1}{36\pi} \times 10^{-9}$ farad/meter
Velocity of propagation c	2.998×10^8 meters/sec
Intrinsic impedance Z_0	$376.7 \approx 120\pi$ ohms

PROBLEMS

Chapter 1

1.1. (*a*) Find the sum and difference of the following two vectors:

$$\mathbf{A} = 4\mathbf{a}_x + 2\mathbf{a}_y - 2\mathbf{a}_z \qquad \mathbf{B} = 2\mathbf{a}_x - 5\mathbf{a}_y - \mathbf{a}_z$$

(*b*) Show that the two vectors in part *a* are orthogonal.

1.2. Derive the law of cosines by squaring both sides of the equation $\mathbf{C} = \mathbf{A} + \mathbf{B}$.

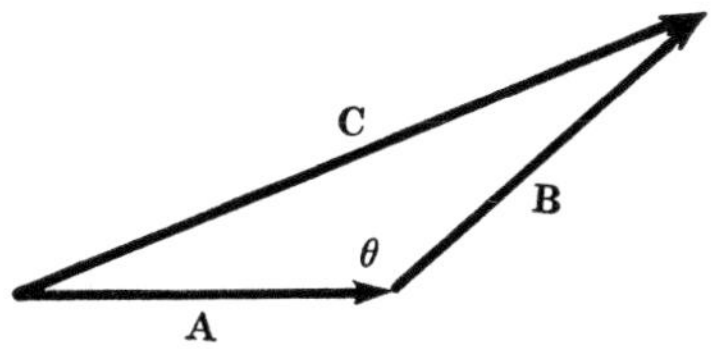

FIG. P 1.2

1.3. (*a*) Show that the direction cosines between each of the unit vectors $\mathbf{a}_x$, $\mathbf{a}_y$, $\mathbf{a}_z$ and the unit vectors ($\mathbf{a}_r$, $\mathbf{a}_\phi$, $\mathbf{a}_z$) in a cylindrical coordinate system are ($\cos\phi$, $-\sin\phi$, 0), ($\sin\phi$, $\cos\phi$, 0), and (0,0,1), respectively.

HINT: Note that the direction cosine between $\mathbf{a}_x$ and $\mathbf{a}_r$ is given by $\mathbf{a}_x \cdot \mathbf{a}_r$, etc.

(*b*) Show that the direction cosines between each of the unit vectors $\mathbf{a}_x$, $\mathbf{a}_y$, $\mathbf{a}_z$ and the unit vectors ($\mathbf{a}_r$, $\mathbf{a}_\theta$, $\mathbf{a}_\phi$) in a spherical coordinate system are ($\sin\theta\cos\phi$, $\cos\theta\cos\phi$, $-\sin\phi$), ($\sin\theta\sin\phi$, $\cos\theta\sin\phi$, $\cos\phi$), and ($\cos\theta$, $-\sin\theta$, 0), respectively, where θ is the polar angle measured from the z axis.

HINT: Find the projection of $\mathbf{a}_r$ and $\mathbf{a}_\theta$ on the xy plane first.

1.4. Consider a force $\mathbf{F}$ acting at a point which is specified by the position vector $\mathbf{r}$. Show that the torque about an axis defined by the unit vector $\mathbf{a}$ is given by $T = (\mathbf{r} \times \mathbf{F}) \cdot \mathbf{a}$.

1.5. Find the components of the following vector along the coordinate directions in a cylindrical and spherical coordinate system, $\mathbf{A} = 2\mathbf{a}_x + \mathbf{a}_y - 3\mathbf{a}_z$.

HINT: The component A_ϕ along the unit vector $\mathbf{a}_\phi$ in a cylindrical coordinate system is the projection of $\mathbf{A}$ on $\mathbf{a}_\phi$; that is, $A_\phi = \mathbf{A} \cdot \mathbf{a}_\phi$, etc. To evaluate the dot products use the results of Prob. 1.3.

1.6. Show that the total vector surface of a closed surface is zero.

HINT: Consider a small plane area, and first show that its projection on any coordinate plane is the component of the vector surface on the axis perpendicular to the coordinate plane. Then by superposition for an arbitrary curved surface, the net component of its vector surface along any coordinate direction is the projection of the surface on the coordinate plane normal to that direction.

1.7. Express **A**, **B**, and **C** in rectangular components, and verify that $\mathbf{A} \times (\mathbf{B} \times \mathbf{C}) = \mathbf{B}(\mathbf{A} \cdot \mathbf{C}) - \mathbf{C}(\mathbf{A} \cdot \mathbf{B})$.

1.8. A five-sided prism as illustrated has its corners at (0,0,0), (2,0,0), (0,2,0), (0,2,3), (0,0,3), and (2,0,3). Evaluate the vector area of each side, and show that the total vector area is zero.

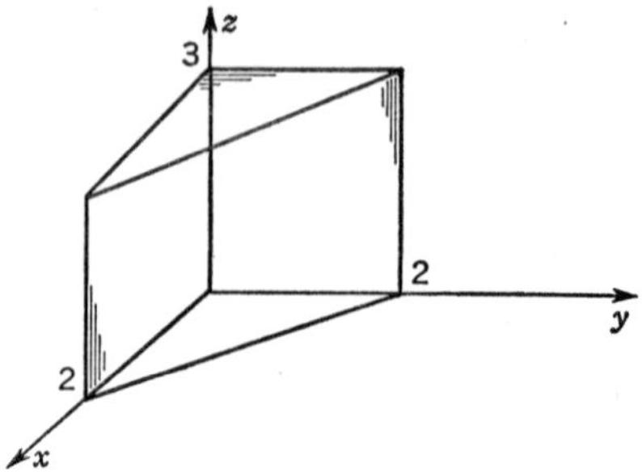

FIG. P 1.8

1.9. Let **A**, **B**, **C** represent position vectors from the origin to three arbitrary points A, B, and C. Prove that the vector

$$\mathbf{A} \times \mathbf{B} + \mathbf{B} \times \mathbf{C} + \mathbf{C} \times \mathbf{A}$$

is orthogonal to the plane determined by A, B, C.

HINT: Note that the vector $\mathbf{A} - \mathbf{B}$, $\mathbf{B} - \mathbf{C}$, or $\mathbf{A} - \mathbf{C}$ lies in the plane determined by the points A, B, C.

1.10. The curl of **F** will be a proper vector function if its form is independent of the choice of axes. Establish this fact by showing that

$$\begin{vmatrix} \mathbf{a}_x & \mathbf{a}_y & \mathbf{a}_z \\ \dfrac{\partial}{\partial x} & \dfrac{\partial}{\partial y} & \dfrac{\partial}{\partial z} \\ F_x & F_y & F_z \end{vmatrix} = \begin{vmatrix} \mathbf{a}_{x'} & \mathbf{a}_{y'} & \mathbf{a}_{z'} \\ \dfrac{\partial}{\partial x'} & \dfrac{\partial}{\partial y'} & \dfrac{\partial}{\partial z'} \\ F_{x'} & F_{y'} & F_{z'} \end{vmatrix}$$

where the primed and unprimed rectangular coordinate systems are arbitrarily oriented.

HINT: **F** itself is tacitly assumed to be a proper vector function, so that

$$\mathbf{F} = \mathbf{a}_x F_x + \mathbf{a}_y F_y + \mathbf{a}_z F_z = \mathbf{a}_{x'} F_{x'} + \mathbf{a}_{y'} F_{y'} + \mathbf{a}_{z'} F_z$$

It is then sufficient to show that

$$\mathbf{a}_{x'} \frac{\partial \Phi}{\partial x'} + \mathbf{a}_{y'} \frac{\partial \Phi}{\partial y'} + \mathbf{a}_{z'} \frac{\partial \Phi}{\partial z'} = \mathbf{a}_x \frac{\partial \Phi}{\partial x} + \mathbf{a}_y \frac{\partial \Phi}{\partial y} + \mathbf{a}_z \frac{\partial \Phi}{\partial z}$$

since then

$$\mathbf{a}_{x'} \frac{\partial}{\partial x'} + \mathbf{a}_{y'} \frac{\partial}{\partial y'} + \mathbf{a}_{z'} \frac{\partial}{\partial z'} = \mathbf{a}_x \frac{\partial}{\partial x} + \mathbf{a}_y \frac{\partial}{\partial y} + \mathbf{a}_z \frac{\partial}{\partial z}$$

This will be facilitated by noting the transformation of a point from one system to the other; i.e.,

$$\begin{aligned} x' &= l_{11}x + l_{12}y + l_{13}z & \qquad x &= l_{11}x' + l_{21}y' + l_{31}z' \\ y' &= l_{21}x + l_{22}y + l_{23}z & \qquad y &= l_{12}x' + l_{22}y' + l_{32}z' \\ z' &= l_{31}x + l_{32}y + l_{33}z & \qquad z &= l_{13}x' + l_{23}y' + l_{33}z' \end{aligned}$$

where l_{11}, l_{12}, l_{13} are the direction cosines of x' relative to x, y, z, etc.

1.11. By direct differentiation show that $\nabla^2(1/R) = 0$ at all points $R \neq 0$, where

$$R = [(x - x')^2 + (y - y')^2 + (z - z')^2]^{1/2}$$

1.12. Find the gradient of the function $\Psi = x^2yz$ and also the directional derivative of Ψ in the direction specified by the unit vector $(3/\sqrt{50})\mathbf{a}_x + (4/\sqrt{50})\mathbf{a}_y + (5/\sqrt{50})\mathbf{a}_z$, at the point $x = 2$, $y = 3$, $z = 1$.

1.13. By direct differentiation show that $\nabla(1/r) = -\nabla'(1/r)$, where $r = [(x - x')^2 + (y - y')^2 + (z - z')^2]^{1/2}$ and ∇' indicates differentiation with respect to x', y', z'. Also show that, for any function $f(r)$, $\nabla f(r) = -\nabla' f(r)$.

HINT: Note that

$$\frac{\partial f}{\partial x} = \frac{df}{dr}\frac{\partial r}{\partial x} \quad \text{and} \quad \frac{\partial f}{\partial x'} = \frac{df}{dr}\frac{\partial r}{\partial x'} \quad \text{etc.}$$

1.14. Find the divergence of the vector function $\mathbf{A} = x^2\mathbf{a}_x + (xy)^2\mathbf{a}_y + 24x^2y^2z^3\mathbf{a}_z$. Evaluate the volume integral of $\nabla \cdot \mathbf{A}$ throughout the volume of a unit cube centered at the origin. Also evaluate the total outward flux of $\mathbf{A}$ over the surface of the cube and thus verify Gauss' law for this particular example.

1.15. Show that the following functions satisfy Laplace's equation in their respective coordinate systems: $\sin kx \sin ly\, e^{-hz}$, where $h^2 = l^2 + k^2$; $r^n(\cos n\phi + A \sin n\phi)$, $r^{-n} \cos n\phi$ (cylindrical coordinates); $r \cos \theta$, $r^{-2} \cos \theta$ (spherical coordinates, no azimuth variation).

1.16. Evaluate the line integral of the vector function $\mathbf{F} = x\mathbf{a}_x + x^2z\mathbf{a}_y + y^2z\mathbf{a}_z$ around the rectangular contour C in the xy plane as illustrated. Also integrate the $\nabla \times \mathbf{F}$ over the surface bounded by C and thus verify that Stokes' law holds for this example.

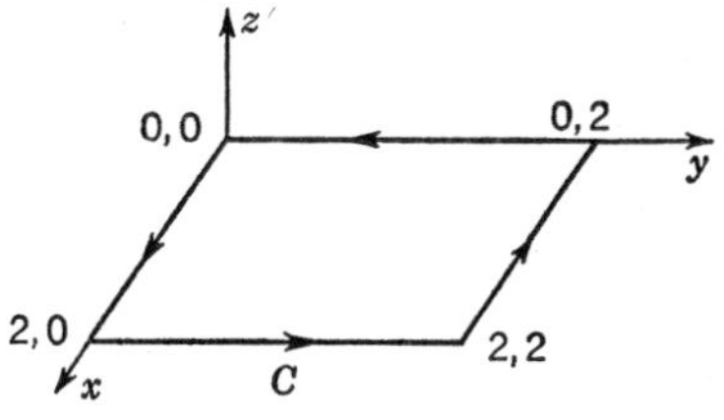

FIG. P 1.16

1.17. Prove the following vector identities: $\nabla \times \nabla\psi = 0$, $\nabla \cdot \nabla \times \mathbf{F} = 0$, $\nabla \times \psi\mathbf{F} = (\nabla\psi) \times \mathbf{F} + \psi\nabla \times \mathbf{F}$, $\nabla \cdot \psi\mathbf{F} = \mathbf{F} \cdot \nabla\psi + \psi\nabla \cdot \mathbf{F}$, where ψ is an arbitrary scalar function and $\mathbf{F}$ is an arbitrary vector function.

1.18. Prove that $\int_V \psi\nabla \cdot \mathbf{F}\, dV = \oint_S \psi\mathbf{F} \cdot \mathbf{n}\, dS - \int_V \mathbf{F} \cdot \nabla\psi\, dV$. This is the vector equivalent of integration by parts where $\mathbf{n}$ is a unit normal to S.

1.19. Evaluate the line integral of the vector function $\mathbf{F} = x^2\mathbf{a}_x + xy^2\mathbf{a}_y$ around the circle $x^2 + y^2 = a^2$. Repeat, making use of Stokes' theorem.

1.20. Prove the following:

$$\nabla \cdot \mathbf{r} = 3 \qquad \nabla \times \mathbf{r} = 0 \qquad \nabla(\mathbf{A} \cdot \mathbf{r}) = \mathbf{A}$$

where $\mathbf{r} = \mathbf{a}_x x + \mathbf{a}_y y + \mathbf{a}_z z$, and $\mathbf{A}$ is a constant vector.

1.21. Show that $(1/F)(\mathbf{F} \cdot \nabla)(\mathbf{F}/F)$ gives the curvature of the flux lines of the vector field $\mathbf{F}$.

HINT: Note that $(\mathbf{F}/F)$ represents a unit tangent to the lines of flux of $\mathbf{F}$.

1.22. Consider an incompressible fluid of density ρ and having a velocity $\mathbf{v}(x,y,z)$. Prove the continuity equation $\rho\nabla \cdot \mathbf{v} = -(\partial\rho/\partial t)$.

HINT: The total mass of fluid flowing out through a closed surface S is given by $\oint \rho\mathbf{v} \cdot d\mathbf{S}$ and must equal the rate at which the enclosed mass of fluid is decreasing, i.e., must equal $-(\partial/\partial t)\int_V \rho\, dV$. Use the divergence theorem (Gauss' law) to convert the surface integral to a volume integral. The results hold for any arbitrary volume, and hence the integrands may be equated.

1.23. Water flowing along a channel with sides along $x = 0$, a has a velocity distribution $\mathbf{v}(x,y) = \mathbf{a}_y(x - a/2)^2z^2$. A small freely rotating paddle wheel with its axis parallel to the z axis is inserted into the fluid as illustrated. Will the paddle wheel rotate? What are the relative rates of rotation at the points $x = a/4$, $z = 1$; $x = a/2$, $z = 1$; $x = 3a/4$, $z = 1$? Will the paddle wheel rotate if its axis is parallel to the x axis or y axis?

HINT: The paddle wheel will rotate provided the fluid is curling or rotating at the point in question. The rate of rotation will be proportional to the z component of the curl of the fluid velocity. The small paddle wheel could form the basis of a curl meter to measure the curl of the fluid velocity.

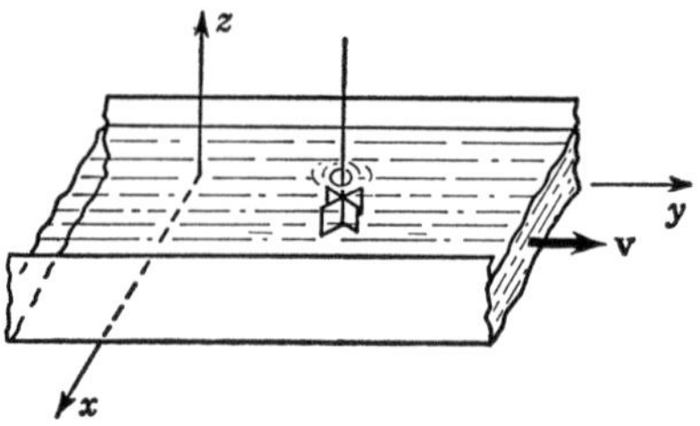

FIG. P 1.23

1.24. Prove that for an arbitrary vector function $\mathbf{A}(x,y,z)$ that is continuous at the point (x',y',z'),

$$\int_V \mathbf{A}(x,y,z)\nabla^2\left(\frac{1}{r}\right) dx\, dy\, dz = \begin{cases} -4\pi\mathbf{A}(x',y',z') & (x',y',z') \text{ inside } V \\ 0 & (x',y',z') \text{ outside volume } V \end{cases}$$

HINT: See Sec. 1.18 (integration of Poisson's equation).

1.25. (*a*) Consider the following vector fields **A**, **B**, **C**, and state which may be completely derived from the gradient of a scalar function and which from the curl of a vector function.

(*b*) Describe a possible source distribution that could set up the field.

$$\begin{aligned} \mathbf{A} &= \sin\theta\cos\phi\mathbf{a}_r + \cos\phi\cos\theta\mathbf{a}_\theta - \sin\phi\mathbf{a}_\phi \\ \mathbf{B} &= z^2\sin\phi\mathbf{a}_r + z^2\cos\phi\mathbf{a}_\phi + 2rz\sin\phi\mathbf{a}_z \\ \mathbf{C} &= (3y^2 - 2x)\mathbf{a}_x + x^2\mathbf{a}_y + 2z\mathbf{a}_z \end{aligned}$$

Chapter 2

2.1. Find the electric field at the point $x = 4$, $y = z = 0$, due to point charges $Q_1 = 8$ coulombs, $Q_2 = -4$ coulombs, and located at $z = 4$ on the z axis and $y = 4$ on the y axis, respectively.

2.2. Positive point charges of magnitude 4, 2, and 2 coulombs are located in the yz plane at $y = 0$, $z = 0$; $y = 1$, $z = 1$; and $y = -1$, $z = -1$; respectively. Find the force acting on a unit negative point charge located at $x = 6$ on the x axis.

HINT: Evaluate the vector force from each charge first, and then add up the partial forces vectorially.

2.3. Find the potential at an arbitrary point (x,y,z) from the three positive charges specified in Prob. 2.2. From the potential function find the electric field and the force exerted on a unit negative charge at $x = 6$ on the x axis. This problem is an example of evaluating the force on a charge by means of the field concept.

2.4. Consider two infinite positive line charges of density q coulombs per meter and located at $y = \pm 1$, $x = 0$, $-\infty \leq z \leq \infty$, together with two similar negative line charges located at $x = \pm 1$, $y = 0$. Derive an analytical expression for the constant potential curves in the xy plane. Sketch the constant potential curves and the electric-field flux lines.

2.5. Find the force per meter exerted on the positive line charges by the negative line charges of Prob. 2.4. Use both the field approach and Coulomb's law. Note that the use of the field approach is much simpler.

HINT: When Coulomb's law is used, it is necessary to integrate over the total length of the negative line charges. Begin by considering the force exerted on a length dl of the positive line charge centered on $z = 0$ by charge elements $-q_l\,dz$ at $\pm z$, $x = \pm 1$. By symmetry this force is in the y direction only.

2.6. Consider two infinite line charges of density q_l coulombs per meter, parallel to the z axis, and located at $x = \pm x_0$, $y = 0$. Consider two particular lines of flux, with an angle θ between them, leaving one line charge, and show that at infinity the angle between these same two flux lines is $\theta/2$.

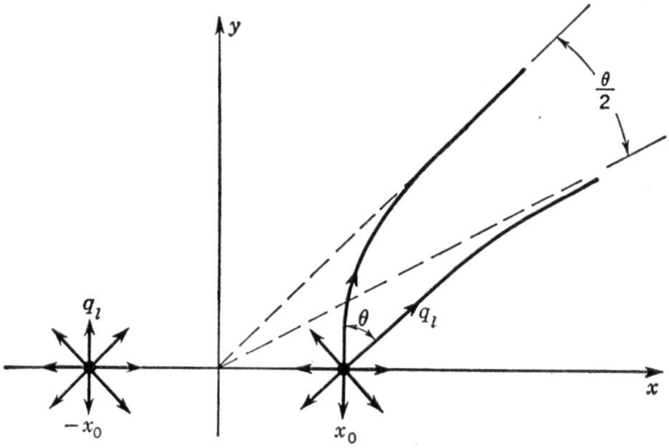

FIG. P 2.6

HINT: Note that very near the line source the flux lines are radial and equally spaced around the line source. (The equal spacing represents the uniformity of the field with azimuth, a condition that must hold very close to the line source.) The number of lines is equal to the charge contained within a cylinder surrounding the line source. At infinity the lines of flux from the two line charges must appear to arise from the sum of the line charges as if concentrated along a line which is their center of gravity. Consequently, they are radial and equispaced. The total lines of flux at infinity is equal to the total charge contained within a cylinder surrounding both line charges. By noting that the total flux in any given flux tube does not change, the relation called for above can be established by setting up a proportionality between the angular spacing of the flux lines near the line charge and at infinity and the charge contained within cylinders surrounding one line source and both line sources, respectively.

2.7. What relation must be satisfied, analogous to that developed in Prob. 2.6, if the line charges are replaced by two positive point charges located at $x = \pm x_0$, $y = z = 0$?

HINT: Note that the flux tubes have rotational symmetry about the x axis.

2.8. A positive line charge Nq_l coulombs per meter and a negative line charge $-q_l$ coulombs per meter, parallel to the z axis, are located at $x = \pm x_0$, $y = 0$. With reference to Fig. P 2.8, determine the maximum angle θ_m for which the lines of flux from the positive line charge will extend out to infinity and not terminate on the negative line charge.

HINT: Apply arguments similar to those needed for Prob. 2.6.

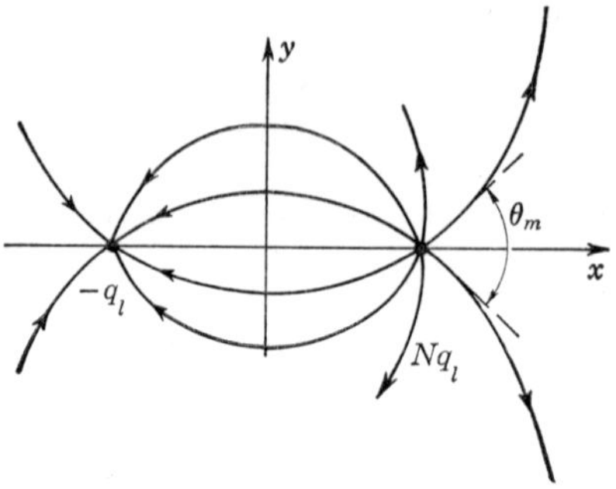

FIG. P 2.8

2.9. Consider two infinitely long concentric coaxial cylinders of radius a and b, as illustrated. The inner cylinder carries a uniform surface charge of density σ_1 coulombs per square meter, while the outer cylinder carries a uniform charge of density σ_2 coulombs per square meter. Use Gauss' law to find the electric field in the three regions $r < a$, $a < r < b$, and $r > b$. What is the potential difference between the two cylinders?

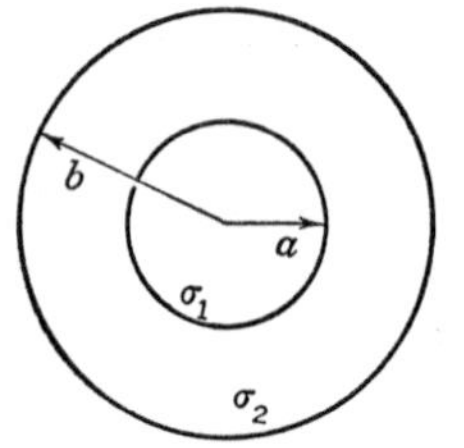

FIG. P 2.9

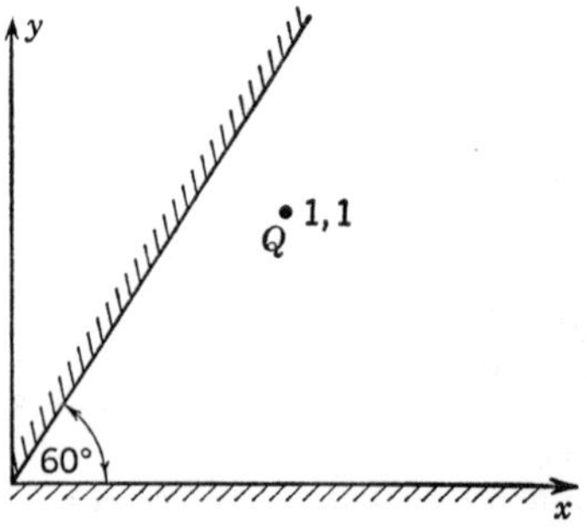

FIG. P 2.10

2.10. A point charge of Q coulombs is located at $x = 1$, $y = 1$ in the space between an infinite 60° conducting wedge. Find the location and sign of all the image charges. What is the potential at the point $x = 2$, $y = 1$?

2.11. A small body of mass m carries a charge Q. The body is placed h meters below an infinite conducting plane. Use image theory to find the required charge Q in order that the electrostatic force may be just sufficient to overcome the force of gravity. Assume $m = 1$ gram, $h = 2$ centimeters.

2.12. The electric field produced by a sphere of charge with a density $\rho(r)$ is given by

$$E_r = \begin{cases} r^3 + Ar^2 & r \leq a \\ (a^5 + Aa^4)r^{-2} & r \geq a \end{cases}$$

Find the charge distribution $\rho(r)$.

2.13. A charge distribution $\rho(r)$ is placed inside a conducting sphere of radius a. The electric field is given by

$$E_r = \begin{cases} Ar^4 & r \leq a \\ Ar^{-2} & r > a \end{cases}$$

Find the charge distribution $\rho(r)$ within the sphere and the surface charge ρ_s on the surface of the sphere.

2.14. Two concentric spheres of radii a and b ($a < b$) are uniformly charged with charge densities ρ_{s1} and ρ_{s2} per square meter. Use Gauss' law to find the electric field for all values of r. If $\rho_{s1} = -\rho_{s2}$, find the potential difference between the spheres.

2.15. (*a*) A conducting sphere of radius a is placed in a uniform field $\mathbf{E}_0$ directed along the z axis. Positive and negative charges are induced on the sphere, which in turn sets up an induced field $\mathbf{E}_i$ such that the total field $\mathbf{E}_0 + \mathbf{E}_i$ vanishes in the interior of the sphere and has a zero tangential component along the surface of the sphere. Find the induced field $\mathbf{E}_i$ inside and outside the sphere. Show that outside the sphere the induced field is the same as that produced by an electric dipole of moment $P = 4\pi a^3\epsilon_0 E_0$ located at the origin.

HINT: The field $\mathbf{E}_0$ may be found from the function $-\nabla\Phi_0$, where $\Phi_0 = -zE_0 = -E_0 r \cos\theta$. Let the induced potential be Φ_1 for $r < a$ and Φ_2 for $r > a$. Both Φ_1 and Φ_2 are solutions of Laplace's equation and must vary with θ according to $\cos\theta$ since Φ_0 does. From Prob. 1.15 appropriate solutions for the induced potentials are found to be $\Phi_1 = Ar\cos\theta$ ($r < a$), $\Phi_2 = Br^{-2}\cos\theta$ ($r > a$), since Φ_1 must remain finite at $r = 0$ and Φ_2 must vanish at $r = \infty$. To find the coefficients A and B, impose the boundary conditions that at $r = a$ the total potential is continuous across the surface $r = a$ and inside the sphere the total field vanishes.

(*b*) Find the charge distribution on the surface of the sphere.

2.16. A point charge Q is located a distance $a < b$ from the center of a conducting sphere of radius b. Find the charge distribution on the inner surface of the sphere. Obtain an expression for the force exerted on Q. The sphere is initially uncharged. Does the force depend on whether the sphere is grounded or not?

2.17. A point charge Q is located at a distance R from the center of an insulated conducting sphere of radius $b < R$. The sphere is ungrounded and initially uncharged. Show that the force attracting Q to the sphere is

$$F = \frac{Q^2 b^3}{4\pi\epsilon_0 R^3} \frac{2R^2 - b^2}{(R^2 - b^2)^2}$$

If the sphere is grounded, what is the force on Q?

2.18. A conducting sphere carries a total charge Q_0. A point charge q is brought into the vicinity of the ungrounded charged sphere. Obtain an expression for the distance from the center of the sphere for which the force on q is zero.

2.19. The entire xz plane is charged with a charge distribution $\rho_s(x,z)$. There is no charge in the region $|y| > 0$. Which of the following potential functions are a valid solution, in the half space $y > 0$, for the problem, and what is the corresponding charge distribution $\rho_s(x,z)$ on the xz plane?

$$\Phi_1 = e^{-y}\cosh x$$
$$\Phi_2 = e^{-y}\cos x$$
$$\Phi_3 = e^{-\sqrt{2}y}\cos x \sin z$$
$$\Phi_4 = \sin x \sin y \sin z$$

2.20. (*a*) A small electric dipole of moment $\mathbf{P}$ is placed in a uniform electric field $\mathbf{E}_0$. Show that the torque acting on the dipole is given by

$$\mathbf{T} = \mathbf{P} \times \mathbf{E}_0$$

(*b*) If the field $\mathbf{E}_0$ varies throughout space, show that the dipole is also subjected to a force given by

$$\mathbf{F} = (\mathbf{P} \cdot \nabla)\mathbf{E}_0$$

and that the torque about an arbitrary origin is then $\mathbf{r} \times (\mathbf{P} \cdot \nabla)\mathbf{E}_0 + \mathbf{P} \times \mathbf{E}_0$.

HINT: Let $d\mathbf{l}$ be the dipole vector length, and expand $\mathbf{E}_0$ in a Taylor series about the negative charge to find the first-order change in $\mathbf{E}_0$ at the positive charge, i.e., to obtain

$$\Delta\mathbf{E}_0 = \mathbf{a}_x \frac{\partial E_{0x}}{\partial l}\, dl + \mathbf{a}_y \frac{\partial E_{0y}}{\partial l}\, dl + \mathbf{a}_z \frac{\partial E_{0z}}{\partial l}\, dl = \frac{\partial \mathbf{E}_0}{\partial l}\, dl$$

Next note that the expression $\boldsymbol{\tau} \cdot \nabla$ gives the derivative in the direction of $\boldsymbol{\tau}$. The net force is the sum of the forces acting on the positive and negative charges.

2.21. For the four point charges located in the xz plane, as in Fig. P 2.21, show that for $r \gg d$ the potential Φ is given by

$$\Phi = \frac{-3Qd^2xz}{4\pi\epsilon_0 r^5} = \frac{-3Qd^2}{4\pi\epsilon_0 r^3} \sin\theta\cos\theta\cos\phi$$

The four charges constitute an electric quadrupole (double dipole).

HINT: Superpose the potential from each charge, and expand the radial distance from each point charge to the field point in a binomial series, retaining the first three terms in the expansion; e.g.,

$$[x^2 + y^2 + (z-d)^2]^{-\frac{1}{2}} \approx r^{-\frac{1}{2}}\left(1 - \frac{2zd}{r^2}\right)^{-\frac{1}{2}} = r^{-\frac{1}{2}}\left[1 + \frac{zd}{r^2} + \frac{3}{2}\left(\frac{zd}{r^2}\right)^2 + \cdots\right]$$

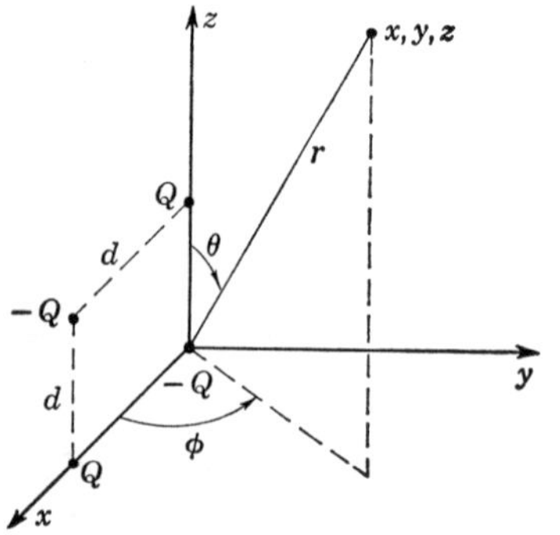

FIG. P 2.21

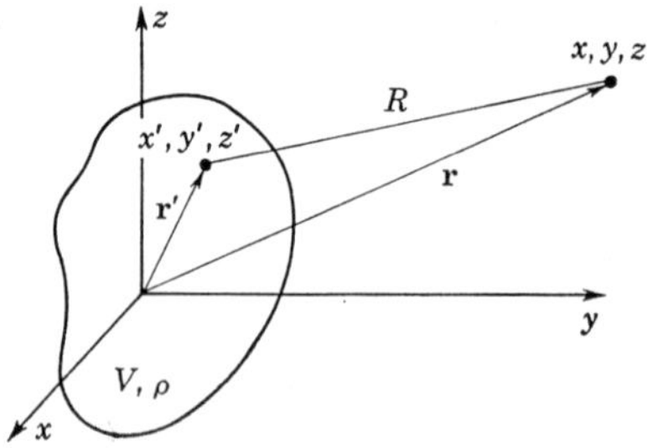

FIG. P 2.22

2.22. Consider a charge distribution $\rho(x',y',z')$ located within a volume V near the origin of an xyz coordinate system, as in Fig. P 2.22. Expand R^{-1} in a Taylor series with respect to x', y', z' about the origin to show that the first three terms of the multipole expansion for the potential Φ is given by

$$\Phi(x,y,z) = \int_V \frac{\rho(x',y',z')}{4\pi\epsilon_0 R}\, dV' = \Phi_1 + \Phi_2 + \Phi_3 + \cdots$$

where $$\Phi_1 = \frac{1}{4\pi\epsilon_0 r}\int_V \rho(x',y',z')\, dV' \qquad \text{coulomb potential}$$

$$\Phi_2 = \frac{-1}{4\pi\epsilon_0}\left(r_x{}^{-1}\int_V \rho x'\,dV' + r_y{}^{-1}\int_V \rho y'\,dV' + r_z{}^{-1}\int_V \rho z'\,dV'\right)$$

$$= \frac{-1}{4\pi\epsilon_0}\nabla\left(\frac{1}{r}\right)\cdot\int_V \mathbf{r}'\rho\,dV' \qquad \text{dipole potential}$$

$$\Phi_3 = \frac{1}{8\pi\epsilon_0}\left(r_{xx}^{-1}\int_V x'^2\rho\,dV' + r_{yy}^{-1}\int_V y'^2\rho\,dV' + r_{zz}^{-1}\int_V z'^2\rho\,dV' + 2r_{xy}^{-1}\int_V x'y'\rho\,dV'\right.$$

$$\left. + 2r_{xz}^{-1}\int_V x'z'\rho\,dV' + 2r_{yz}^{-1}\int_V y'z'\rho\,dV'\right) \qquad \text{quadrupole potential}$$

and $r_x{}^{-1} = \partial r^{-1}/\partial x$, $r_{xy}^{-1} = \partial^2 r^{-1}/\partial x\,\partial y$, etc.

Use this expansion to verify the results of Prob. 2.21.

HINT: Taylor's expansion may be written symbolically as

$$\frac{1}{R} = \frac{1}{r} + \left[\left(x'\frac{\partial}{\partial x'} + y'\frac{\partial}{\partial y'} + z'\frac{\partial}{\partial z'}\right) + \tfrac{1}{2}\left(x'\frac{\partial}{\partial x'} + y'\frac{\partial}{\partial y'} + z'\frac{\partial}{\partial z'}\right)^2 + \cdots\right]\frac{1}{R}$$

where all derivatives are evaluated at the origin. Note that $\partial R^{-1}/\partial x' = -\partial r^{-1}/\partial x$, $\partial^2 R^{-1}/\partial x'\,\partial y' = \partial^2 r^{-1}/\partial x\,\partial y$, etc., for $x' = y' = z' = 0$, that is, at the origin.

2.23. Consider a ring of charge of radius a, center at the origin, and lying in the xy plane. The charge distribution around the ring is given by

$$q_l = q_1\cos\phi + q_2\sin 2\phi \qquad \text{coulombs per meter}$$

Use the multipole expansion of Prob. 2.22 to obtain the first three terms in the expression for Φ for $r \gg a$.

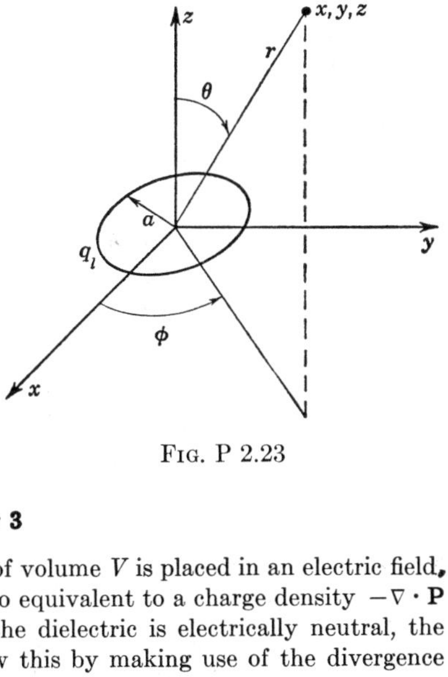

FIG. P 2.23

Chapter 3

3.1. (*a*) If an arbitrarily shaped dielectric of volume V is placed in an electric field, a dielectric polarization $\mathbf{P}$ results which is also equivalent to a charge density $-\nabla\cdot\mathbf{P}$ and a surface charge density $\mathbf{P}\cdot\mathbf{n}$. Since the dielectric is electrically neutral, the total induced charge must equal zero. Show this by making use of the divergence theorem.

(*b*) Consider a specific example where the body is a rectangular parallelepiped whose axis extends from $z = -l/2$ to $z = l/2$ and with a cross-sectional area A. Given that $\mathbf{P} = (Az^2 + B)\mathbf{a}_z$, determine the volume and surface charge density and show explicitly that the total charge is zero.

3.2. Consider a parallel-plate capacitor with sides a, b, and spacing d. The capacitor is half-filled with dielectric (0 to $a/2$, 0 to b) of relative dielectric constant κ. A potential V exists between the plates. Calculate the charge density on the plates and also the equivalent surface polarization charge on the dielectric surfaces. Neglect fringing effects.

3.3. A solid dielectric cylinder of length L and radius a is uniformly polarized with polarization $\mathbf{P}$, where $\mathbf{P}$ is directed axially. Determine the electric field along the cylinder axis both within and outside the cylinder.

3.4. For a field strength of 3×10^6 volts per meter determine the relative displacement and nucleus and electron cloud for He and Ne. Compare with the radius of the atom.

3.5. An infinite dielectric slab of thickness t is placed in a uniform external field $\mathbf{E}_0$. The slab is inclined at an angle θ_1 to the field $\mathbf{E}_0$. Find the angle θ_1 such that the electric lines of flux in the slab make an angle $\theta_2 = \pi/4$ with the sides of the slab, i.e., so that $\theta_2 = \pi/4$. The dielectric constant is $\kappa = 4$. Find the density of surface polarization charge on the two faces of the slab.

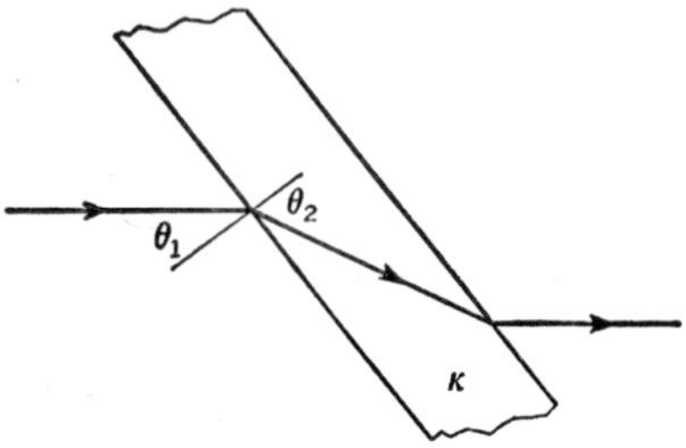

FIG. P 3.5

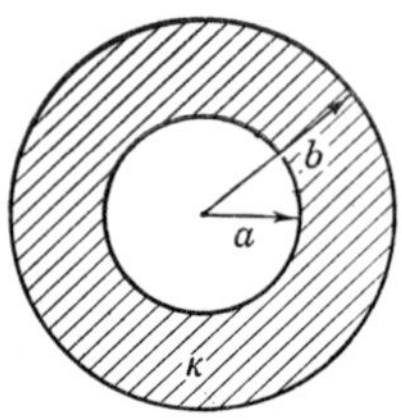

FIG. P 3.6

3.6. (*a*) Consider two coaxial cylinders of inner radius a and outer radius b, as illustrated. The space between cylinders is filled with a dielectric with permittivity $\epsilon = \kappa\epsilon_0$. Find the capacitance per meter length of cylinders.

(*b*) For a given difference of potential between the inner and outer conductors determine the magnitude of b/a that causes the greatest value of $\mathbf{E}$ to be a minimum under the restriction that the radius b is fixed. ($b/a = e$)

3.7. Consider two coaxial cylinders, the inner having a radius a and the outer a radius c. The space $a < r < b$, where $b < c$, is filled with a dielectric with a dielectric constant κ. Find the capacitance per unit length. Show that the capacitance C is equal to the series capacitance of C_1 and C_2, where C_1 is given in Prob. 3.6 and C_2 is the capacitance of an air-filled coaxial cylinder capacitor with inner radius b and outer radius c.

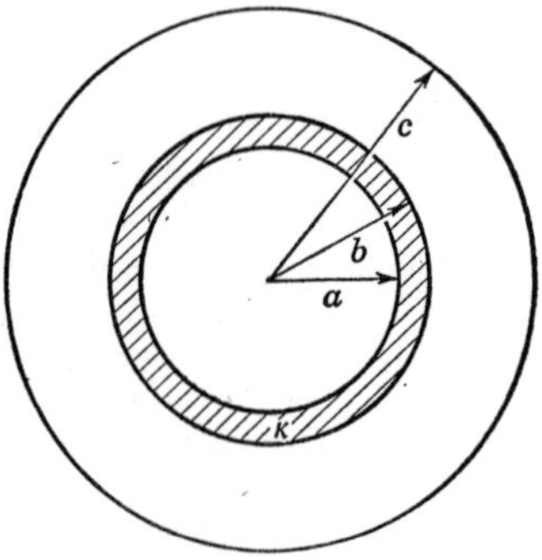

FIG. P 3.7

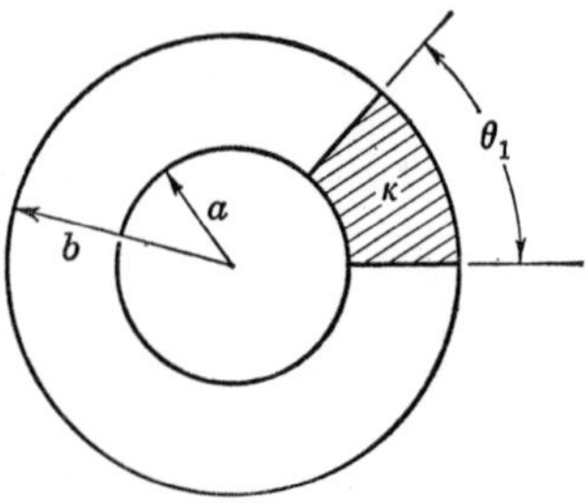

FIG. P 3.8

3.8. Consider two coaxial cylinders with the intervening space $0 < \theta < \theta_1$ filled with a dielectric with a dielectric constant κ. Find the capacitance per unit length, and show that this capacitance is equal to the parallel combination of the capacitances of the air-filled section and the dielectric-filled section.

HINT: Note that the field E_r is independent of θ and depends only on the potential difference between the cylinders.

3.9. A coaxial cable consists of a copper inner conductor of 0.5 centimeter radius, a solid dielectric of permittivity 4.8 to a radius r_0, and a layer of oil ($\kappa = 2.4$) from r_0 to a radius of 2.0 centimeters. The dielectric strength of the solid dielectric is $E_d = 40 \times 10^3$ kilovolts per meter, while that for the oil is $E_0 = 30 \times 10^3$ kilovolts per meter. Find the value of r_0 and the value of potential that will result in maximum stress in both media.

3.10. Show that

$$\int_V \nabla E^2 \, dv = \oint_S E^2\mathbf{n} \, dS$$

HINT: Consider $\int_V \nabla \cdot (\Psi\mathbf{A}) \, dv$, where $\mathbf{A}$ is a constant vector and $\Psi = E^2$, and utilize the divergence theorem.

3.11. Prove that the following relationship exists between the polarization $\mathbf{P}$ and equivalent volume and surface charge densities ρ_p and ρ_{sp} for an arbitrary volume V:

$$\int_V \mathbf{P} \, dv = \int_V \rho_p\mathbf{r} \, dv + \int_S \rho_{sp}\mathbf{r} \, dS$$

In the above, $\mathbf{r}$ is the position vector from an arbitrary origin, and $\mathbf{r} = \mathbf{a}_x x + \mathbf{a}_y y + \mathbf{a}_z z$.

HINT: Note that $\int_V x \, \nabla \cdot \mathbf{P} \, dv = \int_V \nabla \cdot \mathbf{P}x \, dv - \int_V \mathbf{P} \cdot \nabla x \, dv$, and similarly for y and z.

3.12. Consider an arbitrary distribution of point charges q_i ($i = 1, 2, \ldots, n$), and let the potential at the ith charge due to all others be given by Φ_i; that is,

$$\Phi_i = \sum_{\substack{j=1 \\ j \neq i}}^{j=n} \frac{q_j}{4\pi\epsilon_0 r_{ij}}$$

where r_{ij} is the separation of the ith and jth charge. Consider another system of charges located at the same points but with magnitudes q_i' ($i = 1, 2, 3 \ldots, n$) and corresponding to which the potential at the ith point is Φ_i', where

$$\Phi_i' = \sum_{\substack{j=1 \\ j \neq i}}^{j=n} \frac{q_j'}{4\pi\epsilon_0 r_{ij}}$$

Show that

$$\sum_{i=1}^{n} q_i\Phi_i' = \sum_{i=1}^{N} q_i'\Phi_i$$

This result is known as Green's reciprocation theorem.

3.13. (*a*) Given N conducting bodies where under one set of conditions the charge on the ith body is Q_i and for another it is Q_i'. Corresponding to each charge system the potential of the ith body is Φ_i and Φ_i', respectively. Show that

$$\sum_{i=1}^{N} Q_i\Phi_i' = \sum_{i=1}^{N} Q_i'\Phi_i$$

HINT: Subdivide each body into infinitesimal elements, and apply Green's reciprocation theorem developed in Prob. 3.12.

(*b*) Using the result in (*a*) and with $Q_j' = 1$, $Q_i' = 0$ $(i \neq j)$, confirm the formulation of (3.42). What physical significance can be attached to the coefficients of potential?

3.14. A spherical charge distribution of radius R is uniform and has a charge density ρ_0. Calculate the self-energy from (3.60) and from (3.64).

3.15. A dielectric spherical shell has an inner radius r_1, an outer radius r_2, and a permittivity ϵ. What net energy is required to move it from infinity to a point where it is concentric with a point charge Q?

3.16. A variable capacitor is constructed from two coaxial cylinders as illustrated. The inner cylinder is solid and free to slide in an insulating bushing. Find the magnitude and direction of the force acting to displace the center cylinder when a potential difference V is maintained between the cylinders. What is the magnitude of the force when $b = 2a = 1$ centimeter and $V = 1{,}000$ volts?

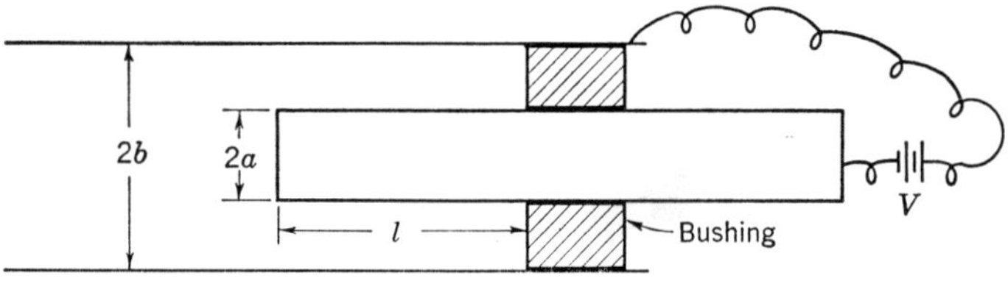

FIG. P 3.16

HINT: Find the rate of change of the energy stored in the capacitor when l is varied.

3.17. In Prob. 3.16 let the variable capacitor be charged so that the total charge on the inner conductor is Q. The battery is now disconnected. What are the magnitude and direction of the force in this case?

3.18. The capacitance of an air capacitor changes linearly from 25 to 350 micromicrofarads during a rotation from 0 to 180°. When set at θ $(0 < \theta < \pi)$, what is the electrostatic torque if a voltage of 400 volts is applied between the plates?

3.19. A homogeneous field **E** exists within an infinite dielectric medium of permittivity ϵ. Find **E** and **D** in a hollow cavity within the dielectric if the cavity shape is

(*a*) A long thin cylinder parallel to **E**.

(*b*) A thin flat plate whose broad dimension is perpendicular to **E**.

(*c*) A sphere.

3.20. The dielectric constant of hydrogen gas at 0°C and atmospheric pressure is 1.000264.

(*a*) Compute the polarizability of the hydrogen molecule.

(*b*) Assuming ideal-gas laws apply, recompute for a pressure of 15 atmospheres and −150°C.

3.21. A dielectric sphere of radius a is placed in a uniform electric field. Find the resultant field inside and outside of the sphere.

HINT: The scheme outlined for the conducting sphere given for Prob. 2.15 may be utilized here. Continuity of Φ and normal **D** across the boundary will serve to determine the coefficients.

3.22. Consider two conducting bodies with an initial charge Q_1, difference of potential V, and capacitance C_1. If the two bodies are displaced so that the new capacitance is C_2, compute the work done on the bodies by the field if

(*a*) V is kept constant (i.e., consider a battery connected between the two bodies).

(*b*) Q_1 is kept constant (bodies isolated, no battery).

In each case account for the change in energy stored in the field. Note that the above process is not a virtual one, but involves finite displacements and finite-energy interchange.

3.23. Consider N conducting bodies at potential Φ_i, with total charge Q_i on the ith body, $i = 1, 2, \ldots, N$. Prove that the charge distributes itself in such a way that the energy stored in the electrostatic field is a minimum (Thomson's theorem).

HINT: Let Φ be the potential for the correct charge distribution and let $\Delta\Phi$ be the change in potential when the charge distribution is perturbed by a small amount on each body, and consider the integral

$$W_e = \frac{\epsilon_0}{2}\int_V [\nabla(\Phi + \Delta\Phi)]^2\, dV$$

Show that

$$\int_V (\nabla\Phi)\cdot(\nabla\Delta\Phi)\, dV$$

vanishes since each body is an equipotential surface and the total charge is kept constant. Thus W_e is a minimum if the positive term

$$\int_V [\nabla(\Delta\Phi)]^2\, dV$$

is zero, i.e., if $\Delta\Phi = 0$.

3.24. A uniform line charge of strength q_l coulombs per meter is placed a distance h above the plane surface of a semi-infinite dielectric of permittivity ϵ. Compute the field everywhere in space.

HINT: Confirm and utilize the following image technique. For field points in the free-space region the source is q_l, plus a line charge (image) q_l' at the image point, both taken as lying in an infinite free-space medium. For points within the dielectric we have q_l, plus q_l'' superposed, both taken as existing in an infinite medium of permittivity ϵ. The relation between q_l' and q_l'' can be found from the boundary conditions at the interface.

Chapter 4

4.1. (*a*) Consider a rectangular parallelepiped bounded by the planes $x = 0, a$; $y = 0, b$; and $z = 0, c$; as in Fig. P 4.1. The side at $y = b$ is kept at a constant potential V_0, while the other sides are kept at a zero potential. Find the solution for the potential Φ inside the enclosure.

(*b*) Let the sides at $y = 0$ and $y = b$ be held at a constant potential V_0, while the other sides are at zero potential. Find the solution for Φ.

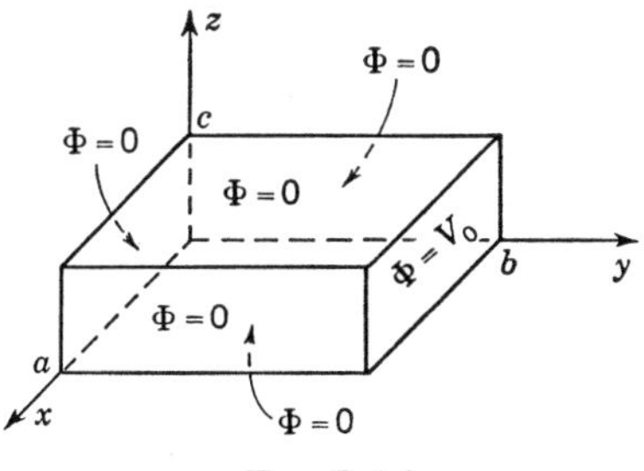

FIG. P 4.1

4.2. Figure P 4.2 illustrates two infinite planes with a very thin conducting septum extending from $y = d$ to $y = b$ $(-\infty < x < \infty)$. The upper plate and septum are kept at a constant potential V_0, while the lower plate is at zero potential. Find a solution for the potential Φ between the planes. Assume that in the plane of the septum the potential $\Phi(y)$ varies linearly from 0 to V_0, that is, $\Phi(y, z = 0) = V_0 y/d$ $(y \leq d)$.

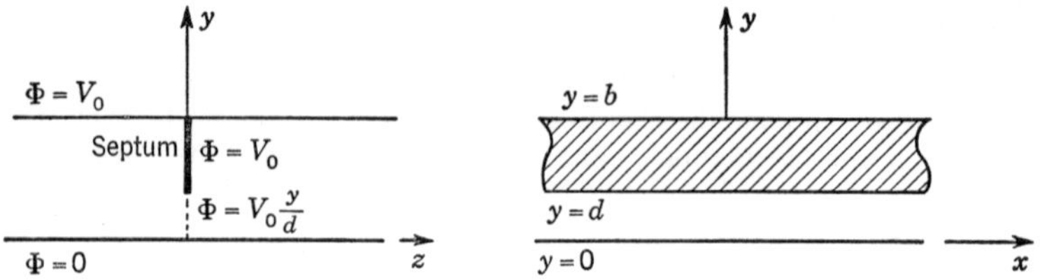

FIG. P 4.2

HINT: Note that there is no variation with x. Find a solution for Φ in the two regions $z > 0$ and $z < 0$. Note that Φ must remain finite as $|z| \to \infty$.

4.3. For Prob. 4.2 evaluate the total energy stored in the electric field contributed by all the terms in the solution except the term $V_0 y/b$. This energy may be used to define the fringing capacitance produced by the septum as follows:

$$C_f = \frac{2We}{V_0^2}$$

Obtain the solution for the fringing capacitance C_f. Note that the total energy stored in the electric field is infinite, a result consistent with the fact that the parallel-plate capacitance for two infinite planes is infinite.

4.4. A rectangular conducting channel has a small slit in the center of the end face (at $x = a/2$). The upper section is kept at a potential V_0 relative to the lower section, as in Fig. P 4.4. The structure is infinite in the y direction. Obtain a solution for the potential Φ in the region $0 < x < a$, $z > 0$.

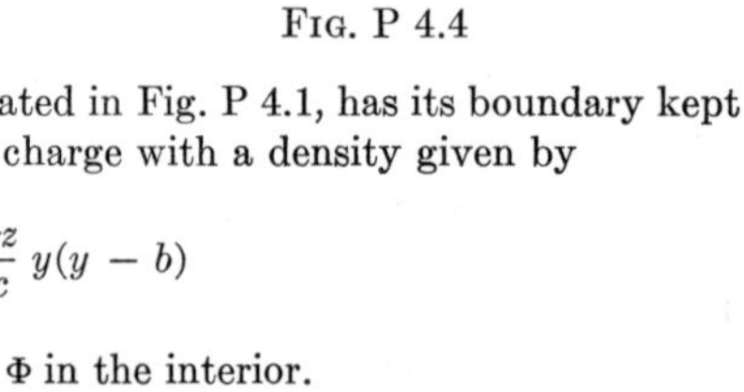

FIG. P 4.4

4.5. A rectangular parallelepiped, as illustrated in Fig. P 4.1, has its boundary kept at zero potential. The interior is filled with charge with a density given by

$$\rho = \sin \frac{\pi x}{a} \sin \frac{\pi z}{c} y(y - b)$$

Find a solution for the potential distribution Φ in the interior.

HINT: Since Φ is not required to satisfy Laplace's equation (it satisfies Poisson's equation since $\rho \neq 0$), assume that Φ can be represented by a general three-dimensional Fourier series

$$\Phi = \sum_{n=1}^{\infty} \sum_{m=1}^{\infty} \sum_{s=1}^{\infty} A_{nms} \sin \frac{n\pi x}{a} \sin \frac{m\pi y}{b} \sin \frac{s\pi z}{c}$$

Expand ρ into a similar three-dimensional Fourier series. Substitute these expansions into Poisson's equation $\nabla^2\Phi = -\rho/\epsilon_0$, and use the orthogonality properties of the sine functions to relate the coefficients A_{nms} to the corresponding coefficients in the expansion of ρ.

4.6. Consider the two-dimensional rectangular region illustrated in Fig. P 4.6. The structure is infinite in extent along the z axis. A line source of strength q coulombs per meter is located at x_0, y_0 and parallel to the z axis. The sides of the enclosure are kept at zero potential. Find a solution for the potential Φ in the interior.

HINT: See comments in Prob. 4.5. The line source may be represented symbolically as the product of two delta functions [analogous to unit impulse function $\delta(t)$ used in circuit theory] $\rho = q\delta(x - x_0)\delta(y - y_0)$, where $\delta(x - x_0) = 0$, $x \neq x_0$; $\int_{x_0-\alpha}^{x_0+\alpha} \delta(x - x_0)\, dx = 1$, and similarly for $\delta(y - y_0)$.

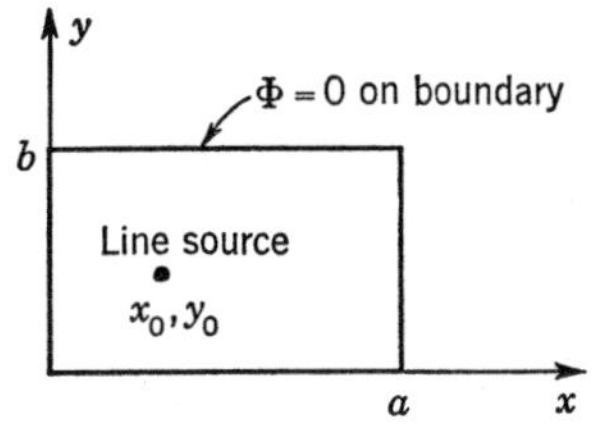

FIG. P 4.6

For any function of x and y, say $f(x,y)$, the following property also holds:

$$\int_{x_0-\alpha}^{x_0+\alpha} \int_{y_0-\beta}^{y_0+\beta} f(x,y)\delta(x - x_0)\delta(y - y_0)\, dx\, dy = f(x_0,y_0)$$

This property may be used to obtain a double Fourier series expansion of the line-charge source function ρ. The solution for Φ may then be constructed as in Prob. 4.5. Note that there is no z variation.

4.7. Consider an infinite dielectric medium with a permittivity ϵ. An infinitely long z-directed cylindrical cavity of radius a is cut out of the material. A uniform electric field E_0 is applied along the x axis as in Fig. P 4.7. Obtain a solution for the total resultant electric field within and outside the cylindrical cavity.

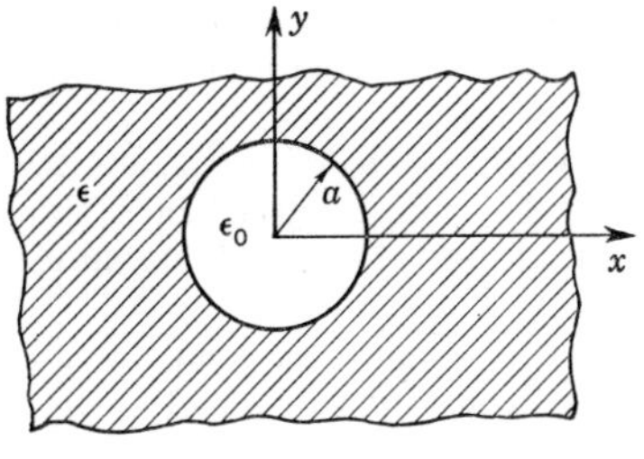

FIG. P 4.7

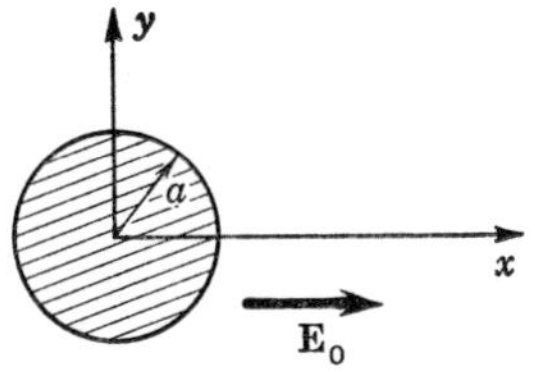

FIG. P 4.8

4.8. An infinitely long z-directed conducting cylinder of radius a is placed in a uniform electric field $E_0\mathbf{a}_x$, as in Fig. P 4.8. Obtain a solution for the induced electric field $\mathbf{E}_i$ in the region internal and external to the cylinder. What is the charge distribution on the surface of the cylinder?

4.9. Consider a very long slab of conducting material (length L, width W, thickness t, conductivity σ). A small cylindrical hole of radius a is cut in the center of the slab. A potential V_0 is applied across the two ends as in Fig. P 4.9 and establishes an electric field in the medium. Find a solution for the current flow in the conductor, and sketch the flow lines in the vicinity of the hole.

HINT: Assume that the electric field consists of a uniform field plus an induced field. The current $\mathbf{J} = \sigma\mathbf{E}$, where $\mathbf{E}$ is the total electric field.

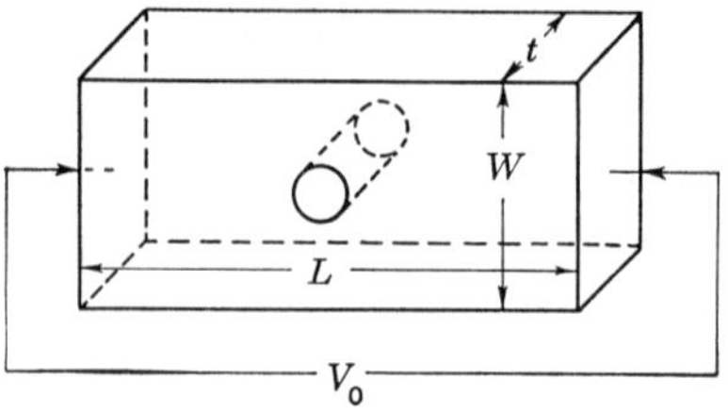

FIG. P 4.9

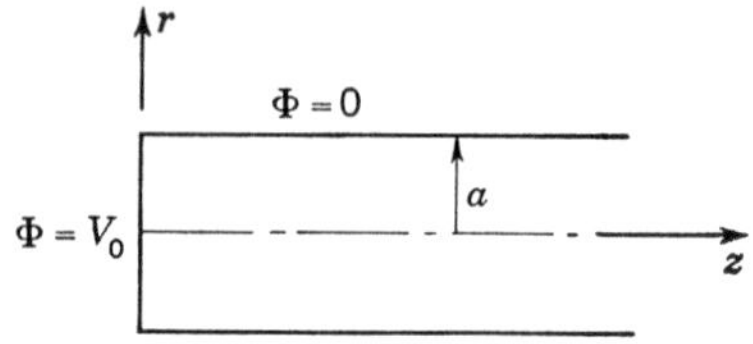

FIG. P 4.10

4.10. Consider a semi-infinitely long cylinder of radius a. The side of the cylinder is kept at zero potential, while the end face at $z = 0$ is kept at a constant potential V_0, as in Fig. P 4.10. Find a solution for Φ within the cylinder.

4.11. A cylinder of length b, radius a, has a small centered ring section of width d kept at a potential V_0. The remainder of the cylinder and the end walls are kept at zero potential, as in Fig. P 4.11. Find a solution for Φ inside the enclosure.

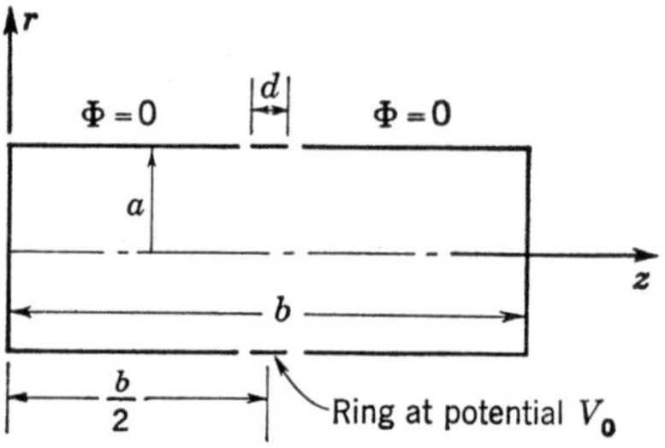

FIG. P 4.11

4.12. A spherical cavity of radius a is cut in an infinite dielectric body of permittivity ϵ. A uniform field $E_0\mathbf{a}_z$ is applied. Find the induced field inside and outside the spherical cavity.

4.13. A conducting sphere of radius a is surrounded by a dielectric with permittivity ϵ over the region $a < r < b$. The dielectric-coated sphere is placed in a uniform applied field $E_0\mathbf{a}_z$. Find the solution for the resultant total electric field for all values of r.

4.14. A hollow conducting sphere of radius a has a small air gap cut around the equatorial plane. The upper hemisphere is kept at a constant potential V_0, while the lower hemisphere is kept at zero potential. Obtain the solution for the potential distribution within the sphere.

4.15. A thin circular ring of charge, radius a and charge density ρ_l coulombs per meter, is located in the xy plane with the center at the origin. By direct integration, find the potential along the z axis. By expanding this potential in a power series in z and comparing with the Legendre function expansion of the potential, determine the solution for the potential for all positions around the ring of charge.

4.16. A conducting sphere (uncharged) is placed in a potential field $\Phi_0 = r^2 \sin 2\theta \cos \phi$. Find the resultant total potential inside and outside the conducting sphere.

4.17. A line charge of density q_l coulombs per meter is located at $x = y = 1$ inside a 90° conducting wedge, as illustrated. By a suitable conformal mapping, obtain the solution for the potential Φ. Compare with the solution obtained by the method of images.

HINT: Note that a transformation of the form $W = Z^\alpha$ or in polar form (with $W = Re^{j\phi}$ and $Z = re^{j\theta}$) $Re^{j\phi} = r^\alpha e^{j\alpha\theta}$ will map the boundary of the wedge into an infinite flat plane for a suitable choice of the parameter α. In the W plane the solution for $\Phi(u,v)$ may be found at once by using image theory (a line source imaged in an infinite plane). By transforming back to the Z plane, the potential inside the wedge region is found. It is simpler to express the solution in terms of the variables r, θ after the solution in the W plane has been expressed in terms of R and ϕ.

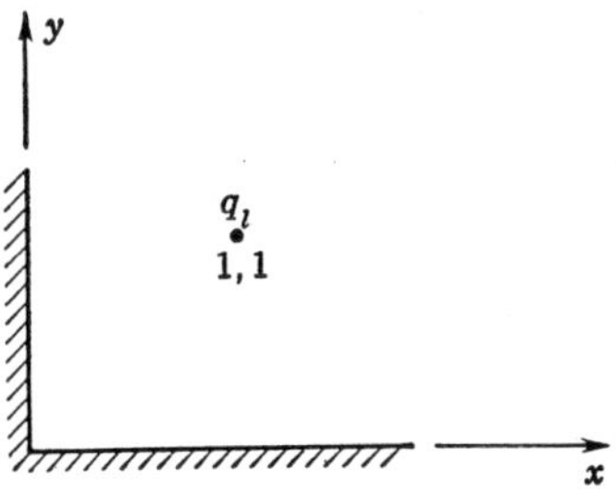

FIG. P 4.17

4.18. For a line charge of density q_l coulombs per meter located at an arbitrary point (x_0,y_0) in the inside of a conducting wedge of arbitrary angle θ_0, find the potential distribution Φ by means of a conformal mapping. Note that the conformal-mapping method will work for any wedge angle while the direct method of images will work only if π/θ_0 is an integer.

HINT: See Prob. 4.17.

4.19. A line charge is located at x_0, y_0 in the vicinity of a semi-infinite plane extending from $x = 0$ to ∞, along $y = 0$, as illustrated. Find the solution for the potential Φ. What is the surface charge distribution on the plane when $y_0 = 0$, $x_0 = -a$?

HINT: The half plane may be considered as a wedge of internal angle 2π.

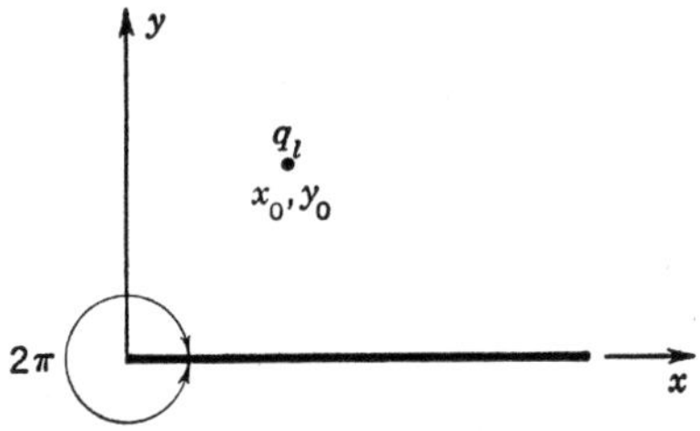

FIG. P 4.19

4.20. Show that the transformation $W = \sin Z$ maps the semi-infinite strip of width π in the Z plane into the upper half of the W plane, as illustrated. Use this transformation to find the potential due to a unit-strength line source located at $x = 0$, $y = y_0$ in the region indicated. The boundary is kept at zero potential.

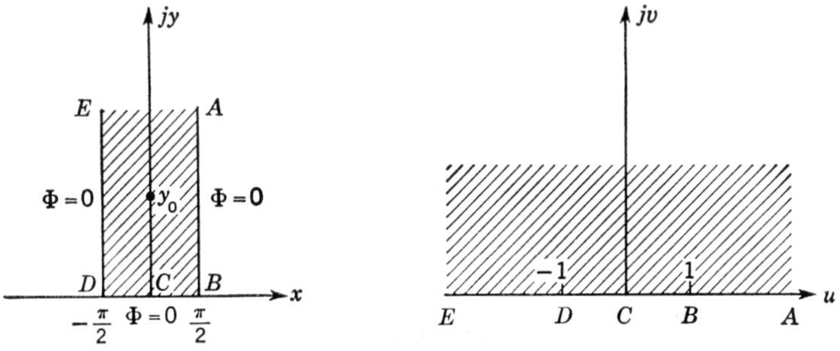

FIG. P 4.20

4.21. Find a Schwarz-Christoffel transformation that will map the region illustrated in the W plane into the x axis in the Z plane. Note that the rectangle is obtained from a triangle as the point u_0 tends to $-\infty$. The external angles of the polygon are those of the triangle as u_0 tends to $-\infty$. Compare this transformation with that in Prob. 4.20.

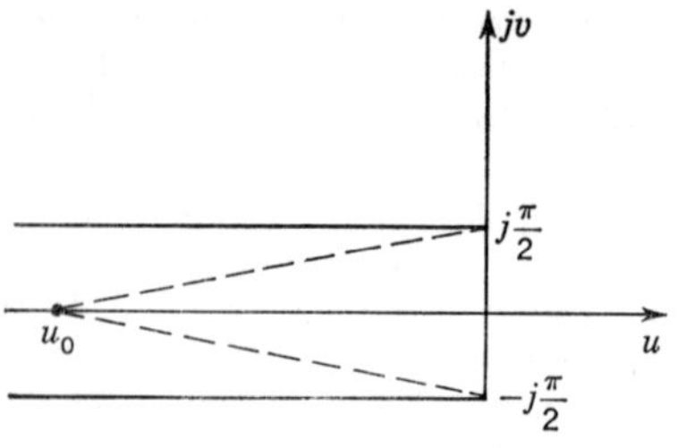

FIG. P 4.21

Chapter 5

5.1. A potential difference V is maintained across two very large coaxial conducting cylinders with radii r_1 and r_2 $(r_2 > r_1)$.

(*a*) If the medium between the cylinders has a conductivity σ, calculate the leakage conductance G per unit length.

(*b*) If the medium were nonconducting with a dielectric constant ϵ, calculate the capacitance C per unit length from basic definitions.

(*c*) Show explicitly from (*a*) and (*b*) that $RC = \epsilon/\sigma$.

(*d*) For (*a*) compute $\mathbf{E}$ and $\mathbf{J}$ in the conducting medium.

5.2. (*a*) For the accompanying half ring, which is composed of material of conductivity σ, compute the total resistance between A and B by a rigorous treatment (the inner radius is R, and the outer radius is $R + d$).

(*b*) Find the total resistance by taking the accompanying figure to be straight with total length equal to π times the mean radius. For $R \gg d$ show that (*a*) reduces to (*b*).

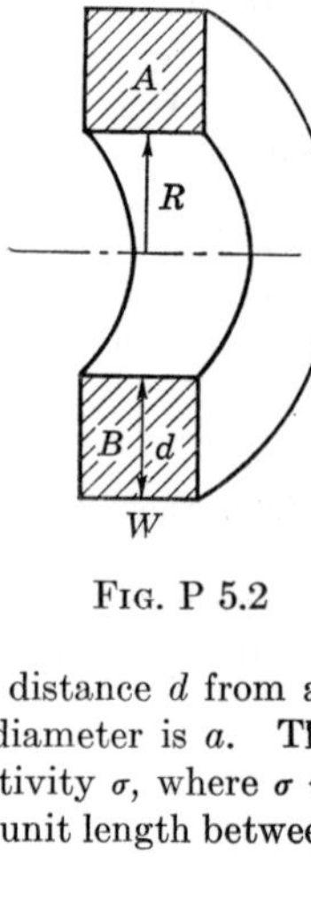

FIG. P 5.2

5.3. A long, highly conducting cylindrical wire is placed at a distance d from an infinite conducting plane and parallel to the plane. The wire diameter is a. The given conductors lie in a uniform conducting medium of conductivity σ, where $\sigma \ll$ conductivity of wire or plane. Show that the total resistance per unit length between the wire and the plane is

$$R = \frac{1}{2\pi\sigma} \cosh^{-1} \frac{d}{a}$$

5.4. Using the results of Prob. 5.3, calculate the total resistance per meter between a copper wire of radius 2 millimeters and an infinite conducting plane where the separation between the two is (*a*) 4, (*b*) 6, and (*c*) 10 millimeters. The conductivity is 10^4 mhos per meter.

5.5. Repeat Prob. 5.4, but determine the resistance per meter by means of flux plotting.

5.6. Map the flux lines and equipotentials for the adjoining figure (shown in cross section) given that (*a*) $b/d = 0.7$, (*b*) $b/d = 0.5$. Take the axial length and width to be essentially infinite, and continue the plot away from the step only to the point where the field becomes essentially uniform.

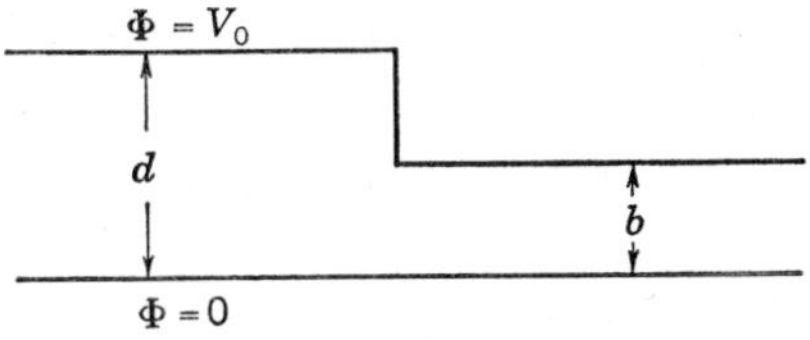

FIG. P 5.6

5.7. For Prob. 5.6 obtain a measure of the total charge per unit width and length at a sufficient distance from the step so that the field is essentially uniform. With these values determine the excess charge that is associated with the discontinuity. Evaluate the capacitance per unit square for each region and also the discontinuity capacitance.

5.8. The accompanying coaxial cable has a two-layered concentric cylindrical insulation with dielectric constants ϵ_1 and ϵ_2. The latter insulating materials have a leakage conductance σ_1 and σ_2, respectively. If the outer conductor is maintained at a potential V relative to the inner, what is the steady-state electric field in the dielectric and the surface charge at the dielectric interface?

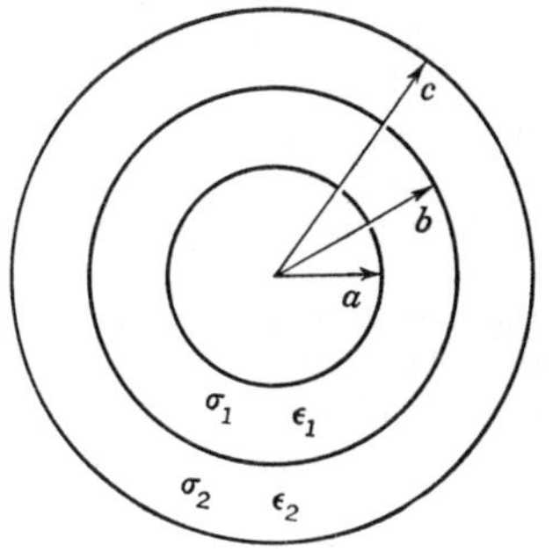

FIG. P 5.8

5.9. (*a*) Determine the transient time constant for the cable described in Prob. 5.8. (*b*) How long would it take to reach steady state if the inner dielectric is mica ($\epsilon_1 = 6.0\epsilon_0$, $\sigma_1 = 10^{-13}$) and the outer dielectric is oil ($\epsilon_2 = 2.5\epsilon_0$, $\sigma_2 = 10^{-14}$)? The cable dimensions are $a = 1$ centimeter, $b = 1.3$ centimeters, $c = 1.8$ centimeters.

5.10. In order to obtain a good ground connection, a hemispherical conductor is embedded in the earth so that its base lies in the earth's surface. Assuming the resistivity of the ground to be 2×10^5 ohm-meters, find the total resistance to ground. (The radius of the hemisphere is 0.15 meter.)

5.11. The equation $\rho(x,y,z,t) = \rho_0(x,y,z)e^{-\sigma t/\epsilon}$ [Eq. (5.25)] required for its derivation only that σ and ϵ were uniform in the region under consideration. Suppose that at $t = 0$ a quantity of charge is localized in a small sphere in an otherwise uncharged medium. According to the above equation, at any $t > 0$, the region immediately surrounding the sphere of charge, and which was originally uncharged, remains neutral, even though the charge in the sphere is disappearing, only to reappear at the surface. A similar situation in heat flow would be quite different; the heat would flow out into the surrounding medium, instead of fading away where it stands, like the electric charge. Explain the reason for this difference.

5.12. A large sphere with uniform conductivity σ and permittivity κ has a radius R. At $t = 0$, a charge Q is placed uniformly over a small concentric spherical surface of radius a, where $a < R$. Calculate the Joule losses during the transient, and show that it is equal to the decrease in stored electric energy.

5.13. A steady current is distributed in a resistive medium which is not homogeneous (that is, σ and ϵ are functions of position). In this case a volume charge density will be set up in a similar way to that whereby a surface charge density accumulates at a lossy dielectric interface. Show that

$$\rho = \frac{-1}{\sigma} (\sigma \nabla \epsilon - \epsilon \nabla \sigma) \cdot \nabla \Phi$$

where Φ is the scalar potential for the electric field.

5.14. Show that for a problem involving N essentially perfectly conducting bodies embedded in a poorly conducting medium, where the medium is homogeneous, the current may be obtained from a scalar function Φ, and Φ is unique provided that

(*a*) $\Phi = \Phi_i$ ($i = 1, 2, \ldots, M$), where Φ_i is a constant specified potential on the ith body.

(*b*) $I_j = -\oint_{S_j} \sigma(\partial\Phi/\partial n)\, dS$, where I_j is a specified total current from the jth body ($j = M + 1, M + 2, \ldots, N$).

(*c*) $\Phi \propto 1/R$, $\partial\Phi/\partial n \propto 1/R^2$ for $R \to \infty$ (that is, Φ is regular at infinity).

5.15. Consider a conducting region V in which a current is caused to flow as a consequence of emf sources in an adjoining region V' (for example, V may encompass an arbitrary circuit and V' a battery). Prove that the current density in the conductor occupying V distributes itself in such a way that the generation of heat is less for the actual distribution than for any other provided the total current supplied by the sources is constant. Note that, since there are no sources in V, the current $\mathbf{J}$ is given by $-\sigma \nabla\Phi$. This problem is similar to Thomson's theorem, and the hints given in Prob. 3.23 are applicable.

5.16. A total of N conducting bodies lie in a conducting medium. If each body is at a constant potential $\Phi_1, \Phi_2, \ldots, \Phi_N$ and the total current from each is $I_1, I_2, \ldots, I_N$, show that the total Joule heat is

$$W_j = \sum_{i=1}^{N} \Phi_i I_i$$

5.17. A steady current flows into a thin conducting spherical shell at one pole and leaves at the other pole. Given the radius to be a, the conductivity σ, and the total current I, determine the potential and current density over the sphere.

5.18. A sphere of uniform conducting material of conductivity σ is placed in a potential field which is capable of maintaining a potential of $\Phi_0 \cos\theta$ over the spherical surface. (This implies a source of emf, of course.) Determine the current density $\mathbf{J}$ within the sphere.

5.19. Consider an infinite-plane conducting sheet, and let current enter at the origin and leave over a body contour at infinity. If a circular hole is cut anywhere in the sheet, not including the origin, show that the difference of potential between any two diametrically opposite points on the circumference of the hole is twice what it is prior to cutting the hole.

HINT: Consider the properties of the inversion of the original potential within the circle to the region outside the circle as discussed for the double-layer electrolytic tank. Note that the required condition at the edge of the hole can be expressed by $\partial\Phi/\partial n = 0$.

5.20. Show that, if the flux-plotting technique of Sec. 5.10 is followed, a correct and unique solution is obtained (subject, of course, to the approximation due to finite-size grids).

HINT: The graphical construction yields a family of equipotential curves $\Phi(x,y) = A_i$ and a family of flow curves $\Psi(x,y) = B_i$. The functions Φ and Ψ are related in that they intersect at right angles and form a square rather than a rectangle. From this it is possible to show that $\partial\Phi/\partial x = C\,\partial\Psi/\partial y$ and $\partial\Phi/\partial y = -C\,\partial\Psi/\partial x$, so that Φ and Ψ satisfy Laplace's equation.

Chapter 6

6.1. Use the Biot-Savart law [Eq. (6.5)] to find the field **B** set up by two infinitely long line currents located at $x = \pm 1$, $y = 0$ and parallel to the z axis. The currents flowing in the line sources are I and $-I$.

6.2. For the line sources in Prob. 6.1, find an equation for the lines of flux and show that these are the same as the constant-potential contours around two line charges of opposite sign.

6.3. Use Eq. (6.14) to find the field **B** at the center of a square current loop with sides d and current I.

6.4. Consider the rectangular U-shaped conductor illustrated. The circuit is completed by means of a sliding bar. When a current I flows in the circuit, what is the force acting on the sliding bar? When $a = 4$ centimeters, $b = 10$ centimeters, and $I = 5$ amperes, what is the value of the force?

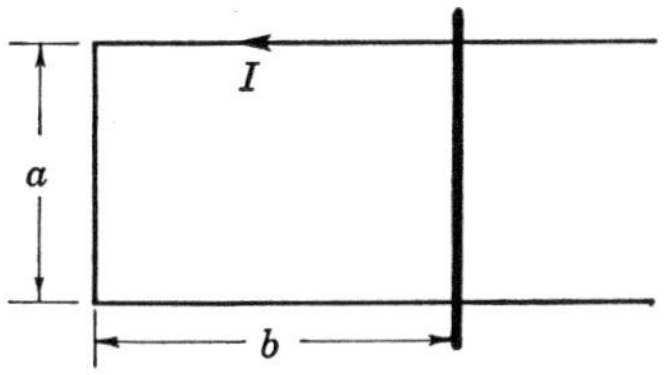

FIG. P 6.4

6.5. Consider two square loops with sides d and equal currents I. One loop is located a distance h above the other loop, as illustrated. Find the force acting on one loop due to the other loop.

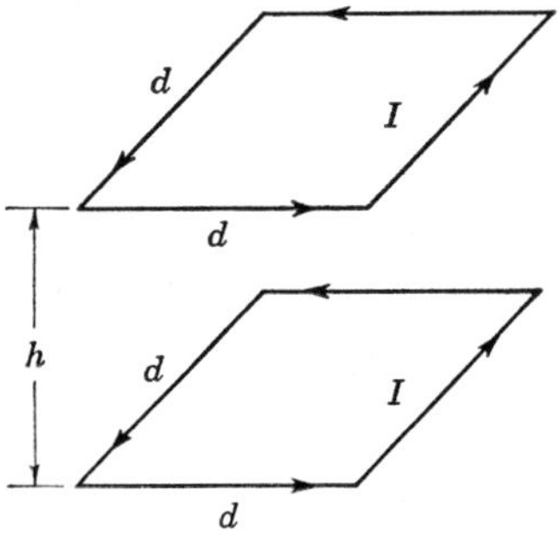

FIG. P 6.5

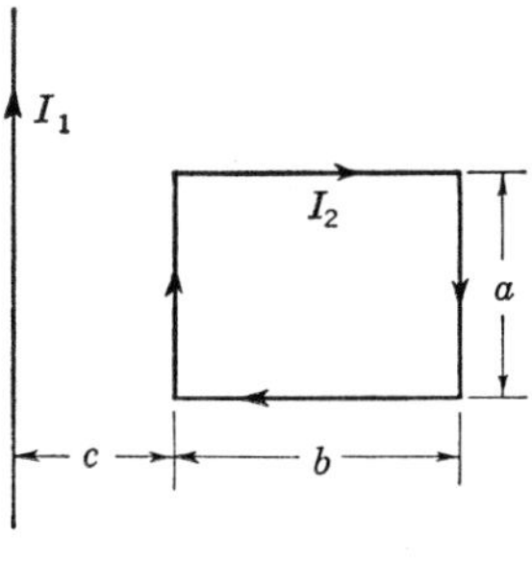

FIG. P 6.6

6.6. A rectangular loop is located near a current line source as illustrated. Find the force acting on the loop.

6.7. Find **B** at any point along the axis of a circular current loop of radius a and current I.

6.8. A solenoid of length $L \gg a$, where a is the radius, has n turns per meter. A current I flows in the winding. Find the field **B** along the axis.

6.9. A regular polygon of N sides has a current I flowing in it. Show that at the center

$$B = \frac{\mu_0 N I}{2\pi d} \tan \frac{\pi}{N}$$

where d is the radius of the circle circumscribing the polygon. Show that as N becomes large, the result reduces to that called for in Prob. 6.7.

6.10. Use Ampère's circuital law to find the field due to the two line sources specified in Prob. 6.1.

6.11. A z-directed current distribution is given by

$$J_z = r^2 + 4r \qquad r \leq a$$

Find **B** by means of Ampère's circuital law.

6.12. The vector potential due to a certain current distribution is given by

$$\mathbf{A} = x^2y\mathbf{a}_x + y^2x\mathbf{a}_y - 4xyz\mathbf{a}_z$$

Find the field **B**.

6.13. A current distribution is given by

$$\mathbf{J} = \mathbf{a}_z J_0 r \qquad r \leq a$$

where r is the radial coordinate in a cylindrical coordinate system. Find the vector potential **A** and the field **B**. Also find **B** directly by using Ampère's circuital law.

HINT: Solve the differential equation for **A** in cylindrical coordinates in the two regions $r < a$ and $r > a$. The arbitrary constants of integration may be found from the condition that **A** is continuous at $r = a$, equals zero at $r = 0$, and for $r \to \infty$ must be asymptotic to $C \ln r$, where C is a suitable constant.

6.14. A square loop with sides d and current I_1 is free to rotate about the axis illustrated. If the plane of the loop makes an angle θ with respect to an infinite line current I_2, find the torque acting to rotate the loop.

HINT: Consider the loop to be made up of infinitesimal dipoles of moment $d\mathbf{M} = I_1\, d\mathbf{S}$.

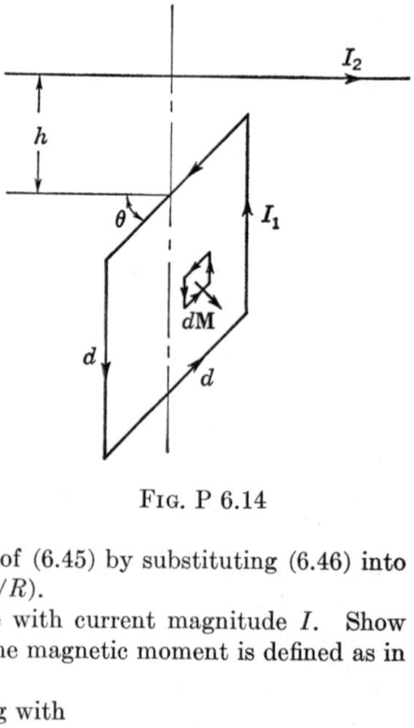

FIG. P 6.14

6.15. Show that Eq. (6.46) for **A** is a solution of (6.45) by substituting (6.46) into (6.45) and using the singularity property of $\nabla^2(1/R)$.

6.16. Given a current loop of arbitrary shape with current magnitude I. Show that (6.26) correctly gives the magnetic field if the magnetic moment is defined as in (6.27).

HINT: First find the vector potential **A** starting with

$$\mathbf{A} = \frac{\mu_0 I}{4\pi} \oint \frac{d\mathbf{l}'}{|\mathbf{r} - \mathbf{r}'|}$$

where **r** is the position vector of the field point and **r**′ that of the source point. Make use of the approximation

$$|\mathbf{r} - \mathbf{r}'|^{-1} = (r^2 + r'^2 - 2\mathbf{r} \cdot \mathbf{r}')^{-\frac{1}{2}} \approx \frac{1}{r}\left(1 + \frac{\mathbf{r} \cdot \mathbf{r}'}{2}\right)$$

and show that

$$d\mathbf{r}' \, (\mathbf{r}' \cdot \mathbf{r}) = \tfrac{1}{2}(\mathbf{r}' \times d\mathbf{r}') \times \mathbf{r} + \tfrac{1}{2}d[\mathbf{r}'(\mathbf{r} \cdot \mathbf{r}')]$$

to eliminate all terms except

$$\mathbf{A} = \frac{\mu_0 I}{4\pi}\left(\tfrac{1}{2}\oint \mathbf{r}' \times d\mathbf{r}'\right) \times \frac{\mathbf{r}}{r^3}$$

Note that $\frac{1}{2}\oint \mathbf{r}' \times d\mathbf{r}'$ equals the vector area of the loop and that the integral of $d[\mathbf{r}'(\mathbf{r} \cdot \mathbf{r}')]$ around a closed loop is zero since the latter is a complete differential.

Chapter 7

7.1. A permeable sphere of radius a is magnetized so that $\mathbf{M}$ within the sphere is uniform.

(*a*) What is the distribution of magnetization current in and on the sphere?

(*b*) In Prob. 7.15 we show that for this case $\mathbf{B}$ is also uniform within the sphere. From this information design a current winding that will set up a uniform $\mathbf{B}$ field over a spherical region of space.

7.2. A permeable sphere of radius a, whose center is at the origin of a system of coordinates, is magnetized so that

$$\mathbf{M} = (Az^2 + B)\mathbf{a}_z$$

Determine the equivalent magnetization currents and charges.

7.3. The magnetic moment of a magnetized body is given by the integral $\int_V \mathbf{M}\, dV$ taken over the body. If the body is placed in a uniform $\mathbf{B}$ field, determine the total torque in terms of the total moment.

7.4. A spherical shell of magnetic material is uniformly magnetized so that $\mathbf{M} = M_0\mathbf{a}_z$ in the shell. Find the scalar potential produced along the polar axis inside and outside of the shell. The inside radius of the shell is R_i, and the outside radius is R_0.

7.5. A very thin cylindrical iron rod and a thin (compared with radius) iron disk are placed in a magnetic field $\mathbf{B}_0$ with their axes parallel to the field. Find $\mathbf{B}$ and $\mathbf{H}$ internal to the iron specimens. Calculate the values of $\mathbf{M}$ in each case, given that $\mathbf{B}_0$ is 1.0 weber per square meter and $\mu = 5{,}000\mu_0$.

7.6. An infinitely long straight copper wire of circular cross section has a radius of 1 centimeter. It is surrounded coaxially by a permeable hollow cylinder which extends from a radius of 2 to 3 centimeters and whose relative permeability is 2,000. A current of 25 amperes flows in the wire.

(*a*) Calculate the total flux in the magnetic material per unit meter.

(*b*) Calculate the magnetization $\mathbf{M}$ in the permeable material.

(*c*) Find the induced magnetization currents in the magnetic medium.

(*d*) Show that the field for $r > 3$ centimeters is the same as it would be in the absence of the magnetic material by considering the net effect of the magnetization currents.

7.7. Find the field produced by a line current I located parallel to and above the plane interface of a magnetic material occupying the half space below the line current. The permeability of the material is μ.

HINT: The problem may be solved by an image method. Show that all boundary conditions can be satisfied if the field in the air is calculated from I and a current I' at a mirror-image position (assuming that both I and I' lie in free space), while the field in the magnetic material is due to I and I'', where I'' is located at I, and I and I'' are assumed to lie in an infinite material with permeability μ. The values of I' and I'' can be determined from the boundary conditions at the interface. Show in general that, for $\mu \to \infty$, lines of **B** on the air side must be perpendicular to the interface, while **H** in the iron goes to zero, and confirm that the above solution reduces to these results. Sketch the field lines.

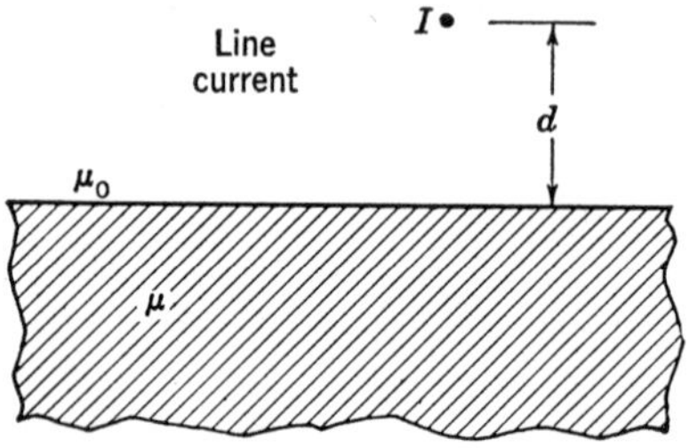

FIG. P 7.7

Answer

$$I' = \frac{\mu - \mu_0}{\mu + \mu_0} I$$
$$I'' = -I'$$

7.8. (*a*) Consider an arbitrary current loop as illustrated. Show that the magnetic scalar potential Φ_m at an arbitrary field point P can be expressed as

$$\Phi_m = -\frac{I\Omega}{4\pi}$$

where Ω is the solid angle subtended at P by an arbitrary surface whose periphery is C; that is,

$$\Omega = \int_S \frac{\mathbf{r} \cdot d\mathbf{S}}{r^3}$$

HINT: Divide the surface into a large number of small circulating current loops as was done in Fig. 6.13.

(*b*) Consider the surface (with periphery C) chosen in (*a*) and assume that a uniform electric dipole layer lies on this surface. ($\mathbf{P}_s$ is the dipole moment per unit area such that $\mathbf{P}_s \cdot d\mathbf{S}$ is the electric dipole moment of a differential surface $d\mathbf{S}$. We assume $\mathbf{P}_s$ to be uniform and normal to the surface.) Show that an electric scalar potential at P can be expressed as

$$\Phi = -\frac{|\mathbf{P}_s|\Omega}{4\pi\epsilon_0}$$

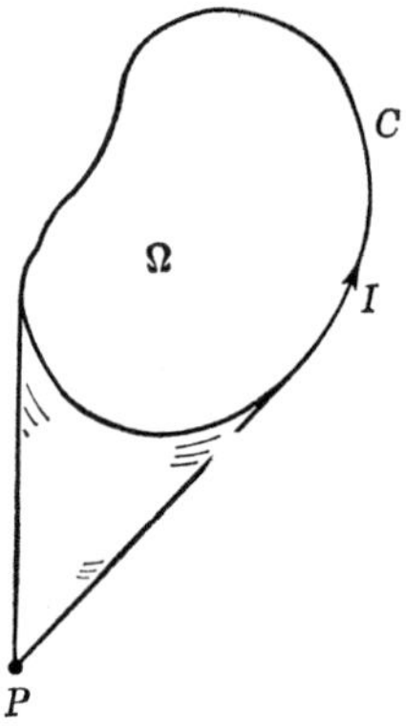

FIG. P 7.8

Note that the potential is discontinuous as the dipole layer is crossed. In the case of the current loop, the discontinuity is not associated with a physical surface, but may be any surface whose periphery is C. If one integrates the field (either $\mathbf{E} = -\nabla\Phi$ or $\mathbf{H} = -\nabla\Phi_m$) over a closed path that intersects this surface, a discontinuity in Ω of 4π results such that for the magnetic field

$$\oint \mathbf{H} \cdot d\mathbf{l} = I$$

as we expect. The multivalued electric field that results, while mathematically correct, violates the known conservative nature of the **E** field and cautions us that a true electric double layer cannot be achieved physically.

7.9. A toroid has the dimensions of 15 centimeters mean radius and 2 centimeters radius of circular cross section and is wound with 1,000 turns of wire. The toroid material is iron, with an effective relative permeability of 1,400 when the current is 0.7 ampere.

(*a*) Calculate the total flux.

(*b*) If a narrow air gap of length x is introduced, determine how the total flux depends on x (assuming $x \ll 2$ centimeters and μ remains the same in the iron). Compute total flux for $x = 0.1, 1.0, 5.0$ millimeters.

7.10. A ferromagnetic material is formed into the illustrated shape, where the cross section is everywhere square and 2 centimeters on a side, and the remaining mean dimensions are illustrated. If the winding carries 500 turns and a current of 0.3 ampere and if $\mu = 2{,}500\mu_0$, calculate the total flux in the central and right-hand leg. (Neglect leakage.)

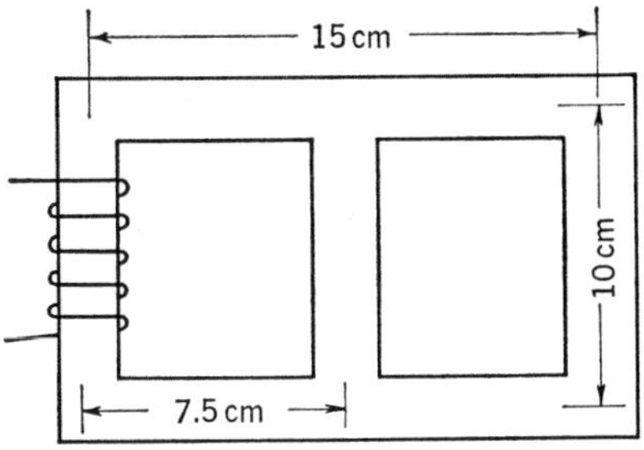

FIG. P 7.10

7.11. Repeat Prob. 7.10, but assume that a 2-millimeter air gap is cut in the central leg. Neglect leakage at air gap, and assume that μ remains constant.

7.12. Repeat Prob. 7.9, except that instead of assuming μ to be constant, the actual *B-H* relationship will be utilized. For this purpose the following data are available:

H, amp-turns/m.	50	100	150	200	250	300	400	500	600	800
B, webers/sq m.	0.07	0.23	0.60	0.85	1.00	1.07	1.18	1.25	1.30	1.33

7.13. A region V contains a permanent magnet of arbitrary shape. No additional sources of magnetic field exist. Show that

$$\tfrac{1}{2}\int \mathbf{B} \cdot \mathbf{H}\, dV = 0$$

where the integral is over all space. (In Chap. 8 we show that this integral evaluates the stored magnetic energy.)

7.14. The magnetic circuit shown consists of a permanent magnet of length 8 centimeters, two lengths of soft iron of 10 centimeters each, and an air gap of 1 centimeter.

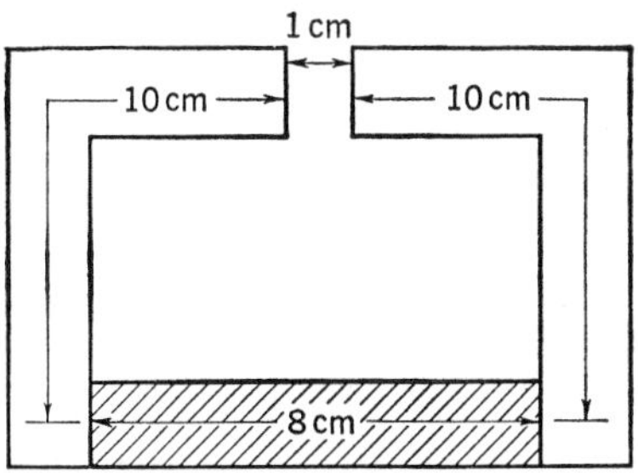

FIG. P 7.14

The cross sections of each are the same, and fringing is to be neglected. The B-H data for the magnet are given below, and $\mu = 5{,}000\mu_0$ for the soft iron.

(a) What is the flux density in the air gap?

(b) Sketch the **B** and **H** fields in the magnetic circuit.

H, amp-turns/m.....	0	−10,000	−20,000	−30,000	−35,000	−40,000	−45,000
B, webers/sq m......	1.25	1.22	1.18	1.08	1.00	0.80	0.00

7.15. A permeable (magnetizable) sphere of radius a is placed in a uniform external magnetic field $\mathbf{a}_z B_0 = \mathbf{a}_z \mu_0 H_0$. Introduce a scalar magnetic potential Φ_m, and find the induced magnetic field intensity $\mathbf{H}_i$ inside and outside of the sphere, as in Prob. 2.15. Show that for $r > a$, $\mathbf{H}_i$ is a dipole field, while for $r < a$, $\mathbf{H}_i$ is uniform. In the interior of the sphere, show that the relation between the magnetic polarization $\mathbf{M}$ and the applied field $\mathbf{H}_0$ is

$$\mathbf{M} = 3\,\frac{\mu - \mu_0}{\mu + 2\mu_0}\,\mathbf{H}_0$$

by using the relation $\mathbf{B} = \mu\mathbf{H} = \mu_0(\mathbf{H} + \mathbf{M})$, where $\mathbf{B}$ and $\mathbf{H}$ are the total fields in the sphere. Next show that the field $\mathbf{H}$ is also given by

$$\mathbf{H} = \mathbf{H}_0 - \frac{\mathbf{M}}{3} \qquad r < a$$

Eliminate $\mathbf{M}$ from the relation $\mathbf{B} = \mu_0(\mathbf{H} + \mathbf{M})$ to get $\mathbf{B} = -2\mu_0\mathbf{H} + 3\mu_0\mathbf{H}_0$. This is the equation of the shearing line. Plot this equation on a B-H plane, and indicate by the point of intersection with the B-H curve (sketch a suitable B-H curve for the purpose) what the remanent flux density B_r in the sphere will be when H_0 is reduced to zero.

7.16. The only shapes for which a rigorous solution such as that in Prob. 7.15 can be found are the ellipsoids and their degenerate forms such as spheroids and the sphere. For these bodies the interior field is uniform and the distant induced field is a dipole field when the external field is applied along an axis. The solution is of the form $\mathbf{H} = \mathbf{H}_0 - D\mathbf{M}$, and D is called the demagnetization factor. Show that in order to obtain a high remanent flux when the applied field $\mathbf{H}_0$ is reduced to zero, D should be as small as possible. In the interior of the body show that

$$\mathbf{B} = \frac{\mu\mathbf{H}_0}{1 + D(\mu/\mu_0 - 1)}$$

by using the relations $\mathbf{B} = \mu\mathbf{H} = \mu_0(\mathbf{H} + \mathbf{M})$ and $\mathbf{H} = \mathbf{H}_0 - D\mathbf{M}$. Thus the effective permeability of the body may be considered to be

$$\mu_e = \frac{\mu}{1 + D(\mu/\mu_0 - 1)}$$

since the flux density $\mathbf{B}$ is increased by μ_e/μ_0 in the body relative to its value $\mu_0\mathbf{H}_0$ for the applied field.

Chapter 8

8.1. (*a*) A rectangular loop is located near an infinite line current, as illustrated. If the current in the linear conductor is $I_0 \cos \omega t$, find the induced voltage in the rectangular loop.

(*b*) If the current in the linear conductor is a constant I_0, find the induced voltage in the rectangular loop as a function of x when the loop moves with a velocity v away from the line current.

(*c*) If the loop moves with velocity v away from the linear conductor and the current in the linear conductor is $I_0 \cos \omega t$, what is the induced voltage in the loop?

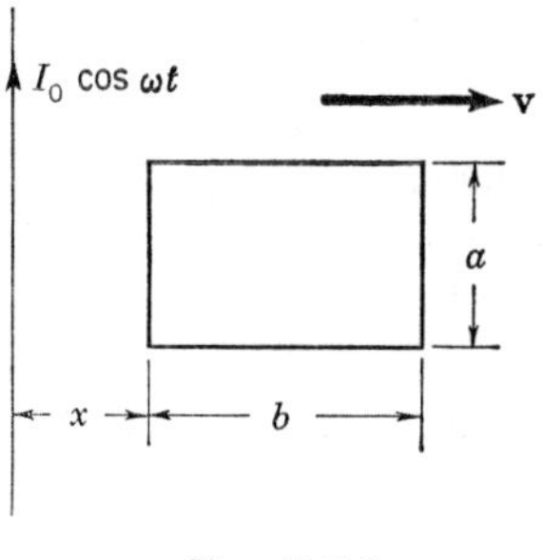

FIG. P 8.1

8.2. A conducting sphere of radius a moves with a constant velocity $v\mathbf{a}_x$ through a uniform magnetic field **B** directed along the y axis. Show that an electric dipole field given by

$$\mathbf{E} = \frac{vBa^3}{r^3}(2\cos\theta\,\mathbf{a}_r + \sin\theta\,\mathbf{a}_\theta)$$

exists around the sphere.

8.3. A large conducting sheet of copper ($\sigma = 5.8 \times 10^7$ mhos per meter) of thickness t falls with a velocity v through a uniform magnetic field B, as illustrated. Show that a force $F = \sigma v t B^2$ per unit area resisting the motion of the conductor exists.

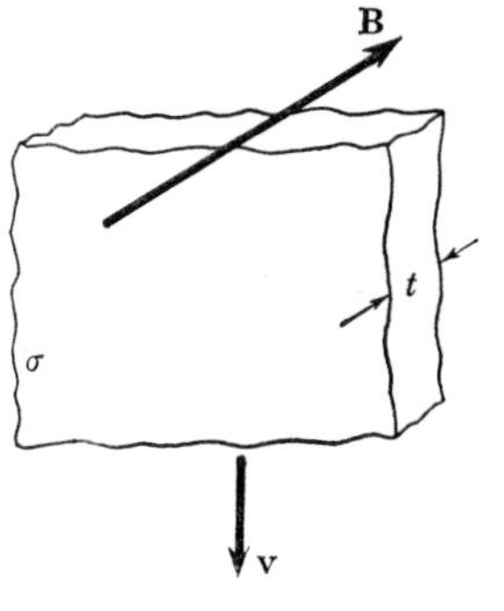

FIG. P 8.3

8.4. A thin conducting spherical shell (radius a, thickness $t_0 \ll a$, conductivity σ) rotates about a diameter (z axis) at the rate ω in the presence of a constant magnetic field **B** directed normal to the axis of rotation (along y axis). Find the resultant current flow in the spherical shell. Assume that the self-inductance of the sphere is negligible so that the current is determined only by the induced electric field and the conductivity σ. Since $t_0 \ll a$, each portion of the spherical surface may be considered to be a plane surface locally.

HINT: Express the velocity of an arbitrary point on the sphere's surface and the field **B** in spherical components, and find the induced electric field along the surface.

Answer. For a stationary observer,

$$I_\theta = \omega a \sigma t_0 B \sin^2\theta \sin\phi$$
$$I_\phi = 3\omega a \sigma t_0 B \sin^2\theta \cos\theta \cos\phi$$

For an observer rotating with the sphere,

$$I_\theta = \omega a\sigma t_0 B \sin^2\theta \sin\omega t$$

Note that the divergence equation $\nabla \cdot \mathbf{J} = -(\partial\rho/\partial t)$ must hold on the sphere. This condition determines Q on the sphere, and hence I_ϕ in the stationary coordinate frame.

8.5. For Prob. 8.4, show that $\cos\phi \sin^3\theta = C$ is the equation for the current flow lines on the sphere. C is the constant which determines a particular member of the family of lines.

8.6. For Prob. 8.4 let the magnetic field **B** be applied parallel to the axis of rotation, and find the resultant current flowing in the spherical shell.

8.7. A circular conducting loop rotates about a diameter at an angular rate ω in the presence of a constant magnetic field **B** normal to the axis of rotation, as illustrated.

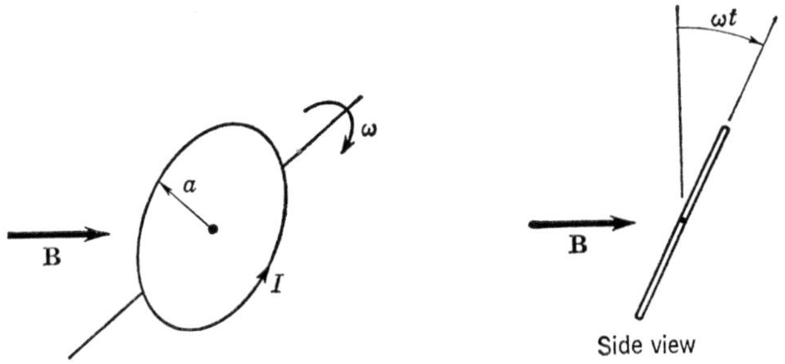

FIG. P 8.7

By making use of Faraday's law and the definition of self-inductance, show that the current flowing in the loop is given by

$$I = \frac{\pi a^2 B\omega \sin(\omega t - \phi)}{[R^2 + (\omega L)^2]^{1/2}}$$

where a is the radius of the loop, R is its resistance, L is its self-inductance, and $\tan\phi = \omega L/R$.

8.8. Show that in Prob. 8.7 the average power dissipated in the resistance R is

$$P = \frac{(\pi a^2 B\omega)^2}{R^2 + (\omega L)^2}\frac{R}{2} \qquad \text{joules/sec}$$

Show that a torque T resisting the rotation of the loop exists, where

$$T = \frac{(\pi a^2)^2 B^2\omega}{[R^2 + (\omega L)^2]^{1/2}} \sin(\omega t - \phi)\sin\omega t$$

HINT: Consider the magnetic dipole moment of the loop.

Show that the average rate of doing work on the loop in keeping it rotating is equal to the average rate at which energy is dissipated in the resistance R. As R goes to zero, show that the peak value of the current I approaches a constant independent of ω and that for $R = 0$ the average resisting torque vanishes.

8.9. A dielectric slab of thickness t moves with a velocity $\mathbf{v}$ normal to an applied uniform magnetic field $\mathbf{B}$, as illustrated. Find the induced polarization charge within and on the surface of the dielectric slab.

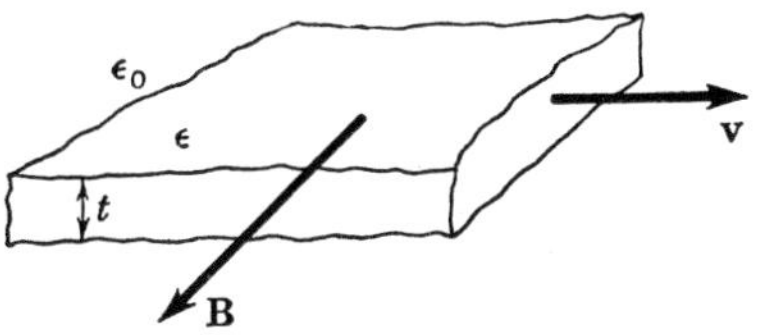

FIG. P 8.9

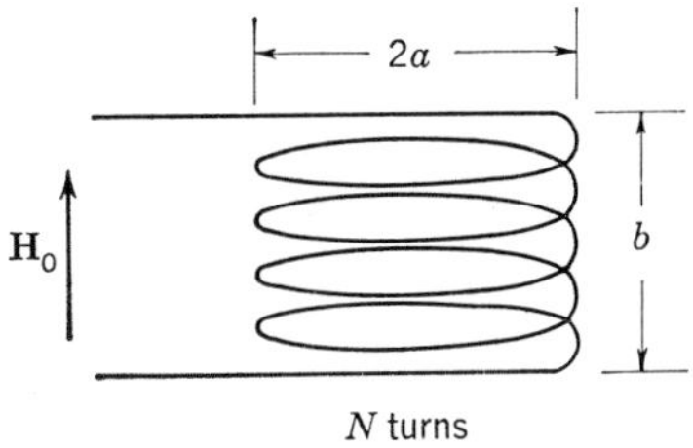

FIG. P 8.10

8.10. A small loop antenna for use in a portable radio receiver consists of N turns wound into a circular coil of radius a and height $b \ll a$. The input circuit to the radio receiver requires that the inductance of the loop be 250 microhenrys. Using the formula

$$L = \frac{0.008a^2n^2}{b} \qquad \text{microhenrys}$$

(dimensions in centimeters, n = number of turns), find the number of turns N required when $a = 6$ centimeters and $b = 1$ centimeter.

What is the voltage induced in the loop when a field $H_0 = 0.1$ microampere per meter is present and the frequency is 1 megacycle per second? Assume H_0 normal to the plane of the loop.

8.11. In place of an air-core loop as in Prob. 8.10, a ferrite rod antenna may be used as illustrated. The rod is in the shape of a prolate spheroid with a length $2d$ and a cross-section radius $a = 0.5$ centimeter. The permeability of the ferrite is $\mu = 200\mu_0$. The demagnetization factor D (see Probs. 7.15 and 7.16) is given by

$$D = \frac{a^2}{2d^2}\left(\ln\frac{4d^2}{a^2} - 2\right) \qquad d \gg a$$

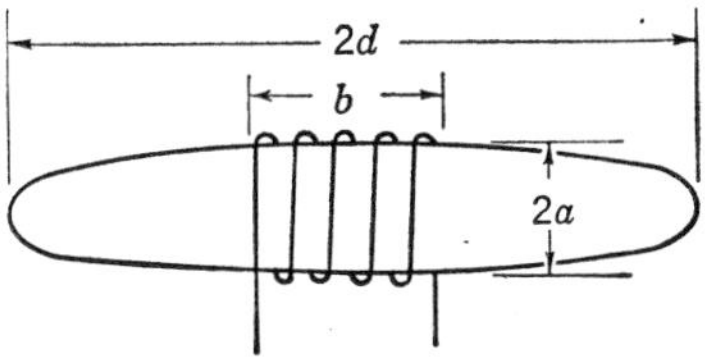

FIG. P 8.11

Find the number of turns N to give an inductance of 250 microhenrys when $d = 6$ centimeters and $b = 4$ centimeters. Use the formula

$$L = \frac{\mu_e}{\mu_0} 0.039 \frac{n^2a^2}{b} \qquad \text{microhenrys} \qquad b \gg a$$

where μ_e is given in Prob. 7.16 and a, b are in centimeters. What is the voltage induced in the coil for the same applied field as in Prob. 8.10? Note that the flux density in the core will be $\mu_e H_0$. What do you conclude regarding the merits of the ferrite rod antenna vs. the air-core loop? The ferrite rod antenna is much smaller, a factor of considerable importance for a portable radio receiver.

8.12. For the infinitely long thin conductor and the rectangular loop arranged as illustrated, show that the mutual inductance is given by

$$M = -\frac{\mu_0 a}{2\pi} \ln \frac{R}{[2b(R^2 - c^2)^{1/2} + b^2 + R^2]^{1/2}}$$

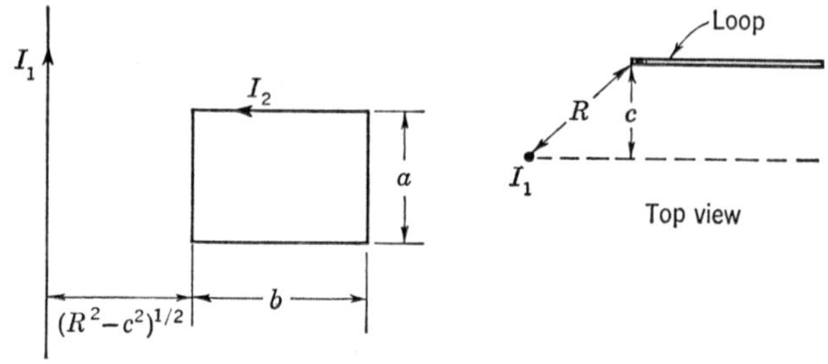

Fig. P 8.12

8.13. For Prob. 8.12 show that the component of force acting on the loop in the direction of R increasing is given by

$$F = \frac{\mu_0 a I_1 I_2}{2\pi R(R^2 - c^2)^{1/2}} \frac{bR^2 - 2bc^2 + b^2(R^2 - c^2)^{1/2}}{2b(R^2 - c^2)^{1/2} + b^2 + R^2}$$

when c is held constant. What is the component of force acting when R is held constant and c is allowed to vary?

8.14. A toroidal coil of mean radius b and cross-sectional radius a consists of N closely wound turns. Show that its self-inductance is given by

$$L = \frac{\mu_0 N^2 a^2}{2b}$$

when $b \gg a$. If the variation in B over the cross section is taken into account, show that

$$L = \mu_0 N^2[b - (b^2 - a^2)^{1/2}]$$

8.15. Two circular loops of radii r_1 and r_2 carry currents I_1 and I_2. The loops are coplanar and separated by a large distance R. By using a dipole approximation for the magnetic field set up by one loop at the position of the second loop, obtain an expression for the mutual inductance between the two loops. What is the force existing between the two loops?

8.16. Design a winding for a U-shaped electromagnet capable of lifting a 1,000-kilogram mass. The cross-section area of each leg is to be 20 square centimeters. The air gap between the electromagnet and the lower bar is 0.1 millimeter, as illustrated. What is the required number of ampere-turns when the magnet mean length is 30 centimeters and its effective relative permeability is 4,000? Assume that the reluctance of the iron bar is negligible.

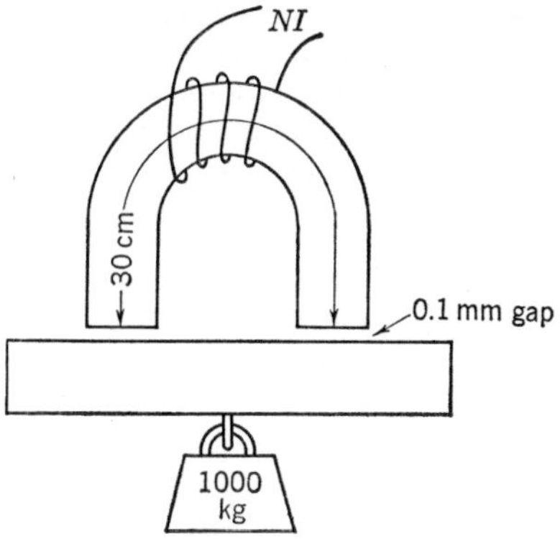

FIG. P 8.16

8.17. A round conductor of radius r_0 is bent into a circular loop of mean radius a. A current I flows in the circuit. Determine if a compressional or tension force acts on the conductor, and also find the magnitude of the force.

Chapter 9

9.1. A cylindrical capacitor has an inner radius of a, an outer radius of b, and a length L. A sinusoidal voltage $V \sin \omega t$ is applied and $2\pi c/\omega \ll L$, so that the electric field distribution is that for static conditions. Calculate the displacement current density in the dielectric and also the total displacement current crossing a cylindrical surface of radius r $(a < r < b)$. Show that the latter equals the conduction current in the leads to the capacitor.

9.2. Repeat Prob. 9.1, but for a spherical capacitor with inner radius a and outer radius b, and find the total displacement current crossing a spherical surface of radius r $(a < r < b)$.

9.3. (*a*) Confirm that the one-dimensional wave equation of (9.11) possesses a general solution $E_x = f(z - ct) + f'(z + ct)$, where f and f' are arbitrary functions.

(*b*) For the specific wave

$$\mathbf{E} = E_0 \cos k_0(z - ct)\,\mathbf{a}_x + E_0 \sin k_0(z - ct)\,\mathbf{a}_y$$

where $k_0 = 2\pi/\lambda_0 = \omega/c$, compute $\mathbf{H}$ and the Poynting vector.

(*c*) For a given z, determine the locus of $\mathbf{E}$ as a function of time. This wave is called circular-polarized.

9.4. The plates of the accompanying parallel-plate capacitor are circular and have a radius R_0, and the medium between the plates is conductive with conductivity σ. A battery with emf V_0 is connected to the capacitor plates (assumed perfectly conducting) as shown.

(*a*) Compute $\mathbf{E}$ and $\mathbf{H}$ in the conducting media.

(*b*) Calculate the Poynting vector, and verify that it correctly evaluates the power flow from the battery.

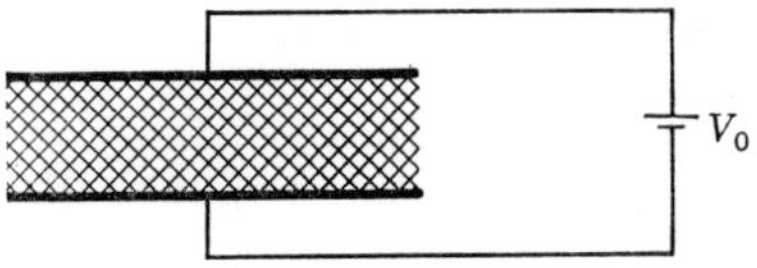

FIG. P 9.4

9.5. A shorted coaxial line has a resistive inner conductor and a perfectly conducting outer conductor. A d-c battery with emf V_0 is connected at the input end. The inner conductor has a radius of a, and the inside of the outer conductor a radius of b. The length is L, and the total current is I_0.

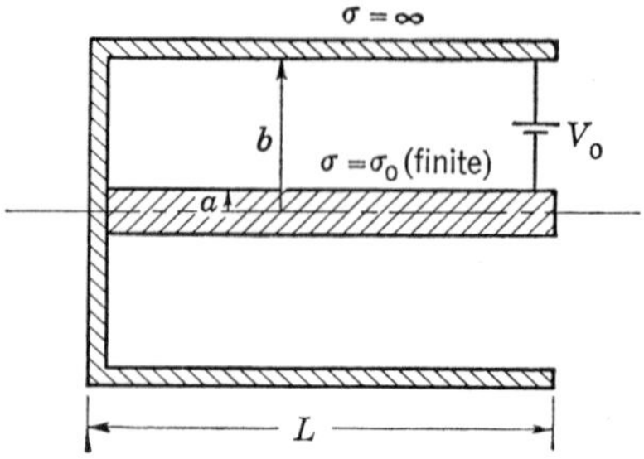

FIG. P 9.5

(a) Derive an expression for E and H in the dielectric region of the coaxial line. Assume that the battery is arranged to set up a potential that varies as $\ln r$ on the input surface. (If this is not done, is the problem uniquely specified?)

(b) Evaluate the Poynting vector, and compute the net power flow into the center conductor. Compare with a conventional circuit analysis.

9.6. An electrostatic field due to static charges and a magnetic field due to a permanent magnet are set up in the same region. In this way, a finite $\mathbf{E} \times \mathbf{H} = \mathbf{P}$ can exist, but no net power flow is taking place. Confirm that $\oint \mathbf{P} \cdot d\mathbf{S} = 0$ for any surface.

9.7. For a plane wave normally incident on an infinitely conducting infinite-plane reflector, the **E** and **H** fields as a function of the distance z from the reflector (with the given polarization) turn out to be

$$\mathbf{E} = jE_0 \sin k_0 z \, \mathbf{a}_x$$
$$\mathbf{H} = \left(\frac{\epsilon_0}{\mu_0}\right)^{1/2} E_0 \cos k_0 z \, \mathbf{a}_y$$

where $k_0 = 2\pi/\lambda_0 = \omega/c$.

(a) What is the instantaneous Poynting vector at $z = 0$, $\lambda_0/8$, $\lambda_0/4$?

(b) Compute the time-average Poynting vector at the above positions.

9.8. For Prob. 9.7, calculate the currents and charges set up on the conducting screen.

9.9. Given two uniform plane waves with electric fields as follows:

$$\mathbf{E}_1 = E_1 e^{-j\omega_1 z/c} \mathbf{a}_x$$
$$\mathbf{E}_2 = E_2 e^{-j\omega_2 z/c} \mathbf{a}_x$$

Both propagate in the same medium simultaneously. If $\omega_1 \neq \omega_2$, prove that the net time-average power flow equals the sum of the individual time-average power flows.

9.10. Find the skin depth for the following common materials, whose conductivity (mhos per meter) is given, at $f = 1$, 10^4, 10^6, 10^{10} cycles per second.

Silver	6.17×10^7
Copper	5.80×10^7
Aluminum	3.72×10^7
Sea water	4.5

9.11. At a frequency $f = 10^8$ cycles per second fused quartz has a relative permittivity of 3.8 and a loss tangent equal to 10^{-4}. Calculate the phase velocity, intrinsic impedance, and attenuation constant for a uniform plane wave propagating in this medium.

9.12. For the following media plot the attenuation of a uniform plane wave vs. frequency over the range 0 to 10^{10} cycles per second (σ in mhos per meter):

Sea water	$\sigma = 4.5$,	$\kappa = 80$
Good ground	$\sigma = 10^{-2}$,	$\kappa = 15$
Poor ground	$\sigma = 10^{-3}$,	$\kappa = 5$

9.13. A long cylindrical conductor of radius a is uniformly excited by an electromagnetic field so that a current flows in the axial direction and has no circumferential variations. We may take the current density at the surface to be equal to J, and for the purposes of this problem it is possible to neglect the axial variations. If the conductivity is σ and $\sigma \gg \omega\epsilon$,

(*a*) Show that

$$\nabla_t^2 J_z = \frac{1}{r^2}\frac{d}{dr}\left(r^2 \frac{dJ_z}{dr}\right) = j\omega\mu\sigma J_z$$

where the axial current density is in the z direction and a function of r only.

(*b*) Solve (*a*) for J_z.

Answer. Solutions to (*b*) come out in the form of a Bessel function $J_0(j^{-1/2}x)$, for which tables are available. Often new functions are defined as

$$\text{Ber } x = \text{Re } J_0(j^{-1/2}x)$$
$$\text{Bei } x = \text{Im } J_0(j^{-1/2}x)$$

in terms of which the solution can be written

$$J_z(r) = J\,\frac{\text{Ber }(2r/\delta) + j\,\text{Bei }(2r/\delta)}{\text{Ber }(2a/\delta) + j\,\text{Bei }(2a/\delta)}$$

where $\delta = 1/\sqrt{\pi f \mu_0 \sigma}$ is the skin depth and J is the current density at $r = a$.

9.14. The current distribution in the conductor of Prob. 9.13 may be found approximately if a/δ is large. In this case we may think of the conductor as if it were flat with a uniform plane wave incident. The diffusion of current into the conductor should be in the form of (9.43), except that the radial variable replaces the coordinate normal to the surface; that is, since **J** and **E** are related by the constant σ, we have

$$J_z = Je^{-(1+j)(a-r)/\delta}$$

By using the asymptotic expressions for Bessel functions of large argument, show that the result of Prob. 9.13 reduces to the above expression.

9.15. Show that in a source-free region of space where $\nabla \cdot \mathbf{E} = 0$, the electric and magnetic fields may be found from a magnetic-type vector potential $\mathbf{A}_m$ by means of the equations

$$\mathbf{E} = \nabla \times \mathbf{A}_m$$
$$\mathbf{H} = j\omega\epsilon\mathbf{A}_m - \frac{\nabla\nabla \cdot \mathbf{A}_m}{j\omega\mu}$$

and $\mathbf{A}_m$ is a solution of

$$\nabla^2\mathbf{A}_m + \omega^2\mu\epsilon\mathbf{A}_m = 0$$

The derivation is similar to that for the electric-type vector potential **A**.

9.16. In a region of space where the only source for the electromagnetic field is a volume dielectric polarization of density **P**, show that the electric and magnetic fields may be found from a vector potential $\mathbf{\Pi}_e$ by means of the equations

$$\mathbf{H} = j\omega\epsilon_0 \nabla \times \mathbf{\Pi}_e$$
$$\mathbf{E} = k_0^2\mathbf{\Pi}_e + \nabla\nabla \cdot \mathbf{\Pi}_e = \nabla \times \nabla \times \mathbf{\Pi}_e - \frac{\mathbf{P}}{\epsilon_0}$$

and $\mathbf{\Pi}_e$ satisfies the equation

$$\nabla^2\mathbf{\Pi}_e + k_0^2\mathbf{\Pi}_e = -\frac{\mathbf{P}}{\epsilon_0}$$

where $k_0^2 = \omega^2\mu_0\epsilon_0$. The derivation is similar to that for the vector potential **A**, and a Lorentz-type condition must be invoked. The polarization **P** is introduced by replacing $\mathbf{D}$ by $\epsilon_0\mathbf{E} + \mathbf{P}$ in Maxwell's equations. The potential function $\mathbf{\Pi}_e$ is known as the electric-type Hertzian potential.

The above result may be extended to arbitrary source distributions through the relations

$$\rho = -\nabla \cdot \mathbf{P}$$
$$\mathbf{J} = \frac{\partial \mathbf{P}}{\partial t}$$

9.17. If the only source for the electromagnetic field is the magnetic polarization $\mathbf{M}$ per unit volume, show that the fields found from a magnetic Hertzian vector potential $\mathbf{\Pi}_m$ by means of the equations

$$\begin{aligned} \mathbf{E} &= -j\omega\mu_0 \nabla \times \mathbf{\Pi}_m \\ \mathbf{H} &= k_0^2\mathbf{\Pi}_m + \nabla\nabla \cdot \mathbf{\Pi}_m \\ &= \nabla \times \nabla \times \mathbf{\Pi}_m - \mathbf{M} \end{aligned}$$

satisfy Maxwell's equations, provided $\mathbf{\Pi}_m$ is a solution of

$$\nabla^2\mathbf{\Pi}_m + k_0^2\mathbf{\Pi}_m = -\mathbf{M}$$

9.18. In an idealized velocity-modulated electron tube, the electron stream can be taken to consist of d-c convection current of amplitude i_0 and an a-c current of the form

$$\mathbf{i}_{\text{a-c}} = i_1 e^{j\omega t - \gamma z}\mathbf{a}_z$$

where i_0 and i_1 are in amperes per unit area of cross section. Note that the a-c current is in the form of a wave and that a geometry is implied where there are no variations with x or y.

The a-c convection current sets up an electromagnetic field most easily found by first computing the vector potential $\mathbf{A}$. If, as we assume, $i_1 \ll i_0$, then all a-c quantities have the space-time dependence $e^{j\omega t - \gamma z}$. (The d-c current can be ignored in calculating time-varying fields.)

(*a*) Find the a-c charge density.

(*b*) Write and solve the differential equation for vector potential $\mathbf{A}$.

(*c*) From (*b*) find the electric field $\mathbf{E}$.

9.19. It is virtually impossible to obtain solutions to the vector wave equation if the fields are written in terms of their spherical components and if spherical coordinates are used. Yet for boundary conditions imposed on spherical boundaries, it is equally difficult to utilize rectangular coordinates since the boundary is not a natural one. It turns out, however, that the vector

$$\mathbf{M} = \mathbf{r} \times \nabla\psi$$

where $\mathbf{r}$ is the radius vector, satisfies the vector wave equation provided that ψ satisfies the scalar wave equation

$$\nabla^2\psi + \omega^2\mu\epsilon\psi = 0$$

Another solution is

$$\mathbf{N} = \frac{1}{\omega\sqrt{\mu\epsilon}} \nabla \times \mathbf{M}$$

Note that $\mathbf{M}$ and $\mathbf{N}$ may be identified with the $\mathbf{E}$ and $\mathbf{H}$ field, or vice versa. In view of the fact that **M** is transverse to spherical surfaces, spherical boundary-value problems may be readily formulated.

Confirm that **N** and **M** do indeed satisfy the vector wave equations

$$\nabla^2\mathbf{A} + \omega^2\mu\epsilon\mathbf{A} = -\nabla \times \nabla \times \mathbf{A} + \nabla\nabla \cdot \mathbf{A} + \omega^2\mu\epsilon\mathbf{A} = 0$$

provided that

$$\nabla^2\psi + \omega^2\mu\epsilon\psi = 0$$

9.20. An isotropic dielectric medium is nonuniform, so that ϵ is a function of position. Show that **E** satisfies

$$\nabla^2\mathbf{E} + k^2\mathbf{E} = -\nabla\left(\mathbf{E} \cdot \frac{\nabla\epsilon}{\epsilon}\right)$$

9.21. If the gauge is chosen so that $\nabla \cdot \mathbf{A} = 0$, confirm that the following equations for Φ and **A** result:

$$\nabla^2\Phi = \frac{-\rho}{\epsilon_0}$$

$$\nabla \times \nabla \times \mathbf{A} - \omega^2\mu\epsilon\mathbf{A} = \mu\mathbf{J} - j\omega\epsilon\mu\,\nabla\Phi$$

where **E** and **H** remain related to Φ and **A** as in (9.62) and (9.63).

What is the relationship between the Φ and **A** in this gauge as compared with that for which the Lorentz condition is satisfied?

9.22. Confirm that $\psi = (1/4\pi R)e^{-jk_0R}$ satisfies the scalar Helmholtz equation $\nabla^2\psi + k_0^2\psi = 0$, provided $R \neq 0$. Show further that for a spherical (ΔV) volume of vanishing radius surrounding the origin, $\int_{\Delta V} (\nabla^2\psi + k_0^2\psi)\,dV$ is finite and actually equals -1.

Hint: Write $\nabla^2\psi = \nabla \cdot \nabla\psi$, and use the divergence theorem.

As a consequence of the above, $\psi = (1/4\pi R)e^{-jk_0R}$ may be said to be a solution of the following inhomogeneous Helmholtz equation:

$$\nabla^2\psi + k_0^2\psi = -\delta(\mathbf{r} - \mathbf{r}')$$

where $R\mathbf{a}_R = \mathbf{r} - \mathbf{r}'$ and $\mathbf{r}'$ is the location of a unit (delta) source. The delta function has the property that

$$\int_V \delta(\mathbf{r} - \mathbf{r}')\,dV = \begin{cases} 0 & \text{if } \mathbf{r}' \text{ is not contained in } V \\ 1 & \text{if } \mathbf{r}' \text{ is contained in } V \end{cases}$$

9.23. Confirm the statement that appears in the text that if, at a perfect conductor, we satisfy $\mathbf{J}_s = \mathbf{n} \times \mathbf{H}$, then $\rho_s = \mathbf{n} \cdot \mathbf{D}$.

Chapter 10

10.1. A parallel-polarized TEM wave is incident on a dielectric interface at an angle θ_i. Find the total reflected and transmitted fields and verify Eqs. (10.46). With reference to Fig. 10.3, let the incident electric field be

$$\mathbf{E} = A_1(\mathbf{a}_x \cos\theta_i - \mathbf{a}_z \sin\theta_i)e^{-jk_0(x\sin\theta_i + z\cos\theta_i)}$$

10.2. The region $z > 0$ is occupied by a dielectric medium with dielectric constant κ_2. In front of this medium a slab of material of thickness d and dielectric constant κ_1 is placed. For a TEM wave incident normally on this structure from the left show that no reflection occurs if $\kappa_1 = \kappa_2^{1/2}$ and $d = \frac{1}{4}\lambda_0\kappa_1^{-1/2}$, where λ_0 is the free-space wavelength. The intermediate dielectric slab forms a quarter-wave transformer which matches the dielectric medium for $z > 0$ to free space. This technique is used in optics to reduce reflections from lenses and is known as lens blooming.

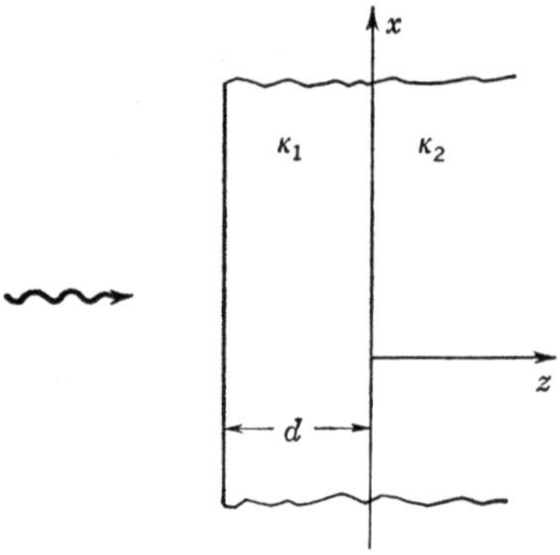

FIG. P 10.2

10.3. When a perpendicular-polarized TEM wave is incident on the structure of Fig. P 10.2, at an angle θ_i with respect to the interface normal, show that the parameters of the quarter-wave matching layer are given by

$$\kappa_1 - \sin^2 \theta_1 = (\kappa_2 - \sin^2 \theta_i)^{1/2} \cos \theta_i$$
$$d = \frac{\lambda_0}{4} (\kappa_1 - \sin^2 \theta_i)^{-1/2}$$

Note that the requirement for a match (no reflection) is that the wave impedance of the intermediate layer be the geometric mean of the wave impedances of free space and the medium to be matched and that the thickness be equal to one-fourth of the effective wavelength in the z direction in the matching layer.

For a parallel-polarized wave show that the equation for d is the same as for perpendicular polarization but that κ_1 is given by

$$\frac{\kappa_1 - \sin^2 \theta_i}{\kappa_1^2} = \frac{(\kappa_2 - \sin^2 \theta_i)^{1/2} \cos \theta_i}{\kappa_2}$$

Can both polarizations be matched simultaneously?

10.4. A perpendicular-polarized wave is incident on a magnetic material at an angle θ_i, as in Fig. 10.3. The electrical parameters of the medium for $z > 0$ are $\epsilon = \epsilon_0$, $\mu = \kappa_m\mu_0$. Show that a Brewster angle exists such that no reflection takes place. Show also that in the present case a similar phenomenon does not occur for the parallel-polarized wave.

10.5. A perpendicular-polarized wave is incident on a dielectric-air interface at an angle θ (relative to the interface normal) from the dielectric side. Show that a critical angle θ_c exists such that the emerging ray on the air side just grazes the surface. For angles of incidence greater than θ_c show that the angle of propagation on the air side must be complex and that the field is exponentially damped in a direction normal and away from the interface on the air side. Note that no energy is transmitted past the interface since the modulus of the reflection coefficient is unity. However, the fields on the air side are not zero. Contrast this rigorous solution with that based on ray optics.

10.6. An infinite dielectric slab of thickness d is placed above a perfectly conducting plane, as illustrated. Along such a structure a field known as a "surface wave" may propagate. This wave consists essentially of a TEM wave propagating along a zigzag path within the dielectric and at an angle greater than the critical angle so that the field is exponentially damped away from the surface on the air side (see Prob. 10.5).

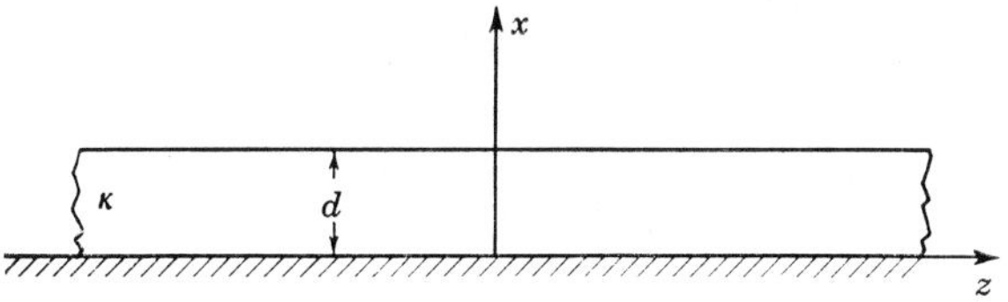

FIG. P 10.6

Find the solutions for a TM wave along this structure. Assume that the field is independent of y. The axial electric field may be taken to be of the form

$$E_z = \begin{cases} A_1 e^{-j\beta z} \sin l_1 x & 0 \le x \le d \\ A_2 e^{-j\beta z} e^{-l_2 x} & x \ge d \end{cases}$$

where the separation constants l_1 and l_2 are given by

$$l_1^2 = k_{c1}^2 = k_0^2 - \beta^2$$
$$l_2^2 = -k_{c2}^2 = \beta^2 - k_0^2$$

since E_z must satisfy Helmholtz's equation. Show that l_1 and l_2 are determined by the equations

$$\kappa l_2 = l_1 \tan l_1 d$$
$$l_1^2 + l_2^2 = (\kappa - 1)k_0^2$$

10.7. A surface wave of the TM type may be guided along an imperfectly conducting plane. Assume that the conducting plane is in the yz plane and that above the plane the axial component of the electric field is

$$E_z = Ae^{-lx-j\beta z}$$

where $l^2 = \beta^2 - k_0^2$. To determine l and β note that at the conducting plane, $E_z = Z_m J_z$, where Z_m is the surface impedance of the plane and J_z is the surface current density as found from the tangential magnetic field at the surface of the conductor. Obtain the solutions for l and β.

10.8. For Prob. 10.7, evaluate the power loss per unit area of the conducting plane and find the attenuation of the wave from the relation

$$\alpha = \frac{P_l}{2P}$$

where P_l is the power loss per unit area and P is the power flow along the plane per unit width. Compare this value of α with that obtained in Prob. 10.7 from the imaginary part of β.

10.9. For a coaxial line with inner radius a and outer radius b, show that its parameters are given by (air-filled)

$$L = \frac{\mu_0}{2\pi} \ln \frac{b}{a}$$
$$C = \frac{2\pi\epsilon_0}{\ln (b/a)}$$
$$Z_c = 60 \ln \frac{b}{a}$$
$$\alpha = \frac{R_m}{4\pi Z_c}\left(\frac{1}{a} + \frac{1}{b}\right)$$

10.10. A 50-ohm transmission line is terminated in a load $Z_L = 20 + j10$. Find the reflection coefficient and the standing-wave ratio. At what distance l from the load is the input impedance equal to $Z_{in} = 50 + jX$? At this point a reactance $-jX$ may be added in series with the line in order to match the load to the line.

10.11. For a rectangular guide with inner dimensions $a = 0.9$ inch, $b = 0.4$ inch, evaluate the parameters λ_0, λ_g, v_p, v_g, α when the frequency is 10^{10} cycles per second. Assume a copper guide with $\sigma = 5.8 \times 10^7$ mhos per meter.

10.12. A rectangular guide of dimensions a, b is filled with a dielectric material with $\epsilon = \kappa\epsilon_0$. Obtain the solution for a TE_{10} mode. Show that the guide wavelength is given by

$$\lambda_g = \frac{\lambda_0}{[\kappa - (\lambda_0/2a)^2]^{1/2}}$$

10.13. A rectangular guide is filled with a dielectric medium ($\epsilon = \kappa\epsilon_0$) for $z \geq 0$. A TE_{10} mode is incident from the region $z < 0$. Find the reflected and transmitted fields.

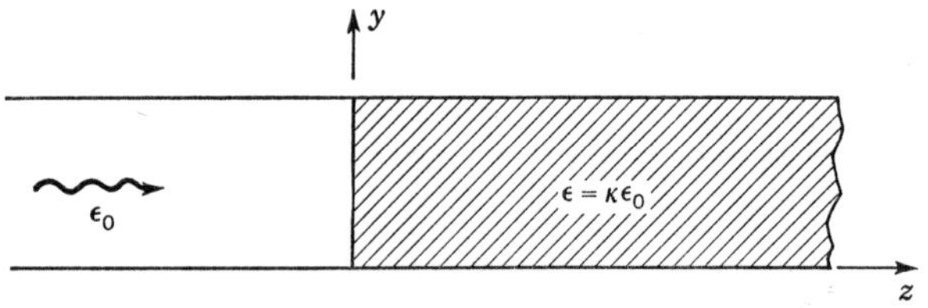

FIG. P 10.13

10.14. For Prob. 10.13 find the thickness and dielectric constant of a quarter-wave dielectric matching layer that will match the empty guide to the dielectric-filled guide.

10.15. A rectangular pulse of width 1 microsecond (frequency components up to 1 megacycle) is used to modulate a carrier of frequency 10^{10} cycles per second. This signal is transmitted through a rectangular guide with $a = 2b = 2.5$ centimeters as a TE_{10} mode. What length of guide is required to produce a signal delay of 2 microseconds?

10.16. Find the solutions for a TE_{111} mode in a rectangular cavity of sides a, b, d. Obtain expressions for the resonant frequency Q and the decay constant α. Evaluate the Q for the case when $a = b = c = 3$ centimeters and $\sigma = 5.8 \times 10^7$ mhos per meter.

10.17. Obtain the solutions for axially symmetric TE_{n0m} modes in a cylindrical cavity of height d and radius a.

10.18. Prove that k_c^2 is always real for TE and TM modes in an arbitrary perfectly conducting waveguide.

HINT: Start with the two-dimensional divergence theorem taken over the waveguide cross section:

$$\int_{cs} (\nabla_t \cdot \mathbf{A})\, dS = \oint \mathbf{A} \cdot \mathbf{n}\, dl$$

and let

$$\mathbf{A} = E_z^* \nabla_t E_z \; (= H_z^* \nabla_t H_z)$$

Chapter 11

11.1. Calculate the far-zone field of a linear antenna whose length is one wavelength under the following two conditions:

(*a*) The antenna is parasitically excited, and the current distribution may be assumed to be $I = I_0 \sin k_0 z$. (The center of the antenna is at $z = 0$.)

(*b*) The antenna is driven at the center so that the current distribution is given by $I = I_0|\sin k_0 z|$.

11.2. The current distribution on a particular half-wave linear antenna is found to be more accurately given by $I = I_0 \cos^2 k_0 z$. Write an expression for the far-zone field, and compute the radiation resistance.

11.3. A two-wire transmission line of length l and separation d is terminated in its characteristic impedance and carries a traveling TEM wave. Calculate the far-zone field, and plot as a function of polar angle, taking the direction of the transmission line along the polar axis with conductors at $x = \pm d/2$. (Neglect radiation from the load itself.)

11.4. (*a*) Show that a linear current radiator normal to and above a perfectly conducting ground plane radiates a field which can be calculated by removing the ground and replacing its effect by a mirror-image current element directed in the same direction (as illustrated).

(*b*) Repeat (*a*) but with a horizontal current element. The image in this case is oppositely directed (see illustration).

HINT: It is sufficient to show that the total field due to antenna and image satisfies Maxwell's equations and that tangential **E** is zero over the surface of the conductor. That this is sufficient follows from a uniqueness proof analogous to the one developed for Laplace's equation.†

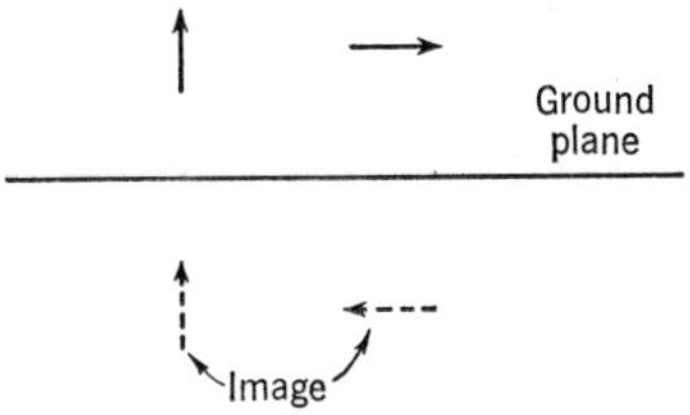

FIG. P 11.4

11.5. Find the field radiated from a small current loop of radius a assumed to carry a current $I = I_0 e^{j\omega t}$. The loop is located at the origin and in the xy plane. Show that the radiation resistance of the loop is $R = 320\pi^4(\pi a^2/\lambda_0^2)^2$. See Prob. 9.17 for fields set up by a magnetic dipole; i.e., consider the loop as a small magnetic dipole.

11.6. At high frequencies the magnetic flux linking a circular current loop C assumed to carry a current $I_0 e^{j\omega t}$ is given by

$$\oint_C \mathbf{A} \cdot d\mathbf{l} = \oint_C \oint_C \frac{I_0}{4\pi\mu_0} \frac{e^{-jk_0 r}}{r} \, d\mathbf{l} \cdot d\mathbf{l}'$$

where $d\mathbf{l}$ and $d\mathbf{l}'$ are two elements of arc length along C separated by a distance r. Expand $e^{-jk_0 r}$ in the form $1 - jk_0 r - \frac{1}{2}k_0^2 r^2 + jk_0^3 r^3/6 + \cdots$, and show that the term involving k_0^3 gives rise to an induced voltage in the loop which is 180° out of phase with the current. The applied voltage must overcome this induced voltage and in so doing does work on the current. Thus this term gives rise to an input resistance to the loop. Show that the applied voltage required to overcome the induced voltage due to this part of the flux linkage is $\frac{1}{6} Z_0 k_0^4 \pi a^4 I_0$ and that the input resistance computed by this method is equal to the radiation resistance of the loop, as found in Prob. 11.5.

11.7. In general, the directivity of an array of elements does not equal the array directivity times the element directivity. Prove this. Under what conditions is the directivity of the array approximately that of the antenna?

11.8. A three-dimensional array of isotropic radiators has sources of equal amplitude

† See J. A. Stratton, "Electromagnetic Theory," p. 486, McGraw-Hill Book Company, Inc., New York, 1941.

and phase located at $\mathbf{r}_0 = na\mathbf{a}_x + mb\mathbf{a}_y + lc\mathbf{a}_z$, where

$$n = 0, 1, 2, \ldots, N$$
$$m = 0, 1, 2, \ldots, M$$
$$l = 0, 1, 2, \ldots, L$$

Find the resultant radiation pattern, and show that it is the product of three array factors and the factor $e^{-jk_0r}/4\pi r$. How many main beams are produced if a and b are equal to $\lambda_0/2$ and $c = \lambda_0$?

11.9. An interferometer consists of two identical antennas, A_1 and A_2, spaced by an amount $L = N\lambda_0$. (See Fig. P 11.9.)

(*a*) Show that the resultant radiation pattern is a multilobed pattern. How many lobes are there?

(*b*) If the relative phase angle of the signals received by antennas A_1 and A_2 can be measured to within an accuracy of 1°, what is the resultant angular accuracy that is obtained in measuring the position of the source that emits the signal? Assume that it is always possible to know which lobe the source is located in and that the source is located in the xy plane.

(*c*) For $N = 1{,}000$ and the source making a true angle of $\theta = 0°, 30°, 60°, 90°$ with the x axis, compute the angular error, under conditions of (*b*).

Interferometers of this type are widely used in radio astronomy and in missile tracking systems. Usually several closer-spaced antennas are also used in order to determine which lobe the source of the received signal is located in. An additional interferometer with base line along the y axis would also be required in order to obtain the direction to an arbitrarily located source.

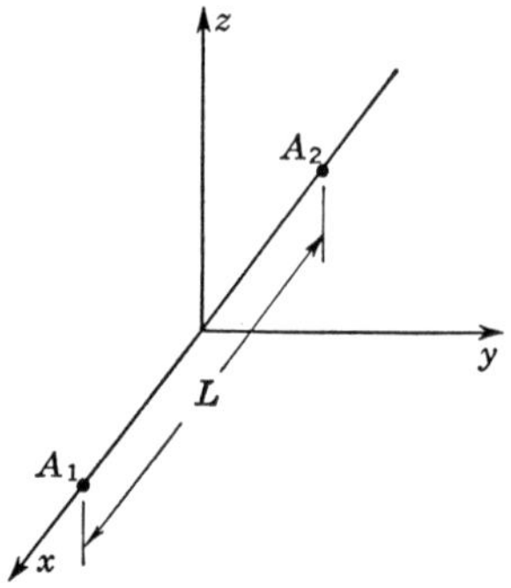

FIG. P 11.9

11.10. An N-element, uniformly spaced, arbitrarily excited linear array has an array factor given by

$$A = \sum_{n=1}^{N} a_n e^{j\psi_n} e^{jk_0 nd\cos\theta}$$

where the line of the array is along the polar axis, θ is the polar angle, and d the element separation. The nth element has an excitation magnitude a_n and relative phase ψ_n.

(*a*) Show that by letting the complex variable $Z = e^{jk_0d\cos\theta}$ and the complex number $A_n = a_n e^{j\psi_n}$ such an array can be represented by a polynomial, and vice versa.

(*b*) What is the locus of the complex variable Z in part a, and what is the range of variation of Z that corresponds to a real pattern (i.e., that corresponds to $0 < \theta < \pi$)?

(*c*) By factoring the polynomial in (*a*), the array factor can be given by the product of line segments from each root to the value of Z corresponding to a given θ. Confirm and describe this in greater detail. Where must at least one polynomial root lie if the pattern is to have a null?

(*d*) The pattern of a two-element array spaced $\lambda_0/4$ apart and with a 90° relative phase is given by $|1 + e^{j\pi(\cos\theta - 1)/2}|$. This pattern is a cardioid, and note that it has no sidelobes, although its main lobe is extremely broad.

If we form the pattern $|1 + e^{j\pi(\cos\theta-1)/2}|^n$ for large n, then the resulting beamwidth should be relatively narrow, as might be visualized by the process of successive multiplication of a cardioid pattern. Interestingly, this pattern remains without any sidelobes. Making use of the polynomial representation, determine the array size and excitation that produce a pattern corresponding to $N = 5$. Plot the pattern. This array is known as the binomial array. (Why?)

11.11. Making use of Prob. 9.22 and Green's second theorem (1.68) with Φ a solution of $\nabla^2\Phi + k^2\Phi = -g(x,y,z)$, and $\psi = (1/4\pi R)e^{-jkR}$, show that the field at any point due to the sources $g(x,y,z)$ may be written

$$\Phi = \frac{1}{4\pi}\int_V g(x,y,z)\,\frac{e^{-jkR}}{R}\,dV + \frac{1}{4\pi}\oint_S \left[\frac{e^{-jkR}}{R}\frac{\partial\Phi}{\partial n} - \Phi\frac{\partial}{\partial n}\left(\frac{e^{-jkR}}{R}\right)\right] dS$$

where n is the outward normal to S. (The surface integrals take into account the presence of sources which lie outside the chosen finite region V. If $V \to \infty$, then all sources are included in the volume integral and the surface integral contribution can be shown to vanish.)

11.12. If the vector wave equation for the electric field in a finite charge-free region is written out, we have

$$\nabla^2\mathbf{E} + k^2\mathbf{E} = 0$$

Note that each component satisfies an equation of the form denoted by the scalar Φ in Prob. 11.11. Show that by summing the components, a vector relationship is obtained of the form

$$\mathbf{E} = \frac{1}{4\pi}\oint_S \left[\frac{e^{-jkR}}{R}\frac{\partial\mathbf{E}}{\partial n} - \mathbf{E}\frac{\partial}{\partial n}\left(\frac{e^{-jkR}}{R}\right)\right] dS$$

11.13. The equation developed in Prob. 11.12 shows how the field in a source-free region of space is obtainable from given values on a bounding surface. This is an electromagnetic formulation of Huygens' principle.

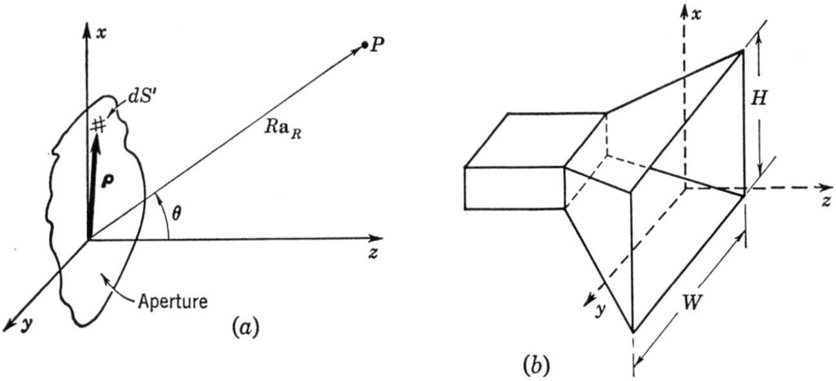

FIG. P 11.13

The equation in Prob. 11.12 can be used to determine the field radiated from a plane aperture since the region beyond the aperture is source-free, and the fields in the aperture may be thought of as a secondary source. Using this as a starting point, Silver† shows that the radiated field may be approximated by the scalar relationship

$$u_p = \frac{jk_0}{4\pi R}\,e^{-jk_0R}\,(1 + \cos\theta)\int_{\text{aperture}} u^{-jk_0\mathbf{r}'\cdot\mathbf{a}_R}\,dS'$$

† S. Silver, "Microwave Antenna Theory and Design," sec. 5.14, McGraw-Hill Book Company, Inc., New York, 1949.

In this expression u_p, $u = E_x$ or E_y, $\mathbf{r}'$ is the position vector of a source point, and $\mathbf{a}_R$ is a unit vector from the origin to the field point. We also have

$$\mathbf{a}_R \cdot \mathbf{r}' = (x' \cos \phi + y' \sin \phi) \sin \theta$$
$$dS' = dx'\, dy'$$

where x', y' are aperture coordinates.

(*a*) Use these results to compute the radiated field from a pyramidal horn of height H and width W. Assume H_{10} mode excitation, so that the aperture field may be taken as $E_x = E_0 \cos (\pi y/W)$.

(*b*) Find the beamwidth between nulls in the xz and yz planes. Compare with that for a large broadside array.

Answer

xz plane $\qquad BW = \dfrac{2\lambda_0}{H}$ radians

yz plane $\qquad BW = \dfrac{3\lambda_0}{W}$ radians

11.14. Show that for a rectangular aperture the expression for u_p in Prob. 11.13 may be thought of as a double Fourier transform of $u(x',y') = f(x')g(y')$ (assumes a separable aperture distribution).

HINT: Introduce new variables $k_0 \sin \theta \cos \phi = \xi$, $k_0 \sin \theta \sin \phi = \eta$, and assume that $\theta \ll 1$ for the main lobe so that $1 + \cos \theta \approx 2$. Note that $u = 0$ outside a certain range, but this does not invalidate the transform relationships.

11.15. A horizontal dipole is located at a height h_1 above a perfectly conducting ground, as illustrated. (Dipole is normal to plane of propagation.) Using the image theory stated in Prob. 11.4, show that the magnitude of the field at P equals the free-space field of the dipole alone times a factor $2 \sin (k_0 h_1 h_2/r)$, if h_1 and h_2 are much less than r.

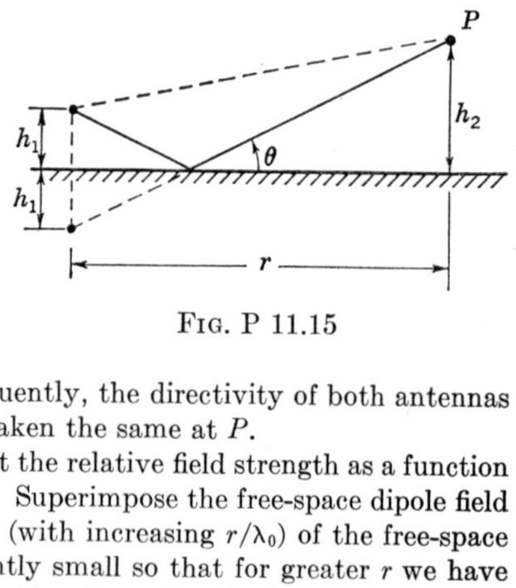

FIG. P 11.15

HINT: Find the difference in path length from antenna and image to the field point to first-order approximation. Note also that since $r \gg h_1$, $r \gg h_2$, both antenna and image may be assumed radiating along the same angle (θ) to the field point P. Consequently, the directivity of both antennas and the polarization of both signals may be taken the same at P.

11.16. For the conditions in Prob. 11.15 plot the relative field strength as a function of r/λ_0 for $h_1 = 4\lambda_0$, $h_2 = 50\lambda_0$, $0 < \theta < 15°$. Superimpose the free-space dipole field on this plot. Note that the last intersection (with increasing r/λ_0) of the free-space field and the actual field occurs for θ sufficiently small so that for greater r we have approximately that $h_2 = r \tan \theta \approx r\theta$. Show, therefore, that in this region the field strength falls off as $1/r^2$ rather than $1/r$. What is the value of r at the "last intersection"? This is a measure of the effective system range in view of the rapid $[O(1/r^2)]$ decrease of field strength for increasing r.

11.17. For practical earth conditions the image technique illustrated by Prob. 11.15 may be a poor approximation. With reference to Fig. P 11.15, the field at the receiving antenna may be considered as arising from a direct ray from the transmitting dipole and a reflected wave from the ground which follows a path determined by ray optics. If we neglect the effect of directivity on the direct and reflected signals, then the relative magnitude of the two signals depends on their relative phase of arrival and on the ground-reflection coefficient; that is, assuming $h_1 \ll r$, $h_2 \ll r$, let the direct distance be r_1 and the ground-reflected path be r_2. The received signal is

$$E_{rec} = E_0|1 + \zeta e^{jk_0(r_2 - r_1)}|$$

and ζ is the ground-reflection coefficient, while E_0 is the free-space field magnitude. (Note that the assumed conditions allow for the approximation that the electric field from the direct and the reflected paths is in the same direction.)

(*a*) Calculate ζ for earth constants $\sigma = 10^{-5}$, $\kappa = 6$, by utilizing (10.42*a*) and replacing ϵ_0 by complex $\epsilon = \kappa\epsilon_0 - j\sigma/\omega$. Plot ζ as a function of θ for $0 < \theta < 90°$. The frequency is 1.0 megacycle per second.

(*b*) For the geometry of Prob. 11.15, repeat, using the more accurate representation in terms of earth conditions and with the specific ζ in (*a*).

11.18. Repeat Prob. 11.17, but assume parallel polarization. For part *b* the antenna may be visualized as being a vertical dipole. The reflection coefficient may be obtained from (10.44).

11.19. Calculate the absorption cross section of an elementary dipole of length l and a half-wave antenna, both under conditions for maximum absorption of power.

11.20. A half-wave receiving antenna provides the input to a receiver at a frequency of 500 kilocycles and with a receiver bandwidth of 10 kilocycles. The receiver has adequate gain; the limitation on its ability to detect a usable signal arises mainly from the presence of noise in the input. The input noise power is given by $P_n = kTB$, and it accompanies the input signal power under matched conditions. In this formula k is Boltzmann's constant (1.38×10^{-23}), T is the "antenna temperature" and may be taken as ambient (290°K), and B is the bandwidth in cycles per second.

It is desired that the output signal-to-noise power ratio equal 20 decibels, and in view of the additional noise introduced within the receiver itself (noise figure of 6 decibels), an input signal-to-noise power ratio of 26 decibels is required. Assuming antenna orientation for maximum signal, what maximum field strength is required? If the transmitter is 200 miles away and the transmitting antenna has a gain of 5 in the direction of the receiver, what transmitting power is required? (Assume free-space conditions. Neglect the effect of the ground, and neglect antenna losses.)

11.21. A communication system is to be designed which will allow for reception of signals from a satellite. A description of its parameters follows:

Satellite transmitter

1. Power radiated, 100 watts effective
2. Antenna pattern omnidirectional (this is necessitated by probable inability to ensure proper orientation of satellite such as would be necessary with a high-gain antenna)

Receiving system

1. Paraboloidal antenna of diameter, 80 feet (the receiving cross section may be approximated as 0.5 times the aperture area)
2. Output signal-to-noise ratio, 10 decibels
3. Noise figure of receiver, 2 decibels (this means that an input signal-to-noise ratio of 12 decibels is required)
4. Losses from antenna to receiver (polarization, transmission lines, etc.), 6 decibels
5. Bandwidth, 10 cycles per second
6. Noise temperature, 290°K [under an assumed matched condition the input noise power P_n is given by $P_n = kTB = 4 \times 10^{-21}B$ (see Prob. 11.20)]
7. Frequency, 400 megacycles per second

(*a*) What is the range of the above system?

(*b*) How does the range depend on the receiving-antenna diameter? What is the range if the diameter is 600 feet?

(*c*) How does range depend on frequency? What is the range for $f = 1{,}000$ megacycles per second. (Note that at higher frequencies satisfactory operation of equipment becomes more difficult.)

11.22. The relation (11.78) may be derived in an alternative way to that used in the text. In (11.67) let the surface S be the sphere at infinity, and a surface enclosing the receiving antenna similar to that in Fig. 11.16. Now note that (11.67) becomes

$$\int_V \mathbf{E}_2 \cdot \mathbf{J}_1 \, dV = \int_{S_1} (\mathbf{E}_1 \times \mathbf{H}_2 - \mathbf{E}_2 \times \mathbf{H}_1) \cdot d\mathbf{S}$$

since the source $\mathbf{J}_2$ is not present in the volume under consideration. When the antenna is used as a transmitting antenna (source $\mathbf{J}_2$ inside $S_1 + S_{10}$), then the field $\mathbf{E}_2$, $\mathbf{H}_2$ in the coaxial line is a TEM wave, $H_\phi = I_0/2\pi r$, $E_r = Z_0 H_\phi$. The current element $I_e \, \Delta l$ also sets up a TEM wave ($\mathbf{E}_1$, $\mathbf{H}_1$) in the coaxial line, given by $H_\phi = -I_r/2\pi r$, $E_r = -Z_0 H_\phi = Z_0 I_r/2\pi r$, where I_r is the received current. The change in sign for H_ϕ arises because of the difference in the direction of propagation of the two TEM waves. Show that (11.67) gives

$$E_t I_e \, \Delta l \cos \alpha = 2 Z_c I_0 I_r = 2 I_0 V_r$$

in place of (11.70). In place of (11.74) show that the appropriate relation to use now is $P_t = \frac{1}{2} V_0 I_0 = \frac{1}{2} I_0^2 Z_c = \frac{1}{2} I_0^2 R_{11}$ for $Z_c = R_{11}$. Using these relations together with (11.71) to (11.73) and (11.75), show that (11.78) follows.

Chapter 12

12.1. Consider two infinite parallel plates with a separation d and with the right-side plate at a potential $-V$ relative to the left-side plate. An electron enters the region between the two plates through a small hole in the left-side plate. What initial velocity must the electron have in order to just reach the plate at potential $-V$? Determine the electron's position as a function of time.

12.2. Consider an electron approaching a stationary point charge Q. At infinity the electron has an initial velocity v_0. In the absence of the positive charge the electron would pass within a distance p of the position of the positive charge. Show that the positive charge causes the electron to follow a hyperbolic orbit.

12.3. Show that circles and ellipses are stable orbits for an electron rotating about a positive charge Q. What restriction, if any, is there on the electron's velocity?

12.4. A magnetron consists of a cylindrical cathode of radius a and an anode of radius b at a potential V relative to the cathode. A uniform magnetic field B is applied along the cylindrical axis. Determine the resultant orbit followed by an electron leaving the cathode with zero initial velocity. For what value of B will the electron just reach the anode?

12.5. Consider a region of space in which uniform fields $\mathbf{a}_z B$ and $E_y \mathbf{a}_y + E_z \mathbf{a}_z$ exist. Express the velocity of the electron as a component along $B\mathbf{a}_z$ plus a component perpendicular to the z axis (a component in xy plane); thus $\mathbf{v} = v_z \mathbf{a}_z + \mathbf{v}_0$, where $\mathbf{a}_z \cdot \mathbf{v}_0 = 0$. Next let $\mathbf{v}_0 = (\mathbf{E} \times \mathbf{B})/B^2 + \mathbf{v}_1$, and show that from the force equation the resultant equation for $\mathbf{v}_1$ is just the equation describing the motion of an electron in a uniform magnetic field. The latter is simply a gyration about the magnetic lines of force. The component $(\mathbf{E} \times \mathbf{B})/B^2$ of the velocity is called the velocity of the guiding center; i.e., it is the average drift velocity of the electron across the magnetic lines of force. Superimposed on this drift or guiding-center velocity is the electron gyration plus the velocity along **B**. The concept of the guiding center is useful since it reduces the problem to one of solving for the motion that takes place in an **E** field or **B** field alone.

12.6. N electrons (mass m, charge $-e$) and N ions (mass M, charge e) per unit volume are entering a region of space where uniform fields $B\mathbf{a}_y$ and $E\mathbf{a}_x$ exist. The initial velocity of both electrons and ions is zero. Show that the resultant motion of

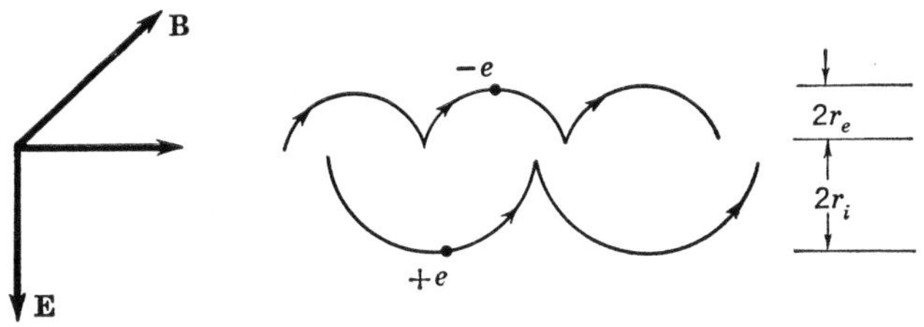

FIG. P 12.6

the charges leads to an average charge separation with a resultant electric dipole polarization

$$\mathbf{P} = \frac{N(m + M)}{B^2}\mathbf{E}$$

being established.

HINT: Show first that the orbits followed are cycloids with different radii for the two types of particles, as illustrated. The dipole moment per particle pair is then $e(r_e + r_i)$.

12.7. Show that the magnetic moment of a charged particle gyrating in a uniform magnetic field **B** with a velocity **v** is

$$\mathbf{M} = -\tfrac{1}{2}mv^2\frac{\mathbf{B}}{B^2}$$

where m is the mass of the particle.

12.8. Consider uniform fields $B_1\mathbf{a}_y$ for $z > 0$ and $B_2\mathbf{a}_y$ for $z < 0$. A particle of mass m and charge q approaches the boundary plane $z = 0$ with a velocity **v** at an angle θ relative to the x axis; that is, $\mathbf{v} = v\mathbf{a}_x \cos\theta + v\mathbf{a}_z \sin\theta$. Show that the resultant motion of the particle is such as to cause it to creep along the boundary with an average velocity

$$v_c = \frac{2v(B_1 - B_2)\sin\theta}{(2\pi - 2\theta)B_2 + 2\theta B_1}$$

in the x direction.

12.9. A stream of electrons leaves the $x = 0$ plane with an initial velocity v_1. At $x = 0$ the potential Φ and $d\Phi/dx$ equal zero. For $x > 0$, steady-state conditions exist so that the current density $J = -\rho_1 v_1 = -\rho v$ is constant, where $-\rho_1$ is the charge density at $x = 0$, and $-\rho$, v are the charge density and velocity at x. By using the energy-conservation equation $e\Phi = mv^2/2$, Poisson's equation, and the relation $E = -d\Phi/dx$, show that $E^2 = (2m/e\epsilon_0)J(v - v_1)$. From the energy-conservation equation show that $mv\,dv/dx = -eE$, and hence show that

$$x^2 = \frac{2\epsilon_0 m v_1^3}{9eJ}\frac{v - v_1}{v_1}\left(\frac{v + 2v_1}{v_1}\right)^2$$

If when $x = d$, $\Phi = V_0$, find the corresponding value of v, and show that the current density J is given by

$$J = \frac{2\epsilon_0 m}{9ed^2}\left(\frac{2eV_c}{m} + v_1^2\right)^{1/2}\left[\left(\frac{2eV_0}{m} + v_1^2\right)^{1/2} + 2v_1\right]^2$$

Compare with Eq. (12.65) when $v_1 = 0$.

12.10. A planar diode has a cathode and anode area of 2 square centimeters each. The cathode-anode separation is 2 millimeters. Assuming space-charge-limited operation and a cathode-anode potential of 100 volts, find the total steady-state current that will flow.

12.11. For the diode described in Prob. 12.10 an a-c voltage 10 cos ωt with $\omega = 2\pi f$ and $f = 10^8$ cycles per second is applied between cathode and anode in addition to the d-c voltage of 100 volts. Find the a-c current that will flow.

12.12. An electron beam with velocity v_0, charge density $-\rho_0$ completely fills a square guide with sides of width a. A large axial magnetic field is applied. Obtain solutions for space-charge waves corresponding to a TM_{11} mode in this beam-filled guide.

12.13. Derive Eq. (12.111) by replacing the term $\gamma_m v_0$ by $\Gamma_m v_0$ in (12.110) and noting that $\Gamma_m^2 = q_m^2 - k_0^2$ and that $\gamma_m^2 + k_0^2$ may be replaced by q_m^2 (approximately) in the small term involving ω_0^2.

12.14. A rectangular coordinate model of a traveling-wave tube that is similar in behavior to the cylindrical helix model consists of two infinitely long and infinitely wide parallel conducting plates. In the plates a large number of closely spaced narrow slits are cut at an angle ψ with respect to the y axis, as illustrated. Current must thus flow along the slits, and the conductivity $\sigma_{\parallel}$ parallel to the slits may be considered infinite, while that normal to the slits, that is, $\sigma_{\perp}$, may be taken as zero.

(*a*) For the above structure find the solution for the lowest-order mode of propagation in the z direction. Note that both TM and TE modes are required to satisfy the boundary conditions. Assume that there is no variation with y.

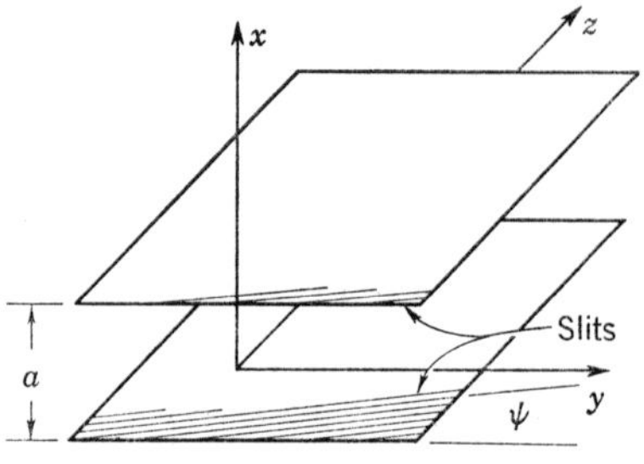

FIG. P 12.14

(*b*) When an electron beam is present (charge density $-\rho_0$, d-c velocity v_0) and a large d-c axial magnetic field is applied, find the corresponding solution. Obtain expressions for the propagation constants of the waves that may exist (use appropriate approximations). What is the power gain per meter for the exponentially growing wave?

(*c*) For part *a* evaluate the phase velocity and propagation constant when $\lambda_0 =$ 3.14 centimeters, $a = 1$ centimeter, $\psi = 5°$.

(*d*) For part *b* evaluate the power gain in decibels per meter for parameters as in part *c*. In addition, take v_0 equal to the phase velocity found in part *c*. Assume that the d-c beam current is 10 milliamperes per square centimeter in order to evaluate the electron plasma frequency ω_0.

HINT: Assume that between the plates E_z is of the form $e^{-\gamma z}\cosh lx$ while H_z is of the form $e^{-\gamma z}\sinh lx$.

12.15. By adding and subtracting positive- and negative-circular-polarized waves, show that in general the relation between the electric field **E** and the displacement flux **D** in an ionized medium with electron particle density N can be represented in matrix form as

$$\begin{bmatrix} D_x \\ D_y \\ D_z \end{bmatrix} = \begin{bmatrix} \epsilon_1 & -j\epsilon_2 & 0 \\ j\epsilon_2 & \epsilon_1 & 0 \\ 0 & 0 & \epsilon_3 \end{bmatrix} \begin{bmatrix} E_x \\ E_y \\ E_z \end{bmatrix}$$

where

$$\epsilon_1 = \epsilon_0 - \frac{Ne^2}{w\omega^2} \frac{\omega^2}{\omega^2 - \omega_g^2}$$

$$\epsilon_2 = \frac{Ne^2}{w\omega^2} \frac{\omega\omega_g}{\omega^2 - \omega_g^2}$$

$$\epsilon_3 = \epsilon_0 - \frac{Ne^2}{w\omega^2}$$

and the d-c magnetic field is applied along the z axis.

12.16. The region $z \geq 0$ is occupied by an ionized gas with a d-c magnetic field B_0 applied in the z direction. A TEM wave with field components $E_x = E_0 e^{-jk_0 z}$, $H_y = Y_0 E_0 e^{-jk_0 z}$ is incident from $z < 0$. Find the field transmitted into the ionized medium and also the reflected field. What is the polarization of the reflected field?

HINT: Decompose the incident field into positive and negative-circular-polarized fields at $z = 0$, and treat each polarization separately.

12.17. Repeat Prob. 12.16 when the medium for $z > 0$ is a ferrite with a z-directed d-c magnetic field B_0 applied.

12.18. Consider an initially neutral plasma (equal number of electrons and ions randomly distributed). If all the electrons are displaced by an amount u in the x direction where the amount of the displacement u varies with position x, show that a charge density

$$\rho = -Ne \frac{\partial u}{\partial x}$$

is produced. This charge density gives rise to an electric field in the x direction. By using the divergence equation $\nabla \cdot \mathbf{E} = \rho/\epsilon_0$, show that $E_x = Neu/\epsilon_0$. This electric field will tend to restore the electrons to their original position, and since the restoring force is proportional to the displacement, simple harmonic oscillations (electron plasma oscillations) occur. Assuming that the heavy ions do not move, show that the equation of motion for the electrons is

$$w \frac{d^2u}{dt^2} + \frac{Ne^2}{\epsilon_0} u = 0$$

and that the electron-plasma-oscillation frequency is given by $\omega_0^2 = Ne^2/w\epsilon_0$, where w is the electron mass.

12.19. Show that for finite conductivity σ, Alfvén waves are attenuated with an attenuation constant which, for small attenuation, is given approximately by $\omega^2/2\sigma\mu_0 u^3$, where the phase velocity u is given by (12.220).

12.20. Consider a volume element of incompressible fluid having a velocity v_x in the x direction only, as illustrated. If v_x is a function of y and z, shearing forces exist on the four faces 1, 2, 3, 4 of the volume element because the velocities of the adjacent fluid layers are slightly different. These forces are proportional to the area and the velocity gradient normal to the surface. Hence on face 4 the force in the x direction due to viscosity is

$$-\eta m_v \frac{\partial v_x}{\partial y} \Delta x \, \Delta z$$

where ηm_v is the coefficient of viscosity (note that η is the coefficient of "kinematic" viscosity). Show that the force on face 2 is given by

$$\eta m_v \frac{\partial v_x}{\partial y} \Delta x \, \Delta z + \eta m_v \frac{\partial^2 v_x}{\partial y^2} \Delta y \, \Delta x \, \Delta z$$

and that the net force on the four faces is given by

$$\eta m_v \left(\frac{\partial^2 v_x}{\partial y^2} + \frac{\partial^2 v_x}{\partial z^2} \right) \Delta x \, \Delta y \, \Delta z$$

Since the fluid is incompressible, $\nabla \cdot \mathbf{v} = 0$; so $\partial v_x/\partial x = 0$, and hence a term $\partial^2 v_x/\partial x^2$

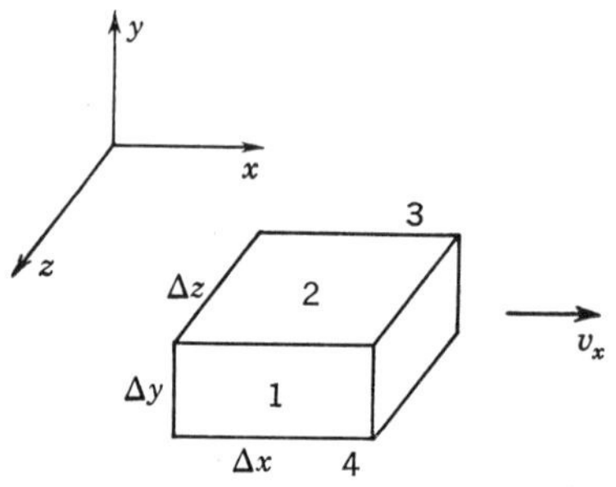

FIG. P 12.20

may be added so that the force per unit volume may be expressed as $\eta m_v \nabla^2 v_x$. In the general case it is clear that the viscous force for an incompressible fluid may then be expressed by $\eta m_v \nabla^2 \mathbf{v}$.

GENERAL BIBLIOGRAPHY

Introductory electromagnetic theory

Fano, R. M., L. J. Chu, and R. B. Adler: "Electromagnetic Fields, Energy, and Forces," John Wiley & Sons, Inc., New York, 1960.

Hayt, W. H., Jr.: "Engineering Electromagnetics," McGraw-Hill Book Company, Inc., New York, 1958.

Krauss, J. D.: "Electromagnetics," McGraw-Hill Book Company, Inc., New York, 1953.

Ramo, S., and J. R. Whinnery: "Fields and Waves in Modern Radio," 2d ed., John Wiley & Sons, Inc., New York, 1953.

Reitz, J. R., and F. J. Milford: "Foundations of Electromagnetic Theory," Addison-Wesley Publishing Company, Reading, Mass., 1960.

Rogers, W. E.: "Introduction to Electric Fields," McGraw-Hill Book Company, Inc., New York, 1954.

Seely, S.: "Introduction to Electromagnetic Fields," McGraw-Hill Book Company, Inc., New York, 1958.

Skilling, H. H.: "Fundamentals of Electric Waves," 2d ed., John Wiley & Sons, Inc., New York, 1948.

Toraldo di Francia, G.: "Electromagnetic Waves," Interscience Publishers, Inc., New York, 1956.

Advanced electromagnetic theory

Abraham, M., and R. Becker: "The Classical Theory of Electricity and Magnetism," 2d ed., Blackie & Son, Ltd., Glasgow, 1950.

Landau, L., and E. Lifshitz: "The Classical Theory of Fields," Addison-Wesley Publishing Company, Reading, Mass., 1951.

Panofsky, W. K. H., and M. Phillips: "Classical Electricity and Magnetism," Addison-Wesley Publishing Company, Reading, Mass., 1955.

Schelkunoff, S. A.: "Electromagnetic Waves," D. Van Nostrand Company, Inc., Princeton, N.J., 1943.

Smythe, W. R.: "Static and Dynamic Electricity," 2d ed., McGraw-Hill Book Company, Inc., New York, 1950.

Stratton, J. A.: "Electromagnetic Theory," McGraw-Hill Book Company, Inc., New York, 1941.

Electromagnetic property of materials

Dekker, A. J.: "Solid State Physics," Prentice-Hall, Inc., Englewood Cliffs, N.J., 1957.

———: "Electrical Engineering Materials," Prentice-Hall, Inc., Englewood Cliffs, N.J., 1959.

Van der Ziel, A.: "Solid State Physical Electronics," Prentice-Hall, Inc., Englewood Cliffs, N.J., 1957.

Von Hippel, A. R.: "Dielectrics and Waves," John Wiley & Sons, Inc., New York, 1954.

———: "Molecular Science and Molecular Engineering," John Wiley & Sons, Inc., New York, 1959.

INDEX

Absorption cross section, 423, 538
Action at a distance, 43, 44
Admittance, electron-beam, 459, 460, 462, 463
  gap, 458, 459
  intrinsic, 308
  triode, 446, 448
  wave, 346, 348
Alfvén waves, 489
Ampère, A.-M., 198
Ampère's circuital law, 218–223
Ampère's law of force, 199–202
Analytic functions, 146–148
Angle, Brewster, 355
  critical, 530
  of incidence, 351
  precession, 483
  of refraction, 352
Angular moment vector, 482
Anisotropic dielectric, 87
Antenna, aperture, 413, 414, 535, 536
    continuous, 412–414
  dipole (*see* Dipole antenna)
  directivity, 397, 398
  effective area, 423
  equivalent circuits, 417, 418
  gain, 398
  half-wave, 398–401
  horn, 413, 414, 535, 536
  loop, 533
  network formulation, 413–418
  parabolic, 413, 414
  pattern, 397, 398
  receiving, 419–423
Antiferromagnetic, 258
Aperture antennas, 412–414, 535, 536
Aristotle, 198
Array factor, 403, 404
Arrays, 401ff.
  beamwidth, 409
  binomial, 535
  broadside, 407, 409
  end-fire, 407, 409
  linear, 407–409
Arrays, representation by polynomials, 534, 535
  two-dimensional, 409–412
  two-element, 402–407
  uniform, universal pattern for, 408
D'Arsonval movement, 215
Associated Legendre polynomial, 140, 141
Atomic currents, 226
Attenuation constant, 358–361, 376, 379, 380

B vector (*see* Magnetic field)
B-H curve, 233–235
Barkhausen effect, 263
Beamwidth, antenna, 409
Bei function, 527
Benham-Müller-Llewellyn equation, 439
Ber function, 527
Bessel functions, 133–136
  asymptotic forms, 134, 135
  differentiation, 135
  imaginary argument, 134, 135
  modified, 134, 135
  orthogonality, 135–137
  recurrence formulas for, 135
  roots, 382, 385
Binomial array, 535
Biot-Savart law, 202
Bohr magneton, 258, 482
Boltzmann distribution law, 82, 83
Boothroyd, A. R., 197
Boundary conditions, in electrostatics, 88–93
  in magnetostatics, 235–240
  at perfect conductor, 55–57, 319–321
  for steady currents, 177–179
  for TE waves, 373
  in time-varying fields, 317–321
  for TM waves, 379
Boundary-value problems, 119ff.
Bowen, A. E., 444
Brewster angle, 355
Buncher grids, 456–459
Bunching, electron stream, 457, 465

Capacitance, coefficients, 97
of concentric spheres, 95, 96
definition, 94, 95, 106
fringing, 160–162, 508
in parallel-plate diode, 160–163
for general circuit, 336, 339
of parallel plate, 95
stray, 331
Capacitor, characteristics, 332, 333
energy in, 104, 105
force on, 109–111
lossy dielectric, 179–183
parallel-plate (*see* Parallel-plate capacitor)
Catcher grids, 457–459
Cathode-ray tube, electrostatic deflection, 427–429
magnetostatic deflection, 433, 434
Cauchy-Riemann equations, 148
Cavity resonator, 385–392
coupling to, 391, 392
Q of, 389–391
as resonant transmission line, 386, 387
Characteristic impedance, 363, 365
Charge density, at dielectric interface, 183
equivalent, 75, 85, 86, 245
line, 46
related to **E** at conductor, 56
surface, 46
volume, 45
Charged particle, in electric field, 427–431
in electric and magnetic field, 434–437
in magnetic field, 431–434
motion (*see* Motion of charged particle)
Charged surface, field behavior at, 46, 47, 55–57, 76–78
Cherry, E. C., 197
Child-Langmuir equation, 441
Circuit analysis, field analysis, 326–341, 364–367
Circuital law, 218–223
Circuits, equivalent (*see* Equivalent circuits)
high-frequency, 334–336
Circular-loop antenna, 533
Circular polarization, 474, 475
Circular waveguide, 380–385
cutoff for, 383, 384
table of properties, 384
Circulation integral, 15
Clausius-Mossotti equation, 116
Coaxial cylinders, electrostatic field, 57, 58
inductance, 276–278
magnetic field, 221, 222
Coefficient, of capacitance, 97
of coupling, 283
of induction, 97
of potential, 97
Coenergy, 293, 294
Coercivity, 235, 251
Commutative law, of addition, 2
of multiplication, 7
Complex conjugate, 147
Complex dielectric constant, 311, 312, 366, 367
Complex function theory, 146–148
Complex permittivity, 311, 312, 366, 367
Complex phasor notation, 310
Complex Poynting vector, 315–317
Conductance, 166
Conduction current, 164–166
Conductivity, 164, 166
table, 493
Conductors, 55, 164, 171
boundary conditions at, 55–57, 319–321
power loss at, 359–361
Conformal mapping, 146ff.
Conservation of charge, 169, 170
Continuity equation, 169, 170
for a fluid, 487
and Lorentz condition, 322–324, 395
Convection current, 188, 189
Coordinate system, 23
Coordinate transformation, 149–152
Coordinates, orthogonal curvilinear, 23–26, 149–151
Cosine integral, 400
Coulomb's law, 40, 41
Coupling, to cavity, 391, 392
coefficient of, 283
probe, 391, 392
Critical angle, 530
Cross product, 5
Curie constant, 261, 263
Curie temperature, 263
Curie-Weiss law, 263
Curl, 16
in orthogonal curvilinear coordinates, 25, 26
Current, conduction, 164–166
density, 164, 165
convection, 188, 189
displacement (*see* Displacement current)
distribution, in plane conductor, 319–321
in round wire, 527
due to equivalent magnetic dipole, 228–231
flow at conducting boundary, 177–179
magnetic field, 203

Current loop, 208–211, 533
  energy, 280–283
Curvilinear-squares method, 189–193
Cutoff, for circular guide, 383, 384
  for rectangular guide, 371
  of TE waves, 374, 380
  of TM waves, 380
Cyclotron frequency (gyrofrequency), 435, 476
Cylindrical coordinates, 130, 131

D vector, 85–91
Damping constant, 389
Deflection, electron beam (*see* Electron beam)
Del operator, 12
Demagnetization factor, 520
Density modulation, 465
Depth of penetration, 314, 315, 319
Diamagnetic materials, 227, 233, 258–260
Diamagnetic susceptibility, 260
Dielectric, anisotropic, 87
Dielectric boundaries, 88–93, 317, 318
Dielectric coated conductor, 530, 531
Dielectric constant, 42, 87
  complex, 311, 312, 366, 367
  table, 89
  tensor, 541
Dielectric cylinder in applied field, 132, 133
Dielectric losses, conduction, 179–183
  polarization damping, 312, 316
Dielectric sphere in applied field, 142, 143
Dielectric strength, 87, 93, 94
  table, 87
Diffusion equation, 314
Diode, a-c signal, 441–444
  d-c conditions, 439–441
  field distribution, 440, 441
  load impedance, 442–444
Dipole antenna, half-wave, 398–400
  infinitesimal, 393–398
  magnetic, 533
Dipole electrostatic field, 74
Dipole moment, definition, 72
  per unit volume, electric, 74–76, 83–85
    magnetic, 227, 228
Dipole potential, 28, 29, 73
Directional derivative, 12
Dirichlet boundary condition, 119
Discontinuity, at charged surface, 54, 76–78
  current sheet, 239, 240
  dipole layer, 518
  volume, 74–76
Disk dynamo, 272
Dispersion, 377, 378
Displacement, electric, 85–91
Displacement current, 300
  in conductors, 314
  in parallel-plate capacitor, 302, 303
Distributed constant circuits, 363–370
Distributive law of multiplication, 6, 7
Divergence, 13–15
  in orthogonal curvilinear coordinates, 24, 25
Divergence theorem, 21
Domain wall, 262, 263
Dominant mode, 374, 385
Dot product, 6
Drift space, 457
Drift velocity, 165
Duality, electric and magnetic circuits, 256, 257
  electromagnetic, 347
  between **J** and **D**, 183–186

E field (*see* Electric field)
E waves (*see* Transverse magnetic waves)
Effective area, 423, 538
Eigenfunction, 123, 373, 383
Eigenvalue, 123, 373, 382
Electric charge, 40, 41
Electric dipole, 72–74
Electric dipole moment per unit volume, 83–87
Electric displacement, 85–91
Electric field, 44–46
  of charged plane, 46, 47, 55–57, 76–78
  of charged sphere, 49, 50
  between coaxial cylinders, 57, 58
  at conductor, 55–57
  due to motion through magnetic field, 268–270
  due to time-varying magnetic field, 266
  energy in, 101–104, 425–427
  induced, 366–371
  of line charge, 48, 49
    in conducting wedge, 510, 511
  molecular, 115–118
  nonconservative, 167–169, 268–271
  of point charge, 45
Electric flux, 47–49
Electric flux density, 85–91
Electric permittivity, 42
Electric stress tensor, 114
Electric susceptibility, 85
Electrolytic tank, 192–197
  double-sheet, 194–197
Electromagnetic duality, 347
Electromagnetic induction, 266–271
Electromagnetic lens, 413

Electromotive force (emf), 167–169
  motional, 270–272
Electron beam, admittance, 459, 460, 462, 463
  deflection, in electric field, 427–429
    in electric and magnetic fields, 436, 437
    in magnetic field, 433, 434
Electron optics, 430, 431
Electronic polarizability, 79–81
Electrostatic energy, 101–104
Electrostatic force, 106–109
Electrostatic lens, 430, 431
Electrostatic shielding, 100, 101
Electrostatics, boundary conditions in, 88, 93
  reciprocity in, 97, 98
  virtual work in, 106–109
Elliptic integrals, 280
Elliptical cylinders, 153–155
Emf, 167–169, 270–272
Energy, in capacitor, 104, 105
  of current loops, 280–283
  in electric fields, 101–104, 425–427
  of inductor, 285, 286
  in magnetic fields, 281–286
  in plane wave, 308–310
  in resonator, 388–390
Energy density, electric, 104
  joule loss, 187
  magnetic, 286
Equipotential surface, 9
Equivalent charge, due to electric polarization, 75, 85, 86
  due to magnetic polarization, 245
Equivalent circuits, antenna, 417, 418
  general, 336–341
  transmission line, 363–370
  wave propagation, 353–356
Equivalent currents, magnetization, 228–231
Evanescent waves, 374
External inductance, 278

Far zone, 396, 402, 403
Faraday, M., 40, 266
Faraday disk dynamo, 272
Faraday rotation, 480–482
Faraday's law, 266, 267
  in differential form, 267
Fermat's law, 430
Ferrite media, effective permeability, 484–486
  propagation in, 482–486
Ferroelectrics, 85
Ferromagnetic domains, 262, 263
Ferromagnetism, 228, 261–264
Field, local, 396
  molecular, 115–118
  nonconservative, 167–169
  polarizing, 115–118
  solenoidal, 19, 23, 30, 31
  (*See also* Electric field; Magnetic field)
Field concept, 44, 102, 103, 222, 223
Field point–source point nomenclature, 26, 27
Field theory–circuit theory relation, 326–341, 364–367
Flow lines, 10, 11
Flux, electric, 47–49
  magnetic, 216–218
Flux concept, 11, 14, 15
Flux linkage, 266, 268
  inductance from, 274–276
Flux plotting, 189–192
Flux tubes, 191
Focusing, magnetic, 431–433
Force, electromotive, 167–169
  electrostatic, 106–109
    between charges, 43
    on conductor, 106–111
    density, 112, 113
    surface, 112–115
    at surface of dielectric, 111
    from virtual work, 106–109
  graphical solution for, 43
  magnetostatic, 286–289
    between current elements, 200, 295–297
    between current loops, 199, 201
    on currents, 201
    of magnets, 290–294
    on moving charge, 202
    between two infinite line currents, 207, 208, 289
    from virtual work, 286–289
Fourier series, 125, 126
  Bessel series, 135–137
  Legendre series, 141
Fourier transforms in antenna synthesis, 536
Fresnel equations, parallel polarization, 355
  perpendicular polarization, 353
Fringing capacitance, 160–163, 508

Gain, 398
Gap admittance, 458, 459
Gap coefficients, 460
Gauge transformation, 225, 529
Gauss' flux theorem, 47, 48
Gauss' law, 21

Gaussian surface, 49
Gilbert, W., 198
Gradient, 11–13
  in orthogonal curvilinear coordinates, 24
Graphical flux plotting, 189–192
Graphical solution for force, 43
Green's reciprocation theorem, 505, 506
Green's theorem, 21, 22
Grounded cathode, 448
Grounded sphere, 68, 69
Group velocity, 371, 377, 378
Guard ring, 95
Guiding center, 539
Gyrofrequency, 435, 476
Gyromagnetic ratio, 482
Gyrotropic media, 473–486

H vector, 231–233
H waves (*see* Transverse electric waves)
Hahn, W. C., 449
Half-wave dipole, 398–401
Half-wave transmission line, 370
Harmonic function, 151
Helix, 469, 470
Helmholtz's equation, in dielectric media, 314
  in free space, 312
  inhomogeneous, 321, 529
Helmholtz's theorem, 29–36
Henry, J., 266
Hertzian potential, 527, 528
Horns, electromagnetic, 413, 414, 535, 536
Huygens' principle, 413, 535
Hysteresis, 234, 235, 262
Hysteresis loop, 234, 235
  minor, 235, 252
Hysteresis loss, 297, 298

Identities, vector, 20, 38
Images, electromagnetic, 533
  electrostatic, 63–72
    in cylinder, 69–72
    in intersecting planes, 67
    in plane, 64–66
    in semi-infinite dielectric, 507
    in sphere, 67–69
Impedance, characteristic, 363, 365
  of diode, 442–444
  intrinsic, 308, 358
  surface, 356–358, 360, 361
  of transmission line, 370
  wave, TE, 346
    TM, 348
Impedance concept, 353–356
Impedance function with double-sheet tank, 197
Incidence angle, 351
Incremental permeability, 235
Index of refraction, 314
Induced electric field, 266–271
Induced voltage, 266–269, 271, 272
Inductance, 273ff.
  of circular loop, 279, 280
  of coaxial cable, 276–279
  from energy, 285, 286
  external, 278
  from flux linkage, 274
  for general circuit, 336, 339
  internal, 278
  of long wire, 278, 279
  mutual, 274–276
  Neumann formula for, 275
  self-, 274, 278–280
Induction, coefficient of, 97
  electromagnetic, 266–271
Inductor, characteristics, 329–332
  energy, 285, 286
Insulators, 171
Interaction fields, electrostatic, 115–118
  magnetostatic, 261, 262
Internal inductance, 278
Intrinsic impedance, 308, 358
Inversion, in cylinder, 69–72
  in sphere, 67–69
Ionosphere, 475–479
  effective permittivity, 478
  electron density in, 479
Irrotational nature of electrostatic field, 52
Irrotational vector function, 16, 29, 30

Joule's law, 186–188

Kirchhoff's law, 169, 256, 257
Klystron, 456–463

Langevin, P., 83, 84
Laplace's equation, 59
  solutions, 60, 119ff.
    in cylindrical coordinates, 130–132
    in rectangular coordinates, 120–122
    in spherical coordinates, 138–141
    in two dimensions, 126–130
Laplacian, 19
  in generalized coordinates, 26
  of vector, 20
Least-action principle, 430

Legendre polynomials, 140, 141
Legendre's equation, 140
Lens, electromagnetic, 413
  electrostatic, 430, 431
Lenz's law, 273
Line integral, 15
Linear arrays, 407–409
Linear polarization, 351, 352, 421, 474
Lines of flow, 10, 11
Llewellyn, F. B., 444
Llewellyn-Peterson equations, 445–449, 457, 458
  table of coefficients for, 447
Local field, 396
Loop, steady current, 208–211
Loop antenna, 533
Lorentz condition, 322
Lorentz force, 202
Lorentz reciprocity theorem, 414
Loss tangent, 312
Losses, in cavities, 389
  at conducting surface, 359–361
  in transmission lines, 368, 369
  in waveguides, 376
Lossy dielectric, conduction in, 179–183
  polarization damping in, 312, 316
Lumped-parameter circuits, 326–328

Magnet, permanent, 242, 243, 247, 248
Magnetic circuits, 252–257
  iron-core transformer, 255, 256
Magnetic dipole, 211–215
  of arbitrary volume current, 214
  comparison with electric dipole, 212
  torque on, 214, 215
Magnetic dipole moment, 211
  atomic, 258
  per unit volume, 227, 228
Magnetic energy, 281–286
Magnetic field, 201–204
  of circular loop, 208–211
  in coaxial line, 221, 222
  of conducting ribbon, 206, 207
  of cut toroid, 248–252
  due to time-varying electric field, 300
  outside long solenoid, 241, 242
  outside long wire, 205, 206, 221
  of magnetic dipole, 212
  of permanent magnet, 242, 243, 247, 248
  of steady current distributions, 201–203
Magnetic field intensity, 231–233
Magnetic flux linkage, 266, 268
Magnetic focusing, 431–433
Magnetic force, 286–289
Magnetic induction, 266–271
Magnetic moment, spin, 258
Magnetic scalar potential, 243–248
Magnetic stress tensor, 294, 295
Magnetic susceptibility, 232
Magnetic vector potential, 204
Magnetohydrodynamics, 486–490
Magnetomotive force, 252, 254
Magnetostatic potential, 243–248
Magnetostatics, boundary conditions in, 235–240
  virtual work in, 286–289
Makar, R., 197
Mawardi, O. K., 294
Maxwell stress tensor, 114
Maxwell's equations, axial and transverse components, 344
  differential form, 301
  integral form, 301
  time-harmonic form, 311
Microscopic theory, for electrical properties, 79–85, 115–118
  for magnetic properties, 257–264
Modes, 373, 374
Modified Bessel functions, 134, 135
Molecular field, 115–118
Moment of sources, 29
Moment vector, angular, 482
Motion of charged particle, in electric field, 427–431
  in electric and magnetic field, 434–437
  in magnetic field, 431–434
  in traveling-wave magnetron, 437
Motional emf, 270–272
Multicapacitor system, 96–100, 105, 106
Multipole expansion, 503
Mutual inductance, from flux linkages, 274–276
  reciprocity for, 274–276
  from vector potential, 276

Near zone, 396
Net circulation integral, 15
Neumann boundary condition, 119
Neumann formula for inductance, 275
Newton's law, 426
Newton's third law, 200, 201
Nonconservative field, 167–169
Nonreciprocal device, 481, 482

Oblique incidence of wave, on conducting surface, 360, 361
  on dielectric interface, 351–356
Oersted, H. C., 198
Ohm's law, 164–166

Orientational polarizability, 81–85
Orthogonal curvilinear coordinates, 23–26, 149–151
Orthogonality property, of Bessel functions, 135–137
  general demonstration, 125
  of Legendre functions, 141
  of trigonometric functions, 124

Paraboloidal reflectors, 413, 414
Parallel-plane diode, conditions in, a-c, 441–444
    d-c, 439–441
Parallel-plane triode, 446–449
Parallel-plate capacitor, 94, 104, 105
  displacement current in, 302, 303
  fringing capacitance in, 160–163
Parallel polarized plane wave, 352, 355, 356
Parallel-wire line, electrostatic field, 71, 72
  magnetic force from steady currents in, 207, 208
Parallelogram law, 3
Paramagnetic materials, 227, 233, 260, 261
Paramagnetic susceptibility, 261
Partial-flux linkage, 277, 278, 285, 286
Particle motion (*see* Motion of charged particle)
Penetration depth, 314, 315, 319
Permanent magnets, 242, 243, 245–248
Permeability, 235
  of free space, 199, 200
  incremental, 235
  tensor, 486
Permittivity, complex, 311, 312, 366, 367
  of free space, 42
Perpendicular polarized wave, 351–355
Peterson, L. C., 445
Phase velocity, 314, 371
Planck's constant, 258
Plane, of incidence, 351
  of polarization, 421
Plane wave, arbitrary direction of propagation, 348–351
  energy relations for, 308–310
  in free space, 308, 309
  of sinusoidal form, 312–315
  superposition, 378, 379
Plasma, 473
Plasma-electron frequency, 452, 454
Point source, 26–29
Poisson equation, integration, 33–35
  scalar, 30, 59
Poisson equation, solution in rectangular parallelpiped, 508, 509
  vector, 31, 224, 225
Polarizability, 79ff.
  electronic, 79–81
  ionic, 81
  orientational, 81–84
Polarization, electric, 83–87
  magnetic, 227–228
  plane of, 421
  of wave, circular, 474, 475
    linear, 351, 352, 421, 474
Polarization charge, electric, 75, 76, 85, 86
  magnetic, 245
Polarization currents, 228–231
Polarizing field, 115–118
Potential, coefficient of, 97
  dipole, 28, 29, 73
  electrostatic, 50–53
    on axis of charged disk, 53, 54
    from charged disk, 144–146
    from charged ribbon, 155–158
    between elliptic cylinders, 152–155
    from line charge, 54
    scalar (*see* Scalar potential)
  magnetostatic, 243–248
  quadrupole, 503
  for time-varying fields, 322–325
    quasi-static, 325, 326, 328, 329
    retarded, 326
  vector (*see* Vector potential)
Power flow, 306, 307
Power loss in plane conductor, 359–361
Poynting vector, complex, 315–317
  instantaneous, 307
  time-average, 313, 316
Precession angle, 483
Precession frequency, 483
Probe coupling, 391, 392
Propagation of wave, in dielectric, 305
  in ferrite medium, 482–486
  in good conductor, 314, 315
  in gyrotropic medium, 473–486
Propagation constant, 343
  for circular waveguide, 383
  for rectangular waveguide, 374
  for traveling-wave tube, 471, 472
Proper function, 8, 17, 496

Q (quality factor), of circuit, 389
  of resonator, 390
Quadrupole potential, 503
Quarter-wave line, 370
Quarter-wave transformer, 530
Quasi-static potentials, 325, 326, 328, 329

Radiated power, from dipole, 395, 396
from half-wave antenna, 400
Radiation, from current element, 393–398
from half-wave dipole, 398–401
Radiation intensity, 397
Radiation resistance, 396, 400
Radiation zone, 396
Ramo, S., 449
Receiving antennas, 419–423
Reciprocity, for electromagnetic field, 414, 417, 419, 420, 538
in electrostatics, 97, 98
for mutual inductance, 274–276
Rectangular coordinates, 120–122
Rectangular resonator, 385–392
Rectangular waveguide, 370–380
cutoff for, 371
Reflection, from conducting plane, 356–361
at dielectric interface, elimination, 355
equivalent circuit for, 353–355
parallel polarization, 355, 356
perpendicular polarization, 351–355
Refraction, angle of, 352
of current flow lines, 178
electric flux lines, 91
magnetic flux lines, 238, 239
of wave at dielectric interface, 351–356
Relaxation time, 55, 170, 171
Reluctance, 252–257
Remanence, 234, 235
Resistance, 166, 177
of arbitrary circuit, 336, 339
of arbitrary shaped conductor, 171–175
radiation, 396, 400
of spherical section, 175–177
Resistivity, 166
surface, 356–358
Resistor characteristics, 333, 334
Resonant cavities, 385–392
Resonator energy, 388–390
Retarded potentials, 326
Retentivity, 235, 251
Rowland, H. A., 40

Saturation magnetization, 262, 264
Scalar, 1
Scalar field, 8–11
Scalar potential, 16, 30
electrostatic, 50–53
magnetostatic, 243–248
time-varying, 322, 323
Scalar product, 6
Scale factors, 24
Schwarz-Christoffel transformation, 158–160
Self-inductance, 274, 278–280
Separation of variables, 120, 121, 130, 131, 138–140
Shearing line, 251, 252
Shielding, electrostatic, 100, 101
Sidelobes, 408
Skin depth, 315
Skin effect, 319, 359, 527
Sky wave, 479
Slow wave, 464, 470
Slow-wave structure, 473
Snell's law, 352
Solenoid, 241, 242
Solenoidal field, 19, 23, 30, 31
Source, vortex, 16, 30, 32
Source point–field point nomenclature, 26, 27
Source strength, 13, 14, 30, 32
Space-charge limited conditions, 440
Space-charge reduction factor, 455, 456
Space-charge theory, 437–445
Space-charge wavelength, 463
Space-charge waves, 449–456
for Klystron, 461, 462
Sphere in applied field, conducting, 501
dielectric, 142, 143
permeable, 520
Spherical coordinates, 138, 139
Spin magnetic moment, 258
Spinning electron, 258
Spontaneous magnetization, 264
Standing-wave ratio, 369, 370
Stokes' theorem, 22, 23
Stray capacitance, 331
Stress, surface, 112–115
Stress tensor, electric, 114
magnetic, 294, 295
Superposition, 42
of plane wave, 378, 379
Surface, charged, 46, 47, 55–57, 76–78
equipotential, 9
Surface charge at dielectric interface, 183
Surface charge density, 46
Surface impedance, 356–358, 360, 361
Surface resistivity, 356–358
Surface sources, 32
Surface stress, 112–115
Surface vector, 4
Surface waves, 530, 531
Susceptibility, diamagnetic, 260
electric, 85
magnetic, 232
paramagnetic, 261

TE waves, 342, 345–347
cutoff, 374, 380
impedance, 346
TE_{10} mode, 375–379
Tensor permeability, 486
Thales, 39, 198
Thompson's theorem, 507, 514
Time constant, 170, 171, 180
TM waves, 343
cutoff, 380
impedance, 348
Toroid, magnetic field, 248–252
Torque, on D'Arsonval movement, 215, 216
on magnetic dipole, 214, 215
on rectangular current loop, 290
Total reflection, 530
Transadmittance, electron beam, 459, 460, 462, 463
Transconductance of triode, 448
Transformation, coordinate, 149–152
gauge, 225
Transit time, 442, 443
Transmission coefficient, 353
Transmission-line parameters, 363–365
Transmission lines, 71, 72, 361–370
analogy, 353–356
attenuation constant, 364, 368, 369
characteristic impedance, 363, 365
equivalent circuit, 364
half-wave, 370
input impedance, 370
lossless, 365, 366
lossy, 367–369
phase constant, 364
propagation constant, 364
standing waves on, 369, 370
Transmitting-receiving system, 413–417, 419–423
Transverse electric (TE) waves, 342, 345–347
in circular waveguide, 383–385
in rectangular waveguide, 371–374
Transverse electromagnetic waves, 342, 344, 345
Transverse-field components, 343
Transverse magnetic (TM) waves, 343
in circular waveguide, 381–383
in rectangular waveguide, 379, 380
Traveling-wave magnetron, 437
Traveling-wave tube, 463–473
amplification, 472
propagation constant, 471, 472
qualitative description, 463–465
Triode, 446–449
Triple scalar product, 8
Triple vector product, 8
Two-conductor line, electrostatic, 71, 72
Two-dimensional arrays, 409–412
Two-element arrays, 402–407

Uniqueness theorem, static, 61–63
time-varying, 321
Unit vector, 3
Units, 41, 42

Variables, separation, 120, 121, 130, 131, 138–140
Vector, 1
addition, 2
components, 3
operator del, 12
Poynting (*see* Poynting vector)
subtraction, 2
surface, 4
unit, 3
Vector field, 8–11
general, 31
irrotational, 16
solenoidal, 19
Vector identities, 20, 38
Vector potential, 30, 204
differential equation for, 224, 225
mutual inductance from, 276
quasi-static, 325, 326
retarded, 326
time-varying, 322–325
Vector product, 5
Vector summary, 36ff.
Velocity, of energy, 310, 377
group, 371, 377, 378
of light, 305
phase, 314, 371, 377
Velocity modulation, 449, 456, 457, 465
Virtual work, in electrostatics, 106–109
in magnetostatics, 286–289
Volt, 52
Volta, A., 198
Voltage, induced, 266–269, 271, 272
Voltage standing-wave ratio, 369, 370
Volume charge density, 45
Volume distribution of dipoles, 74–76
Vortex source, 16, 30, 32

Wave admittance, 346, 348
Wave equation, 303–306
inhomogeneous, 304, 535
Wave impedance, TE, 346
TM, 348

Wave number, 312
Wave propagation (*see* Propagation of wave)
Waveguides, circular, 380–385
  losses in, 376
  rectangular, 370–380
Wavelength, 313, 314
  cutoff, 371, 374
Wavelength, guide, 371
Waves, in conducting materials, 357–361
  in imperfect dielectrics, 366, 367
  surface, 530, 531
Weiss, P., 262

Zone of radiation, 396